Quantum Mechanics

Quantum Mechanics

Non-Relativistic and Relativistic Theory

Lukong Cornelius Fai

CRC Press
Taylor & Francis Group
Boca Raton London New York

CRC Press is an imprint of the
Taylor & Francis Group, an **informa** business

First edition published [2022]
by CRC Press
6000 Broken Sound Parkway NW, Suite 300, Boca Raton, FL 33487-2742

and by CRC Press
4 Park Square, Milton Park, Abingdon, Oxon, OX14 4RN

CRC Press is an imprint of Taylor & Francis Group, LLC

ISBN: 978-1-032-22146-5 (hbk)
ISBN: 978-1-032-22559-3 (pbk)
ISBN: 978-1-003-27307-3 (ebk)

DOI: 10.1201/9781003273073

Typeset in Minion
by KnowledgeWorks Global Ltd.

Contents

Preface

This book is a set of my lecture notes delivered since 1998 at the Universities of Dschang and Bamenda, Cameroon. The scope covers non-relativistic and relativistic quantum mechanics and presents a more in-depth comprehensive course of quantum mechanics. This book serves as a reference textbook that is highly recommended for all undergraduate and graduate students, as well as researchers in theoretical physics, quantum mechanics, condensed matter, mathematical physics, quantum chemistry and electronic students.

This student-friendly and self-contained structured textbook covers, in detail, typical topics in a core undergraduate program as well as more advanced topics in a graduate course (likely to stimulate students' interests and curiosity) with an elegant mathematical rigor, contemporary style and rejuvenated approach. The balance between theory and pertinent worked examples represents an integral part of each chapter that reinforces the readers' understanding of the fundamental concepts. The analytical methods employed in this book describe physical situations written with mathematical rigor and more in-depth clarity emphasizing the essential understanding of the subject matter without need for prior knowledge of classical mechanics, electromagnetic theory, atomic structure and differential equations.

Chapters 1–13 of the book cover non-relativistic quantum mechanics with Chapter 1 devoted to inadequacies of classical mechanics followed by the Richard Feynman famous double-slit experiment that introduces the fundamental notion of the wavefunction and the superposition principle. The book then proceeds with the mathematical framework and terminology of Hilbert spaces, linear self-adjoint (Hermitian) operators after introducing eigenfunctions and observables in addition to the limiting transition from quantum mechanics to classical mechanics. Chapters 2, 3, 4, 5 and 6 review the properties of the Schrödinger equation, dynamical operators and their expected values; translation, momentum, angular momentum and square of angular momentum operators, Heisenberg uncertainty relation, momentum representation, infinitesimal symmetry transformation generator, total angular momentum justification, spherical spinors, spin of a system of two particles, rotation operator and irreducible tensor operators, general principles of a one-dimensional motion particularly, potential well and potential barrier, linear harmonic oscillator, motion in a central and coulombic fields. Chapter 7 reformulates quantum mechanics in the framework of only operators, while Chapter 8 covers approximation methods in quantum mechanics in addition to the perturbation theory, while Chapter 9 treats symmetry considerations. Chapters 10 and 11 treat, respectively, approximate methods for the helium atom and hydrogen molecule. Chapter 12 is devoted to the scattering theory with emphasis on, scattering cross section and elastic scattering amplitude, method of partial waves, s-scattering of slow particles, resonance scattering, the unitary scattering conditions, time-reversal symmetry, Schrödinger equation green's function, born approximation, elastic and inelastic collisions, Wentzel-Kramer-Brillouin (WKB) method and scattering of indistinguishable particles. Chapter 13, devoted to the polaron theory, covers the Lee-Low-Pines (LLP) technique, polaron in a quantum wire, polaronic exciton and Haken exciton.

Chapters 14–21 are devoted to the relativistic theory with emphasis on the relativistic quantum mechanics and covering topics such as case of an electron, Klein-Gordon equation, Dirac equation, probability and current densities, electron spin in the Dirac theory, free electron state with defined momentum-positronium motion and motion in a magnetic field. Detailed algebra on relevant special functions helpful in gaining a deeper understanding and command of the subject matter is covered in Chapters 22–25.

I will like to acknowledge those who have helped at various stages of the elaboration and writing of this book, through discussions, criticism and especially, encouragement and support. I am very thankful to my wife, Prof. Dr. Mrs. Fai Patricia Bi, for all her support and encouragement and to my four children (Fai Fanyuy Nyuydze, Fai Fondzeyuv Nyuytari, Fai Ntumfon Tiysiy and Fai Jinyuy Nyuydzefon) for their understanding and moral support during the writing of this book. I acknowledge with gratitude the library support received from the Abdus Salam International Centre for Theoretical Physics (ICTP), Trieste, Italy.

About the Author

Lukong Cornelius Fai is professor of theoretical physics at the Department of Physics, Faculty of Sciences, University of Dschang. He is Head of Condensed Matter and Nanomaterials as well as Mesoscopic and Multilayer Structures Laboratory. He was formerly a senior associate at the Abdus Salam International Centre for Theoretical Physics (ICTP), Italy. He holds a Masters of Science degree in Physics and Mathematics (June 1991) as well as a Doctor of Science degree in Physics and Mathematics (February 1997) from Moldova State University. He is an author of over 170 scientific publications and five textbooks.

I

Non-Relativistic Theory

1

Quantum Mechanics Basic Concepts

1.1 Inadequacies of Classical Mechanics

Quantum mechanics is the theory of physical phenomena at the atomic and subatomic levels. Historically, quantum mechanics began and took the modern form in the first quarter of the 20th century. The appearance of quantum mechanics was due to the fact that at the end of the last and beginning of our century, there arose a series of experimental facts that could not only get explanations in classical physics but were in contradiction with it:

1. **Planetary model [1–4]**: The existence of the planetary atom itself (Rutherford-Bohr model presented by Niels Bohr and Ernest Rutherford in 1913) was in contradiction with classical mechanics and electrodynamics. After the experiments of Rutherford on the planetary model of the atom (the electron revolves about the nucleus), it was necessary to examine the experimental facts on the one hand and theory on the other. From classical mechanics, an electron moving in a definite orbit around the nucleus is constantly accelerated. Knowledge of a classical charged particle shows that when it is accelerated, it emits radiation. From electrodynamics, it is well known that any charged particle with an accelerated motion should emit an electromagnetic wave. So an electron moving around the nucleus would continuously emit radiation and consequently become less energetic. This is because as the electron emits the electromagnetic wave, it loses energy and, consequently, its kinetic energy as well as its velocity reduces. Consequently, from the laws of classical physics and electrodynamics, the radius of the orbit would constantly shrink in size until the electron falls into the atomic nucleus. But this does not take place.
2. **Blackbody radiation [5–8]**: Theoretical calculations of blackbody radiation spectrum assume electromagnetic radiation to be a wave and so can exchange energy in any continuous amount and yields a result that does not match with experimental facts. The assumption of Planck shows that electromagnetic radiation of frequency ν to exchange energy only in units of $\hbar\nu$. Planck's calculations agree very well with experimental facts. The Planck constant has the value:

$$h = 2\pi\hbar = 2\pi \times 1.054 \times 10^{-27} \text{ erg} - \text{sec} \tag{1.1}$$

3. **Photo-electric effect**: The release of **electrons** from a metal by irradiating it with electromagnetic waves or light is called **photo-electric effect** [9–11]. To free an electron from a metal, the light radiation should have a frequency greater than a critical frequency, a characteristic of the metal, while light radiation frequency lower than this critical frequency would not lead to the photo-electric effect. This was shown by Einstein to be consistent with Planck's hypothesis, considering light with frequency ν to only exchange energy in the amount $h\nu$. In order for the electrons to be released from the metal, the following energy is needed:

$$E = h\nu = h\nu_0 + T_{\text{kin}},\ T_{\text{kin}} = h(\nu - \nu_0) \geq 0 \tag{1.2}$$

DOI: 10.1201/9781003273073-2

Here, $h\nu_0$ is the binding energy for the given metal and T_{kin} is the kinetic energy of the electrons. From the above relation, it is not possible to have any emission of the electrons unless

$$\nu \geq \nu_0 \tag{1.3}$$

From the blackbody radiation and photo-electric effect, it is seen that though classically light is a wave, it has a particle nature which is confirmed by the Compton effect. All these can only be explained by quantum mechanics.

Compton effect: The scattering of light by an electron can only be explained if a photon of frequency ν is considered as a particle moving with energy, E, and momentum, p:

$$E = h\nu,\ p = |\vec{p}| = \frac{E}{c} = \frac{h\nu}{c} = \frac{h}{\lambda} \tag{1.4}$$

Here, c and λ are, respectively, the speed and wavelength of light wave. From relativistic mechanics, mass and energy may be considered as a manifestation of the same thing. The energy of a particle can be expressed through its momentum as:

$$E = \sqrt{p^2c^2 + m^2c^4},\ \vec{p} = m(\vec{v})\vec{v} = \frac{E\vec{v}}{c^2},\ m(\vec{v}) = \frac{m}{\sqrt{1 - \frac{v^2}{c^2}}} \tag{1.5}$$

Here, m with $\vec{v}$ are, respectively, the rest mass and velocity of the particle. For the case of the photon, which is a particle moving with the speed of light, c, if it equals 0, we have $m(\vec{v}) = \frac{0}{0}$. The photon momentum has the value

$$p = \frac{Ec}{c^2} = \frac{E}{c} = \frac{h}{\lambda} \tag{1.6}$$

So if we consider a photon as a particle, its rest mass should vanish (see details in references [12–14]). It is obvious from here that electromagnetic radiation has the property of dualism, i.e., it possesses the corpuscular and wave nature.

4. **Davisson-Germer experiment**: The notion of the trajectory does not coincide with the experimental fact of the diffraction of an electron. The diffraction of the electron was theoretically foreseen by Louis de Broglie in 1924 (wave character of material particles) and experimentally founded by the American physicist Clinton J. Davisson [2, 4, 15] and the Austrian physicist Erwin Schrödinger (1887–1961) in 1927. Davisson and Germer in 1925 and also the British physicist G. P. Thomson (son of J. J. Thomson, discoverer of the electron) in 1926 showed scattered electrons from crystals and found diffraction patterns consistent with the interference of electrons having the de Broglie wavelength and similar to light interacting with a diffraction grating. G. P. Thomson [16] found diffraction rings produced by electrons passing through thin foils. Considering an electronic beam incident on a crystal lattice, we observe a diffraction pattern. As diffraction is a wave phenomenon, the electron having a corpuscular nature sometimes has a wave nature. Such observations lead to a general conclusion that all objects have both corpuscular and wave nature. The question on the wavelength associated with the particle was given by Louis de Broglie's hypothesis through the relation:

$$\lambda = \frac{h}{p} \tag{1.7}$$

It is instructive to note that experimentation on atomic systems shows that measured quantities achieve quantized values unlike classical mechanics' predictions, where observables achieve continuous values.

5. **Experimentation on dualism**: For the dual nature of the particles, sometimes the particles have a corpuscular as well as a wave nature. Classically, particles follow definite trajectories and so do not show interference, whereas waves show interference.

    These and other experimental facts gave birth to quantum mechanics.

1.2 Wave Function

The state of a classical system of particles can be represented by their coordinates and momenta that form a complete set of observables. Suppose, such a system has f degrees of freedom with generalized coordinates $q_1, q_2, \ldots, q_f$ or $\{q\}$ and generalized momenta $p_1, p_2, \ldots, p_f$ or $\{p\}$. Then the state of the system at the moment, t, should be defined by the values $q(t)$ and $p(t)$. Any other observable, say A, can be expressed as a function of coordinates and momenta, which is a **purely classical description**. This classical description is connected with the **motion in a trajectory**. If at an initial time moment, t_0, the initial coordinate, $q(t_0)$, and momentum, $p(t_0)$, are known, then their values $q(t)$ and $p(t)$ at any later time, t, can be determined by the Hamilton canonical equation. On the other hand, in many situations, all the values of coordinates and momenta are not able to be determined precisely, and so the need for a **statistical description**. We ask ourselves what we call the trajectory in the configuration or phase space. The explanation can be given by the famous Richard Feynman thought double-slit diffraction experiment in 1965, which is the cornerstone of quantum mechanics and illustrates key features of quantum mechanics and in particular interference and particle-wave duality of matter [17, 18]. This experiment is depicted in Figure 1.1, where we examine the flux of low-intensity beam of electrons coming from an electronic source, S_e, and allowed to pass through double slits a and b on a diaphragm D and detected behind the slits at a detector screen P facing the source (Figure 1.1). Each electron that hits the detector screen P is detected at a single location. So there is no doubt the electrons should be particles. If the slit b is closed, then we measure on P_1 the intensity J_1 of the electronic flux, and if similarly slit a is closed, then we measure on P_2 the intensity J_2 of the electronic flux, and when slits a and b are both opened, then we measure on P_3 the intensity:

$$J \neq J_1 + J_2 \tag{1.8}$$

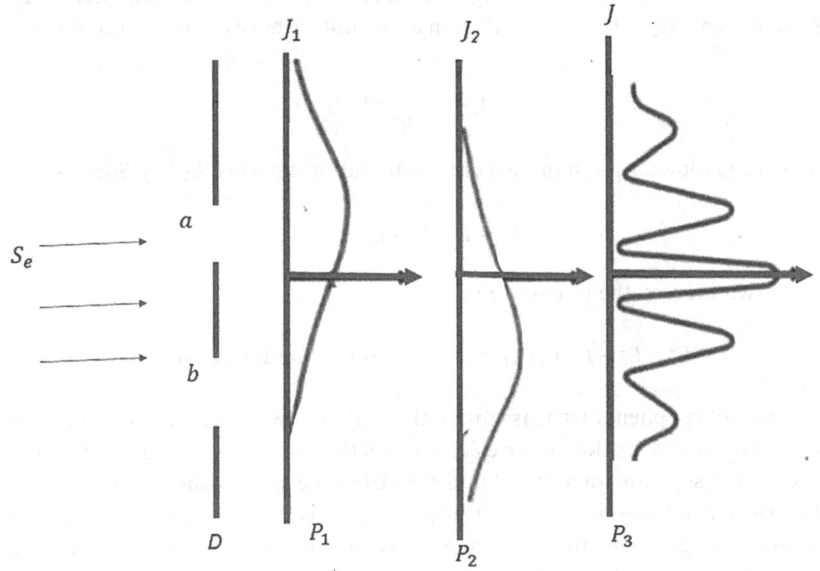

FIGURE 1.1 The double-slit experiment with an electronic beam coming from an electronic source S_e, and passing via slits a and b on a diaphragm D and measured on the sensitive screen P facing the source [17].

So the diffraction pattern from two slits is satisfied (Equation 1.8), which is the general property of all diffraction phenomena. This experiment

- Illustrates the quantum mechanical superposition principle, i.e., the wave properties
- Suggests the electrons, to have combine particle and wave characteristics, the **famous wave-particle duality** of quantum mechanics
- Suggests observing the measurement of a quantum system has a profound effect on the system

We observe from the double-slit experiment that even if the beam intensity is so low that on average only a single electron at the time is in flight and so an interference pattern is built up on the detector screen P after tens of thousands of counts have been detected. This is in clear violation of classical physics since the fluxes at the sensitive screen P in the three cases mentioned above are expected to be the sum:

$$J = J_1 + J_2 \tag{1.9}$$

However, in reality, generally an interference term J_{int} (due to two waves passing via slits a and b on a diaphragm D and measured on the sensitive screen P) is added to it confirming Equation 1.8.

So by controlling the electronic flux through the individual slits of a double slit, we observe the diffraction patterns from slit a, slit b and both on the screens P_1, P_2 and P_3, respectively, with intensities, J_1, J_2 and J, hence observing the wave properties of electrons. On measuring the single electron detection events diffracting through a double slit, we can build up a diffraction pattern and so observe the particle properties of electrons. So the double-slit experiment illustrates the phenomena of wave-particle duality. This is a postulate that all particles exhibit both wave and particle properties.

We now imagine shining a light of a single color, i.e., light of a single wavelength on the diaphragm D with the two slits, a and b (where the distance between the slits is roughly the same as the wavelength of light). When the light wave passes through both slits, it essentially splits into two new waves, each spreading out from each slit. These two waves then interfere with each other and at some points, where a peak meets a trough, they will cancel each other out in a destructive interference phenomenon. At other points, where a peak meets a peak the waves reinforce each other in a constructive interference phenomenon giving the brightest light. Consider the light as an electromagnetic wave described by the vectors $\vec{E}(\vec{r},t)$ (electric field strength) and $\vec{H}(\vec{r},t)$ (magnetic field strength) being its components. If we denote by Ψ any of the components $\vec{E}(\vec{r},t)$ or $\vec{H}(\vec{r},t)$, then in a vacuum we have the wave equation:

$$\Delta\Psi - \frac{1}{c^2}\frac{\partial^2}{\partial t^2}\Psi = 0 \tag{1.10}$$

We may solve this equation for given initial conditions and we have say, $\vec{E}(\vec{r},t)$. Suppose

$$\vec{E} = \vec{E}_1 + \vec{E}_2 \tag{1.11}$$

is the field due to two sources, the intensity J is:

$$E^2 = E_1^2 + E_2^2 + 2\left(\vec{E}_1, \vec{E}_2\right) \cong J = J_1 + J_2 + \text{interference term} \tag{1.12}$$

This justifies many optical phenomena, assuming that light is a wave, as shown also by Young, Fresnel and Maxwell. So, for the description of the electromagnetic wave (or other waves), the resultant intensity J is not the sum of separate intensities but differs from the sum by the interference term that may as a result of phase dependence be positive or negative. This is the principle of superposition of fields. The superposition principle is the fundamental property of quantum systems with its formulation providing the possibility to associate with two wave functions corresponding to the two pure states of the system, the third function is an arbitrary linear combination of the two wave functions. This third function also corresponds to a physical state of the system. We observe that in classical mechanics, the particle state is determined by its position and momentum. In the case of fluctuations, due to the motion in environment, this particle state is determined by the probability density in the particle phase space, while in quantum mechanics, the particle state is determined either by the wave function or by the density operator.

1.3 Wave Function Statistical Interpretation

For the mathematical description of the interference phenomena, it is necessary to introduce the wave function – the function dependent on the position, $\vec{r}$, and time, t. From quantum mechanics, the probability amplitude (square of the probability function or wave function, Ψ) can as well be determined from the Schrödinger equation as it describes the wave properties of the particles. It is instructive to note that each possible physical state of a given system in quantum mechanics is a univalent correspondence with a subspace of a complex linear vector space (Hilbert space). From the double-slit experiment, the probability that a system observed initially in state $|\Psi\rangle$ will be observed finally in state $|\Phi\rangle$:

$$J \approx |\langle\Phi|\Psi\rangle|^2 \tag{1.13}$$

Here, the complex number $\langle\Phi|\Psi\rangle$ is the so-called **probability amplitude**, which can be added and multiplied in the same manner as classical probabilities.

The probability amplitude for an electron at the initial state $|i\rangle$ (electronic source) to have a flight via slit a and be detected at the final state $|f\rangle$ on screen P_1 should be $\langle f|a\rangle\langle a|i\rangle$ and similarly, the amplitude for the electron initially in state $|i\rangle$ with flight via slit b and be detected by state $|f\rangle$ should be $\langle f|b\rangle\langle b|i\rangle$ [17]. The probability amplitude for an electron initially in state $|i\rangle$ (source) to be detected in the final state $|f\rangle$ (screen) can be obtained by adding the amplitudes due to two sources via slits, a and b. So

$$J \cong (\langle f|a\rangle\langle a|i\rangle + \langle f|b\rangle\langle b|i\rangle)^2 = (\langle f|a\rangle\langle a|i\rangle)^2 + (\langle f|b\rangle\langle b|i\rangle)^2 + 2\mathrm{Re}(\langle f|a\rangle\langle a|i\rangle\langle f|b\rangle\langle b|i\rangle) \tag{1.14}$$

We observe that each electron takes both possible paths simultaneously, and the interference between the two paths gives the observed interference pattern. Other paths than the straight lines are canceled by destructive interference. From this result, should we say the wave behavior implies an electron is

- A delocalized object?
- Passing through both slits?

This obviously is not the case and when detected as point-like object on the sensitive screen P. Verifying by a detector set at slits a and b where it went through, we find that it either went through slit a or b. So J should be interpreted as a probability density for a point-like particle position. The particle/wave duality is not contradictory from the indetermination principle: **Attempting to detect the alternative path followed by the electron destroys interference, or any determination of the alternative achieved by a process capable of following more than one alternative destroys the interference between the alternatives [17].** So what adds up is the probability amplitude and not the probability density. The difference between classical and quantum composition of probabilities is given by the interference between classically distinct trajectories.

From the double-slit experiment example, the flight of an electron via a defined slit should be at the price of destroying quantum interference. This indeed presents an alternative, but fully equivalent, method to compute the probability amplitude as should above. Here, the role of the trajectory of a point-like particle, formally, is revived in a way compatible with the indetermination principle.

Compton's formula establishes an electromagnetic wave that can behave like a particle of light when interacting with matter [12]. In 1924, Louis de Broglie proposed the hypothesis that electrons and other particles of matter behave like waves known today as de Broglie's hypothesis of matter waves. In 1926, de Broglie's hypothesis, together with Bohr's early quantum theory, developed the new theory of wave quantum mechanics to describe the physics of atoms and subatomic particles. According to de Broglie's hypothesis, massless photons and also massive particles must satisfy one common set of relations that connect the energy E with the frequency ν and the linear momentum p with the wavelength λ. We have seen earlier these relations for photons in the framework of Compton's effect. We recall them now in a more general framework and so any particle that has energy and momentum is a de Broglie wave of frequency ν and wavelength λ, where $E = h\nu$ and $\lambda = \frac{h}{p}$ taking into account the particle-wave *duality*. Here, E and p are, respectively, the relativistic energy and momentum of a particle. The de Broglie's relations usually are expressed in terms of the wave vector, $\vec{\kappa}$, $\kappa = \frac{2\pi}{\lambda}$, and the wave frequency, $\omega = 2\pi\nu$, as we typically do for waves where

$$E = \hbar\omega,\ \vec{p} = \hbar\vec{\kappa} \tag{1.15}$$

From the wave theory, a wave carries its energy, E, with the group velocity, v. For matter waves, this group velocity, v, is the velocity of the particle. The de Broglie's hypothesis also takes both relativity (Equations 1.5 and 1.6) and quantum mechanics. The derivation given by de Broglie, based on special relativity, is quite different from that given in modern textbooks [12, 13].

It is instructive to note that all microscopic particles, whether massless, like photons, or having mass, like electrons, have a wave-like nature and so the relationship between momentum and wavelength is fundamental for all particles. It is worth noting that de Broglie's work was a watershed for the development of quantum mechanics. de Broglie was awarded the Nobel Prize in 1929 for his idea and also Davisson and G. P. Thomson in 1937 for their experimental verification of de Broglie's hypothesis. Physicists from many countries took this idea and, in particular, Schrödinger developed de Broglie's qualitative idea by writing down the non-relativistic Schrödinger equation that the wave must satisfy. Several scientists and others proposed the relativistic version of this equation in 1926. In 1926, M. Bohr proposed that the diffraction picture of the electrons is due to the probability wave. In the given experiment, it is impossible to say exactly where an electron falls onto a screen. If many electrons fall onto the screen, then we have an interference picture for which

$$\left|\Psi(\vec{r},t)\right|^2 d\vec{r} \tag{1.16}$$

represents the probability that the particle lies in the volume element, $d\vec{r} \equiv dV$, located at $\vec{r}$ at time t and the quantity, $\left|\Psi(\vec{r},t)\right|^2$, is the probability density and the spatial wave function of wave mechanics, $\Psi(\vec{r},t)$, is the probability amplitude that completely specifies the state of the quantum mechanical system and is dependent on the coordinate, $\vec{r}$, of the particle(s) and on time, t. The wave function must satisfy certain mathematical conditions due to its probabilistic interpretation, and for the case of a single particle, the net probability of finding the given particle at some point in space can be obtained by integrating Equation 1.16 over the volume V or entire space that results in unity, leading to the **normalization condition**:

$$\int_{\text{entire space}} \left|\Psi(\vec{r},t)\right|^2 d\vec{r} = 1 \tag{1.17}$$

This enforces that the probability of finding the particle somewhere within a given volume at any particular time moment t in which the particle is known to exist is unity. Equation 1.17 is the condition of normalization of the wave function $\Psi(\vec{r},t)$. In order for Equation 1.17 to be satisfied, it is necessary that the wave function $\Psi(\vec{r},t)$ should have the following constraints:

- **Squarable integrable**
- **Continuous**
- **Twice differentiable**

- **Normalizable**
- **Finite or bounded**

Let us have a system of f degrees of freedom for which we have f generalized coordinates, $q_1, q_2, \dots, q_f = \{q\}$ in the configuration space. The system is described by the wave function $\Psi(q,t)$ and $|\Psi(q,t)|^2 dq$ is the probability of finding the particle in the element dq of the configuration space at the moment t. The wave function must satisfy certain mathematical conditions because of the given probabilistic interpretation, and for the case of a single particle, the probability of finding it somewhere is 1, so that we have the following normalization condition:

$$\int |\Psi(q,t)|^2 dq = 1 \tag{1.18}$$

It is also expected to normalize many-particle wave functions to 1.

1.4 Uncertainty of Two Types of Measurements

Quantum mechanics generally, to the best of our knowledge, is regarded as the physical theory, which is a candidate for a fundamental and universal description of the physical world. The conceptual framework employed by quantum mechanics differs drastically from that of classical physics. In particular, the wave properties of an electron are not in conformism with the motion of the electron and so for the description of the electronic state, we introduce the wave function which is a quantum mechanical concept. One of the outstanding aspects that marks the difference between classical and quantum physics is where classical mechanics presupposes exact simultaneous values which can be assigned to all physical quantities, while quantum mechanics denies this possibility with the prime example being the position and momentum of a particle. From quantum mechanics, the more precisely the position of a particle is given, the less precisely can one say what its momentum is and vice versa. This is a simplistic and preliminary formulation of the quantum mechanical uncertainty principle for position and momentum that in 1927 was formulated by the German physicist Werner Heisenberg (1901–1976) and the mathematical relation is called **Heisenberg's relation**. So for the electronic state, one cannot at a given time moment, t, assign exact simultaneous values to the position and momentum. Hence, these quantities can only be determined with some characteristic **uncertainties** that cannot become arbitrarily small simultaneously. This applies also to other pairs of quantities such as any function of the coordinate and momentum and also the potential and kinetic energies. So, from Heisenberg's uncertainty relation, it is therefore very possible to use the coordinate or the momentum or both to define the state of motion with the main object of the restriction being the trajectory, $q(t)$. Knowledge of the trajectory of a particle by continuously monitoring its location will permit knowledge of the coordinate and momentum simultaneously.

The notion of **uncertainty** sometimes may refer to lack of knowledge of a quantity by an observer, or to the experimental inaccuracy with which a quantity is measured. It may be some ambiguity in the definition of a quantity, or to a statistical spread in an ensemble of similarly prepared systems, say, the measurement of some physical quantity A. So it is impossible to prepare states in which position and momentum are simultaneously arbitrarily well localized. Hence, the knowledge of the spread of the physical quantity, A, is inherent in the probability (wave) function, $\Psi(q,t)$ we forecast the probability of measurement.

For the measurement of the quantity A in an ideal experiment, we neglect errors. Suppose in the state described by the wave function Ψ_1, we measure A_1, and in that described by Ψ_2, we measure A_2 and so on. We can say that for the state described by Ψ_1, the physical quantity A has a defined value, A_1. For the state described by $\Psi(q,t)$, we have different physical quantities such as A_1, A_2 and so on. **The values of the physical quantity A that, for a given system, we obtain when they are measured are called eigenvalues or observables.** Eigen here is a German word, meaning self or own. If these eigenvalues or observables are a countable set, then we say that the physical quantity A for the given system has a **discrete spectrum of eigenvalues**, for example, the energy of an electron in the hydrogen atom. For a physical quantity with a continuous **spectrum** of values, we have any of the values, $A_1 \leq A \leq A_2$. So it is impossible either theoretically or experimentally to forecast the values of A that can be measured. We can only talk about the probability of measurement of A in the state Ψ. One of the problems of the quantum theory is to establish the methods of the definition of this probability which relates already to the generalized formulation of the uncertainty principle.

1.5 Superposition Principle Generalized Formulation

The double-slit experiment [17, 18] entails observing probability distributions in three scenarios i.e., electron flux through

1. **Slit a with slit b closed and with intensity, J_1, on the screen P_1**
2. **Slit b with slit a closed and with intensity, J_2, on the screen P_2**
3. **Both slits and with intensity, J, on the screen P_3**

These scenarios show the quantum mechanical superposition principle, i.e., the wave properties. In this case of the double-slit experiment, suppose we have two states, i.e., $|a\rangle$ for the flux through slit a and $|b\rangle$ for the flux through slit b. For the flux through both slits where we observe on the screen P_3, the state $|f\rangle$ will then be the linear combination of the given states $|a\rangle$ and $|b\rangle$:

$$|f\rangle = C_1|a\rangle + C_2|b\rangle \tag{1.19}$$

with C_1 and C_2 being the expansion coefficients of the state $|f\rangle$ in terms of the states $|a\rangle$ and $|b\rangle$. **So the probability amplitude for an electron initially in a given state (source) to be detected in the final state (screen) is obtained by adding the amplitudes due to two sources through slits a and b (superposition principle).**

Let us examine the case when the physical quantity A has a series of discrete values:

$$A: A_1, A_2, \ldots, A_n \tag{1.20}$$

Let us denote by $\Psi_1, \Psi_2, \ldots, \Psi_n$ the wave functions of some system that describe states with defined values of A. This implies we define the state function as a superposition of eigenstates, $\Psi_1, \Psi_2, \ldots, \Psi_n$, of the physical quantity A:

$$\Psi = \sum_n C_n \Psi_n \tag{1.21}$$

The coefficient, C_n, of Ψ_n in the expansion relates to the probability, $|C_n|^2$, of finding the eigenvalue A_n when A is measured, i.e., the larger the contribution of Ψ_n in the expansion – indicated by the magnitude of C_n – the larger the probability of measuring that eigenvalue. So in the state Ψ, the physical quantity A does not have a defined value and so we can only talk about its probability of measurement, $|C_n|^2$:

$$\sum_n |C_n|^2 = 1 \tag{1.22}$$

This is the probability of one of the independent events which is the sum of probabilities.

Suppose the function Ψ is squarable integrable, such a function can be normalized to unity, i.e.,

$$\int |\Psi(q)|^2 \, dq = 1 \tag{1.23}$$

If we compare Equations 1.22 and 1.23, then we have

$$\int \Psi^*(q)\Psi(q)dq = \sum_n C_n^* C_n \tag{1.24}$$

Take the complex conjugate of Equation 1.21, then multiply the resultant by $\Psi(q)$ and, consequently, integrate over

$$\int \Psi(q)\Psi^*(q)dq = \sum_n C_n^* \int \Psi(q)\Psi_n^*(q)dq \tag{1.25}$$

Comparing Equation 1.24 with Equation 1.25 while considering Equation 1.21 and swapping the index n for m, the identity is as follows:

$$C_n = \int \Psi(q)\Psi_n^*(q)dq = \sum_m C_m \int \Psi_m(q)\Psi_n^*(q)dq \tag{1.26}$$

In order for the given equality to be satisfied, it is necessary that

$$\int \Psi_m(q)\Psi_n^*(q)dq = \delta_{nm} = \begin{cases} 1, & m = n \\ 0, & m \neq n \end{cases} \tag{1.27}$$

where δ_{mn} is the symmetric Kronecker δ-symbol.

This property expresses the **orthogonality** of the functions Ψ_m and Ψ_n for $m \neq n$. The orthogonality is the notion of the geometry of the functional space in the language that can be explained by the entire quantum mechanics. **The eigenfunctions of states with different energies are orthogonal.** The integral Equation 1.27 in the functional space, for $m = n$, is the scalar product:

$$\left(\Psi_n^*, \Psi_n\right) = 1 \tag{1.28}$$

The wave function Ψ_n is called the vector of state which forecasts the result of an arbitrary state.

1.6 Operators of Physical Quantities

1.6.1 Expectation Value (Observable) and Operator of a Physical Quantity

The expectation value (equivalently the average or mean value) is the average measurement of an observable. Mathematically, the formulation of quantum mechanics is built upon the concept of an operator where physical states are represented as basis vectors in the complex Hilber space. We show that any observable, A, i.e., any quantity that can be measured in a physical experiment, ought to be associated with a self-adjoint linear operator, say, $\hat{A}$. The operators must yield real eigenvectors, since they are values that result from experimentation. This implies, mathematically, that the operators must be Hermitian. The probability consistent with each eigenvalue relates the projection of the physical state on the subspace related to that eigenvalue.

Suppose, through an exhaustive series of measurements, we find a particular observable, A, of a physical system, that have all real values as follows:

$$A_1, A_2, \ldots, A_n \tag{1.29}$$

For each possible value of A, the system is prepared in a corresponding state as follows:

$$\Psi_1, \Psi_2, \ldots, \Psi_n \tag{1.30}$$

for which the values (Equation 1.29) are obtained with certainty in any $N \to \infty$ measurement of A:

$$N_1 + N_2 + \cdots + N_n = N \to \infty \tag{1.31}$$

Knowledge of the value of the given physical observable of the quantum system implies some kind of measurement has been made. In the process of acquiring this knowledge, the system, after the measurement has been performed, survives the measurement. In addition, if we were to immediately remeasure the same quantity, we would get the same result. However, in reality, few measurements are of this kind, i.e., the system being subject to measurement is physically modified, if not destroyed, by the measurement process. An example is a measurement designed to count the number of photons in a single mode cavity field. Photons are typically counted by photodetectors whose mode of operation is to absorb a photon and create a pulse of current. So we may well be able to count the number of photons in the field and, while doing so, there is no field left behind after the counting is completed.

We can conclude that the state of the cavity field is left in the vacuum state $\langle 0|$ when the measurement is completed. However, nothing for certain is known on the state of the field before the measurement. Suppose an influx of photons into the cavity from an experimental setup, then every time we measure the photon number in the cavity, we will always get the result n and can assign the state $\langle n|$ to the cavity field.

We apply the same procedure to the spin-half particle in the Stern-Gerlach experiment that involves the splitting of an atomic beam into two components (with spin projections, $+\frac{1}{2}$ or $-\frac{1}{2}$, in units of the Planck constant $\hbar$) when passed through a Stern-Gerlach apparatus and subjected to an inhomogeneous magnetic field oriented in the z-axis direction. For the spin projections, $+\frac{1}{2}$ or $-\frac{1}{2}$, we can assign, respectively, the states $\langle\uparrow|$ or $\langle\downarrow|$ that are mutually exclusive. This implies that if in the state $\langle\uparrow|$, the result $-\frac{1}{2}$ is never observed, then we note that the two states cover all possible values of the spin projection in the z-axis direction. So the observation of the atomic spin behavior shows some evidence of both randomness and interference and so any spin state of the atom can be a linear combination of the states $\langle\uparrow|$ and $\langle\downarrow|$. So the states $\langle\uparrow|$ and $\langle\downarrow|$ constitute a complete set of orthonormal basis states for the state space of the given system and hence the situation that applies to the eigenstates and eigenvectors of a Hermitian operator. So the observable, though initially applied to a physical quantity of interest, is also associated with a Hermitian operator.

At this point, we assume our system to exhibit the properties of intrinsic randomness and interference that characterizes quantum systems. This permits the state of the system to be identified as vectors belonging to the state space of the system. So if the system is prepared in the state Ψ_n, and we measure A, then we ever observe only the result A_n, i.e., we never observe the result A_m with $A_m \neq A_n$. So $(\Psi_n^*, \Psi_m) = \delta_{nm}$ and the states $\{\Psi_n\}$, $n = 1, 2, \ldots$ are orthonormal. Hence, the states in Equation 1.30 span all the possibilities for the system and so the given states form a complete set of orthonormal basis states for the state space of the system.

For the system in state Ψ, the probability of obtaining the result A_n on measuring A is:

$$\left.\frac{N_n}{N}\right|_{N \to \infty} = W_n \tag{1.32}$$

So the mean observable value $\bar{A}$ of the quantity, A, in the state Ψ can then be calculated as follows:

$$\bar{A} = \sum_n W_n A_n \tag{1.33}$$

We define the average value in terms of the probability, W_n, of getting one of the eigenvalues, A_n, when A is measured. So when a property is measured, we can only get one of the eigenvalues as a result with no other numbers possible. The Equation 1.32 is also equal to $|C_n|^2$ which is the probability of finding the eigenvalue A_n and so the mean observable value $\bar{A}$ while considering Equation 1.26:

$$\bar{A} = \sum_n |C_n|^2 A_n = \sum_n C_n^* C_n A_n \equiv \sum_n \int \Psi^*(q)\Psi_n(q)dq C_n A_n = \int \Psi^*(q)dq \sum_n C_n A_n \Psi_n(q) \quad (1.34)$$

This is analogous to the mathematical expectation from the theory of probability.

Let us denote by

$$\sum_n C_n A_n \Psi_n(q) = \hat{A}\Psi(q) \quad (1.35)$$

where $\hat{A}$ is the operator of the physical quantity A. From Equations 1.34 and 1.35, we see that the observable A is represented by a Hermitian operator $\hat{A}$ with eigenvalues that are the possible results of the measurement of A as in Equation 1.29 and the associated eigenstates are the states in Equation 1.30. The appellation observable is often applied to the operator $\hat{A}$ itself. So in quantum mechanics, for any observable, A, there is an operator, $\hat{A}$, that acts on the wave function so that, if a system is in a state described by Ψ, the expectation value of $\bar{A}$ is as follows:

$$\bar{A} = \int \Psi^*(q)\hat{A}\Psi(q)dq \quad (1.36)$$

This linking of particular states with particular measured results provides a way that the observable properties of a quantum system can be described in quantum mechanics, in terms of Hermitian operators, $\hat{A}$. Such operators in quantum mechanics arise as nature is described with waves (the wave function) rather than with discrete particles whose motion and dynamics can be described with the deterministic equations of Newtonian physics. If we consider the possible states with m and n, then

$$A_{mn} = \int \Psi_m \hat{A}\Psi_n \, dq \quad (1.37)$$

is called the **matrix element** of the quantity, A.

The quantum mechanical observable differs from the corresponding classical quantity in that the latter quantity represents a single possible value for that observable, while the former is an operator that, through its eigenvalues, carries with it all the corresponding possibly physical quantity. This implies that when a measurement is made of a particular physical property of a quantum system, the outcome can be any of the possible values associated with the observable, even if the experiment is repeated under identical conditions.

1.6.2 Properties of Operators

Let us examine the properties of operators where we consider $\hat{A}, \hat{B}$ and $\hat{C}$ as linear operators.

1. Addition of operators

   Consider we have the following operator which is the sum of two operators:

$$\hat{A} = \hat{B} + \hat{C} \quad (1.38)$$

   If Ψ is a common wave function to all the operators, then

$$\left(\hat{B} + \hat{C}\right)\Psi = \hat{B}\Psi + \hat{C}\Psi \quad (1.39)$$

Suppose the wave function, Ψ, is represented as a linear combination of two functions ϕ and χ:

$$\Psi = a\phi + b\chi,\ \hat{A}\Psi = a\hat{A}\phi + b\hat{A}\chi \tag{1.40}$$

where a and b are constants.

2. Product of operators
   The product of the operators $\hat{A}$ and $\hat{B}$ written as $\hat{A}\hat{B}$ is defined as

$$\left(\hat{A}\hat{B}\right)\Psi = \hat{A}\left(\hat{B}\Psi\right) \tag{1.41}$$

   where $\hat{B}$ first acts on Ψ to give the new function $\hat{B}\Psi$ after which $\hat{A}$ acts on $\hat{B}\Psi$ to give $\hat{A}\left(\hat{B}\Psi\right)$. It should be noted that, in general, $\hat{A}\hat{B} \neq \hat{B}\hat{A}$. Thus, the order of the operators is important. Also,

$$\hat{A}^n = \underbrace{\hat{A}\cdots\hat{A}}_{n-\text{times}} \tag{1.42}$$

3. Commutation of operators
   The commutator $\left[\hat{A},\hat{B}\right]$ of $\hat{A}$ and $\hat{B}$ is defined as follows:

$$\left[\hat{A},\hat{B}\right] = \hat{A}\hat{B} - \hat{B}\hat{A} \tag{1.43}$$

The commutator itself is a (composite) operator. If we are given the function $f(A)$ of the physical quantity A, then we may also have $f\left(\hat{A}\right)$.

Let us consider again the expectation value $\bar{A}$ of the operator $\hat{A}$. Suppose we consider Equation 1.40:

$$\bar{A} = \int \Psi^*(q)\sum_n C_n A_n \Psi_n\, dq = \int \Psi^*(q)\left\{a\sum_n C_n A_n \phi_n + b\sum_n C_n A_n \chi_n\right\}dq \tag{1.44}$$

So if we consider Equation 1.35, then

$$\hat{A}\Psi \equiv \sum_n C_n A_n \Psi_n(q) = a\sum_n C_n A_n \phi_n + b\sum_n C_n A_n \chi_n \equiv a\hat{A}\phi + b\hat{A}\chi \tag{1.45}$$

Similarly, it may be shown that

$$\left(\hat{A}\hat{B}\right)\Psi = \hat{A}\left(\hat{B}\Psi\right),\quad \hat{A}\hat{B} \neq \hat{B}\hat{A} \tag{1.46}$$

implies the linearity of the operators $\hat{A}$ and $\hat{B}$.

1.7 Linear Self-Adjoint (Hermitian) Operators

In quantum mechanics, it is required that all operators are self-adjoint and linear. This requirement is such that the superposition principle holds. Linear operators satisfy this principle. In order to have meaningful and measurable observables with operators, it is necessary that the expectation (mean) values of these operators be real. This property may be guaranteed if our operator is Hermitian (self-adjoint).

Let us find whether there is some reasoning to the complex conjugate of the mean value $\bar{A}$, so

$$\bar{A} = \int \Psi^* \hat{A}\Psi\, dq = \left(\int \Psi^* \hat{A}\Psi\, dq\right)^* = \int \Psi \hat{A}^* \Psi^*\, dq \tag{1.47}$$

For an arbitrary operator $\hat{A}$, we may select its **adjoint** $\tilde{\hat{A}}$ defined as follows:

$$\int \Phi \hat{A} \Psi \, dq = \int \Psi \tilde{\hat{A}} \Phi dq \tag{1.48}$$

where Ψ and Φ are arbitrary complex valued functions of the configuration coordinate. The operator $\hat{A}$ acts on the function from the right with respect to the integration in the configuration space. If we substitute in place of Φ the function Ψ^* in Equation 1.48 on comparison with Equation 1.47, we have

$$\int \Psi \tilde{\hat{A}} \Psi^* \, dq = \int \Psi \hat{A}^* \Psi^* \, dq \tag{1.49}$$

from where

$$\tilde{\hat{A}} = \hat{A}^* \tag{1.50}$$

The operators satisfying Equation 1.50 are **Hermitian. The operators for which the physical quantities are real are Hermitian.** It is instructive to note that though any observable must be represented by a Hermitian operator, the converse may not be true, i.e., a Hermitian operator need not necessarily represent an observable.

We examine the case for which the physical quantities are complex. Let that quantity be A, we may take its complex conjugate A^*. The operator corresponding to the quantity A^* is denoted by $\hat{A}^\dagger$ (**read: operator A dagger**). It is the conjugate operator of $\hat{A}$ and it is however different from the complex conjugate of $\hat{A}^*$. The operator $\hat{A}^\dagger$ has the matrix elements as follows:

$$A^\dagger_{mn} = (A_{nm})^* \tag{1.51}$$

If we want to get the m,n element of $\hat{A}^\dagger$, we go to the n,m element of $\hat{A}$ (the indices are reversed) and then take its complex conjugate.

If we consider our condition

$$\overline{A^*} = (\overline{A})^* \tag{1.52}$$

then it follows that

$$\hat{A}^\dagger = \tilde{\hat{A}}^* \tag{1.53}$$

from where it is obvious that $\hat{A}^\dagger$ however coincides with $\hat{A}^*$. For real physical quantities, i.e., the operator coincides with its conjugate (the Hermitian operator is also called **self-adjoint**). Equation 1.52 is the reason for which the operator of a physical quantity should be **self-adjoint**. The condition in Equation 1.52 is that the mean value of a physical quantity should be real.

So to every observable in classical mechanics, there corresponds a linear, Hermitian operator in quantum mechanics. This postulate comes from the considerations that if the expectation value of an operator, $\hat{A}$, is real, then it follows that $\hat{A}$ must be a Hermitian operator. Hence, the operator that is linear and self-adjoint is Hermitian. So Hermitian operators should have the following special properties:

- **They always have real eigenvalues but, the eigenfunctions, or eigenvectors if the operator is a matrix, might be complex**
- **Their eigenfunctions can always be chosen so that they are normalized and mutually orthogonal, i.e., an orthonormal set**
- **Their eigenfunctions form a complete set, i.e., any function can be written as some linear combination of the eigenfunctions**

When finding the operators $\tilde{\hat{A}}$ and $\hat{A}^\dagger$ from the following relation:

$$\int \Psi^* \hat{A}\Psi\, dq = \int \left(\tilde{\hat{A}}\Psi^*\right)\Psi\, dq = \int \left(\hat{A}^\dagger\Psi\right)^* \Psi\, dq \tag{1.54}$$

it should be noted that the arbitrary function Ψ should satisfy the following condition:

$$\int |\Psi|^2\, dq < \infty \tag{1.55}$$

It is instructive to note that for state spaces of finite dimension, any Hermitian operator has eigenvalues that are discrete with the eigenvectors forming a complete set of basis states. When spaces have infinite dimension, there is a possibility for the Hermitian operator to be constructed so that the eigenstates cannot form a complete set. For this case, such an operator cannot represent an observable property of the system. It is also instructive to note that there is a possibility to construct all types of Hermitian operators associated with any given physical system. Such operators would have all the mathematical properties associated with their being Hermitian. However, it is not necessary that they represent either any readily identifiable physical feature of the system at least in part since it might not be at all apparent how such **observables** could be measured. This is partly applicable classically.

Question: Square the operator $\frac{d}{dx} + x$

Answer:
Use the arbitrary function Ψ:

$$\left(\frac{d}{dx} + x\right)^2 \Psi = \left(\frac{d}{dx} + x\right)\left(\frac{d\Psi}{dx} + x\Psi\right) = \frac{d^2\Psi}{dx^2} + 2x\frac{d\Psi}{dx} + x^2\Psi + \Psi \tag{1.56}$$

So

$$\left(\frac{d}{dx} + x\right)^2 = \frac{d^2}{dx^2} + 2x\frac{d}{dx} + x^2 + 1 \tag{1.57}$$

Question:

Compare the operators $\left(x\frac{d}{dx}\right)^2$ and $\left(\frac{d}{dx}x\right)^2$

Answer:

$$\left(x\frac{d}{dx}\right)^2 \Psi = \left(x\frac{d}{dx}\right) x\frac{d\Psi}{dx} = x\frac{d\Psi}{dx} + x^2\frac{d^2\Psi}{dx^2} \tag{1.58}$$

So

$$\left(x\frac{d}{dx}\right)^2 = \left(x\frac{d}{dx}\right) x\frac{d}{dx} = x\frac{d}{dx} + x^2\frac{d^2}{dx^2} \tag{1.59}$$

$$\left(\frac{d}{dx}x\right)^2 \Psi = \left(\frac{d}{dx}x\right)\frac{d}{dx}x\Psi = \left(\frac{d}{dx}x\right)\left(x\frac{d\Psi}{dx} + \Psi\right) = 2x\frac{d\Psi}{dx} + x^2\frac{d^2\Psi}{dx^2} + \Psi + x\frac{d\Psi}{dx} \tag{1.60}$$

and

$$\left(\frac{d}{dx}x\right)^2 = x^2\frac{d^2}{dx^2} + 3x\frac{d}{dx} + 1 \tag{1.61}$$

Question:
Square the operator $i\hbar\nabla + \vec{A}(\vec{r})$.

Answer:

$$\left(i\hbar\nabla + \vec{A}(\vec{r})\right)^2\Psi = \left(i\hbar\nabla + \vec{A}(\vec{r})\right)\left(i\hbar\nabla\Psi + \vec{A}(\vec{r})\Psi\right) = -\hbar^2\nabla^2\Psi + i\hbar\left(\nabla\cdot\vec{A} + \vec{A}\cdot\nabla\right)\Psi + \vec{A}^2\Psi \tag{1.62}$$

From

$$\nabla\cdot\vec{A}\Psi = \vec{A}\cdot\nabla\Psi + \operatorname{div}\vec{A}\cdot\Psi \tag{1.63}$$

then

$$\left(i\hbar\nabla + \vec{A}\right)^2 = -\hbar^2\Delta + 2i\hbar\vec{A}\cdot\nabla + i\hbar\operatorname{div}\vec{A} + \vec{A}^2 \tag{1.64}$$

Question:
Find the commutator $\frac{d}{dx}x - x\frac{d}{dx}$

Answer:

$$\left(\frac{d}{dx}x - x\frac{d}{dx}\right)\Psi = \frac{d}{dx}(x\Psi) - x\frac{d\Psi}{dx} = x\frac{d\Psi}{dx} + \Psi - x\frac{d\Psi}{dx} = \Psi \tag{1.65}$$

So

$$\frac{d}{dx}x - x\frac{d}{dx} = 1 \tag{1.66}$$

Question:
For the operator $\hat{A}$ (defined below), find the operator for which with respect to it is self-adjoint, Hermitian conjugate:

$$\hat{A} = i\frac{d}{dx}, \quad -\infty < x < +\infty; \ \hat{A} = i\frac{\partial}{\partial r}, \quad 0 \le r < \infty \tag{1.67}$$

where r is the radial variable of the spherical system of coordinates.

Answer:
From the fact that Ψ and Φ tend to zero as $x \to \pm\infty$, it follows from integration by parts:

$$\int_{-\infty}^{\infty}\Phi^*\left(i\frac{d}{dx}\Psi\right)dx = i\Phi^*\Psi\Big|_{-\infty}^{\infty} - \int_{-\infty}^{\infty}\left(i\frac{d}{dx}\Phi^*\right)\Psi dx = \int_{-\infty}^{\infty}\left(i\frac{d}{dx}\Phi\right)^*\Psi dx \equiv$$
$$\int_{-\infty}^{\infty}\left(i\frac{\tilde{d}}{dx}\Phi^*\right)\Psi dx = -\int_{-\infty}^{\infty}\left[\left(i\frac{d}{dx}\right)^\dagger\Phi\right]^*\Psi dx \tag{1.68}$$

from where

$$i\frac{\tilde{d}}{dx} = -i\frac{d}{dx} \tag{1.69}$$

and so

$$\left(i\frac{d}{dx}\right)^\dagger = i\frac{d}{dx} \tag{1.70}$$

This implies that the operator $i\frac{d}{dx}$ is Hermitian. **So an operator is Hermitian if it is equal to its adjoint, i.e., $\hat{A}=\hat{A}^{\dagger}$.**

In order to find $i\frac{\tilde{\partial}}{\partial x}$, we consider that

$$dq=r^2 dr d\Omega \equiv dV \tag{1.71}$$

where Ω is the solid angle and dV is the volume element:

$$\int \Phi^*\left(\frac{\partial}{\partial r}\Psi\right)r^2\,dr\,d\Omega=\int\left(\Phi^* r^2 d\Omega\right)\frac{\partial}{\partial r}\Psi\,dr=\int\left(\Phi^*\Psi r^2\right)\Big|_{r=0}^{r=\infty}d\Omega-\int\Psi d\left(\Phi^* r^2\right)d\Omega=$$
$$-\int\Psi\frac{\partial}{\partial r}\left(\Phi^* r^2\right)dr\,d\Omega=-\int\left[\left(\frac{2}{r}+\frac{\partial}{\partial r}\right)\Phi^*\right]\Psi\,dV\equiv\int_{-\infty}^{\infty}\left(\frac{\tilde{\partial}}{\partial r}\Phi^*\right)\Psi dV \tag{1.72}$$

i.e.,

$$\frac{\tilde{\partial}}{\partial r}=-\frac{2}{r}-\frac{\partial}{\partial r} \tag{1.73}$$

So

$$\left(i\frac{\partial}{\partial r}\right)^*=-i\frac{\partial}{\partial r} \tag{1.74}$$

and

$$\left(i\frac{\partial}{\partial r}\right)^{\dagger}=i\frac{\partial}{\partial r}+\frac{2i}{r} \tag{1.75}$$

It follows that the operator $i\frac{\partial}{\partial r}$ is not Hermitian.

1.7.1 Translation Operator

The observation of a physical quantity is always done in a certain coordinate system, say $\vec{r}$, and satisfies the following eigenvalue equation:

$$\hat{\vec{r}}\Psi(\vec{r})=\vec{r}\,\Psi(\vec{r}) \tag{1.76}$$

The position vector, $\vec{r}$, is understood as the position of an observer and $\Psi(\vec{r})$ describes the state of an object, O, as seen by an observer, o, located at the point, $\vec{r}=0$. Let us describe now the same object in terms of the state used by an observer, say o', who is displaced with respect to o. Suppose the origin of o' is displaced from the origin of o by a vector, say $\vec{a}$, pointing from o' to o (see Figure 1.2) and o' describes the object through the eigenfunction, $\Psi'(\vec{r})$, and from the fact that the physical attributes of the object should be invariant under the translation of the observer, we infer the translation operator, $\hat{T}_{\vec{a}}$:

$$\hat{T}_{\vec{a}}\Psi(\vec{r})=\Psi'(\vec{r})=\Psi(\vec{r}+\vec{a}) \tag{1.77}$$

We Taylor series expand $\Psi(\vec{r}+\vec{a})$ with respect to $\vec{a}$ as follows:

$$\Psi(\vec{r}+\vec{a})=\Psi(\vec{r})+\vec{a}\frac{\partial\Psi}{\partial\vec{r}}+\frac{\vec{a}^2}{2!}\frac{\partial^2\Psi}{\partial\vec{r}^2}+\cdots=\sum_{n=0}^{\infty}\frac{\vec{a}^n}{n!}\frac{\partial^n}{\partial\vec{r}^n}\Psi=\exp\{\vec{a}\nabla\}\Psi(\vec{r})\equiv\hat{T}_{\vec{a}}\Psi(\vec{r}) \tag{1.78}$$

We find the Hermitian-conjugate of the operator $\hat{T}_{\vec{a}}$. For this, the operator $\hat{T}_{\vec{a}}^{\dagger}$ is defined as follows:

$$\int \Psi^*(\vec{r})\hat{T}_{\vec{a}}\Psi(\vec{r})dq = \int \Psi^*(\vec{r})\Psi(\vec{r}+\vec{a})dq \equiv \int \Psi(\vec{r})\left[\hat{T}_{\vec{a}}^{\dagger}\Psi(\vec{r})\right]^* dq \tag{1.79}$$

From the change of variable,

$$\vec{r}+\vec{a} \equiv \vec{r}' \tag{1.80}$$

we have

$$\int \Psi^*(\vec{r})\Psi(\vec{r}+\vec{a})dq = \int \Psi^*(\vec{r}'-\vec{a})\Psi(\vec{r}')dq = \int \hat{T}_{-\vec{a}}\Psi^*(\vec{r}')\Psi(\vec{r}')dq = \int \Psi(\vec{r})\left[\hat{T}_{-\vec{a}}\Psi(\vec{r})\right]^* dq \tag{1.81}$$

So

$$\hat{T}_{\vec{a}}^{\dagger} = \hat{T}_{-\vec{a}} \tag{1.82}$$

Question:
Find the operator that is Hermitian conjugate of the product of the operators $\hat{A}$ and $\hat{B}$.

Answer:
From our definition and the fact that

$$\hat{B}\Phi = X \tag{1.83}$$

then

$$\int \Psi^*\hat{A}\hat{B}\Phi dq = \int \Psi^*\hat{A}X dq = \int X\left(\hat{A}^{\dagger}\Psi\right)^* dq \equiv \int X\left(\hat{A}^{\dagger}\Psi \equiv \psi\right)^* dq = \int \psi^*\hat{B}\Phi dq =$$
$$\int \Phi\left(\hat{B}^{\dagger}\psi\right)^* dq = \int \Phi\left(\hat{B}^{\dagger}\hat{A}^{\dagger}\Psi\right)^* dq \tag{1.84}$$

So

$$\left(\hat{A}\hat{B}\right)^{\dagger} = \hat{B}^{\dagger}\hat{A}^{\dagger} \tag{1.85}$$

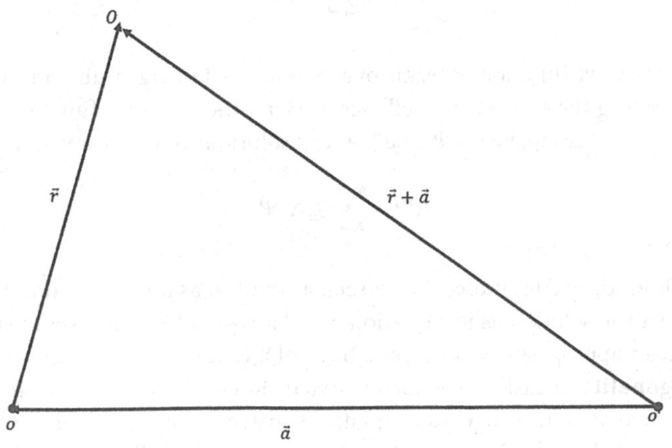

FIGURE 1.2 The position of an object O as seen by an observer o' displaced by $\vec{a}$ with respect to an observer, o.

Question:

Find the operator that is the Hermitian conjugate of the operator $\frac{\partial}{\partial x}$.

Answer:

We consider that the integrals $\int_{-\infty}^{\infty} |\Psi|^2 dx$ and $\int_{-\infty}^{\infty} |\Phi|^2 dx$ exist, and as $x \to \pm\infty$, the functions Ψ and Φ tend to zero, then from the definition of the Hermitian conjugate of an operator, we write the following:

$$\int_{-\infty}^{\infty} \Phi^* \frac{d\Psi}{dx} dx = \Phi^*\Psi\Big|_{-\infty}^{\infty} - \int_{-\infty}^{\infty} \Psi \frac{d\Phi^*}{dx} dx = \int_{-\infty}^{\infty} \Psi\left(-\frac{d\Phi}{dx}\right)^* dx \equiv \int_{-\infty}^{\infty} \Psi\left[\left(\frac{d}{dx}\right)^{\dagger} \Phi\right]^* dx \tag{1.86}$$

So

$$\left(\frac{d}{dx}\right)^{\dagger} = -\frac{d}{dx} \tag{1.87}$$

Question:

Find the operator that is the Hermitian conjugate of the operator $\frac{\partial^n}{\partial x^n}$.

Answer:

Similarly

$$\left(\frac{\partial^n}{\partial x^n}\right)^{\dagger} = (-1)^n \frac{\partial^n}{\partial x^n} \tag{1.88}$$

1.8 Eigenfunction and Eigenvalue

We examine a class of problems having eigenfunctions with similar properties, the **so-called Sturm-Liouville Eigenvalue Problems**, and involve self-adjoint (differential) operators playing an important role in the spectral theory of linear operators and the existence of the eigenfunctions. Many problems in physics arise in the form of boundary value problems involving second order ordinary differential equations. We can write such an equation in an operator form by defining the differential operator, say $\hat{A}$. We can solve some equations using the following eigenvalue expansions:

$$\Psi = \sum_n C_n \Psi_n \tag{1.89}$$

where Ψ is an arbitrary wave function in terms of a complete set of eigenfunctions of $\hat{A}$ with the eigenfunctions Ψ_n and C_n being the expansion coefficients. Here, the arbitrary function, Ψ, obeys the same boundary conditions as the complete set Ψ_n. So we seek solutions to the eigenvalue problem as follows:

$$\hat{A}\Psi = \sum_n C_n A_n \Psi_n \tag{1.90}$$

with homogeneous boundary conditions. So we seek a solution as an expansion of the eigenfunctions where, formally, we let the solution as in Equation 1.89. However, for a nice set of eigenfunctions to be guaranteed, we need an appropriate set to form a basis of a Hilbert space. In addition, it would be nice to **guarantee orthogonality** to easily solve for the expansion coefficients otherwise, we would have to solve an infinite coupled system of algebraic equations instead of an uncoupled and diagonal system. Hence, the eigenfunctions Ψ_n constitute an orthonormal basis of a Hilbert space. So any linear second-order operator can be turned into an **operator possessing self-adjointedness** to carry out the given procedure. The resulting operator in this case is referred to as a **Sturm-Liouville operator** with properties already given above.

Let us examine a particular case when

$$\Psi = \Psi_m, \text{ then } C_n = \delta_{nm} \tag{1.91}$$

Let us substitute $C_n = \delta_{nm}$ into Equation 1.90, for any configuration coordinate, the Sturm-Liouville eigenvalue problem is given by the following differential equation:

$$\hat{A}\Psi_m = A_m \Psi_m \tag{1.92}$$

This postulate captures the central point of quantum mechanics where the values of dynamical variables can be quantized. However, it is still possible to have a continuum of eigenvalues in the case of unbound states. If the system is in an eigenstate of $\hat{A}$ with eigenvalue A_m, then any measurement of the quantity A will yield A_m. So after the measurement in the state Ψ, we have some eigenvalue A_m where the wave function immediately collapses into the corresponding eigenstate Ψ_m (in the case where A_m is degenerate, Ψ_m becomes the projection of Ψ onto the degenerate subspace). So the measurement affects the state of the system with this used in many elaborate experimental tests of quantum mechanics. **So if an operator of a physical quantity acts on an eigenfunction of that quantity (i.e., on the wave function of that state for which the physical quantity has a defined value say A_m), then the result is the product of the eigenvalue and eigenfunction of that quantity. From the Correspondence Principle, for every observable property of a system, there is a corresponding quantum mechanical operator [19]. The converse is not true, for example, say spin and parity.**

Let us examine the following equation:

$$\hat{A}\Psi = A\Psi \tag{1.93}$$

On the left-hand side of Equation 1.93 is an unknown function Ψ on which operator $\hat{A}$ acts, and on the right-hand side is the unknown function Ψ multiplied by a parameter (number) or eigenvalue, A. Suppose we solve the Equation 1.93 and count that its solution should have a physical sense (i.e., Ψ should be a wave function). The physical solution of the equation should satisfy the following conditions in the configuration space: The wave function should be as follows:

- **Finite (bounded)**
- **Continuous**
- **Univalent (single-valued)**

It may appear that these three conditions may be satisfied for some values of the parameter A that are called **eigenvalues or the proper values of the operator** $\hat{A}$. They may form a countable set of **discrete spectra of eigenvalues.** We may also have a **continuous spectrum of eigenvalues.**

Consider the discrete spectrum of eigenvalues. Let us select A_m that corresponds to Ψ_m and we substitute into Equation 1.93. We get an identity that has the same form as Equation 1.92. But in Equation 1.92, A_m and Ψ_m have the sense of the eigenvalue and eigenfunction, respectively, of the physical quantity A (in an experiment) and here A_m is the eigenvalue of the operator $\hat{A}$ and also Ψ_m is eigenfunction of the operator $\hat{A}$. The function Ψ_m in Equation 1.92 is the wave function of the state in which A has a defined value and Ψ_m is the eigenfunction of the solution of Equation 1.93.

1.8.1 Conclusion

The observable value of a physical quantity is the eigenvalue of the operator of that quantity. The corresponding eigenfunction of the operator of the physical quantity is the wave function of that state for which that quantity has a defined value. If we know the operator of the physical quantity, then we may answer a series of questions concerning it and the experiment:

- **What is the possibility of the observable values of the given quantity? For that, it is necessary to write and solve Equation 1.93 applying the constraints of the wave function.**
- **What is the mean value of the quantity A in the arbitrary state Ψ? It is necessary to write the formula of the mean value through the operator and evaluate it.**

If we know the operator, then we solve the Sturm-Liouville problem, and from Equation 1.89, we find C_n and their probability $|C_n|^2$ that is the result of measurement. It follows that the operator gives the answer to all sensitive questions on the given physical quantity. The eigenfunction enables us to answer the question on the transition probabilities.

Question:
Prove that the eigenvalue of a Hermitian operator is real.

Question:
Solve the Sturm-Liouville eigenvalue equation:

$$\hat{A}\Psi = A\Psi,\ \hat{A} = i\frac{\partial}{\partial\phi} \tag{1.94}$$

and Ψ is the eigenfunction and A the eigenvalue of the operator $\hat{A}$.

Answer:
Considering the operator in Equation 1.94:

$$i\frac{\partial}{\partial\phi}\Psi = A\Psi \tag{1.95}$$

and

$$\frac{d\Psi}{\Psi} = -iAd\phi \tag{1.96}$$

from where

$$\Psi(\phi) = \Psi_0 \exp\{-iA\phi\} \tag{1.97}$$

If we consider the fact that the wave function should be

1. Bounded, it follows that A should be real
2. Continuous
3. Single-valued, it follows that

$$\Psi(\phi) = \Psi(\phi + 2\pi) = \Psi_0 \exp\{-iA(\phi + 2\pi)\} \tag{1.98}$$

and

$$\exp\{-2i\pi A\} = 1,\ A = 0, \pm 1, \ldots, \pm m \tag{1.99}$$

It shows that we have a discrete spectrum of eigenvalues.

Hence

$$\Psi = \Psi_0 \exp\{-im\phi\} \tag{1.100}$$

From the normalization condition,

$$1=\int\left|\Psi(\phi)\right|^{2}d\phi=\left|\Psi_{0}\right|^{2}\int_{0}^{2\pi}d\phi=2\pi\left|\Psi_{0}\right|^{2} \tag{1.101}$$

So

$$\Psi=\frac{1}{\sqrt{2\pi}}\exp\{-im\phi\} \tag{1.102}$$

If Ψ_0 is a complex number, then

$$\Psi_{0}=\left|\Psi_{0}\right|\exp\{-i\delta\} \tag{1.103}$$

where δ is a phase factor. This does not influence the physical sense of the problem.

Question:
Find the eigenfunctions and eigenvalues of the operator:

$$\hat{A}=\sin\frac{d}{d\phi} \tag{1.104}$$

Answer: We solve the Sturm-Liouville equation as follows:

$$\hat{A}\Psi=A\Psi \tag{1.105}$$

where the operator, $\hat{A}$, is defined in Equation 1.105. Let us Taylor series expand the operator $\hat{A}$ as follows:

$$\sin\frac{d}{d\phi}\Psi=\left[\frac{d}{d\phi}-\frac{1}{3!}\frac{d^{3}}{d\phi^{3}}+\frac{1}{5!}\frac{d^{5}}{d\phi^{5}}-\dots\right]\Psi=\sum_{k=0}^{\infty}\frac{(-1)^{k}}{(2k+1)!}\frac{d^{2k+1}\Psi}{d\phi^{2k+1}} \tag{1.106}$$

If we consider this and the Sturm-Liouville equation, then we may find the solution as follows:

$$\Psi=\exp\{-\lambda\phi\} \tag{1.107}$$

From here, considering the condition of the single-valuedness of the function, we may take

$$\lambda=im,\quad m=0,\pm1,\dots \tag{1.108}$$

So

$$\sin\frac{d}{d\phi}\Psi=\sum_{k=0}^{\infty}\frac{(-1)^{k}(im)^{k}}{(2k+1)!}\Psi=\sin(im)\Psi,\ A=\sin(im),\quad m=0,\pm1,\dots \tag{1.109}$$

1.9 Properties of Eigenfunctions of Hermitian Operators

From the Hermiticity of a quantum mechanical operator, $\hat{A}$, the eigenvalues of a quantum mechanical operator corresponding to measurable quantities must be real. Consider discrete spectra for the functions Ψ_n and Ψ_m correspond to the eigenvalues A_n and A_m, respectively:

$$\hat{A}\Psi_{m}=A_{m}\Psi_{m},\ \hat{A}\Psi_{n}=A_{n}\Psi_{n} \tag{1.110}$$

Let us take the complex conjugate of Equation 1.110:

$$\hat{A}^*\Psi_m^* = A_m\Psi_m^*, \ \hat{A}^*\Psi_n^* = A_n\Psi_n^* \tag{1.111}$$

we multiply the first equality in Equation 1.111 by Ψ_n and then the second equality of Equation 1.110 by Ψ_m^* and take the difference of the results and integrate over the configuration coordinate q:

$$\int \Psi_n \hat{A}^* \Psi_m^* dq - \int \Psi_m^* \hat{A} \Psi_n dq = (A_m - A_n) \int \Psi_m^* \Psi_n dq \tag{1.112}$$

From the condition that the operator $\hat{A}$ is self-conjugate (Hermitian), it follows that

$$(A_m - A_n) \int \Psi_m^* \Psi_n dq = 0 \tag{1.113}$$

So the eigenfunctions of a Hermitian operator are **orthogonal**:

$$\int \Psi_m^* \Psi_n dq = 0 \tag{1.114}$$

if they have different eigenvalues: $A_m \neq A_n$.

So those eigenfunctions for which we have different eigenvalues are orthogonal. From this theorem, we easily identify orthogonal functions without having to integrate or conduct an analysis based on symmetry or other considerations. This result shows that **non-degenerate eigenfunctions of the same operator are orthogonal.**

If the eigenvalues of n_f eigenfunctions are the same, then the functions, Ψ_{nf}, are said to be **degenerate:**

$$A_n \to \Psi_{n1}, \Psi_{n2}, \ldots, \Psi_{nf} \tag{1.115}$$

The number, f, is the **degree (or multiplicity) of degeneracy**. Let us again consider Equation 1.114. If Ψ_n and Ψ_m are understood to be normalized functions, then

$$(A_m - A_n) = 0 \tag{1.116}$$

and for now, we may not have something to say about the integral $\int \Psi_m^* \Psi_n dq$.

The wave functions

$$\Psi_{n1}, \Psi_{n2}, \ldots, \Psi_{nf} \tag{1.117}$$

may or may not be orthogonal.

Consider degenerate wave functions Φ_n for which we have a linear combination:

$$\Phi_n = \sum_{\alpha=1}^{f} C_{n\alpha} \Psi_{n\alpha} \tag{1.118}$$

where Φ_n is also the eigenfunction of the operator $\hat{A}$ that has the eigenvalue A_n:

$$\hat{A}\Phi_n \equiv A_n \Phi_n \tag{1.119}$$

(The reader should prove this independently).

Any linear combination of degenerate functions is also the eigenfunction of that operator that has same eigenvalue as the degenerate functions. The notion of the degree (or multiplicity) of degeneracy is as follows: The number of linear independent eigenfunctions for which we have one eigenvalue.

If the degenerate functions in Equation 1.117 are not orthogonal, then from *f* linear probabilistic functions, we may represent *f* linear combinations:

$$\Phi_{n1}, \Phi_{n2}, \ldots, \Phi_{nf} \tag{1.120}$$

of the type in Equation 1.118 such that the new functions may be orthogonal. This follows the **orthogonalization of the probabilistic functions.** If we suppose that this operation is satisfied, then it may be assumed that Equation 1.114 is true for any two functions. For the condition of **orthonormalization** of the eigenfunctions of the operator $\hat{A}$, we have

$$\int \Psi_m^* \Psi_n \, dq = \delta_{mn} \tag{1.121}$$

So **the eigenfunctions of a Hermitian operator form an orthonormal system.** In principle, it is proven that such a system has the property of completeness or closeness: **An arbitrary state, Ψ, can be expanded in the complete set of eigenvectors, Ψ_n, of $\hat{A}$.**

$$\Psi(q) = \sum_n C_n \Psi_n(q), \; C_n = \int \Psi(q) \Psi_n^*(q) dq \tag{1.122}$$

where *n* may go to infinity. In this case, we only know the measurement of A will yield one of the values of A_n, but we do not know which one. However, we do know the probability that the eigenvalue A_n will occur is the absolute value squared of the coefficient, $|C_n|^2$.

So far, we have developed basic mathematical concepts consistent for the description of the state of a system where any physical quantity, say A, achieves only a finite number of discrete values (discrete spectrum). However, there exists many physical quantities capable of achieving a continuous range of values:

$$\nu_1 \le \nu \le \nu_2 \tag{1.123}$$

The eigenfunction of the operator $\hat{A}$ is dependent not only on the index but also on the continuous changing parameter ν. Then Equation 1.122 may be written in this case as follows:

$$\Psi(q) = \int_{\nu_1}^{\nu_2} C(\nu) \Psi(q, \nu) d\nu \tag{1.124}$$

which is the generalization of Equation 1.122 in our case. Let us examine the physical sense of $C(\nu)$. In Equation 1.122, $|C_n|^2$ is the probability of the result of measurements. In Equation 1.124, $|C(\nu)|^2 \, d\nu$ is the probability that the measurement of the physical quantity gives its value in the interval $d\nu$. From the law of summation of probabilities:

$$\int |C(\nu)|^2 d\nu \equiv \int |\Psi(q)|^2 dq = 1 \tag{1.125}$$

then substituting $\Psi^*(q)$ from Equation 1.124:

$$\int \Psi^*(q) \Psi(q) dq = \int \Psi(q) dq \int C^*(\nu) \Psi^*(q, \nu) d\nu = \int C^*(\nu) C(\nu) d\nu \tag{1.126}$$

and if we change the order of integration:

$$\int C^*(\nu) \left\{ \int \Psi(q) \Psi^*(q, \nu) dq \right\} d\nu = \int C^*(\nu) C(\nu) d\nu \tag{1.127}$$

then

$$C(\nu)=\int\Psi(q)\Psi^*(q,\nu)dq \tag{1.128}$$

Let us for $\Psi(q)$ in Equation 1.124, we swap $\nu\to\nu'$:

$$C(\nu)=\int C(\nu')d\nu'\int\Psi(q,\nu')\Psi^*(q,\nu)dq \tag{1.129}$$

and in order for this to be true, it is necessary that

$$\int\Psi(q,\nu')\Psi^*(q,\nu)dq=\delta(\nu-\nu') \tag{1.130}$$

For a continuous spectrum, the wave function is normalized to Dirac's delta function $\delta(\nu-\nu')$ as follows:

$$\int C(\nu')\delta(\nu'-\nu)d\nu'=C(\nu) \tag{1.131}$$

The Dirac delta function $\delta(\nu)$ **can also be studied in references** [12, 13]. **Equation** 1.130 **is the condition of orthonormalization of the wave function of a continuous spectrum** – the normalization by the delta-function. The difference between the orthonormalization of the functions of a discrete spectrum and continuous spectrum is that in place of the Kronecker delta-symbol δ_{mn} for the discrete spectrum, we have the Dirac's delta-function for the continuous spectrum. The property of the completeness of the eigenfunctions may be written in the following form:

$$\sum_n\Psi_n^*(q)\Psi_n(q')=\delta(q-q') \tag{1.132}$$

where q is understood as the totality of the phase coordinates.

Homework: **The reader should prove for the following case:**

$$\Psi=\sum_n C_n\Psi_n \tag{1.133}$$

It should be noted that

$$\delta(q-q')=\delta(q_1-q_1')\delta(q_2-q_2')\cdots\delta(q_f-q_f') \tag{1.134}$$

If the Hermitian operator has a continuous spectrum of eigenvalues, then the condition of completeness is as follows:

$$\int\Psi_n^*(\nu,q)\Psi_n(\nu,q')d\nu=\delta(q-q') \tag{1.135}$$

There are some operators that have a mixed spectrum, i.e., in the domain, say D, the spectrum is discrete and, in the domain, say D', it is continuous. Here the wave function of the discrete and that of the continuous spectrum separately do not form a complete system, and for a complete system we have their totality as follows:

$$\Psi(q)=\sum_n C_n\Psi_n(q)+\int_{\nu_1}^{\nu_2}C(\nu)\Psi(\nu,q)d\nu \tag{1.136}$$

We may see that the system of eigenfunctions of a self-conjugate operator is orthonormal and complete, i.e., Equation 1.121 for a discrete spectrum and Equation 1.130 for a continuous spectrum and the function $\Psi(q)$ may be expanded relative to the eigenfunction of the operator $\hat{A}$ as in Equation 1.135. From the conditions of orthonormalization in Equations 1.121 and 1.136 and also in Equation 1.122 for C_n and in Equation 1.128 for $C(\nu)$,

$$\sum_n |C_n|^2 + \int_{\nu_1}^{\nu_2} |C(\nu)|^2 \, d\nu = 1 \tag{1.137}$$

The expectation value $\bar{\nu}$ may then be evaluated as follows:

$$\bar{\nu} = \sum_n |C_n|^2 \nu_n + \int_{\nu_1}^{\nu_2} |C(\nu)|^2 \nu d\nu \tag{1.138}$$

It follows that $|C_n|^2$ and $|C(\nu)|^2 d\nu$ equal correspondingly the probabilities of finding in the state $\Psi(q)$ the value of the quantity $\nu = \nu_n$ (in the discrete spectrum) or value of the quantity ν in the interval ν to $\nu + d\nu$ (in the case of the continuous spectrum).

Question:
Find the eigenfunctions and eigenvalues of the operators:

$$\hat{A} = \frac{d}{dx}; \; \hat{A} = i\frac{d}{dx} \tag{1.139}$$

Answer:
The Sturm-Liouville equation for the eigenfunctions using the operator in Equation 1.139 is

$$\frac{d\Psi}{dx} = A\Psi \tag{1.140}$$

Its solution is

$$\Psi = C\exp\{Ax\} \tag{1.141}$$

where C is the normalization constant. In order for the wave function $\Psi(x)$ to be finite as $x \to \pm\infty$, we select

$$A = i\lambda \tag{1.142}$$

where λ is a real number and so

$$\Psi = C\exp\{i\lambda x\} \tag{1.143}$$

From the normalization condition of the continuous spectrum,

$$\int \Psi^*(x,\lambda)\Psi(x,\lambda')dx = \delta(\lambda - \lambda') = C^2 2\pi\delta(\lambda - \lambda') \tag{1.144}$$

It then follows that the wave function of the continuous spectrum $\Psi(x)$:

$$\Psi = \frac{1}{\sqrt{2\pi}}\exp\{i\lambda x\} \tag{1.145}$$

Answer:
Similarly, for the second operator in Equation 1.139, we have

$$\Psi = \frac{1}{\sqrt{2\pi}}\exp\{-i\lambda x\} \tag{1.146}$$

where λ is a real number. In this case, we also have a continuous spectrum of eigenvalues.

Question:
Find the eigenfunctions and eigenvalues of the operator:

$$\hat{A} = x + \frac{d}{dx} \tag{1.147}$$

Answer:
We write the Sturm-Liouville equation as follows:

$$\left(x + \frac{d}{dx}\right)\Psi = A\Psi \tag{1.148}$$

then

$$\frac{d\Psi}{\Psi} = (A - x)dx \tag{1.149}$$

and

$$\Psi(x) = C\exp\left\{Ax - \frac{x^2}{2}\right\} \tag{1.150}$$

This function should be:

1. **Finite**
2. **Continuous**
3. **Univalent (single-valued) for any A that may be real as well as complex (continuous spectrum)**

From the normalization condition of the function $\Psi(x)$ and from

$$\int_{-\infty}^{\infty}\exp\{-p^2x^2 \pm qx\}dx = \frac{\sqrt{\pi}}{p}\exp\left\{\frac{q^2}{4p^2}\right\}, \quad p > 0 \tag{1.151}$$

then

$$C^2\int_{-\infty}^{\infty}\exp\{2Ax - x^2\}dx = C^2\frac{\sqrt{\pi}}{1}\exp\left\{\frac{4A^2}{4}\right\} = 1 \tag{1.152}$$

and

$$\Psi(x) = \frac{1}{\pi^{\frac{1}{4}}}\exp\left\{Ax - \frac{x^2 + A^2}{2}\right\} \tag{1.153}$$

Question:
Find the eigenfunctions and eigenvalues of the operator:

$$\hat{A} = \frac{d}{d\phi} \tag{1.154}$$

Answer:
We solve again the Sturm-Liouville equation as follows:

$$\frac{d\Psi}{d\phi} = A\Psi \tag{1.155}$$

then

$$\Psi(\phi) = C\exp\{A\phi\} \tag{1.156}$$

Consider that the wave function $\Psi(\phi)$ should be single-valued:

$$\Psi(\phi) = \Psi(\phi + 2\pi) \tag{1.157}$$

Applying this to the wave function $\Psi(\phi)$, we have

$$\exp\{A2\pi\} = 1,\ A = im,\quad m = 0, \pm1, \pm2, \ldots \tag{1.158}$$

Question:
Find the eigenfunctions and eigenvalues of the operator $\hat{A}$:

$$\hat{A} = \frac{d^2}{dx^2} + \frac{2}{x}\frac{d}{dx} \tag{1.159}$$

Answer:
In order to bring the Sturm-Liouville problem to a differential equation easily solvable, we introduce a change of variable as follows:

$$\Psi = x\psi \tag{1.160}$$

$$\left(\frac{d^2}{dx^2} + \frac{2}{x}\frac{d}{dx}\right)\frac{\Psi}{x} = A\frac{\Psi}{x} \tag{1.161}$$

From

$$\frac{d}{dx}\left(\frac{\Psi}{x}\right) = -\frac{1}{x^2}\Psi + \frac{1}{x}\frac{d\Psi}{dx},\ \frac{d^2}{dx^2}\left(\frac{\Psi}{x}\right) = -\frac{2}{x^3}\Psi - \frac{2}{x^2}\frac{d\Psi}{dx} + \frac{1}{x}\frac{d^2\Psi}{dx^2} \tag{1.162}$$

then

$$\frac{d^2\Psi}{dx^2} = A\Psi \tag{1.163}$$

The characteristic equation:

$$\lambda^2 = A \tag{1.164}$$

The solution is therefore

$$\Psi_1 = C_1\exp\left\{x\sqrt{A}\right\},\quad \Psi_2 = C_2\exp\left\{-x\sqrt{A}\right\} \tag{1.165}$$

For

$$A=-\lambda^2<0 \tag{1.166}$$

In this case, the given wave functions will be finite as $x \to \pm\infty$. For the function $\Psi(x)$ to be finite at the point $x=0$, it is necessary to take the linear combination of both functions:

$$\Psi(x)=\frac{C_1\exp\{i\lambda x\}+C_2\exp\{-i\lambda x\}}{x} \tag{1.167}$$

At $x=0$:

$$C_1+C_2=0 \tag{1.168}$$

and

$$\Psi(x)=C\frac{\sin\lambda x}{x} \tag{1.169}$$

where λ is any real number. From the condition of normalization of the wave function $\Psi(x)$,

$$|C|^2\lambda\int_{-\infty}^{\infty}\frac{\sin^2\lambda x}{\lambda x^2}dx=|C|^2\lambda\pi=1 \tag{1.170}$$

and

$$\Psi(x)=\frac{1}{\sqrt{\lambda\pi}}\frac{\sin\lambda x}{x} \tag{1.171}$$

1.10 Theorem on the Commutation of Operators and Their Physical Application

In the state described by the wave function $\Psi(q)$, some physical quantity A may have discrete eigenvalues:

$$A_1, A_2, \ldots, A_n \tag{1.172}$$

and

$$\Psi_1, \Psi_2, \ldots, \Psi_n \tag{1.173}$$

corresponding eigenfunctions. Let us also be interested on the quantity B. In such a situation, these physical quantities may simultaneously have a defined value at one and the same state of the system. It is necessary to define for which conditions the operators $\hat{A}$ and $\hat{B}$ have a common eigenfunction: **The answer is given by the following theorem:**

A wave function can be simultaneously an eigenfunction of two different operators if those operators commute. So in order for the operators $\hat{A}$ and $\hat{B}$ to have a common system of eigenfunctions, it is necessary that these operators should commute with each other:

$$\hat{A}\hat{B}=\hat{B}\hat{A} \tag{1.174}$$

The operators acting on one and same variables are a trivial case.

Proof

Let $\Psi_n(q)$ has eigenvalues defined by the following equation:

$$\hat{A}\Psi_n(q) = A_n\Psi_n(q),\ \hat{B}\Psi_n(q) = B_n\Psi_n(q) \tag{1.175}$$

Necessity Condition

Let $\Psi_n(q)$ be taken arbitrarily and we use the property of completeness as follows:

$$\Psi(q) = \sum_n C_n\Psi_n(q) \tag{1.176}$$

We act the operator $\hat{A}\hat{B}$ on the wave function in Equation 1.176:

$$\hat{A}\hat{B}\Psi(q) = \hat{A}\sum_n C_n\hat{B}\Psi_n(q) \tag{1.177}$$

If we use the property of linearity of the operator $\hat{B}$, then we have

$$\hat{A}\sum_n C_n\hat{B}\Psi_n(q) = \hat{A}\sum_n C_nB_n\Psi_n(q) = \sum_n C_nB_n\hat{A}\Psi_n(q) = \sum_n C_nB_nA_n\Psi_n(q) \tag{1.178}$$

Let the given operators be acted in the reverse order:

$$\hat{B}\hat{A}\Psi(q) = \hat{B}\sum_n C_n\hat{A}\Psi_n(q) = \hat{B}\sum_n C_nA_n\Psi_n(q) = \hat{A}\sum_n C_nA_n\hat{B}\Psi_n(q) = \sum_n C_nA_nB_n\Psi_n(q) \tag{1.179}$$

But as A_n and B_n are numbers, it follows that

$$A_nB_n = B_nA_n \tag{1.180}$$

from where it follows that

$$\hat{B}\hat{A}\Psi(q) = \hat{A}\hat{B}\Psi(q) \tag{1.181}$$

and

$$\hat{B}\hat{A} = \hat{A}\hat{B} \tag{1.182}$$

Sufficiency Condition

Let us prove that if $\hat{B}\hat{A} = \hat{A}\hat{B}$, $\hat{A}$ and $\hat{B}$ have same eigenfunction. Suppose that

$$\hat{A}\Psi_n(q) = A_n\Psi_n(q) \tag{1.183}$$

We act from the left-hand side of Equation 1.183 on the operator $\hat{B}$ as follows:

$$\hat{B}\hat{A}\Psi_n(q) = A_n\hat{B}\Psi_n(q) \tag{1.184}$$

From the condition of commutation defined in Equation 1.182, it then follows:

$$\hat{A}\hat{B}\Psi_n(q) = A_n\hat{B}\Psi_n(q) \tag{1.185}$$

Let us make the denotation:

$$\hat{B}\Psi_n = \Phi_n \tag{1.186}$$

then it follows that

$$\hat{A}\Phi_n = A_n\Phi_n \tag{1.187}$$

and if we compare with the initial equality, we see that the function Φ_n is the eigenfunction of the operator $\hat{A}$ with the eigenvalue A_n that is also same for Ψ_n. This follows that we have two eigenfunctions for the operator $\hat{A}$ that have one and the same eigenvalue A_n. Consider:

1. A_n is non-degenerate, i.e., the eigenvalue has only one eigenfunction. But we have two eigenfunctions. This follows that our eigenfunctions differ only by a factor denoted by B_n:

$$\hat{B}\Psi_n = B_n\Psi_n \tag{1.188}$$

This follows the proof of the theorem.

If the operators $\hat{A}$ and $\hat{B}$ commute, then any non-degenerate function of the operator $\hat{A}$ is the eigenfunction of the operator $\hat{B}$.

2. If the eigenvalue A_n belongs to the eigenfunctions $\Psi_{n_1}, \Psi_{n_2}, \ldots, \Psi_{n_f}$, then we have $f+1$ degenerate functions since $\hat{B}\Psi_{n\alpha}$ is the linear combination of functions with f linear independent functions:

$$\hat{B}\Psi_{n\alpha} = \sum_{\beta} B_{\beta\alpha}\Psi_{n\beta} \tag{1.189}$$

This follows that $\Psi_{n\alpha}$ is not the eigenfunction of the operator $\hat{B}$. So the eigenfunctions of the operator $\hat{A}$ that belong to a set of degenerate functions may not be the eigenfunctions of the commutable operator $\hat{B}$.

Consider again

$$\hat{B}\Psi_{n\alpha} = \sum_{\beta} B_{\beta\alpha}\Psi_{n\beta} \tag{1.190}$$

and we multiply it by $\Psi^*_{n\gamma}$ and integrate over dq as follows:

$$\int \Psi^*_{n\gamma}\hat{B}\Psi_{n\alpha}\,dq = \sum_{\beta}\int \Psi^*_{n\gamma}B_{\beta\alpha}\Psi_{n\beta}\,dq = \sum_{\beta} B_{\beta\alpha}\int \Psi^*_{n\gamma}\Psi_{n\beta}\,dq \tag{1.191}$$

If we orthogonalize the set $\Psi_{n_1}, \Psi_{n_2}, \ldots, \Psi_{n_f}$, then

$$\int \Psi^*_{n\gamma}\Psi_{n\alpha}\,dq = \delta_{\gamma\beta} \tag{1.192}$$

and

$$\int \Psi^*_{n\gamma}\hat{B}\Psi_{n\alpha}\,dq = \sum_{\beta} B_{\beta\alpha}\delta_{\gamma\beta} = B_{\gamma\alpha} \tag{1.193}$$

If we write all the coefficients $B_{\gamma\alpha}$ in the form of a table, then we get the matrix, $\mathbb{B}$, of the operator $\hat{B}$ constructed on the eigenfunctions of the operator $\hat{A}$:

$$\mathbb{B} = \begin{bmatrix} B_{11} & B_{12} & \cdots & B_{1f} \\ B_{21} & B_{22} & \cdots & B_{2f} \\ \cdots & \cdots & \cdots & \cdots \\ B_{f1} & B_{f2} & \cdots & B_{ff} \end{bmatrix} \tag{1.194}$$

In the place of Φ, we select their linear combination X:

$$X = \sum_{\alpha} C_{\alpha} \Psi_{n\alpha} \tag{1.195}$$

and

$$\hat{B}X = BX \tag{1.196}$$

Then from Equation 1.196 considering 1.195, we have

$$\hat{B}X = \hat{B}\sum_{\alpha} C_{\alpha}\Psi_{n\alpha} = \sum_{\alpha} C_{\alpha}\hat{B}\Psi_{n\alpha} = B\sum_{\alpha} C_{\alpha}\Psi_{n\alpha} \tag{1.197}$$

If we multiply Equation 1.195 by $\Psi^{*}_{n\beta}$ and integrate over dq:

$$\sum_{\alpha} C_{\alpha}\int \Psi^{*}_{n\beta}\hat{B}\Psi_{n\alpha}\,dq = \sum_{\alpha} C_{\alpha}B_{\alpha\beta} = B\sum_{\alpha} C_{\alpha}\int \Psi^{*}_{n\beta}\Psi_{n\alpha}\,dq = B\sum_{\alpha} C_{\alpha}\delta_{\beta\alpha} = BC_{\beta} \tag{1.198}$$

or

$$\sum_{\alpha} C_{\alpha}\left[B_{\alpha\beta} - B\delta_{\beta\alpha}\right] = 0 \tag{1.199}$$

which is the system of equations for which we may find C_{α} where $\beta = 1,2,\ldots,f$.

The system in Equation 1.199 may also be expanded in the following form:

$$\begin{cases} C_1(B_{11} - B) + C_2B_{12} + \cdots + C_fB_{1f} = 0 \\ C_1B_{21} + C_2(B_{22} - B) + \cdots + C_fB_{2f} = 0 \\ \cdots \\ C_1B_{f1} + C_2B_{f2} + \cdots + C_f(B_{ff} - B) = 0 \end{cases} \tag{1.200}$$

This system has non-trivial solutions if its determinant is equal to zero (characteristic equation for the operator $\hat{B}$):

$$\begin{vmatrix} B_{11} - B & B_{12} & \cdots & B_{1f} \\ B_{21} & B_{22} - B & \cdots & B_{2f} \\ \cdots & \cdots & \cdots & \cdots \\ B_{f1} & B_{f2} & \cdots & B_{ff} - B \end{vmatrix} = 0 \tag{1.201}$$

If $B^{*}_{\beta\alpha} = B_{\alpha\beta}$, then we have f real roots that we substitute in Equation 1.199 in order to find C_{α}. The roots of this equation are the eigenvalues of the operator $\hat{B}$.

If $\hat{A}$ and $\hat{B}$ commute and the eigenfunction of $\hat{A}$ belongs to a degenerate set, then, in general, it is not the eigenfunction of $\hat{B}$. Let $\hat{A}$, $\hat{B}$ and $\hat{C}$ be three linear operators that, in general, do not commute with each other. So the following relations are true:

$$\left[\hat{A},\hat{B}\right] = -\left[\hat{B},\hat{A}\right], \left[\hat{A}\hat{B},\hat{C}\right] = \hat{A}\hat{B}\hat{C} - \hat{C}\hat{A}\hat{B} = \hat{A}\hat{B}\hat{C} - \hat{A}\hat{C}\hat{B} + \hat{A}\hat{C}\hat{B} - \hat{C}\hat{A}\hat{B} = \hat{A}\left[\hat{B},\hat{C}\right] + \left[\hat{A},\hat{C}\right]\hat{B} \tag{1.202}$$

$$\left[\hat{A},\hat{B}\hat{C}\right] = \left[\hat{B},\hat{A}\right]\hat{C} + \hat{B}\left[\hat{A},\hat{C}\right], \left[\hat{A}+\hat{B},\hat{C}\right] = \left[\hat{A},\hat{C}\right] + \left[\hat{B},\hat{C}\right], \left[\hat{A},\left[\hat{B},\hat{C}\right]\right] + \left[\hat{C},\left[\hat{A},\hat{B}\right]\right] + \left[\hat{B},\left[\hat{C},\hat{A}\right]\right] = 0 \tag{1.203}$$

1.11 Heisenberg Uncertainty Relations for Arbitrary Observables

The basic concept for any measurement theory is the uncertainty in the experimental realization knowledge of an observable when the system is in a given state. Under this chapter, we consider the uncertainty relations in a more general way where we consider two non-commuting Hermitian operators $\hat{A}$ and $\hat{B}$ representing two observables satisfying the following relation:

$$\left[\hat{A},\hat{B}\right]=i\hat{C} \tag{1.204}$$

Here, $\hat{C}$ is a **Hermitian operator,** called the **remainder of commutation (commutation rest**). From Equation 1.204, we have the following fundamental uncertainty relation in a more general way where all the expectation values are related to one and the same arbitrary state of the given system:

$$\overline{\left(\hat{A}-\overline{A}\right)^2}\,\overline{\left(\hat{B}-\overline{B}\right)^2}\geq\frac{\overline{\hat{C}^2}}{4} \tag{1.205}$$

This can easily be proven when examining the following integral that is positive definite and dependent on the real parameter α:

$$J(\alpha)=\int\left|\left(\alpha\hat{A}_1-i\hat{B}_1\right)\Psi\right|^2dq\equiv\int\left(\alpha\hat{A}_1^*+i\hat{B}_1^*\right)\Psi^*\left(\alpha\hat{A}_1-i\hat{B}_1\right)\Psi\,dq=\int\Psi^*\left(\alpha\hat{A}_1+\right.$$
$$\left.i\hat{B}_1\right)\left(\alpha\hat{A}_1-i\hat{B}_1\right)\Psi dq=\int\Psi^*\left(\alpha^2\hat{A}_1^2-i\alpha\left[\hat{A}_1,\hat{B}_1\right]+\hat{B}_1^2\right)\Psi\,dq=\alpha^2\overline{\hat{A}_1^2}-i\alpha\overline{\left[\hat{A}_1,\hat{B}_1\right]}+\overline{\hat{B}_1^2}\geq0 \tag{1.206}$$

We introduce the following abbreviations for the deviation Hermitian operators by the expectation value:

$$\hat{A}_1=\hat{A}-\overline{A},\quad B_1=\hat{B}-\overline{B} \tag{1.207}$$

Since $\overline{A}$ and $\overline{B}$ are just constants, these operators in Equation 1.207 represent two observables and obey the same commutation relations as the operators $\hat{A}$ and $\hat{B}$:

$$\left[\hat{A}_1,\hat{B}_1\right]=i\hat{C} \tag{1.208}$$

From here and considering Equation 1.208, we have **the Heisenberg† uncertainty principle in its most general form. So for all physical quantities with non-commuting operators, the uncertainties obey the inequality in Equation** 1.205. This result shows that for any pair of conjugate variables, there is a minimum of uncertainty associated with their measurement. For commuting operators, i.e., when, $\hat{C}=0$, we have no uncertainty relation for the corresponding physical quantities and so the quantities can be exactly measured simultaneously.

It is instructive to note that the uncertainty in Equation 1.205 should refer to simultaneous measurement of two observables, where the object is in a single state. This may possibly apply, say, to the momentum $\hat{p}_x$ and position $\hat{x}$ of a single object. For this, we consider the following:

$$\left(\alpha\hat{A}_1-i\hat{B}_1\right)\Psi=0 \tag{1.209}$$

So for the operators,

$$\hat{A}=\hat{x},\hat{B}=\hat{p}_x,\hat{C}=\hbar \tag{1.210}$$

we have the first-order differential equation:

$$\left(\alpha x - \hbar\frac{d}{dx}\right)\Psi = f\Psi,\ f = \alpha\bar{A} - i\bar{B} \tag{1.211}$$

and

$$\Psi_f = \Psi_0 \exp\left\{-\frac{f^2}{2\alpha\hbar}\right\}\exp\left\{\frac{(\alpha x - f)^2}{2\alpha\hbar}\right\} \equiv C_0 \exp\left\{\frac{(\alpha x - f)^2}{2\alpha\hbar}\right\} \tag{1.212}$$

We observe from the solution that the eigenfunction describes, however, a Gaussian probability distribution around the value f as the center. If $\alpha > 0$, for $x \to \pm\infty$, then the eigenfunction, Ψ_f, increases for any eigenvalue f. This type of eigenfunction has no physical sense. So the operator $\alpha x - \hbar\frac{d}{dx}$ does not have an eigenfunction. If $\alpha < 0$, then the eigenfunction Ψ_f decreases for $x \to \pm\infty$. In this case,

$$\Psi_f = C_0 \exp\left\{-\frac{(\alpha x - f)^2}{2|\alpha|\hbar}\right\} \equiv \tilde{C}_0 \exp\left\{\frac{ip_0 x}{\hbar} - \frac{(x - x_0)^2}{2\alpha^2}\right\} \tag{1.213}$$

and eigenfunction describes a Gaussian probability distribution around the value x_0 as the center. This eigenfunction is the explicit form of the wave function, which minimizes the uncertainty relation for the operator of the momentum and the coordinate.

It is useful to note that not all pairs of canonical variables in classical mechanics represented in quantum mechanics by the operators satisfying Equation 1.204 are subject to the uncertainty in Equation 1.205. It is instructive to be very prudent as the uncertainty in Equation 1.205 refers to simultaneous measurement of two observables with the object being in one single state. This may possibly apply, to say the momentum and position of a single object. In another sense, we may treat the time variable and energy of an object in classical mechanics as canonical variables. In this case, it would be wrong to infer from this that we cannot determine the energy, E, of an object exactly at a given time moment, t:

$$t\Delta E \geq \hbar \tag{1.214}$$

This refers to the measurements where the object is in different states, consistent with different time moments. This implies that the general proof of the uncertainty relation seen above is not applicable in this case. So the uncertainty relation for the energy should have a different origin.

1.12 Limiting Transition from Quantum Mechanics to Classical Mechanics

We examine under this heading the transition from quantum mechanics to classical mechanics and the conditions for this transition. In quantum mechanics, the state of an electron is described by a wave function, while in classical mechanics, it is completely determined by the equation of motion. Similarly, the link between quantum and classical mechanics in electrodynamics can relate wave optics and geometrical optics. For wave optics, the electromagnetic waves are described by the electric and magnetic field strength vectors satisfying the Maxwell equations while for geometric optics we consider the propagation of light rays through definite trajectories. This already gives an insight on the limiting transition from quantum to classical mechanics and similarly the passage from wave to geometrical optics. Such an analogy permits the passage from quantum to the classical mechanics limit to imitate the passage from wave to geometrical optics.

Though quantum mechanics contains classical mechanics as a limiting case, it requires at the same time the given limiting case for its formulation. Relating the principal role played by optics in quantum mechanics, we consider a particle with energy E moving in the potential field $U(\vec{r})$. One of the mathematical methods for the description of this classical motion is the **Jacobi-Hamilton equation**:

$$E = \frac{p^2}{2m} + U(\vec{r}),\ p(\vec{r}) = \sqrt{2m\left(E - U(\vec{r})\right)} \tag{1.215}$$

where $p(\vec{r})$ is the momentum of the given particle and m is its mass. Suppose $S_0(\vec{r})$ is the classical action of the given particle, the momentum vector can be calculated as follows:

$$\vec{p} = \nabla S_0 \tag{1.216}$$

From Equation 1.220, the **Jacobi-Hamilton equation** is as follows:

$$\left(\nabla S_0\right)^2 = p^2 = 2m\left(E - U(\vec{r})\right) \tag{1.217}$$

This is one of the most elegant approaches to Lagrangian systems such as geometrical optics and classical mechanics. It establishes the duality between trajectories and waves and naturally paves the way for quantum mechanics. From the Jacobi–Hamilton equation (Equation 1.217), we have

$$S_0(\vec{r}) = C = \text{const} \tag{1.218}$$

If we make the variation of C, then we have the totality of curves that form a totality of trajectories. The gradient of S_0 at a given point, i.e.,

$$\text{grad}S_0 = \nabla S_0 \tag{1.219}$$

is in the direction of the normal to the surface $f = \text{const}$.

Let us examine the geometrical optics of a medium with a refractive index $n(\vec{r})$. Considering a wave with a cyclical frequency ω and wave number $\kappa = \frac{\omega}{c}$, we examine the following wave equation for the given medium:

$$\Delta\Psi(\vec{r}) + n^2(\vec{r})\frac{\omega^2}{c^2}\Psi(\vec{r}) = 0 \tag{1.220}$$

We find its solution in the following form:

$$\Psi(\vec{r}) = \exp\left\{i\kappa\Theta(\vec{r})\right\} \tag{1.221}$$

then

$$\frac{\partial\Psi}{\partial x} = i\kappa\frac{\partial\Theta}{\partial x}\Psi,\quad \frac{\partial^2\Psi}{\partial x^2} = i\kappa\frac{\partial^2\Theta}{\partial x^2}\Psi - \kappa^2\left(\frac{\partial\Theta}{\partial x}\right)^2\Psi \tag{1.222}$$

and if we substitute this into Equation 1.220, then we have

$$i\kappa\Delta\Theta - \kappa^2\left(\nabla\Theta\right)^2 + n^2\kappa^2 = 0 \tag{1.223}$$

This is an exact equation and for the transition to geometrical optics, we take the limit $\kappa \to \infty$ from where it follows from Equation 1.223 that

$$(\nabla\Theta)^2 = n^2 \tag{1.224}$$

This is the main differential equation of geometrical optics – the **eikonal equation**. An eikonal is an unknown function, $\Theta(\vec{r})$. Equation 1.217 is equivalent to Equation 1.224. After the solution of Equation 1.224, we plot the family of curves of the eikonal and it gives us a family of rays like in Figure 1.3.

We show that we have a family of rays. In the expansion of $\Theta(\vec{r})$, we assume that the terms from the second derivative are infinitesimal:

$$\Theta(\vec{r}) = \Theta(\vec{r}_0) + (\vec{r} - \vec{r}_0, \nabla\Theta) + \cdots \tag{1.225}$$

Substituting Equation 1.224 into 1.221, we have

$$\begin{aligned}\Psi(\vec{r}) &= \exp\{i\kappa[\Theta(\vec{r}_0) + (\vec{r} - \vec{r}_0, \nabla\Theta)]\} = \exp\{i\kappa[\Theta(\vec{r}_0) - (\vec{r}_0, \nabla\Theta)]\}\exp\{i(\kappa\nabla\Theta, \vec{r})\} \\ &= C\exp\{i(\kappa\nabla\Theta, \vec{r})\} = C\exp\{i(\vec{\kappa}, \vec{r})\}\end{aligned} \tag{1.226}$$

The wave vector $\vec{\kappa}$ contains

$$\nabla\Theta = n(\vec{r}) \tag{1.227}$$

Equation 1.226 is the equation of a plane wave, as we expected. So it follows the notion of the ray.

The wave vector $\vec{\kappa}$ gives the direction of the ray as it is the direction of the normal. On the other hand,

$$\vec{\kappa} = \frac{\omega}{c}\nabla\Theta \tag{1.228}$$

which follows that $\vec{\kappa}$ is in the direction of the normal to the surface:

$$\Theta = \text{const} \tag{1.229}$$

Similarly, this is between the trajectories of a particle moving with respect to the law of classical mechanics and the distribution of the ray in geometrical optics – the optical-mechanical analogy. So geometrical optics is a particular case of wave optics and we may conclude that classical mechanics is the limiting case of wave and not quantum mechanics.

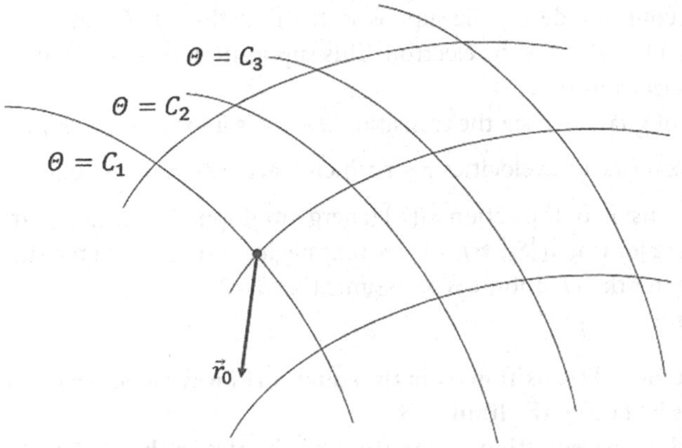

FIGURE 1.3 The family of curves of the eikonal giving a family of rays.

Consider a monochromatic wave distributed in a non-homogenous medium described by the following equation:

$$\Delta\Psi(\vec{r})+\kappa^2 n^2(\vec{r})\Psi(\vec{r})=0,\quad \kappa=\frac{\omega}{c} \tag{1.230}$$

and

$$\Psi=\exp\{i\kappa\Theta\} \tag{1.231}$$

where $\Theta(\vec{r})$ is the eikonal. For the limiting case, $\kappa\to\infty$, we have Equation 1.223 which is the eikonal equation. This is analogous to the equation of the trajectory in the form of the Jacobi-Hamilton equation in Equation 1.217. de Broglie and Schrödinger made a similar analogy as geometrical or wave optics is a limiting case of wave mechanics. Thus, the superposition – classical mechanics – is the limiting case of wave mechanics.

If from the eikonal equation, we move to the wave equation, then we suppose that it has the following form:

$$\Delta\Psi(\vec{r})+\kappa^2\Psi(\vec{r})=0 \tag{1.232}$$

where $\vec{\kappa}$ is the de Broglie wave vector. The limiting transition corresponds to the plane wave:

$$\Psi(\vec{r})\cong C\exp\{i(\vec{\kappa},\vec{r})\} \tag{1.233}$$

Vector $\vec{\kappa}$ is parallel to $\nabla\Theta_0$. The transition from the wave to the classical mechanics corresponds to the wave in Equation 1.226, where $\vec{\kappa}$ is parallel to $\vec{p}=\nabla S_0$. This follows that

$$\vec{\kappa}=\frac{\vec{p}}{\hbar} \tag{1.234}$$

where $\hbar$ is a constant which is a coefficient of proportionality between the wave vector $\vec{\kappa}$ of the de Broglie wave and the momentum $\vec{p}$ of the particle (the link between wave and corpuscular properties). Equation 1.234, which is true for the electron, is also true for a photon where $\vec{p}$ is the momentum of the photon and $\vec{\kappa}$ is the wave vector of the electromagnetic wave. This formula was obtained by Einstein with

$$\hbar=\frac{h}{2\pi} \tag{1.235}$$

where h is the Planck constant. de Broglie supposed that Equations 1.234 and 1.235 are universally true not only for the photon but also for the electron. This supposition was confirmed by the experiment of the diffraction of the electrons [15].

So the unique way of guaranteeing the transition $\kappa\to\infty$ is $h\to 0$: (Similarly as in mechanics $c\to\infty$) this implies that we examine the velocities v for which $v\ll c$ where in the equations $\frac{v}{c}\to 0$. The quantity h here has the dimension of the action $S_0(\vec{r})$ (energy multiplied by time), and $h\to 0$ implies that we should examine the case for which $|S_0|\gg h$ and we may neglect h as a characteristic of the given problem, the action is much greater than h and we may assume that $h=0$.

So we have the result:

1. **We may make a formal transition from the equation of wave mechanics to the equation of classical mechanics by taking the limit $h\to 0$.**
2. **We may write the wave equation in quantum mechanics in the explicit form by replacing $\vec{\kappa}$ in Equation 1.234 by $\vec{p}$ and thus**

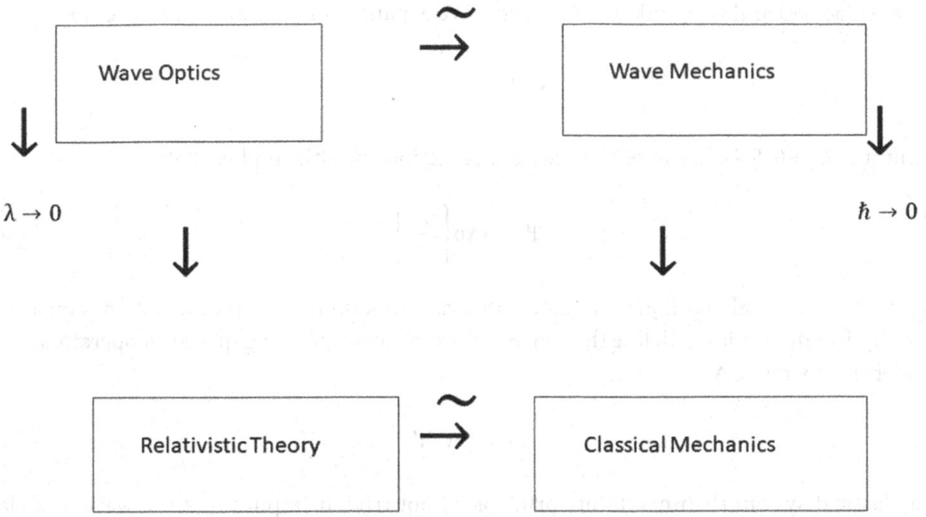

FIGURE 1.4 The transition from wave optics to wave mechanics and relativistic theory and then wave mechanics to classical mechanics.

$$\Delta\Psi + \frac{2m\left(E - U(\vec{r})\right)}{\hbar^2}\Psi = 0 \tag{1.236}$$

This is the **time-independent Schrödinger equation published for the first time by *E.* Schrödinger [20].** This equation is also satisfied by the wave function of the electron, which moves in the potential field, $U(\vec{r})$. Equation 1.236 only makes sense when $U(\vec{r})$ is independent of time. This implies that the force field is constant in time. The development of Equation 1.241 appeared for the first time after the results of de Broglie (Equations 1.234 and 1.235). But Equations 1.234, 1.235 and

$$\Psi(\vec{r}) = C\exp\left\{i(\vec{\kappa},\vec{r})\right\} \tag{1.237}$$

are related to the motion of the particle that was described by de Broglie, which is the result of the Schrödinger equation for a free particle. Equation 1.236 was not developed but written from analogy with the help of Figure 1.4.

It follows that Equation 1.236 is a postulate that should be verified by comparing the effect with the experiment. Further, we write Equation 1.236 as follows:

$$-\frac{\hbar^2}{2m}\Delta\Psi + U(\vec{r})\Psi = E\Psi \tag{1.238}$$

In such a form, it is a particular case of the Sturm-Liouville equation.

We suppose that the transition from quantum mechanics to classical mechanics is such that $h \to 0$ and the Schrödinger equation is transformed to the Jacobi-Hamilton equation. Quantum mechanics is also wave mechanics. Factually, we examine how we can make a transition from the Schrödinger equation to the Jacobi-Hamilton equation. The solution of the Schrödinger equation can be found in the following form:

$$\Psi = \exp\left\{i\frac{S}{\hbar}\right\} \tag{1.239}$$

Here, S replaces Θ and is an unknown function. We expand S in a series with respect to $\hbar$:

$$S = S_0 + \frac{\hbar}{i}\sigma_1 + \cdots \quad (1.240)$$

For the limit as $\hbar \to 0$, $S \to S_0$, where S_0 is the classical action and this implies that

$$\Psi_{cl} = \exp\left\{ i\frac{S_0}{\hbar} \right\} \quad (1.241)$$

where cl denotes classical and Equation 1.241 is the wave function of a classical system. Equation 1.241 provides physical quantities to define the concrete form of corresponding quantum operators.

Consider the operator, $\hat{A}$:

$$\hat{A}\Psi_{cl} = A_{cl}\Psi_{cl} \quad (1.242)$$

As for a classical system the uncertainty principle is not true, it implies that the wave of a classical function is a proper (eigen) function for all quantities A_{cl}. There are systems that approximate to the classical type, i.e.,

$$\hbar < |S_0| \quad (1.243)$$

but for which $\hbar$ is not so less than $|S_0|$ that we may neglect quantum effects. We call such systems **quasi-classical.** For such systems, we may get solutions that are proportional to the solution of the Schrödinger equation. So we consider the next terms after S_0 in the expansion of S in Equation 1.240. The quasi-classical operators for the limiting case lead simply to the multiplication by the corresponding physical quantity.

2 Schrödinger Equation

2.1 Stationary States

In the previous chapter, we did not examine the change of a state with time t. It has been assumed that we have been examining a state at a given moment t. In the general case, the state of a system changes with time t. For example, say the motion of an electron in a potential field, its wave function depends on time t: $\Psi = \Psi(q,t)$. Such a dependence has to be described by a differential equation in which we may have $\frac{\partial \Psi}{\partial t}, \frac{\partial^2 \Psi}{\partial t^2}, \ldots$.

Considering the postulates of quantum mechanics, we evaluate the form of the differential equation and we use:

1. **Principle of reason**. It may be formulated: **The wave function of a system completely describes its properties including its properties at a given moment t as the properties during the evolution of the system with time.**

   If we are given the initial state $\Psi(q,t_0)$ at the initial moment t_0, then the state $\Psi(q,t)$ of the system at the moment t should be defined univalently. The equation that we sort for should have such a structure such that from it, it should be possible to find univalently the state $\Psi(q,t)$. This follows that it should contain the derivative over time to the first order. From the principle of reason, the wave function or state function of a system evolves in time t according to the time-dependent Schrödinger equation that is the first-order differential equation with respect to time t:

$$i\hbar \frac{\partial \Psi}{\partial t} = \hat{H}\Psi \tag{2.1}$$

   where $\hat{H}$ is an operator acting on the wave function Ψ, and in order to find it, we should note that Equation 2.1 should be linear, considering:

2. **Principle of superposition**: If Ψ is the solution of Equation 2.1, then $C\Psi$ is also a solution and $C = \text{const}$. It follows that the linear operator $\hat{H}$ may be written in its complex conjugate as follows:

$$-i\hbar \frac{\partial \Psi^*}{\partial t} = \hat{H}^*\Psi^* \tag{2.2}$$

Let us multiply Equation 2.1 by Ψ^* and Equation 2.2 by Ψ and take the difference of the results and integrate over the configuration coordinates q of the system:

$$i\hbar \frac{\partial}{\partial t}\int \Psi\Psi^* \, dq = i\hbar \frac{\partial}{\partial t} \mathrm{I} = \int \Psi^* \hat{H}\Psi \, dq - \int \Psi \hat{H}^* \Psi^* \, dq \tag{2.3}$$

DOI: 10.1201/9781003273073-3

As the integral I is independent of time, then

$$\int \Psi^* \hat{H}\Psi \, dq = \int \Psi \hat{H}^* \Psi^* \, dq \tag{2.4}$$

That is the condition for which $\hat{H}$ is self-conjugated. It follows that $\hat{H}$ is a Hermitian operator. It follows also for the operator $\hat{H}$.

3. **Principle of the limiting transition.**

As $\hbar \to 0$, we have the principle of correspondence for which considering Equation 2.1 we have

$$\Psi \to \Psi_{cl} \equiv \exp\left\{ i\frac{S_0(q,t)}{\hbar} \right\} \tag{2.5}$$

and

$$-\frac{\partial S_0}{\partial t}\Psi_{cl} = \hat{H}\Psi_{cl} \tag{2.6}$$

From classical mechanics, it follows that

$$\frac{\partial S_0}{\partial t} = -\hat{H}_{cl} \tag{2.7}$$

That is, the classical Hamilton function for the given system. If we substitute Equation 2.7 into 2.6:

$$\hat{H}\Psi_{cl} = H_{cl}\Psi_{cl} \tag{2.8}$$

The operator $\hat{H}$ in classical mechanics corresponds to the Hamilton function and in quantum mechanics is the total energy operator or the so-called Hamiltonian operator that consists of the kinetic plus the potential energy operator. The Hamiltonian operator, $\hat{H}$, is named after the Irish mathematician William Hamilton. This comes from his formulation of classical mechanics based on the total energy. In most of the cases, only the kinetic energy of the particles and the electrostatic or Coulomb potential energy due to their charges are considered. However, generally all terms contributing to the energy appear in the Hamiltonian.

So we have established the form of the differential equation for the wave function $\Psi(q,t)$. Equation 2.1 is called the **time-dependent Schrödinger equation**. In the general case, it is a complex differential equation as $\hat{H}$ is a differential operator.

Let us examine the particular case when the Hamiltonian $\hat{H}$ of a physical system is not dependent explicitly on time. We call such a system **conservative**. In such a case, the solution of the time Schrödinger equation may be found in the form of the method of separation of variables, i.e., in the following form:

$$\Psi(q,t) = \Psi(q)\Phi(t) \tag{2.9}$$

If we substitute Equation 2.9 into 2.1:

$$i\hbar\Psi\frac{\partial \Phi}{\partial t} = \Phi\hat{H}\Psi \tag{2.10}$$

From where, it follows that

$$\frac{i\hbar}{\Phi}\frac{\partial \Phi}{\partial t} = \frac{\hat{H}\Psi}{\Psi} = E \tag{2.11}$$

Here $\frac{i\hbar}{\Phi}\frac{\partial\Phi}{\partial t}$ is a function of only the time t and $\frac{\hat{H}\Psi}{\Psi}$ is the function of only the coordinate. It follows that Equation 2.11 that

$$\hat{H}\Psi = E\Psi \tag{2.12}$$

This is the Sturm-Liouville equation for the energy E and $\hat{H}$ is the operator of the energy (Hamiltonian) or the Hamilton function that is the energy expressed as a function of the coordinate and the momentum. It follows that E is the eigenvalue of the energy of the system for the energy operator $\hat{H}$. Equation 2.12 is called the time-independent Schrödinger equation.

A physical sense has those solutions for which the wave function:

1. **Finite**
2. **Continuous**
3. **Univalent in the entire space**
4. **Twice differentiable**

We denote by $E_1, E_2, \ldots, E_n$ those values of E for which relation in Equation 2.12 is satisfied. These are the eigenvalues of the energy of the system. If we consider the equation:

$$\frac{d\Phi}{\Phi} = -i\frac{E}{\hbar}dt \tag{2.13}$$

then

$$\Phi = \exp\left\{-i\frac{E}{\hbar}t\right\} \tag{2.14}$$

If the conservative system has a discrete energy spectrum, Equation 2.1 has a solution in the following form:

$$\Psi_n(q,t) = \Psi_n(q)\exp\left\{-i\frac{E_n}{\hbar}t\right\} \tag{2.15}$$

where $\Psi_n(q)$ is the solution of the stationary Schrödinger equation (Equation 2.12) or simply the Schrödinger equation whose state is described by a stationary state. The wave function of a stationary state periodically depends on time t. The probability density

$$|\Psi_n(q,t)|^2 = |\Psi_n(q)|^2 \tag{2.16}$$

is independent of time t.

From the point of view of classical electrodynamics, the orbital model (planetary model) of an atom may not exist.

From Equation 2.12, we have

$$\hat{H} = -\frac{\hbar^2}{2m}\Delta + U(\vec{r}) \tag{2.17}$$

which is the explicit form of the operator $\hat{H}$. So, considering Equation 2.17, then Equation 2.12 describes the motion of a single particle in a potential field $U(\vec{r})$. Hence, Equation 2.12 is a general equation for a stationary motion.

It should not be taken that the conservative system should necessary be in the stationary form. From mathematics, Equation 2.15 is a particular solution of the Schrödinger equation. The general solution is the following superposition:

$$\Psi(q,t)=\sum_n C_n\Psi_n(q)\exp\left\{-i\frac{E_n}{\hbar}t\right\} \tag{2.18}$$

where C_n is independent of time t. As $\Psi_n(q)$ is the eigenfunction of the operator of the energy $\hat{H}$, then it follows that Equation 2.15 describes the state in which the energy has a defined value. However, the conservative system in the general case is in the state consistent with Equation 2.17 in which its energy does not have a defined value. The quantity $|C_n|^2$ considering the principle of superposition is the probability of the result of measurement of the energy. It is the probability that if in the state Ψ we measure the energy, then we have E_n. It follows that for the conservative system, the probability of measurement of the energy does not change with time t and its expectation (mean) value:

$$\bar{E}=\sum_n |C_n|^2 E_n \tag{2.19}$$

which is the mathematical expectation. It is the formulation of the law of conservation of energy in quantum mechanics.

It should be noted that C_n might be dependent on time t for the case

$$\hat{H}=\hat{H}_0+\hat{H}_1(t) \tag{2.20}$$

Here, $\hat{H}_0$ is independent of time t and $\hat{H}_1(t)$ is dependent on time t. For such a case, the system is not conservative. The system for which we have $\hat{H}_0$ is conservative.

We examine how the wave function of the classical system depends on time t:

$$\Psi_{cl}\equiv\exp\left\{i\frac{S_0(q,t)}{\hbar}\right\},\ S_0(q,t)=\overline{S_0(q)}-Et \tag{2.21}$$

So, we have exactly the same form as in Equation 2.18:

$$\Psi_{cl}(q,t)=\Psi(q)\exp\left\{-i\frac{E}{\hbar}t\right\} \tag{2.22}$$

2.1.1 Particle in an Infinite Deep Potential Well

We investigate the problem of an infinite deep potential well due to its importance in quantum mechanics. Here, we examine the solutions for a particle of mass m confined to move along the x-axis between 0 and a. At the ends 0 and a of the interval, there are hard walls that prevent the particle from going to $x>a$ and $x<0$. This can be achieved by making the potential 0 between $x=0$ and $x=a$ and $U(x)=\infty$ for $x<0$ and $x>a$ (see Figure 2.1a, 2.1b and 2.1c), i.e.,

$$U(x)=\begin{cases}0, & 0\le x\le a\\ \infty, & x<0,\ \ x>a\end{cases} \tag{2.23}$$

Such a model in quantum mechanics is referred to as particle in a box of length a.

Before solving the Schrödinger equation, we apply de Broglie approach to the given problem where de Broglie associates a wave with every material particle traveling with momentum p and wavelength:

$$\lambda=\frac{2\pi\hbar}{p} \tag{2.24}$$

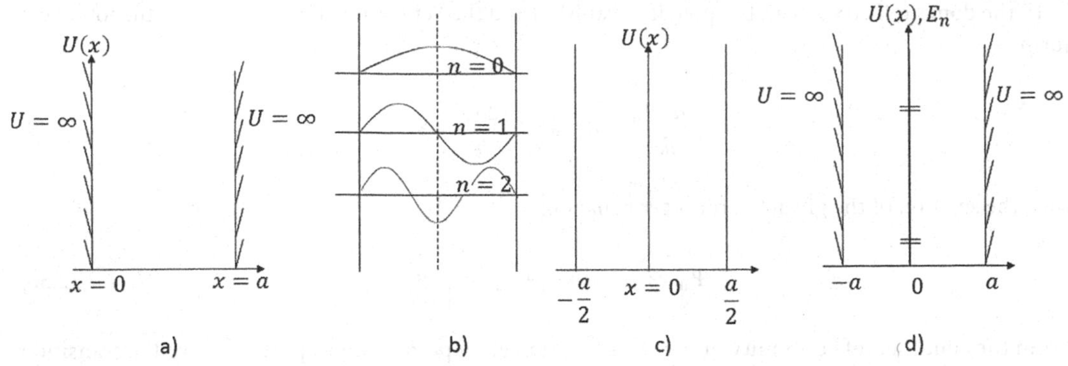

FIGURE 2.1 (a) The infinite square well shifted to the left to make it symmetric about the origin; (b) the three lowest energy eigenstates for the infinite square well potential, while (c) the infinite square well shifted to the left to make it symmetric about the origin; (d) energy levels of a particle in an infinitely narrow and tall potential barrier.

Classically, a string that is fixed at both ends imitates a particle in a box and plucking such a string, the amplitude of the oscillations at the fixed ends achieves the value zero. This implies that an integer number of half-wavelengths must fit in the length of the box:

$$n\frac{\lambda}{2}=n\frac{2\pi\hbar}{2p}=a,\ p=n\frac{2\pi\hbar}{2a},\ n=1,2,\ldots \tag{2.25}$$

Since the particle feels no potential energy, all the energy is in the kinetic energy:

$$E=\frac{p^2}{2m}=\frac{n^2\pi^2\hbar^2}{2ma^2} \tag{2.26}$$

We now analyze this problem quantum mechanically by finding the energy levels and the normalized wave functions of the stationary states of the particle of mass, m, in an infinite deep potential well of breadth a. We now examine the situation where $U=0$ for $\mathrm{II}(0\le x\le a)$ and the domains $\mathrm{I}(x<0)$ and $\mathrm{III}(x>a)$ for which $U=\infty$. The points $x=0$ and $x=a$ should be perfectly rigid impermeable walls classically. We show that the wave function must vanish at these points since the potential energy is infinite at these points. Considering the Schrödinger equation in these domains:

$$\hat{\mathrm{H}}\Psi=E\Psi,\quad \hat{\mathrm{H}}=-\frac{\hbar^2}{2m}\Delta+U(x) \tag{2.27}$$

then

$$\frac{1}{\Psi_{\mathrm{I}}}\frac{d^2\Psi_{\mathrm{I}}}{dx^2}=\frac{2m}{\hbar^2}(U-E) \tag{2.28}$$

We see that for $U=\infty$, the function Ψ_{I} should tend to zero. So, we observe that the wave function must vanish in the region where the potential is infinite. Classically, any region where the potential exceeds the energy of the particle is forbidden; likewise, in quantum mechanics, due to the Tunnel effect, it is not so.

In the domain II $(0 \le x \le a)$, the potential vanishes and the Schrödinger equation takes the following form:

$$\frac{d^2\Psi_{II}}{dx^2} + \kappa^2\Psi_{II} = 0, \frac{2mE}{\hbar^2} = \kappa^2 \tag{2.29}$$

then the solution of the given Schrödinger equation:

$$\Psi_{II} = C_1 \exp\{i\kappa x\} + C_2 \exp\{-i\kappa x\} \tag{2.30}$$

From the condition of continuity of the wave function at all points and in particular for the transition from domain I to II $(x = 0)$ and from II to III $(x = a)$, as the wave function must be continuous, it should vanish at $x = 0$ and at $x = a$:

$$\Psi_{I}(0) = \Psi_{II}(0), \quad \Psi_{II}(a) = \Psi_{III}(a) \tag{2.31}$$

But

$$\Psi_{I} = \Psi_{III} = 0 \tag{2.32}$$

as in those domains there is no particle. These are boundary conditions for the given problem. We look at the problem about the continuity of the first derivative $\Psi'(x)$. This derivative vanishes outside the interval $[0,a]$. The continuity would say that $\Psi'(x)$ vanish at 0 and at a which is impossible. A solution of the Schrödinger equation where both the wave function and its derivative vanish at a point is identically zero. If such a solution exists, then $\Psi'(x)$ should have discontinuities at an infinite wall. So we do not impose any boundary condition on $\Psi'(x)$ and the above two conditions will suffice to find a solution, where $\Psi'(x)$ is discontinuous at the endpoints.

It should also be recollected that for $U = \infty$, the wave function tends to zero:

$$\Psi_{I}(0) = \Psi_{II}(0) \equiv C_1 + C_2 = 0, \Psi_{II}(a) = \Psi_{III}(a) \equiv C_1 \exp\{i\kappa a\} + C_2 \exp\{-i\kappa a\} = 0 \tag{2.33}$$

then

$$C_1 \exp\{i\kappa a\} + C_2 \exp\{-i\kappa a\} = C(\exp\{i\kappa a\} - \exp\{-i\kappa a\}) = 2iC \sin \kappa a = 0 \tag{2.34}$$

From here,

$$\kappa_n a = n\pi, \kappa_n^2 = \frac{2mE_n}{\hbar^2}, E_n = \frac{n^2\pi^2\hbar^2}{2ma^2} \tag{2.35}$$

It is evident that $n = 0$ gives the physically uninteresting result $\Psi_{II} = 0$ or, in other words, there is no particle. So we do the following transformation $n \to n' + 1$. It will now be possible to set $n = 0$ that will give the physical meaning at the ground state. For the given transformation, $n - 1$ coincides with the number of zeros (nodes) of the n eigenfunction $\Psi_n(x)$. This does not include the zeros for $x \to \pm\infty$ or at the potential walls such as the points $x = 0$ and $x = a$. So the stationary states of the particle correspond to the energy:

$$E_n = \frac{(n+1)^2 \pi^2\hbar^2}{2ma^2}, \quad n = 0,1,2,\ldots \tag{2.36}$$

This energy has dimensional quantization and as

$$\Psi_{\text{II}}(x)=C'\sin\frac{(n+1)\pi x}{a} \tag{2.37}$$

then from the condition of normalization of the wave function, we have

$$\int_0^a |\Psi_{\text{II}}(x)|^2\,dx=|C'|^2\int_0^a \sin^2\frac{(n+1)\pi x}{a}dx=1 \tag{2.38}$$

and the wave functions:

$$\Psi_n(x)=\begin{cases}\sqrt{\dfrac{2}{a}}\sin\dfrac{(n+1)\pi x}{a}, & 0\le x\le a\\ 0, \quad x<0, \quad x>a & \end{cases} \tag{2.39}$$

It is instructive to note that the energies obtained by solving the Schrödinger equation coincide with that obtained via qualitative analysis with the help of the de Broglie principle. We observe that each value of n yields a different energy, implying no degeneracies in the energy spectrum for the one-dimensional infinite square well potential. The ground state – the lowest energy state – corresponds to $n=0$ and has non-zero energy,

$$E_0=\frac{\pi^2\hbar^2}{2ma^2} \tag{2.40}$$

which is the so-called **zero-point energy**. This implies even when the system is in the ground state, it has a perpetual motion. The zero-point energy is a result predicted by the uncertainty principle.

With the help of the distribution function,

$$dW_n(x)=|\Psi_n(x)|^2\,dx \tag{2.41}$$

we find

$$\overline{x}=\frac{a}{2},\ \overline{x^2}=a^2\left[\frac{1}{3}-\frac{1}{2\pi^2(n+1)^2}\right],\ \overline{(\Delta x)^2}=a^2\left[\frac{1}{12}-\frac{1}{2\pi^2(n+1)^2}\right] \tag{2.42}$$

We also find

$$\overline{p}=\int_{-\infty}^{\infty}\Psi_n^*(x)\hat{p}\Psi_n(x)dx=0,\ \overline{(\Delta p)^2}=\overline{p^2}=\int_{-\infty}^{\infty}|\hat{p}\Psi_n(x)|^2\,dx=\frac{\hbar^2\pi^2(n+1)^2}{a^2} \tag{2.43}$$

This permits us to have the following uncertainty relation:

$$\overline{(\Delta x)^2}\,\overline{(\Delta p)^2}=\hbar^2\left[\frac{\pi^2(n+1)^2}{12}-\frac{1}{2}\right] \tag{2.44}$$

For the inversion of the coordinate relative to the center of the well, the wave function is transformed:

$$\Psi_n'(x) \equiv \Psi_n(x') = \Psi_n(-x+a) = \sin\kappa_n(-x+a) = \sin(\kappa_n a - \kappa_n x) = \sin\left[(n+1)\pi - \kappa_n x\right]$$
$$= -\cos(n+1)\pi \sin\kappa_n x = (-1)^n \sin\kappa_n x = (-1)^n \Psi_n(x) \quad (2.45)$$

The wave function has a defined parity equal to $(-1)^n$. Figure 2.1b shows the first three solutions to the one-dimensional infinite square well, labeled from $n=0$ to $n=2$, where we consider the following properties:

1. The solutions are alternately symmetric and anti-symmetric about the midpoint $x=a$. The ground state is symmetric under reflection about $x=\frac{a}{2}$, while the first excited state, anti-symmetric with node, equals $\frac{a}{2}$. The second excited state again is symmetric and so symmetry and anti-symmetry continue to alternate. Symmetry in the given case is not accidental and, in general, holds for potentials $U(x)$ that are even functions of x which is not the case for our potential.
2. The ground state $n=0$ has no node which is a zero of the wave function, including the ends of the interval, $[0,a]$. So the zeroes at $x=0$ and $x=a$ do not count as nodes. Since $\Psi_0(x)$ does not vanish anywhere at the interior of $[0,a]$, it has no nodes confirming any normalizable ground state of a one-dimensional potential does not have nodes. The first excited state, $n=1$, has one node at $x=a$ (midpoint of the interval), while the second excited state, $n=2$, has two nodes. The sequence continues till the n-th excited state with n nodes.
3. The wave functions $\Psi_n(x)$ form a complete set and can be used to expand any function that vanishes at endpoints of the interval, $[0,a]$.

We find the wave function in the so-called momentum representation:

$$C(p) = \int_0^a \Psi_n(x)\Psi_p^*(x)dx = -\frac{2\sqrt{\pi a\hbar^2}(n+1)}{p^2a^2 - \pi\hbar^2(n+1)^2}\exp\left\{-\frac{ipa}{2\hbar}\right\}\begin{cases}\cos\left(\frac{pa}{2\hbar}\right), & n-\text{even}\\ i\sin\left(\frac{pa}{2\hbar}\right), & n-\text{odd}\end{cases} \quad (2.46)$$

This is sometimes referred to as the momentum wave function and for the state with energy $E_1 = \frac{4\pi^2\hbar^2}{2ma^2}$, the momentum wave function, $C(p)$:

$$C(p) = -\frac{4\sqrt{\pi a\hbar^2}}{p^2a^2 - 4\pi\hbar^2} i\sin\left(\frac{pa}{2\hbar}\right)\exp\left\{-\frac{ipa}{2\hbar}\right\} = \frac{2\sqrt{\pi a\hbar^2}}{p^2a^2 - 4\pi\hbar^2}\left(\exp\left\{-\frac{ipa}{2\hbar}\right\} - 1\right) \quad (2.47)$$

The probability (distribution function) of the particle is observed to have a momentum in the range p to $p+dp$:

$$dW(p) = |C(p)|^2 dp = \frac{4\pi a\hbar^2 \sin^2\left(\frac{pa}{2\hbar}\right)}{\left(4\pi\hbar^2 - p^2a^2\right)^2}dp \quad (2.48)$$

2.1.2 A Particle in an Infinitely High Potential Well

Apply an infinite deep potential well to the case of a particle of mass, m, captured in a box with potential, U:

$$U = \begin{cases} 0, & 0 \le x \le a, \ 0 \le y \le b, \ 0 \le z \le c \\ \infty, & \text{elsewhere} \end{cases} \quad (2.49)$$

Inside the box, we have $U = 0$. Considering the method of separation of variables, the wave function relative to each direction of the coordinate axis has the following form:

$$\Psi(x,y,z) = \Psi_1(x)\Psi_2(y)\Psi_3(z) \tag{2.50}$$

Each of the wave functions $\Psi_1(x)$, $\Psi_2(y)$ and $\Psi_3(z)$ has the form as in Equation 2.49. Outside the potential well, the wave function must vanish as we can draw an analogy for the one-dimensional case. This is due to the fact that the potential U is infinitely large ($U = \infty$). As the wave functions have to be smooth, we match the wave functions at the boundary of the potential well as for the one-dimensional case:

$$\Psi_1(0) = \Psi_1(a) = 0, \quad \Psi_2(0) = \Psi_2(b) = 0, \quad \Psi_3(0) = \Psi_3(c) = 0 \tag{2.51}$$

This gives the following quantization condition considering the three coordinate axes:

$$\kappa_1 = n_1\frac{\pi}{a}, \quad \kappa_2 = n_2\frac{\pi}{b}, \quad \kappa_3 = n_3\frac{\pi}{c} \tag{2.52}$$

Here $|n_i| = 1, 2, \ldots$ are independent quantum numbers. If we consider the one-dimensional case, the possibility of choosing $n_i = 0$ must be excluded. This is because the corresponding wave function would vanish everywhere. From this argument, we arrive at the total energy with only discrete values, namely

$$E_{n_1n_2n_2} = \frac{\hbar^2}{2m}\left[\left((n_1+1)\frac{\pi}{a}\right)^2 + \left((n_2+1)\frac{\pi}{b}\right)^2 + \left((n_3+1)\frac{\pi}{c}\right)^2\right], |n_i| = 0,1,2,\ldots \tag{2.53}$$

We may convert this discrete energy spectrum into a **quasi-continuum** when the mass m or the extension of the box becomes very large. The lowest energy value E_{000} is not zero. This is expected classically and is an example of the **non-vanishing zero-point energy** (we also see the harmonic oscillator).

The solution inside the box yields the following wave function for $U = 0$:

$$\Psi = C\sin(n+1)\frac{\pi}{a}x\sin(n+1)\frac{\pi}{b}y\sin(n+1)\frac{\pi}{c}z, |C| = \sqrt{\frac{8}{abc}} \tag{2.54}$$

At the rest of space, $\Psi = 0$. If we consider the fact that $a < b < c$, then the level E_{100} is energetically higher than the levels E_{010} and E_{001}. The relation

$$E_{100} > E_{010} > E_{001} \tag{2.55}$$

holds. Suppose a, b and c do not differ too much, then the given levels are close together. In this case, we have a **triplet** (in general, a **multiplet**) of states. If $a = b = c$, then the particle moves in a cube. Here all states belonging to a triplet are degenerate. So

$$E_{100} = E_{010} = E_{001} \tag{2.56}$$

The wave functions that belong to the degenerate states are as follows:

$$\Psi_{100}, \Psi_{010}, \Psi_{001} \tag{2.57}$$

Suppose the degeneracy is broken slightly, our volume approximates that of a cube and the three levels are close together in energy as we have just seen. For the states of higher energy, we observe an equivalent phenomenon. For instance, we have two triplets $\Psi_{110}, \Psi_{011}, \Psi_{101}$ and $\Psi_{200}, \Psi_{020}, \Psi_{002}$ (due to the slight break in the cube's symmetry) close together followed by a singlet Ψ_{111}. We identify such multiplet structures with **shells**. The **shell models** explaining shell structures are very important in atomic and nuclear physics. In nuclear physics, for example, all nucleons in a nucleus are supposed to be in a potential well. This potential is of course spherical symmetric. For small nuclei, a box-like potential is accepted as an approximation. Consider the spin of the proton and the neutron and the Pauli exclusion principle, then only two protons and two neutrons can occupy each level. The lowest energy levels are occupied first. This is due to the fact that the system prefers the state of lowest energy. In this case, the **last** particle determines the most **visible** properties. Suppose this last state is inside a multiplet, a small excitation energy suffices to lift that particle into a higher energy state. The nucleus in this case is easily excitable.

Consider a nucleus to contain just the number of protons and neutrons to fill a shell. In this case, much energy is required to excite a neutron into the first excited state. Particularly such nuclei are stable. This is because they can only be strongly excited if large energy gaps are overcome. This is the so-called **magic nuclei** (comparable with filled electron shells in atoms of inert gases) or double **magic nuclei**.

2.1.3 Coordinate Representation Delta Potential

We investigate the energy levels of a particle in an infinitely narrow and tall potential barrier (see Figure 2.1d):

$$U(x)=\begin{cases}\alpha\delta(x), & \alpha>0, \ |x|<a \\ \infty, & |x|>a\end{cases} \tag{2.58}$$

where α and a are real parameters. The Schrödinger equation for the given problem can be written in the following form:

$$\left[-\frac{\hbar^2}{2m}\frac{d^2}{dx^2}+U(x)\right]\Psi(x)=E\Psi(x) \tag{2.59}$$

From this Schrödinger equation, we can see the continuity of the wave function at the origin, $x=0$, as well as the discontinuity (jump) of the derivative of the wave function at that point. We partition the problem into domains I and II represented, respectively, by $|x|<a$ and $|x|>a$. Let us consider domain II:

$$\frac{1}{\Psi_{\text{II}}}\left[E\Psi_{\text{II}}+\frac{\hbar^2}{2m}\frac{d^2\Psi_{\text{II}}}{dx^2}\right]=U(x)=\infty \tag{2.60}$$

For this case, only $\Psi_{\text{II}}=0$ for $|x|>a$.

We consider now domain I where if we consider the solution away from the origin, the differential Equation 2.59 becomes

$$\frac{d^2}{dx^2}\Psi_{\text{I}}+\kappa^2\Psi_{\text{I}}=0,\ \kappa^2=\frac{2mE}{\hbar^2} \tag{2.61}$$

from where

$$\Psi_{\text{I}}=A_1\exp\{-i\kappa x\}+A_2\exp\{i\kappa x\} \tag{2.62}$$

Consider the continuity or matching of the wave function Ψ at the point $x = \pm a$:

$$\Psi(a) = \Psi(-a) = 0 \tag{2.63}$$

It is instructive to note that our potential is an even function:

$$\delta(-x) = \delta(x) \tag{2.64}$$

So if there is a ground state, it must be even and with no nodes. Since the wave function is even, then the solution of Equation 2.61 can be represented in the following form:

$$\Psi(x) = A\sin\left[\kappa\left(|x| - a\right)\right], \quad 0 < |x| \le a \tag{2.65}$$

This is the wave function of the even energy levels.

We investigate Equation 2.59 at the origin $x = 0$. The potential $U(x)$ is rather singular. Here, it vanishes for all x except for $x = 0$ at which point it has infinite strength. This implies that the potential is delta-function localized at $x = 0$. It is instructive to note that $\Psi'(x)$ is not continuous at $x = 0$ which is a consequence of the potential being infinite at the given point. This permits to integrate both sides of the Schrödinger equation (Equation 2.59) with respect to x over an infinitesimally small region around the delta potential, say from $-\varepsilon$ to $+\varepsilon$, and then letting $\varepsilon \to 0$:

$$-\frac{\hbar^2}{2m}\Delta\Psi'(\varepsilon) + \alpha\Psi(0) = E\int_{-\varepsilon}^{\varepsilon}\Psi(x)dx, \; \Delta\Psi'(\varepsilon) = \Psi'(\varepsilon) - \Psi'(-\varepsilon) \tag{2.66}$$

Here, ε is an arbitrary small positive number which becomes by letting $\varepsilon \to 0$, the right-hand side of Equation 2.66 goes to zero since Ψ is finite for all x while the domain of integration is contracting away. We integrate over a zero breadth and this gives the following:

$$\Delta\Psi'(\varepsilon) - \frac{2m\alpha}{\hbar^2}\Psi(0) = -\frac{2mE}{\hbar^2}\int_{-\varepsilon}^{\varepsilon}\Psi(x)dx \approx -\frac{2mE}{\hbar^2}2\varepsilon\Psi(0), \; \Psi(+0) = \Psi(-0) \tag{2.67}$$

It is instructive to note that the discontinuity of Ψ' at the delta function position is proportional to the value of the wave function at that point. However, at a node, the delta function would have no effect and Ψ' is continuous.

We apply the condition of matching of the wave function in Equation 2.65 at the point $x = 0$ by substituting Equation 2.65 into 2.67 and taking the limit as $\varepsilon \to +0$:

$$\begin{cases} \Psi'(x) = \kappa A\cos\left[\kappa\left(|x| - a\right)\right]\operatorname{sgn} x \\ \Delta\Psi'(0) = \kappa A\left[\cos\kappa a - (-\cos\kappa a)\right] = 2\kappa A\cos\kappa a \\ \Psi(0) = -A\sin\kappa a \end{cases} \tag{2.68}$$

So

$$-\tan\kappa a = 2\kappa\frac{\hbar^2}{2m\alpha} = \frac{\kappa a}{\frac{m\alpha a}{\hbar^2}} = \frac{\kappa a}{\xi}, \quad \xi = \frac{m\alpha a}{\hbar^2} \tag{2.69}$$

This defines the spectrum of the even levels. For $\xi \gg 1$ (also for $\kappa a \ll \xi$), we have lower levels.

So from Equation 2.69, we have

$$-\tan\kappa_n a \approx \frac{\kappa_n a}{\xi} \approx \frac{n\pi}{\xi},\ \kappa_n^{(+)} \approx \frac{n\pi}{a}\left(1-\frac{1}{\xi}\right),\ \ E_n^{(+)} \approx \frac{\hbar^2\pi^2 n^2}{2ma^2}\left(1-\frac{2}{\xi}\right) \tag{2.70}$$

The superscript "+" denotes the positive parity of levels.

For the odd levels, the wave function in the domain $|x|<a$ has the following form:

$$\Psi(x) = B\sin\kappa x \tag{2.71}$$

The condition

$$\Psi(a) = 0 \tag{2.72}$$

defines the spectrum of odd levels:

$$\kappa_n^{(-)} = n\pi,\ \ E_n^{(-)} = \frac{\hbar^2\pi^2 n^2}{2ma^2},\ \ n = 1,2,\ldots \tag{2.73}$$

It should be noted that for the odd state, the particle **does not feel** the presence of the delta-functional potential:

$$U(x) = \alpha\delta(x) \tag{2.74}$$

The odd energy levels in Equation 2.73 have dimensional quantization. From Equation 2.73, the distance between the even and odd levels is

$$\delta E_n \approx \frac{2E_n^{(\pm)}}{\xi} \ll \Delta E \equiv |E_{n\pm1} - E_n| \approx \frac{E_n^{(\pm)}}{n} \tag{2.75}$$

Let us again examine the spectrum of the even energy levels in the domain

$$\kappa a \gg \xi \tag{2.76}$$

For example, say an excited state:

$$-\tan\kappa_n a \approx \frac{\kappa_n a}{\xi} \approx \left(n-\frac{1}{2}\right)\frac{\pi}{\xi},\ \kappa_n^{(+)} a \approx \left(n-\frac{1}{2}\right)\pi + \frac{\xi}{\left(n-\frac{1}{2}\right)\pi},\ E_n^{(+)} \approx \frac{\hbar^2\pi^2(2n-1)^2}{8ma^2} + \frac{\hbar^2\xi}{ma^2} \tag{2.77}$$

The first summand in the expression of $E_n^{(+)}$ describes the spectrum of even levels in an infinite deep potential well and the second is the displacement of these levels due to the action of the delta-potential $\alpha\delta(x)$.

We consider the case, $\alpha < 0$, and investigate if the given potential admits bound states, i.e., when $\xi \gg 1$.

For a bound state, the energy, E, must be negative. This guarantees that all of $x \neq 0$ is classically forbidden, and the wave function decays rapidly allowing a normalized solution. The delta function may be thought of as approximated by a finite square well in the limit when the breadth of the well goes to zero and the depth goes to infinity in such a way that their product, representing the **area** is finite (the delta function is a function with unit area). So for the bound states, we consider

$$\kappa \to i\chi = i\sqrt{-\frac{2mE}{\hbar^2}},\ \ \chi > 0 \tag{2.78}$$

where the wave function, $\Psi(x)$, is constrained by the time-independent Schrödinger equation (Equation 2.61):

$$\Psi(x) = \mathrm{A}\exp\{-\chi x\}\theta(x>0) + \mathrm{B}\exp\{\chi x\}\theta(x<0) \tag{2.79}$$

Since the potential, $\delta(x)$, is even, we expect the ground state to be even and with no nodes as earlier indicated. If there is an excited state, then it must be odd and thus has a node at $x = 0$. The only odd solution can be built with the above exponential functions as $\sinh \chi x$. However, a wave function of the type $\Psi \approx \sinh \chi x$ cannot be normalized and peaks at $x = \pm\infty$. So there cannot be an excited state in the delta function potential, and if there are bound states, there is just one of them.

Considering Equation 2.79, we should recall that the wave function must be continuous to have meaning as a probability amplitude. For the given regions in Equation 2.79, the wave function must have same value at zero. So using Equation 2.67 as $\varepsilon \to 0$, we have $\mathrm{A} = \mathrm{B}$ and

$$\chi = \chi_0 = \frac{|\alpha| m}{\hbar^2}, \; E_0 = -\frac{m\alpha^2}{2\hbar^2} \tag{2.80}$$

So, there is one, and only one energy eigenvalue, E_0.

This implies that we have only one state of a discrete spectrum. This implies, there is one and only one bound state, and one energy eigenvalue. The normalization constant A is found from the following:

$$\int_{-\infty}^{\infty} \Psi^2 dx = 2|\mathrm{A}|^2 \int_0^{\infty} \exp\{-2\chi x\} dx = \frac{2|\mathrm{A}|^2}{2\chi} \int_0^{\infty} \exp\{-t\} dt = \frac{|\mathrm{A}|^2}{\chi} = 1 \tag{2.81}$$

The wave function in the coordinate representation now takes the form:

$$\Psi_0(x) = \sqrt{\chi_0}\exp\{-\chi_0 |x|\} \tag{2.82}$$

This function as expected is an even function. It may be seen that the average potential and kinetic energies take the form:

$$\bar{U} = -|\alpha| \int_{-\infty}^{\infty} \delta(x)\Psi_0^2(x) dx = -\frac{m|\alpha|^2}{\hbar^2} = 2E_0, \; \bar{T} = \frac{1}{2m}\int_{-\infty}^{\infty} \left|\hat{p}\Psi_0(x)\right|^2 dx = \frac{m|\alpha|^2}{2\hbar^2} = -E_0 \tag{2.83}$$

and the total energy, E, which is a constant, and E_0 is the sum of the potential and kinetic energies.

2.2 Time-Dependent Operators

2.2.1 Classical Equation of Motion

We relate under this section the classical equation of motion with the quantum equation of motion. For that, we first examine the temporal derivative of some arbitrary phase space function, $\mathrm{A} = \mathrm{A}(p_i, q_i)$, that does not depend explicitly on time:

$$\frac{d\mathrm{A}}{dt} = \frac{\partial \mathrm{A}}{\partial t} + \sum_{i=1}^{f} \left(\frac{\partial \mathrm{A}}{\partial q^i}\dot{q}^i + \frac{\partial \mathrm{A}}{\partial p_i}\dot{p}_i \right) \tag{2.84}$$

Here, p_i and q_i are the generalized momenta and coordinates, respectively, and f is the number of degrees of freedom. Considering the Hamilton equation of motion:

$$\dot{p}_i = -\frac{\partial \mathrm{H}}{\partial q^i}, \quad \dot{q}^i = \frac{\partial \mathrm{H}}{\partial p_i} \tag{2.85}$$

we have the equation of motion for the function A of the dynamical variables:

$$\frac{d\mathrm{A}}{dt} \equiv \frac{\partial \mathrm{A}}{\partial t} + \{\mathrm{A},\mathrm{H}\} \tag{2.86}$$

Here, the curly bracket denotes the **classical Poisson bracket** of the two functions A and H with respect to the canonical variables (q,p):

$$\{\mathrm{A},\mathrm{H}\} \equiv \sum_{i=1}^{f} \left(\frac{\partial \mathrm{A}}{\partial q^i}\frac{\partial \mathrm{H}}{\partial p_i} - \frac{\partial \mathrm{A}}{\partial p_i}\frac{\partial \mathrm{H}}{\partial q^i} \right) \tag{2.87}$$

The left-hand side of Equation 2.86 is the total time convective derivative of A along a moving phase point. The first summand on the right-hand side gives the explicit dependence of A on time t and the second summand is the change in A due to the motion of the phase point for which A is evaluated. The Poisson bracket has the following properties:

- It is anti-symmetric:

$$\{\mathrm{A},\mathrm{B}\} = -\{\mathrm{B},\mathrm{A}\} \tag{2.88}$$

- It satisfies Leibnitz's rule:

$$\{\mathrm{A},\mathrm{B}C\} = \{\mathrm{A},\mathrm{B}\}C + \mathrm{B}\{\mathrm{A},C\}, \{\mathrm{A}\mathrm{B},C\} = \{\mathrm{A},C\}\mathrm{B} + \mathrm{A}\{\mathrm{B},C\} \tag{2.89}$$

- The Jacobi identity also follows:

$$\{\{\mathrm{A},\mathrm{B}\},C\} + \{\{\mathrm{B},C\},\mathrm{A}\} + \{\{C,\mathrm{A}\}\mathrm{B}\} = 0 \tag{2.90}$$

for arbitrary functions A ,B and C. Also,

$$\{\mathrm{A},C\} = 0, C = \mathrm{const} \tag{2.91}$$

Let us examine Equation 2.86 again. If the function A is explicitly independent of time, its evolution is defined by the equation:

$$\dot{\mathrm{A}} = \{\mathrm{A},\mathrm{H}\} \tag{2.92}$$

Let us substitute in the place of A the canonical coordinates and the momenta, we have

$$\dot{p}_i = \{p_i,\mathrm{H}\}, \quad \dot{q}^i = \{q_i,\mathrm{H}\} \tag{2.93}$$

The quantity A is the integral of motion if

$$\frac{dA}{dt}=\frac{\partial A}{\partial t}+\{A,H\}=0 \tag{2.94}$$

If A is not explicitly time dependent, i.e.,

$$\frac{\partial A}{\partial t}=0 \tag{2.95}$$

it follows that

$$\{A,H\}=0 \tag{2.96}$$

The Poisson bracket for the canonical coordinates and momenta satisfy the following relation:

$$\{q^i,p_j\}=\delta^i_j \tag{2.97}$$

Let us examine the functions:

$$Q^i=Q^i(p,q),\quad P_i=P_i(p,q) \tag{2.98}$$

such that

$$\{Q^i,P_j\}=\delta^i_j \tag{2.99}$$

We say that the functions in Equation 2.98 define the canonical transformation. Here, Q^i and P_i are, respectively, new generalized coordinates and momenta. Let us substitute in Equation 2.87 the following:

$$\frac{\partial}{\partial q^i}=\frac{\partial Q^j}{\partial q^i}\frac{\partial}{\partial Q^j}+\frac{\partial P_j}{\partial q^i}\frac{\partial}{\partial P_j},\quad \frac{\partial}{\partial p_i}=\frac{\partial Q^j}{\partial p_i}\frac{\partial}{\partial Q^j}+\frac{\partial P_j}{\partial p_i}\frac{\partial}{\partial P_j} \tag{2.100}$$

then from Equation 2.99, we have

$$\{A,H\}\equiv\sum_{i=1}^{f}\left(\frac{\partial A}{\partial Q^i}\frac{\partial H}{\partial P_i}-\frac{\partial A}{\partial P_i}\frac{\partial H}{\partial Q^i}\right) \tag{2.101}$$

So, the canonical transformation does not change the Poisson's bracket. This has a very important corollary and in particular the Hamilton equation of motion is invariant relative to the canonical transformation:

$$\dot{P}_i=\{P_i,\tilde{H}\}=-\frac{\partial\tilde{H}}{\partial Q^i},\quad \dot{Q}^i=\{Q^i,\tilde{H}\}=\frac{\partial\tilde{H}}{\partial P_i} \tag{2.102}$$

where

$$\tilde{H}=\tilde{H}(P,Q)=\tilde{H}(p(P,Q),q(P,Q)) \tag{2.103}$$

is the Hamiltonian in the new variables. If we change the variables:

$$q^i = q^i(Q) \text{ or } Q^i = Q^i(q) \tag{2.104}$$

This may be one of the simplest examples of the canonical transformation. The transformation in Equation 2.104 is called point-like. In order to find the new canonical momenta, we use Equation 2.99:

$$\{Q^i, P_j\} = \frac{\partial Q^i}{\partial q^k}\frac{\partial P_j}{\partial p_k} = \delta^i_j \tag{2.105}$$

Furthermore, the following equations also hold:

$$\{P_j, P_j\} = 0, \{Q^i, Q^i\} = 0 \tag{2.106}$$

And Equation 2.105 yields

$$P_j(q,p) = p_f \frac{\partial q^f}{\partial Q^j} = p_f \left\| \frac{Q^j}{\partial q^f} \right\|^{-1} \tag{2.107}$$

Here, $\|A\|^{-1}$ represents the inverse matrix to matrix A. The equality in Equation 2.106 defines the new canonical momenta to the approximation of

$$p_f \to p_f + \frac{\partial}{\partial q^f} f(q) \tag{2.108}$$

Here, $f(q)$ is an arbitrary function of the coordinate.

2.2.2 Quantum-Mechanical Poisson Bracket and Quantum Correspondence Principle

We introduce the so-called **quantum-mechanical Poisson bracket** $[\hat{A}, \hat{H}]$, where $i[\hat{A}, \hat{H}]$ is a Hermitian operator if $\hat{A}$ and $\hat{H}$ are Hermitian operators:

$$\left(i[\hat{A}, \hat{H}]\right)^\dagger = i[\hat{A}, \hat{H}] \tag{2.109}$$

According to Dirac, we have the following correspondence of the classical and quantum mechanical Poisson brackets:

$$\{\hat{A}, \hat{H}\} \to \text{const} \times i[\hat{A}, \hat{H}] \tag{2.110}$$

where "$\to$" implies a correspondence. For the concordance with experimentation, it is necessary to let $\text{const} = \hbar^{-1}$ and thus the **quantum correspondence principle**:

$$[\hat{A}, \hat{H}] \to -i\hbar\{\hat{A}, \hat{H}\} \tag{2.111}$$

We observe from here that the quantum commutator of two operators is $-i\hbar$ times the value of their classical Poisson bracket. For the case of macroscopic systems, where the effects of the order of $\hbar$ are assumed negligible, so the commutator can also be neglected and hence the order of quantities is immaterial. So the Planck constant, $\hbar$, measures the non-classical nature of the system. Hence, classical mechanics can be recovered in the limit, $\hbar \to 0$, as seen earlier.

From the correspondence in Equation 2.111, we find that the transition from classical mechanics to quantum mechanics can be performed by the transition to operators and the replacement of the Poisson bracket $\left\{\hat{A},\hat{H}\right\}$ by the commutator $\frac{1}{i\hbar}\left[\hat{A},\hat{H}\right]$. Suppose $\hat{A}$ and $\hat{B}$ are Hermitian operators:

$$\left[\hat{A},\hat{B}\right]\to -i\hbar\left\{\hat{A},\hat{B}\right\} \tag{2.112}$$

If the Poisson bracket is a number (say $\{p_x,x\}=1$), then the correspondence is obtained from

$$\left[\hat{A},\hat{B}\right]=-i\hbar\left\{\hat{A},\hat{B}\right\} \tag{2.113}$$

If the Poisson's bracket of two quantities A and B are equal to some third quantity C:

$$\left\{\hat{A},\hat{B}\right\}=\hat{C} \tag{2.114}$$

then it follows that

$$\left[\hat{A},\hat{B}\right]=-i\hbar\left\{\hat{A},\hat{B}\right\}=-i\hbar\hat{C} \tag{2.115}$$

Equation 2.113 is the driving principle for the definition of operators of physical quantities and guarantees the transition to classical mechanics to the domain of its application. So from the above, classical mechanics laws need some amendments to be applied to microscopic systems. These amendments are in such a manner that when considering the given macroscopic system, we should recover the familiar predictions of classical mechanics.

2.2.3 Quantum Mechanical Equation of Motion

Any dynamics may aid to predict experimental outcomes. Dynamical equations describing any physical quantity of a system permit us to compute the outcome of a complete measurement at time moment, t, when the outcome of the given measurement at initial time moment, t_0, is known. This is on condition if no other observables of the system are known within the time moments, t and t_0, that may change the initial state of the observable at time moment, t_0. Consider some physical quantity A that depends explicitly on time t, and a conservative system with the wave function that depends on time t. For the physical quantity A in that state, its mean value is

$$\bar{A}=\int \Psi^*(q,t)\hat{A}(t)\Psi(q,t)dq \tag{2.116}$$

where $\hat{A}$ is the operator of the quantity A dependent explicitly on time, t. The wave function of the system is dependent on time t. If the state of the system is stationary, then the dependence on time t in Equation 2.116 cancels out as the dependence on time of Ψ and its conjugate Ψ^* compensates each other. It follows that the dependence of the wave function on time t in Equation 2.116 is particularly for non-stationary states. So we have $\bar{A}(t)$ for $\hat{A}=\hat{A}(t)$ and $\Psi=\Psi(t)$ (for a non-stationary state). We see that if the operator $\hat{A}$ and the wave function Ψ are time dependent, then the mean value $\bar{A}(t)$, in general, is time dependent too. If we evaluate the temporal variation of $\hat{A}$, we can exchange differentiation and integration:

$$\frac{d\bar{A}}{dt}=\int \Psi^*\frac{\partial\hat{A}}{\partial t}\Psi dq+\int\frac{\partial\Psi^*}{\partial t}\hat{A}\Psi dq+\int\Psi^*\hat{A}\frac{\partial\Psi}{\partial t}dq \tag{2.117}$$

In Equation 2.117, the first integral represents the mean value of the partial temporal derivative of the operator $\hat{A}$. We simplify the second and third terms with the aid of the time-dependent Schrödinger equation as follows:

$$\frac{\partial \Psi}{\partial t} = \frac{1}{i\hbar}\hat{H}\Psi, \quad \frac{\partial \Psi^*}{\partial t} = -\frac{1}{i\hbar}\hat{H}^*\Psi^* \tag{2.118}$$

We substitute Equation 2.118 into 2.117 considering the fact that the operator $\hat{H}^*$ only acts on Ψ^*:

$$\frac{d\bar{A}}{dt} = \int \Psi^* \frac{\partial \hat{A}}{\partial t}\Psi\, dq - \frac{1}{i\hbar}\int \left(\hat{H}^*\Psi^*\right)\hat{A}\Psi\, dq + \frac{1}{i\hbar}\int \Psi^*\hat{A}\left(\hat{H}\Psi\right) dq \tag{2.119}$$

We consider the fact that the operator $\hat{H}$ is self-conjugated or Hermitian, it follows that

$$\int \left(\hat{H}^*\Psi^*\right)\hat{A}\Psi\, dq = \int \left(\hat{A}\Psi\right)\hat{H}^*\Psi^*\, dq = \int \Psi^*\hat{H}\hat{A}\Psi\, dq \tag{2.120}$$

We substitute Equation 2.120 into 2.119, this yields

$$\frac{d\bar{A}}{dt} = \int \Psi^* \left\{ \frac{\partial \hat{A}}{\partial t} + \frac{1}{i\hbar}\left[\hat{A}, \hat{H}\right] \right\} \Psi\, dq, \quad \left[\hat{A}, \hat{H}\right] = \hat{A}\hat{H} - \hat{H}\hat{A} \tag{2.121}$$

If we compare Equation 2.116 with 2.121, then the integral in Equation 2.121 is of the type of the mean value:

$$\frac{d\bar{A}}{dt} = \overline{\frac{\partial \hat{A}}{\partial t}} + \frac{1}{i\hbar}\overline{\left[\hat{A}, \hat{H}\right]} \tag{2.122}$$

If we consider Equation 2.122 as a basis, then we may without hesitation define the total temporal derivative of the operator $\frac{d\hat{A}}{dt}$:

$$\frac{d\hat{A}}{dt} = \frac{\partial \hat{A}}{\partial t} + \frac{1}{i\hbar}\left[\hat{A}, \hat{H}\right] \tag{2.123}$$

It follows from this definition that the temporal derivative of the mean value of $\hat{A}$ is equal to the mean value of $\frac{d\hat{A}}{dt}$:

$$\frac{d\overline{\hat{A}}}{dt} = \overline{\frac{d\hat{A}}{dt}} \tag{2.124}$$

This operator is made up of two summands: The change of the physical quantity with respect to time is described by the operator dependent on time and may be related to two situations:

1. The physical quantity may explicitly be dependent on time t, for example, say the potential energy of a particle in a variable (with respect to time) electric field
2. At the time moment, t, the state of the system changes

The commutator in Equation 2.123 is different from zero if the operator $\hat{A}$ of the physical quantity A does not commute with the Hamiltonian $\hat{H}$ of the system, i.e., $\left[\hat{A},\hat{H}\right]\neq 0$. If now $\hat{A}$ does not depend explicitly on time t and commutes with $\hat{H}$, then both summands in Equation 2.123 are equal to zero, i.e., $\frac{\partial\hat{A}}{\partial t}=0$ and $\left[\hat{A},\hat{H}\right]=0$. This follows that $\frac{d\hat{A}}{dt}=0$. The integral in Equation 2.121 and the mean value of the physical quantity appear to be independent of time. Such quantities are called **integrals of motion (constants of motion). The integral of motion for a given system is called a physical quantity, the operator of which does not depend explicitly on time** ***t*** **and commutes with the Hamiltonian of the system.** The mean value of such a quantity for the given system is a constant which is independent of time t. For example, say the energy of the conservative system E for which its operator is $\hat{H}$:

$$\frac{\partial\hat{H}}{\partial t}=0 \tag{2.125}$$

From the definition of a conservative system,

$$\left[\hat{H},\hat{H}\right]=0 \tag{2.126}$$

which is the law of conservation of energy. So the energy E is an integral (constant) of motion. If we consider $\frac{d\overline{\hat{A}}}{dt}=0$, then the mean value of $\hat{A}$ in the state Ψ does not evolve with time and the term **constant of motion.**

2.2.4 Postulates of Quantum Mechanics

For quantum mechanics to be a good description of a physical system, it should reduce to classical mechanics for the case of a macroscopic system. In classical mechanics, the order of measurements and therefore the order of observable is immaterial. When given a classical Hamiltonian system, it is possible to make a transition to the quantum description through the following postulates:

1. A classical mechanics system for a given time moment, t, is described by the coordinate $q_i(t)$ and momentum $p_i(t)$, while in a quantum mechanical system, the state of a system for a given time moment, t, is described by the wave function, $\Psi(t)$, which is a basis vector in the Hilbert space.
2. Each dynamical quantity $A=A(p_i,q_i)$ in classical mechanics is a function of phase space with variables being the generalized coordinates q_i and generalized momenta p_i, while in quantum mechanics, the observables, q_i and p_i, are transformed to the Hermitian operators, $\hat{q}_i$ and $\hat{p}_i$, satisfying the following commutation relation:

$$\left[\hat{q}^i,\hat{p}_j\right]=i\hbar\delta^i_j \tag{2.127}$$

3. If a classical mechanics system is in the state with coordinate $q_i(t)$ and momentum $p_i(t)$, then the measurement of the observable, $A=A(p_i,q_i)$, yields a unique value where the system will be unaffected by the process of measurement while in quantum mechanics, we have probabilistic results. Suppose the wave function, $\Psi(t)$ describes the state of a system, the operator of the physical system $\hat{A}$ acting on the given state yields an eigenvalue A with probability $P(A)$.
4. In classical mechanics, the state of dynamical variables changes according to the Hamilton equation of motion in Equation 2.85, while in quantum mechanics, the wave function evolves according to the time-dependent Schrödinger equation in Equation 2.118 and temporal variation of $\hat{A}$ according to Equation 2.123.

2.2.5 Velocity and Acceleration of a Charged Particle in an Electromagnetic Field

We find the formula for the derivatives of the product of the operators $\hat{A}$ and $\hat{B}$ that will help to find the operator of the acceleration $\vec{a}$ in an electromagnetic field. The Hamiltonian for the charged particle in an electromagnetic field is

$$\hat{H}=\frac{1}{2m}\left(\hat{p}-\frac{e}{c}\vec{A}(\vec{r},t)\right)^2+e\Phi(\vec{r},t) \tag{2.128}$$

where $\hat{p}=-i\hbar\nabla$ is the operator of the momentum, $\vec{A}(\vec{r},t)$ and $\Phi(\vec{r},t)$ are the vector and scalar potentials, respectively, and e is the electronic charge. From

$$\left[\hat{H},\hat{A}\hat{B}\right]=\left[\hat{H},\hat{A}\right]\hat{B}+\hat{A}\left[\hat{H},\hat{B}\right] \tag{2.129}$$

and

$$\frac{d\hat{A}}{dt}=\frac{\partial\hat{A}}{\partial t}+\frac{1}{i\hbar}\left[A,\hat{H}\right] \tag{2.130}$$

So

$$\frac{d\left(\hat{A}\hat{B}\right)}{dt}=\frac{\partial\left(\hat{A}\hat{B}\right)}{\partial t}+\frac{1}{i\hbar}\left[\hat{A}\hat{B},\hat{H}\right] \tag{2.131}$$

We evaluate

$$\frac{1}{i\hbar}\left[\hat{A}\hat{B},\hat{H}\right]=-\frac{1}{i\hbar}\left\{\left[\hat{H},\hat{A}\right]\hat{B}+\hat{A}\left[\hat{H},\hat{B}\right]\right\}=\frac{1}{i\hbar}\left\{\left[\hat{A},\hat{H}\right]\hat{B}+\hat{A}\left[\hat{B},\hat{H}\right]\right\} \tag{2.132}$$

From Equation 2.131:

$$\frac{d\left(\hat{A}\hat{B}\right)}{dt}=\frac{d\hat{A}}{dt}\hat{B}+\hat{A}\frac{d\hat{B}}{dt} \tag{2.133}$$

The operator of the velocity $\vec{v}$ is as follows:

$$\vec{v}=\frac{d\vec{r}}{dt}=\frac{\partial\vec{r}}{\partial t}+\frac{1}{i\hbar}\left[\vec{r},\hat{H}\right]=\frac{1}{m}\left(\hat{p}-\frac{e}{c}\hat{\vec{A}}\right),\ \vec{a}=\frac{d\vec{v}}{dt}=\frac{\partial\vec{v}}{\partial t}+\frac{1}{i\hbar}\left[\vec{v},\hat{H}\right]=\frac{e}{m}\vec{E}+\frac{e}{2cm}\left\{\left[\hat{\vec{v}},\vec{H}\right]-\left[\vec{H},\hat{\vec{v}}\right]\right\} \tag{2.134}$$

The second equation in Equation 2.134 is in agreement with the classical analogue:

$$m\vec{a}\equiv m\ddot{\vec{r}}=\vec{F}=e\vec{E}+\frac{e}{c}\left[\vec{v},\vec{H}\right] \tag{2.135}$$

where the electric and magnetic field strengths are, respectively,

$$\vec{E}=-\nabla\Phi-\frac{1}{c}\frac{\partial\vec{A}}{\partial t},\quad \vec{H}=\mathrm{rot}\vec{A} \tag{2.136}$$

Question:
Suppose the operator $\hat{A}$ of a physical quantity A does not depend explicitly on time. Show that the average value of the operator of the physical quantity in a stationary state of a discrete spectrum is equal to zero and, hence, show that the average force acting on a particle in a stationary state of a discrete spectrum is equal to zero.

Answer: We average the operator $\frac{d\hat{A}}{dt}$:

$$\overline{\frac{dA}{dt}} = \int \Psi_m^* \frac{d\hat{A}}{dt}\Psi_n dq = \frac{1}{i\hbar}\int \Psi_m^*\left(\hat{A}\hat{H} - \hat{H}\hat{A}\right)\Psi_n dq = 0 \tag{2.137}$$

then, from

$$\hat{H}\Psi_n = \left(\frac{\hat{p}^2}{2m} + U\right)\Psi_n = E_n\Psi_n \tag{2.138}$$

we have

$$\overline{F} = -\overline{\nabla U} \equiv -\frac{1}{i\hbar}\overline{\left[\hat{H}, \hat{p}\right]} = \overline{\hat{\dot{p}}} = 0 \tag{2.139}$$

Alternatively, the average value is in the following form, $\overline{F}$:

$$\overline{F} = -\int \Psi_n^*(\text{grad}U)\Psi_n dq = -\int \nabla\left(\Psi_n^* U\Psi_n\right)dq + \int\left\{\left(\text{grad}\Psi_n^*\right)U\Psi_n + \Psi_n^* U\text{grad}\Psi_n\right\}dq \tag{2.140}$$

or

$$\overline{F} = -\int \nabla\left(\Psi_n^* U\Psi_n\right)dq + E_n\int \nabla\left(\Psi_n^*\Psi_n\right)dq + \frac{\hbar^2}{2m}\int\left\{\left(\nabla\Psi_n^*\right)(\Delta\Psi_n) + \left(\Delta\Psi_n^*\right)(\nabla\Psi_n)\right\}dq \tag{2.141}$$

The last integral in Equation 2.141 is transformed considering the commutation of the operators ∇ and Δ and the Hermiticity of the operator:

$$\int\left\{\left(\nabla\Psi_n^*\right)(\Delta\Psi_n) + \left(\Delta\Psi_n^*\right)(\nabla\Psi_n)\right\}dq = \int \nabla\left(\Psi_n^*\Delta\Psi_n\right)dq = 0 \tag{2.142}$$

So, from here and Equation 2.142:

$$\overline{F} = \int \nabla\left(\Psi_n^*\left[E_n - U\right]\Psi_n\right)dq = \int \nabla\left(\Psi_n^*\frac{\hat{p}^2}{2m}\Psi_n\right)dq = -\frac{\hbar^2}{2m}\int \nabla\left(\Psi_n^*\nabla^2\Psi_n\right) = 0 \tag{2.143}$$

2.2.6 Probability Density and Probability Current Density

Consider the Hamiltonian, $\hat{H}$, of a particle in a potential field $U(\vec{r})$ for which we write the time Schrödinger equation:

$$i\hbar\frac{\partial\Psi}{\partial t} = -\frac{\hbar^2}{2m}\Delta\Psi + U\Psi,\ -i\hbar\frac{\partial\Psi^*}{\partial t} = -\frac{\hbar^2}{2m}\Delta\Psi^* + U\Psi^* \tag{2.144}$$

If we multiply the equations in Equation 2.144 correspondingly by Ψ^* and Ψ after which we take the difference of the resultants, then we have

$$i\hbar\frac{\partial}{\partial t}\Psi^*\Psi = -\frac{\hbar^2}{2m}\left(\Psi^*\Delta\Psi - \Psi\Delta\Psi^*\right) \tag{2.145}$$

But

$$\text{div}\left[\Psi^*\nabla\Psi - \Psi\nabla\Psi^*\right] = \left(\nabla\Psi^*\right)\left(\nabla\Psi\right) + \Psi^*\left(\nabla^2\Psi\right) - \left(\nabla\Psi\right)\left(\nabla\Psi^*\right) - \Psi\left(\nabla^2\Psi^*\right) = \Psi^*\Delta\Psi - \Psi\Delta\Psi^* \tag{2.146}$$

and

$$\vec{j} = \frac{i\hbar}{2m}\left[\Psi^*\nabla\Psi - \Psi\nabla\Psi^*\right] \tag{2.147}$$

with

$$\rho = \Psi^*\Psi \tag{2.148}$$

then from Equations 2.145 and 2.146:

$$\frac{\partial\rho}{\partial t} + \text{div}\vec{j} = 0 \tag{2.149}$$

This is a **continuity equation** that expresses some conservation law. Here, $\rho(\vec{r},t)$ is the **probability density** and $\vec{j}$ is the **flux of the probability current density**. The notion of the flux of the current density enables us to introduce another way of normalization of a wave function:

$$\int\rho dq = \int\Psi\Psi^* dq = 1 \tag{2.150}$$

The existence of the equation in Equation 2.149 leads immediately to the physical interpretation of the vector $\vec{j}$ as the probability current density which is just as in the case of a free particle. This implies that $\vec{j}d\vec{\sigma}dt$ is the probability that a particle passes through the element of the surface $d\vec{\sigma}$ within the time interval dt in the direction of the positive normal with less probability that it passes in the opposite direction. The quantity, ρ, may imply that the particle no longer is localized at a defined point of space but is a continuum in the entire space according to the function ρ. Suppose this particle has a charge e, we may say that this charge is a continuum in the entire space in a definite manner. So we have a continuous **charge cloud** with **charge density** $e\rho$ **and current density** $e\vec{j}$.

For $\vec{j}$ being the flux density of a fluid:

$$\vec{j}d\vec{\sigma} = j_n d\sigma \tag{2.151}$$

is the quantity of a liquid through a unit surface in unit time. Here, j_n is the normal component of the vector $\vec{j}$. Apply Gauss' law to Equation 2.149:

$$\frac{\partial}{\partial t}\int_V \rho dV + \int_V \text{div}\vec{j}dV = \frac{\partial}{\partial t}\int_V \rho dV + \oint_\sigma j_n d\sigma = 0 \tag{2.152}$$

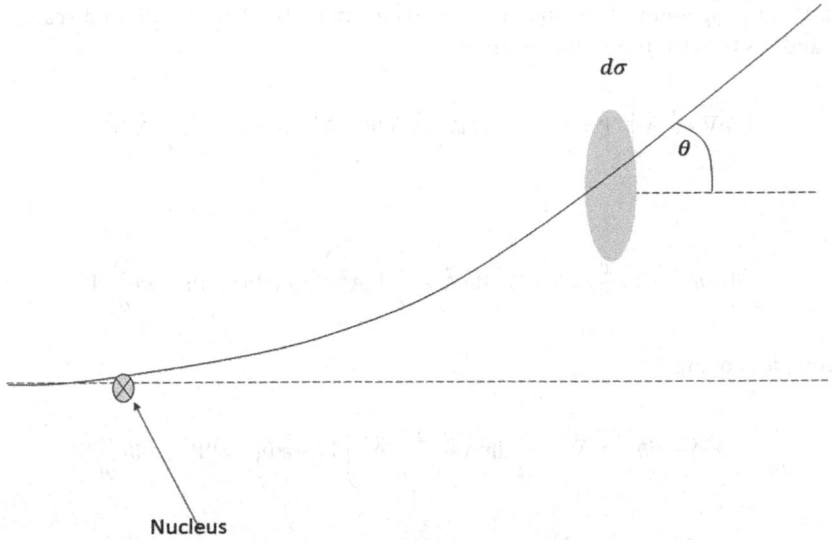

FIGURE 2.2 The angle of deflection θ and $d\sigma$, the surface through which an electron moving in a trajectory with a velocity $\vec{v}$ passes near a positively charged nucleus.

The flux of the particle through the surface σ of a region is equivalent to the variation of the particle density inside the region. From the condition of normalization in Equation 2.150:

$$\oint_{\sigma} j_n d\sigma = 0 \tag{2.153}$$

This implies that the particle current through an infinitely distant surface vanishes. So if we consider Equation 2.153, only those states that are normalized to unity whose current flux through an infinitely distant surface vanishes.

Let us evaluate the mass current density M from the particle current density:

$$\frac{dM}{dt} + \oint_{\sigma} j_n d\sigma = 0,\ M = m\Psi^*\Psi = m\rho \tag{2.154}$$

From Rutherford's experiment where an electron moves in a trajectory with a velocity $\vec{v}$ and passes near a positively charged nucleus (Figure 2.2). We ask ourselves the question whether the particle will pass through the surface $d\sigma$? In quantum mechanics, there exists the probability that the particle may or may not pass through $d\sigma$. The probability that the particle per unit time passes through $d\sigma$ is in Equation 2.151.

2.2.7 Current Density of a Charged Particle in an Electromagnetic Field

The Hamiltonian for the charged particle in an electromagnetic field:

$$\hat{\mathrm{H}} = \frac{1}{2m}\left(\hat{\mathrm{p}} - \frac{e}{c}\vec{\mathrm{A}}(\vec{r},t)\right)^2 + e\Phi(\vec{r},t) \tag{2.155}$$

where $\hat{p} = -i\hbar\nabla$ is the operator of the momentum, $\vec{A}(\vec{r},t)$ and $\Phi(\vec{r},t)$ are vector and scalar potentials, respectively, and e is the electronic charge. From

$$\left(i\hbar\nabla + \frac{e}{c}\vec{A}\right)^2 \Psi = -\hbar^2\Delta\Psi + 2i\hbar\frac{e}{c}\vec{A}\cdot\nabla\Psi + i\hbar\frac{e}{c}\mathrm{div}\vec{A}\Psi + \left(\frac{e}{c}\right)^2 \vec{A}^2\Psi \tag{2.156}$$

then

$$\frac{1}{2m}\left(-\hbar^2\Delta + 2i\hbar\frac{e}{c}\vec{A}\cdot\nabla + i\hbar\frac{e}{c}\mathrm{div}\vec{A} + \left(\frac{e}{c}\right)^2 \vec{A}^2\right)\Psi + e\Phi(\vec{r},t)\Psi = i\hbar\frac{\partial}{\partial t}\Psi \tag{2.157}$$

We take its complex conjugate:

$$\frac{1}{2m}\left(-\hbar^2\Delta - 2i\hbar\frac{e}{c}\vec{A}\cdot\nabla - i\hbar\frac{e}{c}\mathrm{div}\vec{A} + \left(\frac{e}{c}\right)^2 \vec{A}^2\right)\Psi^* + e\Phi(\vec{r},t)\Psi^* = -i\hbar\frac{\partial}{\partial t}\Psi^* \tag{2.158}$$

We multiply Equation 2.157 by Ψ^* and Equation 2.158 by Ψ, then take the difference:

$$\frac{\partial}{\partial t}|\Psi|^2 = \frac{i\hbar}{2m}\left[\Psi^*\Delta\Psi - \Psi\Delta\Psi^*\right] + \frac{e}{mc}\left[\Psi^*\vec{A}\cdot\nabla\Psi + \nabla\Psi^*\cdot\vec{A}\Psi + \mathrm{div}\vec{A}\cdot\Psi^*\cdot\Psi\right] \tag{2.159}$$

From

$$\mathrm{div}\left[\Psi^*\nabla\Psi - \Psi\nabla\Psi^*\right] = \Psi^*\Delta\Psi - \Psi\Delta\Psi^* \tag{2.160}$$

and

$$\Psi^*\vec{A}\cdot\nabla\Psi + \nabla\Psi^*\cdot\vec{A}\Psi + \mathrm{div}\vec{A}\cdot\Psi^*\cdot\Psi = \mathrm{div}\left[\vec{A}\cdot\Psi^*\cdot\Psi\right] \tag{2.161}$$

then

$$\frac{\partial}{\partial t}|\Psi|^2 = \frac{i\hbar}{2m}\mathrm{div}\left[\Psi^*\nabla\Psi - \Psi\nabla\Psi^*\right] + \frac{e}{mc}\mathrm{div}\left[\vec{A}\cdot\Psi^*\cdot\Psi\right] \tag{2.162}$$

then the current density,

$$j = \frac{i\hbar}{2m}\left[\Psi^*\nabla\Psi - \Psi\nabla\Psi^*\right] - \frac{e}{mc}\vec{A}|\Psi|^2 \tag{2.163}$$

and the continuity equation,

$$\frac{\partial}{\partial t}|\Psi|^2 + \mathrm{div}\vec{j} = 0 \tag{2.164}$$

2.2.8 Change with Time of a Wave Packet

We examine the state of a free particle which at time moment $t = 0$ is described by the wave function:

$$\Psi(x,t=0) = A\exp\left\{-\frac{x^2}{2a^2} + \frac{ip_0 x}{\hbar}\right\}, \quad p_0 = mv_0 \tag{2.165}$$

We find the change of state of the particle with time and also the following expectation values:

$$\overline{x(t)}, \overline{p(t)}, \overline{(\Delta x(t))^2}, \overline{(\Delta p(t))^2} \tag{2.166}$$

The normalization constant, A, can be obtained from the normalization condition (the probability of finding the particle in the domain of space is unity):

$$\int_{-\infty}^{\infty} dx \left|\Psi(x,t=0)\right|^2 = |A|^2 \int_{-\infty}^{\infty} dx \exp\left\{-\frac{x^2}{a^2}\right\} = |A|^2 a\sqrt{\pi} = 1 \tag{2.167}$$

In order to find the domain of localization of the particle, we find the initial probability density as follows:

$$\rho(x,t=0) = \left|\Psi(x,t=0)\right|^2 = |A|^2 \exp\left\{-\frac{x^2}{a^2}\right\} \tag{2.168}$$

This function achieves its maximum value at the point, $x = 0$, and decreases faster for $|x| > a$. This shows that the probability of finding the particle at a distance greater than a from the origin is very small compared to the probability of finding the particle within the distance a on either sides of the origin. The breadth of the packet for such a function is obviously a.

From the wave function, $\Psi(x,t=0)$, we assume we have a free particle and so we limit ourselves to the x-dependence. The current density for this initial wave can be evaluated as follows:

$$j_x = \frac{i\hbar}{2m}\left(\Psi(x,0)\frac{\partial}{\partial x}\Psi^*(x,0) - \Psi^*(x,0)\frac{\partial}{\partial x}\Psi(x,0)\right) = \frac{p_0}{m}\rho(x,t=0) \tag{2.169}$$

It is instructive to note that the expression for j_x coincides with its classical value and the factor $\rho(x,t=0)$ is defined only by the real part of the exponent of the wave function, $\Psi(x,t=0)$, and the factor $\frac{p_0}{m}$ by the imaginary part of the given wave function. The quantity, $\frac{p_0}{m}$, is the analogue of the classical velocity.

Initially, the expectation value of the coordinate, $x(t=0)$:

$$\overline{x(t=0)} = \overline{x} = |A|^2 \int_{-\infty}^{\infty} dx\, x \exp\left\{-\frac{x^2}{a^2}\right\} = 0 \tag{2.170}$$

This integral equals zero since the integrand is an odd function and, in addition, we have symmetrical limits. Also,

$$\overline{p} = \int_{-\infty}^{\infty} dx\, \Psi^* \hat{p} \Psi = \frac{p_0}{m},\ \overline{(\Delta x)^2} = |A|^2 \int_{-\infty}^{\infty} dx\, x^2 \exp\left\{-\frac{x^2}{a^2}\right\} = \frac{|A|^2}{2} a^3\sqrt{\pi} = \frac{a^2}{2} \tag{2.171}$$

We examine now the dynamics of the system with time t. For convenience, express the wave function as a packet through the eigenfunction of the operator of the momentum:

$$\Psi(x,t=0) = \int C(p)\Psi_p(x)dp \equiv \frac{1}{\sqrt{2\pi\hbar}} \int C(p)\exp\left\{\frac{ipx}{\hbar}\right\}dp \tag{2.172}$$

From here,

$$C(\mathrm{p})=\int\Psi(x,t=0)\Psi_{\mathrm{p}}^{*}(x)dx=\frac{a\mathrm{A}}{\sqrt{\hbar}}\exp\left\{-\frac{a^2}{2\hbar^2}(\mathrm{p}_0-\mathrm{p})^2\right\} \tag{2.173}$$

The value of $C(\mathrm{p})$ is different from zero at the neighborhood of the point $\mathrm{p}=\mathrm{p}_0$. The quantity

$$\rho(\mathrm{p})d\mathrm{p}=|C(\mathrm{p})|^2 d\mathrm{p}=\frac{a^2|\mathrm{A}|^2}{\hbar}\exp\left\{-\frac{a^2}{\hbar^2}(\mathrm{p}_0-\mathrm{p})^2\right\}d\mathrm{p} \tag{2.174}$$

is proportional to the probability to find the particle with quasi momentum within the interval, $\mathrm{p}\div(\mathrm{p}+d\mathrm{p})$.

The wave function $\Psi(x,t)$ of the state of the particle in an arbitrary time moment can be obtained from the formula substituting the expression for $C(\mathrm{p})$ from Equation 2.173:

$$\Psi(x,t)=\int C(\mathrm{p},t)\Psi_{\mathrm{p}}(x)d\mathrm{p}\equiv\frac{a\mathrm{A}}{\sqrt{2\pi\hbar}}\int\exp\left\{-\frac{i\mathrm{p}^2t}{2m\hbar}+\frac{i\mathrm{p}x}{\hbar}-\frac{a^2}{2\hbar^2}(\mathrm{p}_0-\mathrm{p})^2\right\}d\mathrm{p} \tag{2.175}$$

From definition, the probability density of the particle at the time moment t is a Gaussian:

$$\rho(x,t)=|\Psi(x,t)|^2=\frac{a|\mathrm{A}|^2}{\Delta(t)}\exp\left\{-\frac{\left(x-\frac{\mathrm{p}_0t}{m}\right)^2}{\Delta^2(t)}\right\},\ \Delta(t)\equiv a\left[1+\frac{t^2\hbar^2}{m^2a^4}\right]^{\frac{1}{2}} \tag{2.176}$$

Here, the breadth of Gaussian functions increases with time t as $\Delta(t)$.

From Equation 2.176, we find

$$\overline{x(t)}=\int x|\Psi(x,t)|^2dx=\frac{\mathrm{p}_0}{m}t,\ \sqrt{\overline{(\Delta x(t))^2}}=\sqrt{\overline{\left(x-\overline{x(t)}\right)^2}}=\frac{\Delta(t)}{\sqrt{2}} \tag{2.177}$$

So the breadth of the packet or probability distribution (has the order of the quantity $\sqrt{\overline{(\Delta x(t))^2}}$) increases as t departs from zero in both past and future directions in a phenomenon known as **the spreading of the wave packet** leading to the disappearance of the packet. This disappearance of the packet is related to the fact that the momentum of the particle does not have a defined value. It is instructive to note that it is a general feature of most quantum mechanical systems.

The expectation value of the momentum, $\overline{\mathrm{p}(t)}$, can be evaluated as follows:

$$\overline{\mathrm{p}(t)}=\int\mathrm{p}|C(\mathrm{p},t)|^2d\mathrm{p}=\int\mathrm{p}\left|\exp\left\{-\frac{i\mathrm{p}^2t}{2m\hbar}\right\}C(\mathrm{p})\right|^2d\mathrm{p}=\frac{a^2|\mathrm{A}|^2}{\hbar}\int d\mathrm{p}\,\mathrm{p}\exp\left\{-\frac{a^2}{\hbar^2}(\mathrm{p}_0-\mathrm{p})^2\right\} \tag{2.178}$$

Letting $x=\mathrm{p}-\mathrm{p}_0$:

$$\overline{\mathrm{p}(t)}=\frac{a^2|\mathrm{A}|^2}{\hbar}\mathrm{p}_0\int_{-\infty}^{\infty}dx\exp\left\{-\frac{a^2}{\hbar^2}x^2\right\}=\frac{a^2|\mathrm{A}|^2}{\hbar}\frac{\hbar\mathrm{p}_0}{a}\sqrt{\pi}=\mathrm{p}_0 \tag{2.179}$$

Also,

$$\overline{(\Delta p(t))^2} = \int (p - p_0)^2 |C(p,t)|^2 dp = \frac{a^2|A|^2}{\hbar} \int (p - p_0)^2 \exp\left\{-\frac{a^2}{\hbar^2}(p_0 - p)^2\right\} dp = \frac{\hbar^2}{2a^2} \tag{2.180}$$

From here and Equation 2.177, we establish the following uncertainty relation:

$$\overline{(\Delta x(t))^2}\,\overline{(\Delta p(t))^2} = \frac{a^2}{2}\frac{\hbar^2}{2a^2} = \frac{\hbar^2}{4} \tag{2.181}$$

We observe from here that the average momentum and the momentum dispersion do not vary with time. It is indicative that for a free particle, the momentum is an integral of motion. It is obvious that as the free particle encounters no obstacle, the momentum distribution is not expected to change.

The results in Equations 2.176–2.181 have a simple sense of the distribution function with respect to the coordinates of a particle. Considering Equation 2.175, it has the form of a Gaussian packet with center $\overline{x(t)}$ and moves with the velocity:

$$v_0 = \frac{\overline{p}}{m} \equiv \frac{p_0}{m} \tag{2.182}$$

The current density can be evaluated as follows:

$$j_x = \frac{i\hbar}{2m}\left(\Psi(x,t)\frac{\partial}{\partial x}\Psi^*(x,t) - \Psi^*(x,t)\frac{\partial}{\partial x}\Psi(x,t)\right) = \frac{a^2 p_0}{m\Delta^2(t)}\left(1 + \frac{\hbar^2 xt}{p_0 m a^4}\right)\rho(x,t) \tag{2.183}$$

We make an estimate for the magnitude of breadth of the packet for electrons where $a = 1$ mm. We consider the case where one electron creates a beam of electrons collimated by a slit of breadth approximately one millimeter. For the case when $m = 9\times10^{-28}$ g, we observe that the breadth has doubled about 15 msec. This should be a very short time on a macroscopic scale. However, on a scale consistent to electrons, the time is very long. For example, say the velocity 10^8 cm/sec for an electron is not unusual. Within 15 msec, such an electron travels 1.5×10^6 cm. This implies 15 km, so the electronic beam, initially with a breadth of 1 mm and moving with a velocity of 10^8 cm/sec would travel 15 km before its breadth is doubled. Of course, this is assuming that the electrons are truly free. This implies neglecting collisions, electromagnetic forces, gravitational forces and so on.

3

Momentum Operator

3.1 Translation Operator

In the course of classical and theoretical mechanics, there exists a relation between the laws of conservation (integral or constant of motion) and the property of the symmetry of space and time t. It has been shown that the law of conservation of momentum follows from the isotropy of space and time t. So **any property of space and time symmetry should correspond to a conservable physical quantity.** This theorem is true for quantum mechanics.

The observation of a physical system usually is done in a certain reference frame. This implies that when describing the state of system, we should have information about the position of the observer. So any state must contain in addition to all variable attributes, say α (such as spin, $\vec{S}$ momentum, $\vec{p}$, etc.), the coordinate, $\vec{r}$, which can be considered in the state function.

Consider a system of particles not subjected to an external field where all positions in space of such a system are equivalent. In this case, the Hamiltonian of the system is invariant for any space translation. For this, we consider $\hat{T}_{\delta\vec{r}}$ being an operator of an infinitesimal translation of the coordinate as follows:

$$\hat{T}_{\delta\vec{r}}\Psi(\vec{r}) = \Psi(\vec{r}+\delta\vec{r}) \tag{3.1}$$

This is called the **symmetry translation if the Hamiltonian of the given system is invariant relative to that translation, i.e.,**

$$\hat{T}_{\delta\vec{r}}\hat{H} \to \hat{H} \tag{3.2}$$

where $\delta\vec{r}$ is an infinitesimal translation of the vector $\vec{r}$. We may say that all symmetry translations of the given system form a group. This follows from the group theory. If Equation 3.2 is true, then it follows that the operator $\hat{T}_{\delta\vec{r}}$ commutes with the energy operator $\hat{H}$:

$$\hat{T}_{\delta\vec{r}}\left(\hat{H}\Psi(\vec{r})\right) = \left(\hat{T}_{\delta\vec{r}}\hat{H}\right)\left(\hat{T}_{\delta\vec{r}}\Psi(\vec{r})\right) = \hat{H}\hat{T}_{\delta\vec{r}}\Psi(\vec{r}) \tag{3.3}$$

This follows that $\hat{T}_{\delta\vec{r}}$ and $\hat{H}$ commute. **The symmetry operators of a given system commute with the Hamiltonian of that system and implies from the theory of commutation that the eigenfunctions of the Hamiltonian, i.e., the solution of the Schrödinger equation may be selected such that they are eigenfunctions of the symmetry operator.**

Let us examine the operator $\hat{T}_{\delta\vec{r}}$ of the infinitesimal translation of the system of coordinates: If the system of particles is free in the sense that on it no external force acts, then it is obvious that this is

DOI: 10.1201/9781003273073-4

the symmetry translation considering the homogeneity of space. It follows that $\hat{H}$ is invariant relative to that translation. Consider the right-hand side of Equation 3.1:

$$\hat{T}_{\delta\vec{r}}\Psi(\vec{r})=\Psi(\vec{r})+\delta\vec{r}\,\nabla\Psi(\vec{r})+\cdots \tag{3.4}$$

We expand the right-hand side of Equation 3.1 in a Taylor series with respect to $\delta\vec{r}$ and

$$\delta\vec{r}\,\nabla=\delta x\frac{\partial}{\partial x}+\delta y\frac{\partial}{\partial y}+\delta z\frac{\partial}{\partial z} \tag{3.5}$$

If the right-hand side of Equation 3.4 is represented in the form of an operator, then

$$\hat{T}_{\delta\vec{r}}\cong 1+\delta\vec{r}\,\nabla \tag{3.6}$$

That is, the operator of an infinitesimal translation of the system of coordinates. Equation 3.6 should commute with the Hamiltonian $\hat{H}$ of the given system.

If some operator commutes with $\hat{H}$ and does not depend explicitly on time t, then from definition, the physical quantity corresponding to it is an integral (constant) of motion. Let us find it. The relation in Equation 3.1 has no role to play as it corresponds to no physical quantity. But the operator $\delta\vec{r}\,\nabla$ is not self-conjugate.

Let us act on Equation 3.6 for the wave function of a classical system:

$$\Psi_{cl}=\exp\left\{\frac{iS_0}{\hbar}\right\} \tag{3.7}$$

for which we have $\delta\vec{r}\,\nabla\Psi_{cl}$. It should be noted that $\delta\vec{r}$ is an arbitrary vector and does not influence the property of commutation and so

$$\nabla\Psi_{cl}=\frac{i}{\hbar}\nabla S_0\Psi_{cl}\equiv\frac{i}{\hbar}\vec{p}_{cl}\Psi_{cl} \tag{3.8}$$

From here, we have

$$\hat{p}\Psi_{cl}=\vec{p}_{cl}\Psi_{cl} \tag{3.9}$$

where

$$\hat{\vec{p}}=-i\hbar\nabla \tag{3.10}$$

This is **the momentum operator in the coordinate representation.**

Consider:

1. The principle of homogeneity of space
2. The principle of the limiting translation from classical mechanics to the principle of correspondence

If we consider Equation 3.6, then we have

$$\hat{T}_{\delta\vec{r}}=1+\frac{i}{\hbar}\delta\vec{r}\,\hat{p}+\cdots=\exp\left\{\frac{i}{\hbar}\delta\vec{r}\,\hat{p}\right\} \tag{3.11}$$

This is a unitary operator and is dependent on the infinitesimal translation $\delta\vec{r}$ ($\hat{T}_{\delta\vec{r}}$ is unitary since $\delta\vec{r}$ is real and $\hat{p}$ is a Hermitian operator). By definition and operator, say $\hat{T}_{\delta\vec{r}}$ is unitary if its inverse $\hat{T}_{\delta\vec{r}}^{-1}$ is equal to its adjoint $\hat{T}_{\delta\vec{r}}^{\dagger}$:

$$\hat{T}_{\delta\vec{r}}^{\dagger}\hat{T}_{\delta\vec{r}} = \hat{T}_{\delta\vec{r}}\hat{T}_{\delta\vec{r}}^{\dagger} = 1 \tag{3.12}$$

From Equation 3.12, $\hat{T}_{\delta\vec{r}} \to 1$ as $\delta\vec{r} \to 0$ and follows that the expansion of $\hat{T}_{\delta\vec{r}}$ in a power series in $\delta\vec{r}$:

$$\hat{T}_{\delta\vec{r}} = 1 + \delta\vec{r}\hat{S} + \cdots \tag{3.13}$$

then

$$\hat{T}_{\delta\vec{r}}^{\dagger} = 1 + \delta\vec{r}\hat{S}^{\dagger} + \cdots \tag{3.14}$$

and

$$\hat{T}_{\delta\vec{r}}\hat{T}_{\delta\vec{r}}^{\dagger} = \hat{T}_{\delta\vec{r}}^{\dagger}\hat{T}_{\delta\vec{r}} = 1 + \delta\vec{r}\left(\hat{S} + \hat{S}^{\dagger}\right) + \cdots \tag{3.15}$$

As $\hat{T}_{\delta\vec{r}}$ is a unitary operator, it follows that the first-order terms in $\delta\vec{r}$ on the right-hand side of Equation 3.15 are equal to zero. So

$$\hat{S} + \hat{S}^{\dagger} = 0 \tag{3.16}$$

From here, it follows that the operator $\hat{S}$ is anti-Hermitian and in order for the condition in Equation 3.14 to be satisfied:

$$\hat{S} \to \hat{U} \equiv i\hat{S} \tag{3.17}$$

In that case, Equation 3.14 holds. So the right form of $\hat{T}_{\delta\vec{r}}$ is defined in Equation 3.13 and

$$\hat{U} - \hat{U}^{\dagger} = 0 \tag{3.18}$$

This shows that

$$\hat{U} = \delta\vec{r}\,\frac{\hat{p}}{\hbar} \tag{3.19}$$

is Hermitian. It can be seen that the operator $\hat{T}_{\delta\vec{r}}$ satisfies the condition in Equation 3.14 (unitary operator) as follows:

$$\hat{T}_{\delta\vec{r}}^{\dagger}\hat{T}_{\delta\vec{r}} = \hat{T}_{\delta\vec{r}}\hat{T}_{\delta\vec{r}}^{\dagger} = \exp\left\{-i\hat{U}\right\}\exp\left\{i\hat{U}\right\} = \exp\left\{i\hat{U}\right\}\exp\left\{-i\hat{U}\right\} = 1 \tag{3.20}$$

It is obvious that the operator $-i\hat{U}$ commutes with $i\hat{U}$.

3.2 Momentum Operator

Consider Equation 3.10 where

$$\hat{p}_x = -i\hbar\frac{\partial}{\partial x},\ \hat{p}_y = -i\hbar\frac{\partial}{\partial y},\ \hat{p}_z = -i\hbar\frac{\partial}{\partial z} \tag{3.21}$$

These operators $\hat{p}_x$, $\hat{p}_y$ and $\hat{p}_z$ commute with each other:

$$\hat{p}_x\hat{p}_y - \hat{p}_y\hat{p}_x = 0, \quad \hat{p}_x\hat{p}_z - \hat{p}_z\hat{p}_x = 0, \quad \hat{p}_y\hat{p}_z - \hat{p}_z\hat{p}_y = 0 \tag{3.22}$$

It follows that they have a common system of eigenfunctions that we denote by $\Psi_p(\vec{r})$ and find it in the following form:

$$\Psi_{\vec{p}}(\vec{r}) = \Psi_{p_x}(x)\Psi_{p_y}(y)\Psi_{p_z}(z) \tag{3.23}$$

Consider the Sturm-Liouville equation:

$$\hat{p}_x\Psi = p_x\Psi: \quad -i\hbar\frac{\partial\Psi}{\partial x} = p_x\Psi \tag{3.24}$$

Here p_x is the eigenvalue of the operator $\hat{p}_x$. If we substitute Equation 3.23 into 3.24, then we have

$$-i\hbar\frac{\partial\Psi_{p_x}(x)}{\partial x} = p_x\Psi_{p_x}(x) \tag{3.25}$$

from where, after integration, we have

$$\Psi_{p_x}(x) = C_1\exp\left\{\frac{ip_x x}{\hbar}\right\} \tag{3.26}$$

where C_1 is a constant of integration. If we consider Equation 3.23, then we have

$$\Psi_p(\vec{r}) = C\exp\left\{\frac{i\vec{p}\vec{r}}{\hbar}\right\} \tag{3.27}$$

which is a de Broglie plane wave with the wave vector $\vec{\kappa} = \frac{\vec{p}}{\hbar}$. Hence, de Broglie plane waves are wave functions for which the states have defined momenta $\vec{p}$.

Any formal solution of the Sturm-Liouville equation should be verified for the condition that it should be finite, continuous and univalent. Equation 3.27 is satisfied for any real value of $\vec{p}$. This implies that for any real $\vec{p}$ may be eigenvalue of the operator $\hat{p}$ and this implies that the momentum has a continuous spectrum of real values:

$$-\infty < p_x < \infty \tag{3.28}$$

There arises the question on the definition of C that is the normalization constant. This can be defined from the relation of the condition of normalization. But the wave function does not respect the condition of squarability. It follows that it can be normalized by the Dirac delta-function:

$$\int_{-\infty}^{\infty}\Psi^*_{p'_x}(x)\Psi_{p_x}(x)dx = \delta\left(p'_x - p_x\right) \tag{3.29}$$

where p_x is a parameter.

If we consider Equation 3.26:

$$|C_1|^2 \lim_{a\to\infty} \int_{-a}^{a} \exp\left\{\frac{i(p_x - p_x')}{\hbar}\right\} dx = |C_1|^2 \lim_{a\to\infty} \frac{\sin\frac{(p_x - p_x')}{\hbar}a}{\frac{(p_x - p_x')}{\hbar}} \tag{3.30}$$

and from

$$\lim_{a\to\infty} \frac{\sin xa}{\pi x} = \delta(x) \tag{3.31}$$

$$|C_1|^2 \lim_{a\to\infty} \int_{-a}^{a} \exp\left\{\frac{i(p_x - p_x')}{\hbar}\right\} dx = |C_1|^2 2\pi\delta\left(\frac{p_x - p_x'}{\hbar}\right) = |C_1|^2 2\pi\hbar\delta(p_x - p_x') = \delta(p_x - p_x') \tag{3.32}$$

We consider the fact that

$$\delta(ax) = \frac{1}{a}\delta(x) \tag{3.33}$$

It therefore follows that

$$C_1 = \frac{1}{\sqrt{2\pi\hbar}} \tag{3.34}$$

This is the normalization constant. A possible phase factor $\exp\{i\phi(p_x)\}$ does not affect anything is set equal to unity in Equation 3.34. Hence, the orthonormalized momentum eigenfunction is

$$\Psi_{\vec{p}}(\vec{r}) = \frac{1}{\sqrt{(2\pi\hbar)^3}} \exp\left\{\frac{i\vec{p}\vec{r}}{\hbar}\right\} \tag{3.35}$$

This is the wave function in the **coordinate representation.**

3.3 Heisenberg Uncertainty Relation

Consider the point defined by the radius vector $\vec{r}$:

$$\vec{r} \to \hat{\vec{r}} = \vec{r} \tag{3.36}$$

which shows that **the operator of the coordinate is the coordinate itself.**

Let us find the mean value of $\vec{r}$ in the state $\Psi(\vec{r})$. Considering the statistical interpretation:

$$|\Psi(\vec{r})|^2 d\vec{r}, \quad d\vec{r} = dxdydz \tag{3.37}$$

is the probability that a particle is found in the element of volume $d\vec{r}$. This is the probability that the head of the radius vector $\vec{r}$ lies in the element of volume $d\vec{r}$. So

$$\bar{\vec{r}} = \int \vec{r}\left|\Psi(\vec{r})\right|^2 d\vec{r} \equiv \int \Psi^*(\vec{r})\vec{r}\Psi(\vec{r})d\vec{r} \equiv \int \Psi^*(\vec{r})\hat{\vec{r}}\Psi(\vec{r})d\vec{r} \quad 3.38)$$

Compare the last two expressions of Equation 3.38:

$$\hat{\vec{r}} = \vec{r} \quad (3.39)$$

The physical quantity coincides with the argument of the wave function (the operator of the coordinate is the coordinate itself; i.e., the coordinate representation).

Consider

$$\hat{x}\hat{p}_x\Psi = -i\hbar x\frac{\partial \Psi}{\partial x} \quad (3.40)$$

and

$$\hat{p}_x\hat{x}\Psi = -i\hbar\frac{\partial}{\partial x}(x\Psi) = -i\hbar\Psi - i\hbar x\frac{\partial \Psi}{\partial x} \quad (3.41)$$

and we take the differences of the above equations:

$$\left(\hat{x}\hat{p}_x - \hat{p}_x\hat{x}\right)\Psi = i\hbar\Psi \quad (3.42)$$

from where

$$\hat{x}\hat{p}_x - \hat{p}_x\hat{x} = i\hbar \quad (3.43)$$

It can be seen that

$$\hat{x}\hat{p}_y - \hat{p}_y\hat{x} = 0 \quad (3.44)$$

So, the generalization

$$\left[\hat{x}_i, \hat{p}_j\right] = i\hbar\delta_i^j, \quad i = x, y, z \quad (3.45)$$

Every momentum with other coordinates commutes and it does not commute with its own coordinate. Equation 3.45 is the commutation relation. The fact that they do not commute should have been foreseen long ago. If they should have been commutable, the corresponding physical quantities should have had simultaneously defined values. It follows that we should have constructed a trajectory.

Consider a one-dimensional motion for which we suppose $\Psi(x)$ is the wave function corresponding to a finite motion:

$$\Psi(x)\Big|_{x\to\pm\infty} \to 0 \quad (3.46)$$

In that state, neither x nor p_x has a defined value. The mean value of x is

$$\overline{x} = \int_{-\infty}^{\infty} x\left|\Psi(x)\right|^2 dx \tag{3.47}$$

The momentum p_x at a point does not have a defined value or at that point de Broglie wave does not tend to zero for $x \to \pm\infty$.

Let us consider the mean value of the operator $\hat{p}_x$:

$$\overline{p_x} = \int_{-\infty}^{\infty} \Psi^*(x)\hat{p}_x\Psi(x)dx = -i\hbar\int_{-\infty}^{\infty} \Psi^*(x)\frac{d}{dx}\Psi(x)dx = -\frac{i\hbar}{2}\int_{-\infty}^{\infty} \frac{d}{dx}\left(\Psi(x)\right)^2 dx = 0 \tag{3.48}$$

We may set the mean value $\overline{x}$ to be zero, i.e., we select the origin of the coordinate axis at the point $\overline{x}$. In that state,

$$\overline{p_x} = \overline{x} = 0 \tag{3.49}$$

The deviation of the quantity x from its mean value:

$$\overline{x - \overline{x}} = 0 \tag{3.50}$$

The deviation to the left and right are the same. The measure of the deviation is the **mean-square deviation**:

$$(\Delta x)^2 = \overline{(x-\overline{x})^2} = \overline{x^2} - (\overline{x})^2 = \overline{x^2} \tag{3.51}$$

We select $\overline{x}$ as the origin of the coordinate axis. Similarly,

$$(\Delta p_x)^2 = \overline{p_x^{\,2}} \tag{3.52}$$

Here Equations 3.51 and 3.52 are the uncertainties of the coordinate and momentum in the given state, respectively.

Consider the expression:

$$\int_{-\infty}^{\infty}\left(x\Psi + \alpha\frac{d\Psi}{dx}\right)^2 dx \geq 0 \tag{3.53}$$

where α is a real parameter. We open the brackets in the integrand in Equation 3.53:

$$\int_{-\infty}^{\infty} x^2\Psi^2 dx + 2\alpha\int_{-\infty}^{\infty} x\Psi\frac{d\Psi}{dx}dx + \alpha^2\int_{-\infty}^{\infty}\left(\frac{d\Psi}{dx}\right)^2 dx \geq 0 \tag{3.54}$$

From Equation 3.51, we have

$$\int_{-\infty}^{\infty} x^2 \Psi^2 dx = \overline{x^2} = (\Delta x)^2 \quad (3.55)$$

Consider the second summand in Equation 3.54:

$$\int_{-\infty}^{\infty} x\Psi \frac{d\Psi}{dx} dx = \int_{-\infty}^{\infty} x \frac{1}{2} \frac{d\Psi^2}{dx} dx = \frac{1}{2} \int_{-\infty}^{\infty} x \frac{d\Psi^2}{dx} dx \quad (3.56)$$

We may now use integration by parts:

$$\frac{1}{2} \int_{-\infty}^{\infty} x \frac{d\Psi^2}{dx} dx = \frac{1}{2} x\Psi^2 \Big|_{-\infty}^{\infty} - \frac{1}{2} \int_{-\infty}^{\infty} \Psi^2 dx = -\frac{1}{2} \quad (3.57)$$

Consider the third summand in Equation 3.54:

$$\int_{-\infty}^{\infty} \left(\frac{d\Psi}{dx} \right)^2 dx = \int_{-\infty}^{\infty} \frac{d\Psi}{dx} \frac{d\Psi}{dx} dx \quad (3.58)$$

We do integration by parts considering Equation 3.52:

$$\int_{-\infty}^{\infty} \frac{d\Psi}{dx} \frac{d\Psi}{dx} dx = \Psi \frac{d\Psi}{dx} \Big|_{-\infty}^{\infty} - \int_{-\infty}^{\infty} \Psi \frac{d^2\Psi}{dx^2} dx = \frac{1}{\hbar^2} \int_{-\infty}^{\infty} \Psi \hat{p}_x^2 \Psi dx = \frac{\overline{p_x^2}}{\hbar^2} = \frac{(\Delta p_x)^2}{\hbar^2} \quad (3.59)$$

So Equation 3.54 now becomes

$$(\Delta x)^2 - \alpha + \alpha^2 \frac{(\Delta p_x)^2}{\hbar^2} \geq 0 \quad (3.60)$$

This polynomial of the second order in α is positive definite. According to Equation 3.53, the discriminant must be negative or vanish. The quantity in Equation 3.69 must be positive definite for all α. So the roots of the quadratic equation in Equation 3.60 must be complex. Hence,

$$1 - \frac{4(\Delta x)^2 (\Delta p_x)^2}{\hbar^2} \leq 0 \quad (3.61)$$

or

$$\Delta x \Delta p_x \geq \frac{\hbar}{2} \quad (3.62)$$

This is the **Heisenberg uncertainty relation: In the quantum domain, it is impossible to describe the motion of a particle with the help of the trajectory. There may not be a state for which the coordinates**

of the particle and the projection of the momentum should have simultaneously defined values. It follows that if one of these quantities has a defined value, say $\Delta x = 0$, then the second quantity Δp_x is absolutely undefined. The converse is true.

3.4 Momentum Representation

The momentum representation is preferred in problems where a particle position in space might not be of interest, but rather how fast it goes and in what direction. In the scattering theory, the momentum representation can be applicable where a particle of well-defined momentum is directed towards a scattering center such as an atomic nucleus where interest will be on the direction where a particle is scattered and of the measurement of the momentum or energy of the scattered particle, though the position representation may provide a mental image of the scattering process as waves scattering off an obstacle.

Consider the general eigenfunction of the three operators $\hat{p}_x, \hat{p}_y$ and $\hat{p}_z$ represented in Equation 3.35.

This system of general functions should have the property of completeness: **Any function may be expanded in a series or exchange the expression for the integral (Fourier integral):**

$$\Psi(\vec{r}) = \int C(\vec{p})\Psi_{\vec{p}}(\vec{r})d\vec{p} \tag{3.63}$$

In order to define $C(\vec{p})$, we multiply Equation 3.63 by $\Psi^*_{\vec{p}}(\vec{r})$ and integrate with respect to $\vec{r}$ considering:

$$\int \Psi^*_{\vec{p}'}(\vec{r})\Psi_{\vec{p}}(\vec{r})d\vec{r} = \delta(\vec{p}' - \vec{p}) \tag{3.64}$$

and

$$\int \Psi(\vec{r})\Psi^*_{\vec{p}'}(\vec{r})d\vec{r} = \int C(\vec{p})d\vec{p}\int \Psi_{\vec{p}}(\vec{r})\Psi^*_{\vec{p}'}(\vec{r})d\vec{r} = \int C(\vec{p})\delta(\vec{p}' - \vec{p})d\vec{p} = C(\vec{p}') \tag{3.65}$$

So

$$C(\vec{p}) = \int \Psi(\vec{r})\Psi^*_{\vec{p}}(\vec{r})d\vec{r} \tag{3.66}$$

We find out that the wave function is represented with respect to the momentum eigenstates as basis functions, and that between $\Psi(\vec{r})$ and $C(\vec{p})$, there exists a mutual univalent correspondence and, in particular, if $C(\vec{p})$ is given, then from the function in Equation 3.63, we can define $\Psi(\vec{r})$. The direct and the inverse transformation may be done and considering Equation 3.63, we find out that the momentum wave function, $C(\vec{p})$, and the position wave function, $\Psi(\vec{r})$, are **Fourier transform pairs**.

From the principle of superposition in Equation 3.63, it follows that $\left|C(\vec{p})\right|^2$ is the probability of measurement of the momentum, i.e., at a given state $\Psi(\vec{r})$, the momentum of a particle is found in the element of the volume $d\vec{p}$. The quantity $\left|C(\vec{p})\right|^2$ is the probability density in the momentum representation or $\vec{p}$-representation. We examine how to define operators in the $\vec{p}$-representation. For the operator in the coordinate or $\vec{r}$-representation we place the symbol "∧" above the given physical quantities, say

$$\hat{\vec{r}} = \vec{r},\ \hat{p} = -i\hbar\nabla_{\vec{r}} \tag{3.67}$$

and for the $\vec{p}$-representation, we place above the physical quantity the symbol "∨," say

$$\check{\mathrm{p}} = \vec{\mathrm{p}} \tag{3.68}$$

Then from Equation 3.38, we have

$$\bar{\vec{r}} = \int \Psi^*(\vec{r})\vec{r}\Psi(\vec{r})d\vec{r} = \iint d\vec{\mathrm{p}}'d\vec{\mathrm{p}}d\vec{r}C^*(\vec{\mathrm{p}}')C(\vec{\mathrm{p}})\Psi^*_{\vec{\mathrm{p}}'}(\vec{r})\vec{r}\Psi_{\vec{\mathrm{p}}}(\vec{r}) \tag{3.69}$$

From

$$\vec{r}\Psi_{\vec{\mathrm{p}}}(\vec{r}) = -i\hbar\nabla_{\vec{\mathrm{p}}}\Psi_{\vec{\mathrm{p}}}(\vec{r}) \tag{3.70}$$

and

$$\nabla_{\vec{\mathrm{p}}}(\vec{\mathrm{p}},\vec{r}) = \vec{r} \tag{3.71}$$

then

$$\bar{\vec{r}} = -i\hbar\int d\vec{\mathrm{p}}'d\vec{\mathrm{p}}C^*(\vec{\mathrm{p}}')C(\vec{\mathrm{p}})\nabla_{\vec{\mathrm{p}}}\int \Psi^*_{\vec{\mathrm{p}}'}(\vec{r})\Psi_{\vec{\mathrm{p}}}(\vec{r})d\vec{r} \tag{3.72}$$

From Equation 3.64:

$$\bar{\vec{r}} = -i\hbar\int d\vec{\mathrm{p}}'d\vec{\mathrm{p}}C^*(\vec{\mathrm{p}}')C(\vec{\mathrm{p}})\nabla_{\vec{\mathrm{p}}}\delta(\vec{\mathrm{p}}-\vec{\mathrm{p}}') \tag{3.73}$$

We do integration by parts in the $\vec{p}$-space. We suppose that $C(\vec{p}) = 0$ at infinity as the property is attached to $\Psi(\vec{r})$ and $C(\vec{\mathrm{p}})$, its image. So

$$\begin{aligned}\bar{\vec{r}} &= -i\hbar\int d\vec{\mathrm{p}}'d\vec{\mathrm{p}}C^*(\vec{\mathrm{p}}')C(\vec{\mathrm{p}})\nabla_{\vec{\mathrm{p}}}\delta(\vec{\mathrm{p}}-\vec{\mathrm{p}}') = -i\hbar\int d\vec{\mathrm{p}}'C^*(\vec{\mathrm{p}}')C(\vec{\mathrm{p}})\delta(\vec{\mathrm{p}}-\vec{\mathrm{p}}')\Big|_{-\infty}^{\infty} \\ &+ i\hbar\int d\vec{\mathrm{p}}'d\vec{\mathrm{p}}C^*(\vec{\mathrm{p}}')\delta(\vec{\mathrm{p}}-\vec{\mathrm{p}}')\nabla_{\vec{\mathrm{p}}}C(\vec{\mathrm{p}})\end{aligned} \tag{3.74}$$

or

$$\bar{\vec{r}} = i\hbar\int d\vec{\mathrm{p}}C^*(\vec{\mathrm{p}})\nabla_{\vec{\mathrm{p}}}C(\vec{\mathrm{p}}) = \int d\vec{\mathrm{p}}C^*(\vec{\mathrm{p}})i\hbar\nabla_{\vec{\mathrm{p}}}C(\vec{\mathrm{p}}) \tag{3.75}$$

On the other hand, we use the allegation that $C(\vec{p})$ is the wave function in the $\vec{p}$-representation:

$$\bar{\vec{r}} = \int d\vec{\mathrm{p}}C^*(\vec{\mathrm{p}})\check{\vec{r}}C(\vec{\mathrm{p}}) \tag{3.76}$$

Comparing Equation 3.76 with 3.75, we have

$$\check{\vec{r}} = i\hbar\nabla_{\vec{\mathrm{p}}} \tag{3.77}$$

The $\vec{p}$ or other representations are used to solve problems for convenience.

Homework: Prove that the Heisenberg relations

$$\hat{x}\hat{p}_x - \hat{p}_x\hat{x} = i\hbar,\ \check{x}\check{p}_x - \check{p}_x\check{x} = i\hbar \tag{3.78}$$

are true for the $\vec{p}$-as well as the $\vec{r}$-representations. This is the property of unitary invariance. We may have the energy representation for which the operator and other quantities are represented in the matrix form.

Question:
Find the operator $\check{\vec{r}}$ in the $\vec{p}$-representation, and define its eigenfunctions and spectrum of eigenvalues.

Answer:
In the $\vec{r}$-representation, $\hat{\vec{r}} = \vec{r}$. In order to make a transition to a new representation, we use the expression for the mean value $\bar{\vec{r}}$ of $\vec{r}$ by letting

$$\Psi(\vec{r}) = \int C(\vec{p})\exp\left\{\frac{i\vec{p}\vec{r}}{\hbar}\right\}d\vec{p} \tag{3.79}$$

then

$$\bar{\vec{r}} = \int \Psi^*(\vec{r})\vec{r}\Psi(\vec{r})d\vec{r} = \int \Psi^*(\vec{r})\vec{r}\int C(\vec{p})\exp\left\{\frac{i\vec{p}\vec{r}}{\hbar}\right\}d\vec{p}\,d\vec{r} \tag{3.80}$$

If we change the order of integration over $\vec{p}$ and $\vec{r}$ and consider

$$\vec{r}\exp\left\{\frac{i\vec{p}\vec{r}}{\hbar}\right\} = \frac{\hbar}{i}\frac{\partial}{\partial\vec{p}}\exp\left\{\frac{i\vec{p}\vec{r}}{\hbar}\right\} \tag{3.81}$$

then we do integration by parts over $\vec{p}$ and consider $C(\vec{p}) = 0$ at the boundaries of integration:

$$\int \Psi^*(\vec{r})\int C(\vec{p})\frac{\hbar}{i}\frac{\partial}{\partial\vec{p}}\exp\left\{\frac{i\vec{p}\vec{r}}{\hbar}\right\}d\vec{p}\,d\vec{r} = \int \Psi(\vec{r})\int \exp\left\{\frac{i\vec{p}\vec{r}}{\hbar}\right\}i\hbar\frac{\partial}{\partial\vec{p}}C(\vec{p})d\vec{p}\,d\vec{r} \tag{3.82}$$

But

$$\int \Psi^*(\vec{r})\exp\left\{\frac{i\vec{p}\vec{r}}{\hbar}\right\}d\vec{r} = C^*(\vec{p}) \tag{3.83}$$

then

$$\bar{\vec{r}} = \int C^*(\vec{p})i\hbar\frac{\partial}{\partial\vec{p}}C(\vec{p})d\vec{p} \tag{3.84}$$

i.e.,

$$\check{\vec{r}} = i\hbar\frac{\partial}{\partial\vec{p}} \tag{3.85}$$

and its eigenfunctions are obtained from the following equation:

$$i\hbar\frac{\partial}{\partial\vec{p}}C(\vec{p})=\vec{r}C(\vec{p}) \tag{3.86}$$

The solution of this equation is

$$C(\vec{p})=A\exp\left\{-\frac{i\vec{p}\vec{r}}{\hbar}\right\} \tag{3.87}$$

where A is a normalization constant.

3.4.1 Momentum Representation of Particle in Triangular Potential Well

We find the wave function of the stationary states of a particle in the potential field (**triangular potential well**) $U(x)$:

$$U(x)=\begin{cases}F_0x, & x>0,\ F_0>0\\ \infty, & x<0\end{cases} \tag{3.88}$$

We investigate also the eigenenergy states of this particle with the potential field which is represented in Figure 3.1.

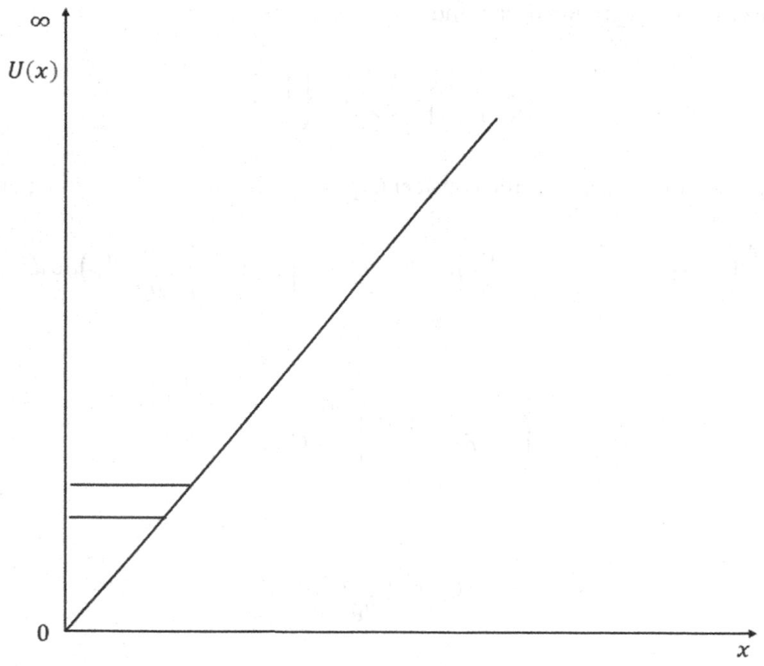

FIGURE 3.1 The triangular potential well.

The Schrödinger equation in the momentum representation:

$$\frac{p^2}{2\mu}C_E(\mathrm{p})+\mathrm{F}_0 i\hbar\frac{d}{d\mathrm{p}}C_E(\mathrm{p})=EC_E(\mathrm{p}) \tag{3.89}$$

The solution of this equation yields the following:

$$C_E(p)=\mathrm{A}(E)\exp\left\{\frac{i}{\mathrm{F}_0\hbar}\left(\frac{\mathrm{p}^3}{6\mu}-E\mathrm{p}\right)\right\} \tag{3.90}$$

We normalize the wave function $C_E(p)$ by the Dirac delta-function with respect to the energy, E:

$$\int_{-\infty}^{\infty}C_E(\mathrm{p})C_{E'}^*(\mathrm{p})d\mathrm{p}=|\mathrm{A}_E|^2\int_{-\infty}^{\infty}d\mathrm{p}\exp\left\{\frac{i}{\mathrm{F}_0\hbar}(E-E')\mathrm{p}\right\}=2\pi|\mathrm{A}_E|^2\mathrm{F}_0\hbar\delta(E-E')=\delta(E-E') \tag{3.91}$$

and

$$2\pi|\mathrm{A}_E|^2\mathrm{F}_0\hbar=1 \tag{3.92}$$

So

$$C_E(p)=\frac{1}{\sqrt{2\pi\mathrm{F}_0\hbar}}\exp\left\{\frac{i}{\mathrm{F}_0\hbar}\left(\frac{p^3}{6\mu}-E\mathrm{p}\right)\right\} \tag{3.93}$$

The boundary condition for the problem for the position space: The wave function in position space must vanish at $x=0$. Transform $C_E(p)$ to the position space through the Fourier integral:

$$\Psi(x)=\frac{1}{\sqrt{2\pi\hbar}}\int_{-\infty}^{\infty}dq\exp\left\{\frac{ipx}{\hbar}\right\}C_E(\mathrm{p})=\frac{1}{2\pi\hbar\mathrm{F}_0^{\frac{1}{2}}}\int_{-\infty}^{\infty}d\mathrm{p}\exp\left\{\frac{i}{\mathrm{F}_0\hbar}\left(\frac{p^3}{6\mu}-E\mathrm{p}\right)+\frac{i\mathrm{p}x}{\hbar}\right\} \tag{3.94}$$

From the condition $\Psi(0)=0$, we obtain the following equation pivotal in energy quantization:

$$\int_{-\infty}^{\infty}d\mathrm{p}\exp\left\{\frac{i}{\mathrm{F}_0\hbar}\left(\frac{p^3}{6\mu}-E\mathrm{p}\right)\right\}=\int_{-\infty}^{\infty}d\mathrm{p}\cos\left[\frac{1}{\mathrm{F}_0\hbar}\left(\frac{p^3}{6\mu}-E\mathrm{p}\right)\right]=0 \tag{3.95}$$

The integral of the sin function (odd function) is equal to zero. Letting

$$q=\frac{p}{(2\mu\mathrm{F}_0\hbar)^{\frac{1}{3}}} \tag{3.96}$$

then the standard mathematical function:

$$\int_{-\infty}^{\infty} dq \cos\left[\frac{q^3}{3} - E\left(\frac{2\mu}{F_0^2\hbar^2}\right)^{\frac{1}{3}} q\right] = \pi A_i\left[-E\left(\frac{2\mu}{F_0^2\hbar^2}\right)^{\frac{1}{3}}\right] = 0 \tag{3.97}$$

Here $A_i(x)$ denotes the Airy function (see details under the appendix of Chapter 24) with zeros:

$$E_n' = -E_n\left(\frac{2\mu}{F_0^2\hbar^2}\right)^{\frac{1}{3}} \tag{3.98}$$

and

$$E_n = -\left(\frac{F_0^2\hbar^2}{2\mu}\right)^{\frac{1}{3}} E_n' \tag{3.99}$$

The first three zeros of the Airy function E_n':

$$E_1' = -2.338, \quad E_2' = -4.088, \quad E_3' = -5.521 \tag{3.100}$$

In our case, the position-space wave function is the Airy function.

3.4.2 Momentum Representation of Particle in Delta Potential Well

In the p-representation, we find the energy level and normalized wave function $C(p)$ of a discrete spectrum in the fields:

$$U(x) = -\alpha\delta(x) \tag{3.101}$$

$$U(r) = -\alpha\delta(r-a) \tag{3.102}$$

In addition, we find the

- Energy when the azimuthal quantum number $l = 0$
- The solution of the Schrödinger equation when $C(p) = 0$ for $p \le p_0$, where $p_0 > 0$
- Discrete energy spectrum for the well of arbitrary depth when the particle is localized in a bounded domain of space
- Binding energy for particle in a shallow well when $\xi \equiv \frac{\mu\alpha a}{\hbar^2} \ll 1$
- Quasi discrete energy levels of the s-state of the particle when $\alpha < 0$ and investigate the system under condition, $\xi \gg 1$

We find the potential in the p-representation:

$$\tilde{U}(p) = \frac{1}{2\pi\hbar}\int U(x)\exp\left\{-\frac{ipx}{\hbar}\right\}dx = -\frac{\alpha}{2\pi\hbar} \tag{3.103}$$

The Schrödinger equation with this potential has the following form:

$$\frac{p^2}{2\mu}C(p)+\int_{-\infty}^{\infty}\tilde{U}(p-p')C(p')dp'=\frac{p^2}{2\mu}C(p)-\frac{\alpha}{2\pi\hbar}\int_{-\infty}^{\infty}C(p)\,dp=EC(p) \tag{3.104}$$

Letting

$$\int_{-\infty}^{\infty}C(p)\,dp=A \tag{3.105}$$

we find the solution in Equation 3.180 in the following form:

$$C(p)=\frac{\alpha\mu}{\pi\hbar}\frac{A}{p^2+p_0^2},\ p_0^2=2\mu|E|,\ E=-|E|<0 \tag{3.106}$$

From Equations 3.105 and 3.106, we have

$$A=\frac{\alpha\mu}{\pi\hbar}A\int_{-\infty}^{\infty}\frac{1}{p^2+p_0^2}dp \tag{3.107}$$

or

$$1=\frac{\alpha\mu}{\pi\hbar}\int_{-\infty}^{\infty}\frac{1}{p^2+p_0^2}dp=\frac{\alpha\mu}{\pi\hbar}\frac{1}{p_0}\tan^{-1}\frac{p}{p_0}\bigg|_{-\infty}^{\infty}=\frac{\alpha\mu}{\hbar p_0}=\frac{\alpha}{\hbar}\sqrt{\frac{\mu}{2|E|}} \tag{3.108}$$

Equation 3.108 represents the energy spectrum of the particle and has only one solution:

$$E_0=-\frac{\mu\alpha^2}{2\hbar^2} \tag{3.109}$$

corresponding to the wave function in Equation 3.106 when

$$A=\sqrt{\frac{2\pi\mu\alpha}{\hbar}} \tag{3.110}$$

We consider the case in Equation 3.102 and so the Schrödinger equation in the *p*-representation:

$$\frac{p^2}{2\mu}C(p)+\int\tilde{U}(p-p')C(p')\,dp=EC(p) \tag{3.111}$$

where

$$\tilde{U}(q)=\frac{1}{(2\pi\hbar)^3}\int U(r)\exp\left\{-\frac{i\vec{q}\vec{r}}{\hbar}\right\}dV=-\frac{\alpha a}{2\pi^2\hbar^2}\frac{1}{q}\sin\left(\frac{aq}{\hbar}\right) \tag{3.112}$$

From the state with $l = 0$, when the wave function is independent of the angular variables, the Schrödinger equation in the *p*-representation becomes

$$\left(\frac{p^2}{2\mu} - E\right)C(p) = \frac{\alpha a}{\pi\hbar^2}\int \frac{\sin\left(\frac{a}{\hbar}\sqrt{p^2 + p'^2 - 2pp'\cos\theta}\right)}{\sqrt{p^2 + p'^2 - 2pp'\cos\theta}} C(p')p'^2\, dp' \sin\theta d\theta$$
$$= \frac{2\alpha}{p\pi\hbar}\sin\left(\frac{pa}{\hbar}\right)\int_0^\infty \sin\left(\frac{p'a}{\hbar}\right)C(p')p' dp' = 0 \quad (3.113)$$

Letting

$$A \equiv \int_0^\infty \sin\left(\frac{p'a}{\hbar}\right)C(p')p' dp' \quad (3.114)$$

the solution of Equation 3.113 becomes:

$$C(p) = \frac{4\mu\alpha A}{\pi\hbar p\left(p^2 - 2\mu E\right)}\sin\left(\frac{pa}{\hbar}\right) \quad (3.115)$$

Considering Equations 3.115 and 3.114:

$$A \equiv \int_0^\infty \sin^2\left(\frac{pa}{\hbar}\right)\frac{4\mu\alpha A}{\pi\hbar\left(p^2 - 2\mu E\right)} dp \quad (3.116)$$

This describes the energy spectrum:

$$E = -\frac{\hbar^2\chi^2}{2\mu} < 0 \quad (3.117)$$

We rewrite Equation 3.116 in the following form:

$$\frac{\mu\alpha}{\pi\hbar}\int_0^\infty \frac{1 - \cos\left(\frac{pa}{\hbar}\right)}{p^2 + \hbar^2\chi^2} dp = 1 \quad (3.118)$$

From the residue theorem and from

$$\cos(x) = \frac{1}{2}\left(\exp\{x\} + \exp\{-x\}\right) \quad (3.119)$$

then

$$\frac{\mu\alpha}{\pi\hbar^2}\left(1 - \exp\{-2a\chi\}\right) = a\chi \quad (3.120)$$

This relation describes the discrete energy spectrum of the s-state and for

$$\xi > \frac{1}{2} \tag{3.121}$$

we have one discrete s-state, while for

$$\xi < \frac{1}{2} \tag{3.122}$$

no discrete state exists. This implies that the potential well cannot have a particle.

From the boundary condition when $C(p) = 0$ for $p \leq p_0$, where $p_0 > 0$, we consider Equations 3.111, 3.113 and 3.118:

$$A \equiv \int_{p_0}^{\infty} \sin^2\left(\frac{pa}{\hbar}\right) \frac{4\mu\alpha A}{\pi\hbar\left(p^2 - 2\mu E\right)} dp \tag{3.123}$$

For this case, the wave function $C(p)$ in the p-representation is obtained:

$$C(p) = \frac{4\mu\alpha A}{\pi\hbar p\left(p^2 - 2\mu E\right)} \sin\left(\frac{pa}{\hbar}\right) \tag{3.124}$$

The bound state of a discrete spectrum of the particle has the energy:

$$E < E_0 = \frac{p_0^2}{2\mu} \tag{3.125}$$

It is instructive to note that the energy is not $E < 0$ as expected since E_0 is the minimum energy of the free particle. So the discrete energy spectrum of the particle can be obtained from

$$A \equiv \int_{p_0}^{\infty} \sin^2\left(\frac{pa}{\hbar}\right) \frac{4\mu\alpha A}{\pi\hbar\left(p^2 - 2\mu E\right)} dp \tag{3.126}$$

The energy E in Equation 3.126 satisfies the condition in Equation 3.125. Equation 3.126 has only one monotone solution for the energy, E. We find the binding energy of the particle under condition $\xi \ll 1$ for the case of a shallow well:

$$\varepsilon = E_0 - E > 0 \tag{3.127}$$

So for the limiting value $\xi \to 0$, from Equation 3.126, we have $E \to E_0$ or $\varepsilon \to 0$. For this case, the integral in Equation 3.126 is divergent considering the lower limit. The value of this integral can be calculated when the domain of definition of p is at the neighborhood of the lower limit. For this reason, for $\varepsilon \to 0$:

$$\int_{0}^{\infty} \sin^2\left(\frac{pa}{\hbar}\right) \frac{1}{p^2 - p_0^2 + 2\mu\varepsilon} dp \cong \sin^2\left(\frac{p_0 a}{\hbar}\right) \int_{p_0}^{\infty} \frac{1}{p^2 - p_0^2 + 2\mu\varepsilon} dp \cong -\frac{1}{2p_0} \sin^2\left(\frac{p_0 a}{\hbar}\right) \ln\frac{\varepsilon}{4E_0} \tag{3.128}$$

From here and considering Equation 3.126 with condition $\xi \ll 1$, we obtain the binding energy of the particle as follows:

$$\varepsilon = 4E_0 \exp\left\{-\frac{\pi p_0 a}{2\hbar\xi \sin^2\left(\frac{p_0 a}{\hbar}\right)}\right\} \tag{3.129}$$

It should be noted that for $\xi \to 0$, the binding energy ϵ exponentially tends to zero.

3.4.3 Momentum Representation of an Operator in Matrix Form

Let us examine again Equation 3.80 taking into consideration Equation 3.72, we have

$$\begin{aligned}\bar{\vec{r}}_{\vec{p}'\vec{p}} &= \int \Psi^*_{\vec{p}'}(\vec{r})\vec{r}\Psi_{\vec{p}}(\vec{r})d\vec{r} = \frac{1}{(2\pi\hbar)^3}\int \exp\left\{\frac{-i\vec{p}\vec{r}}{\hbar}\right\}\vec{r}\exp\left\{\frac{i\vec{p}'\vec{r}}{\hbar}\right\}d\vec{r} \\ &= \frac{\hbar}{i}\frac{\partial}{\partial \vec{p}'}\frac{1}{(2\pi\hbar)^3}\int \exp\left\{\frac{-i(\vec{p}'-\vec{p})\vec{r}}{\hbar}\right\}d\vec{r} = \frac{\hbar}{i}\frac{\partial}{\partial \vec{p}'}\delta(\vec{p}'-\vec{p})\end{aligned} \tag{3.130}$$

or

$$\bar{\vec{r}}_{\vec{p}'\vec{p}} = \frac{\hbar}{i}\frac{\partial}{\partial \vec{p}'}\delta(\vec{p}'-\vec{p}) \tag{3.131}$$

This is the **momentum representation of the operator of the r coordinate in the matrix representation.**

3.5 Particle Hamiltonian in a Potential Field

Consider the quantities described in classical mechanics by functions of the coordinate and momentum:

$$F = F(q,p) \tag{3.132}$$

Here, q is the totality of the configuration coordinates and p is the totality of the momenta. Then in the operator representation:

$$\hat{F} = F(\hat{q},\hat{p}) \tag{3.133}$$

which is a function of the operator and has sense only when Equation 3.133 is related to these quantities such that we may expand in a series with respect to $\hat{p}$. For the construction of the operator $F(\hat{A})$, it should be recollected that $F(\hat{A})$ is Hermitian. Let us construct the operator $\hat{A}\hat{B}$ in classical physics. It is instructive to note that there is an ambiguity of the operator ordering in the definition of such a product of operators, and to resolve this ambiguity of two non-commuting operators, we symmetrize them. So the operator ordering of two Hermitian operators may be done as follows:

$$AB = BA = \frac{AB+BA}{2} \tag{3.134}$$

In quantum mechanics $\left[\hat{A},\hat{B}\right] \neq 0$. Thus,

$$\widehat{AB} = \frac{1}{2}\left(\hat{A}\hat{B} + \hat{B}\hat{A}\right) = \frac{1}{2}\left\{\hat{A},\hat{B}\right\} \tag{3.135}$$

This is a symmetric rule to guarantee Hermiticity.

3.5.1 Hamilton Function Operator and Ehrenfest Theorem

Consider now a zero-spin particle moving in the potential field $U(\vec{r})$:

$$\hat{H} = \hat{T} + U(\vec{r}) \equiv \frac{\hat{p}^2}{2m} + U(\vec{r}) = -\frac{\hbar^2}{2m}\Delta + U(\vec{r}) \tag{3.136}$$

Consider the kinetic energy operator, $\hat{T}$ commutes with the momentum $\hat{\vec{p}}$:

$$\left[\hat{T},\hat{p}\right] = 0 \tag{3.137}$$

The eigenfunction $\Psi_{\vec{p}}$ of the operator of the momentum is also the eigenfunction of the operator of the kinetic energy:

$$\hat{T}\Psi_{\vec{p}} = \frac{\hat{p}^2}{2m}\Psi_{\vec{p}} \tag{3.138}$$

where $\Psi_{\vec{p}}$ is the eigenfunction of the operator $\hat{T}$. The eigenfunction of the operator of the momentum is also the eigenfunction of the Hamiltonian of a particle.

Consider the velocity operator $\hat{\dot{x}}$ for which we apply the rule of the derivative of an operator with respect to time t:

$$\hat{\dot{x}} = \frac{i}{\hbar}\left[\hat{H},\hat{x}\right] \tag{3.139}$$

where the partial derivative $\frac{\partial}{\partial t}$ is absent. We consider $\hat{H}$ and evaluate $\left[\hat{H},\hat{x}\right]$:

$$\hat{\dot{x}} = \frac{i}{\hbar}\left[\frac{\hat{p}_x^2}{2m},\hat{x}\right] = \frac{i}{2m\hbar}\left(\hat{p}_x^2\hat{x} - \hat{x}\hat{p}_x^2\right) \tag{3.140}$$

which may have been easily evaluated using the commutation relation:

$$\hat{x}\hat{p}_x - \hat{p}_x\hat{x} = i\hbar \tag{3.141}$$

from where it follows that

$$\hat{x}\hat{p}_x = i\hbar + \hat{p}_x\hat{x} \tag{3.142}$$

and thus

$$\hat{p}_x^2\hat{x} - \hat{x}\hat{p}_x^2 = \hat{p}_x^2\hat{x} - \hat{x}\hat{p}_x\hat{p}_x = \hat{p}_x^2\hat{x} - \left(i\hbar + \hat{p}_x\hat{x}\right)\hat{p}_x = \hat{p}_x^2\hat{x} - 2i\hbar\hat{p}_x - \hat{p}_x^2\hat{x} \tag{3.143}$$

or

$$\hat{p}_x^2\hat{x} - \hat{x}\hat{p}_x^2 = -2i\hbar\hat{p}_x \tag{3.144}$$

This follows that

$$\hat{\dot{x}} = \frac{\hat{p}_x}{m} \tag{3.145}$$

We find similarly $\hat{\dot{p}}_x$ in classical mechanics the operator of the force:

$$\hat{\dot{p}}_x = \frac{i}{\hbar}\left[\hat{H},\hat{p}_x\right] \tag{3.146}$$

The operator of the kinetic energy commutes with $\hat{p}_x$:

$$\hat{\dot{p}}_x = \frac{i}{\hbar}\left[\hat{U},\hat{p}_x\right] \tag{3.147}$$

Consider $\left[\hat{U},\hat{p}_x\right]$:

$$\left[\hat{U},\hat{p}_x\right]\Psi = \left(\hat{U}\hat{p}_x - \hat{p}_x\hat{U}\right)\Psi = \hat{U}\left(\hat{p}_x\Psi\right) - \hat{p}_x\left(\hat{U}\Psi\right) = \hat{U}\left(\hat{p}_x\Psi\right) - \hat{U}\left(\hat{p}_x\Psi\right) - \Psi\left(\hat{p}_x\hat{U}\right) = i\hbar\Psi\frac{\partial}{\partial x}\hat{U} \tag{3.148}$$

So

$$\hat{\dot{p}}_x = -\frac{\partial\hat{U}}{\partial x} \tag{3.149}$$

which is the well-known Newton's second law.

If we take the expectation value of Equations 3.145 and 3.149, then we have **the time evolution of the expectation value, respectively, of the momentum and coordinate operators (Ehrenfest theorem)**:

$$\overline{\hat{\dot{p}}_x} = -\overline{\frac{\partial\hat{U}}{\partial x}} \tag{3.150}$$

$$\overline{\hat{\dot{x}}} = \frac{\overline{\hat{p}_x}}{m} \tag{3.151}$$

Quantum mechanical expectation values satisfying the law of classical mechanics.

From

$$\dot{\bar{p}}_x = m\ddot{\bar{x}} \tag{3.152}$$

then considering Equation 3.150, we have

$$\ddot{\bar{x}} = \frac{\dot{\bar{p}}_x}{m} \tag{3.153}$$

It follows from Equations 3.150 and 3.153 that

$$\dot{\vec{p}} = -\nabla U, \ddot{\vec{r}} = \frac{\dot{\vec{p}}}{m} \tag{3.154}$$

It can be seen that Ehrenfest's theorem provides an example of correspondence principle. This is because in it is inherent the motion of a wave packet as a classical particle. Equation 3.154 shows the connection with Newtonian equations of motion derived by Ehrenfest.

3.6 Angular Momentum Operator

3.6.1 Infinitesimal Rotation Operator

For the momentum conservation law, we use the property of homogeneity as well as isotropy of space. From space isotropy, the Hamiltonian of the closed system is invariant relative to the rotation of the system through an arbitrary angle about an arbitrary axis. We proceed with this for an infinitesimal translation operator $\hat{T}_{\delta\vec{r}}$ of the coordinate and an infinitesimal rotation operator $\hat{T}_{\delta\vec{\phi}}$ of a system through an angle $\delta\phi$ which is the magnitude of the vector $\delta\vec{\phi}$ of an infinitesimal rotation directed along the axis about which the rotation takes place:

$$\hat{T}_{\delta\vec{r}} \cong 1 + \delta\vec{r}\,\nabla \tag{3.155}$$

Here, $\delta\vec{r}$ is the infinitesimal translation of the radius vector, $\vec{r}$, for an infinitesimal rotation:

$$\delta\vec{r} = \left[\delta\vec{\phi}, \vec{r}\right] \tag{3.156}$$

then

$$\hat{T}_{\delta\vec{\phi}} \cong 1 + \left[\delta\vec{\phi}, \vec{r}\right]\nabla \tag{3.157}$$

And if we consider the cyclical arrangement, then we have

$$\hat{T}_{\delta\vec{\phi}} \cong 1 + \left[\delta\vec{\phi}, \vec{r}\right]\nabla = 1 + \delta\vec{\phi}\left[\vec{r}, \nabla\right] \tag{3.158}$$

From the isotropy of space, Equation 3.158 should be the transformation of the symmetry for the Hamiltonian $\hat{H}$ of a free particle. This implies that $\hat{T}_{\delta\vec{\phi}}$ commutes with $\hat{H}$, and so forms an integral (constant) of motion as is expected for a system that possesses rotation symmetry. Consider the influence of the operator ∇ on the wave function $\exp\left\{\frac{iS_0}{\hbar}\right\}$ of a system where S_0 is the classical action, we have

$$\frac{i}{\hbar}\left[\vec{r}, \nabla S_0\right] = \frac{i}{\hbar}\left[\vec{r}, \vec{p}\right] = \frac{i}{\hbar}\vec{L} \tag{3.159}$$

Here, $\vec{L}$ is the classical angular momentum vector.

The quantization procedure in this case amounts to replacing all quantities in this definition with the corresponding quantum mechanical operators, which are

$$\hat{\vec{p}} = -i\hbar\nabla, \hat{\vec{r}} = \vec{r}, \left[\vec{r}, \hat{\vec{p}}\right] = \hat{\vec{L}} \tag{3.160}$$

So, from Equation 3.158, we arrive at the infinitesimal rotation operator:

$$\hat{T}_{\delta\vec{\phi}} \cong 1 - \frac{i}{\hbar}\delta\phi\hat{L} \tag{3.161}$$

We therefore find the following expression for the operator, $\hat{T}_{\vec{\phi}}$, performing a rotation by an infinitesimal angle $\delta\vec{\phi} \to \vec{\phi}$ around the z-axis:

$$\hat{T}_{\vec{\phi}} = \exp\left\{-\frac{i}{\hbar}\phi\hat{L}\right\} \tag{3.162}$$

This result explains the **rotating frame** name commonly given to the interaction representation.

We arrive at the fact that

- **All rotation operators around a specific axis are exponentials of the angular momentum operator along that axis.**
- **The expressions for the angular momentum operators as well and the conservation laws for the corresponding observables are consequences of the isotropy of space.**
- **All conservation laws stem from symmetries and in particular infinitesimal translation, linear momentum operators and the associated conservation laws appear.**

3.6.2 Angular Momentum Operator

From Equation 3.160, the angular momentum operator, $\hat{L}$, has three components $\hat{L}_x$, $\hat{L}_y$ and $\hat{L}_z$ called the **generators** of the infinitesimal rotations about three coordinate axes through the angles, ϕ_x, ϕ_y and ϕ_z, respectively, while $\hat{\vec{p}}$ is the **generator** of the infinitesimal displacement in space. Considering that the components of $\hat{\vec{p}}$ commute with each other, we arrive at the conclusion that the space-displacement group is abelian. We will find that the components of $\hat{L}$ do not commute with each other and so the rotation group is not abelian. It is instructive to note that the system possesses rotational symmetry or invariance. This implies that $\hat{T}_{\vec{\phi}}$ commutes with the Hamiltonian, $\hat{H}$. So the angular momentum, $\hat{L}$ should be a constant of motion. This should be expected from a system that possesses rotational symmetry. This implies that there is energy degeneracy when $\hat{L}|\eta\rangle$ is linearly independent of $|\eta\rangle$.

We find the explicit form of the angular momentum operator, $\hat{L}$, defined in Equation 3.160 via Cartesian coordinates:

$$\hat{L}_x = y\hat{p}_z - z\hat{p}_y = -i\hbar\left(y\frac{\partial}{\partial z} - z\frac{\partial}{\partial y}\right),\ \hat{L}_y = z\hat{p}_x - x\hat{p}_z = -i\hbar\left(z\frac{\partial}{\partial x} - x\frac{\partial}{\partial z}\right) \tag{3.163}$$

$$\hat{L}_z = x\hat{p}_y - y\hat{p}_x = -i\hbar\left(x\frac{\partial}{\partial y} - y\frac{\partial}{\partial x}\right) \tag{3.164}$$

They do not commute with each other. But $\hat{p}_x$, $\hat{p}_y$ and $\hat{p}_z$ commute with each other. We rewrite the three components of the angular momentum operator in Equations 3.163 and 3.164 as follows:

$$\hat{L}_i = \epsilon_{ijk}\hat{X}_j\hat{p}_k,\quad \hat{X}_i\hat{p}_j - \hat{X}_j\hat{p}_i = \epsilon_{ijk}\hat{L}_k,\ i,j,k = x,y,z \tag{3.165}$$

Here, $\hat{X}$ is the coordinate operator and $\in_{ijk}$ is the **Levi-Civita tensor** which is an anti-symmetric tensor of the third rank. It has the following properties:

$$\in_{ijk}=\begin{cases}+1 & i,j,k \quad \text{is an even permutation of } 1,2,3\\ -1 & i,j,k \quad \text{is an odd permutation of } 1,2,3\\ 0 & i,j,k \quad \text{are not all different}\end{cases} \tag{3.166}$$

3.6.3 Commutation Relations of Angular Momentum Operators

In the above notation, the canonical commutation relations take the form:

$$\left[\hat{X}_i,\hat{p}_j\right]=i\hbar\delta_{ij},\ \left[\hat{X}_i,\hat{X}_j\right]=\left[\hat{p}_i,\hat{p}_j\right]=0 \tag{3.167}$$

Let us establish the commutation relations of the angular momentum components of a spinless particle:

$$\begin{aligned}\left[\hat{L}_x,\hat{L}_y\right]&=\left(y\hat{p}_z-z\hat{p}_y\right)\left(z\hat{p}_x-x\hat{p}_z\right)-\left(z\hat{p}_x-x\hat{p}_z\right)\left(y\hat{p}_z-z\hat{p}_y\right)\\ &=y\hat{p}_x\left(\hat{p}_z z-z\hat{p}_z\right)+x\hat{p}_y\left(z\hat{p}_z-\hat{p}_z z\right)=i\hbar\left(x\hat{p}_y-y\hat{p}_x\right)=i\hbar\hat{L}_z\end{aligned} \tag{3.168}$$

then

$$\begin{aligned}\left[\hat{L}_y,\hat{L}_z\right]&=\left(z\hat{p}_x-x\hat{p}_z\right)\left(x\hat{p}_y-y\hat{p}_x\right)-\left(x\hat{p}_y-y\hat{p}_x\right)\left(z\hat{p}_x-x\hat{p}_z\right)=z\hat{p}_y\left(\hat{p}_x x-x\hat{p}_x\right)-y\hat{p}_z\left(\hat{p}_x x-x\hat{p}_x\right)\\ &=i\hbar\left(y\hat{p}_z-z\hat{p}_y\right)=i\hbar\hat{L}_x\end{aligned} \tag{3.169}$$

Applying a cyclical arrangement of indices, we can obtain $\left[\hat{L}_z,\hat{L}_x\right]$. So

$$\left[\hat{L}_x,\hat{L}_y\right]=i\hbar\hat{L}_z,\left[\hat{L}_z,\hat{L}_x\right]=i\hbar\hat{L}_y,\left[\hat{L}_y,\hat{L}_z\right]=i\hbar\hat{L}_x \tag{3.170}$$

We see from the commutation relation in Equation 3.170 the evidence of the Heisenberg's uncertainty principle which dictates the impossibility to know two components of angular momentum simultaneously, since their commutator does not vanish.

Consider the following relations:

$$\left[\hat{L}_i,\hat{X}_j\right]=i\hbar\in_{ijk}\hat{X}_k \tag{3.171}$$

$$\left[\hat{L}_i,\hat{p}_j\right]=i\hbar\in_{ijk}\hat{p}_k \tag{3.172}$$

We prove the first relation (Equation 3.171) as follows with the help of Equation 3.165:

$$\left[\hat{L}_i,\hat{X}_j\right]=\left[\in_{irs}\hat{X}_r\hat{p}_s,\hat{X}_j\right]=\in_{irs}\hat{X}_r\left[\hat{p}_s,\hat{X}_j\right] \tag{3.173}$$

From the first equation in Equation 3.167:

$$\left[\hat{L}_i,\hat{X}_j\right]=\in_{irs}\hat{X}_r\left[\hat{p}_s,\hat{X}_j\right]=\in_{irs}\hat{X}_r\left(-i\hbar\delta_{sj}\right)=-i\hbar\in_{irj}\hat{X}_r=i\hbar\in_{ijk}\hat{X}_k \tag{3.174}$$

Similarly, we prove the second relation (Equation 3.172):

$$\left[\hat{L}_i,\hat{p}_j\right]=\left[\in_{irs}\hat{X}_r\hat{p}_s,\hat{p}_j\right]=\in_{irs}\left[\hat{X}_r,\hat{p}_j\right]\hat{p}_s=\in_{irs}\left(i\hbar\delta_{rj}\right)\hat{p}_s=i\hbar\in_{ijs}\hat{p}_s=i\hbar\in_{ijk}\hat{p}_k \tag{3.175}$$

The commutation relations (Equations 3.171 and 3.172) show how the coordinate as well as the momentum operators transform by an infinitesimal rotation about the i-axis. Equation 3.170 can be written as follows:

$$\left[\hat{L}_i,\hat{L}_j\right]=i\hbar\in_{ijk}\hat{L}_k,\quad i,j,k=x,y,z \tag{3.176}$$

We show this with the help of Equations 3.165, 3.171 and 3.172:

$$\left[\hat{L}_i,\hat{L}_j\right]=\left[\hat{L}_i,\in_{jmn}\hat{X}_m\hat{p}_n\right]=\in_{jmn}\left(\left[\hat{L}_i,\hat{X}_m\right]\hat{p}_n+\hat{X}_m\left[\hat{L}_i,\hat{p}_n\right]\right) \tag{3.177}$$

Applying Equation 3.171 to the first summand and Equation 3.172 to the second:

$$\left[\hat{L}_i,\hat{L}_j\right]=\in_{jmn}\left(i\hbar\in_{imk}\hat{X}_k\hat{p}_n+\hat{X}_m i\hbar\in_{ink}\hat{p}_k\right) \tag{3.178}$$

From the first relation in Equation 3.167:

$$\begin{aligned}\left[\hat{L}_i,\hat{L}_j\right]&=i\hbar\left(\delta_{ji}\delta_{nk}-\delta_{jk}\delta_{ni}\right)\hat{X}_k\hat{p}_n-i\hbar\left(\delta_{ji}\delta_{mk}-\delta_{jk}\delta_{mi}\right)\hat{X}_m\hat{p}_k=i\hbar\left(\delta_{ji}\hat{X}_k\hat{p}_k-\hat{X}_j\hat{p}_i-\delta_{ji}\hat{X}_k\hat{p}_k+\hat{X}_i\hat{p}_j\right)\\&=i\hbar\left(\hat{X}_i\hat{p}_j-\hat{X}_j\hat{p}_i\right)=i\hbar\in_{ijk}\hat{L}_k\end{aligned} \tag{3.179}$$

This shows how the angular momentum transforms under an infinitesimal rotation and also shows how generators of the rotation along different directions do not commute. From Equation 3.208 for fixed i:

$$\left[\hat{L}_i,\hat{L}_i\right]=0 \tag{3.180}$$

3.6.4 Eigenvalue and Eigenfunction of z-Component Angular Momentum Operator

Since the angular momentum operators obtained above correspond to a three-dimensional rotation, naturally we find their eigenfunctions in spherical coordinates. We do the transformation from Cartesian to spherical coordinates:

$$x=r\sin\theta\cos\phi,\ y=r\sin\theta\sin\phi,\ z=r\cos\theta \tag{3.181}$$

then the components of the total angular momentum operator become

$$\hat{L}_x = i\hbar\left(\sin\phi\frac{\partial}{\partial\theta}+\cot\theta\cos\phi\frac{\partial}{\partial\phi}\right),\ \hat{L}_y = i\hbar\left(-\cos\phi\frac{\partial}{\partial\theta}+\cot\theta\sin\phi\frac{\partial}{\partial\phi}\right),\ \hat{L}_z = -i\hbar\frac{\partial}{\partial\phi} \tag{3.182}$$

and

$$\hat{L}_{\pm} = \hbar\exp\{\pm i\phi\}\left(\pm\frac{\partial}{\partial\theta}+i\cot\theta\cos\phi\frac{\partial}{\partial\phi}\right) \tag{3.183}$$

But $\hat{L}_z$ may be found easily if we use the function relating the angular momentum with the rotation operator $\hat{T}_{\delta\vec{\phi}}$ (about the oz-axis) through an infinitesimally small angle, $\delta\vec{\phi}$:

$$\hat{T}_{\delta\vec{\phi}}\Psi(r,\theta,\phi) = \Psi(r,\theta,\phi+\delta\phi) \cong \Psi(r,\theta,\phi)+\frac{\partial}{\partial\phi}\Psi\delta\phi \equiv \left(1+\frac{i}{\hbar}\hat{L}_z\delta\phi\right)\Psi \tag{3.184}$$

Here, $\hat{L}_z$ is defined in Equation 3.213. It should be noted that $\hat{p}_x$ and x are canonical conjugate variables. For the generalized coordinates, we may select ϕ and then the generalized angular momentum, L. As an analogue, we may do the same for $\hat{L}_x$ and $\hat{L}_y$ but the functions are more complex as in addition to ϕ the angle θ also changes. The Sturm-Liouville equation for $\hat{L}_z$ is as follows:

$$\hat{L}_z\Psi = L_z\Psi \tag{3.185}$$

We find the solution of Equation 3.185 in the following form:

$$\Psi = \mathrm{F}(r,\theta)\Phi(\phi) \tag{3.186}$$

If we substitute Equation 3.186 into 3.185, then $\mathrm{F}(r,\theta)$ falls out as a constant:

$$-i\hbar\frac{d\Phi}{d\phi} = L_z\Phi \tag{3.187}$$

From here,

$$\Phi(\phi) = \exp\left\{\frac{iL_z\phi}{\hbar}\right\} \tag{3.188}$$

The constant of integration may be included in the function $\mathrm{F}(r,\theta)$ that satisfies Equation 3.185 and may be constructed on three conditions: It should be finite, continuous and univalent (single-valued) for any L_z. Considering the cyclical property of ϕ, from the property of univalency, for $\phi \to \phi + 2\pi$, the wave function should be invariant relative to that cyclical property. So

$$\exp\left\{\frac{iL_z(\phi+2\pi)}{\hbar}\right\} = \exp\left\{\frac{iL_z\phi}{\hbar}\right\} \tag{3.189}$$

and from where

$$\exp\left\{\frac{iL_z 2\pi}{\hbar}\right\} = 1 \tag{3.190}$$

This is from the condition of univalency of the wave function. It follows that $\frac{L_z}{\hbar}$ is a whole number, say m:

$$L_z = m\hbar, \quad |m| = 0,1,2,\ldots \tag{3.191}$$

We call m the quantum number of the z-component of the angular momentum or the magnetic quantum number. So, the eigenfunction:

$$\Psi(r,\theta,\phi) = F(r,\theta)\langle\phi|m\rangle \equiv F(r,\theta)\exp\{im\phi\} \tag{3.192}$$

It follows that the projection of the angular momentum onto the oz-axis has a discrete spectrum or is quantized. The eigenvalue of $\hat{L}_z$ is a multiple of $\hbar$. The quantization is obtained from the condition of univalency of the wave function.

It should be noted that, for now, $F(r,\theta)$ is undefined. This will be defined later considering the fact that Ψ is the eigenfunction of again another operator. Equation 3.191 gives the eigenvalue of the projection of $\hat{L}$ in any direction in space. The operators $\hat{L}_x$ and $\hat{L}_y$ do not commute with $\hat{L}_z$ considering the theory of commutation. Then in the state (Equation 3.192) where $\hat{L}_z$ has a defined value, the operators $\hat{L}_x$ and $\hat{L}_y$ do not. It implies that if in a state with a defined axis for which we have $L_z(m\hbar)$, then to measure $\hat{L}_x$, we may not say beforehand that we obtain it but we may say that we obtain a multiple of $\hbar$. We may only define its probability. It follows that in quantum mechanics, $\hat{L}$ is not defined.

3.7 Square of Angular Momentum Operator

We find the square of the angular momentum that corresponds to the squared norm of the total angular momentum:

$$\hat{L}^2 = \hat{L}_x^2 + \hat{L}_y^2 + \hat{L}_z^2 \tag{3.193}$$

and defined through the angular variables θ and ϕ as follows:

$$\hat{L}^2 = -\hbar^2\left\{\frac{1}{\sin\theta}\frac{\partial}{\partial\theta}\left(\sin\theta\frac{\partial}{\partial\theta}\right) + \frac{1}{\sin^2\theta}\frac{\partial^2}{\partial\phi^2}\right\} \equiv -\hbar^2\Delta_{\theta,\phi},\ \Delta_{\theta,\phi} = \frac{1}{\sin\theta}\frac{\partial}{\partial\theta}\left(\sin\theta\frac{\partial}{\partial\theta}\right) + \frac{1}{\sin^2\theta}\frac{\partial^2}{\partial\phi^2} \tag{3.194}$$

In the square of the angular momentum operator, $\hat{L}^2$, the Laplacian, $\Delta_{\theta,\phi}$, acts only on the angular variables θ and ϕ.

3.7.1 Square of Angular Momentum Operator Commutation Relations

We put the square of the angular momentum operator in a compact form:

$$\hat{L}^2 = \hat{L}_i\hat{L}_i = \sum_{i=1}^{3}\hat{L}_i\hat{L}_i \tag{3.195}$$

The operator $\hat{L}^2$ commutes with all components of the angular momentum operator $\hat{L}$:

$$\left[\hat{L}^2,\hat{L}_x\right] = \left[\hat{L}^2,\hat{L}_y\right] = \left[\hat{L}^2,\hat{L}_z\right] = 0 \tag{3.196}$$

In a compact form:

$$\left[\hat{L}^2,\hat{L}_i\right]=0,\quad \mathrm{i}=x,y,z \tag{3.197}$$

From here, it follows that there should exist simultaneous eigenstates of $\hat{L}^2$ and of one of the components, $\hat{L}_i$. Though Heisenberg's uncertainty principle dictates the impossibility to know two components of angular momentum at the same time, since their commutator does not vanish there is a possibility to determine one of the components simultaneously with the total angular momentum, considering Equation 3.196. As an example, we compute the first commutator:

$$\left[\hat{L}^2,\hat{L}_x\right]=\left[\hat{L}_x^2+\hat{L}_y^2+\hat{L}_z^2,\hat{L}_x\right]=\left[\hat{L}_y^2,\hat{L}_x\right]+\left[\hat{L}_z^2,\hat{L}_x\right] \tag{3.198}$$

Here

$$\left[\hat{L}_y^2,\hat{L}_x\right]=\hat{L}_y^2\hat{L}_x-\hat{L}_x\hat{L}_y^2=\hat{L}_y\left(\hat{L}_y\hat{L}_x\right)-\hat{L}_x\hat{L}_y^2=\hat{L}_y\left(-i\hbar\hat{L}_z+\hat{L}_x\hat{L}_y\right)-\hat{L}_x\hat{L}_y^2=-i\hbar\hat{L}_y\hat{L}_z$$
$$+\left(-i\hbar\hat{L}_z+\hat{L}_x\hat{L}_y\right)\hat{L}_y-\hat{L}_x\hat{L}_y^2=-i\hbar\left(\hat{L}_y\hat{L}_z+\hat{L}_z\hat{L}_y\right) \tag{3.199}$$

and

$$\left[\hat{L}_z^2,\hat{L}_x\right]=\hat{L}_z^2\hat{L}_x-\hat{L}_x\hat{L}_z^2=\hat{L}_z\left(\hat{L}_z\hat{L}_x\right)-\hat{L}_x\hat{L}_z^2=\hat{L}_z\left(i\hbar\hat{L}_y+\hat{L}_x\hat{L}_z\right)-\hat{L}_x\hat{L}_z^2=i\hbar\hat{L}_z\hat{L}_y$$
$$+\left(i\hbar\hat{L}_y+\hat{L}_x\hat{L}_z\right)\hat{L}_z-\hat{L}_x\hat{L}_z^2=i\hbar\left(\hat{L}_z\hat{L}_y+\hat{L}_y\hat{L}_z\right) \tag{3.200}$$

If we substitute Equations 3.237 and 3.239 into Equation 3.198, then Equation 3.196 follows and, in particular,

$$\left[\hat{L}^2,\hat{L}_x\right]=0 \tag{3.201}$$

We may provide proof in a similar manner the second and third relation in Equation 3.196.

We prove the commutation of the square of the angular momentum operator with the components of the angular momentum operator in a compact form:

$$\left[\hat{L}^2,\hat{L}_i\right]=\left[\hat{L}_j\hat{L}_j,\hat{L}_i\right]=\hat{L}_j\left[\hat{L}_j,\hat{L}_i\right]+\left[\hat{L}_j,\hat{L}_i\right]\hat{L}_j=\hat{L}_j i\hbar\,\epsilon_{ijk}\,\hat{L}_k+i\hbar\,\epsilon_{ijk}\,\hat{L}_k\hat{L}_j=i\hbar\,\epsilon_{ijk}\left(\hat{L}_j\hat{L}_k+\hat{L}_k\hat{L}_j\right)=0 \tag{3.202}$$

We have used the anti-symmetry of the Levi-Civita tensor, ϵ_{ijk}. This follows that the operators $\hat{L}_x$, $\hat{L}_y$ and $\hat{L}_z$ do not commute with each other but any of them commutes with $\hat{L}^2$ and from the group theory, such an operator is known as the quadratic Casimir operator. It follows that the eigenvalue of the operator $\hat{L}^2$ is degenerate.

It is instructive to note that if the generators commute with the Hamiltonian, then the theory is invariant under rotations and implies

$$\left[\hat{L}_i,\hat{\mathrm{H}}\right]=0 \tag{3.203}$$

For such systems:

$$\left[\hat{L}^2,\hat{H}\right]=\left[\hat{L}_i\hat{L}_i,\hat{H}\right]=\hat{L}_i\left[\hat{L}_i,\hat{H}\right]+\left[\hat{L}_i,\hat{H}\right]\hat{L}_i=0 \tag{3.204}$$

So since different components of the angular momentum operator do not commute with each other, $\hat{H}$, $\hat{L}^2$ and only one component of the angular momentum can simultaneously be diagonalized in a rotationally invariant theory where in this book for convenience we choose to diagonalize $\hat{H}$, $\hat{L}^2$ and $\hat{L}_z$. It implies that the operators are chosen to have the same eigenstates. It is instructive to note that since $\hat{L}$ is expressed in terms of the coordinates and momenta, the quantity, $\hat{L}^2$, is not in general a constant of motion. So the quantum number l need not be well defined and cannot serve to characterize a particle except in very restricted conditions.

3.7.2 Square of Angular Momentum Operator Eigenvalue in the Dirac Representation

Let us find the eigenfunction of the operator $\hat{L}^2$ which is also the eigenfunction of the operator $\hat{L}_z$ and also an eigenfunction of the operator $\hat{L}_x$. We find the eigenvalue and eigenfunction of the operator $\hat{L}^2$. Let L^2 be given and has a defined value. Considering Equation 3.193, it follows that

$$-\sqrt{L^2}\le L_z\le\sqrt{L^2} \tag{3.205}$$

This implies that for a given L^2, there should exist the maximum value relative to the absolute value of L_z defined in Equation 3.191. The absolute value of m has its maximal value, $|m|_{\max}$. We introduce the ladder (or raising and lowering) operators since they shift the projection quantum number of the angular momentum eigenfunction, $|\lambda,m\rangle$, one notch up or down:

$$\hat{L}_\pm=\hat{L}_x\pm i\hat{L}_y \tag{3.206}$$

which are non-Hermitian operators and related by

$$\hat{L}_+^\dagger=\hat{L}_- \tag{3.207}$$

If we consider Equation 3.206, then these operators satisfy the commutation relation:

$$\left[\hat{L}_+,\hat{L}_z\right]=\left[\hat{L}_x,\hat{L}_z\right]+i\left[\hat{L}_y,\hat{L}_z\right]=-i\hbar\hat{L}_y+i^2\hbar\hat{L}_x=i^2\hbar\left(\hat{L}_x+i\hat{L}_y\right)=-\hbar\hat{L}_+ \tag{3.208}$$

Similarly

$$\left[\hat{L}_-,\hat{L}_z\right]=\hbar\hat{L}_- \tag{3.209}$$

It is evident that $\hat{L}_\pm$ commutes with $\hat{L}^2$. It should be noted that

$$\hat{L}_+\hat{L}_z=\left(\hat{L}_x+i\hat{L}_y\right)\hat{L}_z=\hat{L}_z\left(\hat{L}_x+i\hat{L}_y\right)-\hbar\hat{L}_+=\hat{L}_z\hat{L}_+-\hbar\hat{L}_+ \tag{3.210}$$

Suppose we act the operator in Equation 3.210 on the eigenfunction $|\lambda,m\rangle$:

$$\hat{L}_+\hat{L}_z|\lambda,m\rangle=\hat{L}_z\hat{L}_+|\lambda,m\rangle-\hbar\hat{L}_+|\lambda,m\rangle \tag{3.211}$$

and then the operator $\hat{L}_z$ also on $|\lambda,m\rangle$:

$$\hat{L}_z|\lambda,m\rangle = m\hbar|\lambda,m\rangle \tag{3.212}$$

From here, it follows that

$$\hat{L}_z\hat{L}_+|\lambda,m\rangle = \hat{L}_+\hat{L}_z|\lambda,m\rangle + \hbar\hat{L}_+|\lambda,m\rangle = \hat{L}_+m\hbar|\lambda,m\rangle + \hbar\hat{L}_+|\lambda,m\rangle = \hat{L}_+(m+1)\hbar|\lambda,m\rangle \tag{3.213}$$

So

$$\hat{L}_+|\lambda,m\rangle = \Gamma_m|\lambda,m+1\rangle \tag{3.214}$$

It follows that $\hat{L}_+$ is a **raising operator** that increases m by unity (the creation operator). The constant Γ_m will be determined subsequently. If we substitute for $m = l$ in Equation 3.214, then

$$\hat{L}_+|\lambda,l\rangle = 0 \tag{3.215}$$

as l is the maximum value of m:

$$|\lambda,l+1\rangle = 0 \tag{3.216}$$

As the maximum and minimum values of m are l and $-l$, respectively, m has $2l+1$ values.

Consider the action of the operator $\hat{L}_-\hat{L}_+$ on $|\lambda,l\rangle$:

$$\hat{L}_-\hat{L}_+|\lambda,l\rangle = \left(\hat{L}_x - i\hat{L}_y\right)\left(\hat{L}_x + i\hat{L}_y\right)|\lambda,l\rangle = 0 \tag{3.217}$$

then

$$\hat{L}_-\hat{L}_+|\lambda,l\rangle = \left(\hat{L}_x^2 + \hat{L}_y^2 + i\left[\hat{L}_x,\hat{L}_y\right]\right)|\lambda,l\rangle = \left(\hat{L}_x^2 + \hat{L}_y^2 - \hbar\hat{L}_z\right)|\lambda,l\rangle = \left(\hat{L}^2 - \hat{L}_z^2 - \hbar\hat{L}_z\right)|\lambda,l\rangle = 0 \tag{3.218}$$

It follows from here that

$$\hat{L}^2|\lambda,l\rangle = \left(\hat{L}_z^2 + \hbar\hat{L}_z\right)|\lambda,l\rangle = \hat{L}_z\left(\hat{L}_z + \hbar\right)|\lambda,l\rangle = \hat{L}_z(\hbar l + \hbar)|\lambda,l\rangle = \hbar^2 l(l+1)|\lambda,l\rangle \tag{3.219}$$

Here l is a positive whole number.

Question:
Show that for the directional equality of the axes x, y and z, we have

$$\mathrm{L}^2 \equiv l(l+1) \tag{3.220}$$

Answer:
From

$$\mathrm{L}^2 \equiv \overline{\hat{L}^2} = \overline{\left(\hat{L}_x^2 + \hat{L}_y^2 + \hat{L}_z^2\right)} = 3\overline{\hat{L}_z^2} \tag{3.221}$$

and considering the directional equality of different projections of the angular momenta:

$$\overline{\hat{L}_z^2} = (2l+1)^{-1} \sum_{m=-l}^{l} m^2 = \frac{l(l+1)}{3} \tag{3.222}$$

We evaluate the sum in the above equation considering:

$$\sum_{m=0}^{l} m^2 = \left[\frac{d^2}{d\xi^2} \sum_{m=0}^{l} \exp\{\xi m\} \right]_{\xi=0} = \left[\frac{d^2}{d\xi^2} \frac{1-\exp\{\xi(l+1)\}}{1-\exp\{\xi\}} \right]_{\xi=0} = \frac{l(l+1)(2l+1)}{6} \tag{3.223}$$

So, from here and Equation 3.222, we have

$$L^2 \equiv l(l+1) \tag{3.224}$$

EXAMPLE

Examine the addition of the angular momentum for two distinct particles. The operator $\hat{\vec{L}}$ satisfies same commutation relation as the operators $\vec{L}_1$ and $\vec{L}_2$. It can be seen that

$$\left[\hat{L}^2, \hat{L}_x\right] = 0 \tag{3.225}$$

We have shown the total momentum operator commutes with all projection operators. Also

$$\left[\hat{L}^2, \hat{L}_1^2\right] = \left[\hat{L}^2, \hat{L}_2^2\right] = 0 \tag{3.226}$$

If we consider Equations 3.225 and 3.226, then it follows that $\vec{L}$ is quantized by the same rules as L_z and L^2 in Equations 3.191 and 3.219, respectively.

The operators $\hat{L}_1$ and $\hat{L}_2$ are characterized by m_1, l_1 and m_2, l_2, respectively, and we find the link between m, l and m_1, l_1, m_2, l_2 where

$$m = m_1 + m_2 \tag{3.227}$$

If we consider Equation 3.226, then it follows that $\hat{L}_1^2$ and $\hat{L}_2^2$ have a defined value and also $\hat{L}^2$ in that state has a defined value. From the state where $\hat{L}_1^2$ and $\hat{L}_2^2$ are characterized by l_1 and l_2 where such a state is degenerate with the degree of degeneracy:

$$f = (2l_1 + 1)(2l_2 + 1) \tag{3.228}$$

The system may be described such that l_1, l_2 are fixed and m_1, m_2 vary. We have to find l. It should be noted that from definition, l is the maximum value of the quantum number, m. Suppose m_1 and m_2 have maxima l_1 and l_2. If l is given, then it may correspond to $m-1$, i.e., $l_1 + l_2 - 1$ values. We may obtain l through

$$l = \max(m),\ m = l-2;\ l = l_1 + l_2 - 1,\ m = l-2;\ l = l_1 + l_2 - 2,\ m = l \tag{3.229}$$

So Theorem

$$l = l_1 + l_2, l_1 + l_2 - 1, \ldots, |l_1 - l_2| \tag{3.230}$$

It is obvious that the degree of degeneracy of the given state (with given l_1 and l_2) should be independent of the way of description of the system, i.e., with the help of l_1 and l_2 or m_1 and m_2 it follows that if the value of l is limited, i.e., $|l_1 - l_2|$, then we obtain what is necessary. Suppose that $l_1 > l_2$:

$$f = \sum_{l=|l_1-l_2|}^{l_1+l_2} (2l+1) = (2l_1+1)(2l_2+1) \tag{3.231}$$

3.8 Square of Angular Momentum Operator Eigenstates

The construction of the angular momentum eigenstates and determination of their eigenvalues as seen earlier is purely algebraic, and sidesteps any explicit reference to spherical harmonics. We recover their explicit form under this heading by solving the following Sturm-Liouville equation:

$$\hat{L}^2\Psi = L^2\Psi \tag{3.232}$$

We represent $\hat{L}^2$ in a differential form and find the eigenfunction of that operator:

$$\frac{1}{\sin\theta}\frac{\partial}{\partial\theta}\left(\sin\theta\frac{\partial\Psi}{\partial\theta}\right)+\frac{1}{\sin^2\theta}\frac{\partial^2\Psi}{\partial\phi^2}+l(l+1)\Psi=0 \tag{3.233}$$

The solution can be obtained by the method of separation of variables:

$$\Psi(r,\theta,\phi)=\mathrm{F}(r,\theta)\langle\phi|m\rangle=R(r)\Theta(\theta)\langle\phi|m\rangle \tag{3.234}$$

The operator $\hat{L}_z$ commutes with the operator $\hat{L}^2$ and gives us the right to find the eigenfunction of the operator $\hat{L}^2$ which is also the eigenfunction of the operator $\hat{L}_z$. The eigenfunction, $\langle\phi|m\rangle$, of the operator $\hat{L}_z$ is found in Equation 3.192 and the product, $\mathrm{F}(r,\theta)\langle\phi|m\rangle$, is the joint eigenfunction of $\hat{L}^2$ and $\hat{L}_z$.

3.8.1 Legendre Polynomials

Let us again examine Equation 3.206 and substitute Equation 3.234 into 3.233 where the radial function $R(r)$ and the factor $\exp\{im\phi\}$ fall out:

$$\frac{1}{\sin\theta}\frac{d}{d\theta}\left(\sin\theta\frac{d}{d\theta}\Theta\right)+\left[l(l+1)-\frac{m^2}{\sin^2\theta}\right]\Theta=0 \tag{3.235}$$

The variables are changed to $\cos\theta = x$, then we obtain the equation for the **associated Legendre polynomials** $\mathrm{P}_l^m(x)$ of order l and rank m (see appendix, Chapter 25):

$$\Theta(\theta)=\mathrm{P}_l^m(\cos\theta)=\mathrm{P}_l^m(x) \tag{3.236}$$

where

$$\mathrm{P}_l^m(x)=\left(1-x^2\right)^{\frac{m}{2}}\frac{d^m}{dx^m}\mathrm{P}_l(x)=\left(1-x^2\right)^{\frac{m}{2}}\frac{1}{2^l l!}\frac{d^{l+m}}{dx^{l+m}}\left(x^2-1\right)^l \tag{3.237}$$

and the **Legendre polynomials**

$$\mathrm{P}_l(x)=\frac{1}{2^l l!}\frac{d^l}{dx^l}\left(x^2-1\right)^l,\quad l\geq m\geq -l \tag{3.238}$$

are the so-called **Rodriguez recurrence formula**:

$$\begin{aligned}&\mathrm{P}_0(x)=1;\mathrm{P}_1(x)=x;\mathrm{P}_2(x)=\frac{1}{2}\left(3x^2-1\right);\mathrm{P}_3(x)=\frac{1}{2}\left(5x^3-3x\right);\mathrm{P}_4(x)=\frac{1}{8}\left(35x^4-30x^2+3\right);\\&\mathrm{P}_5(x)=\frac{1}{8}\left(63x^5-70x^3+15x\right);\mathrm{P}_6(x)=\frac{1}{16}\left(231x^6-315x^4+105x^2-5\right);\ldots\end{aligned} \tag{3.239}$$

We show the orthogonality of the Legendre polynomials within the interval $(-1,1)$:

$$\mathrm{I}_{mn}\equiv\int_{-1}^{1}\mathrm{P}_m(x)\mathrm{P}_n(x)dx=\frac{1}{2^{m+n}m!n!}\int_{-1}^{1}\left\{\frac{d^m}{dx^m}\left(x^2-1\right)^m\right\}\left\{\frac{d^n}{dx^n}\left(x^2-1\right)^n\right\}dx \tag{3.240}$$

Doing integration by parts, we have

$$\mathrm{I}_{mn}=\frac{(-1)^n}{2^{m+n}m!n!}\int_{-1}^{1}\left\{\frac{d^m}{dx^m}\left(x^2-1\right)^m\right\}\left(x^2-1\right)^n dx \tag{3.241}$$

Consider $m<n$, $\mathrm{I}_{mn}=0$:
or

$$\int_{-1}^{+1}\mathrm{P}_m(x)\mathrm{P}_n(x)dx=0,\ m\neq n \tag{3.242}$$

and if $m=n$, we have

$$\mathrm{I}_{mn}=\frac{(-1)^n}{2^{2n}(n!)^2}\int_{-1}^{1}\left(x^2-1\right)^n\frac{d^{2n}}{dx^{2n}}\left(x^2-1\right)^n dx=\frac{(-1)^n(2n)!}{2^{2n}(n!)^2}\int_{-1}^{1}\left(x^2-1\right)^n dx \tag{3.243}$$

Applying the change of variables $x=2t-1$ to Equation 3.243, we have

$$\mathrm{I}_{mn}=\frac{(-1)^n(2n)!}{(n!)^2}\int_{0}^{1}t^n(t-1)^n dt=\frac{2}{2n+1} \tag{3.244}$$

So the norm of $\mathrm{P}_n(x)$ can also be obtained using the **so-called Jacobi polynomial** $\mathrm{P}_n^{(\lambda,\mu)}(x)$ of order n in Equation 25.29 of the appendix in Chapter 25:

$$\int_{-1}^{+1}\left[\mathrm{P}_n(x)\right]^2 dx=\int_{-1}^{+1}\left[\mathrm{P}_n^{(0,0)}(x)\right]^2 dx=\frac{2}{2n+1} \tag{3.245}$$

From here, considering Equation 3.240, we have

$$\int_{-1}^{1} \mathrm{P}_m(x)\mathrm{P}_n(x)dx = \frac{2}{2n+1}\delta_{mn} \tag{3.246}$$

This is the **orthogonality relation** for Legendre polynomials. We observe that the Legendre polynomials, $\mathrm{P}_n(x)$, do in fact form a complete orthogonal set on $(-1,1)$. From Equation 3.237, we also have for the associated Legendre polynomials (see appendix, Chapter 25):

$$\int_{-1}^{1} \mathrm{P}_l^m(x)\mathrm{P}_{l'}^{m'}(x)dx = \frac{(l+m)!}{(l-m)!}\frac{2}{2l+1}\delta_{ll'}\delta_{mm'} \tag{3.247}$$

From Equation 3.238, the immediate symmetry **of the Legendre polynomials** follows:

$$\mathrm{P}_l(-x) = (-1)^l \mathrm{P}_l(x),\ \mathrm{P}_l(1) = 1,\ \mathrm{P}_l(-1) = (-1)^l \tag{3.248}$$

The Legendre Polynomials, $\mathrm{P}_l(x)$, can be expressed through the **confluent hypergeometric function**:

$$\mathrm{P}_l(x) = \mathrm{F}\left(l+1,-l,1;\frac{1-x}{2}\right) \tag{3.249}$$

Since the differential equation in Equation 3.235 is transformed into itself when m is replaced by $-m$, P_l^{-m} is also a solution of the Legendre differential equation. The given solution is a polynomial in x of order l and is also continuous at the point $x=\pm1$. Therefore, the solutions P_l^m and P_l^{-m} can differ only by a factor for fixed l and $m(0 \le |m| \le l)$:

$$\mathrm{P}_l^{-m}(x) = C\mathrm{P}_l^m(x) \tag{3.250}$$

We may determine the constant C by setting $x=1$, then divide by $\left(1-x^2\right)^{\frac{m}{2}}$:

$$\left(1-x^2\right)^{-m}\frac{d^{l-m}}{dx^{l-m}}\left(x^2-1\right)^l\bigg|_{x=1} = (-1)^m 2^{l-m}\frac{l!}{m!} \tag{3.251}$$

or

$$\left(1-x^2\right)^{-m}\frac{d^{l-m}}{dx^{l-m}}\left(x^2-1\right)^l\bigg|_{x=1} = C2^{l-m}\frac{l!}{m!}\frac{(l+m)!}{(l-m)!} = \frac{d^{l+m}}{dx^{l+m}}\left(x^2-1\right)^l\bigg|_{x=1} \tag{3.252}$$

From here, we have

$$C = (-1)^m\frac{(l-m)!}{(l+1)!} \tag{3.253}$$

So

$$\mathrm{P}_l^{-m}(x) = (-1)^m\frac{(l-m)!}{(l+1)!}\mathrm{P}_l^m(x) \tag{3.254}$$

It is obvious that

$$P_l(-x)=(-1)^l P_l(x),\ P_l(1)=1,\ P_l(-1)=(-1)^l \tag{3.255}$$

The first recurrence relation for the Legendre polynomials can be written as follows:

$$(2n+1)xP_n(x)-(n+1)P_{n+1}(x)-nP_{n-1}(x)=0 \tag{3.256}$$

See appendix in Chapter 25 for other recurrence formulae.

3.8.1.1 Asymptotic Legendre Polynomials

We substitute Equation 3.236 into Equation 3.235 and, for brevity, set $m=0$ and $l=n$:

$$\frac{d^2}{d\theta^2}P_n+\cot\theta\frac{d}{d\theta}P_n+n(n+1)P_n=0 \tag{3.257}$$

Letting

$$P_n(\cos\theta)=\frac{X(\theta)}{\sin^{\frac{1}{2}}\theta} \tag{3.258}$$

then

$$\frac{d}{d\theta}\frac{X(\theta)}{\sin^{\frac{1}{2}}\theta}=\frac{X'(\theta)}{\sin^{\frac{1}{2}}\theta}-\frac{1}{2}\frac{X(\theta)}{\sin^{\frac{3}{2}}\theta}\cos\theta,\ \frac{d^2}{d\theta^2}\frac{X(\theta)}{\sin^{\frac{1}{2}}\theta}=\frac{X''(\theta)}{\sin^{\frac{1}{2}}\theta}-\frac{X'(\theta)}{\sin^{\frac{3}{2}}\theta}\cos\theta+\frac{3}{4}\frac{X(\theta)}{\sin^{\frac{5}{2}}\theta}\cos^2\theta+\frac{1}{2}\frac{X(\theta)}{\sin^{\frac{1}{2}}\theta} \tag{3.259}$$

So

$$X''(\theta)+\left[\left(n+\frac{1}{2}\right)^2+\frac{1}{4\sin^2\theta}\right]X(\theta)=0 \tag{3.260}$$

Find the initial conditions considering Equation 3.258:

$$X'(\theta)=\frac{1}{2}\sin^{-\frac{1}{2}}\theta\cos\theta P_n(\cos\theta)-\sin^{\frac{3}{2}}\theta P_n'(\cos\theta) \tag{3.261}$$

If

$$x=\cos\theta=0 \tag{3.262}$$

so $\theta=\frac{\pi}{2}$ and from Equations 3.258 and 3.261:

$$X\left(\frac{\pi}{2}\right)=A_n(-1)^{\frac{n}{2}}\theta\left((-1)^n\right),\ X'\left(\frac{\pi}{2}\right)=2A_n(-1)^{\frac{l+1}{2}}\theta\left((-1)^{l+1}\right),\ A_n=\frac{\Gamma\left(\frac{n}{2}+\frac{1}{2}\right)}{\sqrt{\pi}\Gamma\left(\frac{n}{2}+1\right)} \tag{3.263}$$

Here, $\theta(x)$ is the Heaviside function of argument x. For quasi-classical approximation $n \gg 1, n\theta \gg 1$ and $n(\pi-\theta) \gg 1$ is fulfilled everywhere except at the small neighborhood of $\theta=0$ and $\theta=\pi$, then $\frac{1}{n\theta} \ll 1$ and

$$\left(n+\frac{1}{2}\right)^2+\frac{1}{4\sin^2\theta}=\left(n+\frac{1}{2}\right)^2\left(1+\frac{1}{4\left(n+\frac{1}{2}\right)^2\sin^2\theta}\right)\approx\left(n+\frac{1}{2}\right)^2 \tag{3.264}$$

So Equation 3.260 becomes

$$X''(\theta)+\left(n+\frac{1}{2}\right)^2X(\theta)=0 \tag{3.265}$$

with the solution

$$X(\theta)=C_1\cos\left[\left(\frac{n}{2}+\frac{1}{2}\right)\theta\right]+C_2\sin\left[\left(\frac{n}{2}+\frac{1}{2}\right)\theta\right] \tag{3.266}$$

The function, $X(\theta)$, satisfies the condition in Equation 3.263 for even and odd n:

$$C_1=A_n\cos\frac{\pi}{4},\ C_2=A_n\sin\frac{\pi}{4} \tag{3.267}$$

So

$$X(\theta)=A_n\sin\left[\left(n+\frac{1}{2}\right)\theta+\frac{\pi}{4}\right] \tag{3.268}$$

If now $\theta \ll 1$, then considering:

$$\cot\theta\approx\frac{1}{\theta},\ n(n+1)\rightarrow\left(n+\frac{1}{2}\right)^2 \tag{3.269}$$

So

$$\frac{d^2}{d\theta^2}P_n+\frac{1}{\theta}\frac{d}{d\theta}P_n+\left(n+\frac{1}{2}\right)^2P_n=0 \tag{3.270}$$

with the solution

$$P_n(\theta)=J_0\left(\left(n+\frac{1}{2}\right)\theta\right) \tag{3.271}$$

Here, $J_0(\theta)$ is the Bessel function of the first kind and of order 0 with argument θ. So, the asymptotic solution of Equation 3.270 takes the following form:

$$P_n(\theta)\approx\sqrt{\frac{2}{\pi n}}\frac{1}{\sqrt{\theta}}\sin\left(\left(n+\frac{1}{2}\right)\theta+\frac{\pi}{4}\right) \tag{3.272}$$

Hence, the quasi-classical expression for $P_n(\theta)$ is as follows:

$$P_n(\theta) \approx \sqrt{\frac{2}{\pi n}} \frac{1}{\sin^{\frac{1}{2}}\theta} \sin\left(\left(n+\frac{1}{2}\right)\theta+\frac{\pi}{4}\right) \tag{3.273}$$

3.8.2 Angular Momentum Eigenstates

So the eigenfunction $\Psi(r,\theta,\phi)$ can be obtained through the method of separation of variables:

$$\Psi(r,\theta,\phi) = R(r)Y_{nm}(\theta,\phi) \tag{3.274}$$

where

$$Y_{lm}(\theta,\phi) = N_{lm}P_l^m(\cos\theta)\langle\phi|m\rangle \equiv \langle\theta,\phi|l,m\rangle \tag{3.275}$$

are spherical functions called the **spherical harmonics** (treated in appendix of Chapter 25). They are eigenfunctions of the square of the angular momentum operator and N_{lm} is the normalization constant for the associated Legendre polynomials, while $|\theta,\phi\rangle$ denotes position basis states on the unit sphere, parameterized in spherical polar coordinates. The states $l = 0,1,2,3,4,5$ are called s,p,d,f,g,h-states, and for $l = 4$, we show the angular momentum quantization in Figure 3.2a.

The eigenvalue of $\hat{L}^2$ is dependent only on one quantum number l and the eigenfunction in Equation 3.274 depends on l and m. For any m, Y_{lm} is the eigenfunction of $\hat{L}^2$ and has same eigenvalue. It implies that the given eigenvalue in Equation 3.220 corresponds to such number of eigenfunctions as the number of quantum numbers m may take for a given quantum number l. So we have

$$f = 2l+1 \tag{3.276}$$

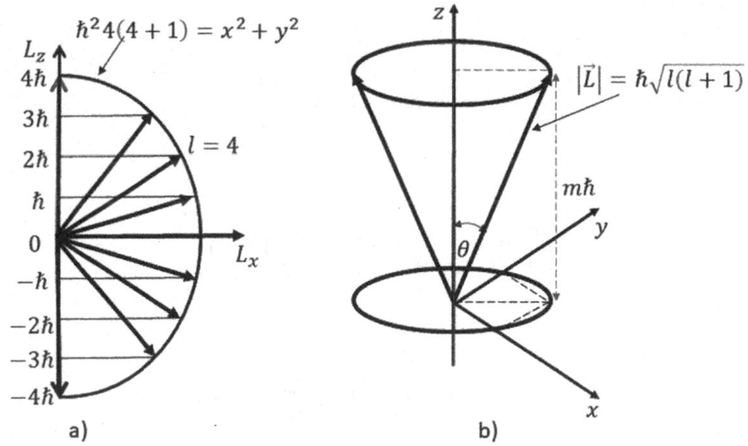

FIGURE 3.2 (a) The quantization of the z-component of the angular momentum. The component in the oxy-plane is not known due to the Heisenberg's uncertainty principle. For $l = 4$, the length of the angular momentum vector is $|\vec{L}| = \hbar\sqrt{20}$ and there are nine possible values of L_z; (b) the angle between the angular momentum and the quantization direction.

degrees of degeneracy for a given quantum number l. These degenerate states differ in the projection of the angular momentum on the oz-axis direction. The length of that vector is a constant. Its projection changes on the oz-axis direction. The number m is called the **magnetic quantum number** that gives the size of the z-component of the angular momentum $\hat{L}_z$ and l is the **orbital (azimuthal) quantum number** that indicates the magnitude of the angular momentum $\hat{L}^2$.

If we are given L^2, i.e., if we are given l or so to say the length of the vector $\left|\vec{L}\right| = \sqrt{L^2}$, then it should be noted that the state for which the vector $\vec{L}$ may have that length is not one but $f = 2l+1$. This is because for every l, there are $f = 2l+1$ eigenfunctions, $Y_{lm}\ (l \geq m \geq -l)$. If we consider Equations 3.233, 3.234 and 3.238, we can infer the z-component $\hat{L}_z$ of the angular momentum $\vec{L}$ with the absolute value:

$$\left|\vec{L}\right| = \sqrt{L^2} = \sqrt{\hbar^2 l(l+1)} = \hbar\sqrt{l(l+1)} \tag{3.277}$$

It takes $f = 2l+1$ different values of $m\hbar$ (see Figure 3.2b). The angle between the angular momentum and the quantization direction is defined from the following relation:

$$\cos\theta = \frac{m}{\sqrt{l(l+1)}} \tag{3.278}$$

This is the quantization of the z-component of the angular momentum $\hat{L}_z$ or sometimes called **quantization direction.**

Let us give a pictorial interpretation of our results (see Figure 3.2b). The angular momentum vector $\vec{L}$ precesses on a cone around the quantization direction (oz-axis). As a consequence, the components L_x and L_y are not constant in time. This shows the uncertainty relations in Equation 3.170 between L_z and L_x and between L_z and L_y. From Equation 3.275 and the appendix of Chapter 25:

$$\int_\Omega Y^*_{l'm'}(\theta,\phi) Y_{lm}(\theta,\phi) d\Omega = |N_{lm}|^2 \frac{4\pi}{2l+1}\frac{(l+m)!}{(l-m)!}\delta_{ll'}\delta_{m'm} = \delta_{ll'}\delta_{m'm},\ N_{lm} = \sqrt{\frac{2l+1}{4\pi}\frac{(l-m)!}{(l+m)!}} \tag{3.279}$$

The eigenfunctions in Equation 3.274 now become

$$Y_{lm}(\theta,\phi) = N_{lm} P_l^m(\cos\theta)\langle\phi|m\rangle \tag{3.280}$$

If we consider Equation 3.280, then we find the symmetry **of the spherical harmonics:**

$$Y_{l-m}(\theta,\phi) = (-1)^m Y^*_{lm}(\theta,\phi) \tag{3.281}$$

Consider the inversion of the coordinates $\vec{r} \to -\vec{r}$ (i.e., reflection through the coordinate origin), the angles θ and ϕ change to $\pi-\theta$ and $\phi+\pi$, respectively. In this case, the function $Y_{lm}(\theta,\phi)$ is multiplied by the factor $(-1)^l$:

$$Y_{lm}(\pi-\theta,\pi+\phi) = (-1)^l Y_{lm}(\theta,\phi) \tag{3.282}$$

The factor $(-1)^l$ is called the **parity of the state with given** l. This implies that a state of odd l has odd parity and that of even l has even parity, regardless of the value of m. We can generally classify all observables as even or odd depending on if their operators do not or do change sign under inversion of the coordinates. We observe from above that the angular solutions (spherical harmonics) are eigenstates of the angular momentum with definite parity eigenvalues.

3.8.3 Dirac Representation Eigenstates

Besides interpreting the angular Schrödinger equation as a differential equation, we solve this equation employing the perspective of operators and vectors with the help of Dirac notation for the eigenstates of the operators, $\hat{L}_z$ and $\hat{L}^2$ denoted by $|\lambda,m\rangle$. From Equation 3.233, we consider the full eigenstate acting on $\hat{L}^2$ as well as all the components of the angular momentum. From Equation 3.214, we find that the constant Γ_+ depends on λ as well as on m.

We note that $\hat{L}_\pm|\lambda,m\rangle$ is still an eigenstate of $\hat{L}_z$ and so examine the action of the product $\hat{L}_z\hat{L}_-$ on $|\lambda,m\rangle$ considering Equation 3.212:

$$\hat{L}_z\hat{L}_-|\lambda,m\rangle = \left(\left[\hat{L}_z,\hat{L}_-\right]+\hat{L}_-\hat{L}_z\right)|\lambda,m\rangle = \left(-\hbar\hat{L}_- + \hat{L}_-\hat{L}_z\right)|\lambda,m\rangle = (m-1)\hbar\hat{L}_-|\lambda,m\rangle \tag{3.283}$$

Also,

$$\hat{L}^2\hat{L}_-|\lambda,m\rangle = \hat{L}_-\hat{L}^2|\lambda,m\rangle = \hbar^2\lambda\hat{L}_-|\lambda,m\rangle \tag{3.284}$$

From the above, it is seen that $\hat{L}_-$ is the lowering operator that acts on $|\lambda,m\rangle$ and lowers the eigenvalue of $\hat{L}_z$ by a unit of $\hbar$ while leaving the eigenvalue of $\hat{L}^2$ invariant. So

$$\hat{L}_-|\lambda,m\rangle = \Gamma'_m|\lambda,m-1\rangle \tag{3.285}$$

We find the physical sense of the constants Γ_m and Γ'_m represented, respectively, in Equations 3.214 and 3.285 by the following relations:

$$\Gamma_m = \langle\lambda,m+1|\hat{L}_+|\lambda,m\rangle,\ \Gamma'_m = \langle\lambda,m-1|\hat{L}_-|\lambda,m\rangle \tag{3.286}$$

So from here, Γ_m and Γ'_m are matrix elements of, respectively, the operators $\hat{L}_+$ and $\hat{L}_-$ in the m-representation. These matrix elements will achieve the values zero when equal to l.

We find the normalized eigenstates of the angular momentum by considering Equations 3.212, 3.266, 3.214 and 3.285 as well as the definition of $\hat{L}_+$ and $\hat{L}_-$ in addition to the Hermicity of $\hat{L}_x$ and $\hat{L}_y$:

$$\left[\langle\lambda,m|\hat{L}_-|\lambda,m+1\rangle\right]^* = \langle\lambda,m+1|\hat{L}_+|\lambda,m\rangle \tag{3.287}$$

Comparing this equation with Equations 3.287 and 3.286:

$$\Gamma'_{m+1} = \Gamma^*_m \tag{3.288}$$

Considering Equations 3.285, 3.214 and 3.288:

$$\hat{L}_x|\lambda,m\rangle = \frac{1}{2}\left[\Gamma_m|\lambda,m+1\rangle + \Gamma^*_{m-1}|\lambda,m-1\rangle\right],\ \hat{L}_y|\lambda,m\rangle = \frac{i}{2}\left[\Gamma^*_{m-1}|\lambda,m-1\rangle + \Gamma_m|\lambda,m+1\rangle\right] \tag{3.289}$$

We find the expression of Γ_m when

$$\langle\lambda,m|\hat{L}_-\hat{L}_+|\lambda,m\rangle = \langle\lambda,m|\left(\hat{L}^2 - \hat{L}_z^2 - \hbar\hat{L}_z\right)|\lambda,m\rangle = \hbar^2\left(\lambda - m(m+1)\right)\langle\lambda,m|\lambda,m\rangle = |\Gamma_m|^2 \tag{3.290}$$

from here

$$\Gamma_m = \Gamma_m^* = \hbar\sqrt{\lambda - m(m+1)} \tag{3.291}$$

This quantity achieves the value zero when it equals l. As Γ_m results from Equation 3.290, it could have been obvious that Γ_m should have a phase factor. It is instructive to know that this phase factor has no physical consequence and so can be safely neglected. So we assume that the quantity Γ_m is real valued and also

$$\langle \lambda, m\pm 1|\hat{L}_x|\lambda, m\rangle = \frac{\hbar}{2}\sqrt{\lambda - m(m\pm 1)}, \langle \lambda, m\pm 1|\hat{L}_y|\lambda, m\rangle = \mp\frac{i}{2}\hbar\sqrt{\lambda - m(m\pm 1)} \tag{3.292}$$

So

$$\hat{L}_+|\lambda, m\rangle = \hbar\sqrt{\lambda - m(m+1)}|\lambda, m+1\rangle, |\lambda, m+1\rangle = \frac{1}{\hbar\sqrt{\lambda - m(m+1)}}\hat{L}_+|\lambda, m\rangle \tag{3.293}$$

Similarly, we find

$$\hat{L}_-|\lambda, m\rangle = \hbar\sqrt{\lambda - m(m-1)}|\lambda, m-1\rangle, |\lambda, m-1\rangle = \frac{1}{\hbar\sqrt{\lambda - m(m-1)}}\hat{L}_-|\lambda, m\rangle \tag{3.294}$$

3.8.4 Matrix Representation and Finite Rotations Eigenstates

We have found the simultaneously eigenstates of $\hat{L}^2$ and $\hat{L}_z$ for a given l. The eigenstates define a sub-space of the total Hilbert space of the angular momentum operators. This implies that the operators $\hat{L}^2$, $\hat{L}_z$, $\hat{L}_+$ and $\hat{L}_-$ take any vector in the given space to another vector of the same space. So, the action of given operators leaves the sub-space of the total Hilbert space invariant. The dimensionality of the given space is $(2l+1)$.

One consequence of operator formalism is the existence of systems with half-integer angular-momentum quantum numbers and provides a formulation for electron spin with no classical analog. We will observe that the operators representing the spin will be described by the Pauli spin algebra and to observe, this we note that the maximum and minimum values of the magnetic quantum number m should be, respectively,

$$|m|_{\max} = l, |m|_{\min} = -l \tag{3.295}$$

Taking the difference of $|m|_{\max}$ and $|m|_{\min}$, this yields an integral value n:

$$|m|_{\max} - |m|_{\min} = 2l = n, n = 0,1,2,\ldots, \tag{3.296}$$

So

$$l = \frac{n}{2}, n = 0,1,2,\ldots, \tag{3.297}$$

We observe from here that l takes integral and half-integral values.

Consider the case when $l=0$, the dimensionality of the representation is $2l+1=1$ with the magnetic quantum number m taking the value 0. For the case when $l=\frac{1}{2}$, we have an eigenstate that is two-fold degenerate, i.e., $2l+1=2$ which is also the dimensionality of the representation with the magnetic quantum number m taking the values $\pm\frac{1}{2}$. The eigenbasis states in the given space are $\left|\frac{1}{2},\frac{1}{2}\right\rangle$ and $\left|\frac{1}{2},-\frac{1}{2}\right\rangle$, representing, respectively, **spin up** and **spin down:**

$$\left|\frac{1}{2},\frac{1}{2}\right\rangle \equiv \begin{bmatrix} 1 \\ 0 \end{bmatrix} \equiv |\uparrow\rangle, \left|\frac{1}{2},-\frac{1}{2}\right\rangle \equiv \begin{bmatrix} 0 \\ 1 \end{bmatrix} \equiv |\downarrow\rangle \tag{3.298}$$

From Equation 3.384:

$$\left\langle \frac{1}{2},\frac{1}{2}\right|\hat{L}_z\left|\frac{1}{2},\frac{1}{2}\right\rangle = -\left\langle \frac{1}{2},-\frac{1}{2}\right|\hat{L}_z\left|\frac{1}{2},-\frac{1}{2}\right\rangle = \frac{\hbar}{2}, \left\langle \frac{1}{2},\frac{1}{2}\right|\hat{L}_z\left|\frac{1}{2},-\frac{1}{2}\right\rangle = \left\langle \frac{1}{2},-\frac{1}{2}\right|\hat{L}_z\left|\frac{1}{2},\frac{1}{2}\right\rangle = 0 \tag{3.299}$$

This follows the matrix representation:

$$\hat{L}_z = \frac{\hbar}{2}\hat{\sigma}_z, \hat{\sigma}_z = \begin{bmatrix} 1 & 0 \\ 0 & -1 \end{bmatrix} \tag{3.300}$$

Here, $\hat{\sigma}_z$ is the component of the Pauli matrix in the z-axis direction. In the same manner, we have for $\hat{L}^2$

$$\langle l,m'|\hat{L}^2|l,m\rangle = \frac{3}{4}\hbar^2\delta_{m'm}, \hat{L}^2 = \frac{3}{4}\hbar^2\hat{\mathbb{I}}, \hat{\mathbb{I}} = \begin{bmatrix} 1 & 0 \\ 0 & 1 \end{bmatrix} \tag{3.301}$$

Here $\hat{\mathbb{I}}$ is a unit matrix. For the raising operator considering Equation 3.287, we have

$$\langle l,m'|\hat{L}_+|l,m\rangle = \Gamma_+\langle l,m'|l,m+1\rangle = \Gamma_+\delta_{m',m+1} = \hbar\sqrt{\frac{3}{4}-m(m+1)}\delta_{m',m+1} \tag{3.302}$$

We write the matrix representation of the given matrix element:

$$\hat{L}_+ = \hbar\begin{bmatrix} 0 & \sqrt{\frac{3}{4}+\frac{1}{4}} \\ 0 & 0 \end{bmatrix} = \hbar\begin{bmatrix} 0 & 1 \\ 0 & 0 \end{bmatrix} \tag{3.303}$$

Similarly,

$$\langle l,m'|\hat{L}_-|l,m\rangle = \Gamma_-\langle l,m'|l,m+1\rangle = \Gamma_-\delta_{m',m-1} = \hbar\sqrt{\frac{3}{4}-m(m-1)}\delta_{m',m-1} \tag{3.304}$$

and represented in the matrix form:

$$\hat{L}_- = \hbar\begin{bmatrix} 0 & 0 \\ \sqrt{\frac{3}{4}+\frac{1}{4}} & 0 \end{bmatrix} = \hbar\begin{bmatrix} 0 & 0 \\ 1 & 0 \end{bmatrix} = \left(\hat{L}_+\right)^\dagger \tag{3.305}$$

From Equations 3.303 and 3.305:

$$\hat{L}_x = \frac{1}{2}\left(\hat{L}_+ + \hat{L}_-\right) = \frac{\hbar}{2}\hat{\sigma}_x,\ \hat{L}_y = \frac{1}{2i}\left(\hat{L}_+ - \hat{L}_-\right) = \frac{\hbar}{2}\hat{\sigma}_y,\ \hat{\sigma}_x = \begin{bmatrix} 0 & 1 \\ 1 & 0 \end{bmatrix},\ \hat{\sigma}_y = \begin{bmatrix} 0 & -i \\ i & 0 \end{bmatrix} \quad (3.306)$$

Here, $\hat{\sigma}_x$ and $\hat{\sigma}_y$ are components of the Pauli matrices for the appropriate direction. We observe that the generators of the angular momentum corresponding to $l = \frac{1}{2}$ are three Pauli matrices up to the multiplicative constants.

Consider the case when $l = 1$, we have a state that is $2l + 1 = 3$ fold degenerate with the magnetic quantum number m taking values $-1, 0, 1$ and the basis states in the given space are $|1,1\rangle$, $|1,0\rangle$ and $|1,-1\rangle$:

$$|1,1\rangle \equiv \mathbb{X}_{1,1} = \begin{bmatrix} 1 \\ 0 \\ 0 \end{bmatrix} \equiv |100\rangle,\ |1,0\rangle \equiv \mathbb{X}_{1,0} = \begin{bmatrix} 0 \\ 1 \\ 0 \end{bmatrix} \equiv |010\rangle,\ |1,-1\rangle \equiv \mathbb{X}_{1,-1} = \begin{bmatrix} 0 \\ 0 \\ 1 \end{bmatrix} \equiv |001\rangle \quad (3.307)$$

while the operators are 3×3 matrices which are the generators:

$$\hat{L}_x = \frac{\hbar}{\sqrt{2}}\begin{bmatrix} 0 & 1 & 0 \\ 1 & 0 & 1 \\ 0 & 1 & 0 \end{bmatrix},\ \hat{L}_y = \frac{\hbar}{\sqrt{2}}\begin{bmatrix} 0 & -i & 0 \\ i & 0 & -i \\ 0 & i & 0 \end{bmatrix},\ \hat{L}_z = \hbar\begin{bmatrix} 1 & 0 & 0 \\ 0 & 0 & 0 \\ 0 & 0 & -1 \end{bmatrix},\ \hat{L}^2 = 2\hbar^2\hat{\mathbb{I}} \quad (3.308)$$

$$\hat{L}_+ = \hat{L}_x + i\hat{L}_y = \hbar\sqrt{2}\begin{bmatrix} 0 & 1 & 0 \\ 0 & 0 & 1 \\ 0 & 0 & 0 \end{bmatrix},\ \hat{L}_- = \hat{L}_x - i\hat{L}_y = \hbar\sqrt{2}\begin{bmatrix} 0 & 0 & 0 \\ 1 & 0 & 0 \\ 0 & 1 & 0 \end{bmatrix},\ \hat{\mathbb{I}} = \begin{bmatrix} 1 & 0 & 0 \\ 0 & 1 & 0 \\ 0 & 0 & 1 \end{bmatrix} \quad (3.309)$$

So, from Equations 3.301 to 3.309:

$$\langle l',m'|\hat{L}_z|l,m\rangle = m\hbar\delta_{m'm}\delta_{l'l},\ \langle l',m'|\hat{L}^2|l,m\rangle = \hbar^2 l(l+1)\delta_{m'm}\delta_{l'l} \quad (3.310)$$

$$\langle l',m'|\hat{L}_+|l,m\rangle = \hbar\sqrt{(l-m)(l+m+1)}\delta_{m',m+1}\delta_{l'l},\ \langle l',m'|\hat{L}_-|l,m\rangle = \hbar\sqrt{(m+l)(l-m+1)}\delta_{m',m-1}\delta_{l'l} \quad (3.311)$$

We find that the generators of the angular momentum have different representations in different spaces and the eigenvectors corresponding to the matrix operator $\hat{L}_z$ satisfy the following eigenvalue equation:

$$\hat{L}_z\mathbb{X} = m\hbar\mathbb{X} \quad (3.312)$$

These eigenvectors satisfy the following normalization condition:

$$\mathbb{X}_{1\alpha}^{\dagger}\mathbb{X}_{1\alpha} = 1 \quad (3.313)$$

The total wave function:

$$\Psi(\vec{r},t,\sigma) = \sum_{m=-\sigma}^{\sigma}\Psi_m(\vec{r},t)\mathbb{X}_{\sigma m} \quad (3.314)$$

Example:
We find from the solution of the eigenvector equation the eigenvector of a particle with $l=1$ and the projection of the moment $l_x=0$. The eigenvector equation has the following form:

$$\hat{L}_x \mathbb{X}_{l_x=0} = \frac{\hbar}{\sqrt{2}} \begin{bmatrix} 0 & 1 & 0 \\ 1 & 0 & 1 \\ 0 & 1 & 0 \end{bmatrix} \mathbb{X}_{l_x=0} \equiv \frac{\hbar}{\sqrt{2}} \begin{bmatrix} 0 & 1 & 0 \\ 1 & 0 & 1 \\ 0 & 1 & 0 \end{bmatrix} \begin{bmatrix} X_1 \\ X_2 \\ X_3 \end{bmatrix} = \frac{\hbar}{\sqrt{2}} \begin{bmatrix} X_2 \\ X_1 + X_3 \\ X_2 \end{bmatrix} = 0 \tag{3.315}$$

This follows from here that

$$X_2 = 0,\ X_1 = -X_3 \tag{3.316}$$

The eigenvector, $\mathbb{X}_{l_x=0}$, has the following form:

$$\mathbb{X}_{l_x=0} \equiv \begin{bmatrix} 0 \\ 0 \\ -1 \end{bmatrix} \tag{3.317}$$

In the complete space, we find the operators, $\hat{L}^2$, $\hat{L}_z$, $\hat{L}_x$ and $\hat{L}_y$ in the following matrix forms:

$$\hat{L}^2 = \begin{bmatrix} |0,0\rangle \\ \left|\frac{1}{2},\frac{1}{2}\right\rangle \\ \left|\frac{1}{2},-\frac{1}{2}\right\rangle \\ |1,1\rangle \\ |1,0\rangle \\ |1,-1\rangle \\ \vdots \end{bmatrix} \begin{bmatrix} 0 & & & & & & \\ & \frac{3\hbar^2}{4} & 0 & & & & \\ & 0 & \frac{3\hbar^2}{4} & & & & \\ & & & 2\hbar^2 & 0 & 0 & \\ & & & 0 & 0 & 0 & \\ & & & 0 & 0 & 2\hbar^2 & \\ & & & & & & \ddots \end{bmatrix} \tag{3.318}$$

$$\hat{L}_z = \begin{bmatrix} |0,0\rangle \\ \left|\frac{1}{2},\frac{1}{2}\right\rangle \\ \left|\frac{1}{2},-\frac{1}{2}\right\rangle \\ |1,1\rangle \\ |1,0\rangle \\ |1,-1\rangle \\ \vdots \end{bmatrix} \begin{bmatrix} 0 & & & & & & \\ & \frac{\hbar}{2} & 0 & & & & \\ & \frac{\hbar}{2} & -\frac{\hbar}{2} & & & & \\ & & & \hbar & 0 & 0 & \\ & & & 0 & 0 & 0 & \\ & & & 0 & 0 & -\hbar & \\ & & & & & & \ddots \end{bmatrix}, \quad \hat{L}_x = \begin{bmatrix} |0,0\rangle \\ \left|\frac{1}{2},\frac{1}{2}\right\rangle \\ \left|\frac{1}{2},-\frac{1}{2}\right\rangle \\ |1,1\rangle \\ |1,0\rangle \\ |1,-1\rangle \\ \vdots \end{bmatrix} \begin{bmatrix} 0 & & & & & & \\ & 0 & \frac{\hbar}{2} & & & & \\ & \frac{\hbar}{2} & 0 & & & & \\ & & & 0 & \frac{\hbar}{\sqrt{2}} & 0 & \\ & & & \frac{\hbar}{\sqrt{2}} & 0 & \frac{\hbar}{\sqrt{2}} & \\ & & & 0 & \frac{\hbar}{\sqrt{2}} & 0 & \\ & & & & & & \ddots \end{bmatrix} \tag{3.319}$$

We do the same for $\hat{L}_y$. We observe that the operators have infinite dimensional representations which are block diagonal. So any product of the operators would also be block diagonal and in particular a finite rotation by an angle ϕ generated by $\hat{\mathrm{T}}_{\vec{\phi}}$ and defined in Equation 3.161 would be a block diagonal matrix:

$$\hat{\mathrm{T}}_{\vec{\phi}} = \exp\left\{-\frac{i}{\hbar}\phi\hat{L}\right\} \tag{3.320}$$

We denote the $2l+1$ dimensional block of $\hat{\mathrm{T}}_{\vec{\phi}}$ for given l by $\hat{\mathbb{D}}(\phi)$ which rotates vectors in the space, $\varepsilon^{(l)}$. This implies that if $|\eta\rangle$ is an arbitrary vector in the subspace spanned by the $2l+1$ vectors $|l,l\rangle, l,l-1\rangle,\ldots,l,-l\rangle$, then $\hat{\mathbb{D}}(\phi)$ should be the rotation operators that would act on it and changing it to another vector, $|\eta'\rangle$:

$$\hat{\mathbb{D}}(\phi)|\eta\rangle = |\eta'\rangle \tag{3.321}$$

4

Total Angular Momentum

4.1 Infinitesimal Symmetry Transformation Generator

The infinitesimal symmetry transformation generator, $\hat{\mathbb{D}}(\phi)$, is recognizable dynamical variable in a simple situation and so can be used to define a dynamical variable in a more complicated situation. We take an example of a particle possessing an internal degree of freedom related to an **intrinsic angular momentum, the so-called spin (pseudo vector),** described by the spin operator, $\hat{S}$. So, the total wave function, considering the definition in Equation 3.314 should be

$$\Psi(\vec{r},\sigma)=\Psi(\vec{r})\mathbb{X}(\sigma) \tag{4.1}$$

where σ is the spin variable. It is instructive to note that when the state is rotated, not only does the eigenfunction $\Psi(\vec{r})$ but also the spin eigenvector $\mathbb{X}$ changes:

$$\hat{\mathbb{D}}(\phi)\Psi \cong \Psi - \frac{i}{\hbar}\phi\hat{L}\Psi - \frac{i}{\hbar}\phi\hat{S}\Psi = \left(1-\frac{i}{\hbar}\phi\,\hat{J}\right)\Psi,\ \hat{J}=\hat{L}+\hat{S} \tag{4.2}$$

or

$$\hat{\mathbb{D}}(\phi)\Psi = \exp\left\{-\frac{i}{\hbar}\phi\hat{J}\right\}\Psi \tag{4.3}$$

Now, the generators of the infinitesimal rotations will be three components of the angular momentum, $\hat{J}$, which is the total angular momentum operator and $\hat{S}$ is the operator of the intrinsic moment (spin) of the orbitary motion. The orbital angular momentum operator, $\hat{L}$, acts only on the $\vec{r}$ dependence of $\Psi(\vec{r})$, while the spin angular momentum, $\hat{S}$, acts only on the σ dependence of the eigenvector, $\mathbb{X}$. So $\hat{L}$ and $\hat{S}$ commute with each other. We will show under Chapter 14 on Relativistic Quantum Mechanics that $\hat{L}$ and $\hat{S}$ separately do not commute with the Hamiltonian, $\hat{\mathrm{H}}$, while $\hat{J}$ does. This implies, physically, that the Hamiltonian, $\hat{\mathrm{H}}$, has terms that couple spin and orbital angular momentum to each other but not to the environment.

4.2 Total Angular Momentum Justification

In order to be convinced that $\hat{J}$ is the total angular momentum, it is necessary to construct the following commutation relations:

$$\left[\hat{S}_i,\hat{S}_j\right]=i\hbar\,\epsilon_{ijk}\,\hat{S}_k \tag{4.4}$$

DOI: 10.1201/9781003273073-5

It is easily seen that

$$\left[\hat{S}_x, \hat{S}_y\right] = i\hbar\hat{S}_z, \left[\hat{S}_z, \hat{S}_x\right] = i\hbar\hat{S}_y, \left[\hat{S}_y, \hat{S}_z\right] = i\hbar\hat{S}_x \tag{4.5}$$

and shows that $\hat{S}$ imitates the commutation relations of the angular momentum operator. However, the operator of the spin, $\hat{S}$, acts in the spin space and $\hat{J}$ is the operator of the total angular momentum which is conserved in the Dirac theory as we see in Relativistic Quantum Mechanics further in Chapter 14.

4.3 Addition of Two Angular Momenta

Let us examine the property of $\hat{J}$ and, for that, we examine a general problem – the development of the rule of addition of the angular momentum. We consider a system consisting of two particles, where each particle is associated with a total angular momentum operator $\hat{J}_1$ and $\hat{J}_2$. All components of total angular momentum operator $\hat{J}_1$ commute with all components of $\hat{J}_2$. We use the subscripts 1 and 2 to label quantities relating to the two subsystems. The eigenvalues of the operators $\hat{J}_1^2$ and $\hat{J}_{1z}$ are, respectively, $\hbar^2 j_1(j_1+1)$ and $\hbar m_1$, while the eigenvalues of the operators $\hat{J}_2^2$ and $\hat{J}_{2z}$ are, respectively, $\hbar^2 j_2(j_2+1)$ and $\hbar m_2$. Here, j_1 and j_2 are the values of the angular momenta for the two particles and take the following ranges:

$$j_1 \geq m_1 \geq -j_1, \ j_2 \geq m_2 \geq -j_2 \tag{4.6}$$

We denote by $\varepsilon^{(j_1)}$ and $\varepsilon^{(j_2)}$ the spaces in which the operators $\hat{J}_1$ and $\hat{J}_2$ act, respectively, so that the total space is a direct product of two spaces:

$$\varepsilon = \varepsilon^{(j_1)} \otimes \varepsilon^{(j_2)} \tag{4.7}$$

This implies we have $(2j_1+1)(2j_2+1)$ kets of the type:

$$\left|j_1 m_1 j_2 m_2\right\rangle = \left|j_1 m_1\right\rangle \otimes \left|j_2 m_2\right\rangle \tag{4.8}$$

These are the common eigenkets of the operators $\hat{J}_1^2$ and $\hat{J}_{1z}$ and also the operators $\hat{J}_2^2$ and $\hat{J}_{2z}$. As

$$\hat{J}_z = \hat{J}_{1z} + \hat{J}_{2z} \tag{4.9}$$

then $\left|j_1 m_1 j_2 m_2\right\rangle$ are also eigenkets of $\hat{J}_z$ with eigenvalues $\hbar(m_1+m_2)$:

$$\hat{J}_z\left|j_1 m_1 j_2 m_2\right\rangle = \hbar(m_1+m_2)\left|j_1 m_1 j_2 m_2\right\rangle \tag{4.10}$$

So we define the total angular momentum operator for the total system as the sum of the individual angular momentum operators $\hat{J}_1$ and $\hat{J}_2$:

$$\hat{J} = \hat{J}_1 + \hat{J}_2 \tag{4.11}$$

From here, we have the square of the total angular momentum operator $\hat{J}$:

$$\hat{J}^2 = \hat{J}_1^2 + \hat{J}_2^2 + 2\left(\hat{J}_{1x}\hat{J}_{2x} + \hat{J}_{1y}\hat{J}_{2y} + \hat{J}_{1z}\hat{J}_{2z}\right) \tag{4.12}$$

The eigenvalues of the operators $\hat{J}^2$ and $\hat{J}_z$ are, respectively, $\hbar^2 j(j+1)$ and $\hbar m$. Our job is to find all possible values of j and m for given j_1, m_1 and j_2, m_2, respectively, and construct the eigenvectors of the operators $\hat{J}^2$ and $\hat{J}_z$. It is instructive to note that in real systems, due to the interaction effect between particles as well as relativistic effects, the operators $\hat{J}_1$ and $\hat{J}_2$ do not commute with the energy operator of the system. Sometimes, this non-commutativity for real systems may be neglected in the first approximation.

4.3.1 Clebsch-Gordan Coefficients

Considering Equation 4.12, $\hat{J}^2$ commutes with $\hat{J}_1^2$ and $\hat{J}_2^2$ as well as Equation 4.9. It is instructive to note that $\hat{J}^2$ does not commute with $\hat{J}_{1z}$ and $\hat{J}_{2z}$ separately. So we should have two ways of describing the states of the system, i.e., $\hat{J}^2, \hat{J}_z, \hat{J}_1^2, \hat{J}_2^2$ and $\hat{J}_1^2, \hat{J}_{1z}, \hat{J}_2^2, \hat{J}_{2z}$. Both of these representations are linked through a unitary transformation. The first case has the eigenvector $|jm\rangle$ and the second $|j_1m_1j_2m_2\rangle$, and we can express each of the basis completely in terms of the other as each defines a complete basis as follows:

$$\sum_{m_1=-j_1}^{j_1}\sum_{m_2=-j_2}^{j_2}|j_1m_1j_2m_2\rangle\langle j_1m_1j_2m_2|jm\rangle = |jm\rangle \tag{4.13}$$

This way of writing $|jm\rangle$ can be suggested by degeneracies of the states $|jm\rangle$ and $|j_1m_1j_2m_2\rangle$ with respect to the quantum numbers m and m_1, m_2, respectively. This implies the dimensionality of the space spanned by $|jm\rangle$ should be the same as that of $|j_1m_1j_2m_2\rangle$. Considering Equation 4.11:

$$m = m_1 + m_2 \tag{4.14}$$

We insert the closure relation:

$$\sum_{m_1=-j_1}^{j_1}\sum_{m_2=-j_2}^{j_2}|j_1m_1j_2m_2\rangle\langle j_1m_1j_2m_2| = 1 \tag{4.15}$$

into the orthogonality relation of the ket vectors $|jm\rangle$:

$$\langle jm|j'm'\rangle = \delta_{jj'}\delta_{mm'} \tag{4.16}$$

then we have the **Clebsch-Gordan (Rudolf F. A. Clebsch and Paul A. Gordan), Wigner or vector-coupling coefficients** [21] $\langle j_1m_1j_2m_2|jm\rangle$ satisfying **the first orthogonality and orthonormality condition**:

$$\sum_{m_1=-j_1}^{j_1}\sum_{m_2=-j_2}^{j_2}\langle j_1m_1j_2m_2|jm\rangle\langle j_1m_1j_2m_2|j'm'\rangle = \delta_{jj'}\delta_{mm'} \tag{4.17}$$

It is instructive to note that the summation is taken over only one index. Similarly, inserting the closure relation:

$$\sum_{j=|j_1-j_2|}^{(j_1+j_2)}\sum_{m=-j}^{j}|jm\rangle\langle jm| = 1 \tag{4.18}$$

into the orthogonality relation of the ket vectors $|j_1m_1 j_2m_2\rangle$, we have

$$\sum_{j=|j_1-j_2|}^{(j_1+j_2)} \sum_{m=-j}^{j} \langle j_1m_1 j_2m_2 | jm\rangle \langle jm | j_1m_1' j_2m_2'\rangle = \delta_{m_1m_1'}\delta_{m_2m_2'} \tag{4.19}$$

or

$$\sum_{j=|j_1-j_2|}^{(j_1+j_2)} \sum_{m=-j}^{j} \langle j_1m_1 j_2m_2 | jm\rangle \langle j_1m_1' j_2m_2' | jm\rangle = \delta_{m_1m_1'}\delta_{m_2m_2'} \tag{4.20}$$

Similarly, as above, the summation is taken over one index. From Equation 4.14, the summation over m is reduced to a single term. The Clebsch-Gordan coefficients $\langle j_1m_1 j_2m_2 | jm\rangle$ can as well be taken as the wave function of the state $|jm\rangle$ in the m_1 and m_2 representations.

4.3.1.1 Other Representation of Clebsch-Gordan Coefficients

From convention, we consider $\langle j_1j_1j_2, j-j_1 | jj\rangle$ real and positive. The Clebsch-Gordan coefficients $\langle j_1m_1 j_2m_2 | jj\rangle$ are fixed after the normalization according to Equation 4.17. If $j=m=0$, then the given procedure yields

$$\langle j_1m_1 j_2m_2 | 00\rangle = \frac{(-1)^{j_1-m_1}}{\sqrt{2j_1+1}}\delta_{j_1j_2}\delta_{m_1,-m_2} \tag{4.21}$$

It is easy to construct an irreducible basis in a higher dimensional product space. For example, say

$$\varepsilon = \varepsilon^{(j_1)} \otimes \varepsilon^{(j_2)} \otimes \varepsilon^{(j_3)} \tag{4.22}$$

and the following transforms as $|jm\rangle$:

$$\sum_{m_1m_2m_3} |j_1m_1\rangle |j_2m_2\rangle |j_3m_3\rangle \langle j_1m_1 j_2m_2 | j_4m_4\rangle \langle j_4m_4 j_3m_3 | jm\rangle = \left|\left[(j_1j_2)j_3\right]jm\right\rangle \tag{4.23}$$

If

$$|jm\rangle = |00\rangle \tag{4.24}$$

and substituting Equation 4.21, then we construct the so-called **invariant** function:

$$\sum_{m_1=-j_1}^{j_1} \sum_{m_2=-j_2}^{j_2} \sum_{m_3=-j_3}^{j_3} |j_1m_1\rangle |j_2m_2\rangle |j_3m_3\rangle \langle j_1m_1 j_2m_2 | j_3,-m_3\rangle \frac{(-1)^{j_3+m_3}}{\sqrt{2j_3+1}} \tag{4.25}$$

This leads us to the definition of the factor called the **3*j* Wigner symbol**:

$$\begin{pmatrix} j_1 & j_2 & j_3 \\ m_1 & m_2 & m_3 \end{pmatrix} \equiv \frac{(-1)^{j_1-j_2-m_2}}{\sqrt{2j_3+1}} \langle j_1m_1 j_2m_2 | j_3,-m_3\rangle \tag{4.26}$$

The phase convention renders the symmetry properties of the $3j$ Wigner symbol simple where permuting two columns or changing all the m_i to $-m_i$, we have an extra factor $(-1)^{j_1+j_2+j_3}$. So the cyclic permutations of the columns leave the $3j$ Wigner symbol invariant:

$$\begin{pmatrix} j_1 & j_2 & j_3 \\ m_1 & m_2 & m_3 \end{pmatrix} = (-1)^{j_1+j_2+j_3} \begin{pmatrix} j_1 & j_2 & j_3 \\ -m_1 & -m_2 & -m_3 \end{pmatrix} = (-1)^{j_1+j_2+j_3} \begin{pmatrix} j_2 & j_1 & j_3 \\ m_2 & m_1 & m_3 \end{pmatrix} \tag{4.27}$$

This can be continued. Considering the inversion relation, we have

$$\langle j_1 m_1 j_2 m_2 | j_3 m_3 \rangle = (-1)^{j_1 - j_2 + m_3} \sqrt{2 j_3 + 1} \begin{pmatrix} j_1 & j_2 & j_3 \\ m_1 & m_2 & -m_3 \end{pmatrix} \tag{4.28}$$

So for a system consisting of two particles, where each particle is associated with a total angular momentum operator $\hat{J}_1$ and $\hat{J}_2$ and $\hat{J}$ the total angular momentum for the entire system, we write the representation of the Clebsch-Gordan coefficients in another form as follows:

$$\Gamma^{jm}_{j_1 m_1 j_2 m_2} = (-1)^{j_1 - j_2 + m} \sqrt{2j+1} \begin{pmatrix} j_1 & j_2 & j \\ m_1 & m_2 & -m \end{pmatrix} \tag{4.29}$$

We write the inverse transformation to Equation 4.13 in the form:

$$\sum_{j=|j_1-j_2|}^{(j_1+j_2)} \sum_{m=-j}^{j} |jm\rangle \langle jm | j_1 m_1 j_2 m_2 \rangle = | j_1 m_1 j_2 m_2 \rangle \tag{4.30}$$

From here, the Clebsch-Gordan coefficients therefore permit us to express the vectors of the old basis $\{|j_1 m_1 j_2 m_2\rangle\}$ through vectors of the new basis, $\{|jm\rangle\}$. The scalar product in Equation 4.30, considering that the Clebsch-Gordan coefficients are selected to be real, can also be represented as follows:

$$\langle jm | j_1 m_1 j_2 m_2 \rangle = \langle j_1 m_1 j_2 m_2 | jm \rangle \tag{4.31}$$

4.3.1.2 Clebsch-Gordan Coefficients Recursion Relations

The Clebsch-Gordan coefficients recursion relations can be obtained by applying the following ladder operators:

$$j_\pm = j_{1\pm} + j_{2\pm} \tag{4.32}$$

to the $|j_1 m_1 j_2 m_2\rangle$ and $|jm\rangle$ forming basis vectors:

$$j_{1\pm} | j_1 m_1 j_2 m_2 \rangle = \sqrt{j_1(j_1+1) - m_1(m_1 \pm 1)} \, | j_1 m_1 \pm 1, j_2 m_2 \rangle \tag{4.33}$$

$$j_{2\pm} | j_1 m_1 j_2 m_2 \rangle = \sqrt{j_2(j_2+1) - m_2(m_2 \pm 1)} \, | j_1 m_1 j_2 m_2 \pm 1 \rangle \tag{4.34}$$

$$j_\pm | jm \rangle = \sqrt{j(j+1) - m(m \pm 1)} \, | jm \pm 1 \rangle \tag{4.35}$$

Considering Equations 4.34, 4.35 and 4.13, for the case of the operator $j_{\pm}$, we have

$$j_{\pm}\sum_{m_1=-j_1}^{j_1}\sum_{m_2=-j_2}^{j_2}\left|j_1m_1j_2m_2\right\rangle\left\langle j_1m_1j_2m_2\right|jm\rangle = j_{\pm}\left|jm\right\rangle \tag{4.36}$$

But $j_{\pm}\left|jm\right\rangle$ can be obtained from Equation 4.35, while $j_{1\pm}\left|j_1m_1j_2m_2\right\rangle$ and $j_{2\pm}\left|j_1m_1j_2m_2\right\rangle$ can be obtained, respectively, from Equations 4.35 and 4.36:

$$\sum_{m_1'=-j_1}^{j_1}\sum_{m_2'=-j_2}^{j_2}\left\langle j_1m_1'j_2m_2'\right|jm\rangle\left[\sqrt{j_1(j_1+1)-m_1'(m_1'\pm1)}\left|j_1m_1'\pm1,j_2m_2'\right\rangle+\sqrt{j_2(j_2+1)-m_2'(m_2'\pm1)}\left|j_1m_1'j_2m_2'\pm1\right\rangle\right]$$
$$=\sqrt{j(j+1)-m(m\pm1)}\left|jm\pm1\right\rangle \tag{4.37}$$

Multiplying the relation by $\left\langle j_1m_1j_2m_2\right|$, we have the recursion relation for the Clebsch-Gordan coefficients:

$$\sqrt{j(j+1)-m(m\pm1)}\left\langle j_1m_1j_2m_2\right|jm\pm1\rangle=\sqrt{j_1(j_1+1)-m_1(m_1\mp1)}\left\langle j_1m_1\mp1,j_2m_2\right|jm\rangle$$
$$+\sqrt{j_2(j_2+1)-m_2(m_2\mp1)}\left\langle j_1m_1j_2m_2\mp1\right|jm\rangle \tag{4.38}$$

Supposing $m=j$, we have

$$\left\langle j_1m_1\mp1,j_2m_2\right|jj\rangle=-\sqrt{\frac{j_2(j_2+1)-m_2(m_2\mp1)}{j_1(j_1+1)-m_1(m_1\mp1)}}\left\langle j_1m_1j_2m_2\mp1\right|jj\rangle \tag{4.39}$$

4.3.2 Triangular Rule

We find the possible values of the quantum numbers j and m given a j_1 and j_2. It is instructive to note that the space spanned by $\left|j_1m_1\right\rangle$ is $(2j_1+1)$ dimensional, considering that m_1 takes $2j_1+1$ values. Similarly, the space spanned by $\left|j_2m_2\right\rangle$ is $(2j_2+1)$ dimensional. So the total space that is the direct product of the two should be $(2j_1+1)(2j_2+1)$ dimensional. This implies we have $(2j_1+1)(2j_2+1)$ states for m_1 and m_2 and for a given j_1 and j_2. This implies different products $\left|j_1m_1\right\rangle\left|j_2m_2\right\rangle$. So the same number $(2j_1+1)(2j_2+1)$ should be equal to the number of the relations in Equation 4.38 for all possible values of j and m.

We also observe that

$$\hat{J}_z\left|j_1m_1j_2m_2\right\rangle=\left(\hat{J}_{1z}+\hat{J}_{2z}\right)\left|j_1m_1j_2m_2\right\rangle=\hat{J}_{1z}\left|j_1m_1\right\rangle\otimes\left|j_2m_2\right\rangle+\left|j_1m_1\right\rangle\otimes\hat{J}_{2z}\left|j_2m_2\right\rangle=\hbar(m_1+m_2)\left|j_1m_1j_2m_2\right\rangle \tag{4.40}$$

So, as seen earlier, the eigenvalues of $\hat{J}_z$ are $\hbar m$ where m is defined in Equation 4.14. But considering Equation 4.6:

$$(j_1+j_2)\ge m\ge-(j_1+j_2) \tag{4.41}$$

It is obvious that a particular m value can be obtained from different values of m_1 and m_2 in order to expect degeneracy of states in the given space. Recalling that the eigenvalues of the operator $\hat{J}^2$ is $\hbar^2 j(j+1)$, m takes the following values for every given j:

$$j\ge m\ge -j \tag{4.42}$$

From the general properties of the angular momentum, the ket vector $\left|jm\right\rangle$ and the Clebsch-Gordan coefficients $\left\langle j_1 m_1 j_2 m_2 \middle| jm\right\rangle$ exist only when Equation 4.42 is satisfied.

From Equation 4.41, we infer that for $m_1 = j_1$ and $m_2 = j_2$, we have one relation corresponding to

$$m_{\max} = j_1 + j_2 \tag{4.43}$$

and so, from Equation 4.42:

$$j_{\max} = j_1 + j_2 \tag{4.44}$$

It is instructive to note that $j_{\max}$ cannot be the only value of j as then the dimensionality of the space would be

$$\left(2 j_{\max} + 1\right) = \left(2 j_1 + 2 j_2 + 1\right) \tag{4.45}$$

But for non-trivial j_1, j_2:

$$\left(2 j_1 + 2 j_2 + 1\right) \neq \left(2 j_1 + 1\right)\left(2 j_2 + 1\right) \tag{4.46}$$

So j must as well take other values. To determine these values, we set the state with

$$m = m_{\max} = j_1 + j_2 \tag{4.47}$$

to be identifiable and unique. For this, we have two states with

$$m_1 = j_1 - 1, m_2 = j_2 \tag{4.48}$$

and

$$m_1 = j_1, m_2 = j_2 - 1 \tag{4.49}$$

Both of these states correspond to one and the same (two-fold degenerate state):

$$m = j_1 + j_2 - 1 \tag{4.50}$$

So there should be two different values of j (this implies two vectors $\left|jm\right\rangle$):

$$j = j_1 + j_2,\ j = j_1 + j_2 - 1 \tag{4.51}$$

We have as well three states with

$$m_1 = j_1 - 2,\ m_2 = j_2;\ m_1 = j_1 - 1,\ m_2 = j_2 - 1;\ m_1 = j_1,\ m_2 = j_2 - 2 \tag{4.52}$$

that correspond to

$$m = j_1 + j_2 - 2 \tag{4.53}$$

So, j has three values:

$$j = j_1 + j_2,\ j = j_1 + j_2 - 1,\ j = j_1 + j_2 - 2 \tag{4.54}$$

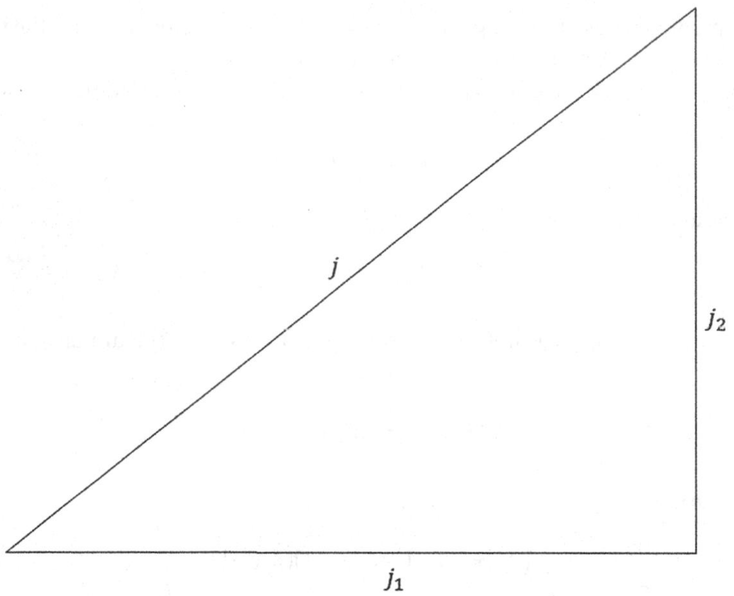

FIGURE 4.1 Triangular rule: The Clebsch-Gordan coefficients $\langle j_1 m_1 j_2 m_2 | jm \rangle$ differ from zero with a triangle with three-line segments of lengths j_1, j_2 and j.

This procedure can be continued up to the minimum value:

$$m_{\min} = |j_1 - j_2| \tag{4.55}$$

So as per the **triangular rule or triangle selection rule**:

$$|j_1 - j_2| \le j \le j_1 + j_2 \tag{4.56}$$

This triangular rule stems from the fact that a triangle can be obtained with three-line segments j_1, j_2 and j (see Figure 4.1). This triangular rule in Equation 4.56 can also be written as follows:

$$|j - j_1| \le j_2 \le j + j_1 \tag{4.57}$$

or

$$|j - j_2| \le j_1 \le j + j_2 \tag{4.58}$$

It is instructive to note that the Clebsch-Gordan-Wigner coefficients in Equation 4.13 are non-trivial only when this triangular rule is satisfied in addition to Equation 4.14. The triangular rule stems from the fact that the inequality imitates the three sides of a triangle. From here, we set

$$j_{\min} = |j_1 - j_2| \tag{4.59}$$

and

$$j_{\max} = j_1 + j_2 \tag{4.60}$$

The minimum value j_{min} has to be such that the dimensionality of the total space should be $(2j_1+1)(2j_2+1)$. For every j value, the dimensionality of the space is $(2j+1)$. So, this constraint yields the total number of available states:

$$\sum_{j_{min}}^{j_{max}}(2j+1)=(2j_1+1)(2j_2+1) \tag{4.61}$$

Hence, the total angular momentum operator achieves the values of j_{max} down to j_{min} decreasing in steps of unity and so the total space is a direct product of two spaces:

$$\varepsilon=\varepsilon^{(j_1)}\otimes\varepsilon^{(j_2)}=\sum_{j_{min}}^{j_{max}}\otimes\varepsilon^{(j)} \tag{4.62}$$

This implies that the $|jm\, j_1m_1\rangle$ basis defines a reducible space where operators take the block diagonal form.

We find the form of normalized eigenvectors when

$$j_1=j_2=1 \tag{4.63}$$

For this, we have a space that is

$$(2j_1+1)(2j_2+1)=3\times3=9 \tag{4.64}$$

fold degenerate. We determine the $\{|jm\rangle\}$ basis of common eigenvectors of j_1^2, j_2^2, j_{1z} and j_{2z} of two particles with $m_1, m_2=1,0,-1$ and $j=2,1,0$. So we construct three families of eigenvectors, $|jm\rangle$.

Case 1

So from

$$j_1=j_2=1,\ m_1=m_2=1 \tag{4.65}$$

and applying the operator j_- to the vector

$$|2,2\rangle=|1,1,1,1\rangle\equiv\begin{bmatrix}1\\0\\0\end{bmatrix}_1\begin{bmatrix}1\\0\\0\end{bmatrix}_2 \tag{4.66}$$

we have

$$|2,1\rangle=\frac{1}{2}j_-|2,2\rangle=\frac{1}{2}(j_{1-}+j_{2-})|1,1,1,1\rangle=\frac{1}{2}\left(\sqrt{2}|1,0,1,1\rangle+\sqrt{2}|1,1,1,1\rangle\right) \tag{4.67}$$

or

$$|2,1\rangle=\frac{1}{\sqrt{2}}\left(|1,1,1,0\rangle+|1,0,1,1\rangle\right)\equiv\frac{1}{\sqrt{2}}\left(\begin{bmatrix}1\\0\\0\end{bmatrix}_1\begin{bmatrix}0\\1\\0\end{bmatrix}_2+\begin{bmatrix}0\\1\\0\end{bmatrix}_1\begin{bmatrix}1\\0\\0\end{bmatrix}_2\right) \tag{4.68}$$

Similarly, we use j_- to evaluate $|2,0\rangle$:

$$|2,0\rangle = \frac{1}{\sqrt{6}}\left(|1,1,1,-1\rangle + 2|1,0,1,0\rangle + |1,-1,1,1\rangle\right) \equiv \frac{1}{\sqrt{6}}\left(\begin{bmatrix}1\\0\\0\end{bmatrix}_1\begin{bmatrix}0\\0\\1\end{bmatrix}_2 + 2\begin{bmatrix}0\\1\\0\end{bmatrix}_1\begin{bmatrix}0\\1\\0\end{bmatrix}_2 + \begin{bmatrix}0\\0\\1\end{bmatrix}_1\begin{bmatrix}1\\0\\0\end{bmatrix}_2\right) \quad (4.69)$$

From here, we find

$$|2,-1\rangle = \frac{1}{\sqrt{2}}\left(|1,0,1,-1\rangle + |1,-1,1,0\rangle\right) \equiv \frac{1}{\sqrt{2}}\left(\begin{bmatrix}0\\1\\0\end{bmatrix}_1\begin{bmatrix}0\\0\\1\end{bmatrix}_2 + \begin{bmatrix}0\\0\\1\end{bmatrix}_1\begin{bmatrix}0\\1\\0\end{bmatrix}_2\right) \quad (4.70)$$

and then

$$|2,-2\rangle = |1,-1,1,-1\rangle \equiv \begin{bmatrix}0\\0\\1\end{bmatrix}_1\begin{bmatrix}0\\0\\1\end{bmatrix}_2 \quad (4.71)$$

So for the total angular momentum, $j = 2$, we have the quintet of states that are symmetric.

Case 2
We now consider the case when

$$j = 1, m = 1 \quad (4.72)$$

In this case, the vector $|1,1\rangle$ should be a linear combination of $|1,1,1,0\rangle$ and $|1,0,1,1\rangle$:

$$|1,1\rangle = c_1|1,1,1,0\rangle + c_2|1,0,1,1\rangle \quad (4.73)$$

But

$$|c_1|^2 + |c_2|^2 = 1 \quad (4.74)$$

For the vector $|1,1\rangle$ to be orthogonal to $|2,1\rangle$, it is necessary that

$$c_1 + c_2 = 0 \quad (4.75)$$

So

$$|1,1\rangle = \frac{1}{\sqrt{2}}\left(|1,1,1,0\rangle - |1,0,1,1\rangle\right) \equiv \frac{1}{\sqrt{2}}\left(\begin{bmatrix}1\\0\\0\end{bmatrix}_1\begin{bmatrix}0\\1\\0\end{bmatrix}_2 - \begin{bmatrix}0\\1\\0\end{bmatrix}_1\begin{bmatrix}1\\0\\0\end{bmatrix}_2\right) \quad (4.76)$$

Applying the operator j_- here, using the same procedure permits us to find $|1,0\rangle$ and $|1,-1\rangle$:

$$|1,0\rangle = \frac{1}{\sqrt{2}}\left(|1,1,1,-1\rangle - |1,-1,1,1\rangle\right) \equiv \frac{1}{\sqrt{2}}\left(\begin{bmatrix}1\\0\\0\end{bmatrix}_1\begin{bmatrix}0\\0\\1\end{bmatrix}_2 - \begin{bmatrix}0\\0\\1\end{bmatrix}_1\begin{bmatrix}1\\0\\0\end{bmatrix}_2\right) \quad (4.77)$$

Here, the vector, $|1,0,1,0\rangle$ is absent though corresponding to $m=0$ since the corresponding Clebsch-Gordan coefficient is zero:

$$\langle 1,0|1,0,1,0\rangle = 0 \tag{4.78}$$

Also, in addition, we have

$$|1,-1\rangle = \frac{1}{\sqrt{2}}\left(|1,0,1,-1\rangle - |1,-1,1,0\rangle\right) \equiv \frac{1}{\sqrt{2}}\left(\begin{bmatrix}0\\1\\0\end{bmatrix}_1\begin{bmatrix}0\\0\\1\end{bmatrix}_2 - \begin{bmatrix}0\\0\\1\end{bmatrix}_1\begin{bmatrix}0\\1\\0\end{bmatrix}_2\right) \tag{4.79}$$

For the total angular momentum, $j=1$, we have the triplet of states that are anti-symmetric.

Case 3

We consider the case when

$$j=0,\ m=0 \tag{4.80}$$

and

$$|0,0\rangle = c_1|1,1,1,-1\rangle + c_2|1,0,1,0\rangle + c_3|1,-1,1,1\rangle \tag{4.81}$$

with

$$|c_1|^2 + |c_2|^2 + |c_3|^2 = 1 \tag{4.82}$$

This should be orthogonal to $|2,0\rangle$ and $|1,0\rangle$:

$$c_1 + 2c_2 + c_3 = 0,\ c_1 - c_3 = 0 \tag{4.83}$$

So

$$c_1 = -c_2 = c_3 \tag{4.84}$$

From here and considering Equation 4.82:

$$|0,0\rangle = \frac{1}{\sqrt{3}}\left(|1,1,1,-1\rangle - |1,0,1,0\rangle + |1,-1,1,1\rangle\right) \equiv \frac{1}{\sqrt{3}}\left(\begin{bmatrix}1\\0\\0\end{bmatrix}_1\begin{bmatrix}0\\0\\1\end{bmatrix}_2 - \begin{bmatrix}0\\1\\0\end{bmatrix}_1\begin{bmatrix}0\\1\\0\end{bmatrix}_2 + \begin{bmatrix}0\\0\\1\end{bmatrix}_1\begin{bmatrix}1\\0\\0\end{bmatrix}_2\right) \tag{4.85}$$

So for $j=0$, we have a singlet state which is symmetric.

Finally, we completed the construction of the $\{|jm\rangle\}$ basis when $j_1 = j_2 = 1$. The subscripts in the eigenvectors indicate the particle or subsystem.

4.4 Spherical Spinors

A **spinor** is a mathematical object which can be Lorentz-transformed while allowing a more general treatment of the notion of invariance under rotation. Spinors play a vital role when seeing electrons and other particles to have an intrinsic angular momentum (spin). Pauli used spinors in representing the electron spin using a two-component complex vector through Pauli spin matrices. For the addition of the spin and orbital moments of a spin-$\frac{1}{2}$ particle, considering Equation 4.41, we have

$$\sum_{\substack{m,\mu \\ (m,\mu=M)}} \langle lm|1/2\mu\rangle \Gamma^{jM}_{1/2\mu lm} = |jM\rangle \tag{4.86}$$

We multiply this relation from the left by $|\theta\phi\mu'\rangle$ considering

$$\langle\theta\phi\mu'|jM\rangle = \Omega_{jlM\mu'} \tag{4.87}$$

and

$$\langle\theta\phi|lm\rangle = \mathrm{Y}_{lm}(\theta,\phi) \tag{4.88}$$

then we have

$$\Omega_{jlM\mu'}(\theta,\phi) = \sum_{m,\mu} \mathrm{Y}_{lm}(\theta,\phi)\delta_{\mu\mu'}\Gamma^{jM}_{1/2\mu lm} \tag{4.89}$$

We represent this result in the form of a two-component wave function-spherical spinor Ω_{jlM}:

$$\Omega_{jlM} = \begin{bmatrix} \mathrm{Y}_{l,M-1/2}(\theta,\phi)\ \Gamma^{jM}_{1/2,1/2,l,M-1/2} \\ \mathrm{Y}_{l,M+1/2}(\theta,\phi)\ \Gamma^{jM}_{1/2,-1/2,l,M+1/2} \end{bmatrix} \tag{4.90}$$

From the triangular rule, for a given j, there are two possible values of l: $j+1/2$ and $j-1/2$. We represent one of them via l and another via l'. It is obvious that

$$l + l' = 2j \tag{4.91}$$

and

$$l - l' = \pm 1 \tag{4.92}$$

Since l and l' differ by unity, the spherical spinors Ω_{jlM} and $\Omega_{jl'M}$ have opposite parities. We write the Clebsch-Gordan coefficients in Equation 4.90 as follows:

$$\Gamma^{jM}_{1/2,1/2,j-1/2,M-1/2} = \sqrt{\frac{j+M}{2j}},\ \Gamma^{jM}_{1/2,-1/2,j-1/2,M+1/2} = \sqrt{\frac{j-M}{2j}} \tag{4.93}$$

$$\Gamma^{jM}_{1/2,1/2,j+1/2,M-1/2} = \sqrt{\frac{j-M+1}{2j+2}},\ \Gamma^{jM}_{1/2,-1/2,j+1/2,M+1/2} = \sqrt{\frac{j+M+1}{2j+2}} \tag{4.94}$$

The spherical spinors Ω_{jlM} and $\Omega_{jl'M}$ are related via

$$\Lambda\Omega_{jlM} = \Omega_{jl'M},\ l' = 2j - l \tag{4.95}$$

We find the expression for Λ. This operator is a 2×2 matrix that may be expressed via the unit matrix and three Pauli matrices $\hat{\sigma}_x$, $\hat{\sigma}_y$ and $\hat{\sigma}_z$. From the properties of the Pauli matrices, we limit the expression of Λ to be linear with respect to the Pauli matrices $\hat{\sigma}_x$, $\hat{\sigma}_y$ and $\hat{\sigma}_z$ with the coefficients of the linear combination being functions of θ and ϕ. The operator Λ is invariant relative to the rotation of the coordinate axes and changes sign due to the inversion of the coordinates since Ω_{jlM} and $\Omega_{jl'M}$ have opposite parities. So the operator Λ should be a pseudo scalar. This implies that it is a scalar product of the pseudo vector (spin) $\hat{S}$ and some vector with components that are dependent on the polar θ and Azimuthal ϕ angles. The only vector that relates Λ and differs only by a constant factor is the unit normal vector $\hat{n}$ dependent on θ and ϕ:

$$\hat{n} = \left(n_x, n_y, n_z\right),\ n_x = \sin\theta\cos\phi,\ n_y = \sin\theta\sin\phi,\ n_z = \cos\theta \tag{4.96}$$

So

$$\Lambda = a\hat{n}\hat{S} \tag{4.97}$$

Here, a is some constant independent of the polar angles θ and ϕ.

4.4.1 Spinor Rotation

A spinor (two-component wave vector), like a vector, can be rotated through the polar angles θ and ϕ with the spinor magnitude fixed. We consider the spin state (spinor)

$$\mathbb{X} = \begin{bmatrix} X_1 \\ X_2 \end{bmatrix} \tag{4.98}$$

For $\mathbb{X}$ to be a unit vector, the scalar product of $\mathbb{X}$ with its Hermitian conjugate, $\mathbb{X}^\dagger$, should be unity, i.e., $\mathbb{X}$ should satisfy the following normalization condition:

$$\mathbb{X}\mathbb{X}^\dagger = |X_1|^2 + |X_2|^2 = 1 \tag{4.99}$$

Here, X_1 and X_2 are functions of the (θ,ϕ)-direction of the spin described by the following spin operator (pseudo vector quantity):

$$\hat{S} = \frac{\hbar}{2}\hat{\sigma},\ \frac{1}{\hbar}\hat{S} = \hat{s} = \frac{1}{2} \tag{4.100}$$

and $\hat{\sigma}$ is the Pauli matrix with components, $\hat{\sigma}_x$, $\hat{\sigma}_y$ and $\hat{\sigma}_z$. When the average value of the spin transforms under rotations of the coordinate system as a vector, one should expect the average $\bar{\sigma}_z$ to be simply the projection of the unit vector in the (θ,ϕ)-direction on the z-axis:

$$\bar{\sigma}_z = \cos\theta \tag{4.101}$$

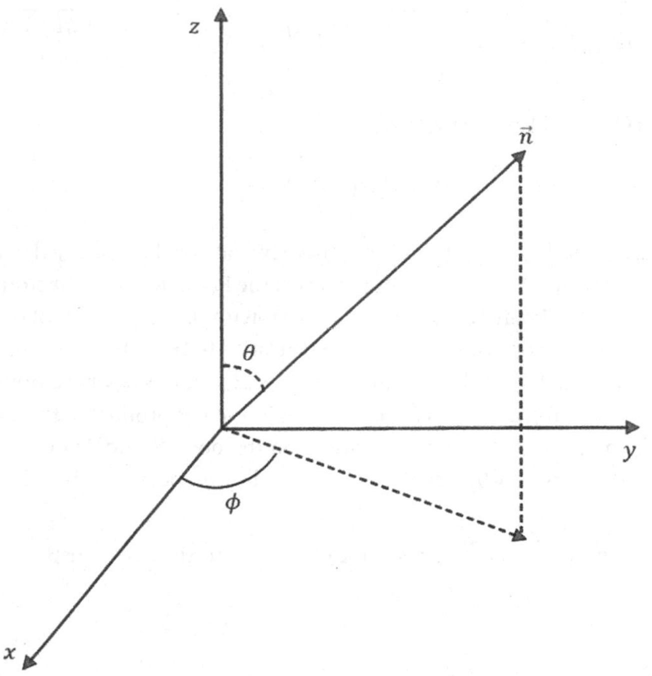

FIGURE 4.2 The polar angles θ and ϕ characterizing a unit vector $\hat{n}$.

Similarly, the average values of $\bar{\sigma}_x$ and $\bar{\sigma}_y$ in the spin state $\mathbb{X}$ are as follows:

$$\bar{\sigma}_x = \sin\theta\cos\phi, \bar{\sigma}_y = \sin\theta\,\sin\phi \tag{4.102}$$

So the operator of the projection of the spin $\hat{s}_n$ along the unit vector $\hat{n}$ direction should be expressed through the components of the spin operator, $\hat{s}_x, \hat{s}_y$ and $\hat{s}_z$. Since the spin operator, $\hat{s}$, is a pseudo vector, the spin projection operator, $\hat{s}_n$, in an arbitrary direction, $\hat{n}$ should be expressed through the components of the spin operator $\hat{s}_x, \hat{s}_y$ and $\hat{s}_z$ as for the case of an ordinary (not operator) vector. So we consider the rotation of the spin vector $\hat{s}$ through the polar angles θ and ϕon the coordinate system as in Figure 4.2 and implies that we are interested in a spin rotation in spin space.

We find the projection s_n of the spin $\hat{s}$ along the unit normal vector $\hat{n}$ characterized by the polar angles θ and ϕ written as follows:

$$s_n = \hat{n}\hat{s} = s_x \sin\theta\cos\phi + s_y \sin\theta\,\sin\phi + s_z\cos\theta, \hat{n}^2 = 1 \tag{4.103}$$

and

$$\hat{s}_n = \frac{1}{2}\hat{\sigma}_{\theta,\phi}, \hat{\sigma}_{\theta,\phi} = \begin{bmatrix} \cos\theta & \sin\theta\exp\{-i\phi\} \\ \sin\theta\exp\{i\phi\} & -\cos\theta \end{bmatrix} \tag{4.104}$$

where the Hermitian operator, $\hat{\sigma}_{\theta,\phi}$, is a 2×2 matrix. It is instructive to note that the operator, $\hat{\sigma}_{\theta,\phi}$, and its Cartesian components $\hat{\sigma}_x$, $\hat{\sigma}_y$ and $\hat{\sigma}_z$ satisfy the same relation which holds between an ordinary vector pointing in the (θ,ϕ)-direction and its Cartesian components:

$$\hat{\sigma}_{\theta,\phi} = \hat{\sigma}_x \sin\theta\cos\phi + \hat{\sigma}_y \sin\theta\,\sin\phi + \hat{\sigma}_z\cos\theta \tag{4.105}$$

We have to find the eigenvalues and eigenvectors of the observables s_x, s_y and s_n via the eigenvalue equation:

$$\hat{s}_n \mathbb{X} = \nu\hbar\mathbb{X} \tag{4.106}$$

where the spinor $\mathbb{X}$ is defined in Equation 4.98.

So

$$\hat{s}_n \mathbb{X} = \frac{1}{2}\hat{\sigma}_{\theta,\phi}\begin{bmatrix} X_1 \\ X_2 \end{bmatrix} = \frac{1}{2}\begin{bmatrix} X_1\cos\theta + X_2\sin\theta\exp\{-i\phi\} \\ X_1\sin\theta\exp\{i\phi\} - X_2\cos\theta \end{bmatrix} \tag{4.107}$$

From here, considering Equation 4.106:

$$X_1(\cos\theta - 2\nu) + X_2\sin\theta\exp\{-i\phi\} = 0,\ X_1\sin\theta\exp\{i\phi\} - X_2(\cos\theta + 2\nu) = 0 \tag{4.108}$$

For the non-trivial solution of these systems of equations in Equation 4.108:

$$\det\begin{bmatrix} \cos\theta - 2\nu & \sin\theta\exp\{-i\phi\} \\ \sin\theta\exp\{i\phi\} & -(\cos\theta + 2\nu) \end{bmatrix} = 0 \tag{4.109}$$

or

$$-\left(\cos^2\theta - 4\nu^2\right) - \sin^2\theta = 0 \tag{4.110}$$

or

$$4\nu^2 - 1 = 0 \tag{4.111}$$

From here:

$$\nu = \pm\frac{1}{2} \tag{4.112}$$

We observe that the operators $\hat{s}_x$, $\hat{s}_y$ and $\hat{s}_n$ have the same eigenvalues $+\frac{1}{2}$ and $-\frac{1}{2}$ as $\hat{s}_z$. If we consider the Stern-Gerlach experiment involving the splitting into two components of an atomic beam in an inhomogeneous magnetic field, then this should be an expected result as it is always possible to rotate the entire Stern-Gerlach apparatus so as to align the axis defined by the magnetic field parallel either with ox, to oy or with $\hat{n}$. It is instructive to note that all directions of space have the same properties and so the phenomenon observed in the Stern-Gerlach apparatus must be invariant under such rotations. So s_x, s_y and s_n should therefore have only one of the results $+\frac{1}{2}$ or $-\frac{1}{2}$ associated with them. We now find the eigenvectors of these operators by solving the system of equations in Equations 4.108 and 4.116:

$$\frac{X_2}{X_1} = \frac{\sin\theta}{\cos\theta + 1}\exp\{i\phi\} = \tan\frac{\theta}{2}\exp\{i\phi\} \tag{4.113}$$

From the normalization condition in Equation 4.99, we have the following spinors that realizes the optimum knowledge attainable about the electronic spin:

$$\mathbb{X}_{\frac{1}{2}} = \begin{bmatrix} \cos\frac{\theta}{2} \\ \sin\frac{\theta}{2}\exp\{i\phi\} \end{bmatrix}, \mathbb{X}_{-\frac{1}{2}} = \begin{bmatrix} -\sin\frac{\theta}{2}\exp\{-i\phi\} \\ \cos\frac{\theta}{2} \end{bmatrix} \tag{4.114}$$

So Equation 4.114 should be the eigenvector associated with the eigenvalue $\pm\frac{1}{2}$ of the component s_n of the spin $\hat{s}$ along the unit vector $\hat{n}$ that is characterized by the polar angles θ and ϕ. Such spin-states can always be prepared by the Stern-Gerlach experiment, say with the external magnetic field in the polar angles θ- and ϕ-direction while allowing only the appropriate component of the split beam to emerge from the apparatus. The spin-states in Equation 4.114 can represent, respectively, the quantum numbers, **spin-up** and **spin-down** in the polar angles θ- and ϕ-direction.

Let us find the average value, $\bar{s}_n$, in the state $\mathbb{X}_{s_z}$ with a defined value of the spin projection, s_z, on the z-axis with $s_z = \frac{1}{2}$:

$$\bar{s}_n = \frac{1}{2}\begin{bmatrix} 1 & 0 \end{bmatrix}\begin{bmatrix} \cos\theta & \sin\theta\exp\{-i\phi\} \\ \sin\theta\exp\{i\phi\} & -\cos\theta \end{bmatrix}\begin{bmatrix} 1 \\ 0 \end{bmatrix} = \frac{1}{2}\cos\theta \tag{4.115}$$

Similarly, with $s_z = -\frac{1}{2}$:

$$\bar{s}_n = -\frac{1}{2}\cos\theta \tag{4.116}$$

These projections correspond, respectively, to the probabilities:

$$W_{\frac{1}{2}} = \frac{1+2s_z\cos\theta}{2} = \cos^2\frac{\theta}{2}, W_{-\frac{1}{2}} = 1 - W_{\frac{1}{2}} = \frac{1-2s_z\cos\theta}{2} = \sin^2\frac{\theta}{2} \tag{4.117}$$

Physical interpretation

From Equation 4.99, the quantities $|X_1|^2$ and $|X_2|^2$ stand for probability interpretations that are consistent with the Stern-Gerlach experiment where an atomic beam with spins $\frac{1}{2}$ is aligned in the (θ,ϕ)-direction and represented by the spinor, $\mathbb{X}$. We perform another Stern-Gerlach experiment with an external magnetic field parallel to the z-axis direction and some anti-parallel to it. In principle, the outcome of the given experiment is only statistically predictable. From Equation 4.99, one of the quantities $|X_1|^2$ and $|X_2|^2$ may be the probability of finding the spin aligned parallel to the z-axis direction and the other probability of finding the spin aligned anti-parallel to the z-axis direction. This motivates the dependence of X_1 and X_2 on the angles θ and ϕ so that $|X_1|^2$ and $|X_2|^2$ can, respectively, be the given probabilities. From here, the states $\mathbb{X}_{\frac{1}{2}} = \begin{bmatrix} 1 \\ 0 \end{bmatrix}$ and $\mathbb{X}_{-\frac{1}{2}} = \begin{bmatrix} 0 \\ 1 \end{bmatrix}$ represent situations where the spin in the z-axis direction has with certainty the values +1 and −1, respectively.

To find the constant a, we find Equation 4.95 for $\theta = \phi = 0$ and then considering

$$Y_{lm}(0,0) = \sqrt{\frac{2l+1}{4\pi}}\delta_{m0} \tag{4.118}$$

we have

$$\Omega_{j,j-1/2,1/2}(0,0)=\begin{bmatrix}\sqrt{\frac{2j}{4\pi}\frac{j+1/2}{2j}}\\ 0\end{bmatrix}=\sqrt{\frac{j+1/2}{4\pi}}\begin{bmatrix}1\\0\end{bmatrix},$$

$$\Omega_{j,j+1/2,1/2}(0,0)=\begin{bmatrix}-\sqrt{\frac{2j+2}{4\pi}\frac{j+1/2}{2j+2}}\\ 0\end{bmatrix}=-\sqrt{\frac{j+1/2}{4\pi}}\begin{bmatrix}1\\0\end{bmatrix} \quad (4.119)$$

Comparing these expressions with Equations 4.95 and 4.97, we find that $a=-1$ and so

$$\Omega_{jl'M}=-\hat{n}\hat{S}\Omega_{jlM} \quad (4.120)$$

4.4.2 Spin Density

We examine the spin density Hermitian matrix $\hat{\rho}$ as follows:

$$\hat{\rho}=\frac{1}{2}\left(\mathbb{I}+P_x\hat{\sigma}_x+P_y\hat{\sigma}_y+P_z\hat{\sigma}_z\right)\equiv\frac{1}{2}\left(\mathbb{I}+\hat{P}\hat{\sigma}\right) \quad (4.121)$$

where $\hat{P}$ is a pseudo vector:

$$P_i=\left\langle\hat{\sigma}_i(\vec{r})\right\rangle, i=x,y,z \quad (4.122)$$

Here P_x, P_y and P_z are components of the pseudo vector $\hat{P}$ that characterizes the spin polarization. This is the **so-called polarization vector.** The idempotence of $\hat{\rho}$ is achieved only if $\hat{P}$ satisfies the following condition:

$$\hat{P}^2=1 \quad (4.123)$$

that leaves for the description of the spinor. From here:

$$\hat{\rho}^2=\frac{1}{4}\left(\mathbb{I}+\hat{P}\hat{\sigma}\right)\left(\mathbb{I}+\hat{P}\hat{\sigma}\right)=\frac{1}{4}\left(\mathbb{I}+2\hat{P}\hat{\sigma}+\hat{P}^2\mathbb{I}\right)=\frac{1}{2}\left(\mathbb{I}+\hat{P}\hat{\sigma}\right)=\hat{\rho} \quad (4.124)$$

Since the spin density $\hat{\rho}$ considering the spinor $\mathbb{X}$ is expected to be

$$\hat{\rho}=\begin{bmatrix}\cos^2\frac{\theta}{2} & \sin\frac{\theta}{2}\cos\frac{\theta}{2}\exp\{-i\phi\}\\ \sin\frac{\theta}{2}\cos\frac{\theta}{2}\exp\{i\phi\} & \sin^2\frac{\theta}{2}\end{bmatrix} \quad (4.125)$$

then

$$\mathrm{Tr}\hat{\rho}=1 \quad (4.126)$$

and is idempotent as in Equation 4.124, and so from Equation 4.123, we have $\bar{\hat{\sigma}}_z$ defined in Equation 4.101. Similarly, the average values of $\bar{\hat{\sigma}}_x$ and $\bar{\hat{\sigma}}_y$ defined in Equation 4.102 for the spin state $\mathbb{X}$:

$$P_x = \sin\theta\cos\phi,\ P_y = \sin\theta\sin\phi,\ P_z = \cos\theta \tag{4.127}$$

So the unit vector $\hat{P}$ points in the (θ,ϕ)-direction and is the **so-called polarization vector of state**. The unit length of $\hat{P}$ which is derived from the idempotence of $\hat{\rho}$ is a consequence of the total polarization of the system in a **pure state**.

It is instructive to note that in the **mixed state**,

$$\hat{P}^2 < 1 \tag{4.128}$$

and in the **absence of polarization**,

$$\hat{P} = 0 \tag{4.129}$$

It is also instructive to note that the ensemble picture of states that are not pure become fictitious models when experimental situations are such that optimum information on subsystems of the entire system are in principle not available. We consider an example of a partially polarized beam of electrons. In the ensemble model, we introduce the density matrix, $\hat{\rho}$:

$$\hat{\rho} = \sum_n W_n\hat{\rho}_n = \sum_n W_n\frac{1}{2}\left(\mathbb{I} + P_n\hat{\sigma}\right) = \frac{1}{2}\left(\mathbb{I} + \bar{P}\hat{\sigma}\right) \tag{4.130}$$

where W_n is the probability for finding the average value $\bar{\hat{\rho}}_n$ in the ensemble. The average polarization

$$\bar{P} = \sum_n W_n P_n \tag{4.131}$$

is no longer a unit vector and $\hat{\rho}$ is no longer idempotent. The absolute value $|\bar{P}|$ represents the degree of polarization of the beam. The relation $|\bar{P}| = 0$ implies the beam is unpolarized and $|\bar{P}| = 1$ implies full polarization. Suppose $\bar{P}$ is the only observable quantity of the system. So the decomposition in Equation 4.131 is fictitious. This is because the full experimental information is inherent in Equation 4.130. This renders the set of electrons each with **definite** orientation of spin, represented by the unit vector $\hat{P}_n$ fictitious. This is undesirable from an operational point of view. So, generally,

$$\bar{\sigma}_i = \mathrm{Tr}\hat{\sigma}_i\hat{\rho} = \frac{1}{2}\mathrm{Tr}\hat{\sigma}_i + \frac{1}{2}\mathrm{Tr}\bar{P}_i + \hat{\sigma}_i\mathrm{Tr}\sum_{\substack{k \\ k\neq i,\ j\neq k,\ i\neq j}} \bar{P}_k\hat{\sigma}_j = \frac{1}{2}\mathrm{Tr}\bar{P}_i = \bar{P}_i \tag{4.132}$$

Question:
Examine the addition of angular momenta j_1 and j_2 of two subsystems with a total angular momentum, j, and described by the eigenfunction:

$$|jm\rangle = \sum_{m_1=-j_1}^{j_1}\sum_{m_2=-j_2}^{j_2} |j_1m_1j_2m_2\rangle\langle j_1m_1j_2m_2|jm\rangle \tag{4.133}$$

where

$$j = j_1 + j_2,\ m = m_1 + m_2 \tag{4.134}$$

$\varepsilon^{(j_1)}$ and $\varepsilon^{(j_2)}$ denote the spaces in which the operators $\hat{J}_1$ and $\hat{J}_2$ act, respectively, so that the total space is a direct product of two spaces:

$$\varepsilon = \varepsilon^{(j_1)} \otimes \varepsilon^{(j_2)} \tag{4.135}$$

This implies there are $(2j_1+1)(2j_2+1)$ kets of the type:

$$\left|j_1 m_1 j_2 m_2\right\rangle = \left|j_1 m_1\right\rangle \otimes \left|j_2 m_2\right\rangle \tag{4.136}$$

and $\left\langle j_1 m_1 j_2 m_2 \middle| jm\right\rangle$ are the Clebsch-Gordan coefficients. The manner of writing $\left|jm\right\rangle$ is suggested by degeneracies of the states $\left|jm\right\rangle$ and $\left|j_1 m_1 j_2 m_2\right\rangle$ with respect to the quantum numbers m and m_1, m_2, respectively. This implies the dimensionality of the space spanned by $\left|jm\right\rangle$ should be the same as that of $\left|j_1 m_1 j_2 m_2\right\rangle$. Find the Clebsch-Gordan coefficients, $\left\langle j_1 m_1 j_2 m_2 \middle| jm\right\rangle$, using the ladder operators, j_-.

Answer:
From the problem:

$$j = j_1 + j_2 \tag{4.137}$$

and for $m = j$, the eigenvector:

$$\left|j_1 j_1 j_2 j_2\right\rangle = \left|j_1 j_1\right\rangle \otimes \left|j_2 j_2\right\rangle \tag{4.138}$$

We consider the ladder operators:

$$j_- = j_{1-} + j_{2-} \tag{4.139}$$

so that

$$j_-\left|jm\right\rangle = \sqrt{j(j+1) - m(m-1)}\left|jm-1\right\rangle \tag{4.140}$$

and

$$\left|jm\right\rangle = \sqrt{\frac{(j+m)!}{(j-m)!(2j)!}}\,(j_-)^{j-m}\left|jj\right\rangle \tag{4.141}$$

Considering Equation 4.139 and the fact that j_{1-} and j_{2-} commute with each other, from the explicit form of the eigenvector in Equation 4.138 and the property in Equation 4.140, we find from Equation 4.141 that

$$\begin{aligned}
\left|jm\right\rangle &= G(j,m)\sum_M C_{j-m}^M (j_{1-})^M (j_{2-})^{j-m-M}\left|j_1 j_1\right\rangle\left|j_2 j_2\right\rangle \\
&= G(j,m)\sum_M C_{j-m}^M G^{-1}(j_1, j_1 - M)G^{-1}(j_2, m + M - j_1)\left|j_1, j_1 - M\right\rangle\left|j_2, j_2 - j + m + M\right\rangle \\
&\equiv \sum_{m_1} \left\langle j_1 m_1 j_2 m_2 \middle| jm\right\rangle \left|j_1 m_1\right\rangle\left|j_2 m_2\right\rangle \equiv \sum_{m_1} \left\langle j_1 m_1 j_2 m_2 \middle| jm\right\rangle \left|j_1 m_1 j_2 m_2\right\rangle
\end{aligned} \tag{4.142}$$

In the above equations, we have the following denotations:

$$G(j,m)=\sqrt{\frac{(j+m)!}{(j-m)!(2j)!}},\ C_j^m=\frac{j!}{m!(j-m)!} \tag{4.143}$$

From Equation 4.155, we find the Clebsch-Gordan coefficients are as follows:

$$\langle j_1m_1 j_2m_2 | jm\rangle \equiv C_{j-m}^{j_1-m_1}G(j,m)G^{-1}(j_1,m_1)G^{-1}(j_2,m_2) \tag{4.144}$$

From here and considering the relations in Equation 4.143, we find

$$\langle j_1m_1 j_2m_2 | jm\rangle \equiv \sqrt{\frac{(2j_1)!(2j_2)!(j+m)!(j-m)!}{(2j)!(j_1+m_1)!(j_1-m_1)!(j_2+m_2)!(j_2-m_2)!}} \tag{4.145}$$

4.5 Spin of a System of Two Particles

We consider the sum of two angular momenta with eigenvalues $j=\frac{1}{2}$ each by using the Clebsch-Gordan coefficients. An example of the two spin-$\frac{1}{2}$ particles is, for example, say an electron and a proton in the ground state of the hydrogen atom with each having spin up or spin down. In this example, we examine the resulting eigenvalues and eigenstates:

$$j_1=\frac{1}{2},\ j_2=\frac{1}{2} \tag{4.146}$$

and so

$$m_1=-\frac{1}{2},\frac{1}{2};\ m_2=-\frac{1}{2},\frac{1}{2} \tag{4.147}$$

The basis states for each of the angular momentum operators are as follows (four possibilities in total):

$$|j_1,m_1\rangle : \left|\frac{1}{2},\frac{1}{2}\right\rangle, \left|\frac{1}{2},-\frac{1}{2}\right\rangle; |j_2,m_2\rangle : \left|\frac{1}{2},\frac{1}{2}\right\rangle, \left|\frac{1}{2},-\frac{1}{2}\right\rangle \tag{4.148}$$

As

$$j_1=j_2=\frac{1}{2} \tag{4.149}$$

then these should be dropped and the states should be denoted by their m quantum numbers only:

$$|j_1m_1 j_2m_2\rangle = |j_1m_1\rangle \otimes |j_2m_2\rangle \tag{4.150}$$

This implies that

$$|m_1,m_2\rangle = |m_1\rangle \otimes |m_2\rangle \tag{4.151}$$

It is obvious that there are four independent basis states in the entire space:

$$\left|\downarrow,\downarrow\right\rangle \equiv v_1v_2; \left|\uparrow,\uparrow\right\rangle \equiv u_1u_2; \left|\downarrow,\uparrow\right\rangle \equiv u_2v_1; \left|\uparrow,\downarrow\right\rangle \equiv u_1v_2 \tag{4.152}$$

From Equations 4.40 and 4.150, we have

$$\hat{J}_z\left|\uparrow,\uparrow\right\rangle = \left(\hat{J}_{1z}+\hat{J}_{2z}\right)\left|\uparrow,\uparrow\right\rangle = \hat{J}_{1z}\left|\uparrow\right\rangle\otimes\left|\uparrow\right\rangle + \left|\uparrow\right\rangle\otimes\hat{J}_{2z}\left|\uparrow\right\rangle = \hbar\left(\frac{1}{2}+\frac{1}{2}\right)\left|\uparrow,\uparrow\right\rangle = \hbar\left|\uparrow,\uparrow\right\rangle \tag{4.153}$$

Similarly, it can be shown that

$$\hat{J}_z\left|\uparrow,\downarrow\right\rangle = 0,\ \hat{J}_z\left|\downarrow,\uparrow\right\rangle = 0,\ \hat{J}_z\left|\downarrow,\downarrow\right\rangle = -\hbar\left|\downarrow,\downarrow\right\rangle \tag{4.154}$$

This implies that, in the product basis, we have

$$\hat{J}_z \rightarrow \begin{matrix} (\uparrow,\uparrow) \\ (\uparrow,\downarrow) \\ (\downarrow,\uparrow) \\ (\downarrow,\downarrow) \end{matrix} \begin{bmatrix} \hbar & 0 & 0 & 0 \\ 0 & 0 & 0 & 0 \\ 0 & 0 & 0 & 0 \\ 0 & 0 & 0 & -\hbar \end{bmatrix} \tag{4.155}$$

So it is clear that the allowed values of the magnetic quantum number m are as follows:

$$m = -1, 0, 1 \tag{4.156}$$

This is consistent with the general result in Equation 4.41 and it is obvious that we have three vector states:

$$|1,-1\rangle, |1,0\rangle, |1,1\rangle \tag{4.157}$$

Consider that

$$\hat{J}^2 = \hat{J}_1^2 + \hat{J}_2^2 + 2\left(\hat{J}_1, \hat{J}_2\right) = \hat{J}_1^2 + \hat{J}_2^2 + 2\left(\hat{J}_{1x}\hat{J}_{2x} + \hat{J}_{1y}\hat{J}_{2y} + \hat{J}_{1z}\hat{J}_{2z}\right) \tag{4.158}$$

or

$$\hat{J}^2 = \hat{J}_1^2 + \hat{J}_2^2 + 2\hat{J}_{1z}\hat{J}_{2z} + \hat{J}_{1+}\hat{J}_{2-} + \hat{J}_{1-}\hat{J}_{2+} \tag{4.159}$$

then we show that, in the product basis,

$$\hat{J}^2 \rightarrow \hbar^2 \begin{bmatrix} 2 & 0 & 0 & 0 \\ 0 & 1 & 1 & 0 \\ 0 & 1 & 1 & 0 \\ 0 & 0 & 0 & 2 \end{bmatrix} \tag{4.160}$$

The operator $\hat{J}^2$ commutes with $\hat{J}_z$ and so has non-zero matrix elements only between eigenvectors of $\hat{J}_z$ associated with the same eigenvalue. From Equation 4.160, it follows that the angular momentum squared, $\hat{J}^2$, is not diagonal in the given basis. Notwithstanding, it is possible to diagonalize this matrix. For that, we consider the sum of the spins $j=1$ for the **triplet state** and $j=0$ for the **singlet state**. We first examine the **triplet state** where it is obvious that

$$|1,-1\rangle = \nu_1\nu_2 \equiv |\downarrow,\downarrow\rangle \equiv \left|\frac{1}{2},-\frac{1}{2}\right\rangle_1 \left|\frac{1}{2},-\frac{1}{2}\right\rangle_2 \equiv \begin{bmatrix} 0 \\ 1 \end{bmatrix}_1 \begin{bmatrix} 0 \\ 1 \end{bmatrix}_2 \tag{4.161}$$

From

$$j_+ = j_{1+} + j_{2+} \tag{4.162}$$

and

$$j_{1+}\nu_1 = u_1,\ j_{2+}\nu_2 = u_2 \tag{4.163}$$

then

$$j_+|1,-1\rangle = (u_1\nu_2 + u_2\nu_1) \equiv \left(|\uparrow,\downarrow\rangle + |\downarrow,\uparrow\rangle\right) \tag{4.164}$$

On the other hand:

$$j_+|1,-1\rangle = \sqrt{j(j+1)-m(m+1)}|1,0\rangle = \sqrt{2}|1,0\rangle \equiv \left(\left|\frac{1}{2},\frac{1}{2}\right\rangle_1 \left|\frac{1}{2},-\frac{1}{2}\right\rangle_2 + \left|\frac{1}{2},-\frac{1}{2}\right\rangle_1 \left|\frac{1}{2},\frac{1}{2}\right\rangle_2\right) \tag{4.165}$$

So

$$|1,0\rangle = \frac{1}{\sqrt{2}} j_+|1,-1\rangle = \frac{1}{\sqrt{2}}(u_1\nu_2 + u_2\nu_1) \equiv \frac{1}{\sqrt{2}}\left(|\uparrow,\downarrow\rangle + |\downarrow,\uparrow\rangle\right)$$
$$\equiv \frac{1}{\sqrt{2}}\left(\begin{bmatrix} 1 \\ 0 \end{bmatrix}_1 \begin{bmatrix} 0 \\ 1 \end{bmatrix}_2 + \begin{bmatrix} 0 \\ 1 \end{bmatrix}_1 \begin{bmatrix} 1 \\ 0 \end{bmatrix}_2\right) \tag{4.166}$$

It is as well as easy to find that

$$|1,1\rangle = u_1u_2 \equiv |\uparrow,\uparrow\rangle \equiv \left|\frac{1}{2},\frac{1}{2}\right\rangle_1 \left|\frac{1}{2},\frac{1}{2}\right\rangle_2 \equiv \begin{bmatrix} 1 \\ 0 \end{bmatrix}_1 \begin{bmatrix} 1 \\ 0 \end{bmatrix}_2 \tag{4.167}$$

Comparing this with Equation 4.123, we have

$$\left\langle \frac{1}{2},\frac{1}{2},\frac{1}{2},-\frac{1}{2}\middle|1,0\right\rangle = \left\langle \frac{1}{2},-\frac{1}{2},\frac{1}{2},\frac{1}{2}\middle|1,0\right\rangle = \frac{1}{\sqrt{2}} \tag{4.168}$$

It is instructive to note that Equation 4.181 describes the state in which with equal probability, the first particle has the spin projection $\frac{1}{2}$ and the second has $-\frac{1}{2}$ and vice versa. This implies that there exists spin delocalization. It is important to note also that the state vectors in Equations 4.161, 4.181 and 4.167

are symmetric with respect to the permutation of two spin indices 1 and 2. This implies that the **triplet states** are **symmetric** with respect to the permutation of two spin indices 1 and 2.

We examine now the **singlet state** that is **anti-symmetric**. In the representation of the total spin, we have only one vector $|0,0\rangle$ that we write in the following form:

$$|0,0\rangle = c_1 u_1 v_2 + c_2 u_2 v_1 \equiv c_1|\uparrow,\downarrow\rangle + c_2|\downarrow,\uparrow\rangle \tag{4.169}$$

We act on this state the operator j_+ in Equation 4.162 considering Equation 4.163 and so

$$j_+|0,0\rangle = 0 \tag{4.170}$$

So

$$0 = (c_1 + c_2)u_1 u_2 \tag{4.171}$$

From the normalization condition, we have

$$c_1 = \frac{1}{\sqrt{2}} \tag{4.172}$$

So

$$\left\langle \frac{1}{2}, \frac{1}{2}, \frac{1}{2}, -\frac{1}{2} \middle| 0,0 \right\rangle = -\left\langle \frac{1}{2}, -\frac{1}{2}, \frac{1}{2}, \frac{1}{2} \middle| 0,0 \right\rangle = \frac{1}{\sqrt{2}} \tag{4.173}$$

and

$$|0,0\rangle = \frac{1}{\sqrt{2}}(u_1 v_2 - u_2 v_1) \equiv \frac{1}{\sqrt{2}}\left(|\uparrow,\downarrow\rangle - |\downarrow,\uparrow\rangle\right) = \frac{1}{\sqrt{2}}\left(\begin{bmatrix} 1 \\ 0 \end{bmatrix}_1 \begin{bmatrix} 0 \\ 1 \end{bmatrix}_2 - \begin{bmatrix} 0 \\ 1 \end{bmatrix}_1 \begin{bmatrix} 1 \\ 0 \end{bmatrix}_2\right) \tag{4.174}$$

It is instructive to note that in the singlet state, there is also spin delocalization. The vector of state $|0,0\rangle$ differs from the triplet state vector by the permutation of the indices 1 and 2 where it is **anti-symmetric**. It is obvious that the four vectors

$$|1,1\rangle, |1,0\rangle, |1,-1\rangle, |0,0\rangle \tag{4.175}$$

form an orthonormal basis. From the above analysis, we write the relation between the two sets of basis states in a matrix form:

$$\begin{bmatrix} |1,1\rangle \\ |1,0\rangle \\ |1,-1\rangle \\ |0,0\rangle \end{bmatrix} = \frac{1}{\sqrt{2}} \begin{bmatrix} \sqrt{2} & 0 & 0 & 0 \\ 0 & 1 & 1 & 0 \\ 0 & 0 & 0 & \sqrt{2} \\ 0 & 1 & 1 & 0 \end{bmatrix} \begin{bmatrix} |\uparrow,\uparrow\rangle \\ |\uparrow,\downarrow\rangle \\ |\downarrow,\uparrow\rangle \\ |\downarrow,\downarrow\rangle \end{bmatrix} \tag{4.176}$$

Note that the elements of the matrix connecting the two sets of basis states are Clebsch-Gordan coefficients for the given problem.

4.6 Rotation Operator

4.6.1 Finite Rotation Operator About Some Angle Along Some Axis

We consider the symmetry operation that involves the rotation in space of a physical system in a state. We already have knowledge of the operator of the total angular momentum, $\hat{J}$, and used in the infinitesimal rotation operator given in Equation 4.3. Interest now will be on the finite rotation operator, $\hat{\mathbb{D}}_i(\alpha)$, about some angle, α, along some axis $i = x, y, z$. Suppose the rotation is about the z-axis through the angle α:

$$j_{x'} = \hat{\mathbb{D}}_z(\alpha) j_x \hat{\mathbb{D}}_z^\dagger(\alpha) = j_x \cos\alpha + j_y \sin\alpha,\ j_{y'} = \hat{\mathbb{D}}_z(\alpha) j_y \hat{\mathbb{D}}_z^\dagger(\alpha) = j_y \cos\alpha - j_x \sin\alpha \tag{4.177}$$

where

$$\hat{\mathbb{D}}_z(\alpha) = \exp\{-ij_z\alpha\},\ J_i = \hbar j_i \tag{4.178}$$

From Equation 4.177, we have

$$\frac{\partial j_{x'}}{\partial\alpha} = -i\hat{\mathbb{D}}_z(\alpha)(j_z j_x - j_x j_z)\hat{\mathbb{D}}_z^\dagger(\alpha) \tag{4.179}$$

Considering that

$$j_z j_x - j_x j_z = ij_y \tag{4.180}$$

then

$$\frac{\partial j_{x'}}{\partial\alpha} = \hat{\mathbb{D}}_z(\alpha) j_y \hat{\mathbb{D}}_z^\dagger(\alpha) = j_{y'} \tag{4.181}$$

Similarly, we have

$$\frac{\partial j_{y'}}{\partial\alpha} = -j_{x'} \tag{4.182}$$

4.6.2 Finite Rotation Operator for Spinor One-Half

We examine the finite rotation operator, $\hat{\mathbb{D}}_n$, in the Pauli scheme with $\hat{n}$, a unit vector pointing along the axis of rotation, and ϕ, the angle of rotation:

$$\hat{\mathbb{D}}_n = \exp\left\{\frac{i\phi\hat{n}\hat{\sigma}}{2}\right\} = \cos\frac{\phi}{2} + i\hat{n}\hat{\sigma}\sin\frac{\phi}{2} \tag{4.183}$$

The vector, $\hat{n}$, can be regarded as a trivial position operator. Considering the Euler relation:

$$\cos\frac{\phi}{2} \pm i\sin\frac{\phi}{2} = \exp\left\{\pm\frac{i\phi}{2}\right\} \tag{4.184}$$

we find the rotation along the x-, y- and z-axes:

$$\hat{\mathbb{D}}_z(\phi)=\begin{bmatrix}\exp\left\{i\frac{\phi}{2}\right\} & 0\\ 0 & \exp\left\{-i\frac{\phi}{2}\right\}\end{bmatrix},\ \hat{\mathbb{D}}_x(\phi)=\begin{bmatrix}\cos\frac{\phi}{2} & i\sin\frac{\phi}{2}\\ i\sin\frac{\phi}{2} & \cos\frac{\phi}{2}\end{bmatrix},\ \hat{\mathbb{D}}_y(\phi)=\begin{bmatrix}\cos\frac{\phi}{2} & \sin\frac{\phi}{2}\\ -\sin\frac{\phi}{2} & \cos\frac{\phi}{2}\end{bmatrix} \tag{4.185}$$

Suppose the angle of rotation is $\phi=2\pi$:

$$\hat{\mathbb{D}}_z(2\pi)=\hat{\mathbb{D}}_x(2\pi)=\hat{\mathbb{D}}_y(2\pi)=-1\times\hat{\mathbb{I}} \tag{4.186}$$

In this case, the geometric rotation $\hat{\mathbb{D}}$ coincides irrespective of the vector $\hat{n}$. Also, we observe that the rotation associated with the given transformation $\hat{\mathbb{D}}$ through the angle 2π is not an identity operator, $\hat{\mathbb{I}}$ but the identity operator multiplied by -1. This implies that the rotation through the angle 2π does not regenerate the original states but the sign change of the original states. So one requires a rotation through the angle 4π to recover the original states. However, this does not represent any inconsistency with observation since rotation through the angles 2π and 4π represent the same physical state. We also observe the rotation through the angle π yields the component of the Pauli matrices in the respective coordinate directions multiplied by i:

$$\hat{\mathbb{D}}_x(\pi)=i\hat{\sigma}_x,\ \hat{\mathbb{D}}_y(\pi)=i\hat{\sigma}_y \tag{4.187}$$

Example

We consider the example of the spin-$\frac{1}{2}$ particle that we examine through the law of transformation of the spinor:

$$\mathbb{X}=\begin{bmatrix}X_1\\ X_2\end{bmatrix},\ \mathbb{Y}=\begin{bmatrix}Y_1\\ Y_2\end{bmatrix} \tag{4.188}$$

when the coordinate system is rotated through the angle, ϕ, relative to the direction defined by the unit vector, $\hat{n}$. We can show the invariant of the quantity relative to the rotation:

$$\mathbb{X}^*\mathbb{Y}=X_1^*Y_1+X_2^*Y_2 \tag{4.189}$$

The transformation of the spinor due to the rotation through the angle, ϕ:

$$\mathbb{X}'=\begin{bmatrix}X_1'\\ X_2'\end{bmatrix}\equiv\exp\left\{\frac{i\phi\hat{n}\hat{\sigma}}{2}\right\}\mathbb{X}=\left(\cos\frac{\phi}{2}+i\hat{n}\hat{\sigma}\sin\frac{\phi}{2}\right)\mathbb{X} \tag{4.190}$$

From here follows the transformation law for the spinor:

$$\mathbb{Y}^*=\begin{bmatrix}Y_1^* & Y_2^*\end{bmatrix},\ \mathbb{Y}^{*\prime}=\mathbb{Y}^*\left(\cos\frac{\phi}{2}-i\hat{n}\,\hat{\sigma}\sin\frac{\phi}{2}\right) \tag{4.191}$$

From here and considering Equation 4.190, we have

$$\mathbb{Y}^{*\prime}\mathbb{X}'=\mathbb{Y}^*\mathbb{X} \tag{4.192}$$

4.6.3 Finite Rotation Operator for Spinor One-Half General Case

We consider the general case when moving from one coordinate system (mobile reference frame K′) to another (fixed reference frame K) by applying Euler rotations. For any rotation of the coordinate the spinor, $\mathbb{X}$ is transformed to

$$\mathbb{X}^{\lambda'} = \left(\hat{\mathbb{D}}\mathbb{X}\right)^{\lambda} \tag{4.193}$$

where the transformation operator $\hat{\mathbb{D}}$ is the following 2×2 matrix:

$$\hat{\mathbb{D}} = \begin{bmatrix} c_{11} & c_{12} \\ c_{21} & c_{22} \end{bmatrix} \tag{4.194}$$

We consider the bilinear form, $\mathbb{X}^1\mathbb{Y}^2 - \mathbb{X}^2\mathbb{Y}^1$, where $\mathbb{X}$ and $\mathbb{Y}$ are two spinors:

$$\mathbb{X}^{1'}\mathbb{Y}^{2'} - \mathbb{X}^{2'}\mathbb{Y}^{1'} = \left(c_{11}c_{22} - c_{12}c_{21}\right)\left(\mathbb{X}^1\mathbb{Y}^2 - \mathbb{X}^2\mathbb{Y}^1\right) \tag{4.195}$$

If

$$c_{11}c_{22} - c_{12}c_{21} = 1 \tag{4.196}$$

then

$$\mathbb{X}^{1'}\mathbb{Y}^{2'} - \mathbb{X}^{2'}\mathbb{Y}^{1'} = \mathbb{X}^1\mathbb{Y}^2 - \mathbb{X}^2\mathbb{Y}^1 \tag{4.197}$$

Considering

$$\mathbb{X}' = \hat{\mathbb{D}}\mathbb{X} \tag{4.198}$$

then

$$\mathbb{X}'^{*} = \hat{\mathbb{D}}^{*}\mathbb{X}^{*} \tag{4.199}$$

From the condition of normalization, we have

$$\int d\vec{r}\,(\mathbb{X}')^{*}\,\mathbb{X} = \int d\vec{r}\,\hat{\mathbb{D}}^{*}\mathbb{X}^{*}\hat{\mathbb{D}}\mathbb{X} = \int d\vec{r}\,\mathbb{X}^{*}\left(\hat{\mathbb{D}}^{*}\right)^{T}\hat{\mathbb{D}}\mathbb{X} = 1 \tag{4.200}$$

From here, it follows that

$$\hat{\mathbb{D}}^{\dagger}\hat{\mathbb{D}} = 1 \tag{4.201}$$

and implies that

$$\hat{\mathbb{D}}^{\dagger} = \hat{\mathbb{D}}^{-1} \tag{4.202}$$

Since the transformation is unitary:

$$\det\hat{\mathbb{D}} = c_{11}c_{22} - c_{12}c_{21} = 1 \tag{4.203}$$

So

$$\hat{\mathbb{D}}^{-1} = \begin{bmatrix} c_{22} & -c_{12} \\ -c_{21} & c_{11} \end{bmatrix}, \hat{\mathbb{D}}^{*} = \begin{bmatrix} c_{11}^{*} & c_{12}^{*} \\ c_{21}^{*} & c_{22}^{*} \end{bmatrix} \tag{4.204}$$

and

$$c_{22} = c_{11}^{*}, -c_{12} = c_{12}^{*}, -c_{21} = c_{21}^{*}, c_{11} = c_{22}^{*} \tag{4.205}$$

Hence,

$$c_{11}c_{22} - c_{12}c_{21} = |c_{11}|^2 + |c_{21}|^2 = 1 \tag{4.206}$$

Let us introduce the covariant, $\mathbb{X}_\lambda$, and contravariant, $\mathbb{X}^\lambda$, vectors, respectively [12, 13]:

$$\mathbb{X}_\lambda = (g_{\lambda\mu})\mathbb{X}^\mu \tag{4.207}$$

where the matric matrix, $(g_{\lambda\mu})$, is as follows:

$$(g_{\lambda\mu}) = \begin{bmatrix} 0 & 1 \\ -1 & 0 \end{bmatrix} = (g^{\lambda\mu}), \; g_{\lambda\mu}g^{\lambda\mu} = \delta_\lambda^\mu \tag{4.208}$$

We examine all the degrees of freedom. Consider the two 2×2 matrices with eight real variables minus one variable consistent with the determinant resulting in seven variables that, if less four real variables consistent with Equation 4.205, yields three degrees of freedom which are three Eulerian angles α, β and γ defining the $x'y'z'$ system in the K′ frame with respect to the xyz system in the K frame. We examine the following rotations about given coordinate axes:

1. Counter-clockwise rotation about the z-axis through the angle α as in Figure 4.3a
2. Counter-clockwise rotation about the y'-axis through the angle β as in Figure 4.3b
3. Counter-clockwise rotation about the z'-axis through the angle γ as in Figure 4.3c

If we consider the rotations about the axes z, y' and then z', then

$$\hat{\mathbb{D}}(\alpha,\beta,\gamma) = \hat{\mathbb{D}}_{z'}(\gamma)\hat{\mathbb{D}}_{y'}(\beta)\hat{\mathbb{D}}_z(\alpha) \tag{4.209}$$

We consider as well the rotations about the axes z through the angle α, y through the angle β and z through the angle γ:

$$\hat{\mathbb{D}}(\alpha,\beta,\gamma) = \hat{\mathbb{D}}_z(\gamma)\hat{\mathbb{D}}_y(\beta)\hat{\mathbb{D}}_z(\alpha) = \begin{bmatrix} \cos\frac{\beta}{2}\exp\left\{i\frac{\alpha+\beta}{2}\right\} & \sin\frac{\beta}{2}\exp\left\{-i\frac{\alpha-\beta}{2}\right\} \\ -\sin\frac{\beta}{2}\exp\left\{i\frac{\alpha-\beta}{2}\right\} & \cos\frac{\beta}{2}\exp\left\{-i\frac{\alpha+\beta}{2}\right\} \end{bmatrix},$$

$$\alpha \in [0,2\pi], \beta \in [0,\pi], \gamma \in [0,2\pi] \tag{4.210}$$

We have found so far the finite rotation operator $\hat{\mathbb{D}}$ for spinor $\frac{1}{2}$ parametrized by three angles α, β and γ. It is instructive to note that the angles α and β coincides with the polar angles θ and ϕ of the

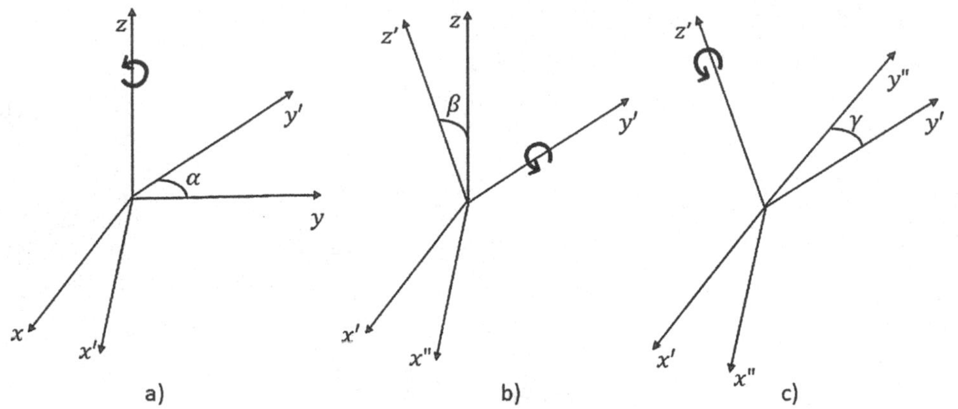

FIGURE 4.3 (4.3a) Counter-clockwise rotation about the z-axis through the angle α; (4.3b) counter-clockwise rotation about the y'-axis through the angle β; (4.3c) counter-clockwise rotation about the z'-axis through the angle γ.

mobile z'-axis of the fixed reference frame K. This implies a rotation of the coordinate in physical space such that the z'-axis points in a direction described by the angles θ and ϕ in the untransformed coordinate system.

4.6.4 Rotation Operator Matrix

We evaluate the rotation operator matrix that is a $(2j+1)\times(2j+1)$ matrix:

$$\mathbb{D}^{(j)}_{mm'} = \left\langle jm' \right| \exp\left\{\frac{i\phi\hat{n}\hat{J}}{\hbar}\right\} \left| jm \right\rangle \tag{4.211}$$

The operator $\hat{\mathbb{D}}$ rotates vectors in the space $\varepsilon^{(j)}$. This implies that if $\left|jm\right\rangle$ is an arbitrary vector in the subspace spanned by $(2j+1)$ vectors, then $\hat{\mathbb{D}}$ is the rotation operator that would act on it. There are only a finite number of independent $(2j+1)$ dimensional square matrices. We examine the action of the operator $\hat{\mathbb{D}}$ on the state vector $\left|jm'\right\rangle$ of the angular momentum:

$$\left|jm'\right\rangle' = \hat{\mathbb{D}}\left|jm'\right\rangle \equiv \sum_{m=-j}^{j} \left|jm\right\rangle\left\langle jm\right|\hat{\mathbb{D}}\left|jm'\right\rangle \tag{4.212}$$

We examine the matrix element $\left\langle jm\right|\hat{\mathbb{D}}\left|jm'\right\rangle$. We consider

$$\hat{\mathbb{D}}_z(\gamma)\left|m'\right\rangle = \exp\{-i\gamma m'\}\left|m'\right\rangle,\ \left\langle m\right|\hat{\mathbb{D}}_z(\alpha) = \left\langle m\right|\exp\{-i\alpha m\} \tag{4.213}$$

and

$$\left\langle jm\right|\hat{\mathbb{D}}(\alpha,\beta,\gamma)\left|jm'\right\rangle = \exp\{-i\alpha m\}\left\langle jm\right|\exp\{-ij_y\beta\}\left|jm'\right\rangle\exp\{-i\gamma m'\} \tag{4.214}$$

while letting

$$\left\langle jm\right|\hat{\mathbb{D}}(\alpha,\beta,\gamma)\left|jm'\right\rangle = \mathbb{D}^{(j)}_{mm'}(\alpha,\beta,\gamma);\ \left\langle jm\right|\exp\{-ij_y\beta\}\left|jm'\right\rangle = d^{(j)}_{mm'}(\beta) \tag{4.215}$$

then from Equation 4.214, the **Wigner function** $\mathbb{D}^{(j)}_{mm'}(\alpha,\beta,\gamma)$ takes the form:

$$\mathbb{D}^{(j)}_{mm'}(\alpha,\beta,\gamma) = \exp\{-i\alpha m\}\mathbb{d}^{(j)}_{mm'}(\beta)\exp\{-i\gamma m'\} \tag{4.216}$$

Considering the unitarity of $\mathbb{D}$ then follows the orthogonality and orthonormality of $\mathbb{D}^{(j)}_{mm'}$:

$$\sum_{m=-j}^{j} \mathbb{D}^{*(j)}_{mm'}\mathbb{D}^{(j)}_{mm''} = \delta_{m'm''} \tag{4.217}$$

From Equation 4.212, the quantity

$$W_{mm'} = \left|\mathbb{D}^{(j)}_{mm'}\right|^2 = \left|\mathbb{d}^{(j)}_{mm'}(\beta)\right|^2 \tag{4.218}$$

is the probability that in the reference frame K, the eigenvalue of j_z equals m with the condition that in the reference frame K′ the eigenvalue of $j_{z'}$ is m'.

For $j=\frac{1}{2}$, we have 2×2 matrices:

$$\hat{j}_z = \frac{\hbar}{2}\begin{bmatrix} 1 & 0 \\ 0 & -1 \end{bmatrix} \equiv \frac{\hbar}{2}\hat{\sigma}_z,\ J_x = \frac{\hbar}{2}\begin{bmatrix} 0 & 1 \\ 1 & 0 \end{bmatrix} \equiv \frac{\hbar}{2}\hat{\sigma}_x,$$

$$\hat{j}_y = \frac{\hbar}{2}\begin{bmatrix} 0 & -i \\ i & 0 \end{bmatrix} \equiv \frac{\hbar}{2}\hat{\sigma}_y,\ J^2 = \frac{3}{4}\hbar^2\begin{bmatrix} 1 & 0 \\ 0 & 1 \end{bmatrix} = \frac{3}{4}\hbar^2\mathbb{I}_2 \tag{4.219}$$

and

$$\exp\left\{-\frac{i\phi\hat{n}\hat{\sigma}}{2}\right\} = \hat{\mathbb{D}}_{z'}(\gamma)\hat{\mathbb{D}}_{y'}(\beta)\hat{\mathbb{D}}_z(\alpha) \tag{4.220}$$

then

$$\mathbb{D}^{\left(\frac{1}{2}\right)}_{mm'} = \langle jm'|\exp\left\{-\frac{i\hat{\sigma}_{z'}\gamma}{2\hbar}\right\}\exp\left\{\frac{i\hat{\sigma}_{y'}\beta}{2\hbar}\right\}\exp\left\{\frac{i\hat{\sigma}_z\alpha}{2\hbar}\right\}|jm\rangle \tag{4.221}$$

or

$$\mathbb{D}^{\left(\frac{1}{2}\right)}_{mm'} = \exp\{im'\gamma\}\langle jm'|\exp\left\{\frac{i\hat{\sigma}_{y'}\beta}{2\hbar}\right\}|jm\rangle\exp\{im\alpha\} = \exp\{im'\gamma\}\mathbb{D}^{\left(\frac{1}{2}\right)}_{mm'}(\beta)\exp\{im\alpha\} \tag{4.222}$$

Here, the probability amplitude is as follows:

$$\mathbb{d}^{\left(\frac{1}{2}\right)}_{mm'}(\beta) = \begin{bmatrix} \cos\frac{\beta}{2} & \sin\frac{\beta}{2} \\ -\sin\frac{\beta}{2} & \cos\frac{\beta}{2} \end{bmatrix},\ \mathbb{d}^{\left(\frac{1}{2}\right)}_{-\frac{1}{2},-\frac{1}{2}}(\beta) = \mathbb{d}^{\left(\frac{1}{2}\right)}_{\frac{1}{2},\frac{1}{2}}(\beta) = \cos\frac{\beta}{2},\ \mathbb{d}^{\left(\frac{1}{2}\right)}_{-\frac{1}{2},\frac{1}{2}}(\beta) = -\mathbb{d}^{\left(\frac{1}{2}\right)}_{\frac{1}{2},-\frac{1}{2}}(\beta) = -\sin\frac{\beta}{2} \tag{4.223}$$

So the probability

$$W_{-\frac{1}{2},-\frac{1}{2}} = W_{\frac{1}{2},\frac{1}{2}} = \cos^2\frac{\beta}{2},\ W_{-\frac{1}{2},\frac{1}{2}} = W_{\frac{1}{2},-\frac{1}{2}} = \sin^2\frac{\beta}{2} \tag{4.224}$$

Considering $J_i = \hbar j_i$ and $J = \hbar j$, where $i = x, y, z$, for $j = 1$, we have 3×3 matrices:

$$j_z = \begin{bmatrix} 1 & 0 & 0 \\ 0 & 0 & 0 \\ 0 & 0 & -1 \end{bmatrix},\ j_x = \frac{1}{\sqrt{2}}\begin{bmatrix} 0 & 1 & 0 \\ 1 & 0 & 1 \\ 0 & 1 & 0 \end{bmatrix},\ j_y = \frac{1}{\sqrt{2}}\begin{bmatrix} 0 & -i & 0 \\ i & 0 & -i \\ 0 & i & 0 \end{bmatrix},\ j^2 = 2\begin{bmatrix} 1 & 0 & 0 \\ 0 & 1 & 0 \\ 0 & 0 & 1 \end{bmatrix} = 2\mathbb{I}_3 \tag{4.225}$$

For $j = \frac{3}{2}$, we have 4×4 matrices:

$$j_x = \frac{1}{2}\begin{bmatrix} 0 & \sqrt{3} & 0 & 0 \\ \sqrt{3} & 0 & 2 & 0 \\ 0 & 2 & 0 & \sqrt{3} \\ 0 & 0 & \sqrt{3} & 0 \end{bmatrix},\ j_y = \frac{1}{2}\begin{bmatrix} 0 & -i\sqrt{3} & 0 & 0 \\ i\sqrt{3} & 0 & -2i & 0 \\ 0 & 2i & 0 & -i\sqrt{3} \\ 0 & 0 & i\sqrt{3} & 0 \end{bmatrix} \tag{4.226}$$

$$j_z = \frac{1}{2}\begin{bmatrix} 3 & 0 & 0 & 0 \\ 0 & 1 & 0 & 0 \\ 0 & 0 & -1 & 0 \\ 0 & 0 & 0 & -3 \end{bmatrix},\ j^2 = \frac{15}{4}\begin{bmatrix} 1 & 0 & 0 & 0 \\ 0 & 1 & 0 & 0 \\ 0 & 0 & 1 & 0 \\ 0 & 0 & 0 & 1 \end{bmatrix} = \frac{15}{4}\mathbb{I}_4 \tag{4.227}$$

Also,

$$j_y^2 = \frac{1}{2}\begin{bmatrix} 0 & -i & 0 \\ i & 0 & -i \\ 0 & i & 0 \end{bmatrix}\begin{bmatrix} 0 & -i & 0 \\ i & 0 & -i \\ 0 & i & 0 \end{bmatrix} = \frac{1}{2}\begin{bmatrix} 1 & 0 & -1 \\ 0 & 2 & 0 \\ -1 & 0 & 1 \end{bmatrix} \tag{4.228}$$

$$j_y^3 = \frac{1}{2\sqrt{2}}\begin{bmatrix} 1 & 0 & -1 \\ 0 & 2 & 0 \\ -1 & 0 & 1 \end{bmatrix}\begin{bmatrix} 0 & -i & 0 \\ i & 0 & -i \\ 0 & i & 0 \end{bmatrix} = \frac{1}{2\sqrt{2}}\begin{bmatrix} 0 & -2i & 0 \\ 2i & 0 & -2i \\ 0 & 2i & 0 \end{bmatrix} = \frac{1}{\sqrt{2}}\begin{bmatrix} 0 & -i & 0 \\ i & 0 & -i \\ 0 & i & 0 \end{bmatrix} = j_y \tag{4.229}$$

$$j_y^4 = \frac{1}{4}\begin{bmatrix} 1 & 0 & -1 \\ 0 & 2 & 0 \\ -1 & 0 & 1 \end{bmatrix}\begin{bmatrix} 1 & 0 & -1 \\ 0 & 2 & 0 \\ -1 & 0 & 1 \end{bmatrix} = \frac{1}{2}\begin{bmatrix} 1 & 0 & -1 \\ 0 & 2 & 0 \\ -1 & 0 & 1 \end{bmatrix} = j_y^2 \tag{4.230}$$

then

$$\exp\{ij_y\beta\} = 1 + ij_y\beta + \frac{(ij_y\beta)^2}{2!} + \frac{(ij_y\beta)^3}{3!} + \frac{(ij_y\beta)^4}{4!} + \cdots = 1 + ij_y\sin\beta + (\cos\beta - 1)j_y^2 \tag{4.231}$$

So

$$\exp\{ij_y\beta\} = \begin{bmatrix} \frac{1}{2}(1+\cos\beta) & \frac{1}{\sqrt{2}}\sin\beta & \frac{1}{2}(1-\cos\beta) \\ -\frac{1}{\sqrt{2}}\sin\beta & \cos\beta & \frac{1}{\sqrt{2}}\sin\beta \\ \frac{1}{2}(1-\cos\beta) & -\frac{1}{\sqrt{2}}\sin\beta & \frac{1}{2}(1+\cos\beta) \end{bmatrix} \equiv \mathbb{d}_{mm'}^{(1)}(\beta) \tag{4.232}$$

It is instructive to note that

$$\mathbb{d}^{(0)}_{mm'}(\beta)=1 \tag{4.233}$$

If we consider Equation 4.216, then

$$\mathbb{D}^{(1)}_{mm'}(\alpha,\beta,\gamma)=\exp\{-i\alpha m\}\,\mathbb{d}^{(1)}_{mm'}(\beta)\exp\{-i\gamma m'\} \tag{4.234}$$

We introduce also a general formula:

$$\mathbb{d}^{(j)}_{mm'}(\beta)=\left[(j+m)!(j-m)!(j+m')!\right]^{\frac{1}{2}}\sum_s \frac{(-1)^{m-m'+s}\left(\cos\frac{\beta}{2}\right)^{2j+m'-m-2s}\left(\sin\frac{\beta}{2}\right)^{m-m'+2s}}{(j+m-s)!s!(m-m'+s)!(j-m-s)!} \tag{4.235}$$

Here, s takes all integral values that do not lead to a negative factorial.

We can derive several symmetry relations from the matrix, $\mathbb{D}$. From the Euler angles of the inverse rotation, we have

$$\mathbb{D}(-\gamma,-\beta,-\alpha)=\mathbb{D}(-\gamma+\pi,\beta,-\alpha-\pi) \tag{4.236}$$

For $\alpha=\gamma=0$,

$$\mathbb{d}^{(j)}_{mm'}(-\beta)=\exp\{-im\pi\}\,\mathbb{d}^{(j)}_{mm'}(\beta)\exp\{im'\pi\}=(-1)^{m-m'}\,\mathbb{d}^{(j)}_{mm'}(\beta) \tag{4.237}$$

It is instructive to note that $m-m'$ must be an integer and so

$$(-1)^{m+m'}=(-1)^{m-m'} \tag{4.238}$$

As $\mathbb{d}^{(j)}_{mm'}(\beta)$ is real:

$$\mathbb{d}^{(j)}_{mm'}(-\beta)=\mathbb{d}^{(j)}_{m'm}(\beta)=(-1)^{m-m'}\,\mathbb{d}^{(j)}_{mm'}(\beta) \tag{4.239}$$

Considering the explicit formula for the matrix, $\mathbb{d}^{(j)}_{mm'}(\beta)$, it follows that

$$\mathbb{d}^{(j)}_{m'm}(\beta)=\mathbb{d}^{(j)}_{-m,-m'}(\beta) \tag{4.240}$$

From the last two equations, we have

$$\mathbb{D}^{(j)*}_{mm'}=(-1)^{m-m'}\,\mathbb{D}^{(j)}_{-m,-m'} \tag{4.241}$$

4.6.4.1 Spherical Harmonics Connection

For the case $j=l$, considering Equation 4.212, we have

$$\sum_m \langle\theta\phi|\,jm\rangle\langle jm|\hat{\mathbb{D}}|\,jm'\rangle=\langle\theta\phi|\hat{\mathbb{D}}|\,jm'\rangle \tag{4.242}$$

From

$$\langle \theta\phi | \hat{\mathbb{D}} = \langle \theta'\phi' | \tag{4.243}$$

then

$$\sum_m \langle \theta\phi | jm \rangle \langle jm | \hat{\mathbb{D}} | jm' \rangle = \langle \theta'\phi' | jm' \rangle \tag{4.244}$$

or

$$\sum_m \mathrm{Y}_{lm}(\theta,\phi)\mathbb{D}^{(j)}_{mm'}(\alpha,\beta,\gamma) = \mathrm{Y}_{lm'}(\theta',\phi') \tag{4.245}$$

From here, we observe that the spherical harmonics can be regarded as constituting a unitary transformation from the angular momentum to the angular coordinate representation.

4.7 Irreducible Tensor Operators

4.7.1 Wigner-Eckart Theorem

It is instructive to note that the space $\varepsilon^{(j)}$ is invariant under rotations. This implies that any state in $\varepsilon^{(j)}$ goes into another state in $\varepsilon^{(j)}$ under rotation and such spaces are called invariant spaces. Invariant sub-spaces are **irreducible** if they cannot be written as sums of invariant spaces of lower dimensionality which are themselves invariant spaces. It is easy to show that $\varepsilon^{(j)}$ does not contain invariant sub-spaces other than itself. On the contrary, a **reducible** space is one that can be written as the sum of spaces of lower dimensionality that are in themselves invariant spaces. For such a case, the symmetry operator $\hat{\mathbb{D}}$ can be further block diagonalized.

We introduce the second rank tensor T_{ik} to explain the **irreducible tensor operator**[1]. Any tensor of the second rank may be represented as a sum of its **symmetric** T^{s}_{ik} and **anti-symmetric** T^{a}_{ik} parts. Note that a (square) matrix is **symmetric** if it is equal to its transpose and **anti-symmetric** if this operation reverses the sign:

$$\mathrm{T}^{a}_{ik} = -\mathrm{T}^{a}_{ki};\ \mathrm{T}^{s}_{ik} = \mathrm{T}^{s}_{ki},\ i,k = 1,2,3 \text{ or } i,k = x,y,z \tag{4.246}$$

We have three independent components for T^{a} and six independent components for T^{s}. The independent components of the anti-symmetric tensor are denoted as $\mathrm{T}^{(1)}_{m}$. For the rotation of the coordinate system, they may be transformed similar to $\mathrm{T}^{(1)}_{m}$ via $\mathbb{D}^{(1)}_{mm'}$:

$$\mathrm{T}^{(1)'}_{m'} = \sum_m \mathrm{T}^{(1)}_{m}\mathbb{D}^{(1)}_{mm'} \tag{4.247}$$

From the symmetric part, we can select the trace $\sum_i \mathrm{T}^{s}_{ii}$ that is an invariant quantity. This implies it is a scalar. The remaining part is the symmetric tensor of the second rank $\mathrm{T}^{(2)}_{m}$ with the trace equal to zero. This tensor has five independent components from where it is not possible to select the totality which is invariant relative to the rotation of the space with less dimensionality. This implies $\mathrm{T}^{(2)}_{m}$ is irreducible and these six quantities are transformed:

$$\mathrm{T}^{(2)'}_{m'} = \sum_m \mathrm{T}^{(2)}_{m}\mathbb{D}^{(2)}_{mm'} \tag{4.248}$$

The general Equations 4.246 and 4.248 define the irreducible tensor operator $T_m^{(j)}$ via the following relation:

$$T_{m'}^{(j)'} = \sum_m T_m^{(j)} \mathbb{D}_{mm'}^{(j)} \tag{4.249}$$

We examine the action of the irreducible tensor operator $T_{m_1}^{(j_1)}$ on the vector of state $\left|j_2 m_2\right\rangle$. We represent the result in the following form:

$$\sum_j \left|jm\right\rangle\left\langle jm\right|T_{m_1}^{(j_1)}\left|j_2 m_2\right\rangle = T_{m_1}^{(j_1)}\left|j_2 m_2\right\rangle \tag{4.250}$$

Compare this with Equation 4.23 and take into consideration that for the rotation of the coordinate system, the tensor operator, $T_{m_1}^{(j_1)}$, is transformed similar to $\left|j_1 m_1\right\rangle$. This implies it is transformed via $\mathbb{D}_{m_1 m_1'}^{(j_1)}$. So it is easy to show that the matrix element $\left\langle jm\right|T_{m_1}^{(j_1)}\left|j_2 m_2\right\rangle$ is proportional to the Clebsch-Gordan coefficients:

$$\left\langle jm \middle| j_1 m_1 j_2 m_2\right\rangle = \Gamma_{j_1 m_1 j_2 m_2}^{jm} \tag{4.251}$$

All the dependence of $\left\langle jm\right|T_{m_1}^{(j_1)}\left|j_2 m_2\right\rangle$ on the quantum numbers m, m_1, and m_2 is described by the coefficient $\Gamma_{j_1 m_1 j_2 m_2}^{jm}$. This is the so-called **Wigner-Eckart theorem.** The proportionality factor is dependent on j, j_1 and j_2 and other quantum numbers if they exist. From here, the selection rule follows:

$$m = m_1 + m_2; \left|j_1 - j_2\right| \leq j \leq j_1 + j_2 \tag{4.252}$$

So the **Wigner-Eckart theorem** can be written as follows:

$$\left\langle jm\right|T_{m_1}^{(j_1)}\left|j_2 m_2\right\rangle = (-1)^{2j_1} \Gamma_{j_1 m_1 j_2 m_2}^{jm} \frac{\left\langle j\right|T^{(j_1)}\left|j_2\right\rangle}{\sqrt{2j+1}} \tag{4.253}$$

Here, the quantity $\left\langle j\right|T^{(j_1)}\left|j_2\right\rangle$ is called the **reduced** or **double-bar matrix element**. It is the same for all projections of m, m_1 and m_2.

Note

1. Irreducible tensor operators are so-called because they have the simplest transformation properties.

5

One-Dimensional Motion General Principles

5.1 One-Dimensional Motion General Principles

Consider the properties of a harmonic oscillator:

1. The motion is bounded: At infinity, the wave function Ψ tends to zero
2. The energy levels are discrete
3. These energy states are non-degenerate, i.e., any energy level corresponds to one eigenfunction
4. The oscillatory theorem: The wave function Ψ of the nth state has n zeros but not considering infinite far removed points

Let us consider now a bounded one-dimensional motion. When $U \rightarrow \infty$, the wave function Ψ tends to zero, and if the converse is true, the total energy becomes infinitely large.

Consider the Schrödinger equations:

$$-\frac{\hbar^2}{2m}\frac{d^2\Psi_1}{dx^2}+U(x)\Psi_1=E\Psi_1,\ -\frac{\hbar^2}{2m}\frac{d^2\Psi_2}{dx^2}+U(x)\Psi_2=E\Psi_2 \tag{5.1}$$

where Ψ_1 and Ψ_2 are two solutions of the Schrödinger equation correspondingly to one and the same energy E and satisfy the boundary condition:

$$\Psi_1(\infty)=\Psi_2(\infty)=0 \tag{5.2}$$

Let us suppose that Ψ_1 and Ψ_2 are real, if we multiply the first equation in Equation 5.1 by Ψ_2 and the second by Ψ_1, then take the difference as follows:

$$\Psi_2\frac{d^2\Psi_1}{dx^2}-\Psi_1\frac{d^2\Psi_2}{dx^2}=\frac{d}{dx}\left(\Psi_2\frac{d\Psi_1}{dx}-\Psi_1\frac{d\Psi_2}{dx}\right)=0 \tag{5.3}$$

with

$$\Psi_2\frac{d\Psi_1}{dx}-\Psi_1\frac{d\Psi_2}{dx}=\text{const} \tag{5.4}$$

Since, at infinity, the wave function tends to zero:

$$\Psi_2\frac{d\Psi_1}{dx}-\Psi_1\frac{d\Psi_2}{dx}=\text{const}\underset{x\rightarrow\infty}{\rightarrow}0 \tag{5.5}$$

DOI: 10.1201/9781003273073-6

and

$$\frac{d}{dx}\left(\frac{\Psi_1}{\Psi_2}\right)=0 \tag{5.6}$$

or

$$\frac{\Psi_1}{\Psi_2}=C=\text{const} \tag{5.7}$$

from where

$$\Psi_1=C\Psi_2 \tag{5.8}$$

This implies that the wave functions Ψ_1 and Ψ_2 differ by a constant factor. In principle, they describe one and the same state and so

$$\Psi_1=\Psi_2 \tag{5.9}$$

So the given energy level corresponds to a single wave function. Hence, we have a non-degenerate state where the wave function Ψ is real.

If some wave function Ψ satisfies the Schrödinger equation, then its complex conjugate Ψ^* satisfies the same Schrödinger equation. But if the level is non-degenerate, then

$$\Psi=\Psi^* \tag{5.10}$$

and implies that Ψ is a real-valued function. The functions Ψ and Ψ^* in any case correspond to one and the same energy level E. If the energy level is degenerate, then Ψ may differ from Ψ^* and following functions correspond to the same eigenvalue:

$$\mathrm{X}=\frac{\Psi+\Psi^*}{\sqrt{2}},\ \Phi=\frac{\Psi-\Psi^*}{\sqrt{2}} \tag{5.11}$$

The simplest example is an unbounded motion.

We consider the example of a free motion with the **Sturm-Liouville** equation:

$$-i\hbar\frac{\partial\Psi}{\partial x}=p_x\Psi \tag{5.12}$$

where p_x is the eigenvalue and the eigenfunction:

$$\Psi=C\exp\left\{\frac{ip_x x}{\hbar}\right\} \tag{5.13}$$

This corresponds to the eigenvalue:

$$E=\frac{p_x^2}{2m} \tag{5.14}$$

But this eigenvalue corresponds also to the eigenfunction:

$$\Psi^* = C\exp\left\{-\frac{ip_x x}{\hbar}\right\} \quad (5.15)$$

So the one-dimensional unbounded motion is degenerate. The eigenfunction Ψ corresponds to the state of motion of a particle from the left to the right and Ψ^* corresponds to the motion from the right to the left.

5.2 Potential Well

Consider the motion in a potential well. An example for such a motion is a harmonic oscillator. We consider a potential well with finite energy E (Figure 5.1b). In this case, the situation is dependent on the sign of the energy E. The sign of the energy is dependent on how the energy reference point is selected. If $E < 0$, then we may write

$$E = -\frac{\hbar^2\chi^2}{2m} \quad (5.16)$$

where χ is some real parameter. The Schrödinger equation has the following form:

$$\frac{d^2\Psi}{dx^2} - \chi^2\Psi = \frac{2m}{\hbar^2}U(x)\Psi \quad (5.17)$$

If we consider the laws of classical mechanics, then the particle should be oscillating between $x = a$ and $x = b$ (Figure 5.1c). It is a periodic motion that may not necessary be simple harmonic. The domain between $x = a$ and $x = b$ is classically permissible contrary to other domains that are classically not permissible. We have to note that the eigenfunction Ψ in the classically not permissible domain is different from zero. Consider the asymptotic behavior of the wave function Ψ at infinity, i.e., we find $\tilde{\Psi}$ ($\tilde{\Psi}$-the asymptotic wave function). At infinity, the right-hand side of Equation 5.17 tends to zero as

$$U(x = \pm\infty) = 0 \quad (5.18)$$

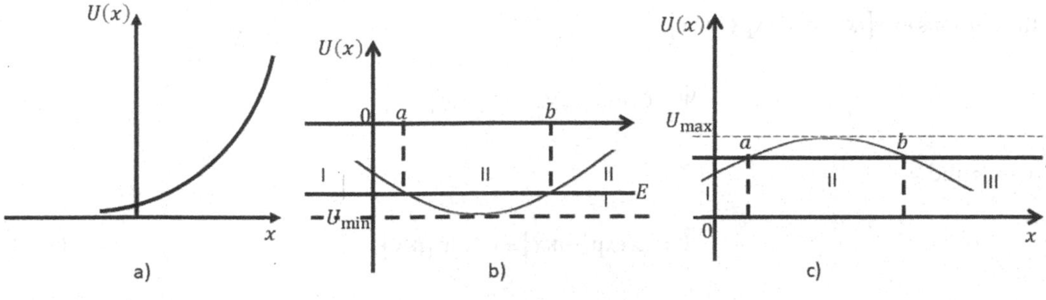

FIGURE 5.1 (5.1a) The potential of a one-dimensional motion unbounded from one side; (5.1b) depicts motion in a potential well with finite energy; (5.1c) depicts the potential barrier for a particle at the initial moment from the left of the point $x = a$ and moving toward the given point $x = a$ where it is then reflected such that it moves in the reverse direction. The particle passes from domain I $(x < a)$ into domain III $(x > b)$ due to the phenomenon called the Tunnel effect.

and we have

$$\frac{d^2\tilde{\Psi}}{dx^2} - \chi^2\tilde{\Psi} = 0 \tag{5.19}$$

with the solution

$$\tilde{\Psi} = C_1 \exp\{-\chi x\} + C_2 \exp\{\chi x\} \tag{5.20}$$

As $x \to \infty$, one of these solutions should be thrown away. If $\chi > 0$, then we throw away the solution $C_2 \exp\{\chi x\}$, considering the condition that the wave function should be bounded at infinity:

$$\Psi_{\text{III}} = C\exp\{-\chi x\} \tag{5.21}$$

On the same basis, we suppose that

$$\Psi_{\text{I}} = C'\exp\{\chi x\} \tag{5.22}$$

But what about Ψ_{III}? However, Ψ_{III} should continuously be transformed to Ψ_{I} on the left and Ψ_{III} on the right. From these conditions, we may define C and C'. But C and C' are defined from the condition of normalization. It follows that we may define Ψ only for some values of $\chi_1, \chi_2, \ldots$ corresponding to $E_1, E_2, \ldots$. So we arrive at a conclusion that the energy spectrum in the domain $E < 0$ should be discrete. The state is non-degenerate, and, at infinity, the wave function tends to zero. There may be a well such that for it there exists no single level. If there exists, then it is discrete.

If $E > 0$, then

$$E = \frac{\hbar^2\kappa^2}{2m} \tag{5.23}$$

and

$$\frac{d^2\tilde{\Psi}}{dx^2} + \kappa^2\tilde{\Psi} = 0 \tag{5.24}$$

with solutions $\exp\{i\kappa x\}$ and $\exp\{-i\kappa x\}$:

$$\tilde{\Psi} = C_1 \exp\{-i\kappa x\} + C_2 \exp\{i\kappa x\} \tag{5.25}$$

as $x \to \infty$ and

$$\tilde{\Psi} = C_3 \exp\{-i\kappa x\} + C_4 \exp\{i\kappa x\} \tag{5.26}$$

as $\to -\infty$.

The energy spectrum is continuous and the state is degenerate. In Equations 5.25 and 5.26, $C_2 \exp\{i\kappa x\}$ is the transmitted wave, $C_3 \exp\{-i\kappa x\}$ is the reflected wave and $C_4 \exp\{i\kappa x\}$ is the incident wave.

If we consider the law of classical mechanics, then the particle with energy $E > 0$ may, with no disturbance, pass above the well, say from the left to right. In Equation 5.25, if $C_1 = 0$, then $\tilde{\Psi} = C_2 \exp\{i\kappa x\}$.

This implies that we have a particle in a force-free region $(U(x)=0)$ that is moving from the left to right in the ox-axis direction with a defined energy E and defined momentum $\hbar\kappa$. However, as $x \to -\infty$, then for the asymptotic eigenfunction $\tilde{\Psi}$, we have two terms that is the superposition of the motion in the positive and negative directions of the ox-axis direction.

If we consider Equations 5.25 and 5.26 taking into consideration $C_1 = 0$, then we evaluate

$$R = \left|\frac{C_3}{C_4}\right|^2, D = \left|\frac{C_2}{C_4}\right|^2 \tag{5.27}$$

where R and D are called **coefficients of reflection** and **transmission,** respectively, and measure the fractions of the incident flux that are reflected and transmitted. From the notion of the conservation of current:

$$R + D = 1 \tag{5.28}$$

For the case of a potential well, the energy levels may be defined as the poles of D or R when $\kappa = i\chi$. So

$$\tilde{\Psi} = C_3 \exp\{\chi x\} + C_4 \exp\{-\chi x\} \tag{5.29}$$

The term $C_4 \exp\{-\chi x\}$ has an exponential increase as $x \to -\infty$ and has no physical sense. For $C_4 = 0$, $\tilde{\Psi} = C_3 \exp\{\chi x\}$ has an exponential decrease as $x \to -\infty$ and corresponds to a bound state. It is obvious that for this, the expression for R and D tends to infinity. Equation 5.29 has no physical sense for imaginary κ as the current density for the wave function $\exp\{\pm\chi x\}$ is equal to zero.

5.3 Particle in a One-Dimensional Finite Square Well Potential

We examine the following finite symmetric square well potential, $U(x)$ (Figure 5.2a):

$$U(x) = \begin{cases} -U_0, & |x| \le a, U_0 > 0 \\ 0, & |x| > a \end{cases} \tag{5.30}$$

The potential energy is zero for $|x| > a$ and equal to $-U_0$ (negative) within the well when $|x| \le a$. The breadth of the well is $2a$.

Interest will be on bound states, namely energy eigenstates, that are normalizable and so the energy, E, of the states must be negative. If $E > 0$, any solutions in the region $|x| > a$ where the potential vanishes would be a plane wave. We will be expected to find the energy levels of the lower part of the spectrum when $U_0 \gg \frac{\hbar^2}{ma^2}$ and define the shift of the energy levels.

We solve the one-dimensional Schrödinger equation for a finite potential well described by the potential $U(x)$ in Equation 5.30, so-called **square well potential** where the motion of the particle is restrained by the reflecting walls that bound the region of constant potential energy. For bound states, $E < 0$, we investigate the Schrödinger equation in the various domains, $|x| < a$ and $|x| > a$. Here the potential is constant and so the necessity to use boundary conditions to match the solutions across the points where the potential is discontinuous. The corresponding Schrödinger equation for these domains is as follows:

$$\left[-\frac{\hbar^2}{2m}\frac{d^2}{dx^2} + U(x)\right]\Psi(x) = E\Psi(x) \tag{5.31}$$

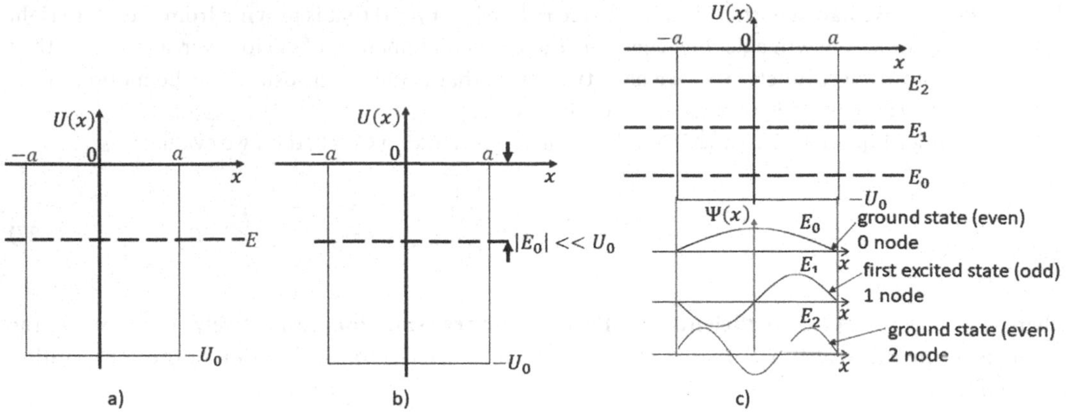

FIGURE 5.2 (5.2a) The square well potential where the motion of the particle is restrained by the reflecting walls that terminate the region of constant potential energy. The potential energy is zero for $|x| > a$ and negative and equal to $-U_0$ in the well for $|x| \leq a$, since U_0 is defined to be positive. The breadth of the well is $2a$. For bound states, the energy of the particle, $E < 0$; (5.2b) The depth of the level in the shallow well where the motion of the particle is restrained by the reflecting walls that terminate the region of constant potential energy; (5.2c) eigenstates of a finite square well potential with eigenenergies, $E_0 < E_1 < E_2$. The eigenstates alternate as even, odd and even with zero, one and two nodes, respectively.

The potential, $U(x)$, for the finite square well is an even function of x:

$$U(-x) = U(x) \tag{5.32}$$

So, as mentioned earlier, for an even potential, the bound states are either symmetric or anti-symmetric. For brevity, we define the following quantities:

$$\kappa^2 = -\frac{2mE}{\hbar^2}, \quad \chi^2 = \frac{2m(E+U_0)}{\hbar^2} \tag{5.33}$$

then if

1. $x < -a$
   so from Equation 5.31, considering Equation 5.33, the differential equation follows as

$$\Psi_{\text{I}}'' - \kappa^2 \Psi_{\text{I}} = 0 \tag{5.34}$$

   with the solution

$$\Psi_{\text{I}} = A_1 \exp\{\kappa x\} + B_1 \exp\{-\kappa x\} \tag{5.35}$$

2. $-a \leq x \leq a$
   Consider Equations 5.31 and 5.33, the differential equation follows as

$$\Psi_{\text{II}}'' - \chi^2 \Psi_{\text{II}} = 0 \tag{5.36}$$

   with the solution

$$\Psi_{\text{II}} = A_2 \cos\chi x + B_2 \sin\chi x \tag{5.37}$$

3. $x > a$

$$\Psi''_{\mathrm{III}} - \kappa^2 \Psi_{\mathrm{III}} = 0 \tag{5.38}$$

This has the solution

$$\Psi_{\mathrm{III}} = A_3 \exp\{\kappa x\} + B_3 \exp\{-\kappa x\} \tag{5.39}$$

The normalization of the bound state requires that the solution should vanish at infinity. This follows that $B_1 = A_3 = 0$. The function $\Psi(x)$ should be continuously differentiable. The function should be continuous together with its first derivatives at the boundary of interface between the inside and outside of the potential well. The second derivative Ψ'' has a jump required by the particular box-type potential of the given Schrödinger equation.

So

$$\Psi_{\mathrm{I}}(-a) = \Psi_{\mathrm{II}}(-a),\quad \Psi_{\mathrm{II}}(a) = \Psi_{\mathrm{III}}(a),\ \Psi'_{\mathrm{I}}(-a) = \Psi'_{\mathrm{II}}(-a),\quad \Psi'_{\mathrm{II}}(a) = \Psi'_{\mathrm{III}}(a) \tag{5.40}$$

Let us find the eigenvalue equations where, considering Equations 5.40 and 5.44, we obtain four linear homogenous equations for the coefficients:

$$\begin{cases} A_1 \exp\{-\kappa a\} + B_2 \sin\chi a - A_2 \cos\chi a = 0 \\ \kappa A_1 \exp\{-\kappa a\} - B_2 \chi \cos\chi a - A_2 \chi \sin\chi a = 0 \\ B_3 \exp\{-\kappa a\} - B_2 \sin\chi a - A_2 \cos\chi a = 0 \\ \kappa B_3 \exp\{-\kappa a\} + B_2 \chi \cos\chi a - A_2 \chi \sin\chi a = 0 \end{cases} \tag{5.41}$$

For the non-trivial solutions, the determinant of the system in Equation 5.41 is equal to zero:

$$\kappa^2 - \chi^2 + 2\kappa\chi \cot 2\chi a = 0 \tag{5.42}$$

If we solve Equation 5.42 relative to κ, then we have two roots where one of them is as follows:

$$\kappa = \chi \tan\chi a \tag{5.43}$$

This is a transcendental equation which, when substituted into the second or last equation in Equation 5.41, yields the following:

$$A_1 = B_3,\ B_2 = 0,\ A_2 = \frac{A_1}{\cos\chi a} \exp\{-\kappa a\} \tag{5.44}$$

and

$$\Psi_{\mathrm{I}} = A_1 \exp\{\kappa x\},\ \Psi_{\mathrm{II}} = A_1 \exp\{-\kappa a\} \frac{\cos\chi x}{\cos\chi a},\ \Psi_{\mathrm{III}} = A_1 \exp\{-\kappa x\} \tag{5.45}$$

This implies that $\Psi(x)$ is an even function:

$$\Psi(x) = \Psi(-x) \tag{5.46}$$

In this case, we speak of **positive parity**. So we observe that Equation 5.43 describes the spectrum of even levels where point $x = -a$ provides the matching condition of the wave functions considering the property in Equation 5.46 (i.e., even function). We now examine one of the roots of Equation 5.42:

$$\kappa = -\chi \cot \chi a \tag{5.47}$$

Substitute this into Equation 5.41:

$$A_2 = 0,\ B_3 = -A_1,\ B_2 = -A_1 \frac{\exp\{-\kappa a\}}{\sin \chi a} \tag{5.48}$$

and

$$\Psi_{\mathrm{I}} = A_1 \exp\{\kappa x\},\ \Psi_{\mathrm{II}} = -A_1 \exp\{-\kappa a\} \frac{\sin \chi x}{\sin \chi a},\ \Psi_{\mathrm{III}} = -A_1 \exp\{-\kappa x\} \tag{5.49}$$

This is an odd function with **negative parity,** i.e.,

$$\Psi(-x) = -\Psi(x) \tag{5.50}$$

The coefficient A_1 is defined from the normalization condition. say for the first solution:

$$\begin{aligned}\int_{-\infty}^{\infty} |\Psi|^2 dx &= |A|^2 \left[\int_{-\infty}^{-a} \exp\{2\kappa x\} dx + \frac{\exp\{-2\kappa a\}}{\cos^2 \chi a} \int_{-a}^{a} \cos^2 \chi x dx + \int_{a}^{\infty} \exp\{-2\kappa x\} dx\right] \\ &= |A|^2 \exp\{-2\kappa a\} \left[\frac{1}{\kappa} + \frac{a}{\cos^2 \chi a} + \frac{\sin 2\chi a}{2\chi \cos^2 \chi a}\right] = 1\end{aligned} \tag{5.51}$$

Apply Equation 5.43 or 5.47 to this:

$$\frac{1}{|A|^2} = a \exp\{-2\kappa a\} \left(\frac{1}{\kappa a} + 1 + \frac{\kappa^2}{\chi^2} + \frac{\kappa}{\chi^2 a}\right) \tag{5.52}$$

Letting

$$\varsigma = \chi a,\ \eta = \kappa a,\ \frac{2mU_0 a^2}{\hbar^2} = \xi^2 \tag{5.53}$$

then for the definition of the energy eigenvalue, we have

$$\varsigma^2 + \eta^2 = \xi^2 \tag{5.54}$$

Here, the quantity ς is a proxy for χ while η is a proxy for κ. Both quantities are dependent on the energy of the bound state. The dimensionless parameter, ξ, characterizes the depth, U_0, and breadth, $2a$, of the well and also the mass, m, of the particle. So, given a potential, the dimensionless parameter, ξ, is known which would be very large. The value of ξ indicates the number of bound states in the square well.

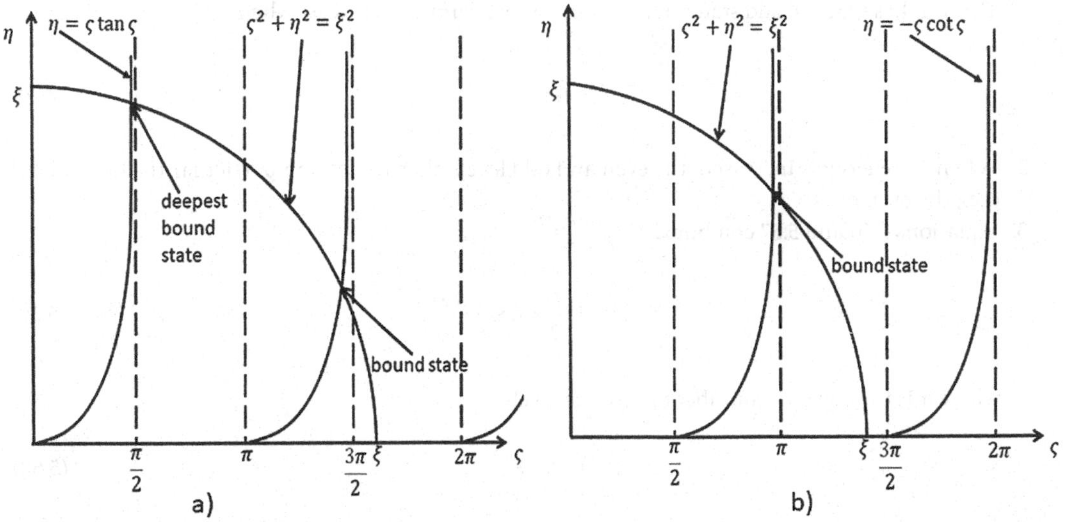

FIGURE 5.3 (5.3a) Graphical representation of the simultaneous Equation 5.55. The intersections of the circle with the $\eta = \varsigma \tan\varsigma$ function represent even bound state solutions in the finite square well potential. The deepest bound state is the one with lowest; (5.3b) graphical representation of Equation 5.55. The intersections of the circle with the curves $\eta = -\varsigma\cot\varsigma$ are odd bound-state solutions in the finite square-well potential. One bound state is displayed in this case. The curve $\eta = -\varsigma\cot\varsigma$ does not appear for $\varsigma < \frac{\pi}{2}$ because η is negative. For $\xi < \frac{\pi}{2}$, there are no odd bound-state solutions, but still there is an even bound state.

From Equation 5.33, we introduce the following:

$$\eta = \varsigma \tan\varsigma, \quad \eta = -\varsigma\cot\varsigma \tag{5.55}$$

Since ς and η are restricted to be positive values, the desired energy values are obtained by constructing the intersection of the given two curves with the circle defined by Equations 5.54 and 5.55 within the (ς,η)-plane (Figure 5.3a and 5.3b). The intersections of the curves determine the energy eigenvalues. For the positive parity at least one solution exists for arbitrary values of the parameter U_0. This is because the tangent function intersects at the origin. For the case of the negative parity, the radius of the circle needs to be larger than a minimum value so that the curves can intersect. The potential should have a certain depth relative to a given size a and a given mass m to permit a solution with negative parity. The number of energy levels increases with U_0, a and mass m.

Let us analyze Equations 5.43 and 5.47: Equation 5.43 for $0 < |E| < U_0$ has n_+ roots (n_+ for even state of a discrete spectrum) defined by the relation:

$$(n_+ - 1)\pi < \xi \le n_+\pi \tag{5.56}$$

This implies that for any parameters of the well, there is at least one even level:

1. The number n_- is the number of roots of the Equation 5.47 defined by the relation:

$$n_-\pi - \frac{\pi}{2} < \xi \le n_-\pi + \frac{\pi}{2} \tag{5.57}$$

This implies that the odd states of the discrete spectrum exist at least when

$$\xi \geq \frac{\pi}{2} \tag{5.58}$$

2. When the energy is increased, the even and odd levels alternate and in particular the lowest level (i.e., the even ones)
3. Equations 5.56 and 5.57 combined:

$$(n-1)\frac{\pi}{2} < \xi \leq n\frac{\pi}{2} \tag{5.59}$$

where n is the common number of discrete levels:

$$n = n_+ + n_- \tag{5.60}$$

4. For the equality signs in Equations 5.56 and 5.57, there appears a new level:
   $(n_+ + 1)^{\text{th}}$ even, $(n_- + 1)^{\text{th}}$ odd or in general $(n+1)^{\text{th}}$ level (see Figure 5.2c) that corresponds to the increase of the well, i.e., for the increase of the parameter $\varsigma_0 = \frac{\xi^2}{2}$. For the case of a very deep well, $\xi \gg 1$:

$$\tan \chi a = \infty \tag{5.61}$$

This corresponds to

$$\chi a = \frac{2n_+ - 1}{2}\pi \tag{5.62}$$

and

$$-\cot \chi a = \infty \tag{5.63}$$

This corresponds to

$$\chi a = n_- \pi \tag{5.64}$$

The combined results yield

$$\chi 2a = n\pi \tag{5.65}$$

with the energy spectrum

$$E_n = \frac{\hbar^2 \pi^2 (n+1)^2}{8ma^2} - U_0 \tag{5.66}$$

When the potential well is increased or increased particle mass m, the difference between two neighboring energy eigenvalues decreases. The lowest state $(n = 0)$ is not at $-U_0$ but a little higher. This difference is called the **zero-point energy**.

For a shallow well, $\xi \ll 1$, we have only one even discrete state which corresponds to the delta-potential well, $U(x) = -\alpha\delta(x)$. Let us find the energy level from Equation 5.43 and consider the argument of the tangent to be small:

$$\chi_0 a \approx \frac{\sqrt{|E_0|}}{\sqrt{U_0 - |E_0|}}, \chi_0^2 = \frac{2m(U_0 - |E_0|)}{\hbar^2} \tag{5.67}$$

since

$$\chi_0 a \ll 1 \tag{5.68}$$

then from Equation 5.67 we have

$$|E_0| \approx \xi^2 U_0 \ll U_0 \tag{5.69}$$

This implies that the depth of the level in the shallow well is much less than the depth of the given well (Figure 5.2b). The approximate form of the ground-state normalized wave function in the shallow well have the form:

$$\Psi_0 = \sqrt{\kappa_0} \exp\{-\kappa_0 |x|\}, \kappa_0^2 = \frac{2m|E_0|}{\hbar^2} \tag{5.70}$$

The average potential and kinetic energies of the particle are, respectively, as follows:

$$\overline{U(x)} \cong -2U_0\xi^2 \cong 2E_0, \bar{T} \cong U_0\xi^2 \cong -E_0 \tag{5.71}$$

From the wave function, Ψ_0 follows the probability

$$|\Psi_0|^2 = \kappa_0 \exp\{-2\kappa_0 |x|\} \tag{5.72}$$

to find the particle outside, the well to decrease with increase distance, $|x|$. So the particle with some probability can be found within the region

$$|x| \cong \frac{1}{\kappa_0} \gg a \tag{5.73}$$

It is obvious that the energy and the wave function for a shallow well are exactly those for the delta-potential well:

$$U(x) = -\alpha\delta(x) \tag{5.74}$$

So any narrow well (or barrier) can be approximated by a delta potential well.

5.4 Potential Barrier

Consider the potential barrier in Figure 5.1c where we examine a particle at the initial moment from the left of the point $x = a$ and moving toward point $x = a$. When the particle arrives at the point $x = a$, it is reflected such that it moves in the reverse direction. There is no basis for which we may consider that the particle arrives at the point $x = b$. The division of the domain in Figure 5.1c into permissible and non-permissible domains is on the basis of the defined values of the potential and kinetic energies. Consider the case for which $U(x)>0$ and $U(x=\pm\infty)$ (Figure 5.1c). Here there is no domain for which $E<0$.

Consider the case $E>0$ where the solution of the Schrödinger equation takes the asymptotic form:

$$\tilde{\Psi}=C_1\exp\{-i\kappa x\}+C_2\exp\{i\kappa x\} \tag{5.75}$$

as $x\to\infty$ and

$$\tilde{\Psi}=C_3\exp\{-i\kappa x\}+C_4\exp\{i\kappa x\} \tag{5.76}$$

as $x\to-\infty$.

For $E<U_{\max}$, the particle may not be found between points $x=a$ and $x=b$, considering classical mechanics. The wave function in this classical not permissible domain is different from zero considering quantum mechanics. In this case, the passage of a particle from domain I $(x<a)$ into domain III $(x>b)$ is feasible. This phenomenon is called the **Tunnel effect** and stems from the fact that the **tail** of the wave function penetrates domain II from I and runs into domain III.

5.5 Particle in a Square Potential Barrier

We find the energy levels and wave functions of a particle in a **square potential barrier** $U(x)$ (Figure 5.4):

$$U(x)=\begin{cases}0, & x<-a\\ U_0, & -a<x<a\\ 0, & x>a\end{cases} \tag{5.77}$$

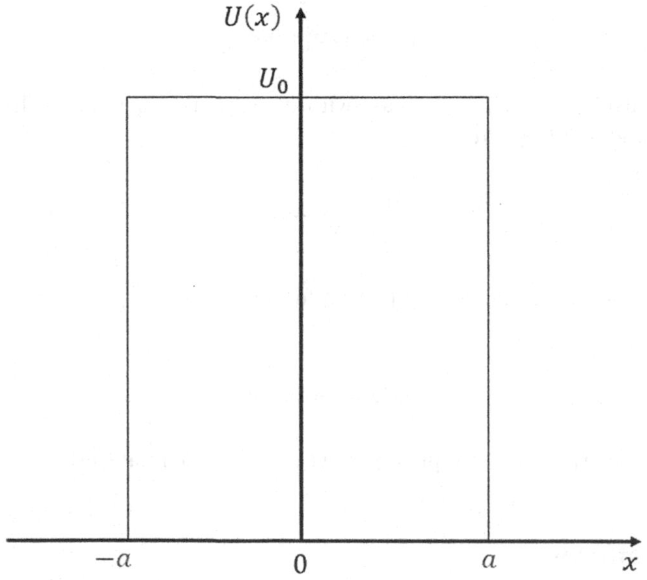

FIGURE 5.4 The potential barrier where we assign domains I, II and III to the regions $x<-a$, $-a<x<a$ and $x>a$, respectively, and consider the case $E>0$. A particle at the initial moment from the left of the point $x=-a$ and moving toward the given point $x=-a$ and arrives at that point is reflected such that it moves in the reverse direction. The particle passes from domain I $(x<-a)$ into domain III $(x>a)$ due to the phenomenon called the Tunnel effect.

Assigning domains I, II and III to the regions $x<-a$, $-a<x<a$ and $x>a$, respectively, we consider the case $E>0$. We consider the particle is incident from the left (I) to right. In this case, the particle is reflected or transmitted by the barrier. If we consider classical mechanics, then the particle is always reflected if $E<U_0$ and is always transmitted if $E>U_0$, where E is the energy of the particle and $U_0=U_{max}$ is the top of the barrier.

From quantum mechanics, it can be seen that both reflection and transmission occur with a probability different from zero for most of the energies of the particle. In this problem, we consider a particle incident from the left with energy $E>0$ and can be reflected by the barrier or transmitted through it.

Consider the case when $E>U_0$ (**resonance**) where, for the resonance energies, the probability of transmission is 1 and that of reflection, 0. In domains I and III, we let

$$\kappa_1^2=\frac{2mE}{\hbar^2}>0 \tag{5.78}$$

then the Schrödinger equation is as follows:

$$\frac{d^2\Psi}{dx^2}+\kappa_1^2\Psi=0 \tag{5.79}$$

with the solution

$$\Psi_{\text{I}}=C_1\exp\{i\kappa_1x\}+C_2\exp\{-i\kappa_1x\},\ \Psi_{\text{III}}=C_5\exp\{i\kappa_1x\}+C_6\exp\{-i\kappa_1x\} \tag{5.80}$$

From domain II:

$$\frac{d^2\Psi}{dx^2}+\kappa_2^2\Psi=0,\ \kappa_2^2=\frac{2m}{\hbar^2}(E-U_0) \tag{5.81}$$

and

$$\Psi_{\text{II}}=C_3\exp\{i\kappa_2x\}+C_4\exp\{-i\kappa_2x\} \tag{5.82}$$

Letting $C_6=0$ in domain III, we have only the transmitted particle (the incident particle is coming from $x=-\infty$). So

$$\Psi=\begin{cases}C_1\exp\{i\kappa_1x\}+C_2\exp\{-i\kappa_1x\}, & x<-a\\ C_3\exp\{i\kappa_2x\}+C_4\exp\{-i\kappa_2x\}, & -a<x<a\\ C_5\exp\{i\kappa_1x\}, & x>a\end{cases} \tag{5.83}$$

The condition of continuity of the wave function in Equation 5.83 and its derivatives at points $x=-a$ and $x=a$:

$$\Psi_{\text{I}}(-a)=\Psi_{\text{II}}(-a),\quad \Psi_{\text{II}}(a)=\Psi_{\text{III}}(a),\left.\frac{d\Psi_I}{dx}\right|_{x=-a}=\left.\frac{d\Psi_{\text{II}}}{dx}\right|_{x=-a},\quad \left.\frac{d\Psi_{\text{II}}}{dx}\right|_{x=a}=\left.\frac{d\Psi_{\text{III}}}{dx}\right|_{x=a} \tag{5.84}$$

$$\begin{cases}C_1\exp\{-i\kappa_1a\}+C_2\exp\{i\kappa_1a\}=C_3\exp\{-i\kappa_2a\}+C_4\exp\{i\kappa_2a\}\\ C_1\kappa_1\exp\{-i\kappa_1a\}-C_2\kappa_1\exp\{i\kappa_1a\}=C_3\kappa_1\exp\{-i\kappa_2a\}-C_4\kappa_1\exp\{i\kappa_2a\}\end{cases} \tag{5.85}$$

$$\begin{cases} C_3 \exp\{i\kappa_2 a\} + C_4 \exp\{-i\kappa_2 a\} = C_5 \exp\{i\kappa_1 a\} \\ C_3\kappa_2 \exp\{i\kappa_2 a\} - C_4\kappa_2 \exp\{-i\kappa_2 a\} = C_5\kappa_1 \exp\{i\kappa_1 a\} \end{cases} \tag{5.86}$$

From these equations:

$$R = \left|\frac{C_2}{C_1}\right|^2 = \frac{\left(\kappa_2^2 - \kappa_1^2\right)^2 \sin^2 2\kappa_2 a}{4\kappa_2^2\kappa_1^2 + \left(\kappa_2^2 - \kappa_1^2\right)^2 \sin^2 2\kappa_2 a}, \quad D = \left|\frac{C_5}{C_1}\right|^2 = \frac{4\kappa_2^2\kappa_1^2}{4\kappa_2^2\kappa_1^2 + \left(\kappa_2^2 - \kappa_1^2\right)^2 \sin^2 2\kappa_2 a}$$

$$= \frac{4E(E-U_0)}{4E(E-U_0) + U_0^2 \sin^2 \xi} = 1 - \frac{U_0^2 \sin^2 \xi}{4E(E-U_0) + U_0^2 \sin^2 \xi}, \quad \frac{2m(E-U_0)a^2}{\hbar^2} \equiv \xi^2 \tag{5.87}$$

where R and D are the coefficients of reflection and transmission. The coefficients essentially measure the fractions of the incident flux that are reflected and transmitted, respectively. For resonance:

$$2\kappa_2 a = n\pi, \quad n = 0,1,2,\ldots \tag{5.88}$$

The transmission coefficient, D, has a periodic oscillation between its minimum $D_{\min}$ and maximum value $D_{\max}$. Resonance is obtained for

$$a = \frac{n\pi}{2\kappa_2}, \quad n = 0,1,2,\ldots \tag{5.89}$$

For $E > U_0$, the reflection of the particle at each potential discontinuity occurs without a phase shift of the wave function. So in domain II, the resonance condition $2\kappa_2 a = n\pi, \quad n = 0,1,2,\ldots$ corresponding to values of a for which we have standing waves. Far from the resonance position, there is destructive interference due to the reflections at $x = -a$ and $x = a$. That is why the value of the wave function is small. For the satisfaction of the resonance condition, the particle spends a relatively long time in domain II. Such a phenomenon is called **resonance scattering** in quantum mechanics. At resonance,

$$\kappa_2^2 = \frac{n^2\pi^2}{4a^2} = \frac{2m}{\hbar^2}(E - U_0) \tag{5.90}$$

and

$$E = \frac{\hbar^2 n^2 \pi^2}{8ma^2} + U_0, \quad n = 0,1,2,\ldots \tag{5.91}$$

This is the energy for the maximum transmission. For the stationary points (at minimum):

$$2\kappa_2 a = \frac{\pi}{2}(2n+1), \quad \kappa_2^2 = \frac{\pi^2(2n+1)^2}{4a^2} = \frac{2m}{\hbar^2}(E - U_0), \quad n = 0,1,2,\ldots \tag{5.92}$$

and

$$E = \frac{\hbar^2\pi^2(2n+1)^2}{32ma^2} + U_0, \quad n = 0,1,2,\ldots \tag{5.93}$$

This is the energy for the minimum transmission.

We consider now the case where $E < U_0$, κ_2 is a pure imaginary quantity:

$$\kappa_2 = i\chi \tag{5.94}$$

In this case, we witness a **Tunnel effect** and

$$D = \frac{4E(E-U_0)}{4E(E-U_0)+U_0^2\sinh^2\xi} = 1 - \frac{U_0^2\sinh^2\xi}{4E(E-U_0)+U_0^2\sinh^2\xi} \tag{5.95}$$

It has already been seen that the wave function in domain II is not zero. It has the behavior of an **evanescent wave.** Since the particle is able to find itself across the potential barrier (**Tunnel effect**), even when the energy of the particle is lower than the height of the barrier, the particle finds itself in domain II due to the **Tunnel effect. Let the reader examine if case $E < 0$ is possible.**

Since the probability is not expected to disappear suddenly, it is easily seen that

$$D + R = 1 \tag{5.96}$$

For

$$2\kappa_2 a = n\pi, \quad n = 0,1,2,\ldots \tag{5.97}$$

the barrier is transparent, i.e., $D=1$ and $R=0$. The solution is true for $U_0 > 0$ and also for $U_0 < 0$. For $U_0 > 0$ and $E < U_0$, considering that $2K_2 a \gg 1$, we have

$$D \approx \frac{16\kappa_2^2\kappa_1^2}{\left(\kappa_2^2-\kappa_1^2\right)^2}\exp\{-4\kappa_2 a\} \tag{5.98}$$

As the exponential function $\exp\{-4\kappa_2 a\}$ rapidly tends to zero as $\kappa_2 a$ tends to infinity, the fraction of particles getting through the barrier is very small where $K_2 a$ is large. Hence, as the energy E becomes smaller compared to U_0 or the breadth $2a$ of the potential barrier becomes greater, fewer particles tunnel through it. It is obvious that in Newtonian mechanics, no particle passes through it.

6

Schrödinger Equation

6.1 Linear Harmonic Equation

A linear harmonic oscillator is a system having 1 degree of freedom and executing in its direction a harmonic oscillation. We devote this heading to a study of an important physical system called the one-dimensional harmonic oscillator. We are concerned with a one-dimensional motion of a point mass that is attracted to a fixed center by a restoring force F_x, which is proportional to the displacement x from that center:

$$F_x = -m\omega^2 x \tag{6.1}$$

where m and ω are the mass and frequency, respectively. This provides one of the fundamental problems of classical dynamics. This study is very important in physics as a large number of systems may always be analyzed in terms of normal modes of motion. Whenever we investigate the behavior of a physical system at the neighborhood of the position of stable equilibrium, we arrive at equations. In the limit of small oscillations, the normal modes are formally equivalent to harmonic oscillators.

The one-dimensional harmonic oscillator is applicable to a series of important physical phenomena such as the vibration of individual atoms of a molecule about their equilibrium position and the oscillation of atoms or ions in crystals. The harmonic oscillator also provides the path in the study of the electromagnetic field. The vibrations of the electromagnetic field in a cavity provide an infinite number of stationary waves, i.e., normal modes of the cavity. We may expand the electromagnetic field in terms of these modes. If we consider Maxwell's equations, it can be seen that each coefficient of expansion conforms to a differential equation. This equation is identical to that of a harmonic oscillator having angular frequency ω associated with a normal mode. This follows that the vibrations of the electromagnetic field in a cavity may be analyzed into harmonic normal modes each of which energy levels is of the oscillator type. Quantizing the given oscillators associated with various normal modes of the cavity may quantize the field. A detail investigation of the problem of the harmonic oscillator in quantum mechanics is very important.

6.2 Harmonic Oscillator Eigenstates and Eigenvalues

Consider a particle of mass m executing a harmonic oscillation in the direction of the ox-axis with a cyclical frequency ω. The force $F = -m\omega^2 x$ may be represented through the potential energy $U(x)$:

$$U(x) = \frac{m\omega^2 x^2}{2} \tag{6.2}$$

DOI: 10.1201/9781003273073-7

where x is the coordinate evaluated from the equilibrium position. The one-dimensional Schrödinger equation for the harmonic oscillator is

$$-\frac{\hbar^2}{2m}\frac{d^2}{dx^2}\Psi+\frac{m\omega^2x^2}{2}\Psi=E\Psi \tag{6.3}$$

Consider Equation 6.3 for the Sturm-Liouville problem where Equation 6.3 is an ordinary differential equation with variable coefficients. The method of its solution is that of power series.

Suppose

$$E=\frac{\hbar\omega\varepsilon}{2} \tag{6.4}$$

where ε is also the dimensionless energy in units of $\hbar\omega$. Let $x=\xi x_0$, where $x_0=\sqrt{\frac{\hbar}{m\omega}}$ and $\frac{\hbar^2}{mx_0^2}=\frac{\hbar^2 m\omega}{m\hbar}=\hbar\omega$. So $m\omega^2x_0^2=\hbar\omega$ and, consequently, from Equation 6.3:

$$\Psi''+\left(\varepsilon-\xi^2\right)\Psi=0 \tag{6.5}$$

The quantity x_0 is the amplitude of the zero oscillations:

$$U(x_0)=\frac{m\omega^2x_0^2}{2}=\frac{\hbar\omega}{2} \tag{6.6}$$

which is the ground-state energy of the oscillator. This energy as we see is different from zero and called the **zero-point energy**. To solve Equation 6.5 easily, we first examine the dominant behavior of Ψ in the asymptotic region $\xi\to\pm\infty$. Note that as ε is finite, then we can consider sufficiently large $\xi\gg\varepsilon$ and Equation 6.5 takes the form:

$$\tilde{\Psi}''-\xi^2\tilde{\Psi}=0 \tag{6.7}$$

where $\tilde{\Psi}$ is the asymptotic solution of Equation 6.5 and from where we have

$$\tilde{\Psi}=\exp\left\{-\frac{\xi^2}{2}\right\}:\tilde{\Psi}'=-\xi\tilde{\Psi},\ \tilde{\Psi}''=-\tilde{\Psi}+\xi^2\tilde{\Psi}=\left(\xi^2-1\right)\tilde{\Psi}\approx\xi^2\tilde{\Psi}\quad\left(\xi\gg1\right) \tag{6.8}$$

So our solution of Equation 6.5 may be selected in the form:

$$\Psi=\exp\left\{-\frac{\xi^2}{2}\right\}U(\xi) \tag{6.9}$$

If we substitute this into Equation 6.5, then

$$U''-2\xi U'+(\varepsilon-1)U=0 \tag{6.10}$$

where primes denote differentiation with respect to ξ. The differential equation in Equation 6.10 matches that of the Hermite polynomials (see appendix, Chapter 25). We find the solution of Equation 6.10 in the form:

$$U=\sum_k a_k\xi^k,\quad a_0\neq0 \tag{6.11}$$

This is necessarily finite for $\varsigma = 0$. Substituting Equation 6.11 into 6.10:

$$\sum_k a_k k(k-1)\xi^{k-2} - 2\sum_k a_k k\xi^k + (\varepsilon - 1)\sum_k a_k \xi^k \equiv 0 \tag{6.12}$$

For the sum of the power series to be zero, it is necessary that the coefficient for all powers of the order k should be equal to zero and so for ξ^k:

$$a_{k+2} = \frac{2k+1-\varepsilon}{(k+2)(k+1)} a_k, \; k = 0,1,2,\ldots \tag{6.13}$$

This is a recurrence formula through which we evaluate all coefficients of the series in Equation 6.11 if a_0 and a_1 are provided. We express through a_0 all even coefficients, a_{2k}, of the given series and through a_1 all odd coefficients, a_{2k+1}. For $a_0 = C_1$ and $a_1 = C_2$:

$$U(\xi) = C_1 U_1(\xi) + C_2 U_2(\xi) \tag{6.14}$$

where U_1 and U_2 are the series corresponding to the even and odd powers of ξ. They separately satisfy Equation 6.10 and, as a consequence, are particular solutions of that equation.

So we examine C_1 and C_2 as arbitrary constants and then Equation 6.14 is the general solution of Equation 6.10. What should be the formal solution of Equation 6.10 which is convergent and is the power series of Equation 6.11, considering Equation 6.13? We examine the condition for convergence, where, from Equation 6.13:

$$\left.\frac{a_{k+2}}{a_k}\right|_{k\to\infty} \approx \frac{2}{k} > 0 \tag{6.15}$$

From the function:

$$\exp\{\xi^2\} = \sum_{k=0}^{\infty} \frac{\xi^{2k}}{k!} \tag{6.16}$$

It is easily seen that for larger k, the ratio of neighboring coefficients leads to $\frac{2}{k}$. This implies that the solution $U(\xi)$ for larger ς behaves as $\exp\{\xi^2\}$. However, $\exp\{\xi^2\}$ and $U(\xi)$ are different series but when ξ is larger, they give equivalent principal contributions. The wave function at infinity should tend to zero. At zero, the condition for the wave function is not satisfied.

Let us find the way out of this crucial problem. We select the parameter ε such that the series in Equation 6.11 is transformed to a polynomial (Hermite polynomial):

$$\varepsilon = 2n+1 \tag{6.17}$$

where n is either even or odd. The polynomial has the parity of n. We select either an even (odd) series. For even n, we have $C_2 = 0$, and for odd n, we have $C_1 = 0$. There is a condition that the wave function should be finite. The polynomial conforms very well since there exists the factor $\exp\left\{-\frac{\xi^2}{2}\right\}$ which ensures the function at infinity is finite. If we consider Equations 6.17 and 6.4, then

$$E_n = \hbar\omega\left(n + \frac{1}{2}\right), \quad n = 0,1,2,\ldots \tag{6.18}$$

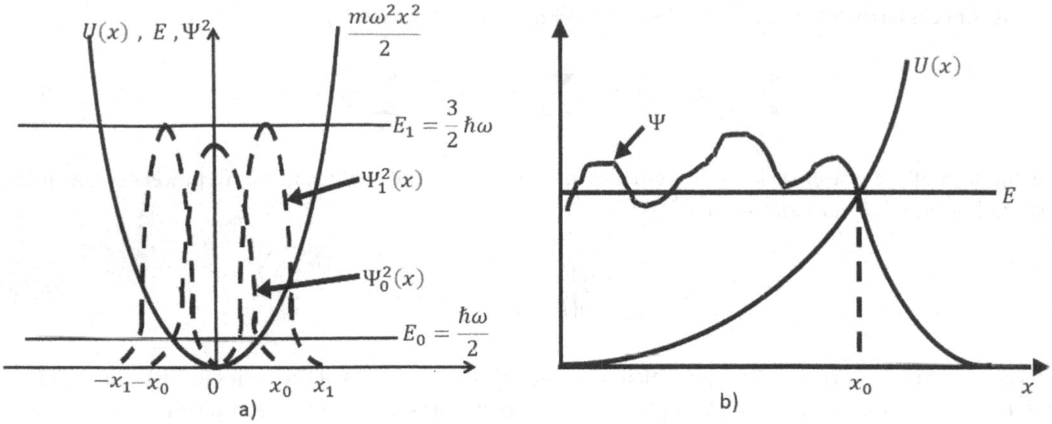

FIGURE 6.1 (6.1a) The ground state and the first excited state of a harmonic oscillator together with their respective eigenstates and eigenenergies; (6.1b) The tail of the wave function (right of the point x_0) going into the classical inaccessible zone (Tunnel effect) [17].

These are the eigenvalues of the energy of the oscillator. The oscillator has a discrete energy spectrum of eigenvalues shown in Equation 6.18. This has been obtained from the mathematical condition that the wave function should be finite at infinity. This brought us to the physical result of the quantization of the energy of the oscillator.

The **energy of the zero oscillation** (**zero-point** or **ground-state energy**) of the oscillator:

$$E_0 = \frac{\hbar\omega}{2} \tag{6.19}$$

Examine the graphs of $U(x)$, E and Ψ^2 (Figure 6.1a). The existence of the ground-state energy is a quantum mechanical result that differs not only from the classical solution but also from Bohr's solution.

If the problem on the oscillator is solved by the method of Bohr's quantization, then

$$\oint p dx = n\hbar\omega \equiv E_n \tag{6.20}$$

This implies that it is not advisable to completely cancel the kinetic energy of an oscillator. Obviously, the origin of the energy is selected to be the potential energy of the oscillator at the equilibrium position. The oscillator has some sought of a **zero-point energy** (**ground-state energy**) that is a characteristic peculiarity of some problems in quantum mechanics. It can be shown that the appearance of E_0 is a result of the Heisenberg uncertainty. It is an observable result. At absolute zero, a system has the possible minimum energy (for the oscillator, it is E_0). The motion does not cease but is frozen in the form of zero oscillations. It follows that the material is not destroyed. The example of a macroscopic system of harmonic oscillators is a solid state. The thermal motion leads to small oscillations of atoms (ions) about the lattice sites of crystalline substances. Introducing normal coordinates, the system of small oscillations may lead to the totality of independent harmonic oscillators.

6.2.1 Hermite Polynomial and Harmonic Oscillator Eigenfunction

6.2.1.1 Hermite Polynomials

We select the eigenfunction of the oscillator in the form:

$$\Psi_n(\xi) = U_n(\xi)\exp\left\{-\frac{\xi^2}{2}\right\} \tag{6.21}$$

where $\xi = \frac{x}{x_0}$ and U_n is a polynomial of the nth order that is an even (odd) function of ξ. Its coefficient may be evaluated using the recurrence formula shown in Equation 6.13. The polynomial U_n should satisfy Equation 6.10 and if Equation 6.17 is considered:

$$U_n'' - 2\xi U_n' + 2nU_n = 0 \tag{6.22}$$

This is the differential equation for the **Tschebycheff-Hermite polynomial** with the solution being the **Hermite polynomial**, $H_n(\xi)$, of order n:

$$U_n(\xi) = H_n(\xi), \quad H_n(\xi) = (-1)^n \exp\{\xi^2\} \frac{d^n}{d\xi^n} \exp\{-\xi^2\} \tag{6.23}$$

where

$$\begin{gathered} H_0(\xi) = 1, H_1(\xi) = 2\xi, H_2(\xi) = 4\xi^2 - 2, H_3(\xi) = 8\xi^3 - 12\xi, \\ H_4(\xi) = 16\xi^4 - 48\xi^2 + 12, H_5(\xi) = 32\xi^5 - 160\xi^3 + 120\xi, \ldots \end{gathered} \tag{6.24}$$

For every n, the solution of Equation 6.22 can be represented as a linear combination of confluent hypergeometric functions for even and odd, n:

$$H_{2n}(\xi) = (-1)^n \frac{(2n)!}{n!} F\left(-n, \frac{1}{2}; \xi^2\right), \; H_{2n+1}(\xi) = (-1)^n \frac{(2n+1)!}{n!} \xi F\left(-n, \frac{3}{2}; \xi^2\right) \tag{6.25}$$

From here:

$$H_{2n}(0) = (-1)^n \frac{(2n)!}{n!}, \; H_{2n+1}(0) = 0 \tag{6.26}$$

We relate the Hermite polynomials with associated Laguerre polynomials, $L_n^{(\lambda)}(\xi)$ for $\lambda = \pm\frac{1}{2}$:

$$H_{2n}(\xi) = C_n L_n^{\left(-\frac{1}{2}\right)}(\xi^2), \; H_{2n+1}(\xi) = C'_n \xi L_n^{\left(\frac{1}{2}\right)}(\xi^2) \tag{6.27}$$

From Equation 6.23:

$$H_{2n}(\xi) = (-1)^{2n} \exp\{\xi^2\} \frac{d^{2n}}{d\xi^{2n}} \exp\{-\xi^2\}; \; L_n^{\left(-\frac{1}{2}\right)}(\xi^2) = \xi \exp\{\xi^2\} \frac{d^n}{d(\xi^2)^n} \left(\exp\{-\xi^2\}\xi^{2n-1}\right);$$

$$H_{2n+1}(\xi) = (-1)^{2n+1} \exp\{\xi^2\} \frac{d^{2n+1}}{d\xi^{2n+1}} \exp\{-\xi^2\}; \; L_n^{\left(\frac{1}{2}\right)}(\xi^2) = \xi^{-1} \exp\{\xi^2\} \frac{d^n}{d(\xi^2)^n} \left(\exp\{-\xi^2\}\xi^{2n+1}\right) \tag{6.28}$$

From Equation 6.27 and using Leibnitz formula, we have

$$C_n = (-1)^n 2^{2n}, \; C'_n = (-1)^n 2^{2n+1} \tag{6.29}$$

So

$$H_{2n}(\xi) = (-1)^n 2^{2n} L_n^{\left(-\frac{1}{2}\right)}(\xi^2), \; H_{2n+1}(\xi) = (-1)^n 2^{2n+1} \xi L_n^{\left(\frac{1}{2}\right)}(\xi^2) \tag{6.30}$$

6.2.1.2 Hermite Polynomials Integral Representation

Considering the Cauchy theorem (residue theorem), we have for any contour, l, enclosing the point $z=\xi$:

$$\oint_l dz\frac{\exp\{-z^2\}}{(z-\xi)^{n+1}}=2\pi i \operatorname*{Res}_{z=\xi}\frac{\exp\{-z^2\}}{(z-\xi)^{n+1}}=\frac{2\pi i}{n!}\lim_{z\to\xi}\frac{d^n}{dz^n}\exp\{-z^2\}=\frac{2\pi i}{n!}\frac{d^n}{dz^n}\exp\{-z^2\}$$
$$=\frac{2\pi i}{n!}(-1)^n\exp\{-\xi^2\}\mathrm{H}_n(\xi)=(-1)^n\exp\{-\xi^2\}\mathrm{H}_n(\xi) \tag{6.31}$$

Letting $z=\xi-t$, for any contour, l_0, enclosing the point $t=0$, we have

$$\mathrm{H}_n(\xi)=\frac{n!}{2\pi i}\oint_{l_0} dt\frac{\exp\{2\xi t-t^2\}}{t^{n+1}} \tag{6.32}$$

6.2.1.3 Harmonic Oscillator Eigenfunction and Normalization Condition

The eigenfunction of the oscillator is obtained considering Equations 6.21 and 6.23:

$$\Psi_n(\varsigma)=C_n\exp\left\{-\frac{\xi^2}{2}\right\}\mathrm{H}_n(\xi) \tag{6.33}$$

where C_n is the normalization constant. We evaluate it from the normalization condition:

$$\int_{-\infty}^{\infty}\Psi_n^2(x)\,dx=1 \tag{6.34}$$

Then, for $x=x_0\xi$, considering Equations 6.23 and 6.33:

$$C_n^2x_0\int_{-\infty}^{\infty}\exp\{-\xi\}\mathrm{H}_n^2(\xi)d\xi=(-1)^nC_n^2x_0\int_{-\infty}^{\infty}\mathrm{H}_n(\xi)\frac{d^n}{d\xi^n}\exp\{-\xi^2\}d\xi \tag{6.35}$$

Using n-fold integration by parts, we have

$$(-1)^nC_n^2x_0\int_{-\infty}^{\infty}\mathrm{H}_n(\xi)\frac{d^n}{d\xi^n}\exp\{-\xi^2\}d\xi=C_n^2x_0\int_{-\infty}^{\infty}\exp\{-\xi^2\}\frac{d^n}{d\xi^n}\mathrm{H}_n(\xi)d\xi \tag{6.36}$$

From Equation 6.23:

$$\int_{-\infty}^{\infty}\exp\{-\xi^2\}\frac{d^n}{d\xi^n}\mathrm{H}_n(\xi)d\xi=2^nn!\int_{-\infty}^{\infty}\exp\{-\xi^2\}d\xi \tag{6.37}$$

So from the Poisson's integral:

$$\int_{-\infty}^{\infty}\exp\{-\xi^2\}d\xi=\sqrt{\pi} \tag{6.38}$$

then

$$C_n^2 x_0 \int_{-\infty}^{\infty} \exp\{-\xi^2\} \frac{d^n}{d\xi^n} H_n(\xi) d\xi = C_n^2 x_0 2^n n! \sqrt{\pi} = 1, \; C_n = \frac{1}{\sqrt{x_0 2^n n! \sqrt{\pi}}} \tag{6.39}$$

6.2.1.4 Hermite Polynomials Orthogonality Condition

We show that the wave functions of two different states, k and n, are orthogonal for $k \neq n$:

$$\int_{-\infty}^{\infty} \Psi_k(x) \Psi_n(x) dx = 0, \quad k \neq n \tag{6.40}$$

Considering, Equations 6.23 and 6.33 for $x = x_0 \varsigma$:

$$\int_{-\infty}^{\infty} \Psi_k(x) \Psi_n(x) dx = C_n C_k x_0 \int_{-\infty}^{\infty} d\xi \exp\{-\xi^2\} H_k(\xi) H_n(\xi) \tag{6.41}$$

We examine the second integral and use integration by parts:

$$\int_{-\infty}^{\infty} d\xi \exp\{-\xi^2\} H_k(\xi) H_n(\xi) = (-1)^n \int_{-\infty}^{\infty} d\xi \, H_k(\xi) \frac{d^n}{d\xi^n} \exp\{-\xi^2\} = -(-1)^n \int_{-\infty}^{\infty} d\xi \, H'_k(\xi) \frac{d^{n-1}}{d\xi^{n-1}} \exp\{-\xi^2\}$$

$$= (-1)^{n-1} 2k \int_{-\infty}^{\infty} d\xi \, H_{k-1}(\xi) \frac{d^{n-1}}{d\xi^{n-1}} \exp\{-\xi^2\} = \cdots = (-1)^{n-k} 2^k k! \int_{-\infty}^{\infty} d\xi \frac{d^{n-k}}{d\xi^{n-k}} \exp\{-\xi^2\} \tag{6.42}$$

From here, we have

$$\int_{-\infty}^{\infty} d\xi \exp\{-\xi^2\} H_k(\xi) H_n(\xi) = (-1)^{n-k} 2^k k! \left[\frac{d^{n-k-1}}{d\xi^{n-k-1}} \exp\{-\xi^2\} \right]_{-\infty}^{\infty} = 0 \tag{6.43}$$

This implies that the Hermite polynomial is orthogonal within the interval $(-\infty, \infty)$ and weighted by $\rho(\xi) = \exp\{-\xi^2\}$. This also implies the orthogonality of the functions, $\Psi_k(x)$ and $\Psi_n(x)$, such that Equation 6.40 is satisfied. We have also shown above that for $k = n$, we find the norm of the Hermite polynomial and, consequently, the normalization condition for the eigenfunction:

$$\int_{-\infty}^{\infty} \exp\{-\xi^2\} H_n^2(\xi) d\xi = 2^n n! \int_{-\infty}^{\infty} \exp\{-\xi^2\} d\varsigma = 2^n n! \sqrt{\pi} \tag{6.44}$$

Taylor series expansion of a function through the Hermite polynomial

We can Taylor expand the function, $f(\xi)$, through the Hermite polynomial, $H_n(\xi)$:

$$f(\xi) = \sum_{n=0}^{\infty} C_n H_n(\xi) \tag{6.45}$$

This function is convergent within the interval $(-\infty, \infty)$ and can be piecewise integrable within the given interval. Considering Equations 6.43 and 6.44, we find the coefficient, C_n:

$$C_n = \frac{1}{2^n n! \sqrt{\pi}} \int_{-\infty}^{\infty} \exp\{-\xi^2\} H_n(\xi) f(\xi) d\xi \tag{6.46}$$

Remarks

Note that the motion is bounded (finite) if the potential energy at infinity is infinite (though from one side). It is obvious that in such a domain of space where the potential energy is infinite, the wave function should tend to zero as the total mean energy:

$$\bar{E} = \bar{T} + \bar{U} \tag{6.47}$$

In the above calculations, we consider that

$$\Psi\big|_{\to\pm\infty} \to 0 \tag{6.48}$$

It follows that we are concerned with a bounded (finite) one-dimensional motion. It is bounded in the sense that the wave function is different from zero for an infinite far removed point from it:

1. The finite motion corresponds to a discrete energy spectrum
2. The energy level is non-degenerate: each energy level corresponds to one wave function
3. The wave function is real-valued

Property 3 follows from 2: The non-degenerate level corresponds to a real-valued function. The functions Ψ and Ψ^* satisfy one and the same energy level for a non-degenerate case. This follows that

$$\Psi = \Psi^* \tag{6.49}$$

are real-valued functions. The ground-state wave function $\Psi_0(x)$:

$$\Psi_0(x) = C_0 \exp\left\{-\frac{x^2}{x_0^2}\right\} \tag{6.50}$$

and $\Psi_0^2(x)$ is the Gaussian function that is the maximum probability about the equilibrium position. The wave function and the probability density for the level $n=1$, respectively, are

$$\Psi_1(x) = C_1 x \exp\left\{-\frac{x^2}{x_0^2}\right\},\ \Psi_1^2(x) = C_1^2 x^2 \exp\left\{-2\frac{x^2}{x_0^2}\right\} \tag{6.51}$$

There exist two maxima for $x = \pm x_0$. Suppose $E = \frac{3}{2}\hbar\omega$, the oscillator should move from $-x_1$ to $+x_1$. It may not be found on the left-hand side of $-x_1$ or on the right-hand side of $+x_1$ that are classical forbidden domains $(U > E)$. In quantum mechanics, there is no such forbidden domain as there exists a probability different from zero that we may find to be an oscillator. This is obvious from the wave function. It does not tend to zero but asymptotically tends to zero (Figure 6.1b). In quantum mechanics, there is no sense in the partition of the total energy E into the kinetic T and the potential U energy. There is no sense at the point, say x for $\mathrm{T} \gtrless U$. There is a sense only in E. As the operators $\hat{\mathrm{T}}$ and $\hat{U}$ are not mutually commutable as $\hat{x}$ and $\hat{p}$ are not mutually commutable, each of them do not commute with the Hamiltonian $\hat{\mathrm{H}}$. It should be noted that $\bar{E} = \bar{\mathrm{T}} + \bar{U}$ is independent of x and $\bar{\mathrm{T}}$ is always greater than zero. If we consider some cases for which $\bar{U}$ is dependent on x, then we have a motion in a central symmetric field.

Question:

Find the energy levels of a three-dimensional oscillator with potential energy:

$$U(x,y,z) = \frac{m\omega_1^2 x^2}{2} + \frac{m\omega_2^2 y^2}{2} + \frac{m\omega_3^2 z^2}{2} \tag{6.52}$$

Answer:
The Schrödinger equation is written in the form:

$$-\frac{\hbar^2}{2m}\frac{d^2}{dx^2}\Psi-\frac{\hbar^2}{2m}\frac{d^2}{dy^2}\Psi-\frac{\hbar^2}{2m}\frac{d^2}{dz^2}\Psi+U(x,y,z)\Psi=E\Psi,\ \Psi(x,y,z)=\psi(x)\psi(y)\psi(z) \tag{6.53}$$

and we have

$$-\frac{\hbar^2}{2m}\frac{d^2}{dx^2}\psi(x)+\frac{m\omega_1^2x^2}{2}\psi(x)=E_1\psi(x),\ -\frac{\hbar^2}{2m}\frac{d^2}{dy^2}\psi(y)+\frac{m\omega_2^2y^2}{2}\psi(y)$$
$$=E_2\psi(y),\ -\frac{\hbar^2}{2m}\frac{d^2}{dz^2}\psi(z)+\frac{m\omega_3^2z^2}{2}\psi(z)=E_3\psi(z) \tag{6.54}$$

The solutions of the above equations are calculated as seen for the linear harmonic oscillator:

$$E_{n_1n_2n_2}=\sum_{i=1}^{3}\hbar\omega_i\left(n_i+\frac{1}{2}\right),\ \Psi_{n_1n_2n_2}(\xi_1,\xi_2,\xi_3)=\prod_{i=1}^{3}C_{n_i}\exp\left\{-\frac{\xi_1^2}{2}\right\}H_{n_i}(\xi_i) \tag{6.55}$$

$$x_i=\xi_ix_{0i},\ C_{n_i}=\frac{1}{\sqrt{x_{0i}2^{n_i}n_i!\sqrt{\pi}}},\ x_{0i}=\sqrt{\frac{\hbar}{m\omega_i}},\ x_1=x,\ x_2=y,\ x_3=z,\ i=1,2,3 \tag{6.56}$$

Question:
Find in the p-representation the stationary Schrödinger equation for a particle found in the field:

$$U(x)=\frac{Kx^2}{2} \tag{6.57}$$

where K is the constant of the elastic coupling. Find the energy spectrum and the normalized wave functions for this case in the p-representation from the Schrödinger equation in that representation.

Answer:
The operator of the kinetic energy in the p-representation has the form:

$$\check{T}=\frac{\check{p}^2}{2m}=\frac{p^2}{2m} \tag{6.58}$$

As the arbitrary linear operator in the general case is an integral operator, we can establish the relation between $L(x,x')$ and $L(p,p')$ that are, respectively, kernels of one and the same operator $\hat{L}$ in the x- and p-representations. The action of the operator $\hat{L}$ on $\Psi(x)$ yields the function $\tilde{\Psi}(x)$ expression through the kernel $L(x,x')$:

$$\tilde{\Psi}(x)\equiv\hat{L}\Psi(x)=\int L(x,x')\Psi(x')dx' \tag{6.59}$$

Here

$$\Psi(x')=\int C(p')\Psi_{p'}^*(x')dp',\ \Psi_p^*(x)=\frac{1}{\sqrt{2\pi\hbar}}\exp\left\{\frac{ipx}{\hbar}\right\} \tag{6.60}$$

From here:

$$\tilde{C}(p) \equiv \hat{L}C(p) = \int L(p,p')C(p')dp' \tag{6.61}$$

where

$$L(p,p') = \int \Psi^*_{p'}(x')\Psi_p(x)L(x,x')dx\,dx',\; L(x,x') = \int \Psi^*_{p'}(x')\Psi_p(x)L(p,p')dp\,dp' \tag{6.62}$$

From here, in the x-representation, the operator of the potential energy is a multiplication operator, and in the p-representation, it is an integral operator with the kernel $U(p,p')$:

$$U(p,p') \equiv \tilde{U}(p-p'),\; U(p) = \frac{1}{2\pi\hbar}\int \tilde{U}(x)\exp\left\{-i\frac{px}{\hbar}\right\}dx' \tag{6.63}$$

From here, the Schrödinger equation in the p-representation:

$$\hat{H}C(p) \equiv \frac{p^2}{2m}C(p) + \int_{-\infty}^{\infty}\tilde{U}(p-p')C(p')dp' = EC(p) \tag{6.64}$$

where

$$\tilde{U}(p-p') = -\frac{K\hbar}{4\pi}\frac{\partial^2}{\partial p^2}\int \exp\left\{-\frac{i}{\hbar}(p-p')x\right\}dx = -\frac{K\hbar^2}{2}\frac{\partial^2}{\partial p^2}\delta(p-p') \tag{6.65}$$

So, the Schrödinger equation:

$$\hat{H}C(p) \equiv \frac{p^2}{2m}C(p) - \frac{K\hbar^2}{2}\frac{\partial^2}{\partial p^2}\int_{-\infty}^{\infty}\delta(p-p')C(p')dp' = \frac{p^2}{2m}C(p) - \frac{K\hbar^2}{2}\frac{\partial^2}{\partial p^2}C(p) = EC(p) \tag{6.66}$$

This is exactly analogous to the Schrödinger equation for a harmonic oscillator in the x-representation. The eigenfunctions of the given equation and the eigenvalues are, respectively:

$$C_n(p) = \frac{\exp\left\{-\frac{p^2}{2m\omega\hbar}\right\}}{\sqrt{2^n n!\sqrt{\pi m\omega\hbar}}}H_n\left(\frac{p}{\sqrt{m\omega\hbar}}\right),\; E_n = \hbar\omega\left(n+\frac{1}{2}\right),\quad n = 0,1,2,\ldots,\quad K = m\omega^2 \tag{6.67}$$

Question:
Find the energy level and wave function of a one-dimensional harmonic oscillator placed in a uniform electric field $\vec{E}$ parallel to the ox-axis. The electric charge is e.

Answer:
The potential energy U of a charged particle placed in a homogenous electric field, $\vec{E}$:

$$U = -e\left|\vec{E}\right|x \tag{6.68}$$

For the case of an ordinary harmonic oscillator placed in a homogenous electric field, the potential

$$\tilde{U} = \frac{m\omega^2 x^2}{2} - e\left|\vec{E}\right| x \equiv \frac{m\omega^2}{2}\left(x - \frac{e\left|\vec{E}\right|}{m\omega^2}\right)^2 - \frac{e^2\left|\vec{E}\right|^2}{2m\omega^2} \tag{6.69}$$

and the Schrödinger equation has the form:

$$-\frac{\hbar^2}{2m}\frac{d^2}{dx^2}\Psi + \frac{m\omega^2}{2}\left(x - \frac{e\left|\vec{E}\right|}{m\omega^2}\right)^2 \Psi = \left(E + \frac{e^2\left|\vec{E}\right|^2}{2m\omega^2}\right)\Psi \tag{6.70}$$

Letting

$$y = x - \frac{e\left|\vec{E}\right|}{m\omega^2}, \quad \tilde{E} = E + \frac{e^2\left|\vec{E}\right|^2}{2m\omega^2}, \quad \xi = \sqrt{\frac{\hbar}{m\omega}}\, y \tag{6.71}$$

we have the standard equation of a harmonic oscillator with the eigenfunctions and eigenvalues, respectively:

$$\Psi_n = C_n \exp\left\{-\frac{\varsigma^2}{2}\right\} \mathrm{H}_n(\xi),\ E_n = \hbar\omega\left(n + \frac{1}{2}\right) - \frac{e^2\left|\vec{E}\right|^2}{2m\omega^2} \tag{6.72}$$

It is seen that for an ordinary harmonic oscillator placed in a homogenous electric field, the entire spectrum is shifted by the quantity $\frac{e^2\left|\vec{E}\right|^2}{2m\omega^2}$. The change of the variable $y = x - \frac{e\left|\vec{E}\right|}{m\omega^2}$ enables us to have a standard problem where $\vec{E} = 0$. It can be seen that the effect of the electric field shifts only the x-origin and the energy origin. It should be noted that in the presence of an electric field, we still have a harmonic oscillator but with a shift in both the coordinate and the energy.

This shift is explained by the fact that the electric field moves the particle away from the origin to the domain where the **elastic** potential energy is greater.

Question:
The one-dimensional oscillator is found in the nth energy level. Find for it $\overline{x^2}$ and the mean potential energy.

Answer:
If we consider the previous problem, the oscillator in the nth quantum state is characterized by the wave function:

$$\Psi_n = C_n \exp\left\{-\frac{\xi^2}{2}\right\} \mathrm{H}_n(\xi) \tag{6.73}$$

From definition:

$$\left\langle x^2 \right\rangle_n = \int_{-\infty}^{\infty} \left|\Psi_n(x)\right|^2 x^2 dx = C_n^2 \left(\frac{\hbar}{m\omega}\right)^{\frac{3}{2}} \int_{-\infty}^{\infty} \exp\left\{-\xi^2\right\} \mathrm{H}_n^2(\xi)\, \xi^2 d\xi \tag{6.74}$$

we change one of $H_n(\xi)$:

$$H_n(\xi)=(-1)^n \exp\{\xi^2\}\frac{d^n}{d\xi^n}\exp\{-\xi^2\} \tag{6.75}$$

then we do n-fold integration by parts:

$$\langle x^2\rangle_n = C_n^2\left(\frac{\hbar}{m\omega}\right)^{\frac{3}{2}}(-1)^n\int_{-\infty}^{\infty}\left(H_n^2(\xi)\xi^2\right)\frac{d^n}{d\xi^n}\exp\{-\xi^2\}d\xi = C_n^2\left(\frac{\hbar}{m\omega}\right)^{\frac{3}{2}}\int_{-\infty}^{\infty}\exp\{-\xi^2\}\frac{d^n}{d\xi^n}\left(H_n^2(\xi)\xi^2\right)d\xi \tag{6.76}$$

It is obvious that

$$\frac{d^n}{d\xi^n}\left(H_n^2(\xi)\xi^2\right)=\frac{d^n}{d\xi^n}\left(a_n\xi^{n+2}+a_{n-2}\xi^n+\ldots\right)=a_n\frac{(n+2)!}{2!}\xi^2+n!a_{n-2} \tag{6.77}$$

If we consider Equation 6.13:

$$a_{k+2}=\frac{2k+1-\varepsilon}{(k+2)(k+1)}a_k \tag{6.78}$$

and

$$H_n(\xi)=(2\xi)^n-\frac{n(n-1)}{1!}(2\xi)^{n-2}+\frac{n(n-1)(n-2)(n-3)}{2!}(2\xi)^{n-4}+\ldots \tag{6.79}$$

Then for $a_n = 2^n$, we have

$$a_{n-2}=-\frac{n(n-1)}{4}a_n \tag{6.80}$$

and from

$$\int_{-\infty}^{\infty}\exp\{-\xi^2\}d\xi=\sqrt{\pi},\quad \int_{-\infty}^{\infty}\xi^2\exp\{-\xi^2\}d\varsigma=\frac{\sqrt{\pi}}{2} \tag{6.81}$$

then we have

$$\langle x^2\rangle_n = C_n^2\left(\frac{\hbar}{m\omega}\right)^{\frac{3}{2}}\left(\frac{(n+2)!}{2}2^n\frac{\sqrt{\pi}}{2}-2^n\frac{n(n-1)}{4}n!\sqrt{\pi}\right)=\frac{m\omega^2}{2}\langle x^2\rangle_n=\frac{1}{2}\hbar\omega\left(n+\frac{1}{2}\right)=\frac{E_n}{2} \tag{6.82}$$

So

$$\langle U\rangle_n=\frac{m\omega^2}{2}\langle x^2\rangle_n=\frac{1}{2}\hbar\omega\left(n+\frac{1}{2}\right)=\frac{E_n}{2} \tag{6.83}$$

The average kinetic energy can be obtained as follows:

$$\bar{T}=\langle T\rangle_n=-\frac{\hbar^2}{2m}\int\Psi^*\Delta\Psi\,dx=\frac{1}{2}E_n \tag{6.84}$$

Question:
Two particles with mass m each move only in the ox-axis direction. The particles are coupled together by an elastic force. In addition, the same type of force but with different elastic constants couples each of them at the point x = 0. Find the energy levels and wave functions for the system.

Answer:
The potential energy $U(x_1, x_2)$:

$$U(x_1,x_2)=\frac{K}{2}\left(x_1^2+x_2^2\right)+\frac{K_1}{2}\left(x_1-x_2\right)^2 \tag{6.85}$$

where K and K_1 are the elastic constants that characterize the coupling of particles at the point x = 0 and with each other, respectively. Introduce the coordinate of the center of mass y and the relative coordinate x:

$$y=\frac{mx_1+mx_2}{m+m}=\frac{x_1+x_2}{2},\quad x=x_1-x_2 \tag{6.86}$$

In order to evaluate the following Schrödinger equation:

$$-\frac{\hbar^2}{2m}\Delta_1\Psi-\frac{\hbar^2}{2m}\Delta_2\Psi+U(x_1,x_2)\Psi=E\Psi \tag{6.87}$$

we consider

$$\frac{\partial^2}{\partial x_1^2}=\frac{1}{4}\frac{\partial^2}{\partial y^2}+\frac{\partial^2}{\partial x^2} \tag{6.88}$$

and

$$\frac{1}{m}\Delta_1+\frac{1}{m}\Delta_2=\frac{1}{m}\left(\frac{1}{4}\frac{\partial^2}{\partial y^2}+\frac{\partial^2}{\partial x^2}\right)+\frac{1}{m}\left(\frac{1}{4}\frac{\partial^2}{\partial y^2}+\frac{\partial^2}{\partial x^2}\right)=\frac{1}{M}\frac{\partial^2}{\partial y^2}+\frac{1}{\tilde{M}}\frac{\partial^2}{\partial x^2} \tag{6.89}$$

where

$$M=2m,\quad \tilde{M}=\frac{m}{2} \tag{6.90}$$

The potential energy may be transformed:

$$U(x_1,x_2)=U(x,y)=\frac{K+2K_1}{2}\frac{x^2}{2}+2K\frac{y^2}{2} \tag{6.91}$$

Then the Schrödinger equation now becomes

$$\left(-\frac{\hbar^2}{2M}\frac{d^2}{dy^2}-\frac{\hbar^2}{2\tilde{M}}\frac{d^2}{dx^2}+\frac{M\omega^2y^2}{2}+\frac{\tilde{M}\omega_1^2x^2}{2}\right)\Psi=E\Psi,\ \omega=\sqrt{\frac{K}{M}},\quad \omega_1=\sqrt{\frac{K+2K_1}{2\tilde{M}}} \tag{6.92}$$

The variables are separable:

$$\Psi=\psi(y)\Phi(x) \tag{6.93}$$

If we substitute this in the Schrödinger equation, it leads to two oscillations:

$$\left(-\frac{\hbar^2}{2M}\frac{d^2}{dy^2}+\frac{M\omega^2 y^2}{2}\right)\psi = E_1\psi,\left(-\frac{\hbar^2}{2\tilde{M}}\frac{d^2}{dx^2}+\frac{\tilde{M}\omega_1^2 x^2}{2}\right)\Phi = E_2\Phi \tag{6.94}$$

then

$$\Psi_{n_1n_2} = \mathrm{A}\exp\left\{-\frac{\xi^2+u^2}{2}\right\}\mathrm{H}_{n_1}(\varsigma)\mathrm{H}_{n_2}(u), \xi=\sqrt{\frac{\hbar}{M\omega}}y,\quad u=\sqrt{\frac{\hbar}{M\omega}}x,\ E_{n_1n_2}=\left(n_1+\frac{1}{2}\right)\hbar\omega+\left(n_2+\frac{1}{2}\right)\hbar\omega_1 \tag{6.95}$$

6.3 Motion in a Central Field

The motion in a central symmetric field is described by the following Schrödinger equation:

$$\hat{\mathrm{H}}\Psi \equiv -\frac{\hbar^2}{2\mu}\Delta\Psi + U(r)\Psi = E\Psi \tag{6.96}$$

where $U(r)$ is the generalized potential and μ the mass of the particle. For the central symmetric field, the potential $U(r)$ depends only on the magnitude of the radius vector $\vec{r}$ from the origin. The potential $U(r)$ is clearly invariant under any rotation about the origin. In this case, the Hamiltonian operator, $\hat{\mathrm{H}}$, commutes with the operator of the angular momentum, $\hat{L}^2$. Obviously, in the Hamiltonian, $\hat{\mathrm{H}}$, the operator of the kinetic energy commutes with $\hat{L}^2$ from definition. This is because the Hamiltonian is invariant relative to the operation of rotation. The operator $\hat{L}^2$ acts only on the function dependent on the angles. But the potential energy in a central symmetric field is independent of the angles and follows that $\hat{L}^2$ commutes with the potential energy. So the solution of the Schrödinger equation (Equation 6.96) should be found such that its eigenfunction should be the eigenfunction of the operator $\hat{L}^2$. From the above argument, we expect the angular momentum to be conserved.

6.3.1 Radial Schrödinger Equation

Write in the spherical coordinates the Schrödinger equation in 6.96 as follows:

$$-\frac{\hbar^2}{2\mu}\frac{1}{r^2}\frac{\partial}{\partial r}\left(r^2\frac{\partial}{\partial r}\Psi\right)+\frac{\hat{L}^2}{2\mu r^2}\Psi+U(r)\Psi = E\Psi \tag{6.97}$$

Substitute the wave function defined in 3.274 into Equation 6.97 and find a solution of the Schrödinger equation which should be the eigenfunction of the operators $\hat{L}^2$ and $\hat{L}_z$, i.e., the eigenfunction of such a state where E, L^2 and L_z have defined values obtained from the following radial equation:

$$-\frac{\hbar^2}{2\mu r^2}\frac{d}{dr}\left(r^2\frac{d}{dr}R\right)+\frac{\hbar^2 l(l+1)}{2\mu r^2}R+U(r)R = ER \tag{6.98}$$

We write the differential equation satisfying the radial function (Sturm-Liouville equation) for the definition of the eigenvalue, E. In order to solve this equation, we substitute for $U(r)$.

Suppose the solution of that equation yields a discrete spectrum of eigenvalues of the energy, E_n, with n being the quantum number. In the Sturm-Liouville equation, the quantum number l (orbital or Azimuthal quantum number) is found as a parameter. This implies the energy eigenvalue, E, should depend also on l. Since there is no m in Equation 6.98, the energy eigenvalue is independent of m. It is therefore obvious that the eigenvalue E in Equation 6.98 has the same degree of degeneracy:

$$f = 2l + 1 \tag{6.99}$$

as the operator of the angular momentum, $\hat{L}^2$.

We may obtain the dependence

$$E = E(m) \tag{6.100}$$

if a magnetic field is introduced in the system that leads to the splitting of the degenerate levels that depend on m. The states with a given l are denoted:

$$\begin{array}{lccccccc} l & 0, & 1 & ,2 & ,3 & ,4 & ,5, & \ldots \\ \text{state} & s & p & d & f & g & h & \ldots \end{array} \tag{6.101}$$

In Equation 6.98, for convenience, we substitute $R(r)$:

$$R(r) = \frac{X(r)}{r} \tag{6.102}$$

which leads to the modified radial wave equation:

$$-\frac{\hbar^2}{2\mu}\frac{d^2}{dr^2}X + \frac{\hbar^2 l(l+1)}{2\mu r^2}X + U(r)X = EX \tag{6.103}$$

Here, X is the modified radial wave function and the radial Equation 6.103 imitates the one-dimensional Schrödinger equation:

$$-\frac{\hbar^2}{2\mu}\frac{d^2}{dx^2}\Psi + U(x)\Psi = E\Psi \tag{6.104}$$

But differs in the additional potential, the so-called **centrifugal potential (or centrifugal barrier)**:

$$\frac{\hbar^2 l(l+1)}{2\mu r^2} > 0 \tag{6.105}$$

This **centrifugal potential** physically relates the angular momentum.

Suppose a classical particle has the angular momentum L about the axis through the origin perpendicular to the plane of its orbit. The particle on this orbit has the angular velocity ω when it has radial distance r from the origin:

$$\omega = \frac{L}{\mu r^2} \tag{6.106}$$

To keep the particle in the given path, an inward force is required:

$$\mu\omega^2 r = \frac{L^2}{\mu r^3} \tag{6.107}$$

This is the **centripetal force** supplied by the potential energy which adds to the potential $U(r)$ appearing in the radial motion an additional **centrifugal potential energy**:

$$\frac{L^2}{2\mu r^2} \equiv \frac{L^2}{2J} \tag{6.108}$$

This is similar to Kepler's problem in classical mechanics where J is the moment of inertia. The force acting on the particle of mass μ always tends to repel it from the force center. The kinetic energy of a material point, rotating about a circle of radius r, should be $\frac{L^2}{2J}$. So our analysis is exactly the type in Equation 6.103 if we let

$$L^2 = \hbar^2 l(l+1) \tag{6.109}$$

In addition to other boundary conditions (three conditions: wave function should be **finite, continuous and single-valued**), X should satisfy the condition:

$$\mathrm{X}(0) = 0 \tag{6.110}$$

This is of course different from the one-dimensional motion on the entire axis $(-\infty, \infty)$.

Consider the effective potential $U_{\text{eff}}(r)$

$$U_{\text{eff}}(r) = U(r) + \frac{\hbar^2 l(l+1)}{2\mu r^2} \tag{6.111}$$

in the central symmetric field, which is the sum of the true potential $U(r)$ and the rotational energy $\frac{L^2}{2\mu r^2}$ as in classical mechanics.

Since $\hat{L}_z$ has an arbitrary dependence on r and θ, $\hat{L}^2$ depends arbitrarily on the radial function R. Since the operator $\hat{L}^2$ acts on the angular variables while $U(r)$ does not, the Hamiltonian, $\hat{\mathrm{H}}$, commutes with $\hat{L}^2$. So the angular momentum is an integral of motion. The parity $(-1)^l$ is also an integral of motion as the operator of inversion commutes with the Hamiltonian of the particle in the central symmetric field.

6.3.2 Radial Wave Function Qualitative Investigation

To obtain the wave function, X, it is necessary to solve Equation 6.103. We have to define the potential, $U(r)$. However, from symmetry, results can be obtained. Consider small distances from the center of the field $(r \to 0)$ (i.e., behavior of the wave function at the origin):

$$\mathrm{X} \to 0 \tag{6.112}$$

If we expand X in series with respect to r, then

$$\lim_{r \to 0} \mathrm{X} \to 0 \tag{6.113}$$

This implies that X should begin from the term $r^s\,(s > 0)$:

$$\mathrm{X}(r) = r^s + C_1 r^{s+1} + \cdots \tag{6.114}$$

Let

$$\lim_{r\to 0} U(r)r^2 \to 0 \tag{6.115}$$

For example, say a Coulombic field. We substitute into Equation 6.103 the expansion of X and multiply the result by r^2:

$$-\frac{\hbar^2}{2\mu}\left[s(s-1)r^s+\cdots\right]+\frac{\hbar^2}{2\mu}l(l+1)\left[r^s+\cdots\right]+U(r)\left[r^{s+2}+\cdots\right]=Er^2\left[r^s+\cdots\right] \tag{6.116}$$

(Here $(\cdots)$ are terms with order greater than *s*.) Divide the above equation by r^s and take limits as $r\to 0$, then we have

$$-\frac{\hbar^2 s(s-1)}{2\mu}+\frac{\hbar^2}{2\mu}l(l+1)=0 \tag{6.117}$$

or

$$s=l+1;\ -l \tag{6.118}$$

and as $s>0$, we neglect the second solution $-l$ and consider $s=l+1$:

$$\lim_{r\to 0} X(r)\to r^{l+1} \tag{6.119}$$

Then there arises the question on the normalization of the wave function Ψ. The radial and angular parts of Ψ are normalized separately. So

$$\int_0^\infty |R(r)|^2 r^2 dr=\int_0^\infty |X(r)|^2 dr=1 \tag{6.120}$$

and

$$\int |Y_{lm}(\theta,\phi)|^2 \sin\theta\, d\theta d\phi=1 \tag{6.121}$$

In Equation 6.120, the quantity $|X(r)|^2$ is the radial probability density that the particle at a given state finds in a spherical layer of thickness dr from $r\to r+dr$. From Equation 6.119 onwards, those terms with an orbital quantum number greater than l have lesser probability to find the particle at the neighborhood of the center of the field. Letting

$$\lim_{r\to 0} U(r)r^2 \to -a\frac{\hbar^2}{2\mu} \tag{6.122}$$

then

$$-\frac{\hbar^2 s(s-1)}{2\mu}+\frac{\hbar^2}{2\mu}l(l+1)-a\frac{\hbar^2}{2\mu}=0 \tag{6.123}$$

For

$$a>l(l+1)+\frac{1}{4} \tag{6.124}$$

the roots of the Equation 6.123 will be complex-valued:

$$s_{1,2}=\frac{1}{2}(1\pm i\gamma),\ \gamma=\sqrt{4a-1-4l(l+1)} \tag{6.125}$$

In this case, either Equation 6.98 or 6.103 will not have a regular solution for $r\to 0$. From Equation 6.125, the radial solutions will behave like $r^{-\frac{1}{2}}\exp\{i\gamma\ln r\}$ and $r^{-\frac{1}{2}}\exp\{-i\gamma\ln r\}$. So the general solution of Equation 6.98 or 6.103 for $r\to 0$ will be their linear combination:

$$R(r)=A\,r^{-\frac{1}{2}}\cos(\gamma\ln r+\alpha) \tag{6.126}$$

This function for $r\to 0$ has infinite number of zeros. This is because as $r\to 0$, $\ln r\to\infty$. However, from experimenting the solution of a one-dimensional problem of a particle in a potential well, it is well known where the number of nodes (zeros) of the wave function coincide with the number of levels considered from the ground state. The significance of an infinite number of zeros implies that the ground-state level is $E=-\infty$. This result corresponds to the classical falling of a particle into the force center.

For larger distances r from the center of the field, $U(r)\to 0$ as $r\to\infty$. Here, 0 is conditional and implies that $U(r)\to$ const as $r\to\infty$, where we select the *const* as the reference point on the energy scale.

In non-relativistic mechanics, the energy ε is given to the approximation of a constant. We find that the asymptotic wave function $\tilde{X}$ for larger r satisfies Equation 6.103 if we neglect $U(r)$ and the centrifugal term:

$$\frac{d^2}{dr^2}\tilde{X}+\frac{2\mu E}{\hbar^2}\tilde{X}=0 \tag{6.127}$$

Suppose $E<0$, i.e.,

$$\frac{2\mu E}{\hbar^2}=-\alpha^2 \tag{6.128}$$

Here α is a real number. Equation 6.127 has two solutions:

$$C_1\exp\{-\alpha r\}\text{ and }C_2\exp\{\alpha r\} \tag{6.129}$$

So

$$\tilde{X}=C_1\exp\{-\alpha r\}+C_2\exp\{\alpha r\} \tag{6.130}$$

If we select the regular solution of Equation 6.103 at $r=0$, then generally all coefficients in Equation 6.130 are different from zero. Since $\tilde{X}$ should be finite, $C_2=0$:

$$\tilde{X}=C_1\exp\{-\alpha r\} \tag{6.131}$$

As a rule, this function should be dependent on the Azimuthal quantum number l and so the radial wave function for this case has the following asymptotic form:

$$R_l=C\frac{1}{r}\exp\{-\alpha r\} \tag{6.132}$$

The motion is finite if $E<0$ and the wave function exponentially tends to zero. So at infinity, the wave function tends to zero and we select it such that

$$\int_0^{\infty}|X(r)|^2\,dr=\text{const} \tag{6.133}$$

Such an expression corresponds to a discrete spectrum.

The general solution of Equation 6.103 imitates that in Equation 6.130 with two arbitrary constants which are independent. The discrete energy levels in E should be dependent on two quantum numbers l and n, i.e., the eigenvalue and eigenfunction should depend on the parameter l in Equation 6.101 as well as m. There is a greater dependence on the **principal** or **Bohr quantum number** of the corresponding state:

$$n=l+1,l+2,\ldots \tag{6.134}$$

This is at least one larger than the Azimuthal quantum number l, and $n-l-1$ is the number of zeros of the radial part of the wave function. The discrete levels for the motion of an electron in the central symmetric field are described by s, p, f-states with no 1p,1d and 2d states (see Figure 6.2).

The wave function $\Psi_{nlm}(r,\theta,\phi)$ for the motion in the central symmetric can be represented as

$$\Psi_{nlm}(r,\theta,\phi)=R_{nl}(r)Y_{lm}(\theta,\phi) \tag{6.135}$$

The energy level E for the motion in a central symmetric field is degenerate. The parameters n,l and m describing the wave function are the principal, orbital (azimuthal) and magnetic quantum

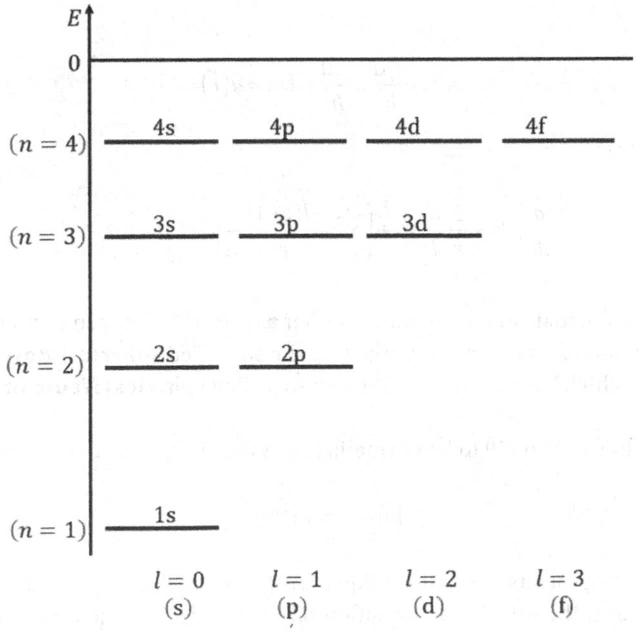

FIGURE 6.2 The discrete levels for the motion of an electron in a central symmetric field described by *s*, *p*, *f*-states.

numbers, respectively. Different values of m and E_{nlm} correspond to different wave functions and m for given l where the degree of degeneracy, f, is defined in Equation 6.99. The physical interpretation of degeneracy: **every number n, l and m corresponds to an energy.** The parameter l is from the square of the angular momentum and m is from the projection of the angular momentum in the z-axis direction. It is not possible to have a two-fold degenerate energy levels since we do not consider the spin of the particle; otherwise, it could have been possible. For the ground state, $l=0, n=1$, and for the first excited state, $l=0, n=2$, the level is non-degenerate or for $l=1, n=2$, the level is three-fold degenerate. The concrete value of the quantum numbers for the first excited state is dependent on the type of potential.

6.3.3 Continuous Spectra Radial Wave Functions

We examine the wave functions of a continuous spectrum which are different from those for the discrete spectrum. For the continuous spectrum, we have an infinite degeneracy relative to the Azimuthal quantum number l. This implies that one and the same energy value corresponds to radial functions for arbitrary l. So the general solution of the Schrödinger equation for $E \geq 0$ can be represented as a linear combination for arbitrary l. A simple example of this linear combination may be the linear combination of plane waves which are very useful in the scattering theory.

We examine the radial wave function for an arbitrary Azimuthal quantum number l described by the Schrödinger equation defined in Equation 6.98 as follows:

$$\frac{1}{r^2}\frac{d}{dr}\left(r^2\frac{d}{dr}R_l\right)-\frac{l(l+1)}{r^2}R_l+\frac{2\mu E}{\hbar^2}R_l=\frac{2\mu}{\hbar^2}U(r)R_l \tag{6.136}$$

We consider

$$\lim_{r\to\infty} rU(r)=0 \tag{6.137}$$

Letting

$$\kappa^2=\frac{2\mu E}{\hbar^2},\ \frac{2\mu}{\hbar^2}U(r)=u(r) \tag{6.138}$$

Equation 6.136 becomes

$$\frac{d^2}{dr^2}R_l+\frac{2}{r}\frac{d}{dr}R_l+\left(\kappa^2-\frac{l(l+1)}{r^2}-u(r)\right)R_l=0 \tag{6.139}$$

As $r\to 0$, the solution of Equation 6.139 behaves either as r^l or r^{-l-1}, as seen earlier. The regular solution at the neighborhood of $r=0$ behaving like r^l is the so-called **spherical Bessel function,** $\mathrm{j}_l(\kappa r)$, of order l and the one which blows up like r^{-l-1} is the so-called **spherical Neumann function,** $\mathrm{n}_l(\kappa r)$, of order l.

For the solution of Equation 6.139 to be normalizable as $r\to 0$, we consider the behavior r^l where

$$\lim_{r\to 0} R_{\kappa l}=\mathrm{j}_l(\kappa r) \tag{6.140}$$

This should be the homogeneous solution of Equation 6.139 which is regular at the origin, i.e., at 0. Note that for any finite r, the solution of Equation 6.139 considering Equation 6.140 is an analytical function of κ.

Examine the asymptotic solution $R_{\kappa l}$ of Equation 6.139 for $r \to \infty$ considering Equation 6.139 together with Equation 6.140. This permits us to write Equation 6.139 in an integral form with the first term being the homogeneous solution and the second being the non-homogeneous solution:

$$R_{\kappa l}(r) = \mathrm{j}_l(\kappa r) + \int_0^r g_l(r,r')u(r')R_{\kappa l}(r')dr' \tag{6.141}$$

Here, $g_l(r,r')$ is Green's function (resolvent) and satisfies the following equation:

$$\frac{d^2}{dr^2}g_l + \frac{2}{r}\frac{d}{dr}g_l + \left(\kappa^2 - \frac{l(l+1)}{r^2}\right)g_l = 0 \tag{6.142}$$

considering the following boundary conditions:

$$g_l(r,r) = 0, \frac{\partial}{\partial r}g_l(r,r')\Big|_{r=r'} = 1 \tag{6.143}$$

This implies that $g_l(r,r')$ is continuous at $r = r'$, while its derivative $\frac{\partial}{\partial r}g_l(r,r')$ is not. The singular solution of the Schrödinger equation (Equation 6.142) at the origin is not considered in the solution in Equation 6.141. So for $r < r'$, Green's function $g_l(r,r')$ must have no singularity at $r = 0$. But $\mathrm{n}_l(\kappa r)$ has the singularity of the type r^{-l-1} at $r = 0$ and so should be excluded from the solution. Green's function, $g_l(r,r')$, therefore should have the form:

$$g_l(r,r') = \frac{\kappa r'^2}{2i}\left(\mathrm{h}_l^{(1)}(\kappa r)\mathrm{h}_l^{(2)}(\kappa r') - \mathrm{h}_l^{(2)}(\kappa r)\mathrm{h}_l^{(1)}(\kappa r')\right),\ \mathrm{h}_l^{(1,2)}(\kappa r) \equiv \mathrm{j}_l(\kappa r) \pm i\mathrm{n}_l(\kappa r) \tag{6.144}$$

Here, $\mathrm{h}_l^{(1)}(\kappa r)$ and $\mathrm{h}_l^{(2)}(\kappa r)$ are called the **spherical Hankel functions** of the first and second kind, respectively (see appendix, Chapter 24). These functions have just the type of behavior at large distances that is expected of a bound state and in particular exponential decrease.

Introduce the **spherical Bessel function of order** l (see appendix, Chapter 25):

$$\mathrm{j}_l(\kappa r) = \frac{1}{2}\left(\mathrm{h}_l^{(1)}(\kappa r) + \mathrm{h}_l^{(2)}(\kappa r)\right) \tag{6.145}$$

then Equation 6.141 is represented in a more symmetric form:

$$R_{\kappa l}(r) = \frac{1}{2}\mathrm{h}_l^{(1)}(\kappa r)\left(1 - i\kappa\int_0^r \mathrm{h}_l^{(2)}(\kappa r')u(r')R_{\kappa l}(r')r'^2dr'\right) + \frac{1}{2}\mathrm{h}_l^{(2)}(\kappa r)\left(1 + i\kappa\int_0^r \mathrm{h}_l^{(1)}(\kappa r')u(r')R_{\kappa l}(r')r'^2dr'\right) \tag{6.146}$$

6.3.3.1 Jost Function

We consider in Equation 6.146 the **spherical Hankel functions** $\mathrm{h}_l^{(1),(2)}(\kappa r)$ which have acceptable behaviors at large distances $(r \to \infty)$ convenient for achieving bound states:

$$\mathrm{h}_l^{(1),(2)}(\kappa r) \sim \pm\frac{1}{i\kappa r}\exp\left\{i\left(\kappa r \pm \frac{l\pi}{2}\right)\right\} \tag{6.147}$$

and so

$$R_{\kappa l}(r) \sim \frac{1}{2i\kappa r}\left(\mathrm{I}_l(-\kappa)\exp\left\{i\left(\kappa r - \frac{l\pi}{2}\right)\right\} - \mathrm{I}_l(\kappa)\exp\left\{-i\left(\kappa r - \frac{l\pi}{2}\right)\right\}\right) \tag{6.148}$$

where the quantity, $\mathrm{I}_l(\kappa)$, is the so-called **Jost function [22]**:

$$\mathrm{I}_l(\kappa) = 1 + i\kappa\int_0^\infty \mathrm{h}_l^{(1)}(\kappa r')u(r')R_{\kappa l}(r')r'^2 dr' \tag{6.149}$$

Since $R_l(r)$ is real-valued:

$$\mathrm{I}_l(-\kappa) = \mathrm{I}_l^*(\kappa),\ \mathrm{I}_l(\kappa) = |\mathrm{I}_l(\kappa)|\exp\{-i\delta_l\} \tag{6.150}$$

and from Equation 6.148, we have the asymptotic radial function:

$$R_{\kappa l}(r) \sim \frac{1}{\kappa r}|\mathrm{I}_l(\kappa)|\sin\left(\kappa r - \frac{l\pi}{2} + \delta_l\right) \tag{6.151}$$

From the normalization condition of $R_{\kappa l}(r)$:

$$\int_0^\infty R_{\kappa' l}(r)R_{\kappa l}(r)r^2 dr = \frac{\pi}{2\kappa^2}|\mathrm{I}_l(\kappa)|^2\delta(E - E')\frac{d}{d\kappa}E \tag{6.152}$$

So normalizing the radial wave function $\langle r|E,l\rangle$ to the Dirac delta-function $\delta(E - E')$:

$$\langle r|E,l\rangle = \sqrt{\frac{2}{\pi}\kappa^2\frac{d\kappa}{dE}}\frac{1}{\mathrm{I}_l(\kappa)}R_{\kappa l}(r) \tag{6.153}$$

6.3.4 Delta Potential Radial Solution

Suppose the particle is subjected to the potential

$$U(r) = -\alpha\delta(r - a) \tag{6.154}$$

then we find the

- Energy when the azimuthal quantum number $l = 0$
- Binding energy for the particle in the shallow well under the condition:

$$\xi \equiv \frac{\mu\alpha a}{\hbar^2} \ll 1 \tag{6.155}$$

- Quasi-discrete energy levels of the s-state of the particle when $\alpha < 0$ and investigate the system under the condition: $\xi \gg 1$

For $\alpha < 0$, we find the wave function of the stationary state of the particle with $l = 0$ in the form:

$$\psi_{\kappa,l=0,m=0} = \frac{R_\kappa(r)}{r},\ \kappa^2 = \frac{2\mu E}{\hbar^2} \tag{6.156}$$

The radial wave function, $R_\kappa(r)$, can be obtained from the following equation:

$$-R''_\kappa + \frac{2\mu\alpha}{\hbar^2}\delta(r-a)R_\kappa = \kappa^2 R_\kappa \tag{6.157}$$

To find the quasi-discrete levels, we find the solution of Equation 6.157. For $r \to \infty$, this solution has the form:

$$R_\kappa(r) \approx \exp\{i\kappa r\} \tag{6.158}$$

This solution is feasible for some complex value of κ:

$$\kappa = \kappa_1 - i\kappa_2,\ E = E_0 - i\frac{\Gamma}{2},\ \Gamma = \frac{2\hbar^2\kappa_1\kappa_2}{\mu},\ \kappa_{1,2} > 0 \tag{6.159}$$

Here, E_0 is the energy level and Γ its breadth. The solution of Equation 6.157 satisfying the right boundary conditions has the form:

$$R_\kappa(r) = C_1\theta(r<a)\sin(i\kappa r) + C_2\theta(r>a)\exp\{i\kappa r\} \tag{6.160}$$

Here, $\theta(r)$ is the Heaviside function. The condition for matching the wave function at the point $r = a$:

$$i\kappa a - \kappa a \cot\kappa a = 2\xi \tag{6.161}$$

This describes the spectrum of quasi-discrete levels for $l = 0$. From $\xi \gg 1$, we get the values of κa for the lower levels from the roots (nodes) of Equation 6.161, when $|\kappa a| \ll \xi$:

$$\kappa_n a = (n+1)\pi + \epsilon_1 - i\epsilon_2,\ n = 0,1,\cdots;\ |\epsilon_{1,2}| \ll 1 \tag{6.162}$$

So from Equation 6.161, we find the following approximate values:

$$\epsilon_1 \cong -\frac{(n+1)\pi}{2\xi},\ \epsilon_2 \cong \epsilon_1^2 \cong \frac{(n+1)^2\pi^2}{4\xi^2},\ (E_0)_n \cong \frac{(n+1)^2\hbar^2\pi^2}{2\mu a^2}\left(1-\frac{1}{\xi}\right),\ \Gamma_n \cong \frac{(n+1)^3\hbar^2\pi^3}{2\mu a^2\xi^2} \tag{6.163}$$

The energy of quasi-discrete levels is closer to the particle's energy levels in an infinite deep potential well of radius a. For $\xi \to \infty$, when the potential barrier becomes non-transparent for the particle, the quasi-discrete levels make a transition to the corresponding levels of the discrete spectrum.

6.4 Motion in a Coulombic Field

6.4.1 Hydrogen Atom (Spherical Coordinates)

Consider the hydrogen atom which is a two-particle system consisting of an electron and a proton. In a simplified picture, the hydrogen atom is a system describing the motion of a lighter electron of charge $-e$ in the Coulomb potential of a heavier proton (which may probably be placed at the origin). This problem is solved exactly in the analytical form.

We represent the potential energy $-\frac{ze^2}{r}$ of two opposite charged particles z_1e and z_2e that are equal to $-z_1z_2$. The Schrödinger equation, for the motion of a particle in a Coulombic field, considers the nucleus of the hydrogen atom with the nuclear charge ze located at the origin of the coordinate axis. The hydrogen atom is the most important example of the motion of a particle in a Coulomb field. The electron and proton attract each other with the force $\frac{ze^2}{r^2}$. This corresponds to the potential energy, $-\frac{ze^2}{r}$. Here, $r=|r_1-r_2|$ is the coordinate of the relative motion or the separation between the proton and the electron. So our problem can be reduced to that of a single-particle motion in the center of mass frame. This permits us to isolate the Hamiltonian associated with the relative motion of two particles. So we consider the proton to have a center of mass which is stationary while the electron describes relative motion. The electron is bound to the nuclear field and the system imitates the classical situation of a particle moving in a Kepler ellipse.

Apart from hydrogen isotopes (deuterium and tritium), there exist hydrogen-type ions such as He^+, Li^{++} and so on. We write the radial function, $R(r)$, in the form:

$$R(r)=\frac{X(r)}{r} \tag{6.164}$$

where $X(r)$ satisfies the Schrödinger equation:

$$-\frac{\hbar^2}{2\mu}\frac{d^2X}{dr^2}+\frac{\hbar^2 l(l+1)}{2\mu r^2}X-\frac{ze^2}{r}X=EX \tag{6.165}$$

We find the solution of this equation in the sense of the Sturm-Liouville problem where the solution should be finite, continuous and univalent. Equation 6.165 is the second-order linear homogenous differential equation with coefficients dependent on r. This equation has no standard form for its solution. However, we select the method of mixed series and consider the transformation where we use the atomic units – the system of units based on the fundamental length scale which is the **Bohr radius** of the hydrogen atom:

$$a=\frac{\hbar^2}{\mu e^2}\cong 0.529\,\text{Å} \tag{6.166}$$

where $1\,\text{Å}=10^{-8}$ cm. The Bohr radius, a, determines the size of the hydrogen atom. So we introduce dimensionless length:

$$r=\rho a \tag{6.167}$$

with ρ being the length in units of the Bohr radius, a. The natural energy scale

$$W=\frac{\mu e^4}{2\hbar^2}=\frac{e^2}{2a}\cong 13.6\text{eV} \tag{6.168}$$

is the **binding energy of the hydrogen atom** in the ground state. It is also called the **Rydberg** and permits to introduce dimensionless energy:

$$E=W\varepsilon \tag{6.169}$$

with ε being the energy in units of W. If we substitute these parameters into Equation 6.165, then

$$X''+\left[\varepsilon-\frac{l(l+1)}{\rho^2}+\frac{2z}{\rho}\right]X=0 \tag{6.170}$$

We investigate the behavior of the solution at infinity, i.e., as $\rho \to \infty$:

$$\tilde{X}'' + \varepsilon \tilde{X} = 0 \tag{6.171}$$

where $\tilde{X}$ is the asymptotic wave function. The positive energy ε corresponds to a continuous spectrum-ionization state. The negative energy $\varepsilon < 0$ corresponds to a discrete spectrum (bound state):

$$\varepsilon = -\alpha^2, \alpha > 0 \tag{6.172}$$

where α is a real parameter. So the partial solution of

$$\tilde{X}'' - \alpha^2 \tilde{X} = 0 \tag{6.173}$$

is

$$\exp\{-\alpha\rho\} \text{ or } \exp\{\alpha\rho\} \tag{6.174}$$

We neglect the second solutions as at infinity it is infinite, i.e.,

$$\lim_{\rho \to \infty} \exp\{\alpha\rho\} \to \infty \tag{6.175}$$

Hence, we find the solution of Equation 6.170 in the form:

$$X(\rho) = \exp\{-\alpha\rho\} V(\rho) \tag{6.176}$$

where $V(\rho)$ is a new unknown function:

$$V'' - 2\alpha V' - \frac{l(l+1)}{\rho^2} V + \frac{2z}{\rho} V = 0 \tag{6.177}$$

In this equation, we find the right behavior of the wave function at infinity. We find the solution in the form of a power series. If

$$\lim_{\rho \to 0} \rho^2 V(\rho) \to 0 \tag{6.178}$$

then the function X behaves as ρ^{l+1}. It follows that we find the power series:

$$V(\rho) = \sum_{\kappa=0}^{\infty} a_\kappa \rho^{\kappa+l+1} \tag{6.179}$$

where a_K is constant coefficients. Substituting Equation 6.179 into 6.177, we find recurrence relations between the coefficients:

$$\sum_\kappa a_\kappa (\kappa+l+1)(\kappa+l)\rho^{\kappa+l+1} - 2\alpha \sum_\kappa a_\kappa (\kappa+l+1)(\kappa+l)\rho^{\kappa+l} - l(l+1)\sum_\kappa a_\kappa \rho^{\kappa+l-1} + 2z \sum_\kappa a_\kappa \rho^{\kappa+l} \equiv 0 \tag{6.180}$$

This holds for any value of ρ. In order for the sum of the power series of ρ to be zero, the coefficients for any power of ρ should be equal to zero. We select powers of $\kappa + l$:

$$a_{\kappa+1}\left[(\kappa+l+2)(\kappa+l+1)-l(l+1)\right]=a_{\kappa}\left[2\alpha(\kappa+l+1)-2z\right] \tag{6.181}$$

or

$$a_{\kappa+1}=\frac{2\left[\alpha(\kappa+l+1)-z\right]}{\left[(\kappa+l+2)(\kappa+l+1)-l(l+1)\right]}a_{\kappa} \tag{6.182}$$

This is the recursion formula for the coefficients of the series in Equation 6.179 and hence the function $V(\rho)$. It is necessary to set the value of a_0 which becomes an overall constant that can eventually be fixed by normalization. Consider, the following questions:

1. **If the series is convergent and for which radius, is it convergent?**
2. **To what is the series convergent?**

We infer the criterion of convergence from the coefficients in Equation 6.182 where if

$$\left.\frac{a_{\kappa+1}}{a_{\kappa}}\right|_{\kappa\to+\infty}=\frac{2}{\kappa}\to 0 \tag{6.183}$$

then the series is convergent for any ρ. We compare this by expanding $\exp\{2\alpha\rho\}$ in a power series:

$$\exp\{2\alpha\rho\}=\sum_{\kappa}b_{\kappa}\rho^{\kappa} \tag{6.184}$$

then we test the convergence from the ratio:

$$\left.\frac{b_{\kappa+1}}{b_{\kappa}}\right|_{\kappa\to\infty}=\frac{2}{\kappa}\to 0 \tag{6.185}$$

It follows that the solutions $V(\rho)$ and $\exp\{2\alpha\rho\}$ have the same behavior at infinity: The function

$$\left.\exp\{2\alpha\rho\}\right|_{\kappa\to\infty}\to\infty \tag{6.186}$$

This follows that $V(\rho)$ has a similar behavior at infinity. So the function $X(\rho)$ should also be divergent at infinity and our solution does not satisfy the condition at infinity, i.e., the condition that the wave function should be finite, which is unacceptable. So this result is unphysical and we need to find one physical result.

6.4.2 Eigenvalue and Eigenfunction

We select the parameters of Equation 6.182 such that for some $\kappa = n_r$, all the coefficients should be equal to zero. So $a_{\kappa} \neq 0, a_{\kappa+1} = 0$ and the series disintegrates and implies that the series is expected to tend to a polynomial which should be the convenient solution. So if in Equation 6.182 we let $\kappa = n_r$ and approach the result to zero, then

$$\alpha=\frac{z}{n_r+l+1},\quad l,n_r=0,1,2,\ldots \tag{6.187}$$

So to obtain the solution in the form of a polynomial, α is selected as in Equation 6.187, permitting the energy spectrum to become discrete.

Let

$$n = n_r + l + 1 \tag{6.188}$$

where n_r and n are, respectively, the **radial** and **principal quantum numbers**:

$$\alpha = \frac{z}{n} \tag{6.189}$$

From here:

$$E_n = -\frac{\mu e^4}{2\hbar^2}\frac{z^2}{n^2} = -\frac{z^2 W}{n^2} \tag{6.190}$$

The energy has a negative sign and implies that it takes a positive amount of work to remove the electron to infinity. When

$$U(r) = -\frac{ze^2}{r} \tag{6.191}$$

we select the zero point as the energy of an electron is located far from the proton. If the electron is close to the proton, its energy is less and below zero. The lowest level (**ground state**) is obtained when $n = 1$, and as n tends to infinity, the energy tends to zero.

The peculiarity of Equation 6.190 is the independence of the energy, E_n, on l, i.e., it has degeneracy with respect to l and also m: One and the same energy corresponds to eigenstates with different values of l. As $n - l - 1 \geq 0$, $0 \leq l \leq n - 1\left(-l \leq m \leq l\right)$. It follows that for a given energy, the total wave function may be represented as linear combinations of the corresponding eigenfunctions:

$$\Psi_n = \sum_{l=0}^{n-1}\sum_{m=-l}^{l} a_{lm} R_{nl}(r) Y_{lm}(\theta,\phi) \tag{6.192}$$

This type of wave function has no defined parity except the ground state $n = 1$ that is non-degenerate. The quantity a_{lm} is the normalization constant. It is expected that linear combinations of the degenerate hydrogen atom eigenfunctions having the same n and different l can be found corresponding to some new choice of coordinate system. This is evident when the hydrogen atom wave function is separated into parabolic coordinates. Generally, degeneracy exists when the wave equation is solved in more than one manner and involves different coordinate systems, or a single coordinate system oriented in different ways. When there is no degeneracy, the wave functions obtained from different coordinate systems would have to be identical except for a multiplicative constant and which is seldom possible. Exceptions occur for a general central symmetric field when $l = 0$ where the wave function is spherical symmetric and has the same form for all orientations of the polar axis so that there should be no degeneracy. The degeneracy with respect to m is characteristic of any central symmetric force field where $U(r)$ depends only on the radial distance r from some point. The l degeneracy is a characteristic of the Coulomb field.

It is obvious from Equation 6.188 that for a given l, the principal quantum number

$$n = l + 1, l = 2, \ldots \tag{6.193}$$

When $l = 0, n = 1$, then $n_r \neq 0$ and $n = 1, 2, \ldots$. The given value of n characterizes the so-called **electron shell**.

For the case of the hydrogen atom, when $l=0, n=1$, then the solutions obtained from the spherical and parabolic separation of the wave equation match. Equation 6.190 coincides exactly with Bohr's formula. If the Schrödinger theory gives the same result as Bohr for E_n, then there arises the question: **What is the advantage of the Schrödinger equation?** The Bohr theory gives only the eigenvalue E_n and the Schrödinger theory also gives, in addition to the eigenvalue, the eigenfunction that has important information and other physical properties of the system. Bohr gave an exact explanation to the spectral lines of the hydrogen atom and the exact value of the Rydberg constant. But the line spectra differ also in their intensity and Bohr could not explain this as it was necessary to have knowledge of the wave function.

6.4.2.1 Hydrogen Atom's Wave Function

A complete wave function is the product:

$$R_{nl}(r)Y_{lm}(\theta,\phi)=C_{nl}\exp\{-\alpha\rho\}\rho^l L_{n-l-1}(\rho)Y_{lm}(\theta,\phi) \tag{6.194}$$

where C_{nl} can be obtained from the condition of normalization. From here, we observe that the wave function depends on three quantum numbers m, n, l. The function $L_{n-l-1}(\rho)$ is called **Laguerre polynomials of order** $n-l-1$ and can be obtained from Equation 6.182 by substituting α from Equation 6.187:

$$R_{nl}(\xi)=C_{nl}\exp\left\{-\frac{\xi}{2}\right\}\xi^l L_{n+l}^{2l+1}(\xi),\ \xi=2\alpha\rho \tag{6.195}$$

where L_{n+l}^{2l+1} are the **associated Laguerre polynomials** of order $n+l$ which are $(2l+1)$-times differentiable over the variable ξ (see appendix, Chapter 25):

$$L_n^{|m|}(\xi)=\xi^{-|m|}\exp\{\xi\}\frac{d^n}{d\xi^n}\left(\xi^{|m|+n}\exp\{-\xi\}\right) \tag{6.196}$$

with

$$L_0^{|m|}(\xi)=1,\ L_1^{|m|}(\xi)=-\xi+(\xi+1),\ L_2^{|m|}(\xi)=\xi^2-2\xi(|m|+2)+(|m|+1)(|m|+2);\ L_3^{|m|}(\xi)$$
$$=-\xi^3+3\xi^2(|m|+3)-3\xi(|m|+2)(|m|+3)+(|m|+1)(|m|+2)(|m|+3);\ldots \tag{6.197}$$

If $|m|=0$, then we have the **Laguerre polynomials,** $L_n(\xi)$:

$$L_n(\xi)=\exp\{\xi\}\frac{d^n}{d\xi^n}\left(\xi^n\exp\{-\xi\}\right) \tag{6.198}$$

and

$$L_0(\xi)=1,\ L_1(\xi)=-\xi+1,\ L_2(\xi)=\xi^2-4\xi+2,\ L_3(\xi)=-\xi^3+9\xi^2-18\xi+6;\cdots \tag{6.199}$$

Apply Leibnitz formula to Equation 6.196:

$$L_n^{|m|}(\xi)=\sum_{k=0}^{n}(-1)^k\frac{\Gamma(n+1)\Gamma(|m|+n+1)}{\Gamma(k+1)\Gamma(n-k+1)\Gamma(|m|+k+1)}\xi^k \tag{6.200}$$

and

$$L_n(\xi) = \sum_{k=0}^{n} (-1)^k \frac{\Gamma^2(n+1)}{\Gamma^2(k+1)\Gamma(n-k+1)} \xi^k \tag{6.201}$$

We observe from here that the confluent hypergeometric functions are related to the associated and Laguerre polynomials, respectively:

$$L_n^{|m|}(\xi) = \frac{\Gamma(|m|+n+1)}{\Gamma(|m|+1)} F(-n,|m|+1;\xi),\ L_n(\xi) = n!F(-n,1;\xi) \tag{6.202}$$

From here, the associated Laguerre polynomials, $L_n^{|m|}(\xi)$, satisfy the equation:

$$\frac{d}{d\xi} L_n^{|m|}(\xi) = -nL_{n-1}^{|m|+1}(\xi) \tag{6.203}$$

Then effecting $(k-1)$ differentiation and from Equation 6.264, we have

$$\frac{d^k}{d\xi^k} L_n^{|m|}(\xi) = \frac{(-1)^k \Gamma(n+1)}{\Gamma(n-k+1)} L_{n-k}^{|m|+k}(\xi) = \sum_{m=0}^{n-k} \frac{(-1)^{k+m} \Gamma(n+1)\Gamma(|m|+n+1)}{\Gamma(m+1)\Gamma(n-k-m+1)\Gamma(|m|+k+m+1)} \xi^m \tag{6.204}$$

From here, if $k=n$, then

$$\frac{d^k}{d\xi^k} L_n^{|m|}(\xi) = (-1)^n n! \tag{6.205}$$

6.4.2.2 Laguerre Polynomials Integral Representation

From the residue theorem and the contour l enclosing the point $z=\xi$, we have the following:

$$\begin{aligned}\oint_l \frac{\exp\{-z\} z^{|m|+n} dz}{(z-\xi)^{n+1}} &= 2\pi i \operatorname*{Res}_{z=\xi} \frac{\exp\{-z\} z^{|m|+n}}{(z-\xi)^{n+1}} = \frac{2\pi i}{n!} \lim_{z\to\xi} \frac{d^n}{dz^n}\left(\exp\{-z\} z^{|m|+n}\right) \\ &= \frac{2\pi i}{n!} \frac{d^n}{d\xi^n}\left(\exp\{-\xi\} \xi^{|m|+n}\right) = \frac{2\pi i}{n!} \exp\{-\xi\} \xi^{|m|} L_n^{|m|}(\xi)\end{aligned} \tag{6.206}$$

From here,

$$\exp\{-\xi\} \xi^{|m|} L_n^{|m|}(\xi) = \frac{n!}{2\pi i} \oint_l \frac{\exp\{-z\} z^{|m|+n} dz}{(z-\xi)^{n+1}} \tag{6.207}$$

Consider the contour l_0 enclosing the point $t=0$ where we have the change of variable:

$$z = \frac{\xi}{1-t} \tag{6.208}$$

then

$$\exp\{-\xi\}\xi^{|m|}L_n^{|m|}(\xi)=\frac{n!}{2\pi i}\oint_{l_0}\frac{\xi^{|m|+n}\exp\left\{-\frac{t\xi}{1-t}\right\}\exp\{-\xi\}(1-t)^{n+1}\xi dt}{(1-t)^{|m|+n}\xi^{n+1}t^{n+1}(1-t)^2}\tag{6.209}$$

or

$$\frac{1}{n!}L_n^{|m|}(\xi)=\frac{1}{2\pi i}\oint_{l_0}\frac{\exp\left\{-\frac{t\xi}{1-t}\right\}dt}{(1-t)^{|m|+1}t^{n+1}}\tag{6.210}$$

6.4.2.3 Eigenvalue and Degeneracy

The eigenvalue, E_n, depends only on n. The shell in the hydrogen atom is characterized by n and is said to be n sub-shells each of which corresponds to one of the values of l. Each sub-shell contains $2l+1$ distinct states associated with $2l+1$ possible values of m for fixed l. So the total degeneracy of the energy level E_n is then

$$f_n=\sum_{l=0}^{n-1}\sum_{m=-l}^{l}m=\sum_{l=0}^{n-1}(2l+1)=2\frac{n(n-1)}{2}+n=n^2\tag{6.211}$$

This is the degree of degeneracy for the given energy eigenstate. If we consider the electron spin, then this is multiplied by a factor of 2 likewise taking into account the proton spin which is equal to that of the electron then we have another factor of 2. The ground state $n=1$ is non-degenerate. The state $n=2$ is degenerate with the degree of degeneracy 4. The eigenfunction and eigenvalue can also be obtained by revisiting again the radial Equation 6.98:

$$-\frac{\hbar^2}{2\mu r^2}\frac{d}{dr}\left(r^2\frac{d}{dr}R\right)+\frac{\hbar^2 l(l+1)}{2\mu r^2}R+U(r)R=ER\tag{6.212}$$

or

$$\rho\frac{d^2}{d\rho^2}R+2\frac{d}{d\rho}R+\left(-\frac{\rho}{4}+\sqrt{-\frac{\mu z^2e^4}{2\hbar^2E}}-\frac{l(l+1)}{\rho}\right)R=0,\ r=\rho\sqrt{-\frac{\hbar^2}{8\mu E}}\tag{6.213}$$

Letting the radial solution as

$$R=\rho^{-\frac{1}{2}}X\tag{6.214}$$

then

$$\rho\frac{d^2}{d\rho^2}X+\frac{d}{d\rho}X+\left(-\frac{\rho}{4}+n-\frac{\lambda^2}{4\rho}\right)X=0,\ \lambda=2l+1,\ n=\frac{\lambda+1}{2}+n_r=\sqrt{-\frac{\mu z^2e^4}{2\hbar^2E}}\tag{6.215}$$

The quantity, n_r, is the radial quantum number:

$$n=n_r+l+1\tag{6.216}$$

So

$$n = l+1, l+2, l+3, \dots \tag{6.217}$$

and the eigenenergy:

$$E_n = -\frac{\mu e^4}{2\hbar^2}\frac{z^2}{n^2} \tag{6.218}$$

and eigenfunction:

$$\Psi_{nlm}(r,\theta,\phi) = C_{nl}\exp\left\{-\frac{\rho}{2}\right\}\rho^l L_{n+l}^{2l+1}(\rho) Y_{lm}(\theta,\phi) \tag{6.219}$$

Remarks

The state has the same n and different l and m (the square of the angular momentum and its projection, respectively). This problem differs in others where E is dependent on n and l. The degeneracy with respect to n is called **accidental degeneracy.** We already know that for fixed l, there exists an infinite number of possible energy values for which each of them is $(2l+1)$-fold degenerate. This is called an **essential degeneracy** that is related with the fact that the radial equation is dependent only on the quantum number l and not on m. The degeneracy with respect to m is a characteristic of any central symmetric field for which the potential $U(r)$ is dependent on the radial distance r from some point. As distinguished from most central symmetric fields, the l degeneracy is characteristic of the Coulomb field. For problems like the motion of the valence electron of an alkali atom, the electron's potential energy is central and only approximates the Coulomb form. This prevents the n energy levels having the same total quantum number n and different l from coinciding and leading to the splitting of the nth hydrogen-type level into n distinct levels. Suppose the system is subjected to some external field such as a magnetic field (that destroys the spherical symmetry), then the $(2l+1)$-fold m degeneracy disappears, leading to the splitting of the nth hydrogen level into n^2 distinct levels.

6.4.3 Hydrogen Atom (Parabolic Coordinates)

For the motion in a central symmetric field, the method of separation of variables is always possible with the Schrödinger equation written in spherical coordinates. We examine the problem of a particle in a Coulomb field where the method of separation of variables is described by parabolic coordinates ξ, η and ϕ expressed by the following equations in terms of the spherical polar coordinates:

$$x = \sqrt{\xi\eta}\cos\phi,\ y = \sqrt{\xi\eta}\sin\phi,\ z = \frac{1}{2}(\xi-\eta),\ r = \sqrt{x^2+y^2+z^2} = \frac{1}{2}(\xi+\eta),\ 0 \le \xi < \infty,\ 0 \le \eta < \infty,\ 0 \le \phi \le 2\pi \tag{6.220}$$

or

$$\xi = r+z = r(1+\cos\theta),\ \eta = r-z = r(1-\cos\theta),\ \phi = \phi = \tan^{-1}\frac{y}{x} \tag{6.221}$$

The surfaces of constant ξ are a set of confocal paraboloids of revolution about the z or polar axes, with focus at the origin and opens in the direction of negative z or $\theta = \pi$, while, similarly, the surfaces of constant η are a set of confocal paraboloids that open in the direction of positive z or $\theta = 0$. The surfaces of constant ϕ are the same in the spherical coordinate system corresponding to planes through the polar axis. The elements of the length and volume are defined, respectively, by the expressions:

$$dl^2 = \frac{\xi+\eta}{4\xi}d\xi^2 + \frac{\xi+\eta}{4\eta}d\eta^2 + \xi\eta d\phi^2,\ dV = \frac{\xi+\eta}{4}d\xi d\eta d\phi \tag{6.222}$$

From Equation 6.222, the Laplace operator in the parabolic coordinates is defined as

$$\Delta = \frac{4}{\xi+\eta}\left[\frac{\partial}{\partial\xi}\left(\xi\frac{\partial}{\partial\xi}\right)+\frac{\partial}{\partial\eta}\left(\eta\frac{\partial}{\partial\eta}\right)\right]+\frac{1}{\xi\eta}\frac{\partial^2}{\partial\phi^2} \tag{6.223}$$

The Coulomb potential energy U in parabolic coordinates is defined as

$$U = -\frac{Ze^2}{r} = -\frac{2Ze^2}{\xi+\eta} \tag{6.224}$$

The Schrödinger equation for the hydrogen atom in parabolic coordinates is defined as

$$-\frac{\hbar^2}{2\mu}\left\{\frac{4}{\xi+\eta}\left[\frac{\partial}{\partial\xi}\left(\xi\frac{\partial}{\partial\xi}\Psi\right)+\frac{\partial}{\partial\eta}\left(\eta\frac{\partial}{\partial\eta}\Psi\right)\right]+\frac{1}{\xi\eta}\frac{\partial^2}{\partial\phi^2}\Psi\right\}-\frac{2Ze^2}{\xi+\eta}\Psi = \mathrm{E}\Psi,\ E<0 \tag{6.225}$$

Using the method of separation of variables:

$$\Psi = f_1(\xi)f_2(\eta)\Phi(\phi) \tag{6.226}$$

where substituting into Equation 6.225, we have

$$\frac{4\xi\eta}{\xi+\eta}\left[\frac{1}{f_1}\frac{d}{d\xi}\left(\xi\frac{d}{d\xi}f_1\right)+\frac{1}{f_2}\frac{d}{d\eta}\left(\eta\frac{d}{d\eta}f_2\right)\right]+\frac{4\mu Ze^2\xi\eta}{\hbar^2(\xi+\eta)}-\frac{2\mu|\mathrm{E}|\xi\eta}{\hbar^2} = \frac{1}{\Phi}\frac{d^2}{d\phi^2}\Phi \tag{6.227}$$

The left-hand side of Equation 6.227 is dependent on ξ and η, while the right-hand side is dependent only on ϕ. Both sides of Equation 6.227 should be equal to the constant m^2 and this permits us to have the function:

$$\Phi_m(\phi) = \frac{1}{\sqrt{2\pi}}\exp\{im\phi\},\ m = 0,\pm1,\pm2,\ldots \tag{6.228}$$

Here, m is the magnetic quantum number. We separate Equation 6.227 to the parts dependent on ξ and η:

$$\frac{1}{f_1}\frac{d}{d\xi}\left(\xi\frac{d}{d\xi}f_1\right)-\frac{\mu|E|}{2\hbar^2}\xi-\frac{m^2}{4\xi}+\frac{\mu Ze^2}{\hbar^2} = -\left[\frac{1}{f_2}\frac{d}{d\eta}\left(\eta\frac{d}{d\eta}f_2\right)-\frac{\mu|E|}{2\hbar^2}\eta-\frac{m^2}{4\eta}\right] = \beta \tag{6.229}$$

The separation constant β can be determined from the boundary conditions, and so the equations for f_1 and f_2 can be rewritten in the form:

$$\frac{\mathrm{d}}{d\xi}\left(\xi\frac{d}{d\xi}f_1\right)-\left(\frac{\mu|E|}{2\hbar^2}\xi+\frac{m^2}{4\xi}-\frac{\mu Ze^2}{\hbar^2}+\beta\right)f_1 = 0,\ \frac{d}{d\eta}\left(\eta\frac{d}{d\eta}f_2\right)-\left(\frac{\mu|E|}{2\hbar^2}\eta+\frac{m^2}{4\eta}-\beta\right)f_2 = 0 \tag{6.230}$$

6.4.3.1 Energy Levels

It is instructive to note that the above two equations are the same and differ only by a constant term. So it is sufficient to solve only one equation and then apply to both. Letting

$$\alpha^2 = \frac{2\mu|E|}{\hbar^2},\ \beta_1 = \frac{1}{\alpha}\left(\frac{\mu Ze^2}{\hbar^2}-\beta\right),\ \rho_1 = \alpha\xi,\ \rho_2 = \alpha\eta,\ \alpha^2 = \frac{2\mu|E|}{\hbar^2},\ \beta_2 = \frac{\beta}{\alpha} \tag{6.231}$$

Equation 6.230 becomes

$$\frac{1}{\rho_1}\frac{d}{d\rho_1}\left(\rho_1\frac{d}{d\rho_1}f_1\right)+\left(\frac{\beta_1}{\rho_1}-\frac{1}{4}-\frac{m^2}{4\rho_1^2}\right)f_1=0,\ \frac{1}{\rho_2}\frac{d}{d\rho_2}\left(\rho_2\frac{d}{d\rho_2}f_2\right)+\left(\frac{\beta_2}{\rho_2}-\frac{1}{4}-\frac{m^2}{4\rho_2^2}\right)f_2=0 \tag{6.232}$$

In the first equation of Equation 6.232, the function f_1 for $\rho_1 \to \infty$ behaves as $\exp\left\{-\frac{\rho_1}{2}\right\}$, and for $\rho_1 \to 0$, it behaves as $\rho_1^{\frac{|m|}{2}}$ and, similarly, in the second equation of Equation 6.232, the solution of Equation 6.232 can be represented in the following form:

$$f_1(\rho_1)=\rho_1^{\frac{|m|}{2}}F_1(\rho_1)\exp\left\{-\frac{\rho_1}{2}\right\},\ f_2(\rho_2)=\rho_2^{\frac{|m|}{2}}F_2(\rho_2)\exp\left\{-\frac{\rho_2}{2}\right\} \tag{6.233}$$

Substituting Equation 6.233 into 6.232, we have the following equation:

$$\rho_1\frac{d^2}{d\rho_1^2}F_1(\rho_1)+(|m|+1-\rho_1)\frac{d}{d\rho_1}F_1(\rho_1)+n_1F_1(\rho_1)=0,\ \rho_2\frac{d^2}{d\rho_2^2}F_2(\rho_2)$$

$$+(|m|+1-\rho_2)\frac{d}{d\rho_2}F_2(\rho_2)+n_2F_2(\rho_2)=0 \tag{6.234}$$

Here, the following are positive integers or zero and are the so-called **parabolic quantum numbers**:

$$n_1=\beta_1-\frac{|m|+1}{2},\ n_2=\beta_2-\frac{|m|+1}{2} \tag{6.235}$$

The solution of Equation 6.234 has two poles with a regular one at $\rho_1=0$ and the other irregular at $\rho_1=\infty$. It is instructive to note that for

$$|m|+1\neq 0,-1,-2,\cdots \tag{6.236}$$

the solution of Equation 6.234 satisfying the finite boundary conditions has the form:

$$F_1(\rho_1)=F(-n_1,|m|+1;\rho_1) \tag{6.237}$$

Here, $F(-n_1,|m|+1;\rho_1)$ is the confluent hypergeometric function where the series from D'Alembert rule is convergent for any of the variables $-n_1, |m|+1$ and ρ_1. Equation 6.312 is solved in a similar manner where the above analysis applies. So three integral quantum numbers n_1, n_2 and m define every stationary state of a discrete spectrum in the parabolic system of coordinates. From Equations 6.235 and 6.314, we have the following principal quantum number (non-zero positive integer) n:

$$\beta_1+\beta_2=n_1+n_2+|m|+1\equiv n \tag{6.238}$$

For given n, the number $|m|$ may take n different values from 0 to $n-1$. For fixed n and $|m|$, the number n_1 takes $n-|m|$ values from 0 to $n-|m|-1$. We also consider that for given $|m|$, we also select the function with $m=\pm|m|$. So we find that for given n, we have always the following number of different states (total degeneracy of the energy level E_n):

$$2\sum_{|m|=1}^{n-1}(n-|m|)+(n-0)=2\left[n(n-1)-\frac{n(n-1)}{2}\right]+(n-0)=n^2 \tag{6.239}$$

The degenerate energy level E_n can be obtained by considering Equations 6.231 and 6.238:

$$E_n = -|E_n| = -\frac{\hbar^2\alpha^2}{2\mu} = -\frac{\mu Z^2 e^4}{2\hbar^2 n^2} \quad (6.240)$$

6.4.3.2 Wave Functions

The wave function $\Psi_{n_1 n_2 m}$ describing the discrete spectrum has the form:

$$\Psi = C_n (\alpha\xi)^{\frac{|m|}{2}} \mathrm{F}(-n_1, |m|+1; \alpha\xi) \exp\left\{-\frac{\alpha\xi}{2}\right\} (\alpha\eta)^{\frac{|m|}{2}} \mathrm{F}(-n_2, |m|+1; \alpha\eta) \exp\left\{-\frac{\alpha\eta}{2}\right\} \frac{1}{\sqrt{2\pi}} \exp\{im\phi\} \quad (6.241)$$

and should be normalized:

$$\int |\Psi_{n_1 n_2 m}|^2 dV = \frac{1}{4} \int_0^\infty \int_0^\infty \int_0^{2\pi} |\Psi_{n_1 n_2 m}|^2 (\xi + \eta) d\xi d\eta d\phi = 1 \quad (6.242)$$

The normalized wave function has the form:

$$\Psi_{n_1 n_2 m} = \frac{\sqrt{2}}{n^2} f_{n_1 m}\left(\frac{\xi}{n}\right) f_{n_2 m}\left(\frac{\eta}{n}\right) \frac{1}{\sqrt{2\pi}} \exp\{im\phi\},$$

$$f_{nm}(\rho) = \frac{1}{|m|!} \sqrt{\frac{(|m|+n)!}{n!}} \exp\left\{-\frac{\rho}{2}\right\} \rho^{\frac{|m|}{2}} \mathrm{F}(-n, |m|+1; \rho), \; n = \frac{1}{\sqrt{-2E}} \quad (6.243)$$

The confluent hypergeometric functions $\mathrm{F}(-n, |m|+1; \rho)$ are related to the associated Laguerre polynomials, $L_n^{|m|}$:

$$L_n^{|m|}(\rho) = C_n \mathrm{F}(-n, |m|+1; \rho) \quad (6.244)$$

Here, C_n is an arbitrary constant that can be found in our case when the wave function is normalized.

For a given energy level, E_n, and magnetic quantum number $(n > |m|)$, the parabolic quantum numbers n_1 and n_2 can be selected such that

$$n_1 + n_2 = n - |m| - 1 \quad (6.245)$$

This implies selection in $n - |m|$ different ways. Similarly, for given n and m, the Azimuthal quantum number l for the spherical solution can be selected:

$$|m| \le l \le n - 1 \quad (6.246)$$

This also implies the selection in $n - |m|$ different ways. So the $n - |m|$ product of the functions of ξ and η are linear combinations of the $n - |m|$ product of the function of r and θ. The ground-state energy shows an intimate connection between the parabolic and spherical solutions, where the parabolic solution is $\exp\left\{-\frac{MZe^2}{2\hbar^2}(\xi + \eta)\right\}$ for

$$n_1 = n_2 = m = 0 \quad (6.247)$$

and the spherical solution is $\exp\left\{-\frac{\mu Ze^2}{\hbar^2}r\right\}$ for

$$n=1, l=m=0 \tag{6.248}$$

It is obvious from Equation 6.221 that the given two solutions are identical for the given geometries.

The wave function of the first excited state for $n=2$ has the form:

$$\Psi_{100}=\frac{1}{\sqrt{16\pi}}\left(1-\frac{\xi}{2}\right)\exp\left\{-\frac{\xi+\eta}{4}\right\} \tag{6.249}$$

In order to find the distribution function of the electron over the z-coordinate, it is convenient to move from the variables ξ, η and ϕ to the cylindrical coordinates ρ, z and ϕ. We then integrate over the plane perpendicular to the z-axis:

$$\begin{aligned}W(z)dz&=\int_0^\infty\int_0^{2\pi}\left|\Psi_{100}\right|^2\rho d\rho d\phi=\frac{dz}{8}\int_0^\infty\rho d\rho\left(1-\frac{\sqrt{\rho^2+z^2}+z}{2}\right)^2\exp\left\{-\sqrt{\rho^2+z^2}\right\}\\&=\frac{\exp\{-|z|\}}{8}\left(\left(\left(\frac{|z|+z}{2}\right)^2+\frac{1}{2}\right)(|z|+1)-|z|\frac{|z|+z}{2}\right)dz\end{aligned} \tag{6.250}$$

With the help of Equation 6.250, we find the probability of finding the electron in the domain $z\geq 0$ and $z\leq 0$:

$$P(z\geq 0)=\int_0^\infty W(z)dz=\frac{7}{8},\ P(z\leq 0)=\frac{1}{8} \tag{6.251}$$

It is instructive to note that the wave functions in the parabolic coordinates are asymmetric relative to the plane $z=0$, so that for $n_1>n_2$ and $n_1<n_2$, there is the probability of localization of the particle, respectively, in the domains $z\leq 0$ and $z\leq 0$.

6.4.4 Spherical Oscillator (Spherical Coordinates)

VARIANT 1

We apply Equation 6.98 to the case of a spherical oscillator:

$$-\frac{\hbar^2}{2\mu r^2}\frac{d}{dr}\left(r^2\frac{d}{dr}R\right)+\frac{\hbar^2 l(l+1)}{2\mu r^2}R+\frac{Kr^2}{2}R=ER \tag{6.252}$$

We introduce the new variable as follows:

$$\rho=\frac{\mu\omega}{\hbar}r^2,\ K=\mu\omega^2 \tag{6.253}$$

then Equation 6.212 becomes

$$\rho\frac{d^2}{d\rho^2}R+\frac{3}{2}\frac{d}{d\rho}R+\left(\frac{\epsilon}{4}-\frac{l(l+1)}{4\rho}-\frac{\rho}{4}\right)R=0,\ E=\frac{\hbar\omega\epsilon}{2} \tag{6.254}$$

Considering the asymptotic solutions for the motion in the central symmetric field (for the cases seen above), the solution of Equation 6.254 can be selected in the form:

$$R(r)=\exp\left\{-\frac{\rho}{2}\right\}\rho^{\frac{l}{2}}V(\rho) \quad (6.255)$$

Substituting Equation 6.255 into 6.254, we have the equation:

$$\rho\frac{d^2}{d\rho^2}V+(\beta-\rho)\frac{d}{d\rho}V-\alpha V=0 \quad (6.256)$$

Here

$$\alpha=-\frac{\epsilon}{4}+\frac{l}{2}+\frac{3}{4},\ \beta=l+\frac{3}{2} \quad (6.257)$$

Equation 6.256 has a regular asymptotic point $\rho\to 0$ and an irregular asymptotic point $\rho\to\infty$. For $\rho\to 0$, the asymptotic wave function R has the form:

$$R(r)\sim\rho^{\frac{l}{2}} \quad (6.258)$$

So

$$\lim_{\rho\to 0}V(\rho)\to \text{const} \quad (6.259)$$

and the solution of Equation 6.256 for $\beta\neq 0,-1,-2,\ldots$ has the form:

$$V(\rho)=CF\,(\alpha,\beta;\rho) \quad (6.260)$$

Here, $F\,(\alpha,\beta;\rho)$ is the confluent hypergeometric function and matches a polynomial for $\rho\to\infty$ only when the following condition is satisfied:

$$\alpha=-n_r,\ n_r=0,1,2,\ldots \quad (6.261)$$

This defines the energy level of the particle:

$$E_{n_r l}=\hbar\omega\left(l+2n_r+\frac{3}{2}\right)\equiv\hbar\omega\left(N+\frac{3}{2}\right),\ N=2n_r+l=0,1,2,\ldots \quad (6.262)$$

The wave function corresponding to this energy level is as follows:

$$\Psi_{n_r lm}(r,\theta,\phi)=R_{n_r}(r)Y_{lm}(\theta,\phi)=C_{n_r l}r^l\exp\left\{-\frac{\mu\omega r^2}{2\hbar}\right\}F\left(-n_r,l+\frac{3}{2};\frac{\mu\omega r^2}{2\hbar}\right)Y_{lm}(\theta,\phi) \quad (6.263)$$

For a given value of N corresponding to the state of the particle with angular moment l, $l=N,N-2,\ldots$. So the degree of degeneracy is then

$$\sum_{l=N,N-2,\ldots}(2l+1)=\frac{(N+1)(N+2)}{2} \quad (6.264)$$

The classification of the eigenvalues and eigenfunctions by the quantum numbers n_r, l and m shows that we are concerned with a problem in a central symmetric field. The ground state is defined by the quantum numbers, $n_r = 0$ and $= 0$. The eigenfunction Ψ_{000} does not have nodes and is spherically symmetric. This implies that it corresponds to zero.

VARIANT 2

The wave function for the spherical symmetric oscillator as any central symmetric problem can be obtained in the form:

$$\Psi(r,\theta,\phi) = R(r)\mathrm{P}_{lm}(\cos\theta)\exp\{im\phi\} \equiv R(r)\mathrm{Y}_{lm}(\theta,\phi) \tag{6.265}$$

Here, $\mathrm{P}_{lm}(\cos\theta)$ is the Legendre polynomials, where

$$l = 0,1,2,\ldots, m = 0,\pm1,\ldots,\pm l \tag{6.266}$$

The function $R(r)$ satisfies the following Schrödinger equation:

$$-\frac{\hbar^2}{2\mu}\left[\frac{d^2}{dr^2} + \frac{2}{r}\frac{d}{dr} - \frac{l(l+1)}{r^2}\right]R + \frac{\mu\omega^2 r^2}{2}R = ER,\ \mathrm{K} = \mu\omega^2 \tag{6.267}$$

We introduce the following radial function $\mathrm{X}(r)$:

$$R(r) = \frac{\mathrm{X}(r)}{r} \tag{6.268}$$

and also do the following change of variable:

$$\rho = \sqrt{\frac{\mu\omega}{\hbar}}r,\ \epsilon = \frac{E}{\hbar\omega} \tag{6.269}$$

then the Schrödinger equation (Equation 6.267) becomes

$$\frac{d^2}{d\rho^2}\mathrm{X} + \left[2\epsilon - \frac{l(l+1)}{\rho^2} - \rho^2\right]\mathrm{X} = 0 \tag{6.270}$$

We find the asymptotic solution at an irregular asymptotic point when $\rho \to \infty$:

$$\tilde{\mathrm{X}} = \exp\left\{-\frac{\rho^2}{2}\right\} \tag{6.271}$$

The behavior at zero which is a regular asymptotic point when $\rho \to 0$:

$$\tilde{\mathrm{X}} = \rho^s \tag{6.272}$$

The value of s can be obtained from the equation:

$$s(s-1) = l(l+1) \tag{6.273}$$

Solving this equation, we have

$$s_1 = l+1,\ s_2 = -l \tag{6.274}$$

So for the behavior when $\rho \to 0$, we have only $s_1 = l+1$:

$$\tilde{X} = \rho^{l+1} \tag{6.275}$$

From the above solutions, we have to find the solution of Equation 6.270 in the form:

$$X = \exp\left\{-\frac{\rho^2}{2}\right\}\rho^{l+1}U(\rho) \tag{6.276}$$

This solution is substituted into Equation 6.270:

$$\frac{d^2}{d\rho^2}U + 2\left(\frac{l(l+1)}{\rho} - \rho\right)\frac{d}{d\rho}U + 2\left(\epsilon - l - \frac{3}{2}\right)U = 0 \tag{6.277}$$

We find the solution in the form:

$$U(\rho) = \sum_{k=k_0}^{\infty} a_k \rho^k \tag{6.278}$$

We equate the coefficients for each power of ρ and obtain the following recurrent relation:

$$a_{k+2} = \frac{2\left(k + l + \frac{3}{2} - \epsilon\right)}{(k+2)(k+2l+3)} a_k \tag{6.279}$$

For convergence, we have

$$\lim_{k\to\infty} \frac{a_{k+2}}{a_k} = \lim_{k\to\infty} \frac{2\left(k + l + \frac{3}{2} - \epsilon\right)}{(k+2)(k+2l+3)} = \frac{2}{k} \tag{6.280}$$

This permits us to find the least power k_0 from the equation:

$$k_0(k_0+1) - 2(l+1)k_0 = 0 \tag{6.281}$$

This follows that the finite value at zero will be the solution with $k_0 = 0$. From Equation 6.280, it follows that when $\rho \to \infty$, this series should be that of the function $\exp\{\rho^2\}$ and $U(\rho) \to \exp\left\{\frac{\rho^2}{2}\right\}$ and implies divergence of the series. This divergence can be resolved when the series is convergent for

$$\epsilon = p + l + \frac{3}{2} \tag{6.282}$$

So if we consider the fact that the series should start from $k_0 = 0$, then p in the given expression should be even:

$$p = 2n_r,\ n_r = 1, 2, \ldots \tag{6.283}$$

Hence

$$\epsilon = 2n_r + l + \frac{3}{2} \tag{6.284}$$

and the energy level:

$$E_{n_r l} = \hbar\omega\left(2n_r + l + \frac{3}{2}\right) \tag{6.285}$$

The wave function has the form:

$$\Psi_{n_r lm}(r,\theta,\phi) = C_{n_r l}\exp\left\{-\frac{\rho^2}{2}\right\}\rho^l Y_{lm}(\theta,\phi)F\left(-n_r, l+\frac{3}{2};\frac{\rho^2}{2}\right) \tag{6.286}$$

6.4.5 Particle in an Infinite Deep Spherical Symmetric Potential Well

We investigate the states of a particle in an infinite deep spherical symmetric potential, $U(r)$, dependent on the radial coordinate:

$$U(r) = \begin{cases} 0, & r < a \\ \infty, & r > a \end{cases} \tag{6.287}$$

As we have seen earlier, the wave function of a particle in a central symmetric field is as follows:

$$\Psi_{n_r lm}(r,\theta,\phi) = R_{n_r l}(r)Y_{lm}(\theta,\phi) \tag{6.288}$$

The radial function, $R_{n_r l}(r)$, satisfies the following Schrödinger equation:

$$-\frac{\hbar^2}{2\mu r^2}\frac{d}{dr}\left(r^2\frac{d}{dr}R_{n_r l}\right) + \frac{\hbar^2 l(l+1)}{2\mu r^2}R_{n_r l} + U(r)R_{n_r l} = ER_{n_r l} \tag{6.289}$$

For $r > a$, the radial solution will be $R_{n_r l} = 0$ since $U = \infty$. So the boundary condition for the solution of Equation 6.289 should be

$$R_{n_r l}(a) = 0 \tag{6.290}$$

We select the solution of Equation 6.289 in the form:

$$R_{n_r l}(r) = \frac{X_{n_r l}(r)}{\sqrt{r}} \tag{6.291}$$

then the Schrödinger equation (Equation 6.289) becomes

$$\frac{d^2}{dr^2}X_{n_r l} + \frac{1}{r}\frac{d}{dr}X_{n_r l} + \left(\chi_{n_r l}^2 - \frac{\left(l+\frac{1}{2}\right)^2}{r^2}\right)X_{n_r l},\ \chi_{n_r l}^2 = \frac{2\mu E_{n_r l}}{\hbar^2} \tag{6.292}$$

The solution of this equation has the form:

$$X_{n_r l} = C_1 j_{l+\frac{1}{2}}(\chi_{n_r l} r) + C_2 n_{l+\frac{1}{2}}(\chi_{n_r l} r) \tag{6.293}$$

Since

$$\lim_{r\to 0} n_{l+\frac{1}{2}}(\chi_{n_r l} r) \to -\infty \tag{6.294}$$

$C_2 = 0$, and for $r \to 0$, the function $X_{n_r l}$ is as follows:

$$X_{n_r l} = C j_{l+\frac{1}{2}}(\chi_{n_r l} r) \to r^{\pm\left(l+\frac{1}{2}\right)} \tag{6.295}$$

So, for the function to be finite for $l = 0$, we have

$$j_{\frac{1}{2}}(\chi_{n_r l} r) = \sqrt{\frac{2}{\pi^3 r}} \sin \chi_{n_r,0} r \tag{6.296}$$

The energy levels can be obtained from the following boundary condition:

$$X_{n_r l}(0) = X_{n_r l}(a) = 0 \tag{6.297}$$

from where, we have

$$E_{n_r l} = \frac{\hbar^2 \alpha^2_{n_r+1,l}}{2\mu a^2}, \ \chi_{n_r l} a = \alpha_{n_r+1,l} \tag{6.298}$$

Here, $\alpha_{n_r l}$ is the nth root in the increasing order and not considering the root for $x = 0$ of the Bessel function $j_{l+\frac{1}{2}}(x)$. This implies $j_{l+\frac{1}{2}}(\alpha_{n_r l}) = 0$. Particularly, the ground state corresponds to $n_r = 0, l = 0$, where $\alpha_{10} = \pi$ and

$$j_{l+\frac{1}{2}}(x) = \sqrt{\frac{2}{\pi x}} \sin x \tag{6.299}$$

From $\alpha_{n_r l} = n_r \pi$:

$$E^{(0)}_{n_r} = \frac{\hbar^2 \pi^2 n_r^2}{2\mu a^2} \tag{6.300}$$

6.4.6 Kepler Problem in Two Dimensions

VARIANT 1

We investigate the energy and wave function of a particle with the potential:

$$U(\rho) = -\frac{\alpha}{\rho}, \ Ze^2 \equiv \alpha, \ \rho = \sqrt{x^2 + y^2} \tag{6.301}$$

The Laplace operator for such a problem:

$$\Delta = \frac{\partial^2}{\partial \rho^2} + \frac{1}{\rho}\frac{\partial}{\partial \rho} + \frac{1}{\rho^2}\frac{\partial^2}{\partial \phi^2} \tag{6.302}$$

Since $U(\rho)$ is independent of the angle ϕ, the variables are separable in such a problem. So the wave function can be selected in the form:

$$\Psi_{n_\rho |m|}(\rho,\phi) = \exp\{im\phi\} R_{n_\rho |m|}(\rho) \tag{6.303}$$

For this problem, the operators of the Hamiltonian, $\hat{\mathrm{H}}$, and the angular momentum, $\hat{L}_z$, in the z-axis direction commute. The Schrödinger equation for the radial function $\mathrm{X}_{n_\rho |m|}(\rho)$ has the form:

$$-\frac{\hbar^2}{2\mu}\left(\frac{d^2}{d\rho^2} + \frac{1}{\rho}\frac{d}{d\rho} - \frac{m^2}{\rho^2}\right) R_{n_\rho |m|} - \frac{\alpha}{\rho} R_{n_\rho |m|} = E_{n_\rho |m|} R_{n_\rho |m|} \tag{6.304}$$

We find the solution when $E < 0$. It should be noted that for $E > 0$, we have a continuous spectrum. Letting

$$R_{n_\rho |m|}(\rho) = \frac{\mathrm{X}_{n_\rho |m|}(\rho)}{\sqrt{\rho}} \tag{6.305}$$

Equation 6.304 becomes

$$-\frac{\hbar^2}{2\mu}\left(\frac{d^2}{d\rho^2} - \frac{m^2 - \frac{1}{4}}{\rho^2}\right) \mathrm{X}_{n_\rho |m|} - \frac{\alpha}{\rho} \mathrm{X}_{n_\rho |m|} = E_{n_\rho |m|} \mathrm{X}_{n_\rho |m|} \tag{6.306}$$

This imitates the equation of the motion in a Coulombic field seen in Equation 6.165. So Equation 6.306 has the following boundary conditions as those of the problem of the Coulombic field:

$$\mathrm{X}_{n_\rho |m|}(0) = 0,\ \mathrm{X}_{n_\rho |m|}(\infty) = 0 \tag{6.307}$$

Comparing Equation 6.306 to that of Coulombic field seen in Equation 6.165:

$$l(l+1) \to m^2 - \frac{1}{4} \tag{6.308}$$

The energy level compared to that from Equation 6.165:

$$E_n = -\frac{\mu\alpha^2}{2\hbar^2 (n_r + l + 1)^2} \to E_{n_\rho |m|} = -\frac{\mu\alpha^2}{2\hbar^2 \left(n_\rho + |m| + \frac{1}{2}\right)^2},\ n_\rho = 0,1,2,\ldots,\ |m| = 0,1,2,\ldots \tag{6.309}$$

From here, we observe an accidental degeneracy since the energy is dependent only on the combination $n_\rho + |m|$ of the quantum numbers n_ρ and $|m|$.

Suppose

$$N = n_\rho + |m| + 1 \tag{6.310}$$

then

$$E_N = -\frac{\mu\alpha^2}{2\hbar^2\left(N-\frac{1}{2}\right)^2},\ N=1,2,\ldots \tag{6.311}$$

The energy level, E_N, has the following degeneracy:

$$g(N)=1+\sum_{|m|=1}^{N-1}2=2N-1 \tag{6.312}$$

Variant 2

Considering Equations 6.304 and 6.305, and letting

$$\rho=\beta\xi,\ \beta=\frac{\hbar^2}{\mu\alpha},\ \frac{2\mu E_{n_\rho|m|}}{\hbar^2}\beta^2=-\chi^2_{n_\rho|m|} \tag{6.313}$$

Equation 6.304 becomes

$$\frac{d^2}{d\xi^2}X_{n_\rho|m|}+\left(\frac{2}{\xi}-\chi^2_{n_\rho|m|}-\frac{m^2-\frac{1}{4}}{\xi^2}\right)X_{n_\rho|m|}=0 \tag{6.314}$$

We find now the asymptotic solution of Equation 6.314 when $\xi\to\infty$:

$$\tilde{X}_{n_\rho|m|}=\exp\left\{\pm\chi_{n_\rho|m|}\xi\right\} \tag{6.315}$$

We consider only the convergent solution when $\xi\to\infty$:

$$\tilde{X}_{n_\rho|m|}=\exp\left\{-\chi_{n_\rho|m|}\xi\right\} \tag{6.316}$$

If $E>0$, then we have a continuous spectrum and $\chi_{n_\rho|m|}=i\lambda_{n_\rho|m|}$. So Equation 6.315 becomes

$$\tilde{X}_{n_\rho|m|}=\exp\left\{\pm i\lambda_{n_\rho|m|}\xi\right\} \tag{6.317}$$

We find the asymptotic solution of Equation 6.314 when $\xi\to 0$:

$$\tilde{X}_{n_\rho|m|}=\xi^s \tag{6.318}$$

This permits us to have

$$s(s-1)=m^2-\frac{1}{4} \tag{6.319}$$

with

$$s_1=m+\frac{1}{2},\ s_2=-\left(m-\frac{1}{2}\right) \tag{6.320}$$

From

$$m = 0, \pm 1, \pm 2, \ldots \tag{6.321}$$

then we select

$$s = |m| + \frac{1}{2} \tag{6.322}$$

This permits us to select the solution of Equation 6.314 in the form:

$$\tilde{X}_{n_\rho|m|} = \xi^{|m|+\frac{1}{2}} \exp\{-\chi_{n_\rho|m|}\xi\} V(\rho) \tag{6.323}$$

We substitute this function into Equation 6.314 and have

$$\xi \frac{d^2}{d\xi^2} V + 2\left(|m| + \frac{1}{2} - \chi_{n_\rho|m|}\xi\right)\frac{d}{d\xi} V + 2\left(1 - \left(|m| + \frac{1}{2}\right)\chi_{n_\rho|m|}\right) V = 0 \tag{6.324}$$

We find the solution in the form of a series:

$$V(\xi) = \sum_{k=k_0}^{\infty} a_k \xi^k \tag{6.325}$$

We equate the coefficient to zero for the least power of ξ, i.e., ξ^{k_0-1} and we have

$$k_0\left(k_0 + 2|m|\right) = 0 \tag{6.326}$$

From here, we have

$$k_0 = 0, \; k_0 = -2|m| \tag{6.327}$$

The second equation for $\xi \to 0$ gives a divergent solution. The coefficient for ξ^k gives the recurrent equation:

$$a_{k+2} = 2\frac{\left(k + |m| + \frac{1}{2}\right)\chi_{n_\rho|m|} - 1}{(k+1)(k+1+2|m|)} a_k \tag{6.328}$$

For convergence, we have

$$\lim_{k\to\infty} \frac{a_{k+2}}{a_k} = \lim_{k\to\infty} 2\frac{\left(k + |m| + \frac{1}{2}\right)\chi_{n_\rho|m|} - 1}{(k+1)(k+1+2|m|)} = \frac{2\chi_{n_\rho|m|}}{k} \tag{6.329}$$

From Equation 6.329, it follows that when $\xi \to \infty$, this series should be that of the function $\exp\{2\chi_{n_\rho|m|}\xi\}$ and $V(\xi) \to \exp\{2\chi_{n_\rho|m|}\xi\}$ and implies divergence of the series. This divergence can be resolved when the series is convergent for

$$\left(n + |m| + \frac{1}{2}\right)\chi_{n_\rho|m|} - 1 = 0 \tag{6.330}$$

From here and considering the third equation in Equation 6.313, the energy spectrum is obtained:

$$E_{n_\rho|m|} = -\frac{\mu Z^2 e^4}{2\hbar^2\left(n+|m|+\frac{1}{2}\right)^2} \tag{6.331}$$

The wave function has the form:

$$\Psi_{n_\rho|m|}(\rho,\phi) = \frac{1}{\sqrt{\beta\xi}}\xi^{|m|+\frac{1}{2}}\exp\left\{-\chi_{n_\rho|m|}\xi\right\}V(\xi) \tag{6.332}$$

7

Representation Theory

Though the wave formulation has proven successful in describing quantum mechanics, some quantum mechanical properties cannot be characterized by a wave-like description. For example, the electron spin degree of freedom cannot be translated by the action of a gradient operator and so the necessity to reformulate quantum mechanics in a framework of only operators. Note that selecting a representation implies selecting an orthonormal basis which can either be discrete or continuous in the given state space. Vectors and operators are then represented in the given basis as **numbers** with components imitating vectors and matrix elements imitating operators. It is worthy to note that the selection of a representation is arbitrary and of course depends on the given problem leading to the simplest algebra.

We have examined the Schrödinger equation in some cases of physical interest and now we examine a different formulation of quantum mechanics where the dynamical variables (i.e., coordinates, momentum components, energy, etc.) of a particle appear explicitly in the equations of motion of the system without their having to multiply or differentiate a wave function. Classical equations have just this structure. Here we might expect more resemblance between the classical and quantum formalism than the Schrödinger theory. The main formal difference here is that the quantum dynamical variables do not obey the commutative law of multiplication. What is more convenient to represent such non-commutative dynamical variables is that the non-commutative dynamical variables which are operators can be represented by matrices.

7.1 Matrix Wave Functions and Operator Representation

The matrix theory provides one more flexible representation since with them: there exist an arbitrary large number of ways of choosing their rows and columns. For convenience, we consider a system with a discrete spectrum, though subsequent relations can be generalized to include the continuous spectrum. We evaluate the mean value of some physical operators $\hat{A}$ in some state, Ψ:

$$\overline{A} = \int \Psi^* \hat{A} \Psi \, dq \tag{7.1}$$

We select another operator $\hat{B}$ which does not commute with the operator $\hat{A}$ and has the eigenfunction $\Psi_n = |n\rangle$ (of a discrete spectrum of stationary states) being also the basis function in the series expansion of the arbitrary function, Ψ:

$$\Psi = \sum_n C_n \Psi_n = \sum_n C_n |n\rangle \tag{7.2}$$

DOI: 10.1201/9781003273073-8

This should be the **energy representation** since the arbitrary wave function Ψ is series expanded in terms of the eigenfunctions of the energy or of the Hamiltonian $\hat{H}$. Substituting Equation 7.2 into 7.1:

$$\bar{A}=\int\sum_{m}C_m^*\Psi_m^*\hat{A}\sum_{n}C_n\Psi_n\,dq=\sum_{m,n}C_m^*\int\Psi_m^*\hat{A}\Psi_n\,dqC_n=\sum_{m,n}C_m^*A_{mn}C_n,\ A_{mn}=\int\Psi_m^*\hat{A}\Psi_n\,dq=\langle m|\hat{A}|n\rangle \quad (7.3)$$

All the quantities A_{mn} can be represented in the form of a table with m denoting rows and n columns

$$\mathbb{A}=\begin{bmatrix} A_{11} & A_{12} & \cdots & A_{1n} \\ A_{21} & A_{22} & \cdots & A_{2n} \\ \cdots & \cdots & \cdots & \cdots \\ A_{m1} & A_{m2} & \cdots & A_{mn} \end{bmatrix} \quad (7.4)$$

which is the matrix of the operator $\hat{A}$ in the $\hat{B}$-representation. **So, we observe that an operator can be written in matrix form to map one basis vector to another and as the operators are linear, the matrix is a linear transformation between bases.**

We observe that for a Hermitian operator, the eigenfunctions corresponding to different eigenvalues are orthogonal. The orthogonality permits a suitable basis set of vectors to represent the state of the quantum system. In matrix form, operators permit real eigenvalues to be obtained, corresponding to measurements. If we consider Equation 7.5, then it is obvious that the quantities are dependent on the selection of the wave functions Ψ_n and Ψ_m which are selected as the eigenfunctions of the operator, $\hat{B}$. **The complete set of eigenfunctions of some operator used for the construction of the matrix is called the basis of the representation.** So choosing a representation implies choosing an orthonormal basis either discrete or continuous in state space. Numbers then represents vectors and operators in this basis: components for vectors, matrix elements for operators.

7.2 Properties of Matrices

The matrix which is the **complex conjugate matrix** of $\mathbb{A}$ is defined as follows:

$$\mathbb{A}^* = \left(\mathbb{A}_{mn}^*\right) \quad (7.5)$$

The **transposed matrix** $\tilde{\mathbb{A}}$ of $\mathbb{A}=\left(\mathbb{A}_{mn}\right)$ is as follows:

$$\tilde{\mathbb{A}}=\tilde{\mathbb{A}}_{mn}=\mathbb{A}_{nm} \quad (7.6)$$

Here the **rows and columns of the matrix are interchanged (transpose matrix) and the complex conjugate taken as follows:**

$$\mathbb{A}_{mn}^\dagger=\tilde{\mathbb{A}}_{mn}^*=\mathbb{A}_{nm}^* \quad (7.7)$$

So

$$\mathbb{A}=\mathbb{A}^\dagger \quad (7.8)$$

and the matrix, $\mathbb{A}$, is **self-adjoint** or **Hermitian.** It is obvious that only square matrices can be Hermitian. We show that a Hermitian matrix represents a Hermitian operator:

$$\mathrm{A}_{mn} = \int \Psi_m^* \hat{\mathrm{A}} \Psi_n \, dq = \int \Psi_n \hat{\mathrm{A}}^* \Psi_m^* \, dq = \left(\int \Psi_n^* \hat{\mathrm{A}} \Psi_m \, dq \right)^* = \mathrm{A}_{nm}^* \tag{7.9}$$

Find the complex conjugate of the matrix element in Equation 7.9:

$$\left(\mathrm{A}^*\right)_{mn} = \int \Psi_n^* \hat{\mathrm{A}}^\dagger \Psi_m \, dq = \int \Psi_n^* \tilde{\hat{\mathrm{A}}}^* \Psi_m \, dq = \int \Psi_m \hat{\mathrm{A}}^* \Psi_n^* \, dq \tag{7.10}$$

So

$$\left(\mathbb{A}^*\right)_{nm} = \left(\mathbb{A}_{mn}\right)^* \tag{7.11}$$

and for real physical quantities, we have

$$\mathbb{A}_{nm} = \mathbb{A}_{mn}^* \tag{7.12}$$

Example
Examine the following matrix:

$$\mathbb{A} = \begin{bmatrix} 0 & i \\ -i & 0 \end{bmatrix} \tag{7.13}$$

then

$$\tilde{\mathbb{A}} = \mathbb{A}^* = \mathbb{A} \tag{7.14}$$

and implying that $\mathbb{A}$ is a Hermitian matrix.

Note that Hermitian matrices represent Hermitian operators where any two elements symmetrical with respect to the principal diagonal are complex conjugates of each other. From linear algebra of real matrices, Hermitian operators are simply symmetric matrices with elements of the principal diagonal different from zero:

$$\mathrm{A}_{mn} = \mathrm{A}_n \delta_{mn} \tag{7.15}$$

Proof
Consider Ψ_m and Ψ_n as eigenfunctions of the operator $\hat{\mathrm{A}}$. So from the Sturm–Liouville equation:

$$\hat{\mathrm{A}} \Psi_n = \mathrm{A}_n \Psi_n \tag{7.16}$$

and A_n is the eigenvalue that helps in calculating the following matrix element:

$$\mathrm{A}_{mn} = \int \Psi_m^* \hat{\mathrm{A}} \Psi_n \, dq = \int \Psi_m^* \mathrm{A}_n \Psi_n \, dq = \mathrm{A}_n \int \Psi_m^* \Psi_n \, dq = \mathrm{A}_n \delta_{mn} \tag{7.17}$$

So the question of the solution for the eigenvalues of an operator transforms its matrix to diagonal form. Hence, the diagonal elements of a Hermitian matrix are always real numbers.

If $\hat{A}$ and $\hat{B}$ are non-commutative operators, then the matrix A in the B-representation is not a diagonal matrix. A unit matrix

$$\mathbb{I} = \begin{bmatrix} 1 & 0 & \cdots & 0 \\ 0 & 1 & \cdots & 0 \\ \cdots & \cdots & \cdots & \cdots \\ 0 & 0 & \cdots & 1 \end{bmatrix} \tag{7.18}$$

is a square matrix where only elements on the principal diagonal are different from zero and each of them is equals to unity. It may be proven that

$$\mathbb{A}\mathbb{I} = \mathbb{A} \tag{7.19}$$

If

$$\mathbb{A}\mathbb{A}^{-1} = \mathbb{I} \tag{7.20}$$

then the matrix $\mathbb{A}^{-1}$ is called the inverse of the matrix $\mathbb{A}$.

7.3 Rule on Matrix Operations

1. Addition Law
   Suppose

$$\hat{A} = \hat{C} + \hat{D} \tag{7.21}$$

   then we show that the matrix corresponding to $\hat{A}$ is the sum of the matrices $\hat{C}$ and $\hat{D}$:

$$A_{mn} = \int \Psi^*_m \hat{A} \Psi_n \, dq = \int \Psi^*_m \left(\hat{C} + \hat{D}\right) \Psi_n \, dq = \int \Psi^*_m \hat{C} \Psi_n \, dq + \int \Psi^*_m \hat{D} \Psi_n \, dq = C_{mn} + D_{mn} \tag{7.22}$$

2. Multiplication Law
   Let

$$\hat{A} = \hat{C}\hat{D} \tag{7.23}$$

   then multiplication of matrices is defined as follows:

$$A_{mn} = \sum_k C_{mk} D_{kn} \tag{7.24}$$

   We show that the matrices of the operators $\hat{A}$, $\hat{C}$ and $\hat{D}$ fulfil the same relation:

$$A_{mn} = \int \Psi^*_m \hat{A} \Psi_n \, dq = \int \Psi^*_m \left(\hat{C}\hat{D}\right) \Psi_n \, dq = \langle m | \hat{C}\hat{D} | n \rangle = \sum_k \langle m | \hat{C} | k \rangle \langle k | \hat{D} | n \rangle = \sum_k C_{mk} D_{kn} \tag{7.25}$$

3. Distributive Law of Multiplication
   The distributive law of multiplication is defined as follows:

$$\hat{A}\left(\hat{B}+\hat{C}\right)=\hat{A}\hat{B}+\hat{A}\hat{C} \tag{7.26}$$

4. Associative Law of Multiplication
   The associative law of multiplication is valid as well:

$$\hat{A}\left(\hat{B}\hat{C}\right)=\left(\hat{A}\hat{B}\right)\hat{C} \tag{7.27}$$

   The product in Equation 7.27 can be written simply as

$$\hat{D}=\hat{A}\hat{B}\hat{C} \tag{7.28}$$

   Then from Equation 7.28, we have

$$D_{mn}=\left(\hat{A}\hat{B}\hat{C}\right)_{mn}=\langle m|\hat{A}\hat{B}\hat{C}|n\rangle=\sum_{k,l}\langle m|\hat{A}|k\rangle\langle k|\hat{B}|l\rangle\langle l|\hat{C}|n\rangle=\sum_{k,l}A_{mk}B_{kl}C_{ln} \tag{7.29}$$

   It is obvious from Equation 7.23 that $\hat{C}\hat{D}$ is not in general equal to $\hat{D}\hat{C}$.

In addition to these matrices (square matrices), an important place in quantum mechanics is played by row or column matrices obtained from the wave function:

$$\Psi(q)=\sum_n C_n\Psi_n(q) \tag{7.30}$$

where the basis function, $\Psi_n(q)$, is the eigenfunction of the operator $\hat{A}$ and

$$C_n=\int\Psi(q)\Psi_n^*(q)dq \tag{7.31}$$

This shows a mutual and single-valued correspondence of the coefficients, C_n, and the wave function $\Psi(q)$. So as the wave function $\Psi(q)$ describes the state of a system, it can be said that the complete selection of the coefficients C_n describes the state of the given system which is the wave function in the A-representation:

$$\Psi=\begin{bmatrix} C_1 \\ C_2 \\ \vdots \\ C_n \end{bmatrix} \tag{7.32}$$

Suppose we have the vector

$$\vec{R}=R_x\vec{\xi}+R_y\vec{\eta}+R_z\vec{\zeta} \tag{7.33}$$

where $\vec{\xi},\vec{\eta},\vec{\zeta}$ are unit vectors in the coordinate axis and $R_{x,yz}$ is the projection of the $\vec{R}$ on the respective coordinate axis:

$$R_x = \left(\vec{R},\vec{\xi}\right) \tag{7.34}$$

In the configuration space for the unit vector $\vec{\xi}$, we have $\left(\vec{\xi},\vec{\xi}\right)=1$, while for the functional space:

$$\int \left|\Psi_n(q)\right|^2 dq = 1 \tag{7.35}$$

so, $\Psi_n(q)$ is a unit vector and C_n is the projection of the vector Ψ on $\Psi_n(q)$. Such is the functional space.

In place of Ψ^*, we introduce the matrix $\Psi^\dagger$:

$$\Psi^\dagger = \begin{bmatrix} C_1^* & C_2^* & \cdots & C_n^* \end{bmatrix} \tag{7.36}$$

Further, on such a matrix, we execute the rule of multiplication if it is assumed to be a square matrix and all the rest of the elements are zero. So

$$\Psi^\dagger\Psi = \begin{bmatrix} \sum_n |C_n|^2 & 0 & 0 & \cdots & 0 \\ 0 & 0 & 0 & \cdots & 0 \\ \vdots & \vdots & \vdots & \vdots & \vdots \\ 0 & 0 & 0 & \cdots & 0 \end{bmatrix} \tag{7.37}$$

It should be noted that if the wave function $\Psi(q)$ is normalized, i.e.,

$$\int \left|\Psi(q)\right|^2 dq = 1 \tag{7.38}$$

then

$$\sum_n |C_n|^2 = 1 \tag{7.39}$$

The condition of normalization of the wave function in the matrix representation:

$$\Psi^\dagger\Psi = \mathbb{I} \tag{7.40}$$

7.4 Action of an Operator on a Wave Function

For the wave function Ψ as well as Φ, there should exist a correspondence, i.e., we expand the function Φ through the orthogonal functions Ψ_n. If we consider

$$\Phi = \hat{A}\Psi \tag{7.41}$$

then we may have expanded the wave functions Ψ and Φ in Equation 7.41 in terms of eigenfunctions of the Hamiltonian, $\hat{H}$:

$$\hat{H}\Psi_n = E_n\Psi_n \tag{7.42}$$

So, we have

$$\Psi=\sum_n C_n\Psi_n=\sum_n C_n|n\rangle,\ \Phi=\sum_m b_m\Psi_m=\sum_m b_m|m\rangle \tag{7.43}$$

The set of the coefficients C_n and b_n give, respectively, the energy representation of the functions Ψ and Φ. Let us find the relation between C_n and b_n. We evaluate the integral below as follows:

$$\int\Psi_m^*\Phi dq=\sum_n b_n\int\Psi_m^*\Psi_n dq=\sum_n b_n\delta_{mn}=b_m \tag{7.44}$$

We find the relation between C_n and b_n through the matrix, $\mathbb{A}_{mn}$, considering Equation 7.41:

$$b_m=\int\Psi_m^*\Phi dq=\int\Psi_m^*\hat{A}\Psi dq=\int\Psi_m^*\hat{A}\sum_n C_n\Psi_n dq=\sum_n\langle m|\hat{A}C_n|n\rangle=\sum_n\langle m|\hat{A}|n\rangle C_n=\sum_n A_{mn}C_n \tag{7.45}$$

This equation is the **energy representation** of Equation 7.41 and the set of A_{mn} is the matrix $\mathbb{A}_{mn}$ constituting the **energy representation** of the operator, $\hat{A}$:

$$\Phi\equiv\begin{bmatrix}b_1\\b_2\\\vdots\\b_n\end{bmatrix}=\begin{bmatrix}A_{11}&A_{12}&\cdots&A_{1n}\\A_{21}&A_{22}&\cdots&A_{2n}\\\cdots&\cdots&\cdots&\cdots\\A_{m1}&A_{m2}&\cdots&A_{mn}\end{bmatrix}\Psi\equiv\mathbb{A}\Psi \tag{7.46}$$

Here, $\boldsymbol{\Phi}$ and $\boldsymbol{\Psi}$ are, respectively, the **matrix representations** of the wave functions Φ and Ψ.

7.5 Mean Value of an Operator

Evaluate the expectation value, $\overline{A}$, of the operator $\hat{A}$ in the Ψ state through the matrix representation:

$$\overline{A}=\int\Psi^*\hat{A}\Psi dq=\sum_{n,k}\int C_n^*\Psi_n^*\hat{A}C_k\Psi_k dq=\sum_{n,k}C_n^*\langle n|\hat{A}|k\rangle C_k=\sum_{n,k}C_n^*A_{nk}C_k \tag{7.47}$$

or from Equations 7.32 and 7.36:

$$\overline{A}=\Psi^\dagger\mathbb{A}\Psi \tag{7.48}$$

It is important to note that in addition to the coordinate and momentum representations, there exist other representations for expressing quantum-mechanical relations. Knowledge of this heading will be more applicable for describing spin and spin systems.

7.6 Eigenstate and Eigenvalue Problem

The eigenfunctions and eigenvalues can be obtained through the Sturm–Liouville problem:

$$\hat{A}\Psi=A\Psi \tag{7.49}$$

and in a matrix form:

$$\hat{A}\Psi = A\Psi \tag{7.50}$$

We represent the eigenfunction, Ψ, of the operator, $\hat{A}$, through the basis function (or eigenfunction) Ψ_n of the operator, $\hat{B}$:

$$\Psi = \sum_n C_n \Psi_n \tag{7.51}$$

So

$$\hat{A}\Psi = \sum_n C_n \hat{A}\Psi_n = A\sum_n C_n \Psi_n \tag{7.52}$$

Determining the eigenvalues and eigenvectors of the operator $\hat{A}$, we multiply Equation 7.52 by Ψ^*_m, and integrate over the configuration coordinate, q:

$$\sum_n C_n \int \Psi^*_m \hat{A}\Psi_n \, dq = A\sum_n C_n \int \Psi^*_m \Psi_n \, dq = A\sum_n C_n \delta_{mn} \equiv \sum_n C_n A_{mn} \tag{7.53}$$

or

$$\sum_n C_n \left(A_{mn} - A\delta_{mn}\right) = 0 \tag{7.54}$$

For the trivial solution, all the coefficients C_n are zero (i.e., when Equation 7.54 is non-singular) and for non-trivial solutions of Equation 7.54, its determinant is equal to zero (characteristic equation for A) (i.e., when Equation 7.54 is singular):

$$\det\left(A_{mn} - A\delta_{mn}\right) = 0 \tag{7.55}$$

or

$$\mathbb{D}_m(A) = \begin{vmatrix} A_{11} - A & A_{12} & \cdots & A_{1m} \\ A_{21} & A_{22} - A & \cdots & A_{2m} \\ \vdots & \ddots & \vdots & \vdots \\ A_{m1} & A_{m2} & \cdots & A_{mm} - A \end{vmatrix} = 0 \tag{7.56}$$

Equation 7.56 is the so-called **characteristic equation** and the polynomial $\mathbb{D}_m(A)$ is the **characteristic polynomial**. The eigenvalues of $\hat{A}$, labelled A_n, are the n roots of the characteristic polynomial which are not necessarily distinct. The **spectrum** of $\hat{A}$ is the set of eigenvalues $\{A_n\}$. The spectrum is said to be **degenerate** if a given eigenvalue is a multiple root of the characteristic polynomial. Note that from algebra, every polynomial of degree n has n roots which are not necessarily distinct in the complex domain. If we substitute the eigenvalues A_n into Equation 7.54, then after solving it, we have a set of coefficients C_n matching each A_n. We may have the **geometric multiplicity** of an eigenvalue, A, where the number of linearly independent eigenvectors are consistent with A while the **algebraic multiplicity** of the given root is the number of times it is repeated in the solution of the characteristic equation.

7.7 Unitary Transformation in State Vector Space

We have seen the eigenfunction of an operator in a given representation can be represented in matrix form. Changing the representation, the same eigenfunction or operator is represented by a different matrix. In a further development, we show how these matrices in different representations relate each other.

Suppose the operator $\hat{A}$ is given in the representation with basis functions, Ψ_n, which are the eigenfunctions of the operator $\hat{B}$, i.e.,

$$\hat{B}\Psi_n = B_n\Psi_n \tag{7.57}$$

then

$$A_{mn} = \int \Psi_m^* \hat{A}\Psi_n \, dq \tag{7.58}$$

is the B-**representation** of the operator $\hat{A}$. We can also have the representation of $\hat{A}$ with eigenfunctions Φ_α of $\hat{C}$, i.e.,

$$\hat{C}\Phi_\alpha = C_\alpha\Phi_\alpha \tag{7.59}$$

and the C-**representation** of the operator $\hat{A}$:

$$A_{\alpha\beta} = \int \Phi_\alpha^* \hat{A}\Phi_\beta \, dq \tag{7.60}$$

We determine the transformation matrix that relates (A_{mn}) with $(A_{\alpha\beta})$. From the B-representation with the eigenfunctions, Ψ_n, and the C-representation with the eigenfunctions, Φ_α, we find the transformation matrices from one representation to the other. We expand Ψ_n with respect to the basis functions of the C-representation:

$$\Psi_n = \sum_\beta S_{\beta n}\Phi_\beta \tag{7.61}$$

Multiply Equation 7.61 by Φ_α^* and the result integrated over the configuration coordinate q:

$$\int \Phi_\alpha^* \Psi_n \, dq = \sum_\beta S_{\beta n} \int \Phi_\alpha^* \Phi_\beta \, dq = \sum_\beta S_{\beta n}\delta_{\alpha\beta} = S_{\alpha n} \tag{7.62}$$

The matrix element $S_{\alpha n}$ is the projection of Φ_α onto the state Ψ_n or the transformation matrix, and from matrix representation, we have the matrix $\mathbb{S}$:

$$\Psi = \mathbb{S}\Phi \tag{7.63}$$

where Ψ is a vector of state in the $\mathbb{B}$-representation and Φ is the vector of the same state in the $\mathbb{C}$-representation, while $\mathbb{S}$ is the transformation matrix which we show further to be a unitary matrix. So Equation 7.63 consists of a change in the basis vectors which, geometrically, corresponds to a rotation of the coordinate frame in the state vector space.

7.7.1 Unitary Matrix

We find the properties of the transformation matrix, $\mathbb{S}$, by considering Equation 7.61 and the orthogonality of the basis functions, Ψ_n and Ψ_m^*:

$$\delta_{mn} = \int \Psi_m^* \Psi_n \, dq = \sum_{\alpha,\beta} S_{\alpha n} S_{\beta m}^* \int \Phi_\beta^* \Phi_\alpha \, dq = \sum_{\alpha,\beta} S_{\alpha n} S_{\beta m}^* \delta_{\beta\alpha} = \sum_{\alpha} S_{\alpha n} S_{\alpha m}^* = \sum_{\alpha} S_{m\alpha}^\dagger S_{\alpha n} \tag{7.64}$$

From Equation 7.64, we observe that Ψ_n form an orthonormal (or orthogonal) basis spanning the state vector space. From the element of the adjoint matrix:

$$\left(\tilde{S}\right)_{\alpha m}^* = \left(S_{m\alpha}^\dagger\right) \tag{7.65}$$

we use the rule of matrix multiplication:

$$\delta_{mn} = \sum_{\alpha} S_{m\alpha}^\dagger S_{\alpha n} = \left(S^\dagger S\right)_{mn} \tag{7.66}$$

where δ_{mn} is the matrix element of a unit matrix and so

$$\mathbb{S}^\dagger \mathbb{S} = \mathbb{I} \tag{7.67}$$

The matrix that has the property in Equation 7.67 **is called a unitary matrix** and the transformation in Equation 7.63 is called **unitary**.

Equation 7.67 may be written in another form:

$$\mathbb{S}\mathbb{S}^{-1} = \mathbb{I} \tag{7.68}$$

where $\mathbb{S}^{-1}$ is an inverse matrix of $\mathbb{S}$. Equation 7.68 is the definition of the inverse matrix and can be rewritten as

$$\mathbb{S}^\dagger = \mathbb{S}^{-1} \tag{7.69}$$

This is the property of the unitary matrix. **So a matrix is unitary if its Hermitian adjoint is equal to its inverse.** From Equation 7.68, $\mathbb{S}$ is said to be **non-singular**, since it possesses an inverse, otherwise it is **singular**.

The physical interpretation of the unitary transformation is the conservation of the probability. Suppose a particle is in the Ψ_n state with probability 1, then in the Φ_β state it can be found with probability $\left|S_{\beta n}\right|^2$. So for the set

$$\left|S_{\beta 1}\right|^2, \ldots, \left|S_{\beta n}\right|^2 \tag{7.70}$$

the probability distribution of the given particle over the Φ_β states permits to obtain

$$\sum_n \left|S_{\beta n}\right|^2 = \sum_n S_{\beta n}^* S_{\beta n} = 1 \tag{7.71}$$

This confirms the unitarity of $\mathbb{S}$ from Equation 7.67.

7.7.2 Matrix Element of a Transformation Operator

We consider again the operator $\hat{A}$ of a physical quantity and relate its matrix elements to the eigenfunctions Ψ_n and Φ_α in given representations and considering Equation 7.58:

$$A_{mn} = \int \Psi_m^* \hat{A} \Psi_n \, dq = \sum_{\alpha,\beta} S_{\alpha m}^* S_{\alpha n} \int \Phi_\beta^* \hat{A} \Phi_\alpha \, dq = \sum_{\alpha,\beta} S_{\beta m}^* S_{\alpha n} A_{\beta\alpha} \tag{7.72}$$

If we use the rule of matrix multiplication, then

$$A_{mn} = \left(S^\dagger \tilde{A} S\right)_{mn} \tag{7.73}$$

where $\tilde{A}$ is the operator A in the C-representation and A is the operator in the $\hat{B}$-representation. So

$$\mathbb{A} = \mathbb{S}^\dagger \tilde{\mathbb{A}} \mathbb{S} \tag{7.74}$$

is the transformation formula for matrices (or operators) for the transition from one representation to another.

It is obvious that the given physically accessible quantities are invariant relative to the transformation. We show that the transformed matrix (operator) $\mathbb{A}$ retains the Hermitian property of $\tilde{\mathbb{A}}$. From Equations 7.73 and 7.74, we have

$$\mathbb{A} = \mathbb{S}^{-1} \tilde{\mathbb{A}} \mathbb{S} \tag{7.75}$$

Considering Equation 7.74 with the help of Equation 7.63, we obtain the formula of the inverse transformation by multiplying Equation 7.63 from the left-hand side by $\mathbb{S}^\dagger$:

$$\mathbb{S}^\dagger \Psi = \mathbb{S}^\dagger \mathbb{S} \Phi = \Psi \tag{7.76}$$

So

$$\Phi = \mathbb{S}^\dagger \Psi = \mathbb{S}^{-1} \Psi \tag{7.77}$$

Multiply Equation 7.74 on the left-hand side by $\mathbb{S}$ and on the right-hand side by $\mathbb{S}^\dagger$:

$$\tilde{\mathbb{A}} = \mathbb{S}\mathbb{A}\mathbb{S}^\dagger \tag{7.78}$$

The unitary transformations are very important as the **quantum mechanical equation has the property of unitary invariance**. Consider the Sturm–Liouville equation in Equation 7.50 and multiply this equation from the right-hand side by $\mathbb{S}^\dagger$ and from the left-hand side by $\mathbb{S}$:

$$\mathbb{S}\mathbb{A}\Psi\mathbb{S}^\dagger = A\mathbb{S}\Psi\mathbb{S}^\dagger \tag{7.79}$$

from where

$$\mathbb{S}\mathbb{A}\mathbb{S}^\dagger\mathbb{S}\Psi\mathbb{S}^\dagger = A\mathbb{S}\Psi\mathbb{S}^\dagger \tag{7.80}$$

But

$$\Phi = \mathbb{S}\Psi \tag{7.81}$$

and

$$\mathbb{S}\mathbb{A}\mathbb{S}^{\dagger} = \tilde{\mathbb{A}} \tag{7.82}$$

then

$$\tilde{\mathbb{A}}\Phi = A\Phi \tag{7.83}$$

which has same form in Equation 7.50. This follows **that the Sturm–Liouville equation is invariant relative to the unitary transformation**. This justifies Equations 7.74 and 7.78, requiring that $\mathbb{S}$ to be a unitary operator as defined in Equation 7.67.

We examine the property of unitary invariance for the commutation relation as follows:

$$[\mathbb{M},\mathbb{N}] = \mathbb{P} \tag{7.84}$$

Multiply the left-hand side by $\mathbb{S}$ and from the right-hand side by $\mathbb{S}^{\dagger}$:

$$\mathbb{S}\mathbb{M}\mathbb{S}^{\dagger}\mathbb{S}\mathbb{N}\mathbb{S}^{\dagger} - \mathbb{S}\mathbb{N}\mathbb{S}^{\dagger}\mathbb{S}\mathbb{M}\mathbb{S}^{\dagger} \equiv \left[\tilde{\mathbb{M}},\tilde{\mathbb{N}}\right] = \mathbb{S}\mathbb{P}\mathbb{S}^{\dagger} \equiv \tilde{\mathbb{P}} \tag{7.85}$$

So **the permutation relation between the operators has the property of unitary invariance. Consequently, quantum mechanical equations have the property of unitary invariance.**

All quantum mechanical problems can be formulated in the language of unitary transformations. Consider Equation 7.78 again and multiply it from the right-hand side by $\mathbb{S}$:

$$\tilde{\mathbb{A}}\mathbb{S} = \mathbb{S}\mathbb{A} \tag{7.86}$$

From where

$$\sum_{\beta} \mathrm{A}_{\alpha\beta}\mathrm{S}_{\beta n} = \mathrm{A}\mathrm{S}_{\alpha n} \tag{7.87}$$

For fix n, we have an equation with unknown quantities $S_{\alpha n}$:

$$\left(\tilde{\mathrm{A}}_{11} - \mathrm{A}\right)\mathrm{S}_{1n} + \mathrm{A}_{12}\mathrm{S}_{2n} + \cdots = 0,\ \mathrm{A}_{21}\mathrm{S}_{1n} + \left(\tilde{\mathrm{A}}_{22} - \mathrm{A}\right)\mathrm{S}_{2n} + \cdots = 0, \cdots \tag{7.88}$$

Letting $\mathrm{S}_{\alpha n} = C_{\alpha}$, we have a system of linear equations for the quantities, C_{α}. This leads us to a secular equation. Here, the role of the wave function is played by the corresponding columns of the matrix, $\mathbb{S}$.

7.7.3 Invariance of the Trace of a Matrix Under Unitary Transformations

Let us evaluate the invariance of the trace of a matrix under unitary transformations. The trace of the matrix $\mathbb{A}$, denoted as $\mathrm{Tr}\mathbb{A}$, is the sum of all diagonal elements:

$$\mathrm{Tr}\mathbb{A}_C = \sum_{\beta} \mathrm{A}_{\beta\beta} = \sum_{\beta}\sum_{n,m} \mathrm{S}^{*}_{n\beta}\mathrm{A}_{nm}\mathrm{S}_{m\beta} = \sum_{\beta,n,m} \mathrm{A}_{nm}\mathrm{S}_{m\beta}\mathrm{S}^{*}_{n\beta} = \sum_{n,m} \mathrm{A}_{nm}\left(\mathrm{S}\mathrm{S}^{\dagger}\right)_{mn} = \sum_{n,m} \mathrm{A}_{nm}\delta_{mn} = \sum_{n} \mathrm{A}_{nn} = \mathrm{Tr}\mathbb{A}_B \tag{7.89}$$

This follows that the Trace of a matrix is independent of the representation. We may also show that the trace of a product of two operators is independent of the order of multiplication, notwithstanding the fact that the operators themselves are not commutable.

7.8 Schrödinger and Heisenberg Representations

Schrödinger Representation

We examine the time-dependent Schrödinger equation:

$$i\hbar\frac{d}{dt}|\Psi_S(q,t)\rangle=\hat{H}|\Psi_S(q,t)\rangle \tag{7.90}$$

Considering the Hamiltonian, $\hat{H}$, to be independent of time t, the solution of Equation 7.90 is an arbitrary eigenvector, $|\Psi_S(q,t)\rangle$, in terms of the stationary state eigenvector written in an operator form:

$$|\Psi_S(q,t)\rangle=\hat{U}(t,t_0)|\Psi_S(q,t_0)\rangle,\ \hat{U}(t,t_0)=\exp\left\{-\frac{i}{\hbar}\hat{H}(t\text{-}t_0)\right\} \tag{7.91}$$

where the evolution operator, $\hat{U}(t,t_0)$, transforms the eigenvector of the system at some initial time moment, t_0, into the eigenvector at an arbitrary time moment, t.

We again consider the dynamical evolution of the operator $\hat{A}$ and take its expectation value, and from the Ehrenfest theorem, we have

$$\frac{d\langle\hat{A}_S\rangle}{dt}=\frac{\partial\langle\hat{A}_S\rangle}{\partial t}+\frac{1}{i\hbar}\langle\hat{A}_S,\hat{H}\rangle \tag{7.92}$$

or

$$\frac{d}{dt}\langle\Psi_S(q,t)|\hat{A}_S|\Psi_S(q,t)\rangle=\langle\Psi_S(q,t)|\frac{\partial\hat{A}_S}{\partial t}|\Psi_S(q,t)\rangle+\frac{1}{i\hbar}\langle\Psi_S(q,t)|\left[\hat{A}_S,\hat{H}\right]|\Psi_S(q,t)\rangle \tag{7.93}$$

If we consider the eigenvector in Equation 7.91, then

$$\begin{aligned}\frac{d}{dt}\langle\Psi_S(q,t_0)|\hat{U}(t,t_0)\hat{A}_S\hat{U}^\dagger(t,t_0)|\Psi_S(q,t_0)\rangle&=\langle\Psi_S(q,t_0)|\hat{U}(t,t_0)\frac{\partial}{\partial t}\hat{A}_S\hat{U}^\dagger(t,t_0)|\Psi_S(q,t_0)\rangle\\&+\frac{1}{i\hbar}\langle\Psi_S(q,t_0)|\left[\hat{U}(t,t_0)\hat{A}_S\hat{U}^\dagger(t,t_0),\hat{H}\right]|\Psi_S(q,t_0)\rangle\end{aligned} \tag{7.94}$$

We observe that though the operators are time-independent, their expectation values, which are physical observables, change with time since the arbitrary eigenvectors in Equation **7.91 bring time-dependence in the so-called Schrödinger representation and so the subscript,** S.

Heisenberg Representation

We now find the description where the **eigenstates are time-independent** while the **operators of the physical quantity are time-dependent** in the **so-called Heisenberg representation**. We find the equation of motion as viewed in the Heisenberg representation:

$$\begin{aligned}&\frac{d}{dt}\langle\Psi_S(q,t_0)|\hat{U}(t,t_0)\hat{A}_S\hat{U}^\dagger(t,t_0)|\Psi_S(q,t_0)\rangle\\&=\langle\Psi_S(q,t_0)|\hat{U}(t,t_0)\frac{\partial\hat{A}_S}{\partial t}\hat{U}^\dagger(t,t_0)|\Psi_S(q,t_0)\rangle\frac{1}{i\hbar}\langle\Psi_S(q,t_0)|\left[\hat{U}(t,t_0)\hat{A}_S\hat{U}^\dagger(t,t_0),\hat{H}\right]|\Psi_S(q,t_0)\rangle\end{aligned} \tag{7.95}$$

In Equation 7.95, we use the fact that $\hat{H}$ commutes with $\hat{U}(t,t_0)$. From Equation 7.91, we have the following state:

$$\left|\Psi_H(q,t_0)\right\rangle = \hat{U}^\dagger(t,t_0)\left|\Psi_S(q,t)\right\rangle = \hat{U}^\dagger(t,t_0)\hat{U}(t,t_0)\left|\Psi_S(q,t_0)\right\rangle = \left|\Psi_S(q,t_0)\right\rangle \tag{7.96}$$

and the expectation value

$$\left\langle \hat{A}_H \right\rangle = \left\langle \Psi_S(q,t_0)\right|\hat{U}(t,t_0)\hat{A}_S\hat{U}^\dagger(t,t_0)\left|\Psi_S(q,t_0)\right\rangle \tag{7.97}$$

then follows the time dynamical variables (operators) in the Heisenberg representation:

$$\hat{A}_H \equiv \hat{U}^\dagger(t,t_0)\hat{A}_S\hat{U}(t,t_0) \tag{7.98}$$

We may now write

$$\left\langle \hat{A}_H \right\rangle = \left\langle \Psi_S(q,t_0)\right|\hat{A}_H\left|\Psi_S(q,t_0)\right\rangle \tag{7.99}$$

In this, all-time dependence lies in the operators while the state vector is independent of time. It is seen that $\hat{A}_H$ generally depends on time even if $\hat{A}_S$ does not. It should be noted that unless $\hat{A}_S$ commutes with $\hat{H}$, the operator $\hat{A}_H$ is dependent on time t even if $\hat{A}_S$ has no explicit time dependence. If the operator $A_S(t)$ commutes with $\hat{H}$, it also commutes with $\hat{U}(t,t_0)$:

$$\hat{A}_H \equiv \hat{U}^\dagger(t,t_0)\hat{A}_S\hat{U}(t,t_0) = \hat{U}^\dagger(t,t_0)\hat{U}(t,t_0)\hat{A}_S = \hat{A}_S \tag{7.100}$$

The operators $\hat{A}_S$ and $\hat{A}_H$ are therefore equal in this case. The time $t = t_0$ at which the kets and operators in the Schrödinger representation are the same as those in the Heisenberg representation is arbitrary and chosen for convenience. In Equation 7.98, $\hat{H}$ is the same in the Schrödinger and the Heisenberg representations. As $\left|\Psi_H(q,t_0)\right\rangle$ does not depend on time, we may omit the explicit mention of time t.

Let us clarify ourselves on the appellation Heisenberg. If we consider Equations 7.95 and 7.98, then the operators of quantum mechanics are dependent on time t through the Heisenberg equation of motion. This is why we call it the Heisenberg representation (picture):

$$\frac{d}{dt}\hat{A}_H = \frac{\partial}{\partial t}\hat{A}_H + \frac{i}{\hbar}\left[\hat{H},\hat{A}_H\right] \tag{7.101}$$

We have seen that in the Schrödinger representation, the dynamical variables are constant in time and in the Heisenberg representation, the state vectors are constant in time and the dynamical variables vary according to Equation 7.101. In between two extremes of the Schrödinger and the Heisenberg representations, there is an intermediate representation called the **interaction representation.**

7.9 Interaction Representation

We may specify the third representation called the interaction representation or the Tamanag-Schrödinger representation if the Hamiltonian $\hat{H}$ of the system splits into two parts:

$$\hat{H} = \hat{H}_0 + \hat{V} \tag{7.102}$$

In Equation 7.102, the quantity $\hat{H}_0$ is independent explicitly on time t and has also a simple structure. This quantity, say $\hat{H}_0$, might be the kinetic energy and $\hat{V}$ the potential energy. This quantity, $\hat{H}_0$, might as well be the Hamiltonian for a relatively simple potential such as the Coulombic field and $\hat{V}$ some additional interaction such as an external electromagnetic field. In the latter, let us consider $\hat{V}$ to be explicitly dependent on time t. We assume $\hat{V}$ as a perturbation and so for the corresponding Hamiltonian, there is no stationary state. This implies the solution of the Schrödinger equation does not have the form $\exp\left\{-\frac{iEt}{\hbar}\right\}|E\rangle$. Note that there exist stationary states for the Hamiltonian, $\hat{H}_0$. The dynamical evolution of the non-stationary state can be described with the parameters of the transition between the stationary states of the operator, $\hat{H}_0$.

Let us define the interaction picture eigenvector by introducing the evolution operator $\hat{U}_0(t,t_0)$:

$$|\Psi_I(q,t)\rangle = \hat{U}_0^\dagger(t,t_0)|\Psi_S(q,t_0)\rangle \tag{7.103}$$

and the interaction picture expectation value by

$$\hat{A}_I = \hat{U}_0^\dagger(t,t_0)\hat{A}_S\hat{U}_0(t,t_0),\ \hat{U}_0(t,t_0) = \exp\left\{-\frac{i\hat{H}_0(t-t_0)}{\hbar}\right\} \tag{7.104}$$

The interaction and Heisenberg representations are the same when $\hat{V} = 0$. Earlier, it is known that $\hat{H}_H = \hat{H}_S$, now from Equation 7.104, it follows that $\hat{H}_{0I} = \hat{H}_{0S}$. Generally, $\hat{H}_I \neq \hat{H}_S$, $\hat{H}_{0H} \neq \hat{H}_{0S}$ and $\hat{V}_S$, $\hat{V}_H$ and V_I are different. Differentiate Equation 7.103 relative to time t, which yields the equation:

$$i\hbar\frac{d}{dt}|\Psi_I(q,t_0)\rangle = \hat{V}_I|\Psi_I(q,t_0)\rangle,\ \hat{V}_I = \hat{U}_0^\dagger(t,t_0)\hat{V}\hat{U}_0(t,t_0) \tag{7.105}$$

It may be seen that in the interaction representation, both the state vectors and the operators are time-dependent. In this representation, the state vector changes in accordance with $\hat{V}$ and the operator in accordance with $\hat{H}_0$.

Also differentiate Equation 7.104 relative to time t, which yields the equation of motion for $\hat{A}_I(t)$:

$$\frac{d\hat{A}_I}{dt} = \frac{\partial\hat{A}_I}{\partial t} + \frac{i}{\hbar}\left[\hat{H}_{0I}, \hat{A}_I\right] \tag{7.106}$$

7.10 Energy Representation

The energy representation is one in which the basis representation is the Hamiltonian $\hat{H}$ of some conservative system (i.e., the Hamiltonian $\hat{H}$ is independent of time t). The energy representation is specified by the kets $|\Psi_n(q,t)\rangle$ which are eigenstates of the full Hamiltonian $\hat{H}$ with the eigenvalues E_n. It is interesting to get the dependence of the matrix elements on time t. There exist two types of energy representation:

1. Schrödinger
2. Heisenberg

Let us start with (1). The bases of this representation are kets $|\Psi_n(q,t)\rangle$ dependent on time t of systems with the Hamiltonian, $\hat{\mathrm{H}}$. Consider the stationary Schrödinger equation:

$$\hat{\mathrm{H}}|\Psi_n(q)\rangle = E_n|\Psi_n(q)\rangle \tag{7.107}$$

and from Equation 7.90, we have

$$|\Psi_n^S(\mathrm{q,t})\rangle = |\Psi_n^S(q)\rangle \exp\left\{-\frac{\mathrm{i}\mathrm{E}_n}{\hbar}t\right\} \tag{7.108}$$

We can use this equation and its adjoint to evaluate the time dependence of the matrix elements in the Schrödinger picture of the operator, $\hat{\mathrm{A}}^S$, that has no explicit time dependence:

$$\hat{\mathrm{A}}_{\mathrm{mn}}^S(t) = \langle\Psi_m^S(\mathrm{q,t})|\hat{\mathrm{A}}^S|\Psi_n^S(\mathrm{q,t})\rangle = \langle\Psi_m^S(q)|\hat{\mathrm{A}}^S|\Psi_n^S(q)\rangle \exp\{i\omega_{\mathrm{mn}}t\},\ \omega_{\mathrm{mn}} = \frac{E_m - E_n}{\hbar} \tag{7.109}$$

where ω_{mn} is the Bohr's frequency for the transition between the energy levels E_m and E_n. So from Equation 7.109, we have

$$\hat{\mathrm{A}}_{\mathrm{mn}}^S(t) = \hat{\mathrm{A}}_{\mathrm{mn}}^S(0)\exp\{i\omega_{\mathrm{mn}}t\} \tag{7.110}$$

Hence, in the representation (1), the matrix elements are periodic functions of time t with the frequency which is equal to Bohr's transition frequency. If we differentiate Equation 7.109 with respect to time t:

$$\dot{\hat{\mathrm{A}}}_{\mathrm{mn}}^S(t) = i\omega_{\mathrm{mn}}\hat{\mathrm{A}}_{\mathrm{mn}}^S(t) \tag{7.111}$$

In the Schrödinger representation, the operator $\hat{\mathrm{A}}^S$ is independent of time t but $\Psi(q,t)$ is a function of time.

Consider representation (2). The ket of the Heisenberg representation (Heisenberg picture) is independent of time$t$, i.e., the state vector $|\Psi(q)\rangle$ of the Hamiltonian $\hat{\mathrm{H}}$ is independent of time t. This does not imply that the matrix elements of the operator $\hat{\mathrm{A}}_{\mathrm{H}}$ are also independent of time t, although there is no explicit time dependence. Time-dependent operators describe the dynamical evolution. The operators of quantum mechanics are dependent on time t through the equation of motion:

$$\frac{\mathrm{d}\hat{\mathrm{A}}_{\mathrm{H}}}{\mathrm{dt}} = \frac{\partial\hat{\mathrm{A}}_{\mathrm{H}}}{\partial t} + \frac{i}{\hbar}\left[\hat{\mathrm{H}},\hat{\mathrm{A}}_{\mathrm{H}}\right] \tag{7.112}$$

If we take the matrix element of this equation, then we have

$$\dot{\mathrm{A}}_{\mathrm{m,n}}^{\mathrm{H}} = \frac{i}{\hbar}\sum_k\left(\mathrm{H}_{\mathrm{m,\,k}}^{\mathrm{H}}\mathrm{A}_{\mathrm{kn}}^{\mathrm{H}} - \mathrm{A}_{\mathrm{m,\,k}}^{\mathrm{H}}\mathrm{H}_{\mathrm{kn}}^{\mathrm{H}}\right) \tag{7.113}$$

In this representation, the diagonal matrix represents the operator, $\mathrm{H}_{\mathrm{mk}} \neq 0$ when $n = m$, $\mathrm{H}_{\mathrm{km}} = E_k$.

So

$$\frac{i}{\hbar}\left(\mathrm{H}_{\mathrm{m,\,k}}^{\mathrm{H}}\mathrm{A}_{\mathrm{kn}}^{\mathrm{H}} - \mathrm{A}_{\mathrm{m,\,k}}^{\mathrm{H}}\mathrm{H}_{\mathrm{kn}}^{\mathrm{H}}\right) = \mathrm{A}_{\mathrm{mn}}^{\mathrm{H}}\frac{i}{\hbar}\left(E_m\text{-}\mathrm{E}_n\right) = i\omega_{\mathrm{mn}}\mathrm{A}_{\mathrm{mn}}^{\mathrm{H}} \tag{7.114}$$

and

$$\dot{\mathrm{A}}_{\mathrm{mn}}^{\mathrm{H}} = i\omega_{\mathrm{mn}}\mathrm{A}_{\mathrm{mn}}^{\mathrm{H}} \tag{7.115}$$

7.10.1 Evolution Operator

We consider the temporal evolution of a system described by a series of unitary transformations. The operator of this time-evolution transformation is $\hat{U}(t,t_0)$. Let us show some of the properties of $\hat{U}(t,t_0)$. We substitute Equation 7.91 into the time-dependent Schrödinger Equation 7.91, we have

$$i\hbar\frac{\partial\hat{U}}{\partial t}=\hat{H}\hat{U} \tag{7.116}$$

From the initial condition, $t = t_0$, the differential Equation 7.116 completely defines $\hat{U}(t,t_0)$. We write Equation 7.92 in the form:

$$\hat{U}(t,t_0)=1-\frac{i}{\hbar}\int_{t_0}^{t}\hat{H}(t')\hat{U}(t',t_0)dt' \tag{7.117}$$

Considering the parameter t_0 as a variable just like t, Equation 7.91 takes the form:

$$\left|\Psi(q,t)\right\rangle=\hat{U}(t,t')\left|\Psi(q,t')\right\rangle \tag{7.118}$$

The state vector $\left|\Psi(q,t')\right\rangle$ may be obtained from the relation:

$$\left|\Psi(q,t')\right\rangle=\hat{U}(t',t'')\left|\Psi(q,t'')\right\rangle \tag{7.119}$$

Substituting Equation 7.118 into 7.117 yields

$$\left|\Psi(q,t)\right\rangle=\hat{U}(t,t')\hat{U}(t',t'')\left|\Psi(q,t'')\right\rangle \tag{7.120}$$

Also,

$$\left|\Psi(q,t)\right\rangle=\hat{U}(t,t'')\left|\Psi(q,t'')\right\rangle,\ \hat{U}(t,t'')=\hat{U}(t,t')\hat{U}(t',t'') \tag{7.121}$$

If this procedure is continued, then we have

$$\hat{U}(t_n,t_1)=\hat{U}(t_n,t_{n-1})\ldots\hat{U}(t_3,t_2)\hat{U}(t_2,t_1) \tag{7.122}$$

Here $t_1,t_2,t_3,\ldots,t_{n-1},t_n$ are arbitrary and we assume that the system progresses with increase in time:

$$t_1<t_2<t_3<\cdots<t_n \tag{7.123}$$

Examine the significance of Equation 7.122. In order to go from t_1 to t_n, the system progresses from t_1 to t_2 then from t_2 to t_3, and so on then finally from t_{n-1} to t_n. Suppose in Equation 7.121 we let $t''=t$:

$$\hat{\mathbb{I}}=\hat{U}(t,t')\hat{U}(t',t)=\hat{U}(t',t)\hat{U}(t,t') \tag{7.124}$$

From here

$$\hat{U}(t,t')=\hat{U}^{-1}(t',t) \tag{7.125}$$

We evaluate the evolution operator between two time intervals that are separated by dt. This may be done if the Schrödinger equation in Equation 7.90 takes the form:

$$d\left|\Psi(q,t)\right\rangle=\left|\Psi(q,t+dt)\right\rangle-\left|\Psi(q,t)\right\rangle=-\frac{i}{\hbar}\hat{H}(t)\left|\Psi(q,t)\right\rangle dt \tag{7.126}$$

i.e.,

$$\left|\Psi(q,t+dt)\right\rangle=\left(1-\frac{i}{\hbar}\hat{H}(t)dt\right)\left|\Psi(q,t)\right\rangle \tag{7.127}$$

So, from Equation 7.127, follows the quantity

$$\hat{U}(t+dt,t)=1-\frac{i}{\hbar}\hat{H}(t)dt \tag{7.128}$$

called the **infinitesimal evolution operator**. Since $\hat{H}(t)$ is Hermitian, $\hat{U}(t+dt,t)$ is unitary. So $\hat{U}(t,t')$ is also unitary since the interval $[t,t']$ may be partitioned into a large number of infinitesimal intervals. Then from Equation 7.122, $\hat{U}(t,t')$ represents a product of unitary operators and hence a unitary operator itself. This follows that Equation 7.125 may be rewritten as follows:

$$\hat{U}(t,t')^{\dagger}=\hat{U}^{-1}(t,t')=\hat{U}(t',t) \tag{7.129}$$

It has been seen above that the operator $\hat{U}(t,t')$ conserves the norm of the state vectors on which it acts, i.e., the norm of the state vector does not change over time.

7.10.2 Oscillator in the Energy Representation

We solve this problem using the matrix method where we construct the matrix of the coordinate of the harmonic oscillator in the energy representation. This implies, considering the general theorem of the one-dimensional motion which is satisfied for any representation and knowing that the oscillator has a discrete spectrum E_n with non-degenerate eigenfunctions Ψ_n. We use Ψ_n as the basis for the evaluation of that matrix.

7.10.2.1 Matrix Element of the Oscillator Coordinate

The operators of physical quantities in quantum mechanics satisfy the same equations of motion as the same physical quantities in classical mechanics. We consider the Hamiltonian $\hat{H}$ of the harmonic oscillator written in the form:

$$\hat{H}=\frac{\hat{P}^2}{2M}+\frac{M\omega^2q^2}{2} \tag{7.130}$$

where M, ω, q and $\hat{P}$are, respectively, the mass, oscillatory frequency, coordinate and the operator of the momentum. The equation of motion for the oscillator

$$\ddot{q}+\omega^2q=0 \tag{7.131}$$

is a classical equation of motion. The operator, $\hat{q}$, in the quantum mechanical equation of motion:

$$\ddot{\hat{q}}+\omega^2\hat{q}=0 \tag{7.132}$$

Here, ω is the cyclical frequency. Considering the matrix elements of Equation 7.132:

$$\langle n|\ddot{\hat{q}}+\omega^2\hat{q}|n\rangle = 0 \tag{7.133}$$

From Equation 7.131, we write

$$\ddot{q}_{mn} = i\omega_{mn}\dot{q}_{mn} = -\omega_{mn}^2 q_{mn} \tag{7.134}$$

Consider the dependence of time of all the matrix elements, and consider Equation 7.132 again:

$$q_{mn}\left(\omega^2 - \omega_{mn}^2\right) = 0 \tag{7.135}$$

It is seen that $q_{mn} \neq 0$ only if

$$\omega_{mn} = \pm\omega \tag{7.136}$$

This is the Bohr's frequency and is the eigenfrequency of the oscillator. From here, the oscillator has no other levels, i.e., the minimum frequency between levels should be ω. This implies that we assume Equation 7.136 is satisfied if

$$m = n \pm 1 \tag{7.137}$$

The transition of the oscillator from level n to level $n+1$ or $n-1$ corresponds to the creation or destruction of a quantum of energy $\hbar\omega$. Equations 7.136 and 7.137 give the selection rule. However, it is impossible to define the value of q_{mn} from Equation 7.135. Develop in the coordinate representation the Heisenberg relation:

$$\hat{q}\hat{P}_q - \hat{P}_q\hat{q} = i\hbar,\ \hat{P}_q = -i\frac{\partial}{\partial q} \tag{7.138}$$

From Ehrenfest theorem:

$$\hat{P}_q = M\hat{\dot{q}}^2 \tag{7.139}$$

and substituting into Equation 7.138:

$$\hat{q}\hat{\dot{q}} - \hat{\dot{q}}\hat{q} = \frac{i\hbar}{M} \tag{7.140}$$

Find the matrix element of this relation:

$$\sum_m \left(q_{nm}\dot{q}_{mn} - \dot{q}_{nm}q_{mn}\right) = \frac{i\hbar}{M} \tag{7.141}$$

From relation:

$$-\omega_{mn} = \omega_{nm} \equiv \frac{E_n - E_m}{\hbar} \tag{7.142}$$

and Equation 7.134:

$$\sum_m \omega_{mn} q_{nm} q_{mn} = \frac{\hbar}{2M} \tag{7.143}$$

Since the coordinate maps a Hermitian matrix:

$$q_{nm} = q^*_{mn} \tag{7.144}$$

and considering Equation 7.143, we have

$$q_{nm} q_{mn} = |q_{nm}|^2 \tag{7.145}$$

From Equation 7.137 and considering

$$q_{mn} \neq 0 \tag{7.146}$$

then

$$|q_{n+1,n}|^2 - |q_{n,n-1}|^2 = \frac{\hbar}{2M\omega},\ n = 0,1,2,\ldots \tag{7.147}$$

If we consider in Equation 7.147, $n+1 \to n$:

$$|q_{n,n-1}|^2 - |q_{n-1,n-2}|^2 = \frac{\hbar}{2M\omega} \tag{7.148}$$

This procedure is continued until we have

$$|q_{1,0}|^2 = \frac{\hbar}{2M\omega} \tag{7.149}$$

We the arrive at $(n+1)$ equations which, after simplification, reduce to

$$|q_{n+1,n}|^2 = \frac{\hbar}{2M\omega}(n+1) \tag{7.150}$$

So we have evaluated the first coordinate matrix.

If we let $n \to n-1$, then

$$|q_{n,n-1}|^2 = \frac{\hbar}{2M\omega} n \tag{7.151}$$

Then we arrive at the second coordinate matrix element.

Letting the amplitude of the zero oscillations which has the dimension of length:

$$\sqrt{\frac{\hbar}{M\omega}} = q_0 \tag{7.152}$$

then

$$\left|q_{n+1,n}\right| = q_0\sqrt{\frac{n+1}{2}} \tag{7.153}$$

This is the matrix element of the oscillator in the energy representation. It should be noted that

$$q = |q|\exp\{i\alpha\} \tag{7.154}$$

where $\exp\{i\alpha\}$ is a phase factor.

7.10.2.2 Hamiltonian Operator Eigenvalue

We take the matrix element of the Hamiltonian in Equation 7.130:

$$E_n = \mathrm{H}_{nn} = \frac{M}{2}\sum_m\left(\dot{q}_{nm}\dot{q}_{mn} + \omega^2 q_{nm}q_{mn}\right) =_{7.133} \frac{M}{2}\sum_m\left(\omega_{mn}^2 + \omega^2\right)\left|q_{mn}\right|^2 \tag{7.155}$$

From the selection rule where $q_{mn} \neq 0$, we have $m = n \pm 1$ and $\omega_{mn}^2 = \omega^2$. So

$$E_n = M\omega^2\left(\left|q_{n+1,n}\right|^2 + \left|q_{n-1,n}{}^2\right|\right) = M\omega^2 q_0^2\left(\frac{n+1}{2} + \frac{n}{2}\right) = M\omega^2 q_0^2\left(n + \frac{1}{2}\right) \tag{7.156}$$

Considering Equation 7.152:

$$E_n = \hbar\omega\left(n + \frac{1}{2}\right), n = 0,1,2,\ldots \tag{7.157}$$

We observe that the energy is quantized with the quantum of energy being, $\hbar\omega$, in contrast to the classical oscillator which can have a continuous energy. The ground-state or zero-point energy has the value, $E_0 = \frac{\hbar\omega}{2}$. This is the energy of zero oscillations.

7.10.2.3 Harmonic Oscillator Ground-State Eigenfunction

We find the wave function of the harmonic oscillator by considering Equation 7.130:

$$\hat{\mathrm{H}} = \frac{M}{2}\left(\hat{\dot{q}} + i\omega\hat{q}\right)\left(\hat{\dot{q}} - i\omega\hat{q}\right) \tag{7.158}$$

then

$$\langle n+1|\hat{\dot{q}} + i\omega\hat{q}|n\rangle = 2i\omega q_{n+1,n} = 2i\omega q_0\sqrt{n+1}, \langle n-1|\hat{\dot{q}} + i\omega\hat{q}|n\rangle = 0 \tag{7.159}$$

Similarly,

$$\langle n+1|\hat{\dot{q}} - i\omega\hat{q}|n\rangle = 0, \langle n-1|\hat{\dot{q}} - i\omega\hat{q}|n\rangle = -2i\omega q_0\sqrt{n} \tag{7.160}$$

This implies that the matrix element of $\left(\hat{\dot{q}} \pm i\omega\hat{q}\right)$ is different from zero for the transition between such states of the oscillator where its energy may either increase or decrease by $\hbar\omega$. But

$$q_{n-1,n} = q_{n,n-1} = \sqrt{\frac{n\hbar}{2\mu\omega}}, \dot{q}_{n-1,n} = -i\omega q_{n-1,n} \tag{7.161}$$

This follows that $\left(\hat{\dot{q}}+i\omega\hat{q}\right)$ and $\left(\hat{\dot{q}}-i\omega\hat{q}\right)$ represent, respectively, the operator of annihilation and creation.

Consider the operator $\hat{A}$:

$$\hat{A}\Psi_n = \Phi_n,\quad \Phi_n = \sum_m C_{mn}\Psi_m \tag{7.162}$$

then

$$C_{mn} = \int \Psi_m^* \hat{A}\Psi_n \, dq = A_{mn} \tag{7.163}$$

and

$$\hat{A}\Psi_n = \sum_m A_{mn}\Psi_m \tag{7.164}$$

So

$$\left(\hat{\dot{q}}-i\omega\hat{q}\right)\Psi_0 = \sum_m \left(\hat{\dot{q}}-i\omega\hat{q}\right)_{m,0}\Psi_m \tag{7.165}$$

and

$$\left(\hat{\dot{q}}-i\omega\hat{q}\right)\Psi_0 = \left(\hat{\dot{q}}-i\omega\hat{q}\right)_{-1,0}\Psi_{-1} = 0 \tag{7.166}$$

From

$$\hat{\dot{q}} = \frac{\hat{P}_q}{M} = -i\frac{\hbar}{M}\frac{d}{dq} \tag{7.167}$$

then

$$\left(-i\frac{\hbar}{M}\frac{d}{dq} - i\omega q\right)\Psi_0 = 0 \tag{7.168}$$

from where:

$$\Psi_0(q) = C_0 \exp\left\{-\frac{M\omega q^2}{2\hbar}\right\} \tag{7.169}$$

Considering the normalization condition:

$$1 = \int_{-\infty}^{+\infty} \Psi_0^* \Psi_0 dq \tag{7.170}$$

then

$$C_0 = \sqrt[4]{\frac{M\omega}{\pi\hbar}} \tag{7.171}$$

and the normalized ground-state wave function is as follows:

$$\Psi_0(q) = \sqrt[4]{\frac{M\omega}{\pi\hbar}} \exp\left\{-\frac{M\omega q^2}{2\hbar}\right\} \tag{7.172}$$

7.10.2.4 Quantization of Operators

Let us consider again the Hamiltonian in Equation 7.130 and introduce dimensionless quantities:

$$\hat{\xi} = \left(\frac{M\omega}{\hbar}\right)^{\frac{1}{2}} \hat{q} \tag{7.173}$$

which permit us to define the new operator as follows:

$$\hat{\text{P}}_q = \sqrt{M\omega\hbar}\,\hat{\text{P}}_\xi,\ \hat{\text{P}}_\xi = -i\frac{\partial}{\partial\xi} \tag{7.174}$$

So the Hamiltonian in the new operators is as follows:

$$\hat{\text{H}} = \frac{\hbar\omega}{2}\left(\hat{\xi}^2 + \hat{\text{P}}_\xi^2\right) \tag{7.175}$$

Introduce the following dimensionless creation $\hat{a}^\dagger$ and annihilation $\hat{a}$ operators, respectively:

$$\hat{a}^\dagger = \frac{1}{\sqrt{2}}\left(\hat{\xi} - i\hat{\text{P}}_\xi\right),\ \hat{a} = \frac{1}{\sqrt{2}}\left(\hat{\xi} + i\hat{\text{P}}_\xi\right) \tag{7.176}$$

Note that as

$$\left[\hat{\xi}, \hat{\text{P}}_\xi\right] = i \tag{7.177}$$

then

$$\left[\hat{a}, \hat{a}^\dagger\right] = \frac{1}{2}\left[\hat{\xi} + i\hat{\text{P}}_\xi, \hat{\xi} - i\hat{\text{P}}_\xi\right] = \frac{i}{2}\left[\hat{\text{P}}_\xi, \hat{\xi}\right] - \frac{i}{2}\left[\hat{\xi}, \hat{\text{P}}_\xi\right] = -\frac{i^2}{2} - \frac{i^2}{2} = 1 \tag{7.178}$$

If we consider Equations 7.175–7.178, then we have

$$\hat{\text{H}} = \frac{\hbar\omega}{2}\left\{\left(\hat{\xi} - i\hat{\text{P}}_\xi\right)\left(\hat{\xi} + i\hat{\text{P}}_\xi\right) + \left(\hat{\xi} + i\hat{\text{P}}_\xi\right)\left(\hat{\xi} - i\hat{\text{P}}_\xi\right)\right\} = \frac{\hbar\omega}{2}\left(\hat{a}^\dagger\hat{a} + \hat{a}\hat{a}^\dagger\right) = \frac{\hbar\omega}{2}\left(2\hat{a}^\dagger\hat{a} + 1\right) = \hbar\omega\left(\hat{\text{N}} + \frac{1}{2}\right),\ \hat{\text{N}} = \hat{a}^\dagger\hat{a} \tag{7.179}$$

Here $\hat{N}$ is the phonon number operator. So

$$\hat{N}\Psi_n = n\Psi_n \quad \left(\hat{N}|n\rangle = n|n\rangle\right) \tag{7.180}$$

It is instructive to note from our results that as the Hamiltonian operator $\hat{H}$ in Equation 7.175 is a positive quantity, its eigenvalues may not be negative. So the spectrum of $\hat{H}$ is of course discrete and bounded from below. From Equation 7.179, the operator $\hat{N}$ has a similar property. So if Ψ_n is the eigenfunction of the operator $\hat{N}$, then from Equation 7.179, we have

$$\hat{H}\Psi_n = \hbar\omega\left(\hat{N} + \frac{1}{2}\right)\Psi_n = \hbar\omega\left(n + \frac{1}{2}\right)\Psi_n \tag{7.181}$$

It may be seen that the eigenvectors of $\hat{H}$ are also eigenvectors of $\hat{N}$ and vice versa.

Let us consider the property of the Hermite polynomial:

$$\xi\Psi_n = \sqrt{\frac{n}{2}}\Psi_{n-1} + \sqrt{\frac{n+1}{2}}\Psi_{n+1}, \quad \frac{\partial\Psi_n}{\partial\xi} = \sqrt{\frac{n}{2}}\Psi_{n-1} - \sqrt{\frac{n+1}{2}}\Psi_{n+1} \tag{7.182}$$

Then, considering Equation 7.176, we have

$$\hat{a}\Psi_n = \sqrt{n}\Psi_{n-1}, \quad \hat{a}^{\dagger}\Psi_n = \sqrt{n+1}\Psi_{n+1} \tag{7.183}$$

It is obvious that it is convenient to solve our problem in a representation where $\hat{N}$ represents a diagonal matrix. It can be seen that when the operator $\hat{a}\left(\hat{a}^{\dagger}\right)$ acts on the phonon state $|n\rangle = \Psi_n$, it reduces (increases) the $|n\rangle$ state by 1 quantum number of oscillations (Figure 7.1). So

$$\hat{a}|n\rangle = \sqrt{n}|n-1\rangle, \quad \hat{a}^{\dagger}|n\rangle = \sqrt{n+1}|n+1\rangle \tag{7.184}$$

where $\hat{a}$ and $\hat{a}^{\dagger}$ are called the **annihilation** and **creation** operators, respectively. In addition,

$$\hat{a}\hat{a}^{\dagger}|n\rangle = (n+1)|n\rangle, \quad \hat{a}^{\dagger}\hat{a}|n\rangle = n|n\rangle \tag{7.185}$$

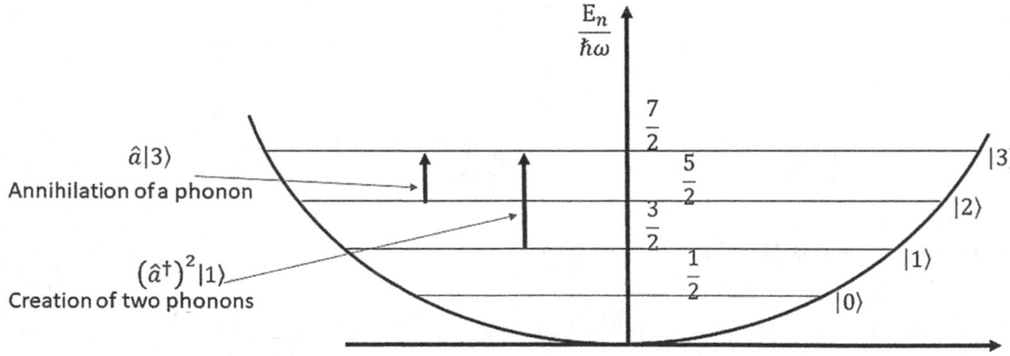

FIGURE 7.1 The energy levels of the harmonic oscillator and the effect of the creation and annihilation operators.

and from here also Equation 7.178. The matrix elements of the operators $\hat{a}$, $\hat{a}^\dagger$, $\hat{\xi}$ and $\hat{P}_\xi$ are, respectively, as follows:

$$\langle n'|\hat{a}|n\rangle = \sqrt{n}\delta_{n',n-1},\ \langle n'|\hat{a}^\dagger|n\rangle = \sqrt{n+1}\delta_{n',n+1} \tag{7.186}$$

$$\langle n'|\hat{\xi}|n\rangle = \sqrt{\frac{\hbar}{2M\omega}}\left(\sqrt{n+1}\delta_{n',n+1} + \sqrt{n}\delta_{n',n-1}\right),\ \langle n'|\hat{P}_\xi|n\rangle = i\sqrt{\frac{M\omega\hbar}{2}}\left(\sqrt{n+1}\delta_{n',n+1} - \sqrt{n}\delta_{n',n-1}\right) \tag{7.187}$$

The introduction of the phonon representation is called second quantization. The quanta of the field of the oscillator are exactly the phonons. The matrices for $\hat{a}$ and $\hat{a}^\dagger$ are, respectively, as follows:

$$\hat{a} = \begin{bmatrix} 0 & \sqrt{1} & 0 & \cdots & 0 \\ 0 & 0 & \sqrt{2} & \cdots & 0 \\ 0 & 0 & 0 & \sqrt{3} & 0 \\ \vdots & \vdots & \vdots & \ddots & \vdots \\ 0 & 0 & 0 & 0 & \ddots \end{bmatrix},\ \hat{a}^\dagger = \begin{bmatrix} 0 & 0 & 0 & \cdots & 0 \\ \sqrt{1} & 0 & 0 & \cdots & 0 \\ 0 & \sqrt{2} & 0 & \cdots & 0 \\ \vdots & \vdots & \ddots & 0 & \vdots \\ 0 & 0 & 0 & 0 & \ddots \end{bmatrix} \tag{7.188}$$

As we can see from the explicit expressions of $\hat{a}$ and $\hat{a}^\dagger$, the matrices are Hermitian conjugates of each other. Let us now find the eigenfunction $\Psi_n = |n\rangle$ in the coordinate representation where the ground-state eigenvector is defined from the relation:

$$\hat{a}|0\rangle = 0 \tag{7.189}$$

So with the help of the second equality in Equation 7.184, we can construct a recurrence relation for all eigenfunctions:

$$\hat{a}^\dagger|0\rangle = \sqrt{1}|1\rangle,\ \hat{a}^\dagger|1\rangle = \sqrt{2}|2\rangle,\ \hat{a}^\dagger|2\rangle = \sqrt{3}|3\rangle, \ldots \tag{7.190}$$

From here, we have

$$|n\rangle = \frac{\left(\hat{a}^\dagger\right)^n}{\sqrt{n!}}|0\rangle \tag{7.191}$$

Moving to the coordinate representation, we consider Equation 7.172:

$$\langle q|0\rangle \equiv \Psi_0(q) = \sqrt[4]{\frac{M\omega}{\pi\hbar}}\exp\left\{-\frac{M\omega q^2}{2\hbar}\right\} \tag{7.192}$$

as well as Equation 7.173–7.176:

$$\langle q|n\rangle = \sqrt[4]{\frac{M\omega}{\pi\hbar}}\frac{1}{\sqrt{2^n n!}}\left(\left(\hat{\xi} - i\hat{P}_\xi\right)\right)^n \exp\left\{-\frac{\xi^2}{2}\right\} \equiv \sqrt[4]{\frac{M\omega}{\pi\hbar}}\frac{1}{\sqrt{2^n n!}}\exp\left\{-\frac{\xi^2}{2}\right\}H_n(\xi) \tag{7.193}$$

Here, $H_n(\xi)$ is the Hermite polynomial.

We examine the coherent state that is a special form of a non-stationary state which, for time $t=0$, is an eigenvector of the annihilation operator, $\hat{a}$ (see the details in reference [23]):

$$\hat{a}|\alpha\rangle = \alpha|\alpha\rangle \tag{7.194}$$

Considering the coordinate representation, we have

$$\hat{a}\langle\xi|\alpha\rangle = \frac{1}{\sqrt{2}}\left(\hat{\xi} + i\hat{P}_{\xi}\right)\langle\xi|\alpha\rangle = \alpha\langle\xi|\alpha\rangle \tag{7.195}$$

The solution of this equation gives

$$\langle\xi|\alpha\rangle = C\exp\left\{-\frac{1}{2}\left(\xi-\bar{\xi}\right)^2\right\},\ \bar{\xi} = \alpha\sqrt{2} \tag{7.196}$$

For $t>0$, the eigenvector $|\alpha\rangle$ becomes $\exp\left\{-\frac{i\hat{H}t}{\hbar}\right\}|\alpha\rangle$ so that the wave function becomes

$$\Psi(\xi,t) = \left\langle \xi\right|\exp\left\{-\frac{i\hat{H}t}{\hbar}\right\}\left|\alpha\right\rangle = \sum_{nm}\langle\xi|n\rangle\left\langle n\right|\exp\left\{-\frac{i\hat{H}t}{\hbar}\right\}\left|m\right\rangle\langle m|\alpha\rangle = \sum_{n}\langle\xi|n\rangle\exp\left\{-in\omega t - \frac{i}{2}\omega t\right\}\langle n|\alpha\rangle \tag{7.197}$$

We show that

$$|\Psi|^2 = |C|^2\exp\left\{-\left(\xi-\bar{\xi}\cos\omega t\right)^2\right\} \tag{7.198}$$

and find first $\langle n|\alpha\rangle$:

$$\langle n|\alpha\rangle = \langle\alpha|n\rangle^* = \left\langle \alpha\right|\frac{\left(\hat{a}^\dagger\right)^n}{\sqrt{n!}}\left|0\right\rangle^* \equiv \left\langle 0\right|\frac{\left(\hat{a}\right)^n}{\sqrt{n!}}\left|\alpha\right\rangle = \frac{\alpha^n}{\sqrt{n!}}\langle 0|\alpha\rangle \tag{7.199}$$

From the normalization condition:

$$\sum_n \langle\alpha|n\rangle\langle n|\alpha\rangle = 1 \tag{7.200}$$

and follows

$$\left|\langle 0|\alpha\rangle\right|^2\exp\left\{\alpha^2\right\} = 1 \tag{7.201}$$

from where:

$$\langle n|\alpha\rangle = \frac{\alpha^n}{\sqrt{n!}}\exp\left\{-\frac{\alpha^2}{2}\right\} \tag{7.202}$$

Substitute 7.202 into 7.197 then

$$\Psi(\xi,t)=\exp\left\{-\frac{\alpha^2}{2}-\frac{i\omega t}{2}\right\}\sum_n \langle\xi|n\rangle\frac{(\alpha\exp\{-i\omega t\})^n}{\sqrt{n!}} \tag{7.203}$$

From Equation 7.196:

$$\Psi(\xi,t)=C\exp\left\{-\frac{\overline{\xi}^2}{4}\exp\{-2i\omega t\}-\frac{\xi^2}{4}-\frac{i\omega t}{2}-\frac{1}{2}\left(\xi-\overline{\xi}\exp\{-i\omega t\}\right)^2\right\} \tag{7.204}$$

Equation 7.198 can therefore be easily obtained.

The distribution over the number of particles in the given state is described by the eigenfunction of the operator, $\hat{a}$:

$$|\langle n|\alpha\rangle|^2=\frac{|\alpha|^{2n}}{n!}\exp\left\{-|\alpha|^2\right\} \tag{7.205}$$

This is the Poisson distribution.

8

Quantum Mechanics Approximate Methods

In quantum mechanics, there exist a lot of problems from the domain of atomic, molecular, solid-state physics, physics of magnetic phenomena, nuclear physics, etc. The number of problems which may be solved exactly are few such as some one-dimensional problems, problems in the rectangular well (barrier), the oscillator, problems of motion in a central field such as motion in a Coulomb field. These methods are well formulated and are used in evaluations to a greater approximation. However, in some cases, these may create problems that are very difficult, if not impossible, to be solved exactly. In such problems, the Hamiltonian cannot be exactly diagonalized in a simple manner. For such problems, we resort to approximate methods for finding the eigenvalues and eigenfunctions of the Hamiltonian. Besides, these approximate methods yield results closer to experimental ones. Amongst the approximate methods, we examine the variational method.

8.1 Variational Principle

The variational method is a very appropriate method when interest is on estimates of the ground state or higher energy of complex physical systems. One of the basics of the variational method is the Ritz method.

8.1.1 Ritz Method

Let us examine the least action principle that involves

1. Writing the variational functional
2. Solving the problem for the minimum of the functional
3. Getting the Euler-Lagrange equation which is the equation of motion. In mechanics, it is the Lagrange's equation of the second kind

We consider the Hamiltonian $\hat{H}$ of the given system and assume that the eigenfunction $\Psi(q)$ is the state in which we evaluate the expectation value of the Hamiltonian, $\hat{H}$:

$$\bar{H}[\Psi] = \int \Psi^*(q)\hat{H}\Psi(q)dq \tag{8.1}$$

Here, integration is extended over the entire configuration space, and $\bar{H}[\Psi]$ is the function of the state $\Psi(q)$. This implies that the expectation value $\bar{H}[\Psi]$ is dependent not only on the numerical form

DOI: 10.1201/9781003273073-9

of $\Psi(q)$ but also on its functional form. For a finite motion, the wave function, $\Psi(q)$, is squarable integrable:

$$\int \Psi^*(q)\Psi(q)dq = 1 \tag{8.2}$$

In the Ritz variational method, we determine the ground state without the explicit solution of the Schrödinger equation. This requires that the energy should have the lowest of all possible energies for all possible wave functions. For this, we consider an arbitrary Hamiltonian $\hat{H}$ and demands that its spectrum should have a lower limit. This follows that it should have the lowest non-degenerate energy eigenvalue.

We modify the wave function $\Psi(q)$ infinitesimally by the quantity, $\delta\Psi$ (small variation of $\Psi(q)$), so that we have the variation $\Psi \to \Psi + \delta\Psi$ then for the stationary value of $\bar{H}[\Psi]$:

$$\delta\bar{H} = 0 \tag{8.3}$$

(the variation of the functional $\bar{H}[\Psi]$).

Problem

Find such a $\Psi(q)$ the variation for which gives

$$\delta\bar{H} = 0 \tag{8.4}$$

For the auxiliary condition of normalization of $\Psi(q)$ (see Equation 8.2), this can be obtained by using the method of Lagrange's multipliers: the auxiliary condition for the undefined factors of the energy E:

$$\delta_\Psi\left(\bar{H} - E\int \Psi^*\Psi\, dq\right) = 0 \tag{8.5}$$

In general, considering $\Psi^* \pm \Psi$, we make the variation with respect to Ψ^*. If we substitute Equation 8.1 into 8.5, then

$$\int \delta\Psi^*\left[\hat{H}\Psi - E\Psi\right]dq = 0 \tag{8.6}$$

This is true for arbitrary $\delta\Psi^*$ and follows that the integral is equal to zero.

Note that the wave function Ψ should be finite, continuous and univalent (single-valued) in the entire space. From the Schrödinger equation, E is the energy eigenvalue of the operator $\hat{H}$:

$$\hat{H}\Psi = E\Psi \tag{8.7}$$

Problem Solution

The variational principle in Equation 8.3 with the auxiliary condition in Equation 8.2 leads to the Schrödinger equation. We establish the variational principle in mechanics. We examine the type of extremum (either minimum or maximum) by series expanding the state in the basis of the eigenfunctions (forming a complete orthonormal set), Ψ_n, of the Hamiltonian, $\hat{H}$, with eigenvalues, E_n:

$$\Psi = \sum_n C_n\Psi_n \tag{8.8}$$

then from Equation 8.7, we write

$$\hat{H}\Psi_n = E_n\Psi_n \tag{8.9}$$

We assume the eigenvalues, E_n, to be discrete.

Suppose E_0 is the ground-state energy (zero-order energy) and E is the first perturbation and so on. If we substitute Equation 8.8 into 8.1, considering Equation 8.9:

$$\bar{H} = \sum_n |C_n|^2 E_n \tag{8.10}$$

and if we substitute Equation 8.8 into 8.2, then

$$\sum_n |C_n|^2 = 1 \tag{8.11}$$

Suppose we swap E_0 with E_n, from Equation 8.10:

$$\bar{H} \geq E_0 \sum_n |C_n|^2 = E_0 \tag{8.12}$$

In order for the inequality in Equation 8.12 to become an equality, it is necessary and sufficient that all the coefficients C_n are zero except the coefficient C_0. In this case, Ψ achieves then the ground eigenstate of $\hat{H}$ with the eigenvalue E_0. From the inequality in Equation 8.12, we observe that **the expectation value of the Hamiltonian in an arbitrary state is greater or equal to the ground-state energy**. This is called the **variational method**.

If in Equation 8.1, in the place of Ψ, we have an arbitrary function, then we have a quantity greater than the energy of the ground state of a system. The equal sign corresponds to the function Ψ that maps the exact eigenfunction of the ground state. It follows that the wave function of the ground state corresponds to the absolute minimum of the functional Equation 8.1.

Let us apply on Ψ again an auxiliary condition of orthogonality on the ground-state wave function:

$$\int \Psi\Psi_0^* dq = 0 \tag{8.13}$$

Then in Equation 8.8 (series expansion of orthogonal eigenstates), we have $C_0 = 0$ from where it follows from Equation 8.10 that

$$\bar{H} = \sum_{n\neq 0} |C_n|^2 E_n \geq E_1 \sum_{n\neq 0} |C_n|^2 = E_1 \tag{8.14}$$

where E_1 is the energy of the first excited state. It follows that if a wave function satisfies the variational principle and the auxiliary condition in Equation 8.13, then we have the inequality in Equation 8.14. It follows that $\Psi = \Psi_1$ is the exact eigenfunction of the first excited state. Usually, in the variational calculus, the problem of finding the minimum of an integral leads to the Schrödinger equation that should be solvable.

We observe that the variational method determines the upper bound of the energy on the ground-state energy which is the absolute minimum of the expectation value of the Hamiltonian. So for this, we can also select a wave function (variational wave function) which imitates the ground-state wave function. So for the solution of the Schrödinger equation, we find the minimum of the variational functional with the help of a test function (variational or trial wave function), $\Psi(q;\alpha,\beta,\gamma,\ldots)$ with arbitrary parameters, $\alpha,\beta,\gamma,\ldots$ which help to minimize the expectation value. The test function is normalized and satisfies Equation 8.2. It follows that in Equation 8.1, we evaluate the integral and have $\bar{\mathrm{H}}(\alpha,\beta,\gamma,\ldots)$ and, in particular, we have Equation 8.12:

$$\frac{\partial\bar{\mathrm{H}}}{\partial\alpha}=0,\frac{\partial\bar{\mathrm{H}}}{\partial\beta}=0,\ldots \tag{8.15}$$

We find from Equation 8.15 the numerical values of $\alpha,\beta,\gamma,\ldots$ and substitute them in Ψ and, consequently, $\bar{\mathrm{H}}$; we find $\bar{\mathrm{H}}[\Psi]$ which is then the upper bound of the correct ground-state energy. It is necessary to try a handful of variational functions with more variational parameters and select the ones for which the energy has the least upper bound. When the upper bound can no longer be lowered, then we should have achieved the least upper bound.

The variational method can also be applied to estimate the upper bound of the energy eigenvalues of excited states. The wave function should be selected such that the condition in Equation 8.12 is satisfied. Though the variation method can give good estimates of the energy eigenvalues, the wave function may not be achieved accurately.

8.2 Case of the Hydrogen Atom

We have studied in Chapter 6 the exact solution of the hydrogen atom. We calculate now the ground state of the hydrogen atom using the variational method by using the Hamiltonian, $\hat{\mathrm{H}}$, for the hydrogen atom in the center of mass frame:

$$\hat{\mathrm{H}}=-\frac{\hbar^2}{2M}\Delta-\frac{e^2}{r} \tag{8.16}$$

where M is the electronic mass assuming that the proton mass is infinitely heavy and, otherwise, the reduced mass of the system. Since rotation is the symmetry of the system, the angular momentum is conserved. From Equation 8.16:

$$\bar{\mathrm{H}}=\frac{\hbar^2}{2M}\int|\nabla\Psi|^2\,dq-e^2\int\frac{|\Psi|^2\,dq}{r} \tag{8.17}$$

We find the wave function of the ground state (with one parameter) where Azimuthal quantum number, $l=0$. We select a spherical symmetric variational wave function (for a bound state) independent of θ and ϕ which at infinity should tend to zero:

$$\Psi(r)=C\exp\{-\alpha r\} \tag{8.18}$$

Here, α is a variational parameter and C can be found from the normalization condition:

$$\int\Psi^2(r)\,dq=C^2 4\pi\int_0^\infty\exp\{-2\alpha r\}r^2\,dr=C^2 4\pi\frac{2}{(2\alpha)^3}=\frac{C^2\pi}{\alpha^3}=1 \tag{8.19}$$

For the expectation value of the kinetic energy, $\bar{\mathrm{T}}$:

$$\bar{\mathrm{T}}=\frac{\hbar^2}{2M}\int|\nabla\Psi|^2\,dq=4\pi\int_0^\infty\left(\frac{d\Psi}{dr}\right)^2 r^2dr=\frac{\hbar^2}{2M}4\pi C^2\alpha^2\int_0^\infty\exp\{-2\alpha r\}r^2dr=\frac{\hbar^2}{2M}4\pi C^2\alpha^2\frac{2}{8\alpha^3}=\alpha^2=\frac{\hbar^2\alpha^2}{2M} \quad (8.20)$$

and the expectation value of the potential energy, $\bar{\mathrm{U}}$:

$$\bar{\mathrm{U}}=-e^2 4\pi\int_0^\infty\Psi^2(r)rdr=-e^2 4\pi C^2\int_0^\infty\exp\{-2\alpha r\}rdr=-\frac{e^2 4\pi C^2}{4\alpha^2}=-e^2\alpha \quad (8.21)$$

We find the expectation value of the total energy through

$$\frac{\partial}{\partial\alpha}\bar{\mathrm{H}}=\frac{\hbar^2\alpha}{M}-e^2=0 \quad (8.22)$$

from where the ground-state energy is as follows:

$$E_0=\frac{\hbar^2}{2M}\frac{M^2e^4}{\hbar^4}-\frac{Me^4}{\hbar^2}=-\frac{Me^4}{2\hbar^2},\ \alpha=\frac{Me^2}{\hbar^2}=\frac{1}{a} \quad (8.23)$$

Here, a is the Bohr radius. So

$$\Psi(r)=C\exp\left\{-\frac{r}{a}\right\} \quad (8.24)$$

is the exact wave function of the ground state of the hydrogen atom.

The reader should verify this with another variational wave function:

$$\Psi(r)=C\exp\{-\beta r^2\} \quad (8.25)$$

8.3 Perturbation Theory

8.3.1 Stationary Perturbation Theory – Non-Degenerate Level Case

As earlier seen in quantum mechanics and also classical mechanics, there are relatively few systems of physical interest where the equations of motion can be solved exactly. Approximate methods are thus expected to play an important role in virtually all applications of the theory. For quantum mechanics, an exact solution of the Schrödinger equation exists only for few idealized problems. The perturbation theory is applied to those cases in which the real system can be described by a small change in an easily solvable idealized system. This implies that the perturbation theory should be a method of the evaluation of the correction to the eigenenergies and eigenfunctions for a small change in the Hamiltonian of the system. This method is very useful where it is not possible to calculate the exact spectrum and exact eigenfunctions of the Hamiltonian, $\hat{\mathrm{H}}$. However, in such a case, the exact spectrum and eigenfunctions of the Hamiltonian $\hat{\mathrm{H}}_0$ are known. In addition, the quantity $\hat{\mathrm{H}}-\hat{\mathrm{H}}_0$ should be a small quantity where the criterion of smallness is specified under the next heading. Under this heading, we examine the perturbation approach which is applicable for physical systems when their Hamiltonians are independent of time and thus the name stationary perturbation theory.

Presentation of The Problem

The Hamiltonian, $\hat{H}$, of the system that is of course Hermitian:

$$\hat{H} = \hat{H}^{\dagger} \equiv \hat{H}_0 + \hat{v} \tag{8.26}$$

where the Hamiltonian (Hamiltonian of the unperturbed system), $\hat{H}_0$, has a sufficiently simple structure with its Schrödinger equation exactly solvable. The operator, $\hat{v}$, is small enough and considered as a perturbation on $\hat{H}_0$. The operator, $\hat{H}_0$, is the Hamiltonian of the zero-order approximation.

For the perturbation theory, we find the changes in the discrete and non-degenerate eigenenergy levels and eigenfunctions of the system when a small perturbation is introduced. The exact solution of the following Schrödinger equation,

$$\hat{H}\Psi = E\Psi \tag{8.27}$$

can be obtained in the form of the Taylor series expansion in the parameter $\hat{v}$, where Ψ is the wave function. For the stationary perturbation theory, we find the changes in the discrete energy levels and eigenfunctions of the system when a small perturbation is applied. We solve Equation 8.27 while selecting $\Psi_m^{(0)}$ as the basis eigenfunction of the operation, $\hat{H}_0$. We suppose that the eigenfunctions, $\Psi_m^{(0)}$, correspond to the discrete eigenvalues, $E_m^{(0)}$:

$$\hat{H}_0\Psi_m^{(0)} = E_m^{(0)}\Psi_m^{(0)} \tag{8.28}$$

We first examine the absence of degeneracy. We find the correction to $E_m^{(0)}$ and $\Psi_m^{(0)}$ through series expansion of the desired exact function, Ψ, in terms of the known solution, $\Psi_m^{(0)}$ (complete basis function), of the unperturbed system:

$$\Psi = \sum_m C_m \Psi_m^{(0)} \tag{8.29}$$

where C_m are expansion coefficients. Considering $\hat{H}$ from Equation 8.26 and Ψ from Equation 8.29, we solve Equation 8.27, while considering Equation 8.28:

$$\hat{H}\Psi = \left(\hat{H}_0 + \hat{v}\right)\Psi = \left(\hat{H}_0 + \hat{v}\right)\sum_m C_m \Psi_m^{(0)} = E\sum_m C_m \Psi_m^{(0)} \tag{8.30}$$

From Equation 8.28:

$$\left(\hat{H}_0 + \hat{v}\right)\sum_m C_m \Psi_m^{(0)} = \sum_m C_m \hat{H}_0 \Psi_m^{(0)} + \sum_m C_m \hat{v}\Psi_m^{(0)} = \sum_m C_m E_m^{(0)}\Psi_m^{(0)} + \sum_m C_m \hat{v}\Psi_m^{(0)} \tag{8.31}$$

So

$$\sum_m C_m \left(E_m^{(0)} + \hat{v}\right)\Psi_m^{(0)} = E\sum_m C_m \Psi_m^{(0)} \tag{8.32}$$

If we multiply both sides of this equation by $\Psi_k^{*(0)}$ and integrate over the configuration coordinate q, then from the orthogonality of the basis eigenfunctions, we have

$$C_k\left(E_k^{(0)} - E\right) + \sum_{m\neq k} C_m \mathrm{v}_{km} = 0 \tag{8.33}$$

where

$$\mathrm{v}_{km} = \int \Psi_k^{*(0)} \hat{\mathrm{v}} \Psi_m^{(0)}\, dq \equiv \langle k|\hat{\mathrm{v}}|m\rangle \tag{8.34}$$

is the matrix element of the perturbation operator, $\hat{\mathrm{v}}$, constructed on the basis function of the operator $\hat{H}_0$ and

$$\int \Psi_k^{*(0)} \Psi_m^{(0)}\, dq = \delta_{km} \equiv \langle k|m\rangle \tag{8.35}$$

Equation 8.33 is a system of algebraic equations for the coefficients C_m. To write these equations in an explicit form, we select k and find its solution when the determinant is equal to zero. This yields a secular equation for which, if exactly solvable, the Schrödinger equation is exactly solvable. For this, the perturbation theory is not applicable and otherwise applicable.

So when the perturbation theory is applicable, we solve Equation 8.33 approximately by series expansion both the desired expansion coefficients C_m and the energy eigenvalues, E_k:

$$C_m = C_m^{(0)} + C_m^{(1)} + C_m^{(2)} + \cdots \tag{8.36}$$

and

$$E = E_n^{(0)} + E_n^{(1)} + E_n^{(2)} + \cdots \tag{8.37}$$

Let the exact Hamiltonian $\hat{H}$ differ slightly from $\hat{H}_0$, the exact eigenvalue, E, should differ slightly from the eigenvalue of $\hat{H}_0$. So each eigenvalue, $E_n^{(0)}$, of the Hamiltonian $\hat{H}_0$ slightly differs from the corresponding eigenvalue of $\hat{H}$. The quantity (correction) $E_n^{(0)}$ is the eigenvalue of the energy of the zero approximation while the index, n numbers that state. Considering Equations 8.33, 8.36 and 8.37, we have

$$\left(C_k^{(0)} + C_k^{(1)} + C_k^{(2)} + \cdots\right)\left(E_k^{(0)} - E_n^{(0)} - E_n^{(1)} - E_n^{(2)} - \cdots\right) + \sum_{m\neq k}\left(C_m^{(0)} + C_m^{(1)} + C_m^{(2)} + \cdots\right)\mathrm{v}_{km} = 0 \tag{8.38}$$

For the zero approximation, we have

$$C_m^{(0)} = \delta_{mn} \tag{8.39}$$

This is obvious from Equation 8.29:

$$\Psi_n^{(0)} = \sum_m C_m^{(0)} \Psi_m^{(0)} = \sum_m \delta_{mn} \Psi_m^{(0)} \tag{8.40}$$

To solve the problem in the first approximation, we let in Equation 8.38 $k=n$ and neglect terms of higher order of smallness compared to the first order where $E_n^{(1)} \approx \mathrm{V}_{nn}$. Then for the zero approximation, we have $\mathrm{C}_k^{(0)}$ and from Equation 8.39:

$$\mathrm{C}_k^{(0)} = 1 \tag{8.41}$$

and so

$$1 \times \left(\mathrm{E}_n^{(0)} - \mathrm{E}_n^{(0)} - \mathrm{E}_n^{(1)}\right) + \mathrm{V}_{nn} = 0 \tag{8.42}$$

or

$$\mathrm{E}_n^{(1)} = \mathrm{V}_{nn} \tag{8.43}$$

with

$$\mathrm{E} = \mathrm{E}_n^{(0)} + \mathrm{V}_{nn} \tag{8.44}$$

Equation 8.43 is the correction to the energy of the non-degenerate energy level in the first approximation of the perturbation theory. This correction corresponds to the diagonal matrix element of the perturbation operator which is equal to the expectation value of the perturbation in the non-degenerate state. The given relation does not give the possibility to define the correction of the first order to the wave function. This can be obtained from Equation 8.38 if we substitute for $\mathrm{C}_m^{(1)}$.

Consider again Equation 8.38 for the case when $n \neq k$, the term of the first order of smallness is conserved and

$$C_k = \mathrm{C}_k^{(0)} + \mathrm{C}_k^{(1)} \tag{8.45}$$

From Equation 8.38, it follows that $\mathrm{C}_k^{(0)} = 0$ and so

$$\mathrm{C}_k^{(1)}\left(\mathrm{E}_k^{(0)} - \mathrm{E}_n^{(0)}\right) + \mathrm{V}_{kn} = 0 \tag{8.46}$$

or

$$\mathrm{C}_k^{(1)} = \frac{\mathrm{V}_{kn}}{\mathrm{E}_n^{(0)} - \mathrm{E}_k^{(0)}} = \frac{\langle k|\hat{\mathrm{V}}|n\rangle}{\mathrm{E}_n^{(0)} - \mathrm{E}_k^{(0)}} \tag{8.47}$$

The energy level $E_n^{(0)}$ is non-degenerate, and if it were degenerate, then $E_n^{(0)}$ should have been equal to $E_k^{(0)}$, although $n \neq k$. There arises the question on $\mathrm{C}_k^{(1)}$ which is arbitrary. Let

$$\mathrm{C}_k^{(1)} = 0 \tag{8.48}$$

then we have the normalization condition as follows:

$$\sum_m |C_m|^2 = 1 \tag{8.49}$$

It is necessary to write it to some approximation on which we have

$$\sum_m \left| C_m^{(0)} + C_m^{(1)} \right|^2 = 1 + o\left(v^2\right) \tag{8.50}$$

and if $m = n$, then

$$\left| 1 + C_n^{(1)} \right|^2 = 1 + o\left(v^2\right) \tag{8.51}$$

The terms for which $m \neq n$ are considered to include the infinitesimal function $o\left(v^2\right)$. If $C_n \neq 0$, then we get terms equivalent to $C_n^{(1)}$ which is the quantity of the first order and follows that the condition of normalization should be broken in the first order.

From the correction to the wave function in the first approximation, the eigenfunction

$$\Psi = \Psi_n^{(0)} + \Psi_n^{(1)} \tag{8.52}$$

Here

$$\Psi_n^{(1)} = \sum_{m \neq n} C_m^{(1)} \Psi_m^{(0)} = \sum_{m \neq n} \frac{v_{mn}}{E_n^{(0)} - E_m^{(0)}} \Psi_m^{(0)} \tag{8.53}$$

Already in the first order of the perturbation theory, in the combination of the wave function, the wave function of the zero approximation of all other states is found. The quantity v_{mn} is the measure of the mixing of wave functions in the zero approximation.

From the second order of the perturbation theory:

$$E = E_n^{(0)} + E_n^{(1)} + E_n^{(2)} \tag{8.54}$$

Considering Equation 8.38 and letting $n = k$, as a consequence, the conserved terms are proportional to v^2:

$$-\left(E_n^{(1)} + E_n^{(2)}\right) + v_{nn} + \sum_{m \neq n} C_m^{(1)} v_{nm} = 0 \tag{8.55}$$

If we consider Equation 8.43, then

$$E_n^{(2)} = \sum_{m \neq n} C_m^{(1)} v_{nm} \tag{8.56}$$

From Equation 8.47, we have

$$E_n^{(2)} = \sum_{m \neq n} \frac{\left| v_{nm} \right|^2}{E_n^{(0)} - E_m^{(0)}} \tag{8.57}$$

From Hermiticity,

$$v_{nm} = v_{mn}^* \tag{8.58}$$

So the eigenenergy and the eigenfunctions are as follows:

$$E = \mathrm{E}_n^{(0)} + \mathrm{v}_{nn} + \sum_{m \neq n} \frac{|\mathrm{v}_{nm}|^2}{\mathrm{E}_n^{(0)} - \mathrm{E}_m^{(0)}} \tag{8.59}$$

$$\Psi = \Psi_n^{(0)} + \Psi_n^{(1)} + \Psi_n^{(2)} \tag{8.60}$$

The criterion of the approximation of the perturbation theory is the condition of the convergence of a series. It has been possible to establish the necessary condition: The correction should be less than the quantity that we search for. The physical sense is conserved if, after the correction, the interval between the energies $E_n^{(0)}$ and $E_m^{(0)}$ is considerable:

$$|\mathrm{v}_{nm}| \ll \left|\mathrm{E}_n^{(0)} - \mathrm{E}_m^{(0)}\right| \tag{8.61}$$

for $m \neq n$ and

$$|\mathrm{v}_{mn}| \ll \left|\mathrm{E}_n^{(0)} - \mathrm{E}_m^{(0)}\right| \tag{8.62}$$

This is necessary in order that $\mathrm{C}_n^{(1)} \ll 1$. **Equation 8.62 is the necessary condition for the application of the perturbation theory as well as the criterion of smallness of the perturbation.** If n is the ground state, then **always**

$$\mathrm{E}_n^{(2)} = \sum_{m \neq n} \frac{|\mathrm{v}_{nm}|^2}{\mathrm{E}_n^{(0)} - \mathrm{E}_m^{(0)}} < 0 \tag{8.63}$$

As the energies $E_m^{(0)}$ and $E_n^{(0)}$ are very close to each other for large quantum numbers in the Coulombic field, the perturbation theory can only be applied to the case of strongly bound states. It was required during the derivation of the perturbation formulae that the nature of the spectra not be changed quantitatively. The perturbed states Ψ_n should continuously emerge from the unperturbed states $\Psi_n^{(0)}$ if the perturbation $\hat{\mathrm{v}}$ is turned on. The sufficient condition is a difficult question that can be solved by the quantum theory.

Remark:
Considering Equation 8.57, it is obvious that the correction of the second order to the energy of the ground state is always less than zero.

Anharmonic Oscillator
The Hamiltonian, $\hat{\mathrm{H}}$, of the anharmonic oscillator may be written as follows:

$$\hat{\mathrm{H}} = \frac{\hat{\mathrm{P}}^2}{2M} + \frac{M\omega^2 q^2}{2} + \alpha q^3 \tag{8.64}$$

where the last term αq^3 is a small anharmonism and M, ω, q and $\hat{\mathrm{P}}$ are, respectively, the mass, oscillatory frequency, coordinate and the operator of the momentum. We consider the solution of the harmonic oscillator treated earlier by finding the matrix element of the anharmonism αq^3 in Equation 8.64:

$$\left(q^3\right)_{n,n-3} = q_{n,n-1}q_{n-1,n-2}q_{n-2,n-3} = \left(\frac{\hbar}{M\omega}\right)^{\frac{3}{2}}\left(\frac{n(n-1)(n-2)}{8}\right)^{\frac{1}{2}} = \left(q^3\right)_{n-3,n} \tag{8.65}$$

From here, then we have

$$\left(q^3\right)_{n-1,n} = q_{n-1,n}q_{n,n-1}q_{n-1,n} + q_{n-1,n}q_{n,n+1}q_{n+1,n} + q_{n-1,n-2}q_{n-2,n-1}q_{n-1,n} \tag{8.66}$$

or

$$\left(q^3\right)_{n-1,n} = \left(\frac{\hbar}{8M\omega}\right)^{\frac{3}{2}}\left(n^{\frac{3}{2}} + n^{\frac{1}{2}}(n+1) + n^{\frac{1}{2}}(n-1)\right) = 3\left(\frac{n\hbar}{8M\omega}\right)^{\frac{3}{2}} \tag{8.67}$$

There are no diagonal matrix elements since the correction to the first-order perturbation:

$$E^{(1)} = \alpha\langle n|q^3|n\rangle = 0 \tag{8.68}$$

The correction to the second-order perturbation is as follows:

$$E^{(2)} = \alpha^2\sum_m \frac{\langle n|q^3|m\rangle\langle m|q^3|n\rangle}{E_n^{(0)} - E_m^{(0)}} \tag{8.69}$$

or

$$\begin{aligned} E^{(2)} = {} & \alpha^2\frac{\langle n|q^3|n-3\rangle\langle n-3|q^3|n\rangle}{E_n^{(0)} - E_{n-3}^{(0)}} + \alpha^2\frac{\langle n|q^3|n+3\rangle\langle n+3|q^3|n\rangle}{E_n^{(0)} - E_{n+3}^{(0)}} \\ & + \alpha^2\frac{\langle n|q^3|n-1\rangle\langle n-1|q^3|n\rangle}{E_n^{(0)} - E_{n-1}^{(0)}} + \alpha^2\frac{\langle n|q^3|n+1\rangle\langle n+1|q^3|n\rangle}{E_n^{(0)} - E_{n+1}^{(0)}} \end{aligned} \tag{8.70}$$

or

$$\begin{aligned} E^{(2)} = {} & \alpha^2\frac{\langle n|q^3|n-3\rangle\langle n-3|q^3|n\rangle}{3\hbar\omega} + \alpha^2\frac{\langle n|q^3|n+3\rangle\langle n+3|q^3|n\rangle}{-3\hbar\omega} \\ & + \alpha^2\frac{\langle n|q^3|n-1\rangle\langle n-1|q^3|n\rangle}{\hbar\omega} + \alpha^2\frac{\langle n|q^3|n+1\rangle\langle n+1|q^3|n\rangle}{-\hbar\omega} \end{aligned} \tag{8.71}$$

From Equations 8.65 and 8.67:

$$\left(q^3\right)_{n,n+3} = q_{n,n+1}q_{n+1,n+2}q_{n+2,n+3} = \left(\frac{\hbar}{M\omega}\right)^{\frac{3}{2}}\left(\frac{(n+1)(n+2)(n+3)}{8}\right)^{\frac{1}{2}} = \left(q^3\right)_{n+3,n} \tag{8.72}$$

$$\left(q^3\right)_{n,n-1} = q_{n,n-1}q_{n-1,n}q_{n,n-1} + q_{n,n+1}q_{n+1,n}q_{n,n-1} + q_{n,n-1}q_{n-1,n-2}q_{n-2,n-1} \tag{8.73}$$

or

$$\left(q^3\right)_{n,n-1} = \left(\frac{\hbar}{8M\omega}\right)^{\frac{3}{2}}\left(n^{\frac{3}{2}} + n^{\frac{1}{2}}(n+1) + n^{\frac{1}{2}}(n-1)\right) = 3\left(\frac{n\hbar}{8M\omega}\right)^{\frac{3}{2}} \tag{8.74}$$

and

$$\left(q^3\right)_{n+1,n} = \left(q^3\right)_{n,n+1} = 3\left(\frac{(n+1)\hbar}{8M\omega}\right)^{\frac{3}{2}} \tag{8.75}$$

then

$$E^{(2)} = \frac{\alpha^2}{\hbar\omega}\left(\frac{\hbar}{M\omega}\right)^3\left(\frac{n(n-1)(n-2)}{3\times 8} - \frac{(n+3)(n+2)(n+1)}{3\times 8} + \frac{9n^3}{8} - \frac{9(n+1)^3}{8}\right) \tag{8.76}$$

After simplification, this always yields a negative correction as follows:

$$E^{(2)} = -\frac{15\alpha^2}{4\hbar\omega}\left(\frac{\hbar}{M\omega}\right)^3\left(n^2 + n + \frac{11}{30}\right) \tag{8.77}$$

So the energy of the anharmonic oscillator via the perturbation theory is as follows:

$$E_n = \hbar\omega\left(n + \frac{1}{2}\right) - \frac{15\alpha^2}{4\hbar\omega}\left(\frac{\hbar}{M\omega}\right)^3\left(n^2 + n + \frac{11}{30}\right) \tag{8.78}$$

8.4 Perturbation Theory – Case of a Degenerate Level

The Schrödinger equation in the $\hat{H}_0$-representation in the form of a system of linear algebraic equations:

$$C_k\left(E_k^{(0)} - E_n\right) + \sum_m C_m v_{km} = 0 \tag{8.79}$$

where k index the level and m the terms in the equation. The index n is that of the state of the zero approximation. The energy eigenvalue, $E_n^{(0)}$, has f-fold degeneracy and so the following functions or any of their linear combination have the same energy:

$$\Psi_{n\alpha}^{(0)}, \alpha = 1, 2, \ldots, f \tag{8.80}$$

Now, it is not clear the type of functions (or proper functions) which should be selected for the zero approximation to perform the calculations. However, due to the perturbation, the system chooses to achieve a particular state and when the perturbation is switched off, the state goes into a specific linear combination of f-unperturbed states. This implies the correct unperturbed state that we begin with. So using this, the perturbation calculations will be indeed valid. Choosing a different initial state with the same energy, the terms in the perturbation series will be large. When we infer from the nearly degenerate case, then there is a large mixing and so the states are expected to be further diagonalized. So as for the nearly degenerate case, we select as for the initial states:

$$\Psi_n^{(0)} = \sum_{\alpha} C_{\alpha}^{(0)} \Psi_{n\alpha}^{(0)} \tag{8.81}$$

Here, we assume that all the states, $\Psi_{n\alpha}^{(0)}$, are degenerate and have the energy eigenvalues, $E_n^{(0)}$, while the coefficients $C_{\alpha}^{(0)}$ for now are unknown. Applying in Equation 8.79 the first-order perturbation:

$$E_n = E_n^{(0)} + E^{(1)} \tag{8.82}$$

where $E^{(1)}$ is the correction of the first order. Selecting the equation when $k = \alpha$:

$$C_{\alpha}^{(0)}\left(-E^{(1)}\right) + \sum_{\beta=1}^{f} C_{\beta}^{(0)} v_{\alpha\beta} = 0 \tag{8.83}$$

This is a system of f linear homogenous algebraic equations relative to the coefficients defined in Equation 8.81 and α the number of equations and β the terms of the equations. From Equation 8.83:

$$\begin{array}{l} \alpha = 1: \\ \alpha = 2: \\ \vdots \;: \\ \alpha = f: \end{array} \left\{ \begin{array}{ccccccccc} \left(v_{11} - E^{(1)}\right)C_1^{(0)} & + & v_{12}C_2^{(0)} & + & \cdots & + & v_{1f}C_f^{(0)} & = & 0 \\ v_{21}C_1^{(0)} & + & \left(v_{22} - E^{(1)}\right)C_2^{(0)} & + & \cdots & + & v_{2f}C_f^{(0)} & = & 0 \\ \cdots & \cdots & \cdots & \cdots & \cdots & \cdots & \cdots & \cdots & \cdots \\ v_{f1}C_1^{(0)} & + & v_{f2}C_2^{(0)} & + & \cdots & + & \left(v_{ff} - E^{(1)}\right)C_f^{(0)} & = & 0 \end{array} \right. \tag{8.84}$$

This system has a non-trivial solution when its $f \times f$-determinant equals zero:

$$\begin{vmatrix} v_{11} - E^{(1)} & v_{12} & \cdots & v_{1f} \\ v_{21} & v_{22} - E^{(1)} & \cdots & v_{2f} \\ \cdots & \cdots & \cdots & \cdots \\ v_{f1} & v_{f2} & \cdots & v_{ff} - E^{(1)} \end{vmatrix} = 0 \tag{8.85}$$

This is a secular algebraic equation of the first order relative to $E^{(1)}$, and if we solve it, then we have f roots which are first-order corrections to the energy levels:

$$E_1^{(1)}, E_2^{(1)}, \ldots, E_f^{(1)} \tag{8.86}$$

These roots are real since they are eigenvalues of a Hermitian matrix.

Let the matrix $v_{\alpha\beta}$ has f rows and f columns, then we write the equation of the transformation of the matrix to the diagonal form where eigenvalues are real. We find the correction to the first order of the energy. In the zero approximation, the value of the energy is f-fold degenerate and in the first approximation the degenerate level splits as a result of the perturbation. The perturbation removes the degeneracy completely in the first order when all roots are distinct. Substituting the roots into Equation 8.84, we solve for the coefficients, and considering Equation 8.81, it is seen that each root should correspond to its own function of the zero approximation:

$$\Psi_1^{(0)},\Psi_2^{(0)},\ldots,\Psi_f^{(0)} \tag{8.87}$$

They are defined to the approximation of a constant factor that is defined from the normalization condition. Thus, in our theory, we solve two problems:

1. Finding the correction of the first order to the energy
2. Finding the proper wave functions of the zero approximation – this is also the linear combination of the wave functions of the zero approximation that transforms the matrix operator of the perturbation to the diagonal form or such wave functions that change slightly as a result of the perturbation

Consider the inverse problem and we claim knowledge of the exact solution of the full Schrödinger equation:

$$E_1 \to \Psi_1;\, E_2 \to \Psi_2 \tag{8.88}$$

They all contain the parameter v. Letting $v \to 0$, we get the levels from Equation 8.79 and the functions correspond to the proper wave functions of the zero approximation $\left(\Psi_1 \to \Psi_1^{(0)}\right)$.

For the non-degenerate case, $E_n^{(1)} = v_{nn}$. However, for now, this is not true. Let us guess the wave function of the zero approximation and the secular equation yields

$$\begin{vmatrix} v_{11}-E^{(1)} & 0 & 0 & \cdots & 0 \\ 0 & v_{22}-E^{(1)} & 0 & \cdots & 0 \\ \cdots & \cdots & \cdots & \cdots & \cdots \\ 0 & 0 & 0 & \cdots & v_{ff}-E^{(1)} \end{vmatrix} = 0 \tag{8.89}$$

Consider the case when all the elements $v_{\alpha\beta} = 0$. If we consider Equation 8.85, then

$$E^{(1)} = 0 \tag{8.90}$$

So in the first approximation, there is no splitting of the degenerate level. Since v fails to remove the degeneracy in the first order, it is necessary to return to Equation 8.79 with the correction of the second order of the proper wave function:

$$E_n = E_n^{(0)} + E^{(2)} \tag{8.91}$$

Suppose in Equation 8.83 $k = \alpha$:

$$C_\alpha^{(0)}\left(-E^{(2)}\right) + \sum_{m\neq\alpha} C_m^{(1)} v_{\alpha m} \tag{8.92}$$

Let us substitute $C_m^{(1)}$ when $m \neq \alpha$. For this, we examine Equation 8.79 and let $k = m$:

$$C_m^{(1)}\left(E_m^{(0)} - E_n^{(0)}\right) + \sum_{\beta} C_\beta^{(0)} v_{m\beta} = 0 \tag{8.93}$$

or

$$C_m^{(1)} = \frac{\sum_{\beta} C_\beta^{(0)} v_{m\beta}}{E_n^{(0)} - E_m^{(0)}} \tag{8.94}$$

If we substitute Equation 8.94 into 8.92, then

$$C_\alpha^{(0)}\left(-E^{(2)}\right) + \sum_{\beta} C_\beta^{(0)} \sum_{m \neq \alpha} \frac{v_{\alpha m} v_{m\beta}}{E_n^{(0)} - E_m^{(0)}} = 0 \tag{8.95}$$

Compare Equation 8.95 with 8.83 and denote

$$v_{\alpha\beta}^{(2)} = \sum_{m \neq \alpha} \frac{v_{\alpha m} v_{m\beta}}{E_n^{(0)} - E_m^{(0)}} \tag{8.96}$$

This formally should coincide with Equation 8.83. Further, we have the final solution when we get the second correction to the energy.

Suppose we make the substitution

$$\alpha = \beta = n \tag{8.97}$$

into Equation 8.96, we get the non-degenerate case. The quantity $v_{\alpha\beta}^{(2)}$ is the transition matrix element through the virtual state.

8.4.1 The Stark Effect

We examine an example of the perturbation theory where we calculate the splitting of the energy level of a hydrogen atom subjected to a homogenous electric field. We see under this heading that the effect of the electric field on a hydrogen atom is to split the spectral lines. **Stark** experimentally showed this phenomenon in 1913. When the electron moves in a central symmetric field in the presence of an external homogenous electric field ε (the field is in the direction of the oz-axis), then full Hamiltonian of the system:

$$\hat{H} = \hat{H}_0 - e\varepsilon z \tag{8.98}$$

where $\hat{H}_0$ is the usual Hamiltonian of the electron moving in the central symmetric field, and $e\varepsilon z$ is the extra energy of the nucleus and the electron in the external electric field or energy of interaction of the electron with the external field ε. The quantity

$$\hat{v} = -e\varepsilon z \tag{8.99}$$

is a perturbation and ε is a small parameter.

8.4.1.1 Hydrogen Atom

We study the hydrogen atom that will permit us to bring out important quantum mechanical features. The hydrogen atom is a two-particle system consisting of a negatively charged electron having a Coulomb interaction with a positively charged proton. We examine the Hamiltonian of the hydrogen atom:

$$\hat{H} = -\frac{\hbar^2}{2m_1}\Delta_1 - \frac{\hbar^2}{2m_2}\Delta_2 + U(\vec{r}_2 - \vec{r}_1) \tag{8.100}$$

Here the first and second summands are the kinetic energy operators of the electron and the proton, respectively, and the third summand is the Coulomb potential energy due to the Coulomb interaction of the electron and the proton.

We suppose that the electron and proton interact via a potential dependent only on the relative separation of the two particles. This will lead us to separate the motion of the system of particles into two distinct and decoupled parts, i.e., the part describing the motion of the center of mass of the system and the other describing the relative motion of the two particles. For this, we introduce the coordinate of the center of mass, $\vec{R}$, and that of the relative coordinate, $\vec{r}$, of the system of the electron and the proton:

$$\vec{R} = \frac{m_2\vec{r}_2 + m_1\vec{r}_1}{m_2 + m_1}, \quad \vec{r} = \vec{r}_2 - \vec{r}_1 \tag{8.101}$$

From here:

$$\vec{r}_1 = \vec{R} - \frac{m_2}{m_2 + m_1}\vec{r}, \ \vec{r}_2 = \vec{R} + \frac{m_1}{m_2 + m_1}\vec{r} \tag{8.102}$$

where $\vec{r}_1, m_1$ and $\vec{r}_2, m_2$ are, respectively, the coordinate and mass of the electron and proton.

We find the explicit expressions of the Laplace operators Δ_1 and Δ_2 in the coordinate of the center of mass, $\vec{R}$, and of relative motion, $\vec{r}$:

$$\Delta_1 = \frac{\partial^2}{\partial x_1^2} + \frac{\partial^2}{\partial y_1^2} + \frac{\partial^2}{\partial z_1^2} \tag{8.103}$$

$$\frac{\partial}{\partial \vec{r}_2} = \frac{\partial}{\partial \vec{R}}\frac{\partial \vec{R}}{\partial \vec{r}_2} + \frac{\partial}{\partial \vec{r}}\frac{\partial \vec{r}}{\partial \vec{r}_2} = \frac{m_2}{m_2 + m_1}\frac{\partial}{\partial \vec{R}} + \frac{\partial}{\partial \vec{r}} \tag{8.104}$$

then

$$\frac{\partial^2}{\partial \vec{r}_2^{\,2}} = \frac{\partial}{\partial \vec{r}_2}\left(\frac{m_2}{m_2 + m_1}\frac{\partial}{\partial \vec{R}} + \frac{\partial}{\partial \vec{r}}\right) \tag{8.105}$$

or

$$\frac{\partial^2}{\partial \vec{r}_2^{\,2}} = \frac{\partial}{\partial \vec{R}}\left(\frac{m_2}{m_2 + m_1}\frac{\partial}{\partial \vec{R}} + \frac{\partial}{\partial \vec{r}}\right)\frac{\partial \vec{R}}{\partial \vec{r}_2} + \frac{\partial}{\partial \vec{r}}\left(\frac{m_2}{m_2 + m_1}\frac{\partial}{\partial \vec{R}} + \frac{\partial}{\partial \vec{r}}\right)\frac{\partial \vec{r}}{\partial \vec{r}_2} \tag{8.106}$$

or

$$\frac{\partial^2}{\partial \vec{r}_2^{\,2}}=\left(\frac{m_2}{m_2+m_1}\right)^2\frac{\partial^2}{\partial \vec{R}^2}+\frac{\partial^2}{\partial \vec{r}^{\,2}}+2\frac{m_2}{m_2+m_1}\frac{\partial}{\partial \vec{R}}\frac{\partial}{\partial \vec{r}} \tag{8.107}$$

Also,

$$\frac{\partial}{\partial \vec{r}_1}=\frac{\partial}{\partial \vec{R}}\frac{\partial \vec{R}}{\partial \vec{r}_1}+\frac{\partial}{\partial \vec{r}}\frac{\partial \vec{r}}{\partial \vec{r}_1}=\frac{m_1}{m_2+m_1}\frac{\partial}{\partial \vec{R}}-\frac{\partial}{\partial \vec{r}} \tag{8.108}$$

then

$$\frac{\partial^2}{\partial \vec{r}_1^{\,2}}=\frac{\partial}{\partial \vec{r}_1}\left(\frac{m_1}{m_2+m_1}\frac{\partial}{\partial \vec{R}}-\frac{\partial}{\partial \vec{r}}\right) \tag{8.109}$$

$$\frac{\partial^2}{\partial \vec{r}_1^{\,2}}=\frac{\partial}{\partial \vec{R}}\left(\frac{m_1}{m_2+m_1}\frac{\partial}{\partial \vec{R}}-\frac{\partial}{\partial \vec{r}}\right)\frac{\partial \vec{R}}{\partial \vec{r}_1}+\frac{\partial}{\partial \vec{r}}\left(\frac{m_1}{m_2+m_1}\frac{\partial}{\partial \vec{R}}-\frac{\partial}{\partial \vec{r}}\right)\frac{\partial \vec{r}}{\partial \vec{r}_1} \tag{8.110}$$

or

$$\frac{\partial^2}{\partial \vec{r}_1^{\,2}}=\left(\frac{m_1}{m_2+m_1}\right)^2\frac{\partial^2}{\partial \vec{R}^2}+\frac{\partial^2}{\partial \vec{r}^{\,2}}-2\frac{m_1}{m_2+m_1}\frac{\partial}{\partial \vec{R}}\frac{\partial}{\partial \vec{r}} \tag{8.111}$$

So

$$\begin{aligned}\frac{1}{m_1}\Delta_1+\frac{1}{m_2}\Delta_2=&\frac{1}{m_1}\left[\left(\frac{m_1}{m_2+m_1}\right)^2\frac{\partial^2}{\partial \vec{R}^2}+\frac{\partial^2}{\partial \vec{r}^{\,2}}-2\frac{m_1}{m_2+m_1}\frac{\partial}{\partial \vec{R}}\frac{\partial}{\partial \vec{r}}\right]\\&+\frac{1}{m_2}\left[\left(\frac{m_2}{m_2+m_1}\right)^2\frac{\partial^2}{\partial \vec{R}^2}+\frac{\partial^2}{\partial \vec{r}^{\,2}}+2\frac{m_2}{m_2+m_1}\frac{\partial}{\partial \vec{R}}\frac{\partial}{\partial \vec{r}}\right]\end{aligned} \tag{8.112}$$

or

$$\frac{1}{m_1}\Delta_1+\frac{1}{m_2}\Delta_2=\frac{1}{\mu}\frac{\partial^2}{\partial \vec{r}^{\,2}}+\frac{1}{M}\frac{\partial^2}{\partial \vec{R}^2} \tag{8.113}$$

Here

$$m_1\left(\frac{1}{m_2+m_1}\right)^2+m_2\left(\frac{1}{m_2+m_1}\right)^2=\frac{\mu_1+\mu_2}{M}\equiv\frac{1}{M} \tag{8.114}$$

$$m_2+m_1\equiv M,\ \frac{m_1}{m_2+m_1}\equiv\mu_1,\ \frac{m_2}{m_2+m_1}\equiv\mu_2,\ \frac{1}{m_1}+\frac{1}{m_2}\equiv\frac{1}{\mu} \tag{8.115}$$

Here, M and μ are the total mass and reduced mass of the system, respectively. So the Hamiltonian of the hydrogen atom after the change of variables:

$$\hat{H}=-\frac{\hbar^2}{2\mu}\Delta_{\vec{r}}-\frac{\hbar^2}{2M}\Delta_{\vec{R}}+U(\vec{r}) \tag{8.116}$$

We see from here that the motion of our system can equivalently be described by the motion of two fictitious particles, i.e., one with the total mass M and with the center of mass coordinate, $\vec{R}$, and the other with reduced mass, μ, and relative coordinate, $\vec{r}$. From the Hamiltonian, $\hat{H}$, we observe that the motion of the two particles is uncoupled. From here and considering the Schrödinger equation:

$$\hat{H}\Psi = E\Psi \tag{8.117}$$

the wave function Ψ is expected to be represented in the form of separation of variables:

$$\Psi = X_R \psi(r,\theta,\phi) \tag{8.118}$$

So

$$\left(-\frac{\hbar^2}{2\mu}\Delta_{\vec{r}} - \frac{\hbar^2}{2M}\Delta_{\vec{R}} + U(\vec{r})\right) X_R \psi(r,\theta,\phi) = E X_R \psi(r,\theta,\phi) \tag{8.119}$$

From here:

$$-\frac{\hbar^2}{2M}\Delta_{\vec{R}} X_R = \varepsilon_\kappa X_R,\ \varepsilon_\kappa = \frac{\hbar^2\kappa^2}{2M} \tag{8.120}$$

with the solution

$$X_R = \exp\left\{i\vec{\kappa}\vec{R}\right\} \tag{8.121}$$

This describes the free motion of the centre of mass of the hydrogen atom. The normalization factor for this wave function will be included in the solution of the radial equation which will be obtained from the equation of the relative motion described by the following Schrödinger equation:

$$\left(-\frac{\hbar^2}{2\mu}\Delta_{\vec{r}} + U(\vec{r})\right)\psi(r,\theta,\phi) = \mathcal{E}\psi(r,\theta,\phi) \tag{8.122}$$

8.4.2 Stark Effect (Spherical Coordinates)

We consider now the Stark effect described by the Hamiltonian in Equation 8.98, where

$$\hat{H}_0 = -\frac{\hbar^2}{2\mu}\Delta_{\vec{r}} - \frac{\hbar^2}{2M}\Delta_{\vec{R}} + U(\vec{r}) \tag{8.123}$$

The wave function of the zero approximation is as follows:

$$\psi_{nlm}^{(0)} = R_{nl}(r)Y_{lm}(\theta,\phi) \tag{8.124}$$

The eigenvalue, $E_{nl}^{(0)}$, of the zero approximation is dependent only on n with l and not on m where we have $(2l+1)$ multiplicity of degeneracy. We have to find the splitting of the energy levels due to the field ε. If $l \neq 0$, then the level is degenerate. For the motion in the central field, there exist two integrals of motion: the angular momentum and the parity. The Hamiltonian $\hat{H}$ of the system is invariant relative to inversion:

$$\vec{r} \to -\vec{r} \tag{8.125}$$

For this, the quantity

$$|\vec{r}| = \sqrt{x^2 + y^2 + z^2} \tag{8.126}$$

does not change. The kinetic energy does not change also as it contains the second derivative with respect to the coordinates. The center of the field is the center of inversion. The eigenfunction of the operator $\hat{H}$ should be the eigenfunction of the operator of inversion, i.e., it should have the property

$$\psi(-\vec{r}) = \psi(\vec{r}) \tag{8.127}$$

or

$$\psi(-\vec{r}) = -\psi(\vec{r}) \tag{8.128}$$

Suppose we are given the inversion operator $\hat{I}$:

$$\hat{I}\psi(\vec{r}) = \psi(-\vec{r}) \tag{8.129}$$

$$\hat{I}^2\psi(\vec{r}) = \psi(\vec{r}) \tag{8.130}$$

then, from Equation 8.128:

$$I = \pm 1 \tag{8.131}$$

If we have $I = +1$, then the wave function is even, and if $I = -1$, then it is odd. It can be proven that the parity, P:

$$P = (-1)^l \tag{8.132}$$

i.e., the state with an even Azimuthal quantum number l is even and that with an odd l is odd.

Consider the matrix element of the perturbation operator $\hat{v}$:

$$\langle nlm|z|nlm'\rangle = 0 \tag{8.133}$$

This matrix element is equal to zero considering the reasoning that z is an odd function. We may write the matrix element in the form:

$$\iint dx\,dy\,dz\,\psi_{nlm}(\vec{r})\,z\,\psi_{nlm'}(\vec{r}) \tag{8.134}$$

Suppose the wave function ψ should be even. If we do the change of variable:

$$x \to -x,\ y \to -y,\ z \to -z \tag{8.135}$$

nothing changes except that

$$z \to -z \tag{8.136}$$

The entire integral changes sign and so the integral should be equal to zero. It is same when ψ is odd and also all the matrix elements are equal to zero. This implies that there is no effect that is linear in ε. So the first-order correction to the ground-state energy is zero and follows from the symmetry of the system.

Further, under this heading, we examine effects that are quadratic in ε. However, it will be shown that for the case of the hydrogen atom, the linear effect in ε exists. We may show this in another way: As the operator of the perturbation $\hat{v}$ acts on $R_{nl}(r)$ and $Y_{lm}(\theta,\phi)$, we have to write the variable z through the spherical system of coordinates:

$$z = r\cos\theta \tag{8.137}$$

We examine the most general form of the matrix elements as follows:

$$\langle n'l'm'|z|nlm\rangle = \int_0^\infty R_{n'l'}^*(r)R_{nl}(r)r^3dr\int_0^\pi\int_0^{2\pi} Y_{l'm'}^*(\theta,\phi)Y_{lm}(\theta,\phi)\cos\theta\sin\theta d\theta d\phi \tag{8.138}$$

Here

$$Y_{lm}(\theta,\phi) = P_l^{|m|}(\cos\theta)\exp\{im\phi\} \tag{8.139}$$

From where, considering Equation 8.138:

$$\int_0^{2\pi} \exp\{i(m-m')\phi\}d\phi = 2\pi\delta_{m'm} \tag{8.140}$$

and considering the change of variable, $\cos\theta = \xi$, we have

$$\int_{-1}^{1} P_{l'}^{|m|}(\xi)P_l^{|m|}(\xi)\xi d\varsigma = \begin{cases} \neq 0, & l' = l \pm 1 \\ 0, & l' = l \end{cases} \tag{8.141}$$

This implies that in our case, all the matrix elements of the operator $\hat{v}$ are equal to zero and so we have to evaluate those matrix elements for equal value of l and different m. For different m, they are also equal to zero. It thus follows that the correction of the first order disappears and we resort to the second order:

$$E_n^{(2)} = e^2\varepsilon^2 \sum_{n\neq 1,\, l,m} \frac{|\langle 100|z|nlm\rangle|^2}{E_1 - E_n},\; E_n = -\frac{E_{\text{Ryd}}}{n^2} \tag{8.142}$$

as ε is of small order. Here, the Rydberg

$$E_{\text{Ryd}} = \frac{\mu e^4}{2\hbar^2} = \frac{e^2}{2a} \cong 13.6\text{eV},\; a = \frac{\hbar^2}{\mu e^2} \tag{8.143}$$

and a is the Bohr radius. The difference, $E_1 - E_n$, is always negative and so the ground state is lowered. It follows that for the electron we have a **quadratic Stark effect**. There is no first-order Stark effect. Any system with a center of inversion has this effect. But for the central symmetric field, there is an exception – for the hydrogen-like atom where a **linear Stark effect** (the energy being proportional to ε) is observed. This is connected with the conditional degeneracy. The energy levels of the hydrogen atom (e.g., the Ballmer series) for which fields split are proportional to the first power of the field strength ε. For the motion in a Coulomb field, we have $E_n^{(0)}$ (which is not due to l) for every given

$$n, l = 0, 1, \ldots, n-1 \tag{8.144}$$

Suppose we solve the problem for the first excited state for the hydrogen-like atom:

$$n = 2, l = 0, 1 \tag{8.145}$$

The second matrix element between the $2p$ and $2s$ states should not be equal to zero as in the integrand, we should have $\Psi_l(\vec{r})\Psi_{l'}(\vec{r})$ with one of the eigenfunctions which is an even function and the other odd function while z changes sign. This follows that the integral does not change sign. This implies that the integral is different from zero. So the sum in Equation 8.142 is different from zero as there exist states $|nlm\rangle$ with parity opposite to that of $|100\rangle$.

In the first correction to the energy:

$$E_n^{(1)} \neq 0 \tag{8.146}$$

This is the **linear Stark effect**.

The first level

$$\Psi_{100} \equiv |100\rangle \tag{8.147}$$

of the hydrogen atom is not degenerate. Then for the simplest case we start from the splitting of the level, $n = 2$. The hydrogen eigenstates are n^2-fold degenerate:

$$\sum_{l=0}^{n-1}(2l+1) = n^2 \tag{8.148}$$

So the state is four-fold degenerate:

$$|2,0,0\rangle, |2,1,-1\rangle, |2,1,0\rangle, |2,1,1\rangle \tag{8.149}$$

with all the four states having the same energy eigenvalue:

$$E_n^{(0)} = E_2^{(0)} \tag{8.150}$$

of the unperturbed hydrogen atom. As the parity P of the state defined in Equation 8.132 is dependent only on the Azimuthal quantum number l, all the degenerate states do not have the same parity.

The wave function for the $2s$-state is an even function:

$$\psi_1 = \psi_{200} \equiv |2,0,0\rangle = \frac{1}{\sqrt{2a^3}}\left(1 - \frac{r}{2a}\right)\exp\left\{-\frac{r}{2a}\right\}Y_{00}(\theta,\phi) \tag{8.151}$$

TABLE 8.1 Four States Denoted by Index α and Quantum Number

	2s	2p		
α	1	2	3	4
m	0	0	1	-1

and the $2p$-state are three odd functions:

$$\psi_2 = \psi_{210} \equiv |2,1,0\rangle = \frac{1}{\sqrt{6a^3}} \frac{r}{2a} \exp\left\{-\frac{r}{2a}\right\} Y_{10}(\theta,\phi),\ \Psi_{3,4} = \Psi_{2,1,\pm1} \equiv |2,1,\pm1\rangle$$

$$= \frac{1}{\sqrt{6a^3}} \frac{r}{2a} \exp\left\{-\frac{r}{2a}\right\} Y_{1,\pm1}(\theta,\phi) \tag{8.152}$$

where the above spherical harmonics are as follows:

$$Y_{00}(\theta,\phi) = \frac{1}{\sqrt{4\pi}},\ Y_{10}(\theta,\phi) = i\sqrt{\frac{3}{4\pi}}\cos\theta,\ Y_{1,\pm1}(\theta,\phi) = \mp i\sqrt{\frac{3}{8\pi}}\sin\theta\exp\{\pm i\phi\} \tag{8.153}$$

The four-fold degeneracy is lifted by the appearance of an electric field. For the first excited state, we have four-fold degeneracy. For the 2s, we have $m = 0$, and for the $2p$, we have

$$m = 1,0,-1 \tag{8.154}$$

Let us examine the four states. We assign the index α to them:

Table 8.1 **shows four states denoted by the index α and the quantum number** m.

We examine on the basis of this, the matrix elements again. Here

$$v_{\alpha\alpha} = 0 \tag{8.155}$$

as considering Equation 8.171, the matrix elements that are different from zero may be those for the states for which the l differ from each other by unity. As only

$$v_{12} = v_{21} \neq 0 \tag{8.156}$$

then the secular equation for the first-order Stark effect of the $n = 2$ of hydrogen is as follows:

$$\begin{vmatrix} -E^{(1)} & v_{12} & 0 & 0 \\ v_{21} & -E^{(1)} & 0 & 0 \\ 0 & 0 & -E^{(1)} & 0 \\ 0 & 0 & 0 & -E^{(1)} \end{vmatrix} = 0 \tag{8.157}$$

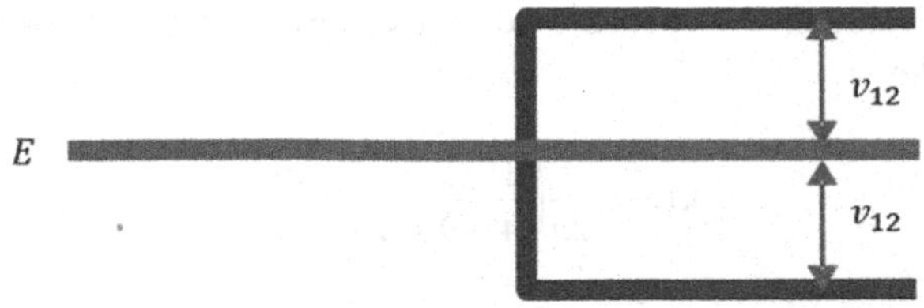

FIGURE 8.1 The first-order corrections to the energy of the first excited state of the hydrogen atom.

or

$$\left[\left(E^{(1)}\right)^2 - v_{12}^2\right]\left(E^{(1)}\right)^2 = 0 \tag{8.158}$$

with roots

$$E_{1,2}^{(1)} = \pm v_{12},\ E_{3,4}^{(1)} = 0 \tag{8.159}$$

that are the first-order corrections to the energy of the first excited state of the hydrogen atom that splits into three sub-levels (Figure 8.1) where two of the levels are non-degenerate and one two-fold degenerate. If we consider the spin of the electron, then the multiplicity of degeneracy is doubled.

This splitting corresponds to the first order of the field ε. This follows that we have a linear Stark effect. We observe that the linear Stark effect leads to two levels of opposite parities and the same energy, i.e., for the 2s and 2p states. This is a singular case for the hydrogen atom due to the l-fold degeneracy of the $n \neq 1$ shells. In this problem, we observe the degeneracy being lifted only partially. This means that the applied electric field breaks the accidental symmetry and lifts the degeneracy in the l-quantum numbers. As the field is applied along the z-direction, there still remains a rotational symmetry about the z-axis which, as a consequence, implies the degeneracy in the m-quantum numbers persist.

If we consider Equation 8.159, we get four roots:

$$|v_{12}|, -|v_{12}|, 0, 0 \tag{8.160}$$

So that half of the four-fold degeneracy is removed in the first order. The last two values of $E_{3,4}^{(1)}$ correspond to any two linearly independent combinations of Ψ_{211} and $\Psi_{2,1,-1}$, and the first and second values correspond to $\frac{1}{\sqrt{2}}(\Psi_{200} - \Psi_{210})$ and $\frac{1}{\sqrt{2}}(\Psi_{200} + \Psi_{210})$, respectively. This implies that a hydrogen atom in its first excited state behaves as though it has a permanent electric dipole moment of magnitude $3ea$ that can be oriented in three different ways: one state parallel to the external field, one state anti-parallel to the field and two states with zero component along the field.

From

$$v_{12} = v_{21} = -\frac{e|\varepsilon|}{12a^4}\int_0^\infty \left(1 - \frac{r}{2a}\right)\exp\left\{-\frac{r}{a}\right\} r^4 dr \int d\Omega |Y_{10}|^2 = -\frac{e|\varepsilon|a}{12}\int_0^\infty \xi^4 d\xi\left(1 - \frac{\xi}{2}\right)\exp\{-\xi\} = 3e|\varepsilon|a,\ \xi \equiv \frac{r}{a} \tag{8.161}$$

This follows that our results are true if ε is sufficiently small as it plays the role of a parameter. From here, it appears that the splitting of the energy levels of the hydrogen atom in an electric field is proportional to the value of the applied field. The criterion of the application of the results in Equation 8.159 is as follows:

$$3e\varepsilon a \ll \frac{e^2}{2a}\left(\frac{1}{4}-\frac{1}{9}\right), \frac{e^2}{2a}=\frac{\mu e^4}{2\hbar^2} \tag{8.162}$$

and

$$\varepsilon \ll \frac{e}{2a^2}\frac{5}{216} \approx \frac{1}{43} \tag{8.163}$$

It is important to note from the above that the Stark-splitting should be exceedingly greater than the fine structure interval and, however, exceedingly less than the difference of the energy levels of the non-degenerate hydrogen atom. It is also instructive to note that states of $n=2$ level are not stable. The lifetime of the 2s state is much longer than that of the 2p state. This is due to the fact that the atom passes easily from the 2p to 1s state by Lyman α-photon spontaneous emission, whereas the decay from the 2s state entails a two-photon emission. So the 2p states are **unstable** while the 2*s* state is **metastable**. Since Equation 8.156 is satisfied, any static or oscillating electric field leads to the **mixing** of the 2s **metastable** state with that of the 2p **unstable** state, thereby leading to the reduction of the 2s state lifetime in the so-called **metastable state quenching** phenomenon.

8.4.3 Stark Effect (Parabolic Coordinates)

First Order

We have earlier used parabolic coordinates for the problem of the hydrogen atom and now for the hydrogen atom subjected to an electric field where the energy due to the electric field (perturbation) is as follows:

$$\hat{v} = -e\varepsilon z \equiv \in z = \frac{\in}{2}(\xi-\eta) \tag{8.164}$$

We are interested in the matrix element for the following transition where n does not change:

$$n_1 n_2 m \rightarrow n_1' n_2' m' \tag{8.165}$$

From the matrix elements, only the diagonal elements will be different from zero:

$$\mathrm{I} = \int |\Psi_{n_1n_2m}|^2 \in z\,dV = \frac{\in}{8}\int_0^\infty\int_0^\infty\int_0^{2\pi} |\Psi_{n_1n_2m}|^2(\xi-\eta)(\xi+\eta)\,d\xi\,d\eta\,d\phi \tag{8.166}$$

The wave function, $\Psi_{n_1n_2m}$, describing the discrete spectrum is defined as follows:

$$\Psi_{n_1n_2m} = \frac{\sqrt{2}}{n^2} f_{n_1m}\left(\frac{\xi}{n}\right) f_{n_2m}\left(\frac{\eta}{n}\right)\frac{1}{\sqrt{2\pi}}\exp\{im\phi\},\ f_{nm}(\rho) = \frac{1}{|m|!}\sqrt{\frac{(|m|+n)!}{n!}}\exp\left\{-\frac{\rho}{2}\right\}\rho^{\frac{|m|}{2}}\mathrm{F}(-n,|m|+1;\rho) \tag{8.167}$$

where $\mathrm{F}(-n,|m|+1;\rho)$ is the **confluent hypergeometric function**. Considering Equation 8.167, Equation 8.166 becomes

$$\mathrm{I} = \int |\Psi_{n_1n_2m}|^2 \in z\,dV = \frac{\in}{4n^4}\int_0^\infty\int_0^\infty f_{n_1m}^2\left(\frac{\xi}{n}\right) f_{n_2m}^2\left(\frac{\eta}{n}\right)(\xi-\eta)(\xi+\eta)\,d\xi\,d\eta \tag{8.168}$$

In order to calculate Equation 8.168, we find first the following integrals by letting $\frac{\xi}{n} \equiv z$:

$$I^0_{n_1|m|} = \int_0^\infty f^2_{n_1 m}\left(\frac{\xi}{n}\right)d\xi \equiv n\int_0^\infty f^2_{n_1 m}(z)dzn\left(\frac{1}{|m|!}\sqrt{\frac{(|m|+n_1)!}{n_1!}}\right)^2 \int_0^\infty dz F^2(-n_1,|m|+1;z)\exp\{-z\}z^{|m|} \tag{8.169}$$

Express the confluent hypergeometric function through the associated Laguerre polynomials:

$$F(-n_1,|m|+1;z) = \frac{1}{C_{n_1}} L^{|m|}_{n_1}(z),\ C_{n_1} = \frac{\Gamma(|m|+n_1+1)}{\Gamma(|m|+1)} \tag{8.170}$$

then the integral in Equation 8.169 now becomes

$$\begin{aligned} I^0_{n_1|m|} &= n\left(\frac{1}{|m|!}\sqrt{\frac{(|m|+n_1)!}{n_1!}}\frac{\Gamma(|m|+1)}{\Gamma(|m|+n_1+1)}\right)^2 \int_0^\infty dz\left[L^{|m|}_{n_1}(z)\right]^2 \exp\{-z\}z^{|m|} \\ &= n\left(\frac{1}{|m|!}\sqrt{\frac{(|m|+n_1)!}{n_1!}}\frac{\Gamma(|m|+1)}{\Gamma(|m|+n_1+1)}\right)^2 n_1!\Gamma(|m|+n_1+1) \end{aligned} \tag{8.171}$$

Likewise,

$$I^0_{n_2|m|} = \int_0^\infty f^2_{n_2 m}\left(\frac{\eta}{n}\right)d\eta = n\left(\frac{1}{|m|!}\sqrt{\frac{(|m|+n_2)!}{n_2!}}\frac{\Gamma(|m|+1)}{\Gamma(|m|+n_2+1)}\right)^2 n_2!\Gamma(|m|+n_2+1) \tag{8.172}$$

From

$$\int_0^\infty \rho^2 f^2_{nm}(\rho)d\rho = \left(\frac{1}{|m|!}\sqrt{\frac{(|m|+n)!}{n!}}\right)^2 \int_0^\infty \exp\{-\rho\}\rho^{|m|+2}F^2(-n,|m|+1;\rho)d\rho = \left(\frac{\Gamma(|m|+n+1)}{|m|!}\right)^2 \tag{8.173}$$

So the energy due to the linear Stark effect:

$$\int |\Psi_{n_1 n_2 m}|^2 \in z\,dV = \frac{3}{4}\in n(n_1-n_2) = E^{(1)} \tag{8.174}$$

or in ordinary units:

$$E^{(1)} = \frac{3}{4}\varepsilon|e|\frac{\hbar^2}{\mu e^2}n(n_1-n_2) \tag{8.175}$$

Second Order

The potential energy in parabolic coordinates has the form:

$$U = -\frac{Ze^2}{r} - e\varepsilon z = \frac{\in}{2}(\xi-\eta) - \frac{2Ze^2}{\xi+\eta} \tag{8.176}$$

The Schrödinger equation in parabolic coordinates has the form:

$$-\frac{\hbar^2}{2\mu}\left\{\frac{4}{\xi+\eta}\left[\frac{\partial}{\partial\xi}\left(\xi\frac{\partial}{\partial\xi}\Psi\right)+\frac{\partial}{\partial\eta}\left(\eta\frac{\partial}{\partial\eta}\Psi\right)\right]+\frac{1}{\xi\eta}\frac{\partial^2}{\partial\phi^2}\Psi\right\}+\left(\frac{\in}{2}(\xi-\eta)-\frac{2Ze^2}{\xi+\eta}\right)\Psi=\mathrm{E}\Psi,\ E<0 \quad (8.177)$$

Using the method of separation of variables:

$$\Psi=f_1(\xi)f_2(\eta)\Phi(\phi) \quad (8.178)$$

This permits us to have the following equations:

$$\frac{d}{d\xi}\left(\xi\frac{d}{d\xi}f_1\right)-\left(\frac{\mu|E|}{2\hbar^2}\xi+\frac{m^2}{4\xi}+\frac{\in}{4}\xi^2+\beta_1\right)f_1=0,\ \frac{d}{d\eta}\left(\eta\frac{d}{d\eta}f_2\right)-\left(\frac{\mu|E|}{2\hbar^2}\eta+\frac{m^2}{4\eta}-\frac{\in}{4}\eta^2-\beta_2\right)f_2=0 \quad (8.179)$$

The total wave function is written so that

$$n_1=\frac{|m|+1}{2}+n\beta_1,\ n_2=-\frac{|m|+1}{2}+n\beta_2 \quad (8.180)$$

Letting

$$\frac{\xi}{n}\equiv\rho_1,\ \frac{\eta}{n}\equiv\rho_2,\ n=\Delta^{-1},\ \Delta=\sqrt{-2E} \quad (8.181)$$

then

$$f_1(\rho)=\sqrt{\Delta}f_{n_1m}(\xi\Delta),\ f_2(\rho)=\sqrt{\Delta}f_{n_2m}(\eta\Delta) \quad (8.182)$$

and the zero-order corrections in the perturbation theory are, respectively, as follows:

$$\beta_1^{(0)}=\left(n_1+\frac{|m|+1}{2}\right)\Delta,\ \beta_2^{(0)}=\left(n_2+\frac{|m|+1}{2}\right)\Delta \quad (8.183)$$

The normalization of the wave functions f_1 and f_2:

$$\int_0^\infty f_1^2(\xi)d\xi=1,\ \int_0^\infty f_2^2(\eta)d\eta=1 \quad (8.184)$$

The perturbation corresponding to the functions f_1 and f_2 are, respectively, $\frac{\in}{4}\xi^2$ and $-\frac{\in}{4}\eta^2$, and the first-order corrections are then, respectively,

$$\beta_1^{(1)}=\frac{\in}{4}\int_0^\infty f_1^2(\xi)\xi^2d\xi=\frac{\in}{4}n^2\left(6n_1^2+6n_2|m|+m^2+6n_1+3|m|+2\right) \quad (8.185)$$

$$\beta_2^{(1)} = -\frac{\in}{4}\int_0^{\infty} f_2^2(\eta)\eta^2 d\eta - = \frac{\in}{4} n^2\left(6n_2^2 + 6n_2|m| + m^2 + 6n_2 + 3|m| + 2\right) \tag{8.186}$$

From the general formula of the perturbation theory, we have the second order of the perturbation theory:

$$\beta_1^{(2)} = \frac{\in^2}{16}\sum_{n_1' \neq n_1} \frac{\left|\left(\xi^2\right)_{n_1' n_1}\right|^2}{\beta_1^{(0)}(n_1') - \beta_1^{(0)}(n_1)} \tag{8.187}$$

Here

$$\left(\xi^2\right)_{n_1' n_1} = \frac{1}{|m|!}\sqrt{\frac{(|m| + n_1)!}{n_1!}}\frac{1}{|m|!}\sqrt{\frac{(|m| + n_1')!}{n_1'!}}\int_0^{\infty} \frac{(\xi\Delta)^2}{\Delta^2}\exp\{-\xi\Delta\}(\xi\Delta)^{|m|} F(-n_1, |m| + 1; \xi\Delta)$$
$$\times F(-n_1', |m| + 1; \xi\Delta) d(\xi\Delta) \tag{8.188}$$

From

$$\beta_1 + \beta_2 = \beta_1^{(0)} + \beta_2^{(0)} + \beta_1^{(1)} + \beta_2^{(1)} + \beta_1^{(2)} + \beta_2^{(2)} \tag{8.189}$$

then the energy considering the second order of the perturbation theory is as follows:

$$E = -\frac{1}{2n^2} + \frac{3}{2}\in n(n_1 - n_2) - \frac{3}{2}\in^2 n^4\left[17n^2 - (n_1 - n_2)^2 - 9m^2 + 19\right] \tag{8.190}$$

8.5 Time-Dependent Perturbation Theory

One of the principal tasks in quantum mechanics is the **evaluation of transition probabilities** from say state Ψ_n to another state Ψ_m. This takes place under the influence of a time-dependent perturbation $v(t)$ that perturbs the system and causes transition between energy eigenstates. When does the transition of a system from one state to another make sense? It makes sense if the cause of the transition, i.e., $v(t)$, acts only within a finite time period, say from $t = t_0$ to $t = T$. With the exception of this time period, the total energy is a **constant of motion** that can be measured.

For the case when the Hamiltonian is dependent on time t, there are no stationary solutions of the Schrödinger equation. Hence, the identification of bound states with discrete energy levels and stationary eigenfunctions must be modified. The **time-dependent perturbation theory** or the so-called **method of variation of constants** assures that. The Hamiltonian, $\hat{H}$, of the system is as follows:

$$\hat{H}(t) = \hat{H}_0 + \hat{v}(t) \tag{8.191}$$

where $\hat{H}_0$ describes a conservative system, i.e., that which is independent of time t. The index, 0, stands for time independence. The perturbation $\hat{v}(t)$ is dependent on time t and has the effect of causing transitions between eigenstates of $\hat{H}_0$ which is stationary in the absence of $\hat{v}(t)$. For convenience, other variables on which $\hat{v}(t)$ is dependent will not be shown.

Instead of decomposing the Hamiltonian in Equation 8.191, the adiabatic approximation assures that $\hat{H}$ has parameters that change very slowly with time. This follows that the system can be described approximately by means of stationary eigenfunctions of the instantaneous Hamiltonian. We finally consider the sudden approximation in which $\hat{H}$ is constant in time except for very short time intervals where it changes from one form to another.

If we have the Hamiltonian in Equation 8.191, then the state of a system should be described by the time Schrödinger equation as follows:

$$i\hbar\frac{\partial}{\partial t}\Psi = \hat{H}\Psi \tag{8.192}$$

Problem

We solve Equation 8.192 considering the perturbation, $\hat{v}(t)$, to be time dependent and so

$$\Psi_m^{(0)}(t) = \Psi_m^{(0)}\exp\left\{-i\frac{E_m^{(0)}t}{\hbar}\right\} \tag{8.193}$$

where $\Psi_m^{(0)}(t)$ is the eigenfunction and $E_m^{(0)}$ is the eigenvalue of the Hamiltonian $\hat{H}_0$. Equation 8.193 is the solution of Equation 8.192 considering the Hamiltonian, $\hat{H}_0$. The functions in Equation 8.193 form a **complete set of eigenfunctions**. The solution of Equation 8.192 may be series expanded in terms of these eigenfunctions with the expansion coefficients, $C_m(t)$, dependent on time:

$$\Psi(t) = \sum_m C_m(t)\Psi_m^{(0)}(t) \tag{8.194}$$

The eigenfunctions, $\Psi_m^{(0)}(t)$, satisfy the following Schrödinger equation:

$$i\hbar\frac{\partial}{\partial t}\Psi_m^{(0)}(t) = \hat{H}_0\Psi_m^{(0)}(t) \tag{8.195}$$

If the Hamiltonian $\hat{H}$ should have not been dependent on time t, then C_m should also be independent of time, t. If we substitute Equation 8.194 into 8.192 and considering Equation 8.191:

$$i\hbar\frac{\partial}{\partial t}\sum_m C_m(t)\Psi_m^{(0)}(t) = \left(\hat{H}_0 + \hat{v}(t)\right)\sum_m C_m(t)\Psi_m^{(0)}(t) \tag{8.196}$$

or

$$i\hbar\sum_m\left(\Psi_m^{(0)}(t)\frac{\partial}{\partial t}C_m(t) + C_m(t)\frac{\partial}{\partial t}\Psi_m^{(0)}(t)\right) = \sum_m\left(\hat{H}_0C_m(t)\Psi_m^{(0)}(t) + \hat{v}(t)C_m(t)\Psi_m^{(0)}(t)\right) \tag{8.197}$$

From Equation 8.195:

$$\sum_m\left(i\hbar\Psi_m^{(0)}(t)\frac{\partial}{\partial t}C_m(t) + C_m(t)\hat{H}_0\Psi_m^{(0)}(t)\right) = \sum_m\left(\hat{H}_0C_m(t)\Psi_m^{(0)}(t) + \hat{v}(t)C_m(t)\Psi_m^{(0)}(t)\right) \tag{8.198}$$

or

$$i\hbar\sum_m\Psi_m^{(0)}(t)\frac{\partial}{\partial t}C_m(t) = \sum_m\hat{v}(t)C_m(t)\Psi_m^{(0)}(t) \tag{8.199}$$

We multiply both sides by $\Psi_k^{*(0)}(t)$ and integrate over the configuration coordinate q, then we have the following equation for the expansion coefficients:

$$i\hbar\frac{\partial}{\partial t}C_k(t)=\sum_m C_m(t)v_{km}(t) \tag{8.200}$$

where $v_{km}(t)$ is the matrix element of the perturbation operator:

$$v_{km}(t)=\int\Psi_k^{*(0)}(t)\hat{v}\Psi_m^{(0)}(t)dq=v_{km}\exp\{i\omega_{km}t\} \tag{8.201}$$

Considering Equation 8.200, we have

$$i\hbar\frac{\partial}{\partial t}C_k(t)=\sum_m C_m(t)v_{km}\exp\{i\omega_{km}t\} \tag{8.202}$$

The matrix element v_{km} may also be dependent on time t if the perturbation itself is dependent explicitly on time t.

Equation 8.202 is the exact equation for the matrix element that is equivalent to the initial time Schrödinger equation in the $\hat{H}_0$-representation. It is the system of equations relative to the coefficients C_m defined in Equation 8.194 where their set is the wave function in the $\hat{H}_0$-representation. Here k numbers the equations and m the terms at the interior of the equations.

Considering C_m, we have a system of linear homogenous equations which are differential equations of the first order (in the case of a stationary system it should be an algebraic equation).

8.5.1 Transition Probability

Assume before the perturbation is switched on, i.e., for $t_0 \to -\infty$, the system is found at the state n of the unperturbed Hamiltonian, i.e.,

$$C_k(t\to-\infty)=\delta_{kn} \tag{8.203}$$

Further, unless otherwise stated, we swap $C_{kn}(t)$ for $C_k(t)$, and so considering the perturbation, v small enough, we write $C_k(t)$ in series expansion relative to the perturbation:

$$C_{kn}(t)=C_{kn}^{(0)}(t)+C_{kn}^{(1)}(t)+C_{kn}^{(2)}(t)+\cdots=\delta_{kn}+C_{kn}^{(1)}(t)+C_{kn}^{(2)}(t)+\cdots \tag{8.204}$$

From here, considering Equation 8.202:

$$\frac{d}{dt}C_{kn}^{(0)}(t)=0 \tag{8.205}$$

This implies that

$$C_{kn}^{(0)}(t)=\text{const}\equiv\delta_{kn} \tag{8.206}$$

From here, substituting Equation 8.204 into 8.202 and equating terms of the same order of the perturbation of both sides of the equality, we have

$$i\hbar\frac{d}{dt}C_{kn}^{(1)}(t)=v_{kn}\exp\{i\omega_{kn}t\},\ i\hbar\frac{\partial}{\partial t}C_{k}^{(2)}(t)=\sum_{m}C_{m}^{(1)}(t)v_{km}\exp\{i\omega_{km}t\} \tag{8.207}$$

From here, considering the initial conditions:

$$C_{kn}^{(1,2)}(t=-\infty)=0 \tag{8.208}$$

then

$$C_{kn}^{(1)}(t)=\frac{1}{i\hbar}\int_{-\infty}^{t}v_{kn}(t')\exp\{i\omega_{kn}t'\}dt' \tag{8.209}$$

$$C_{kn}^{(2)}(t)=-\frac{1}{\hbar^2}\sum_{m}\int_{-\infty}^{t}v_{km}(t')\exp\{i\omega_{km}t'\}dt'\int_{-\infty}^{t'}v_{mn}(t'')\exp\{i\omega_{mn}t''\}dt'' \tag{8.210}$$

The first order in Equation 8.209 describes a direct transition from state n to state k and the second order in Equation 8.210 and higher orders the transitions through intermediate states.

The calculation of the transition probability, $P_{n\to k}$, of the system from the initial n to the final k state when $k\neq n$ after the perturbation is switched off:

$$P_{n\to k}=\left|C_{kn}(t=\infty)\right|^2=\left|C_{kn}^{(1)}(t=\infty)+C_{kn}^{(2)}(t=\infty)+\cdots\right|^2 \tag{8.211}$$

From

$$C_{kn}^{(1)}(t=+\infty)=0 \tag{8.212}$$

the transition probability of the transition of the system from state n to k is as follows:

$$P_{n\to k}^{(2)}=\left|C_{kn}^{(2)}(t=+\infty)\right|^2 \tag{8.213}$$

We calculate the probability $P_{n\to k}$, considering that after when the perturbation is switched on, the state remains invariant. So from Equation 8.204:

$$P_{nn}^{(2)}=\left|C_{nn}(t=\infty)\right|^2=1+\frac{1}{\hbar^2}\left(\int_{-\infty}^{\infty}v_{nn}(t')dt'\right)^2-\frac{1}{\hbar^2}\sum_{m}\int_{-\infty}^{\infty}v_{nm}(t')\exp\{i\omega_{nm}t'\}dt'\int_{-\infty}^{t'}v_{mn}(t'')\exp\{i\omega_{mn}t''\}dt''$$

$$-\frac{1}{\hbar^2}\sum_{m}\int_{-\infty}^{\infty}v_{nm}^{*}(t')\exp\{-i\omega_{nm}t'\}dt'\int_{-\infty}^{t'}v_{nm}^{*}(t'')\exp\{-i\omega_{mn}t''\}dt'' \tag{8.214}$$

Note that the matrix element, $v_{nn}(t)$, is a real quantity and, in addition, $\omega_{kn} = -\omega_{nk}$. From the Hermiticity of the operator, $\hat{v}$, $v^*_{kn}(t) = v_{nk}(t)$ and the following property:

$$\int_{-\infty}^{\infty} dt \int_{-\infty}^{t} dt' \mathrm{F}(t,t') = \int_{-\infty}^{\infty} dt' \int_{t'}^{\infty} dt \mathrm{F}(t,t') \tag{8.215}$$

then the transition probability, $\mathrm{P}^{(2)}_{nn}$, in Equation 8.214 has the form:

$$\mathrm{P}^{(2)}_{nn} = 1 + \frac{1}{\hbar^2}\left|\int_{-\infty}^{\infty} v_{nn}(t)dt\right|^2 - \frac{1}{\hbar^2}\sum_m \left|\int_{-\infty}^{\infty} v_{nm}(t)\exp\{i\omega_{nm}t'\}dt\right|^2 \tag{8.216}$$

or

$$\mathrm{P}^{(2)}_{nn} = 1 - \frac{1}{\hbar^2}\sum_m{}' \left|\int_{-\infty}^{\infty} v_{nm}(t)\exp\{i\omega_{nm}t'\}dt\right|^2 \equiv 1 - \frac{1}{\hbar^2}\sum_m{}' \mathrm{P}^{(1)}_{n\to m} \tag{8.217}$$

The prime in the sum in Equation 8.217 implies the absence of the term with $n = m$. The transition probability, $\mathrm{P}^{(1)}_{n\to m}$, considering the perturbation, v, is small enough, has the form:

$$\mathrm{P}^{(1)}_{n\to m} \equiv \frac{1}{\hbar^2}\left|\int_{-\infty}^{\infty} v_{nm}(t)\exp\{i\omega_{nm}t'\}dt\right|^2 \tag{8.218}$$

This is the probability that after the perturbation is switched on, there is a transition from the n state to the m. The expression in Equation 8.217 expresses the law of conservation of the normalized eigenfunction of the system to the approximation of the terms to the second order when the perturbation is switched on. The probability that after the perturbation is switched on, the state remains at the same state:

$$\mathrm{P}_{n\to n} = 1 + \frac{1}{\hbar^2}\left|\int_{-\infty}^{\infty} v_{nn}(t)dt\right|^2 \tag{8.219}$$

8.5.2 Adiabatic Approximation

In Equation 8.218, the integrand has two factors dependent on time t. Suppose the perturbation, v, which is slowly varying, is switched on, then the time t increases and achieves its maximum value at time τ. If

$$\tau \gg \frac{1}{\omega_{nm}} \tag{8.220}$$

then the transition probability is equal to zero, i.e., when the integrand in Equation 8.218 is a fast-oscillatory function and yields a constant function leading to the so-called adiabatic perturbation switch-on. We have just calculated the transition probability, $P_{n\to k}$, of the system that departs from the initial state, n, to the final, k state when $k \neq n$ after the perturbation is switched off, considering the perturbation to be small enough. There are situations when small changes in the velocity and the perturbation may not be small enough such as small atomic scatterings. In such situations, the adiabatic approximation is used where the solutions of the Schrödinger equation can be approximated through stationary eigenfunctions, $\Psi_m(\vec{r},t)$, of the instantaneous Hamiltonian, $\hat{H}(t)$, with eigenvalues, $E_m(t)$:

$$\hat{H}(t)\Psi_m(\vec{r},t) = E_m(t)\Psi_m(\vec{r},t) \tag{8.221}$$

In this case, a particular eigenfunction at one time evolves continuously into the corresponding eigenfunction at a later time.

We solve Equation 8.221 at each time and consider the system in a discrete non-degenerate state, $\Psi_m(0)$, with energy, $E_m(0)$, at time $t=0$. This implies we find the system probably to be in the state $\Psi_m(t)$ with energy, $E_m(t)$, at time. This is in case when $\hat{H}(t)$ changes slowly enough with time. We expect to estimate the extend to which our expectation is not satisfied in order that other states appear in the series expansion of Ψ of the function, Ψ_m, assumed to be real-valued and normalizable:

$$\frac{d}{dt}\int d\vec{r}\,\Psi_m^2 = 0 \tag{8.222}$$

This implies that

$$\int d\vec{r}\,\Psi_m \frac{\partial}{\partial t}\Psi_m = 0 \tag{8.223}$$

Since the wave function, $\Psi(\vec{r},t)$, satisfies the time-dependent Schrödinger equation in Equation 8.192, we series expand this wave function as follows:

$$\Psi(\vec{r},t) = \sum_m C_m(t)\Psi_m(\vec{r},t)\exp\left\{-\frac{i}{\hbar}\int_{t_0}^{t} E_m(t)dt\right\} \tag{8.224}$$

Here, $C_m(t)$ is the expansion coefficient and is the probability amplitude for finding the system at the moment t and states, Ψ_m, which are orthogonal, discrete and non-degenerate. We substitute Equation 8.224 in the Schrödinger equation in Equation 8.192 and have a system of equations for C_m:

$$\sum_n \left(\frac{d}{dt}C_n(t)\Psi_n + C_n\frac{\partial}{\partial t}\Psi_n\right)\exp\left\{-\frac{i}{\hbar}\int_{t_0}^{t} E_m(t)dt\right\} = 0 \tag{8.225}$$

We multiply this equality from the left-hand side by Ψ_k^* and integrate over the entire space:

$$\frac{d}{dt}C_k(t) = -\sum_{n\neq k} C_m \exp\left\{i\int_{t_0}^{t}\omega_{kn}dt\right\}\int d\vec{r}\Psi_k^*\frac{\partial}{\partial t}\Psi_n,\ \omega_{kn} = \frac{E_k(t) - E_n(t)}{\hbar} \tag{8.226}$$

We compare this with Equation 8.202 and observe that the role of the perturbation provoking the transition is played by the factor, $\int d\vec{r}\Psi_k^* \frac{\partial}{\partial t}\Psi_n$, with the help of Equation 8.221:

$$\frac{\partial \hat{H}(t)}{\partial t}\Psi_n(t)+\hat{H}(t)\frac{\partial \Psi_n(t)}{\partial t}=\frac{\partial E_n(t)}{\partial t}\Psi_n(t)+E_n(t)\frac{\partial \Psi_n(t)}{\partial t} \tag{8.227}$$

Multiply this through from the left-hand side by Ψ_k^* and integrate over the entire space, considering $n \neq k$, then we have

$$\int d\vec{r}\Psi_k^* \frac{\partial}{\partial t}\Psi_n = \frac{1}{E_n(t)-E_k(t)}\int d\vec{r}\Psi_k^* \frac{\partial \hat{H}(t)}{\partial t}\Psi_n \tag{8.228}$$

Substituting this into Equation 8.226:

$$\frac{d}{dt}C_k(t)=\sum_{n\neq k}\frac{C_n}{\hbar\omega_{kn}}\exp\left\{i\int_{t_0}^{t}\omega_{kn}dt\right\}\int d\vec{r}\Psi_k^* \frac{\partial \hat{H}(t)}{\partial t}\Psi_n \tag{8.229}$$

This is the solution of the Schrödinger equation in Equation 8.192 for all k.

We estimate $\frac{d}{dt}C_k(t)$, assuming all the quantities C_k, ω_{kn}, Ψ_n and $\frac{\partial \hat{H}(t)}{\partial t}$, defined on the right-hand side of Equation 8.229 and expected to be slowly changing, are indeed constant in time. Assume also for the system to be in the state m at $t=0$, then we let $C_k = \delta_{nm}$ and so

$$\frac{d}{dt}C_k(t)\approx\frac{1}{\hbar\omega_{kn}}\exp\{i\omega_{km}t\}\int d\vec{r}\Psi_k^* \frac{\partial \hat{H}(t)}{\partial t}\Psi_n,\ k\neq m \tag{8.230}$$

From here:

$$C_k(t)\approx\frac{1}{i\hbar\omega_{km}^2}\left(\exp\{i\omega_{km}t\}-1\right)\int d\vec{r}\Psi_k^* \frac{\partial \hat{H}(t)}{\partial t}\Psi_n,\ k\neq m \tag{8.231}$$

This shows the probability amplitude for a state other than the initial states to oscillate in time. In addition, though $\hat{H}$ changes by a finite quantity, there is no steady increase over exceedingly long periods of time. The magnitude of the given probability amplitude over exceedingly long periods of time is of the order of the ratio of the change in $\hat{H}$ within the Bohr period, $\frac{2\pi}{\omega_{kn}}$, to the energy, $E_k - E_n$ between the states. So there should be the most probable transition between the most probable near energy levels, E_n and E_k, when the matrix element $\Psi_k^* \frac{\partial \hat{H}(t)}{\partial t}\Psi_n$ is not small enough. Hence, the simplest approximation of the system in Equation 8.229 should be two equations responsible for the states between which there exists an enhanced transition while the rest is neglected. So we have the system of equations as follows:

$$\frac{d}{dt}C_1(t)=C_2(t)\Psi_1-\frac{\partial}{\partial t}\Psi_2\exp\left\{i\int_{t_0}^{t}\omega_{12}dt\right\},\ \frac{d}{dt}C_2(t)=C_1(t)\Psi_2-\frac{\partial}{\partial t}\Psi_1\exp\left\{-i\int_{t_0}^{t}\omega_{12}dt\right\} \tag{8.232}$$

For a trivial case,

$$\int d\vec{r}\Psi_1 \frac{\partial}{\partial t}\Psi_2 = 0 \tag{8.233}$$

So, from Equation 8.232 where the expansion coefficients are constant:

$$\Psi(\vec{r},t) = C_1 \exp\left\{-\frac{i}{\hbar}\int_{t_0}^{t} E_1(t)dt\right\} + C_2 \exp\left\{-\frac{i}{\hbar}\int_{t_0}^{t} E_2(t)dt\right\} \tag{8.234}$$

The expansion coefficients C_1 and C_2 can be obtained from the initial conditions.

8.6 Time-Independent Perturbation

We re-examine Equation 8.218 for the time-independent perturbation and do the t integrals from $t = 0$ to $t = T$, and so in the first approximation of the perturbation theory the transition probability from the state with n to that with m, considering the perturbation to be small enough:

$$P_{n\to m} = \frac{1}{\hbar^2}|v_{nm}|^2 \frac{4\sin^2\frac{\omega_{mn}T}{2}}{\omega_{mn}^2} = \frac{1}{\hbar^2} f(\mathrm{T},\omega_{mn})|v_{nm}|^2 \tag{8.235}$$

To understand the main features of the result for the transition probability, we examine the behavior of the function $f(t,\omega)$ for different values of

$$f(t,\omega) = \frac{4\sin^2\frac{\omega t}{2}}{\omega^2} \tag{8.236}$$

8.7 Time and Energy Uncertainty Relation

The function $f(t,\omega)$ shown in Figure 8.2 has basically a large peak (resonant peak), centered at $\omega = 0$, with the height of the peak that increases in proportion to t^2 as proven below:

$$\lim_{\omega\to 0} f(t,\omega) = \lim_{\omega\to 0}\frac{4\sin^2\frac{\omega t}{2}}{\omega^2} = \lim_{\omega\to 0}\frac{\sin^2\frac{\omega t}{2}}{\left(\frac{\omega t}{2}\right)^2}t^2 = t^2 \tag{8.237}$$

and the breadth $\cong \frac{2\pi}{t}$ (decreases inversely to t) with the area under the curve proportional to t. This resonant breadth is the distance between the first zeros on either side of the resonant frequency. The resonant breadth shows that the longer the perturbation acts the narrower the resonance would be. Away from the value of the resonance frequency that equals zero, the function f oscillates with

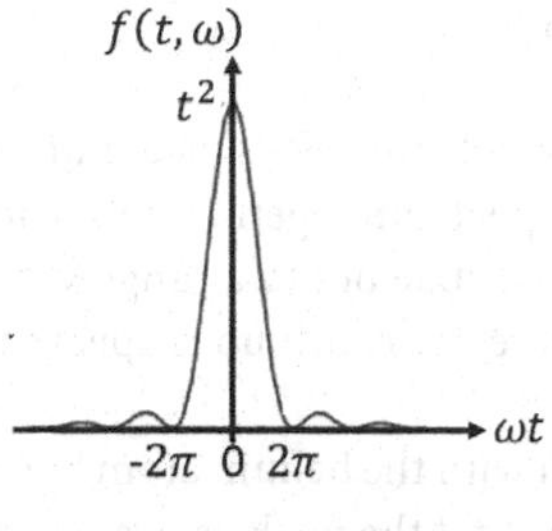

FIGURE 8.2 The dependence of the function $f(t,\omega)$ on the frequency ω.

damped amplitude that imitates a diffraction pattern. So for larger t, there is only an appreciable transition probability for those states whose energy lies in a band of breadth ΔE about the initial energy E_n:

$$t \cong \frac{2\pi}{\omega} = \frac{2\pi\hbar}{|E_m - E_n|} = \frac{2\pi\hbar}{\Delta E} \tag{8.238}$$

So from here:

$$t\Delta E \geq \hbar \tag{8.239}$$

This is the time-energy uncertainty relation and was obtained for the first time by Bohr. For the conservative system, the greater the energy uncertainty, the more rapid the time evolution. For the time t, the perturbation is switched on. The time here is a parameter and not an observable. The perturbation, v is regarded as device permitting measurement of final energy of the system by transferring it to one of the states, say k. The time of measurement is t in order that the uncertainty in energy predicted by Equation 8.238 is of the order $\frac{2\pi}{t}$.

From

$$\lim_{t\to\infty} f(t,\omega) = t\lim_{t\to\infty}\frac{\sin^2\frac{\omega t}{2}}{\left(\frac{\omega}{2}\right)^2 t} = t\pi\delta\left(\frac{\omega}{2}\right) = 2\pi t\delta(\omega) \tag{8.240}$$

and

$$\int_{-\infty}^{\infty}\frac{\sin^2\alpha t}{\alpha^2 t}d\alpha = \pi \tag{8.241}$$

then

$$\int_{-\infty}^{\infty} f(t,\omega)d\omega = 2\pi t \tag{8.242}$$

and $f(t,\omega)$ achieves an infinite peak of area $2\pi t$ as $t \to \infty$ as described by Equation 8.240. So for small t, the peak is broader and implies large energy change. For large t, the peak is narrow (Figure 8.2).

8.8 Density of Final State

We examine the case of a continuous spectrum. We consider one of the cases associated with the operator $\hat{H}_0$ where there exists the discrete spectrum together with the continuous spectrum. So for this, we suppose a transition not to a single final state but to a range ν of final states about the energy E_m. This situation arises in a potential well where the continuous spectrum is situated above the discrete spectrum in the energy scale.

We examine a many-particle system with the helium atom being a good example. The helium nucleus has two protons and two neutrons. Around the nucleus, we have two elections. The two electrons have an attractive force for the nucleus in addition to the repulsion between the two electrons. We assume that the two electrons are distinguishable though in reality not so.

If we consider the spin property of the two electrons, then they are indeed distinguishable. For brevity, we neglect the spin coupling interaction. The helium atom is only slightly more complex than the hydrogen atom and may not be solved analytically. Variational techniques are usually applied to the helium atom that may aid in solving more complex problems. For brevity, we ignore the mutual repulsion between the two electrons then it is possible to talk of the state of each electron separately in the field of the helium nucleus. The system will then be equivalent to two hydrogen atoms. The Hamiltonian for such a problem will then be that of two hydrogen atoms. We can then find the ground-state energy for this system. This ground-state energy is the same as the binding energy or ionization energy that can be easily measured. Comparing the ground-state energy of the hydrogen atom with that of the helium atom, we observe that the mutual repulsion between the two electrons in the helium atom contributes significantly to the ground-state energy. This comparison permits us to put on the same energy scale, the discrete spectrum of the helium atom, the excited states of the two electrons, the continuous spectrum of two-electron system and the ionized helium atom with electron continuous spectrum. From the interaction between electrons, we can show the nonexistence of the one-electron state. For the many-body problem, the discrete spectrum is below the continuous spectrum on the energy scale. The energy of the excited state of the helium atom is the sum of the ground-state energy of the ion and the free electron.

We consider the case associated with the Hamiltonian operator, $\hat{H}_0$, where after switching on the perturbation at zero, there exists a discrete spectrum together with the continuous spectrum and the transition $dP_{n\to\nu}$ is not to a single final state but to a range ν of final states about the energy E_n of the discrete spectrum. This should be a type of approximation mentioned above that permits the solution of the problem. Since the final state is a continuous spectrum, we examine the probability that the final state will have energy in the following interval:

$$E \to E + dE \tag{8.243}$$

In this problem, the total angular moment of the system equals the angular moment of the excited atom. For convenience, we will not write the expressions for the angular momenta without which the material will still be understood. We begin with the transition for the first-order perturbation theory. We suppose that the energy, E, of the continuous spectrum is the unique parameter within the range in Equation 8.243 and so

$$dP_{n\to\nu} = \rho(E_\nu)dE_\nu \tag{8.244}$$

then further we perform integration over the parameter E. This implies evaluating the transition probability:

$$P_{n\to\nu} = \int P_{n\to E}\varrho(E)dE \tag{8.245}$$

In the above relations, $\varrho(E)dE$ is the **number of states within the range** $E \to E + dE$, while the function $\varrho(E)$ is the so-called **density of final states.**

8.8.1 Transition Rate

We assume the range ν to be small enough so that $\varrho(E)$ and $v_{n\nu}$ are constant within the given range. For exceedingly large T,

$$\lim_{T\to\infty} \frac{4\sin^2 \frac{\omega_{\nu n} T}{2}}{\omega_{\nu n}^2} = 2\pi T \delta(\omega_{\nu n}) \tag{8.246}$$

and the only dominant contribution to the integral arises from the energy range that corresponds to the narrow central peak of the function $f(t,\omega)$. So the integration limits can safely be extended to infinity without affecting the given result:

$$\mathrm{P}_{n\to\nu} = \int_{-\infty}^{\infty} \varrho(E) \frac{4\sin^2 \frac{\omega_{\nu n} T}{2}}{\omega_{\nu n}^2} |v_{n\nu}|^2 dE = \frac{|v_{n\nu}|^2 \varrho(E_\nu)}{\hbar^2} \int_{-\infty}^{\infty} \frac{4\sin^2 \frac{\omega_{\nu n} T}{2}}{\omega_{\nu n}^2} \hbar d\omega_{\nu n} = \frac{2\pi}{\hbar} |v_{n\nu}|^2 \varrho(E_\nu) T \tag{8.247}$$

Here, the transition probability $\mathrm{P}_{n\to\nu}$ is proportional to the time T and this permits us to introduce the **transition rate** (or **transition probability per unit time**):

$$\mathrm{W}_\nu = \frac{2\pi}{\hbar} |v_{n\nu}|^2 \varrho(E_\nu) \tag{8.248}$$

This is the so-called **Enrico Fermi celebrated Golden Rule.** It is instructive to note that in the matrix element $v_{n\nu}$, one of the parameters ν from the wave function is due to the continuous spectrum. The normalization by the Dirac delta function $\delta(E - E')$ has the dimension $(\text{energy})^{\frac{1}{2}}$. The above derivations have been possible by considering the condition:

$$\frac{2\pi}{\hbar} |v_{n\nu}|^2 T \ll 1 \tag{8.249}$$

So to know what happens within the following condition:

$$\frac{2\pi}{\hbar} |v_{n\nu}|^2 T \geq 1 \tag{8.250}$$

is impossible.

8.9 Transition Probability-Continuous Spectrum

Considering the Hamiltonian, $\hat{\mathrm{H}}$, to have a discrete as well as continuous spectrum, we modify the function in Equation 8.194:

$$\Psi = \sum_m C_m(t) \Psi_m^{(0)} + \int C_\nu(t) \Psi_\nu^{(0)} d\nu \tag{8.251}$$

where ν is the parameter of the continuous spectrum. We find, considering Equation 8.211, the transition probability. For this, we do some changes for the continuous spectrum, where

$$|C_\nu(\infty)|^2 d\nu \tag{8.252}$$

is the probability finding the system within the interval $d\nu$. So

$$dP_{\nu n} = \frac{1}{\hbar^2}\left|\int_0^T v_{\nu n}(t)\exp\{i\omega_{\nu n}t\}dt\right|^2 d\nu \tag{8.253}$$

is the transition probability of a state within the interval $d\nu$ for the continuous spectrum.

8.9.1 Harmonic Perturbation

For the case of $\hat{v}(t)$ having a harmonic dependence on time except when switched on at one time and off at a later time:

$$\hat{v}(t) = \hat{u}\exp\{i\omega t\} + \hat{w}\exp\{-i\omega t\} \tag{8.254}$$

Here, ω is a constant angular frequency which is positive, $\hat{u}$ and $\hat{w}$ are operators that explicitly do not depend on time t while $\hat{v}$ is Hermitian, i.e., $\hat{v}^\dagger = \hat{v}$ which, from Equation 8.254:

$$\hat{u}\exp\{i\omega t\} + \hat{w}\exp\{-i\omega t\} = \hat{u}^\dagger\exp\{-i\omega t\} + \hat{w}^\dagger\exp\{i\omega t\} \tag{8.255}$$

From here:

$$\hat{u}^\dagger = \hat{w} \tag{8.256}$$

then

$$u_{nm} = u_{mn}^* \tag{8.257}$$

We switch on the perturbation at $t = 0$, switch off at $t = T$ and so, from Equation 8.254:

$$dP_{n\to\nu} = \frac{1}{\hbar^2}\left| u_{\nu n}\frac{\exp\{i(\omega_{\nu n}+\omega)T\}-1}{i(\omega_{\nu n}+\omega)} + u_{\nu n}^*\frac{\exp\{i(\omega_{\nu n}-\omega)T\}-1}{i(\omega_{\nu n}-\omega)}\right|^2 d\nu \tag{8.258}$$

For fixed t, i.e., T, the transition probability, $dP_{n\nu}$, is a function of only the variable ω. For a resonance phenomenon, then one of the two denominators in Equation 8.258 satisfies the following frequencies:

$$\omega_{\nu n} \approx \pm\omega \tag{8.259}$$

and correspond to the energies:

$$E_\nu \approx E_n \pm \hbar\omega \tag{8.260}$$

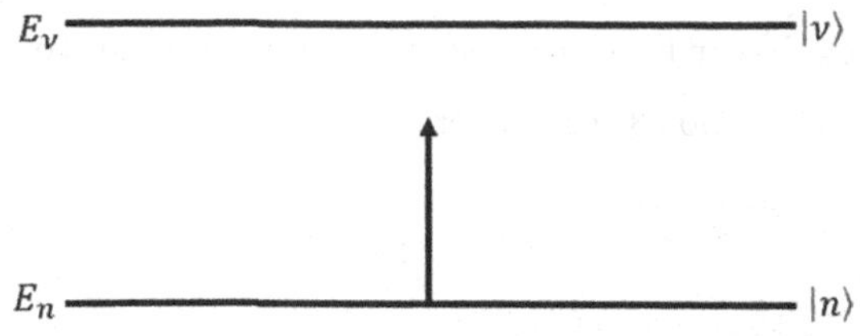

FIGURE 8.3 The resonance condition corresponding to $\omega_{\nu n} > 0$ $(\omega_{\nu n} < 0)$. For $\omega_{\nu n} > 0$, the system goes from the lower energy level E_n to the higher energy level E_ν by the resonant absorption of an energy quantum $\hbar\omega$.

So the first-order effect of the perturbation varying sinusoidally in the time with angular frequency, ω for these two cases match, respectively, with absorption and emission of a quantum of radiation when electromagnetic interactions are involved.

From $\omega \geq 0$, $\omega_{\nu n} = \omega(-\omega)$ gives the resonance conditions corresponding to $\omega_{\nu n} > 0$ $(\omega_{\nu n} < 0)$. For $\omega_{\nu n} > 0$, the system goes from the lower energy level, E_n, to the higher energy level, E_ν, by the resonant absorption of an energy quantum $\hbar\omega$ (Figure 8.3).

For $\omega_{\nu n} < 0$, the resonant perturbation stimulates the passage of a system from the higher energy level, E_ν, to the lower level, E_n. This is accompanied by an induced emission of an energy quantum $\hbar\omega$. Further, we consider the case where $\omega_{\nu n} > 0$ and $\omega > 0$ as the case when $\omega_{\nu n} < 0$ is considered analogously. Consider again Equation 8.258:

$$\mathrm{P}^{+}_{\nu\to n} = \frac{\exp\{i(\omega_{\nu n} + \omega)T\} - 1}{i(\omega_{\nu n} + \omega)} = \frac{\exp\left\{i(\omega_{\nu n} + \omega)\frac{T}{2}\right\}\sin\left[(\omega_{\nu n} + \omega)\frac{T}{2}\right]}{(\omega_{\nu n} + \omega)\frac{T}{2}} \tag{8.261}$$

$$\mathrm{P}^{-}_{\nu\to n} = \frac{\exp\{i(\omega_{\nu n} - \omega)T\} - 1}{i(\omega_{\nu n} + \omega)} = \frac{\exp\left\{i(\omega_{\nu n} - \omega)\frac{T}{2}\right\}\sin\left[(\omega_{\nu n} - \omega)\frac{T}{2}\right]}{(\omega_{\nu n} - \omega)\frac{T}{2}} \tag{8.262}$$

The denominator of $\mathrm{P}^{-}_{\nu\to n}$ is zero when $\omega = \omega_{\nu n}$ and that of $\mathrm{P}^{+}_{\nu\to n}$ is zero when $\omega = -\omega_{\nu n}$. So if we consider ω close to $\omega_{\nu n}$, $\mathrm{P}^{-}_{\nu\to n}$ should be dominant compared to $\mathrm{P}^{+}_{\nu n}$. So $\mathrm{P}^{-}_{\nu\to n}$ is a **resonant term** while $\mathrm{P}^{+}_{\nu n}$ is the **anti-resonant term**. Note that $\mathrm{P}^{+}_{\nu\to n}$ becomes resonant for negative $\omega_{\nu n}$ close to ω and as well close to $-\omega_{\nu n}$. Note also that we deal with a situation where the initial state $|n\rangle$ is a discrete bound state and final state $|\nu\rangle$ is one of a continuous set of dissociated states. Then $E_\nu > E_n$ and we consider only $\mathrm{P}^{-}_{\nu\to n}$ and hence Equation 8.258 becomes

$$d\mathrm{P}_{\nu\to n} = \frac{1}{\hbar^2}|u_{\nu n}|^2 \frac{4\sin^2(\omega_{\nu n} - \omega)\frac{T}{2}}{(\omega_{\nu n} - \omega)^2} d\nu \tag{8.263}$$

Consider the time $T \to \infty$ or exceedingly large T by noting

$$\lim_{T\to\infty} \frac{\sin^2 \alpha T}{\alpha^2 T} = \pi\delta(\alpha) \tag{8.264}$$

The resonance breadth $\Delta\alpha = \frac{2\pi}{T}$ can be obtained at the zeros on both sides of the resonant point and for the transition probability in Equation 8.263, the zeros occur when

$$(\omega_{\nu n} - \omega)\frac{T}{2} = \pm\pi \tag{8.265}$$

Then, from here, the time T is larger if the breadth is smaller. Within the time interval, $[0,T]$, the perturbation performs numerous oscillations which are sinusoidal perturbations (Figure 8.3). If T were small, then the perturbation would not have time to oscillate and this will be equivalent to a perturbation linear in time.

For $T \to \infty$, the maximum of the function tends to infinity and the breadth of the curve becomes smaller and the curve increases in height and leads to a Dirac delta function:

$$\pi T\delta\left(\frac{\omega_{\nu n} - \omega}{2}\right) = 2\pi\hbar T\delta(E_\nu - E_n - \hbar\omega) \tag{8.266}$$

So

$$dW_\nu = \frac{2\pi}{\hbar}|u_{\nu n}|^2 \delta(E_\nu - E_n - \hbar\omega)d\nu \tag{8.267}$$

which is the transition rate. This formula cannot yield a numerical value since it contains a Dirac delta function. In order to compare Equation 8.266 with experimentation, we do away with the Dirac delta function through integration.

Suppose that the energy E is the unique parameter:

$$d\nu = \rho(E_\nu)dE_\nu \tag{8.268}$$

We now integrate Equation 8.267 over the parameter E and this implies evaluating the transition rate for the continuous spectrum:

$$W_\nu = \frac{2\pi}{\hbar}|v_{\nu n}|^2 \rho(E_\nu) \tag{8.269}$$

Here, $\rho(E_\nu)$ is the density of state for the continuous spectrum and

$$E_\nu = E_n + \hbar\omega \tag{8.270}$$

From where it shows that if in the perturbation is due to the absorption of light, then $\hbar\omega$ is the quantum of light and Equation 8.270 is the equation of Einstein photo effect. The transition rate in Equation 8.269 portrays the so-called **Enrico Fermi celebrated Golden Rule** and illustrates the rate of transition from a discrete state to a group in a continuum is a constant independent of time. Note that it is impossible to achieve a constant transition rate when only discrete states are concerned in the transition.

8.10 Transition in a Continuous Spectrum Due to a Constant Perturbation

Consider the case when the perturbation is brought about by a quantum transition independent of time t. Suppose the system in the zero approximation contains the state with a discrete and continuous spectrum, the solution of the Schrödinger equation $\Psi(q,t)$ is as follows:

$$\Psi(q,t) = \sum_m C_m(t)\Psi_m^{(0)} + \int C_\nu(t)\Psi_\nu^{(0)} d\nu \tag{8.271}$$

Substitute this into the time Schrödinger equation:

$$i\hbar\frac{\partial}{\partial t}\Psi=\left(\hat{H}_0+\hat{v}\right)\Psi \tag{8.272}$$

Then we have a system of equations for the coefficients, C_ν:

$$i\hbar\frac{dC_\nu}{dt}=\sum_m C_m v_{\nu m}\exp\{i\omega_{\nu m}t\}+\int C_{\nu'}v_{\nu\nu'}\exp\{i\omega_{\nu\nu'}t\}d\nu' \tag{8.273}$$

where $v_{\nu m}$ and $v_{\nu\nu'}$ are the matrix elements of the perturbation $\hat{v}$ operator constructed on the eigenfunctions of $\hat{H}_0$.

For the initial state when $t=0$ corresponds to the state of a continuous spectrum numbered by the index ν_0, it is easily seen that C_ν and C_m satisfy the initial condition:

$$C_m(0)=0,\quad C_\nu(0)=\delta(\nu-\nu_0) \tag{8.274}$$

Substitute this into Equation 8.271, the summation falls out and we have

$$\Psi(0)=\Psi_{\nu_0}^{(0)} \tag{8.275}$$

We find the solution of the coefficients, C_ν, in the form:

$$C_\nu=C_\nu^{(0)}+C_\nu^{(1)}+\cdots \tag{8.276}$$

It is obvious that

$$C_\nu^{(0)}=\delta(\nu-\nu_0) \tag{8.277}$$

This shows the perturbation in the zero approximation does not exist, and from Equation 8.273, we have

$$i\hbar\frac{d}{dt}C_\nu^{(1)}=v_{\nu\nu_0}\exp\{i\omega_{\nu\nu_0}t\} \tag{8.278}$$

Integrate Equation 8.278 from 0 to t:

$$C_\nu^{(1)}(t)=\frac{1}{i\hbar}v_{\nu\nu_0}\frac{\exp\{i\omega_{\nu\nu_0}t\}-1}{i\omega_{\nu\nu_0}} \tag{8.279}$$

The transition probability of the states within the interval $\nu\to\nu+d\nu$:

$$dP_\nu(t)=\left|C_\nu^{(1)}(t)\right|^2d\nu=\frac{\left|v_{\nu\nu_0}\right|^2}{\hbar^2}\frac{4\sin^2\left(\frac{\omega_{\nu\nu_0}t}{2}\right)}{\omega_{\nu\nu_0}^2}d\nu=\frac{f(t,\omega_{\nu\nu_0})}{\hbar^2}\left|v_{\nu\nu_0}\right|^2d\nu \tag{8.280}$$

This imitates the transition probability of a harmonic perturbation except for the fact that resonance occurs when

$$\omega_{\nu\nu_0}=0 \tag{8.281}$$

From here, we deduce that the resonance phenomenon occurs under constant perturbation when there are degenerate levels.

From Equation 8.280, the function, $f(t, \omega_{\nu\nu_0})$, oscillates appreciably and behaves as delta-function when $T \to \infty$:

$$dW = \frac{2\pi}{\hbar^2} |v_{\nu\nu_0}|^2 \delta\left(E_\nu^{(0)} - E_{\nu_0}^{(0)}\right) d\nu \tag{8.282}$$

This is the transition probability per unit time or the rate of the transition probability.

Problem

Find $C_m^{(1)}$ where the system in Equation 8.273 is supplemented with $C_m^{(1)}$:

$$i\hbar \frac{d}{dt} C_k = \sum_m C_m v_{km} \exp\{i\omega_{\nu m} t\} + \int C_{\nu'} v_{k\nu'} \exp\{i\omega_{k\nu'} t\} d\nu' \tag{8.283}$$

Solution

We solve Equation 8.283 in the first approximation:

$$i\hbar \frac{d}{dt} C_k^{(1)} = v_{k\nu_0} \exp\{i\omega_{k\nu_0} t\} \tag{8.284}$$

and after integration, we have $C_k^{(1)}$:

$$C_k^{(1)} = \frac{v_{k\nu_0}}{i\hbar} \frac{\exp\{i\omega_{k\nu_0} t\} - 1}{i\omega_{k\nu_0}} \tag{8.285}$$

So

$$i\hbar \frac{d}{dt} C_\nu^{(2)} = \sum_m \frac{v_{m\nu_0} v_{\nu m}}{E_{\nu_0}^{(0)} - E_m^{(0)}} \left[\exp\{(i\omega_{m\nu_0} + i\omega_{\nu m})t\} - \exp\{(i\omega_{\nu m})t\}\right] \tag{8.286}$$

Suppose $E_m^{(0)} \neq E_\nu^{(0)}$ is the level of the discrete spectrum that does not fall within the continuous spectrum and implies $\omega_{\nu m} \neq 0$.

So

$$i\hbar \frac{d}{dt} C_\nu^{(2)} = v_{\nu\nu_0}^{(2)} \exp\{i\omega_{m\nu_0} t\} \tag{8.287}$$

Here, the states where we have the following summation are called virtual states:

$$v_{\nu\nu_0}^{(2)} = \sum_m \frac{v_{m\nu_0} v_{\nu m}}{E_{\nu_0}^{(0)} - E_m^{(0)}} \tag{8.288}$$

9

Many-Particle System

Details of the many-particle system can also be read from my book titled "Quantum Field Theory: Feynman Path Integrals and Diagrammatic Techniques in Condensed Matter" [23]. The fundamental difference between classical and quantum mechanics relates the concept of indistinguishability of identical particles where each particle can be equipped with an identifying marker without influencing its behavior in classical mechanics. Besides in classical mechanics, the particle follows its own continuous path in phase space and so it is possible to determine the particular process taking place. So, principally, each particle in a group of identical particles can be identified which is not so in quantum mechanics. It is not possible to mark a particle without influencing its physical state, and worse, if a number of identical particles are brought to the same region in space, their wave functions will rapidly spread out and overlap with one another, thereby soon rendering it impossible to say which particle is where. One of the fundamental assumptions for many-particle systems is therefore that identical particles (i.e., particles characterized by the same quantum numbers such as mass, charge and spin) are, in principle, indistinguishable.

We consider the example of two electrons located in two potential wells (Figure 9.1).

From a classical mechanics point of view, if there is no external perturbation, the two electrons will sit forever in the potential wells A and B. However, in quantum mechanics the situation changes as a result of the tunnel effect. The positions in A and B do not change due to the tunneling effect of the two states. The system of two electrons is already a many-electron system. In particular, its wave function should satisfy the principle of indistinguishability, i.e., the wave function should be anti-symmetric. We consider a two-electron system with spin-independent Hamiltonian.

9.1 System of Indistinguishable Particles

We examine the system of two particles with the wave function $\psi(\xi_1,\xi_2)$, where ξ_1 and ξ_2 take into consideration the system of space coordinates $\vec{r}$ and the spin variable s_z, i.e., $\xi=\{\vec{r},s_z\}$. The essential indistinguishability of particles of the same species is one of the fundamental postulates of quantum mechanics. The principle of indistinguishability relates the fact that the wave functions $\psi(\xi_1,\xi_2)$ and $\psi(\xi_2,\xi_1)$ describe one and the same state. So in quantum mechanics, particles of the same species are regarded as being indistinguishable. This follows that they may not be labeled and thus the wave functions may only differ by a constant factor C:

$$\psi(\xi_1,\xi_2)=C\psi(\xi_2,\xi_1) \tag{9.1}$$

This postulate limits the possibility of selection of these wave functions for the system of two particles. Suppose that we interchange the positions of the particles with the help of the projector operator, $\hat{P}_{12}$:

$$\hat{P}_{12}\psi(\xi_1,\xi_2)=\psi(\xi_2,\xi_1) \tag{9.2}$$

DOI: 10.1201/9781003273073-10

If Equation 9.2 is substituted into Equation 9.1:

$$\hat{P}_{12}\psi(\xi_1,\xi_2)=C\psi(\xi_1,\xi_2) \tag{9.3}$$

then we observe that the wave function $\psi(\xi_1,\xi_2)$ (which is the eigenfunction of the projector operator $\hat{P}_{12}$ and the constant C being the eigenvalue) obeys the principle of indistinguishability.

Since we observe that the two particles are truly indistinguishable, the quantum states (and also the probability of observing the system in a given state) remain invariant after the swap of the particles:

$$\left|\hat{P}_{12}\psi(\xi_1,\xi_2)\right|^2=\left|C\psi(\xi_2,\xi_1)\right|^2 \tag{9.4}$$

To find C, it is sufficient to note that

$$\hat{P}_{12}^2=1 \tag{9.5}$$

from where:

$$C^2=1 \tag{9.6}$$

and so

$$C=\pm 1 \tag{9.7}$$

If $C=1$, then the given vector state has the symmetric property:

$$\psi_s(\xi_1,\xi_2)=\psi_s(\xi_2,\xi_1) \tag{9.8}$$

This implies that it is a symmetric wave function. So **a wave function is symmetric if the permutation of any pair of the particles among its arguments leaves the wave function invariant.**

Consider a gas made up of identical fermions. If $C=-1$, then the function is anti-symmetric:

$$\psi_a(\xi_1,\xi_2)=-\psi_a(\xi_2,\xi_1) \tag{9.9}$$

A wave function is anti-symmetric if the permutation of any pair of particles changes the sign of the wave function.

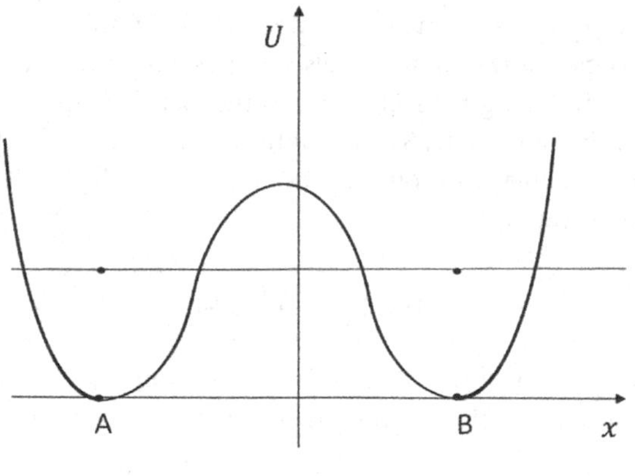

FIGURE 9.1 Two electrons are located in the potential wells A and B.

When the Pauli Exclusion Principle is violated, $\psi_a(\xi_1,\xi_2)=0$. It seems therefore that only symmetric or anti-symmetric many-particle states are encountered in nature. If $\hat{H}$ is the Hamiltonian of a system of particles, then

$$\hat{H}\hat{P}_{12}=\hat{P}_{12}\hat{H} \tag{9.10}$$

Then $\hat{H}$ commutes with $\hat{P}_{12}$. Therefore, $\hat{H}$ and $\hat{P}_{12}$ have a common system of eigenfunctions that may be symmetric or anti-symmetric. Considering that $\xi_1 \neq \xi_2$, then the normalized eigenfunctions of the projector operator, $\hat{P}_{12}$, are as follows:

$$\psi_s(\xi_1,\xi_2)=\frac{1}{\sqrt{2!}}\left[\psi_{n_1}(\xi_1)\psi_{n_2}(\xi_2)+\psi_{n_1}(\xi_2)\psi_{n_2}(\xi_1)\right] \tag{9.11}$$

$$\psi_a(\xi_1,\xi_2)=\frac{1}{\sqrt{2!}}\left[\psi_{n_1}(\xi_1)\psi_{n_2}(\xi_2)-\psi_{n_1}(\xi_2)\psi_{n_2}(\xi_1)\right] \tag{9.12}$$

with eigenvalues +1 and −1, respectively. Though these states normally do not yet correspond to eigenstates of the two-particle Hamiltonian $\hat{H}$, they now have the right symmetry so that eigenstates of $\hat{H}$ will be linear combinations of these (anti-)symmetric two particle states, depending on the identity of the particles involved.

For the case of identical fermions, the Pauli Exclusion Principle results from the requirement that an N-particle state must be anti-symmetrical upon interchange of any two particles. For the case of two particles, this implies that the relevant state is anti-symmetric with eigenfunction defined in Equation 9.12. The given state vanishes when $\xi_1=\xi_2$ and thus takes into account Pauli Exclusion Principle. The symmetric state in Equation 9.11 for two bosons is not yet properly normalized when $\xi_1=\xi_2$, though it demonstrates the possibility a boson can occupy any particle quantum state.

9.2 Interacting System of Particles

We define the (anti-)symmetrization operator $\hat{P}$ by its action on the many-body wave function $\psi(\xi_1,\xi_2,\ldots,\xi_N)$ where we consider, respectively, the symmetric and anti-symmetric wave functions:

$$\psi_s(\xi_1,\xi_2,\ldots,\xi_N)=\frac{1}{\sqrt{N!}}\sum_{P}\hat{P}\psi_{n_1}(\xi_1)\psi_{n_2}(\xi_2)\cdots\psi_{n_N}(\xi_N) \tag{9.13}$$

$$\psi_a(\xi_1,\xi_2,\ldots,\xi_N)=\frac{1}{\sqrt{N!}}\sum_{P}(-1)^P\hat{P}\psi_{n_1}(\xi_1)\psi_{n_2}(\xi_2)\cdots\psi_{n_N}(\xi_N) \tag{9.14}$$

From these two equations, we have

$$\hat{P}_\chi\psi(\xi_1,\xi_2,\ldots,\xi_N)=\frac{1}{N!}\sum_{P}\chi^P\hat{P}\psi(\xi_1,\xi_2,\ldots,\xi_N) \tag{9.15}$$

where

$$\psi(\xi_1,\xi_2,\ldots,\xi_N)\equiv\sqrt{N!}\psi_{n_1}(\xi_1)\psi_{n_2}(\xi_2)\cdots\psi_{n_N}(\xi_N) \tag{9.16}$$

and

$$\hat{P}\psi(\xi_1,\xi_2,\ldots,\xi_N)=\psi\left(\xi_{P(1)},\xi_{P(2)},\ldots,\xi_{P(N)}\right) \quad (9.17)$$

Applying the operator $\hat{P}_\chi$ twice, we set $M \equiv PP'$ as the group composition of P' and P, while considering $\chi^{P+P'}=\chi^{PP'}$, we take summation over P' and P and then swap with summation over M and P:

$$\hat{P}_\chi^2\psi(\xi_1,\xi_2,\ldots,\xi_N)=\frac{1}{N!}\sum_{P}\left(\frac{1}{N!}\sum_{M}\chi^{M}\hat{M}\psi(\xi_1,\xi_2,\ldots,\xi_N)\right)=\frac{1}{N!}\sum_{P}\hat{P}_\chi\psi(\xi_1,\xi_2,\ldots,\xi_N)$$
$$=\hat{P}_\chi\psi(\xi_1,\xi_2,\ldots,\xi_N) \quad (9.18)$$

This equality holds for any wave function ψ and therefore for the operator itself, and the (anti-) symmetrization operator is a projector.

Since in quantum mechanics particles of the same species are regarded as being indistinguishable, this follows that they may not be labeled and thus the eigenstates may only differ by a constant factor χ: $\chi=+1$ for **bosons** and $\chi=-1$ for **fermions**; the symbol χ^P denotes 1 equals unity for bosons and $(-1)^P$ for fermions. It is easy to show that $\hat{P}_\chi^2=\hat{P}_\chi$. For $\chi=+1$, which implies that for **bosons**, several particles can occupy the same one-particle state. This was shown empirically for the first time by the Indian physicist S.N. Bose by proving the relation:

$$\hat{P}\psi(\xi_1,\xi_2,\ldots,\xi_N)=\psi\left(\xi_{P(1)},\xi_{P(2)},\ldots,\xi_{P(N)}\right)=\psi(\xi_1,\xi_2,\ldots,\xi_N) \quad (9.19)$$

From Equation 9.19, bosons are genuinely indistinguishable when enumerating the different possible states of the particles. For $\chi=-1$, the state in Equation 9.14 or 9.15 vanishes if two ξ_is are identical. This implies that any two fermions cannot occupy the same particle state (**Pauli Exclusion Principle**). The **Pauli Exclusion Principle** was developed empirically by the German physicist **Wolfgang Pauli** (1900–1958). It is instructive to note that this principle follows directly the symmetry requirements on vector states. This follows that **Pauli Exclusion Principle** is a corollary to the principle of indistinguishability of particles. The Pauli Exclusion Principle poses a severe constraint on vector of states of many-fermion systems and limits the number of them that are physically admissible. This relates **fermions** due to the Italian physicist Enrico Fermi who first studied the properties of fermion gases.

9.3 System of Two Electrons

Suppose we have a system of two electrons and let X_{s_z} be the normalized spin function of each particle with the spin projection, s_z:

$$s_z=-s,-s+1,\ldots,s \quad (9.20)$$

If we neglect the spin-spin interaction, then the non-symmetrized spin functions have the following form:

$$X=X_{s_z}^{(1)}X_{s_z'}^{(2)} \quad (9.21)$$

where $X_{s_z}^{(i)}$ is the spin function for the ith particle for a defined spin projection, s_z. From here, we examine the following normalized functions:

$$X_{s_z s_z}^{(+)}=X_{s_z}^{(1)}X_{s_z}^{(2)} \quad (9.22)$$

$$X^{(+)}_{s_z s_z'} = \frac{1}{\sqrt{2}}\left(X^{(1)}_{s_z} X^{(2)}_{s_z'} + X^{(1)}_{s_z'} X^{(2)}_{s_z}\right),\ s_z \neq s_z' \tag{9.23}$$

$$X^{(-)}_{s_z s_z'} = \frac{1}{\sqrt{2}}\left(X^{(1)}_{s_z} X^{(2)}_{s_z'} - X^{(1)}_{s_z'} X^{(2)}_{s_z}\right),\ s_z \neq s_z' \tag{9.24}$$

It is obvious that these functions have a defined symmetry relative to the permutation of any pair spin variables. Considering that the permutation of any pair of spin variables leaves the wave function invariant, $X^{(+)}_{s_z s_z}$ and $X^{(+)}_{s_z s_z'}$ are symmetric functions while the wave function, $X^{(-)}_{s_z s_z'}$, is anti-symmetric, since the permutation of any pair of spin variables changes the sign of the wave function. From Equation 9.20, the function, $X^{(+)}_{s_z s_z}$ as well as $X^{(+)}_{s_z s_z'}$ and $X^{(-)}_{s_z s_z'}$ constitutes, respectively, $(2s+1)$ and $\frac{(2s+1)2s}{2!} = (2s+1)s$ independent states. So the total number of symmetric and anti-symmetric states are, respectively, $(s+1)(2s+1)$ and $(2s+1)s$ giving the total number of states, $(2s+1)^2$. From Equation 9.22 to 9.24, the total spin, S, for the symmetric state is as follows:

$$S = 2s, 2s-2, 2s-4, \ldots \tag{9.25}$$

while for the anti-symmetric state:

$$S = 2s-1, 2s-3, 2s-5, \ldots \tag{9.26}$$

Example

We examine an example of the hydrogen atom with an electron and a proton with each having spin-up $\uparrow$ or spin-down $\downarrow$ with four possibilities of spin arrangements:

$$\uparrow\uparrow \quad \uparrow\downarrow \quad \downarrow\uparrow \quad \downarrow\downarrow \tag{9.27}$$

Assuming the first arrow refers to the electron and the second to the proton, the total spin of the hydrogen atom is as follows:

$$\hat{s} = \hat{s}_1 + \hat{s}_2 \tag{9.28}$$

It should be noted that each of the four composite states in Equation 9.27 are eigenstates of $\hat{s}_z$:

$$\begin{aligned} \hat{s}_z X_1 X_2 &= (\hat{s}_{1z} + \hat{s}_{2z}) X_1 X_2 = (\hat{s}_{1z} X_1) X_2 + X_1 (\hat{s}_{2z} X_2) = (m_1 X_1) X_2 + X_1 (m_2 X_2) \\ &= (m_1 + m_2) X_1 X_2 = m X_1 X_2 \end{aligned} \tag{9.29}$$

Here we consider $\hat{s} = \frac{1}{\hbar}\hat{S}$ and $m = m_1 + m_2$ is the quantum number with values 1, 0, 0 and –1 that correspond to four eigenvectors:

$$\begin{cases} \uparrow\uparrow, & m = 1 \\ \uparrow\downarrow, & m = 0 \\ \downarrow\uparrow, & m = 0 \\ \downarrow\downarrow, & m = -1 \end{cases} \tag{9.30}$$

In the $|sm\rangle$ notation, the three states with $s=1$ (**triplet state**) may be represented as follows:

$$|11\rangle = \uparrow\uparrow, |10\rangle = \frac{1}{\sqrt{2}}(\uparrow\downarrow + \downarrow\uparrow), |1(-1)\rangle = \downarrow\downarrow \tag{9.31}$$

These are **symmetric eigenvectors** since the permutation of the spin variables leaves the eigenvectors invariant. The orthogonal state with $m=0$ has $s=0$ (**singlet state**):

$$|00\rangle = \frac{1}{\sqrt{2}}(\uparrow\downarrow - \downarrow\uparrow) \tag{9.32}$$

This is an **anti-symmetric eigenvector** since the permutation of the spin variables causes the eigenvectors to change sign. For the electron or proton:

$$s_1 = s_2 = \frac{1}{2} \tag{9.33}$$

Then the total spin quantum number s is as follows:

$$s = s_1 + s_2 = 1,\ s_1 - s_2 = 0 \tag{9.34}$$

This depends on whether the particles occupy, respectively, the triplet or singlet configuration. We may justify this by proving that the triplet states are eigenvectors of the operator, $\hat{s}^2$, with eigenvalue $2\hbar^2$, and the singlet is an eigenvector of the operator, $\hat{s}^2$, with eigenvalue 0:

$$\hat{s}^2 = (\hat{s}_1 + \hat{s}_2)\cdot(\hat{s}_1 + \hat{s}_2) = (\hat{s}_1)^2 + (\hat{s}_2)^2 + 2\hat{s}_1\cdot\hat{s}_2 \tag{9.35}$$

or

$$\hat{s}^2 = \frac{1}{4}(\hat{\sigma}_1 + \hat{\sigma}_2)^2 = \frac{3}{2} + 2\hat{\sigma}_1\cdot\hat{\sigma}_2,\ \hat{\sigma}_1^2 = \hat{\sigma}_2^2 = 3 \tag{9.36}$$

From here, we observe the eigenvector of the operator $\hat{s}^2$ is also the eigenvector of the operator $\hat{s}_1\cdot\hat{s}_2$. So

$$\begin{aligned}\hat{s}_1\cdot\hat{s}_2(\uparrow\downarrow) &= (\hat{s}_{1x}\uparrow)(\hat{s}_{2x}\downarrow) + (\hat{s}_{1y}\uparrow)(\hat{s}_{2y}\downarrow) + (\hat{s}_{1z}\uparrow)(\hat{s}_{2z}\downarrow)\\ &= \left(\frac{1}{2}\uparrow\right)\left(\frac{1}{2}\downarrow\right) + \left(\frac{i}{2}\uparrow\right)\left(-\frac{i}{2}\downarrow\right) + \left(\frac{1}{2}\uparrow\right)\left(-\frac{1}{2}\downarrow\right)\end{aligned} \tag{9.37}$$

So

$$\hat{s}_1\cdot\hat{s}_2(\uparrow\downarrow) = \frac{1}{4}(2\downarrow\uparrow - \uparrow\downarrow) \tag{9.38}$$

We may also find in a similar manner:

$$\hat{s}_1\cdot\hat{s}_2(\downarrow\uparrow) = \frac{1}{4}(2\uparrow\downarrow - \downarrow\uparrow) \tag{9.39}$$

and

$$\hat{s}_1 \cdot \hat{s}_2 |10\rangle = \frac{1}{4}\frac{1}{\sqrt{2}}(2\downarrow\uparrow - \downarrow\uparrow + 2\uparrow\downarrow - \downarrow\uparrow)|10\rangle \tag{9.40}$$

$$\hat{s}_1 \cdot \hat{s}_2 |00\rangle = \frac{1}{4}\frac{1}{\sqrt{2}}(2\downarrow\uparrow - \downarrow\uparrow - 2\uparrow\downarrow + \downarrow\uparrow)|00\rangle = -\frac{3}{4}|00\rangle \tag{9.41}$$

So

$$\hat{s}^2|10\rangle = \left(\frac{3}{4}+\frac{3}{4}+2\frac{1}{4}\right)|10\rangle = 2|10\rangle,\ \hat{s}^2|00\rangle = \left(\frac{3}{4}+\frac{3}{4}-2\frac{3}{4}\right)|00\rangle = 0 \tag{9.42}$$

The reader may do similarly for $|11\rangle$ and $|1(-1)\rangle$.

We observe that the spin eigenvectors of two electronic systems satisfying the principle of indistinguishability appear to be the eigenvectors of the square of the total spin $\hat{s}^2$. For that, we have three symmetric vectors corresponding to $s=1$ and $m_s=-1,0,1$, respectively, and one anti-symmetric vector corresponding to $s=0$. This follows that $m_s=0$ and, as a result, there arises the following effect: We solve a problem in the non-relativistic approximation without spin-spin and spin-orbital interaction. So it appears our system of two electrons is dependent on the total spin. This dependence relates the fact that the total spin is defined by a symmetric spinor. So the exchange interaction should be considered in the non-relativistic approximation.

In addition to the spin variables, we consider the space function:

$$\Phi(\vec{r}_1,\vec{r}_2) \equiv \psi_n(\vec{r}_1)\psi_n(\vec{r}_2) \tag{9.43}$$

This function is symmetric since the permutation of the pair of space variables for the particles leaves the wave function invariant. If we consider the spin variables as well as the space variables, and form the symmetry character, then the wave function of two indistinguishable particles relative to the permutation of the pair space or spin variables should be symmetric for bosons and anti-symmetric for fermions. So the wave function of possible independent states should have the form:

$$\Psi_{2\text{boson}}(\xi_1,\xi_2) \equiv \Phi(\vec{r}_1,\vec{r}_2)X^{(+)}_{s_z s_z'},\ \Psi_{2\text{fermion}}(\xi_1,\xi_2) \equiv \Phi(\vec{r}_1,\vec{r}_2)X^{(-)}_{s_z s_z'} \tag{9.44}$$

As seen earlier, ξ_1 and ξ_2 denote the variables which determine its spatial position $\vec{r}$, and the spin coordinate, s_z, determining its internal state, i.e., $\xi=(\vec{r},s_z)$. As shown above, $(s+1)(2s+1)$ is the number of bosonic states, while $(2s+1)s$ is the number of fermionic states. For bosons, s is an integral, while for fermions, it is a half-integral.

For $n_1 \neq n_2$, the space wave functions have the form:

$$\Phi^{(\pm)}(\vec{r}_1,\vec{r}_2) = \frac{1}{\sqrt{2!}}\left[\psi_{n_1}(\vec{r}_1)\psi_{n_2}(\vec{r}_2) \pm \psi_{n_1}(\vec{r}_2)\psi_{n_2}(\vec{r}_1)\right] \tag{9.45}$$

Here, $\Phi^{(+)}$ and $\Phi^{(-)}$ are, respectively, symmetric and anti-symmetric functions and their production with spin functions, $X^{(\pm)}$, yields

$$\Psi_{2\text{boson},1}(\xi_1,\xi_2) \equiv \Phi^{(+)}(\vec{r}_1,\vec{r}_2)X^{(+)}_{s_z s_z'},\ \Psi_{2\text{fermion},1}(\xi_1,\xi_2) \equiv \Phi^{(+)}(\vec{r}_1,\vec{r}_2)X^{(-)}_{s_z s_z'} \tag{9.46}$$

$$\Psi_{2\text{boson},2}(\xi_1,\xi_2) \equiv \Phi^{(-)}(\vec{r}_1,\vec{r}_2)X^{(-)}_{s_z s_z'},\ \Psi_{2\text{fermion},2}(\xi_1,\xi_2) \equiv \Phi^{(-)}(\vec{r}_1,\vec{r}_2)X^{(+)}_{s_z s_z'} \tag{9.47}$$

So from here, $(2s+1)^2$ should be the number of bosonic as well as fermionic states. We observe from the above eigenstates that symmetric solutions of the Schrödinger equation require singlet spin states whereas anti-symmetric, triplets. In addition, the coordinate wave function $\Phi(\vec{r}_1,\vec{r}_2)$ may either be anti-symmetric or symmetric since we neglect relativistic effects.

Consider the case of N different single particle states:

$$\psi_{s_{z,1}}(\xi_1,\xi_2,\ldots,\xi_N)=\psi_{n_1,s_{z,1}}(\xi_1)\psi_{n_2,s_{z,2}}(\xi_2)\cdots\cdot\psi_{n_N,s_{z,N}}(\xi_N) \tag{9.48}$$

Any change in the set of $s_{z,i}$ to $s'_{z,i}$ leads to a new set of single particle states, yielding a new state of the system:

$$\psi_{s_{z,2}}(\xi_1,\xi_2,\ldots,\xi_N)=\psi_{n_{1'},s'_{z,1}}(\xi_2)\psi_{n_{2'},s'_{z,2}}(\xi_1)\cdots\cdot\psi_{n_{N'},s'_{z,N}}(\xi_N) \tag{9.49}$$

From here, we observe that $N!$ solutions can be obtained from any one solution with each solution corresponding to one of the $N!$ permutations of the N arguments of the function, ψ. Such solutions correspond to one and the same eigenvalue, i.e., the solutions derived from any ψ through permutations of its arguments are degenerate with the original ψ. This is the so-called **degeneracy with permutation or exchange degeneracy**. Since n_i is different then not any of the functions is either symmetric or anti-symmetric. It is easily seen that if we sum all these functions, we get a symmetric function as follows:

$$\psi_s(\xi_1,\xi_2,\ldots,\xi_N)=\frac{1}{\sqrt{N!}}\sum_{\mathrm{P}}\hat{\mathrm{P}}\psi_{n_1}(\xi_1)\psi_{n_2}(\xi_2)\cdots\cdot\psi_{n_N}(\xi_N) \tag{9.50}$$

Here the sum is over all possible permutations P. For any single particle state, $\psi_{n_i}(\xi_i)$, we also have the anti-symmetric function:

$$\psi_a(\xi_1,\xi_2,\ldots,\xi_N)=\frac{1}{\sqrt{N!}}\sum_{\mathrm{P}}(-1)^{\mathrm{P}}\hat{\mathrm{P}}\psi_{n_1}(\xi_1)\psi_{n_2}(\xi_2)\cdots\cdot\psi_{n_N}(\xi_N) \tag{9.51}$$

If $n_1=n_2$, then for the change of the values of the spin projections, $s_{z,i}$, due to the permutation of any pair $s_{z,1}$ and $s_{z,2}$, we arrive at a new set of single particle states. But every $s_{z,i}$ has $(2s+1)$ values, so the total number of different sets of single particle states should be $(2s+1)^N$ as in Equations 9.48 and 9.49.

In Equation 9.51, half of the terms have the negative sign and half the positive sign. In this case, the anti-symmetric wave function ψ_a should be represented in the form of the Slater's determinant:

$$\psi_a(\xi_1,\xi_2,\ldots,\xi_N)=\frac{1}{\sqrt{N!}}\begin{bmatrix}\psi_{n_1}(\xi_1) & \psi_{n_1}(\xi_2) & \cdots & \psi_{n_1}(\xi_N)\\ \psi_{n_2}(\xi_1) & \psi_{n_2}(\xi_2) & \cdots & \psi_{n_2}(\xi_N)\\ \cdots & \cdots & \cdots & \cdots\\ \psi_{n_N}(\xi_1) & \psi_{n_N}(\xi_2) & \cdots & \psi_{n_N}(\xi_N)\end{bmatrix} \tag{9.52}$$

For example, if $N=2$, then

$$\psi_a(\xi_1,\xi_2)=\frac{1}{\sqrt{2!}}\left[\psi_{n_1}(\xi_1)\psi_{n_2}(\xi_2)-\psi_{n_1}(\xi_2)\psi_{n_2}(\xi_1)\right]=\frac{1}{\sqrt{2!}}\begin{bmatrix}\psi_{n_1}(\xi_1) & \psi_{n_1}(\xi_2)\\ \psi_{n_2}(\xi_1) & \psi_{n_2}(\xi_2)\end{bmatrix} \tag{9.53}$$

In order for the Slater's determinant to be different from zero, it is necessary that the n_i should be different, otherwise it is equal to zero. For fermions in any state, there may not be more than one particle (Pauli Exclusion Principle). This follows that the Pauli Exclusion Principle is a corollary to the principle of indistinguishability of particles.

9.3.1 Exchange Interaction

From the Hamiltonian, $\hat{H}$, we find the potential energy of the electron-electron interaction to be a positive quantity, $u = \frac{e^2}{r_{12}}$. For the evaluation of the eigenvalue, this term has a positive contribution. The anti-symmetric function, $\psi_a(\vec{r}_1, \vec{r}_2) = 0$, for $\vec{r}_1 = \vec{r}_2$. So if the total spin $s = 1$, then $\frac{e^2}{r_{12}}$ plays a less important rule than $s = 0$. The energy corresponding to the function, $\psi_s(\vec{r}_1, \vec{r}_2)X_a$, is greater than that for the function, $\psi_a(\vec{r}_1, \vec{r}_2)X_s$. This effect is called an **exchange interaction** which is the result of the electrostatic Coulomb interaction between electrons that becomes spin-dependent due to the requirement that the wave function of a pair of electrons is anti-symmetric with respect to the interchange of electron coordinates and spins. The ground state should be for the case of $s = 0$ as there is an oscillatory theorem according to which the wave function may not be equal to zero. This follows that ψ_a may not be the wave function of the ground state. Suppose the electron spins are parallel, the coordinate part of the wave function should be anti-symmetric:

$$\psi_{\uparrow\uparrow}(\vec{r}_1, \vec{r}_2) = -\psi_{\uparrow\uparrow}(\vec{r}_2, \vec{r}_1) \tag{9.54}$$

This explains the fact that the probability that two electrons are very close to each other is small compared to the opposite case, when the spins are anti-parallel. So their coordinate wave function is symmetric. It is instructive to note that since electrons with parallel spins are better separated in space, their repulsion is less. Thus, the energy of the electrostatic interaction for parallel spins is lower. If electrons are more than two, then the energy is also dependent on s but the wave function will have a complicated form.

It is instructive to note that there are two categories of exchange interactions: The **intra**-atomic interactions that favor spin alignment and are responsible for the operations of Hund's rules in unfilled atomic shells, and the **inter**-atomic interactions that govern relative spin alignment in neighboring ions or atoms.

9.3.2 Two Electrons in an Infinite Square Potential Well – Heisenberg Exchange Interaction

Consider an example of two electrons in an infinite rectangular potential well of breadth a. We expect that inside the potential well, the electron experiences no force (they are non-interacting). The Schrodinger equation for these particles may be written as follows:

$$\hat{H}\psi = E\psi, \hat{H} = -\frac{\hbar^2}{2m}\Delta \tag{9.55}$$

or

$$\psi'' + \kappa^2\psi, \kappa^2 = \frac{2mE}{\hbar^2} \tag{9.56}$$

and the solution

$$\psi = C_1 \exp\{i\kappa x\} + C_2 \exp\{-i\kappa x\} \tag{9.57}$$

with the symmetric wave function:

$$\psi_s = C\cos(\kappa x),\ \frac{\kappa_s a}{2} = \frac{\pi}{2} + \pi n \tag{9.58}$$

and anti-symmetric wave function:

$$\psi_a = C\sin(\kappa x),\ \frac{\kappa_a a}{2} = \pi + \pi n \tag{9.59}$$

So

$$\kappa_s = \frac{\pi}{a}(2n+1),\ \kappa_a = \frac{2\pi n}{a} \tag{9.60}$$

and energy eigenvalues:

$$E_n = \frac{\hbar^2\kappa^2}{2m} = \frac{\hbar^2\pi^2(n+1)^2}{2ma^2},\ n = 0,1,2,\ldots \tag{9.61}$$

with eigenfunctions:

$$\psi_n = \sqrt{\frac{2}{a}}\begin{cases} \cos\dfrac{(n+1)\pi x}{a}, & n \text{ even} \\ \sin\dfrac{(n+1)\pi x}{a}, & n \text{ odd} \end{cases} \tag{9.62}$$

Suppose both particles have spins $s_1 = s_2 = 0$, the total spin $s = 0$ and eigenfunction, ψ, must be symmetric:

$$\psi(x_1,x_2) = \frac{1}{\sqrt{2}}\left(\phi_1(x_1)\phi_2(x_2) + \phi_1(x_2)\phi_2(x_1)\right) = \frac{2\sqrt{2}}{a}\cos\left(\frac{\pi x_1}{a}\right)\cos\left(\frac{\pi x_2}{a}\right) \tag{9.63}$$

and ground-state energy:

$$E_0 = \frac{\hbar^2\pi^2}{2ma^2} \tag{9.64}$$

The eigenfunctions corresponding to the first excited state are as follows:

$$\phi_1(x) = \sqrt{\frac{2}{a}}\cos\left(\frac{\pi x}{a}\right),\ \phi_2(x) = \sqrt{\frac{2}{a}}\sin\left(\frac{2\pi x}{a}\right) \tag{9.65}$$

and

$$\psi(x_1,x_2) = \sqrt{\frac{2}{a}}\left(\cos\left(\frac{\pi x_1}{a}\right)\sin\left(\frac{2\pi x_2}{a}\right) + \cos\left(\frac{\pi x_2}{a}\right)\sin\left(\frac{2\pi x_1}{a}\right)\right) \tag{9.66}$$

with eigenenergy:

$$E_1 = \frac{\hbar^2\pi^2}{2ma^2} + \frac{\hbar^2\pi^2 4}{2ma^2} = \frac{5\hbar^2\pi^2}{2ma^2} \tag{9.67}$$

The anti-symmetric combination is forbidden by symmetry. From classical mechanics, the particles are distinguishable with the excited sates:

$$\sqrt{\frac{2}{a}}\cos\left(\frac{\pi x_1}{a}\right)\sin\left(\frac{2\pi x_2}{a}\right), \sqrt{\frac{2}{a}}\cos\left(\frac{\pi x_2}{a}\right)\sin\left(\frac{2\pi x_1}{a}\right) \tag{9.68}$$

while from quantum mechanics, it is seen that the particles are indistinguishable. This results, respectively, into symmetric and anti-symmetric eigenfunctions:

$$\psi_s(x_1,x_2) = \frac{1}{\sqrt{2}}\left(\phi_1(x_1)\phi_2(x_2) + \phi_1(x_2)\phi_2(x_1)\right) \tag{9.69}$$

or

$$\psi_a(x_1,x_2) = \frac{1}{\sqrt{2}}\left(\phi_1(x_1)\phi_2(x_2) - \phi_1(x_2)\phi_2(x_1)\right) = \frac{1}{\sqrt{2}}\begin{vmatrix} \phi_1(x_1) & \phi_1(x_2) \\ \phi_2(x_1) & \phi_2(x_2) \end{vmatrix} \tag{9.70}$$

The two-body states corresponding to the eigenfunctions, $\psi_{s,a}(x_1,x_2)$, can be presented as follows:

$$\psi_{s,a}(x_1,x_2) = \frac{1}{\sqrt{2}}\left(\phi_1(x_1)\phi_2(x_2) + \chi\phi_1(x_2)\phi_2(x_1)\right) \tag{9.71}$$

where $\chi = -1$ for fermions and $\chi = 1$ for bosons. The leading factor of $\frac{1}{\sqrt{2}}$ ensures that the new eigenfunctions are normalized. However, ψ_a is not allowed by symmetry since the total eigenfunction must be symmetric.

Suppose both electrons have spins $s_1 = s_2 = \frac{1}{2}$, the total eigenfunction is anti-symmetric. Then for $s_1 + s_2 = 1$, we have a triplet state (symmetric eigenvectors), and for $s_1 + s_2 = 0$, the singlet state (anti-symmetric eigenvector). From here, it can be easily shown that the three vectors $|1m\rangle, m = 1,0,-1$ constitute the triplet and the vector $|00\rangle$ the singlet state. Note that the triplet states are symmetric with respect to the permutation of two spins whereas the singlet state is anti-symmetric. So $\psi(x_1,x_2,s_{1z},s_{2z})$ must be anti-symmetric:

$$\psi(x_1,x_2,s_{1z},s_{2z}) = X_{s_{1z}s_{2z}}\phi(x_1,x_2) \tag{9.72}$$

$$\phi(x_1,x_2) = \frac{2\sqrt{2}}{a}\cos\left(\frac{\pi x_1}{a}\right)\cos\left(\frac{\pi x_2}{a}\right) \tag{9.73}$$

Since

$$\kappa_1 = \kappa_2 = \frac{\pi}{a} \tag{9.74}$$

then the spins must be different (Pauli Exclusion Principle). So

$$\psi(x_1,x_2,s_{1z},s_{2z})=\mathrm{X}_s\frac{2\sqrt{2}}{a}\cos\left(\frac{\pi x_1}{a}\right)\cos\left(\frac{\pi x_2}{a}\right) \tag{9.75}$$

For the first excited state:

$$\phi_1(x)=\sqrt{\frac{2}{a}}\cos\left(\frac{\pi x}{a}\right),\ \phi_2(x)=\sqrt{\frac{2}{a}}\sin\left(\frac{2\pi x}{a}\right) \tag{9.76}$$

we have the following symmetric eigenfunction:

$$\psi(x_1,x_2)=\sqrt{\frac{2}{a}}\left(\cos\left(\frac{\pi x_1}{a}\right)\sin\left(\frac{2\pi x_2}{a}\right)+\cos\left(\frac{\pi x_2}{a}\right)\sin\left(\frac{2\pi x_1}{a}\right)\right) \tag{9.77}$$

The total eigenfunction is as follows:

$$\psi(x_1,x_2,s_{1z},s_{2z})=\mathrm{X}_s\psi(x_1,x_2) \tag{9.78}$$

The anti-symmetric eigenfunctions are as follows:

$$\psi(x_1,x_2)=\sqrt{\frac{2}{a}}\left(\cos\left(\frac{\pi x_1}{a}\right)\sin\left(\frac{2\pi x_2}{a}\right)-\cos\left(\frac{\pi x_2}{a}\right)\sin\left(\frac{2\pi x_1}{a}\right)\right) \tag{9.79}$$

and the total eigenfunction is as follows:

$$\psi(x_1,x_2,s_{1z},s_{2z})=\mathrm{X}_T\psi(x_1,x_2) \tag{9.80}$$

The subscript T denotes the triplet state. Then we have the degenerate energy levels:

$$\overset{\uparrow\downarrow}{\rightarrow} E_{3s}=4\frac{\hbar^2\pi^2}{2ma^2},\ \overset{\uparrow\downarrow}{\rightarrow} E_{2s}=5\frac{\hbar^2\pi^2}{2ma^2},\ \overset{\uparrow\uparrow}{\rightarrow} E_{2T}=5\frac{\hbar^2\pi^2}{2ma^2},\ \overset{\uparrow\downarrow}{\rightarrow} E_{1s}=\frac{\hbar^2\pi^2}{2ma^2} \tag{9.81}$$

and so on.

We introduce an interaction between electrons sitting at level 2:

$$\psi_s(x_1,x_2)=\frac{1}{\sqrt{2}}\left(\phi_1(x_1)\phi_2(x_2)+\phi_1(x_2)\phi_2(x_1)\right) \tag{9.82}$$

The eigenfunctions ψ_s and ψ_a are, respectively, symmetric and anti-symmetric under the permutation of the pair x_1 and x_2.

We consider mutual interaction energy $u(x_1-x_2)$ to be small and spin-independent between the electrons. Then from the perturbation theory:

$$\Delta E_n=\int u(x_1-x_2)|\psi_n|^2\,dx_1\,dx_2=\langle n|u|n\rangle \tag{9.83}$$

$$\Delta E_s = \Delta E_{\text{singlet}} = \mathrm{C} + \mathrm{A},\ \Delta E_a = \Delta E_{\text{Triplet}} = \mathrm{C} - \mathrm{A} \tag{9.84}$$

where

$$\mathrm{C} = \int |\phi_1(x_1)|^2 u(x_1 - x_2) |\phi_2(x_2)|^2 \, dx_1 \, dx_2 \tag{9.85}$$

are the direct

$$\mathrm{A} = \int \phi_1^*(x_1)\phi_2(x_2) u(x_1 - x_2) \phi_2^*(x_2)\phi_1(x_1) dx_1 \, dx_2 \tag{9.86}$$

and exchange integrals, respectively.

Then

$$\overset{\uparrow\downarrow}{\rightarrow} E_{3s} = 4\frac{\hbar^2\pi^2}{2ma^2},\ \overset{\uparrow\downarrow}{\rightarrow} E_{2s} = E_2 + \mathrm{A} = 5\frac{\hbar^2\pi^2}{2ma^2} + \mathrm{A} \tag{9.87}$$

$$\overset{\uparrow\uparrow}{\rightarrow} E_{2T} = E_2 - \mathrm{A} = 5\frac{\hbar^2\pi^2}{2ma^2} - \mathrm{A},\ \overset{\uparrow\downarrow}{\rightarrow} E_{1s} = \frac{\hbar^2\pi^2}{2ma^2} \tag{9.88}$$

Note from above that the explicit symmetrization of the eigenfunctions is necessitated by quantum mechanical **indistinguishability**: for fermions (bosons), the eigenfunction has to be anti-symmetric (symmetric) under particle permutation. Since ψ_s and ψ_a still refer to only spatial symmetry, we consider now the case when the spin is considered.

Let us think about this splitting as effective interaction between $\hat{s}_1$ and $\hat{s}_2$. We find the operator that involves the splitting by considering the interaction that gives rise to a coupling between spin operators $\hat{s}_1$ and $\hat{s}_2$ of two particles:

$$\mathrm{H} \approx \hat{s}_1 \hat{s}_2 \tag{9.89}$$

We define the total spin, $\hat{s}_{\text{total}}$, so that the mutual interaction Hamiltonian for the system has the form:

$$\mathrm{H} = \Delta E_s + \frac{1}{2}(\Delta E_a - \Delta E_s)\hat{s}_{\text{total}}^2 \tag{9.90}$$

The sum of the spins for the two particles,

$$\hat{s}_{\text{total}} = \hat{s}_1 + \hat{s}_2,\ s_{\max} = s_1 + s_2 \tag{9.91}$$

is identifiable and unique. We have two states corresponding to one and the same state (two-fold degenerate state):

$$s = s_1 + s_2 - 1 \tag{9.92}$$

as well three states corresponding to

$$s = s_1 + s_2 - 2 \tag{9.93}$$

This procedure can be continued up to the minimum value:

$$s_{min} = |s_1 - s_2| \tag{9.94}$$

So, the **triangle rule selection** is as follows:

$$|s_1 - s_2| \leq s \leq s_1 + s_2 \tag{9.95}$$

We examine the case for the following examples:

1. $s_1 = s_2 = \frac{1}{2}$ (9.96)

$$s_{max} = s_1 + s_2 = 1,\ s_{min} = |s_1 - s_2| = 0,\ s = 1,0 \tag{9.97}$$

2. $s_1 = 1,\ s_2 = \frac{1}{2},\ s_1 + s_2 = \frac{3}{2},\ |s_1 - s_2| = \frac{1}{2}$ (9.98)
3. $s_1 = 2,\ s_2 = \frac{3}{2},\ s_1 + s_2 = \frac{7}{2},\ |s_1 - s_2| = \frac{3}{2},\ s = \frac{7}{2}, \frac{2}{2}, \frac{3}{2}, \frac{1}{2}$ (9.99)

We introduce the general form for the expectation value of the Coulomb interaction between electrons (two-spin-$\frac{1}{2}$ particles):

$$\langle u \rangle = A\left(a + b\hat{s}_1\hat{s}_2\right) \tag{9.100}$$

We complete the square:

$$\hat{s}_1\hat{s}_2 = \frac{1}{2}\left(\left(\hat{s}_1 + \hat{s}_2\right)^2 - \left(\hat{s}_1\right)^2 - \left(\hat{s}_2\right)^2\right) \tag{9.101}$$

then we replace $\left(\hat{s}_1\right)^2$ by its eigenvalue $s_1(s_1+1)$, and $\left(\hat{s}_2\right)^2$ by $s_2(s_2+1)$ and $\left(\hat{s}_1 + \hat{s}_2\right)$ by the total spin operator $\hat{s}_{total}$, yielding

$$\hat{s}_1\hat{s}_2 = \frac{1}{2}\left(s_{total}\left(s_{total}+1\right) - s_1\left(s_1+1\right) - s_2\left(s_2+1\right)\right) \tag{9.102}$$

with $\hat{s}_{total}$ taking on any possible integral values ranging from s_{min} to s_{max}.

So for $s_{total} = 1$, we have

$$\hat{s}_1\hat{s}_2 = \frac{1}{2}\left(1(1+1) - \frac{1}{2}\left(\frac{1}{2}+1\right) - \frac{1}{2}\left(\frac{1}{2}+1\right)\right) = \frac{1}{4} \tag{9.103}$$

and for $s_{total} = 0$, we have

$$\hat{s}_1\hat{s}_2 = \frac{1}{2}\left(0(0+1) - \frac{1}{2}\left(\frac{1}{2}+1\right) - \frac{1}{2}\left(\frac{1}{2}+1\right)\right) = -\frac{3}{4} \tag{9.104}$$

Then

$$\hat{s}_1\hat{s}_2 = \begin{cases} \frac{1}{4}, & s_{\text{total}} = 1 \quad \text{Triplet} \\ -\frac{3}{4}, & s_{\text{total}} = 0 \quad \text{Singlet} \end{cases} \tag{9.105}$$

For the singlet and triplet states, we have, respectively,

$$\text{A}\left(a - \frac{3}{4}b\right) = \text{A}, \; \text{A}\left(a + \frac{1}{4}b\right) = -\text{A} \tag{9.106}$$

implying

$$a = -\frac{1}{2}, \; b = -2, \; \langle u \rangle = -\frac{1}{2}\left(1 + 4\hat{s}_1\hat{s}_2\right) \tag{9.107}$$

Here, $-2\text{A}\hat{s}_1\hat{s}_2$ denotes the contribution to the energy from a pair of electrons located at sites 1 and 2. This is also the splitting of the energy levels:

$$\Delta\text{H} = 2\text{A}\hat{s}_1\hat{s}_2 \tag{9.108}$$

We have thus taken a spin-independent interaction (say a Coulomb potential) between electrons and imitate it to a spin-spin interaction. Note that the splitting is proportional to the "exchange integral" A which depends on the overlap of the eigenfunctions. Note that the introduction of the interaction term $-2\text{A}\hat{s}_1\hat{s}_2$ is the source of the Weiss internal field which produces ferromagnetism.

10

Approximate Method for the Helium Atom

The helium atom with chemical symbol He is the second element of the periodic table. It provides an example of a quantum-mechanical problem which cannot be solved exactly. However, approximate methods applied to the helium atom can give accurate solutions in perfect agreement with experimental results. The helium atom He constitutes of a nucleus with two protons, two neutrons and two electrons on the 2p orbital. This is the simplest example of the many-electron system. Before the appearance of Schrödinger quantum mechanics, there was an attempt to calculate the energy of the helium atom, He, where the helium atom represents a problem of three bodies with variables that are not separable. So the problem cannot be solved classically. It appears that the Bohr theory is not successful in solving this problem too. The classical and the Bohr theory could not be successful in considering the interaction effect which is of principal importance in quantum mechanics.

As is well known, the hydrogen molecule H_2 constitutes two hydrogen atoms and linked via a covalent bonding which can be explained by an exchange interaction. This implies that the covalent bonding is a pure quantum mechanical phenomenon. The hydrogen molecule relates the helium atom in that it has two electrons and two protons, so the helium Hamiltonian will have the hydrogen-like Hamiltonian.

10.1 The State of the Helium Atom

The non-relativistic Hamiltonian, $\hat{H}$, of the helium atom can be written in the coordinate representation as follows:

$$\hat{H} = \hat{H}_0(1) + \hat{H}_0(2) + \frac{e^2}{r_{12}} \tag{10.1}$$

Here, 1 and 2 index the two electrons moving about the nucleus of charge $Z = +2e$ and $\frac{e^2}{r_{12}}$ represents the Coulomb repulsion between the two electrons in the helium atom which prevents an exact solution of the Schrödinger equation and accounts for much of the complication in the theory; $\hat{H}_0$ is the hydrogen-like Hamiltonian with $\hat{H}_0(1)$ and $\hat{H}_0(2)$ being, respectively, the interaction energy of the first and second electrons with the helium nucleus:

$$\hat{H}_0 = -\frac{\hbar^2}{2\mu}\Delta - \frac{2e^2}{r} \tag{10.2}$$

The last term in Equation 10.2 is the energy of interaction of the electron with the nucleus. In Equation 10.1, the distance between the two electrons is $r_{12} = |\vec{r}_1 - \vec{r}_2|$, while $\vec{r}_1$ and $\vec{r}_2$ are the position vectors of electrons 1 and 2, respectively, with respect to the nucleus as origin and μ denotes the mass of the electron. In addition to the attractive force between electrons and the nucleus, there is also repulsion

DOI: 10.1201/9781003273073-11

between the two electrons. We assume the two electrons to be distinguishable which of course is not so. From the ground state of the two-electron system, the Pauli Exclusion Principle constraints the spins to be in opposite directions. So the electrons can be treated effectively as distinguishable. We also neglect the spin-orbit coupling interaction. From Equation 10.1, the helium atom is slightly complex than the hydrogen atom and yet may not be solved analytically and so the variational or perturbative technique will be more appropriate. The system in Equation 10.1 imitates that of two hydrogen atoms.

From Equation 10.1, the separation of variables seems not feasible. So we find an approximate method for its solution. The most probable is the perturbation theory where we consider a small perturbative parameter. It is obvious that from Equation 10.1, there is no such perturbative small parameter. So it is possible to find the perturbation for such a term in the Hamiltonian $\hat{\mathrm{H}}$ that disturbs the exact solution of such a problem. This perturbative term should be $\frac{e^2}{r_{12}}$ which is a mutual potential energy between the electrons. Note that though this method does not give the exact solution, it permits to study the energy spectra. The Hamiltonian in Equation 10.1 does not have spin variables. However, the total wave function can be represented as the product of the space and spin wave functions. In this way, the wave function has to be anti-symmetric considering the permutation of the states of two electrons. If the Hamiltonian is symmetric during the permutation of space variables, then we select the space function either completely symmetric or completely anti-symmetric. In the former case, we have **a para-state,** while in the latter, an **ortho-state.**

The zero-order wave function of the helium atom in the perturbation theory:

$$\psi_1^{(0)} = \psi_{n_1}(\vec{r}_1)\psi_{n_2}(\vec{r}_2) \tag{10.3}$$

and the zero-order energy:

$$E^{(0)} = E_{n_1} + E_{n_2} \tag{10.4}$$

This also corresponds to the wave function:

$$\psi_2^{(0)} = \psi_{n_1}(\vec{r}_2)\psi_{n_2}(\vec{r}_1) \tag{10.5}$$

From Equations 10.3 and 10.5, we can represent the symmetric and anti-symmetric functions:

$$\psi_s^{(0)} = \frac{1}{\sqrt{2}}\left(\psi_1^{(0)} + \psi_2^{(0)}\right),\ \psi_a^{(0)} = \frac{1}{\sqrt{2}}\left(\psi_1^{(0)} - \psi_2^{(0)}\right) \tag{10.6}$$

We select the zero-order wave function from the principle of indistinguishability. However, here we have to understand that we are talking about the perturbation theory for degenerate levels. Since Equations 10.3 and 10.5 correspond to $E^{(0)}$, this implies two-fold degeneracy. This is of course not a total degeneracy since

$$\psi = \psi(n,m,l) \tag{10.7}$$

and

$$E^{(0)} = E^{(0)}(n) \tag{10.8}$$

Since we need exchange degeneracy, Equation 10.6 should be the correct zero-order wave function approximation:

$$\mathrm{A} = \int \psi_s^*(\vec{r}_1,\vec{r}_2)\psi_a(\vec{r}_1,\vec{r}_2)d\tau_{1,}\, d\tau_2 = 0 \tag{10.9}$$

For this, if we swap the coordinates $\vec{r}_1 \leftrightarrows \vec{r}_2$, then $-A = A$ and it implies $A = 0$. This implies that the wave functions ψ_s^* and ψ_a are orthogonal. But the matrix operator constructed from these wave functions should be diagonal. Also, similarly to Equation 10.9:

$$\int \psi_s^*(\vec{r}_1,\vec{r}_2)\frac{e^2}{r_{12}}\psi_a(\vec{r}_1,\vec{r}_2)d\tau_1, d\tau_2 = 0 \tag{10.10}$$

as $r_{12} = |\vec{r}_1 - \vec{r}_2|$.

So we come to a conclusion that ψ_a and ψ_s are the correct zero-order wave functions. This implies that the first-order correction to the energy can be calculated as a non-degenerate level, i.e.,

$$E_{s(a)}^{(1)} = e^2 \int \frac{|\psi_{s(a)}(\vec{r}_1,\vec{r}_2)|^2}{r_{12}} d\tau_1, d\tau_2 = \frac{e^2}{2} \int \frac{|\psi_{n_1}(1)|^2 |\psi_{n_2}(2)|^2 |\psi_{n_1}(2)|^2 |\psi_{n_2}(1)|^2}{r_{12}} d\tau_1, d\tau_2 \tag{10.11}$$

If we swap the coordinates $\vec{r}_1 \leftrightarrows \vec{r}_2$, then

$$E_{s(a)}^{(1)} = e^2 \int \frac{|\psi_{n_1}(1)|^2 |\psi_{n_2}(2)|^2}{r_{12}} dq_1\, dq_2 \pm e^2 \int \frac{\psi_{n_1}^*(1)\psi_{n_2}(1)\psi_{\psi_{n_2}}^*(2)\psi_{n_1}(2)}{r_{12}} dq_1\, dq_2 \tag{10.12}$$

and letting

$$C_{n_1,n_2} = e^2 \int \frac{|\psi_{n_1}(1)|^2 |\psi_{n_2}(2)|^2}{r_{12}} dq_1\, dq_2 = \int \frac{\rho_{n_1}(1)\rho_{n_2}(2)}{r_{12}} dq_1\, dq_2,\ \rho_n(\vec{r}) = e|\psi_n(\vec{r})|^2 \tag{10.13}$$

Equation 10.13 is due to the electron-electron repulsion and is called the **electrostatic** or **Coulomb integral.** If ρ is the electronic density, then Equation 10.13 formally is called the **electrostatic** or **Coulomb potential energy**. Since the integrand in Equation 10.13 is strictly positive as expected for electron repulsion, this would only raise the average energy of the helium atom.

The **exchange integral** defining the energy distribution of the singlet and triplet terms:

$$A_{n_1,n_2} = e^2 \int \frac{\psi_{n_1}^*(1)\psi_{n_1}(2)\psi_{n_2}^*(2)\psi_{n_2}(1)}{r_{12}} dq_1\, dq_2 \tag{10.14}$$

and implies

$$E_s^{(s)} = C + A,\ E_a^{(s)} = C - A \tag{10.15}$$

In the first order of the perturbation theory, both energies differ by the quantity 2A which is the measure of the exchange interaction and is as well the difference of the energies of the singlet and triplet states.

In the helium atom, there should be principally two types of states with $s = 1$ and $s = 0$ corresponding, respectively, to the **para**- and **ortho**-states, i.e., **para**- and **ortho-helium** (Figure 10.1). At the ground state, we have $n_1 = n_2 = 1$. This implies, any electron, is in a hydrogen-like state. In this case, $\psi^{(0)}$ is a symmetric function. This implies that the exchange interaction does not exist and $\psi_a^{(0)} = 0$. For $s = 0$, the spinor wave function is anti-symmetric.

The ground-state energy is $2E_1 + C_{11}$. We examine the first excited state when

$$n_1 = n_2 = 1 \tag{10.16}$$

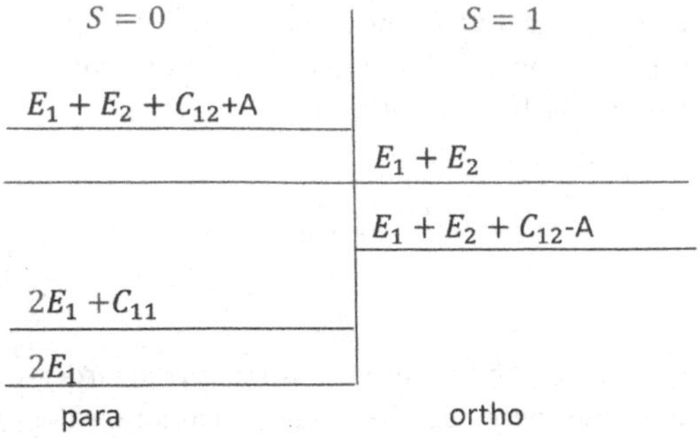

FIGURE 10.1 The two types of states (para- and ortho-helium) of the helium atom with $s=1$ and $s=0$ corresponding, respectively, to the para- and ortho-states.

This implies that there exists exchange interaction. So the first excited state is the ortho-state, i.e., state with spin $s=1$. This implies where there is exchange interaction, the energy level is always lower than for the case $s=0$. The first excited state of helium has an abnormal large life time. $\tau \equiv W^{-1} \approx 10^{-7} - 10^{-8}$ sec, where W is the probability of the spontaneous transitions from the first excited level to the ground state.

10.2 Self-Consistent Field Method

This is one of the methods in quantum mechanical calculations and is the Hartree-Fock method. We examine this method for the helium atom with the Hamiltonian:

$$\hat{H} = \hat{H}_0(1) + \hat{H}_0(2) + \frac{e^2}{r_{12}} \tag{10.17}$$

We select the wave function:

$$\psi_{s,(a)}(\vec{r}_1, \vec{r}_2) = \psi_1(1)\psi_2(2) \pm \psi_1(2)\psi_2(1) \tag{10.18}$$

Here, ψ_1 and ψ_2 are trial variational functions which are not eigenfunctions of the hydrogen-like Hamiltonian, $\hat{H}_0$. The variational principle gives

$$\delta \int \psi^* \left(\hat{H} - E\right) \psi \, dq_1 \, dq_2 = 0 \tag{10.19}$$

and implies that

$$\delta \int \left[\psi_1^*(1)\psi_2^*(2) \pm \psi_1^*(2)\psi_2^*(1)\right]\left(\hat{H} - E\right)\left[\psi_1(1)\psi_2(2) \pm \psi_1(2)\psi_2(1)\right] dq_1 \, dq_2 = 0 \tag{10.20}$$

We calculate ψ_1^* and ψ_2^* independent of the variation, $\delta\psi^*$, which implies

$$\delta\psi^* \int [\cdots] dq_1 \, dq_2 = 0 \tag{10.21}$$

We have to change ψ_1^* in the first and second summands by ψ_1^*. This implies that

$$\int \delta\psi_1^*[\cdots]dq_1\,dq_2 = 0 \tag{10.22}$$

In order that we have the integral

$$\int \delta\psi_1^*[\cdots]dq_1 = 0 \tag{10.23}$$

it is necessary that the coefficient of $\delta\psi_1^*$ should be equal to zero. We find this coefficient while considering that $\hat{H}_1$ does not act on $\psi(2)$. Also,

$$\hat{H}_1\psi_1(1) = E\psi_1(1) \tag{10.24}$$

$$H_{22} = \int \psi_2^*(2)\hat{H}_2\psi_2(2)dq_2 \tag{10.25}$$

Let

$$A_{22}(\vec{r}_1) = e^2 \int \frac{\psi_2^*(2)\psi_2(2)}{r_{12}}dq_2 \tag{10.26}$$

From the first summand of the first bracket and first summand of the second bracket of Equation 10.20, we have

$$\hat{H}_1\psi_1(1) - E\psi_1(1) + H_{22}\psi_1(1) + A_{22}(\vec{r}_1)\psi_1(1) \tag{10.27}$$

We examine now the first and second summand of the brackets in Equation 10.20. From the orthogonality of the eigenfunctions ψ_1 and ψ_2:

$$\int \psi_1^*(\vec{r})\psi_2(\vec{r})dq = 0 \tag{10.28}$$

If

$$H_{21} = \int \psi_2^*(2)\psi_1(2)dq_2 \tag{10.29}$$

which implies we have

$$H_{21}\psi_2(1) + A_{21}(\vec{r}_1)\psi_2(1) \tag{10.30}$$

So

$$\hat{H}_1\psi_1(1) - E\psi_1(1) + H_{22}\psi_1(1) + A_{22}(\vec{r}_1)\psi_1(1) + H_{21}\psi_2(1) + A_{21}(\vec{r}_1)\psi_2(1) = 0 \tag{10.31}$$

Now we do also for the second summand of the first bracket and both summands of the second bracket of Equation 10.20, and so we have

$$\left[\hat{H}_1 + A_{22}(\vec{r}_1)\right]\psi_1(\vec{r}_1) + \left[H_{21} + A_{21}(\vec{r}_1)\right]\psi_2(\vec{r}_1) = E_1\psi_1(\vec{r}_1),\ E_1 = E - H_{22} \tag{10.32}$$

If we do the variation with respect to ψ_1^*, then the same result is obtained by swapping the indices $1 \leftrightarrows 2$ in Equation 10.32:

$$\left[\hat{H}_2 + A_{11}(\vec{r}_2)\right]\psi_2(\vec{r}_2) + \left[H_{12} + A_{12}(\vec{r}_2)\right]\psi_1(\vec{r}_2) = E_2\psi_2(\vec{r}_2),\ E_2 = E - H_{11} \tag{10.33}$$

Equations 10.32 and 10.33 are a system of two integral-differential equations with two independent variables. Such equations are called self-consistent field equations. These equations are also called Fock equations. The equations were for the first time established by Hartree without considering the contribution of the interaction. It is easily seen in this case that the Hartree equation:

$$\left[-\frac{\hbar^2}{2\mu} - \frac{2e^2}{r} + e^2\int \frac{\left|\psi_2(\vec{r}\,')\right|^2}{\left|\vec{r} - \vec{r}\,'\right|} d\tau'\right]\psi_1(\vec{r}) = E_1\psi_1(\vec{r}) \tag{10.34}$$

$$\left[-\frac{\hbar^2}{2\mu} - \frac{2e^2}{r} + e^2\int \frac{\left|\psi_1(\vec{r}\,')\right|^2}{\left|\vec{r} - \vec{r}\,'\right|} d\tau'\right]\psi_2(\vec{r}) = E_2\psi_2(\vec{r}) \tag{10.35}$$

These equations have the standard form of the Schrodinger equation apart from the integral terms in the brackets. These are self-consistent field equations. We observe for example that the first electron moves in the field H_2 with the electrostatic field of the second electron.

11

Approximate Method for the Hydrogen Molecule

We consider Figure 11.1 depicting the electrons for fixed positions of protons. This implies we consider the adiabatic approximation since the mass of the proton μ_p is exceedingly greater than that of the electron μ_e:

$$\mu_p \gg \mu_e \tag{11.1}$$

The exchange integral in Equation 10.14 is also applicable for the case of the hydrogen molecule when calculating the interaction energy of the two atoms in the hydrogen molecule. The nuclei of the two hydrogen atoms at positions a and b are assumed fixed and located at a distance of R from each other in a central symmetric field, r_{12}, distance between electron e_1 and e_2; r_{1b}, distance between electron e_1 and proton b; r_{2b}, distance between electron e_2 and proton b; and r_{2a}, distance between electron e_2 and proton a.

The Hamiltonian, $\hat{\mathrm{H}}$, of the hydrogen molecule is as follows:

$$\hat{\mathrm{H}} = \hat{\mathrm{T}}_1 + \hat{\mathrm{T}}_2 - \frac{e^2}{r_{1a}} - \frac{e^2}{r_{2b}} - \frac{e^2}{r_{2a}} - \frac{e^2}{r_{1b}} + \frac{e^2}{r_{12}} + \frac{e^2}{R} \tag{11.2}$$

where $\hat{\mathrm{T}}_1$ and $\hat{\mathrm{T}}_2$ are, respectively, the kinetic energies of electron e_1 and e_2 while $\frac{e^2}{R}$ is the Coulomb repulsion energy between the two protons a and b. From the Hamiltonian in Equation 11.2, it is obvious that the problem can only be solved approximately. With an increase in R, in the zero-approximation electron, e_1 approaches a and e_2 approaches b. In this case, the wave function of the hydrogen molecule is the product $\psi_a(\vec{r}_1)\psi_b(\vec{r}_2)$ and the energy of the zero approximation is as follows:

$$E^{(0)} = 2E_1 \tag{11.3}$$

Here, E_1 is the ground state energy of the hydrogen atom. If we swap the coordinates $\vec{r}_1 \leftrightarrows \vec{r}_2$, then we have the wave function $\psi_a(\vec{r}_2)\psi_b(\vec{r}_1)$. This permits us to write the symmetric and antisymmetric wave functions, respectively:

$$\psi^{(0)}_{s,(a)}(\vec{r}_1,\vec{r}_2) = C_{s,(a)}\left[\psi_a(\vec{r}_1)\psi_b(\vec{r}_2) \pm \psi_a(\vec{r}_2)\psi_b(\vec{r}_1)\right],\ C_{s,(a)} = \frac{1}{\sqrt{2(1 \pm S^2)}} \tag{11.4}$$

DOI: 10.1201/9781003273073-12

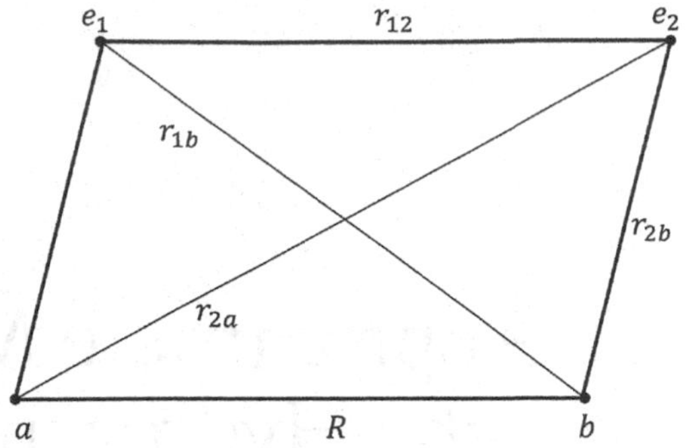

FIGURE 11.1 The covalent bonding of the hydrogen molecule.

where $C_{s,(a)}$ is the normalization coefficient. Here, the **overlap integral** is S since contributions come only from points of space where the atomic wave functions $\psi_a(\vec{r}_1)$ and $\psi_b(\vec{r}_1)$ are both different from zero with the existence of such points when the two atomic orbitals partially overlap:

$$S = \int \psi_a^*(\vec{r}_1)\psi_b(\vec{r}_1)dq,\ 0 \le S \le 1 \tag{11.5}$$

In the above relations, $\psi(\vec{r})$ is the one-electron wave function describing the ground state of the electron in the Coulomb field of the nucleus.

We solve the problem with the help of the perturbation theory by considering

$$E = E^{(0)} + E_s^{(1)},\ E^{(0)} = 2E_1 \tag{11.6}$$

As mentioned earlier, E_1 is the ground-state energy of the hydrogen atom and E can be calculated as in Equation 11.6 if we consider the symmetric wave function:

$$\psi = \psi_s^{(0)} + \psi_s^{(1)} \tag{11.7}$$

So

$$\hat{H}\left(\psi_s^{(0)} + \psi_s^{(1)}\right) = \left(E^{(0)} + E_s^{(1)}\right)\left(\psi_s^{(0)} + \psi_s^{(1)}\right) \tag{11.8}$$

Substituting the wave functions in Equations 11.4 into 11.8:

$$\begin{aligned} &C_s E^{(0)}\psi_a(1)\psi_b(2) + C_s U_1 \psi_a(1)\psi_b(2) + C_s E^{(0)}\psi_a(2)\psi_b(1) + C_s U_2 \psi_a(2)\psi_b(1) + \hat{H}\psi_s^{(1)} \\ &= E^{(0)}\psi_s^{(0)} + E_s^{(1)}\psi_s^{(0)} + E^{(0)}\psi_s^{(1)} + E_s^{(1)}\psi_s^{(1)} \end{aligned} \tag{11.9}$$

Here

$$U_1 = -\frac{e^2}{r_{2a}} - \frac{e^2}{r_{1b}} + \frac{e^2}{r_{12}} + \frac{e^2}{R},\ U_2 = -\frac{e^2}{r_{1a}} - \frac{e^2}{r_{2b}} + \frac{e^2}{r_{12}} + \frac{e^2}{R} \tag{11.10}$$

In Equation 11.10, the quantity $E_s^{(1)}\psi_s^{(1)}$ is an infinitesimal of the second order and we limit ourselves to the first order. This implies that

$$\left(\hat{\text{H}}-E^{(0)}\right)\psi_s^{(1)}=E_s^{(1)}\psi_s^{(0)}-C_sU_1\psi_a(1)\psi_b(2)-C_sU_2\psi_a(2)\psi_b(1) \tag{11.11}$$

This is a non-homogenous equation for $\psi_s^{(1)}$.

If

$$\hat{\text{A}}\psi=\text{B} \tag{11.12}$$

then this equation has a solution if B is orthogonal to the eigenfunction of the operator $\hat{\text{A}}$, i.e.,

$$\int\psi_0^*\text{B}\,dq=0 \tag{11.13}$$

Note on the left-hand side of Equation 11.11 we have $E^{(0)}$, and we multiply Equation 11.11 by $\psi_s^{(0)}$:

$$E_s^{(1)}=C_s\int\psi_s^{(0)}U_1\psi_a(1)\psi_b(2)dq_1\,dq_2+\int\psi_s^{(0)}U_2\psi_a(2)\psi_b(1)dq_1\,dq_2 \tag{11.14}$$

Substituting in this relation Equation 11.4 and considering Equation 11.6, we have

$$E_s^{(1)}=\frac{C+A}{1+S^2}+\frac{e^2}{R} \tag{11.15}$$

Similarly, for the anti-symmetric wave function $\psi_a^{(0)}$ we have:

$$E_a^{(1)}=\frac{C-A}{1-S^2}+\frac{e^2}{R} \tag{11.16}$$

Here, C and A are the **electrostatic and exchange integrals,** respectively.

Obviously, substituting Equations 11.10 and 11.4 into Equation 11.15, we have, respectively, the **Coulomb** and **exchange integrals**:

$$C=e^2\int\frac{|\psi_a(1)|^2|\psi_b(2)|^2}{r_{12}}dq_1\,dq_2-e^2\int\frac{|\psi_b(2)|^2}{r_{2b}}dq_2-e^2\int\frac{|\psi_a(1)|^2}{r_{1b}}dq_1 \tag{11.17}$$

and

$$A=e^2\int\frac{\psi_a^*(1)\psi_a(2)\psi_b^*(2)\psi_b(1)}{r_{12}}dq_1\,dq_2-e^2S\int\frac{\psi_a^*(1)\psi_b(1)}{r_{1b}}dq_1-e^2S\int\frac{\psi_a(2)\psi_b^*(2)}{r_{2a}}dq_2 \tag{11.18}$$

Further, we calculate these integrals considering that they are for the case of the central symmetric field. The calculation may be approximate though there may be an exact solution. In the first approximation

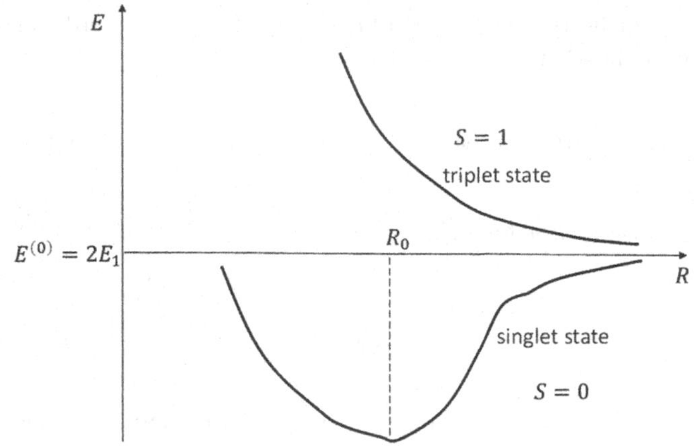

FIGURE 11.2 The dependence of the energy E on the distance R for the triplet and singlet states.

of the perturbation theory, we have $E(R)$. So considering the term $\frac{e^2}{R}$ in Equations 11.15 and 11.16, we observe that $E^{(1)} > 0$ with $\frac{\partial}{\partial R}E_a^{(1)} < 0$. This follows that the force between two hydrogen atoms is positive and implies this force is repulsive. This corresponds to $S=1$ spin state (triplet state). For $E_s^{(1)}$, when $R < R_0$, the force between the hydrogen atoms is positive and for $R > R_0$, the force is negative. This implies that at R_0, the force of attraction equals the force of repulsion. The stable state of the hydrogen molecule is achieved when the symmetric wave function corresponds to the spin antisymmetric wave function. This implies that the spin of the hydrogen molecule $S=0$ (singlet state). This corresponds to the quantum mechanical explanation of the chemical bonding of two atoms with forces of different signs subjected on them. This phenomenon is also true for any diatomic molecule with atoms of the same kind. The dependence of the energy E on the distance R for the triplet and singlet states is shown in Figure 11.2.

We observe that the exchange integral is expressed in a manner such that in the triplet sates, the force between the atoms is repulsive, while in the singlet state, at large distances, there is attraction. This attraction swaps with repulsion for small distances and leads to formation of a bound state – the hydrogen molecule. The energy of the triplet state is observed to be higher than that of the singlet state. From Figure 11.2, the singlet state of the energy as a function of the distance R has the form of a potential well and leads to the bound state of the system. The bonding of two atoms of the same kind in the molecule is called a **homopolar bonding** and cannot be explained by classical physics and best explained by quantum mechanics.

The symmetrization of the coordinate dependence wave function described by Equation 11.4 leads to a defined correlation of the distribution of electrons. The probability of location of the first electron in the element of volume dq_1 and the second electron in the element of volume dq_2 considering Equation 11.4 is equal to:

$$dP = C_{s,(a)}^2\left\{|\psi_a(1)|^2|\psi_b(2)|^2 + |\psi_a(2)|^2|\psi_b(1)|^2 \pm\left[\psi_a(1)\psi_b(2)\psi_a^*(2)\psi_b^*(1) + \psi_a^*(1)\psi_a(2)\psi_b^*(2)\psi_b(1)\right]\right\} \quad (11.19)$$

The first two terms in this expression give the product of probabilities of two statistical independent events, while the third term describes the correlation of statistically non-independent events for electrons in a selected volume. It is instructive to note the correlation:

- Leads to exchange as well as overlap integrals
- Terms decrease faster with increase inter-nuclei R distance and implying the exponential decrease of the electronic wave functions with increase electronic coordinate relative to that of the nuclei

- May be safely neglected for exceedingly large distances R and implying neglecting wave function symmetry properties
- Being appreciable in electronic positions leads to limitation in symmetrization

11.1 Vibrational and Rotational Levels of Diatomic Molecules

The calculations will be based on the adiabatic approximation proposed in 1927 by Born and Oppenheimer. Their idea consists in that for the first atom, the electron is a light particle and moves faster than the nucleus of the atom relative to their velocities that is proportional to $\sqrt{\frac{M}{m}}$ as $m \ll M$, where M and m are, respectively, the mass of the nucleus and that of the electron. So it is possible to examine the state of the electrons for fixed positions of the nuclei. In this way, the electronic energy E will be a function of the distance R between the nuclei:

$$E_e = E_e(R) \tag{11.20}$$

In this case, R will just be a parameter.

In this context, the Schrödinger equation can be solved by the method of separation of variables and we have the eigenvalue equation for the nuclei:

$$\left(\hat{T}_R + E_e(R)\right)\Phi(R) = E\Phi(R) \tag{11.21}$$

In this equation, $\Phi(R)$ is the wave function describing the motion of the nuclei while the electronic energy $E_e(R)$ is a parameter. In the case of diatomic molecules, the electronic energy E_e is independent of R and so we have a situation that imitates motion in a central symmetric field where

$$\Phi(R) = \frac{X(R)}{R} Y_{lm}(\theta,\phi) \tag{11.22}$$

and

$$\frac{\hbar^2}{2\mu^*}\frac{d^2}{dR^2}X(R) + \frac{\hbar^2 l(l+1)}{2\mu^* R^2}X(R) + E_e(R)X(R) = EX(R) \tag{11.23}$$

We examine a situation where the atom is at the neighborhood of the point R_0 so that we Tailor series expand $E_e(R)$ about the point R_0:

$$E_e(R) = E_e(R_0) + \frac{1}{2}\frac{\partial^2}{\partial R^2}E_e(R)\bigg|_{R=R_0}(R-R_0) + \cdots \tag{11.24}$$

Substitute Equation 11.24 into 11.23 and then swap R_0 for R in the centrifugal term in Equation 11.23. So in the left-hand side of Equation 11.23 we observe an equation of a harmonic oscillator where the equilibrium position will be the point R_0. So the energy of the total system is as follows:

$$E_e(R) = E_e(R_0) + \hbar\omega\left(v + \frac{1}{2}\right) + \frac{\hbar^2 l(l+1)}{2\mu^* R_0^2} \tag{11.25}$$

Here, the vibrational energy is as follows:

$$E_{\text{vib}} = \hbar\omega\left(v + \frac{1}{2}\right), \, v = 0,1,2,\ldots \tag{11.26}$$

while the rotational energy is as follows:

$$E_{\text{rot}} = \frac{\hbar^2 l(l+1)}{2M^* R_0^2} \tag{11.27}$$

which are the results of the adiabatic approximation for a diatomic molecule.

12

Scattering Theory

The scattering process is equivalent to the collision process and relates the physics of gaseous states, solid state physics (electrons colliding with phonons in the crystal lattice), plasma physics and so on. When we cannot directly visualize particles, scattering can be a valuable probe for their study. An example is the hydrogen atom that scattering experiments confirm it to have a nucleus with an electron orbiting around it. Scattering experiments also confirm protons to have some constituents called quarks. So the scattering theory is pivotal in understanding of the quantum nature of particles. In principle, we can formulate all problems of quantum mechanics in the language of scattering theory.

12.1 Scattering Cross Section and Elastic Scattering Amplitude

Usually for problems related to the collision of a particle with a force field, $U(r)$, the energy eigenvalues are continuously distributed. Contrary to the problems seen earlier, for the collision problem, the energy is specified in advance where the behavior of the wave function at large distances is obtained in terms of it. The given asymptotic behavior is then related to the amount of scattering of the particle by the force field. For the **elastic scattering**, the scattered particle has the same kinetic energy as the incoming particle. This implies that no energy is transferred to the particle. For the **inelastic scattering**, the electron is scattered by an atom and excites the atom during the scattering process. This implies that the outgoing electron would then have less kinetic energy than the incoming one, less by the amount needed for the excitation of the atom.

Consider Rutherford experiment on the scattering of α-particles. This establishes the structure of the atom and the study of collision of particles is one of the most important sources of information on the structure of atoms and, in particular, their interaction as earlier indicated. The classical representation of the problem entails the following: the particle moves from infinity toward the positive direction of the z-axis and scatters on some particle through the angle θ (Figure 12.1a) and the scattering distance d. On the path of the scattering particle, we introduce an elementary surface S. Experimentation shows that different particles scatter through different scattering angles θ.

This is explained by the spread of the scattering distance, d and $d+\Delta d$, with the scattering angle, θ and $\theta+\Delta\theta$. In quantum mechanics, such a representation of the problem is not adequate as there is no existence of the notion of a trajectory in quantum mechanics. Here, the state of a particle is described by the wave function $\psi(\vec{r})$. We introduce here the notion of the probability density that describes the state with the **probability current density**:

$$\vec{j} = \frac{i\hbar}{2\mu}\left(\psi\Delta\psi^* - \psi^*\Delta\psi\right) \tag{12.1}$$

DOI: 10.1201/9781003273073-13

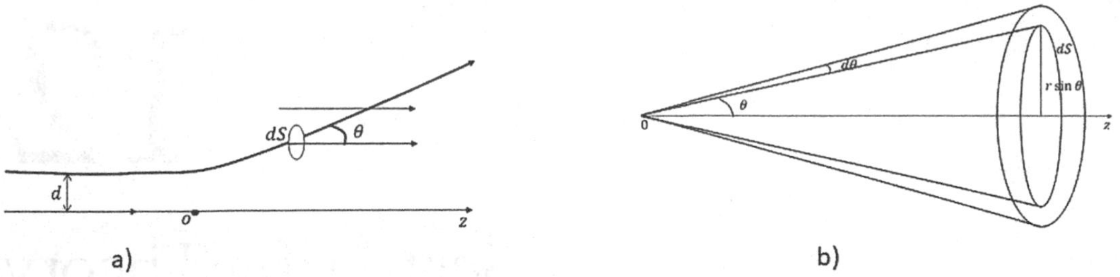

FIGURE 12.1 (12.1a) The classical scattering of a beam of particles in passing near a nucleus; (12.1b) the element of the scatting surface, dS, and the scattering angle, θ.

We examine the physical sense of the probability current density $\vec{j}$. Taking the element of the scattering surface:

$$\overrightarrow{dS} = \vec{n}dS \tag{12.2}$$

$\vec{j}\overrightarrow{dS}$ is the probability that the particle with the state described by the wave function ψ within the period of one second passes through the elementary surface dS. This is the probability of the scattering within a period of one second with the **differential scattering cross section (effective differential scattering cross section)** being

$$\overrightarrow{d\sigma} = \frac{\vec{j}\overrightarrow{dS}}{j_z} \tag{12.3}$$

Here, j_z is the **flux current density** of the incident particles. A similar definition of $\overrightarrow{d\sigma}$ can be seen from classical physics where, instead of the probability density, we have the particle density. We have two ways of looking at this problem, i.e., considering elastic and inelastic scattering. For the elastic scattering, the particles conserve their energies, and for the inelastic, energy is not conserved.

Suppose we examine the **elastic scattering**, in this case, the wave function is non-stationary. In this case, the differential scattering cross section is always defined by the interaction potential energy of the scattering particles. Separating the motion of the center of mass from the problem, we arrive at the motion of the particle in the field of a fixed field center. We assume for now that the interaction of the particle with center is independent of the spin.

Considering the spread over the momentum is small compared to the atomic dimensions, we can describe the initial state by a plane wave:

$$\frac{1}{(2\pi\hbar)^{-\frac{3}{2}}}\exp\{i\vec{\kappa}\vec{r}\} \tag{12.4}$$

We are talking about the non-stationarity in the continuous energy spectrum. This is for the case of the infinite motion where

$$\mathrm{E} = \frac{\hbar^2\kappa^2}{2\mu} \tag{12.5}$$

Here, $\vec{\kappa}$ is the de Broglie wave vector and μ is the mass of the particle. What type of wave function should be put in Equation 12.1, leading to Equations 12.3–12.5 in order to find the vector $\overrightarrow{d\sigma}$? Considering that $\vec{j}\overrightarrow{dS}$ only has sense when we talk about a state at large distance, we assume it to achieve the value considering the interaction with the center of the system. This implies that the particle is moving freely. This implies that in Equation 12.3, we substitute the value of the current density $\vec{j}$ obtained from the asymptotic wave function $\tilde{\psi}$. This implies that the distance is increased where the interaction with the center may be assumed to achieve the value zero. So we suppose that

$$\lim_{\vec{r}\to\infty} U(\vec{r}) \to 0 \tag{12.6}$$

Swapping ψ with $\tilde{\psi}$ is an approximation that arises from its very definition. In the scattering theory, $\tilde{\psi}$ is written in the form:

$$\tilde{\psi} = \exp\{i\kappa z\} + f(\theta)\frac{\exp\{i\kappa r\}}{r} \tag{12.7}$$

This is the standard wave function of the scattering theory. The first summand is the de Broglie plane wave that corresponds to the free motion of a particle in the positive z-axis direction or along the polar axis $\theta = 0$ since $z = r\cos\theta$. This is the asymptotic wave function before scattering, i.e., the incident wave. The second summand, $\frac{\exp\{i\kappa r\}}{r}$, in Equation 12.7 is the wave function of a particle moving radially outward in the form of a divergent spherical wave. This implies the wave moving with a center $f(\theta)$ that is dependent on the scattering angle. This has the dimension of length and is called the **scattering amplitude**. So this scattering amplitude, $f(\theta)$, is dependent on the potential $U(\vec{r})$. This scattering amplitude, $f(\theta)$, measures the ratio of the flux of the scattered particles to that of the incident particles. The scattering amplitude, $f(\theta)$, is of great importance in quantum mechanics as its absolute value squared is what is measured in the scattering experiments. The second term of Equation 12.7 is inversely proportional to r and indicative that the radial flux must fall off as an inverse square of the distance.

We effectively are working in the spherical system of coordinates, where

$$\psi = \psi(r,\theta,\phi) \tag{12.8}$$

We examine the motion in the central symmetric field. However, the angle ϕ describes the scattering about the z-axis. From Figure 12.1b, the element of the surface is

$$dS = 2\pi r^2 \sin\theta \, d\theta \tag{12.9}$$

The asymptotic wave function $\tilde{\psi}$ is normalized by the first probability density of the initial state. In this case,

$$j_z = \frac{i\hbar}{\mu}\left(\psi\nabla\psi^* - \psi^*\nabla\psi\right) = \frac{\hbar\kappa}{\mu} = v \tag{12.10}$$

This is as well the velocity of the particle. After calculation, it is obvious that $\overrightarrow{d\sigma}$ is independent of the normalization constant. But

$$\vec{j}\,\overrightarrow{dS} = j_r dr \tag{12.11}$$

where j_r is the radial component of vector $\vec{j}$. This implies that in Equation 12.1, we have to take the projection of $\vec{j}$ on the radial axis:

$$j_r = \frac{i\hbar}{2\mu}\left|f(\theta)\right|^2 \frac{1}{r^2} 2(-i)\kappa = \frac{\hbar\kappa}{\mu r^2}\left|f(\theta)\right|^2 \tag{12.12}$$

We substitute Equations 12.10 and 12.12 into 12.3, then the differential cross section:

$$d\sigma = \frac{1}{r^2}|f(\theta)|^2 \, dS = |f(\theta)|^2 \, d\Omega \tag{12.13}$$

So the differential cross section $d\sigma$ is simply proportional to the square of the absolute value of the scattering amplitude. Here, $d\Omega$ is the element of the solid angle. Equation 12.13 can be calculated by considering the following:

- Find the exact solution of the Schrödinger equation for the wave function, describing the motion of the scattering particle in the field of the scattering center.
- Find the asymptotic equation for the wave function in the form represented in Equation 12.7. This can be done with help of boundary conditions.
- Substitute in Equation 12.13 the expression of the solid angle $d\Omega$and then integrate and find the total scattering cross section.

12.1.1 Relation Between the Laboratory and Center-of-Mass Systems

It is instructive to note that when the only forces acting on two particles result from their mutual interaction, then the overall motion can be translated into the motion of the center of mass and that of relative motion of the two particles. It could have been interesting to know the relation between the differential cross section and the angles in the laboratory system and that of the center-of-mass system. This can be possible by translating the laboratory system in the direction of the incident particle with sufficient speed that brings the center of the mass to rest. We can relate the differential cross sections and angles in the laboratory system and the center-of-mass system by translating the laboratory system in the direction of the incident particle with sufficient speed to bring the center of mass to rest.

We examine a particle of mass m_1 with initial velocity $\vec{v}$ and scattering on a particle of mass m_2 that was initially at rest or in motion. Figure 12.2 depicts the radius vectors of the collision of the two particles with masses m_1 and m_2:

$$\vec{\rho} = \vec{r}_1 - \vec{r}_2, \; \vec{R} = \frac{m_1\vec{r}_1 + m_2\vec{r}_2}{m_1 + m_2}, \; M = m_1 + m_2 \tag{12.14}$$

Here, $\vec{\rho}$ and $\vec{R}$ are, respectively, the radius vectors of the relative motion and that of the center of mass, while M is the total mass of the two particles:

$$m_1\vec{r}_1 + m_2\vec{r}_2 = M\vec{R} \tag{12.15}$$

We examine the system of the center of mass, when $\vec{R} = 0$:

$$m_1\vec{r}_1 + m_2\vec{r}_2 = 0 \tag{12.16}$$

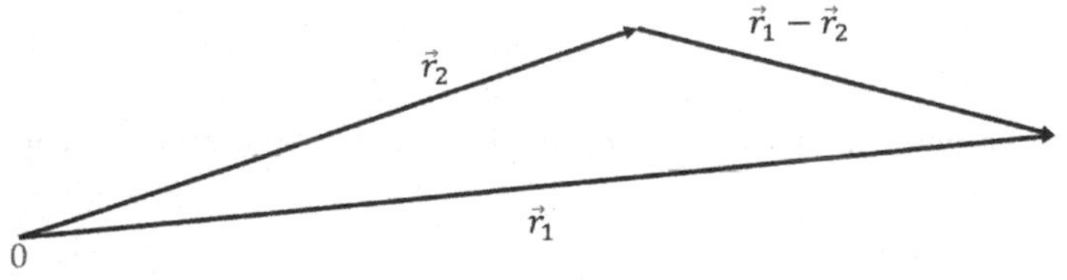

FIGURE 12.2 The radius $\vec{r}_1$ and $\vec{r}_2$ vectors of the collision of the two particles with masses m_1 and m_2, respectively.

So considering the first equation in Equation 12.14, we have

$$\vec{r}_2 = -\frac{m_1\vec{\rho}}{m_1+m_2} \tag{12.17}$$

and from Equation 12.16:

$$\vec{r}_1 = -\frac{m_2}{m_1}\vec{r}_2 = \frac{m_2}{m_1}\frac{m_1\vec{\rho}}{m_1+m_2} = \frac{m_2\vec{\rho}}{m_1+m_2} \tag{12.18}$$

From here:

$$M\vec{r}_2 = -m_1\vec{\rho} \tag{12.19}$$

The velocity of the relative motion of the particles can be obtained as follows:

$$\vec{v} = \frac{d}{dt}\vec{\rho} = \vec{v}_1 - \vec{v}_2 \tag{12.20}$$

From the law of conservation of momentum, we have

$$m_1\vec{v}_{10} + m_2\vec{v}_{20} = 0 \tag{12.21}$$

where the velocities of the particles before collision with masses m_1 and m_2 are, respectively, $\vec{v}_{10}$ and $\vec{v}_{20}$:

$$\vec{v}_{10} = \frac{m_2}{m_1+m_2}\vec{v},\ \vec{v}_{20} = -\frac{m_1}{m_1+m_2}\vec{v} \tag{12.22}$$

The velocities after collision:

$$\vec{v'}_{10} = \frac{m_2}{m_1+m_2}v\vec{n}_0,\ \vec{v}'_{20} = -\frac{m_1}{m_1+m_2}v\vec{n}_0 \tag{12.23}$$

The momentum is conserved as well as the energy. The positive and negative velocities show that particles with masses m_1 and m_2 are moving in opposite directions to each other.

The laboratory system moves with the velocity:

$$\vec{V} = \frac{d}{dt}\vec{R} = \frac{m_1\vec{v}_1 + m_2\vec{v}_2}{m_1+m_2} \tag{12.24}$$

So the velocities of the particles with masses m_1 and m_2 after collision are, respectively, as follows:

$$\vec{V'}_1 = \frac{m_2}{m_1+m_2}v\vec{n}_0 + \frac{m_1\vec{v}_1 + m_2\vec{v}_2}{m_1+m_2},\ \vec{V}'_2 = -\frac{m_1}{m_1+m_2}v\vec{n}_0 + \frac{m_1\vec{v}_1 + m_2\vec{v}_2}{m_1+m_2} \tag{12.25}$$

Then the corresponding momenta after collision:

$$\vec{p'}_1 = \mu v\vec{n}_0 + \frac{m_1}{m_1+m_2}(\vec{p}_1 + \vec{p}_2),\ \vec{p}'_2 = -\mu v\vec{n}_0 + \frac{m_2}{m_1+m_2}(\vec{p}_1 + \vec{p}_2),\ \mu = \frac{m_1 m_2}{m_1+m_2} \tag{12.26}$$

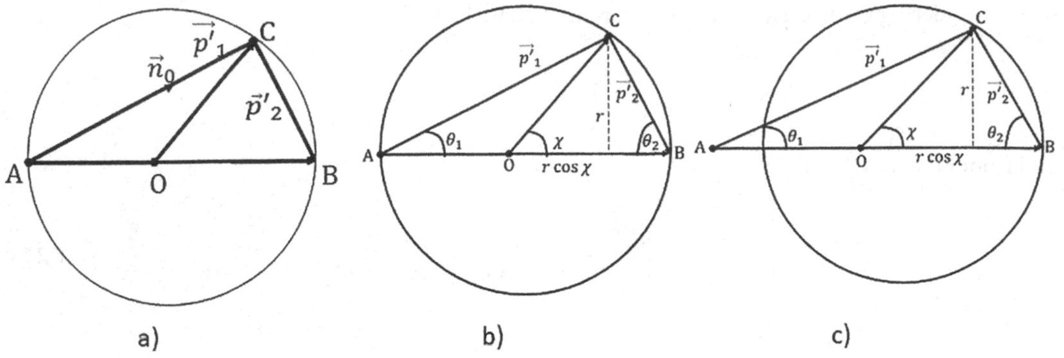

FIGURE 12.3 (12.3a) The momenta of the particles with masses m_1 and m_2 after collision; (12.3b) The momenta of the particles with masses m_1 and m_2 after collision considering, $m_1 < m_2$; (12.3c) The momenta of the particles with masses m_1 and m_2 after collision considering $m_1 > m_2$.

This is depicted geometrically in Figure 12.3a.

$$\overrightarrow{OC} = m\vec{V},\ \overrightarrow{AO} = \frac{m_1}{m_1 + m_2}(\vec{p}_1 + \vec{p}_2),\ \overrightarrow{OB} = \frac{m_2}{m_1 + m_2}(\vec{p}_1 + \vec{p}_2),\ \overrightarrow{AB} = \vec{p}_1 + \vec{p}_2 \tag{12.27}$$

Suppose the particle with mass m_2 is initially at rest then $\vec{p}_2 = 0$ and

$$\overrightarrow{OB} = \frac{m_2}{m_1 + m_2}\vec{p}_1 = \mu v \vec{n}_0 \tag{12.28}$$

For $m_1 < m_2$, we have the diagram in Figure 12.3b, and for $m_1 > m_2$, we have the diagram in Figure 12.3c:

$$\overrightarrow{AO} = \frac{m_1^2}{m_1 + m_2}\vec{v}n = \frac{m_1}{m_2}\mu v \vec{n}_0 \tag{12.29}$$

$$\overrightarrow{AB} = \vec{p}_1,\ \overrightarrow{OC} = \mu \vec{v} \tag{12.30}$$

$$\left|\overrightarrow{AO}\right| = \frac{m_1}{m_2} r,\ h = C\tan\theta_1,\ h = r\sin\chi,\ h = \left(\left|\overrightarrow{AO}\right| + r\cos\chi\right)\tan\theta_1 = r\sin\chi \tag{12.31}$$

$$\tan\theta_1 = \frac{r\sin\chi}{\Gamma r + r\cos\chi} = \frac{m_2\sin\chi}{m_1 + m_2\cos\chi} = \frac{\sin\chi}{\Gamma + \cos\chi},\ \Gamma \equiv \frac{m_1}{m_2},\ \theta_2 = \frac{\pi - \chi}{2} \tag{12.32}$$

Here, χ is the angle of deviation of the first particle after collision. Equation 12.32 can be valid also for inelastic collisions. We apply this to the case of a nuclear reaction where a particle of mass m_1 collides with a particle of mass m_2 initially at rest, and after the collision, particles of masses m_3 and m_4 emerge:

$$m_1 + m_2 = m_3 + m_4 \tag{12.33}$$

Consider an amount of energy, Q, is converted from the internal energy into the kinetic energy of the emergent particles and the particle of mass m_3 is observed, then Equation 12.32 is always valid.

When Q is positive, then we have an exothermic collision, and when negative, then an endothermic collision. In the given case:

$$\Gamma \to \left(\Gamma \frac{m_3}{m_4} \frac{E}{E+Q}\right)^{\frac{1}{2}} \tag{12.34}$$

Here, $E = \frac{\mu v^2}{2}$ is the energy initially associated with the relative motion in the center-of-mass system. It can be easily shown that the total cross section is the same for both laboratory and center-of-mass systems. This is the same for both the outgoing particles as the total number of collisions taking place is independent of the manner in which the process is described.

If $m_1 = m_2$ or $\Gamma = 1$, then

$$\tan\theta_1 = \frac{\sin\chi}{1+\cos\chi} = \tan\frac{\chi}{2} \tag{12.35}$$

$$\chi + \frac{\pi}{2} + \alpha = \pi,\ \theta_2 + \frac{\pi}{2} + \beta = \pi,\ \chi + \theta_2 + \alpha + \beta = \pi,\ \alpha + \beta = \theta_2 \tag{12.36}$$

So

$$\theta_1 = \frac{\chi}{2} \tag{12.37}$$

and varies from 0 to $\frac{\pi}{2}$ as χ varies from 0 to π and then

$$\theta_1 + \theta_2 = \frac{\pi}{2} \tag{12.38}$$

In the given case, no particles appear in the backward hemisphere in the laboratory system.

If $\Gamma < 1$, Equation 12.32 shows θ_1 to have a monotone increase from 0 to π since χ increases from 0 to π. If $\Gamma > 1$, θ_1 increases first from 0 to a maximum value:

$$\sin^{-1}\frac{1}{\Gamma} < \frac{\pi}{2} \tag{12.39}$$

as χ increases from 0 to $\cos^{-1}\frac{1}{\Gamma}$. So θ_1 then decreases to 0 since χ increases further to π. In the given case, no particles appear beyond the maximum θ_1 in the laboratory system. The two values of χ corresponding to the given value of θ_1 between 0 and $\sin^{-1}\frac{1}{\Gamma}$ can be isolated by the energy of the observed particle which is greater for smaller χ.

The use of geometrical relations in the above-mentioned discussion is true for the quantum mechanical systems as well as classical systems. This is due to the fact that they, in essence, relate momentum vectors applied in asymptotic regions where the particles need not necessarily be precisely localized in space and so can have definite momenta. The difference between the laboratory and the center-of-mass system is negligible when we consider the collisions of electrons with atoms due to large mass ratio of the colliding particles. However, nuclear collisions show a significant difference between the two coordinate systems.

12.2 Method of Partial Waves

We examine the case of the motion of a particle in a central symmetric field where the field at infinity tends to zero. For the motion in a central symmetric field, we examine a special case of the central potential, $U(r)$, where the orbital angular momentum, L, of a particle is an integral of motion. So as seen earlier, there should exist stationary states with well-defined angular momenta. This implies the eigenstates common, respectively, to the Hamiltonian, square of the angular momentum and the projection of the angular momentum on the z-axis, $\hat{H}$, $\hat{L}^2$ and $\hat{L}_z$. We call the wave functions consistent to these states **partial waves**. The angular dependence of the eigenstates is expressed usually by the spherical harmonic as seen earlier. The potential, $U(r)$, influences only the radial dependence of the eigenstates. For, $r \to \infty$, the partial waves are close to the common eigenfunctions of the operators, $\hat{H}_0$, $\hat{L}^2$ and $\hat{L}_z$, where $\hat{H}_0$ is the Hamiltonian of a free particle. This will be the motivation of investigating first the stationary states of a free particle and in particular those with well-defined angular momenta. The radial wave function for such a motion in a central symmetric field can be written in the form:

$$R_{\kappa l}(r) = \frac{X_{\kappa l}(r)}{r} \tag{12.40}$$

Here, the modified wave function $X_{\kappa l}(r)$ satisfies the following Schrödinger equation:

$$-\frac{\hbar^2}{2\mu}\frac{d^2}{dr^2}X + \frac{\hbar^2 l(l+1)}{2\mu r^2}X + UX = \frac{\hbar^2\kappa^2}{2\mu}X \tag{12.41}$$

We find the asymptotic solution by considering

$$-\frac{d^2}{dr^2}X + \frac{l(l+1)}{r^2}X + \frac{2\mu}{\hbar^2}UX = \kappa^2 X \tag{12.42}$$

with r being so large that we neglect the potential energy and the centrifugal field:

$$\frac{2\mu}{\hbar^2}U \underset{r\to\infty}{\to} 0,\ \frac{l(l+1)}{r^2} \underset{r\to\infty}{\to} 0 \tag{12.43}$$

then

$$\frac{d^2}{dr^2}\tilde{X} + \kappa^2\tilde{X} = 0 \tag{12.44}$$

and the asymptotic solution $\tilde{X}$ is then

$$\tilde{X} = C\sin\left(\kappa r - \frac{l\pi}{2} + \delta_l\right) \tag{12.45}$$

Here, C and δ_l are, respectively, the normalization constant and the scattering **phase shift** of the l partial wave that can be found by matching the asymptotic and the exact solutions. In the general case, for the motion in the central symmetric field:

$$\psi(r,\theta,\phi) = \sum_{l=0}^{\infty}\sum_{m=-l}^{l} A_{lm} R_{\kappa l}(r) Y_{lm}(\theta,\phi) \tag{12.46}$$

Any partial solution in Equation 12.46 is a solution with a defined momentum. The problem has axial symmetry (relative to the z-axis). So the solution is not dependent on the angle ϕ. As we know,

$$Y_{lm}(\theta,\phi) = P_{lm}^{|m|}(\cos\theta)\exp\{im\phi\} \tag{12.47}$$

Here, $P_{lm}^{|m|}(x)$ is the associate Legendre polynomials with argument x. This implies that we have to take the unique term where $m=0$ and so the asymptotic wave function:

$$\tilde{\psi}(r,\theta)=\sum_{l=0}^{\infty} A_l P_l(\cos\theta)\frac{1}{\kappa r}\sin\left(\kappa r-\frac{l\pi}{2}+\delta_l\right) \tag{12.48}$$

where $P_l(x)$ is the Legendre polynomials with argument x. It is necessary to represent $\tilde{\psi}$ in the form:

$$\tilde{\psi}=\exp\{i\kappa z\}+f(\theta)\frac{\exp\{i\kappa r\}}{r} \tag{12.49}$$

where the first term represents the incoming wave and the second an outgoing wave; $f(\theta)$ is the so-called **scattering amplitude** and measures the ratio of the flux of the scattered particles to that of the incident particles. This scattering amplitude, $f(\theta)$, has the dimensions of length. It is of great importance in quantum mechanics as its absolute square is the quantity measured in experiments.

We can represent the exponential function $\exp\{i\kappa z\}$ (plane wave) in the form:

$$\exp\{i\kappa z\}=\sum_{l=0}^{\infty}(-i)^l(2l+1)P_l(\cos\theta)\left(\frac{r}{\kappa}\right)^l\frac{d^l}{(rdr)^l}\frac{\sin\kappa r}{\kappa r} \tag{12.50}$$

This is a pure mathematical formula for the expansion of plane waves via spherical functions. From the energy, $E=\frac{\hbar^2\kappa^2}{2\mu}$, two states are possible, i.e.,

- The state with a defined momentum and one with
- A defined angular momentum

We can recollect that the operators of the momentum $\hat{p}$ and the angular momentum $\hat{L}$ do not commute. This implies that they do not have a common eigenfunction and, consequently, the state is degenerate. This implies $\exp\left\{i\frac{\vec{p}\vec{r}}{\hbar}\right\}$ that imitates the de Broglie wave corresponds to a degenerate state and of course infinitely degenerate since $\vec{p}$ does not quantize. The momentum $\vec{p}$ may take any direction. On the other hand, we can write this formula as $f(\vec{L})$. This means the function, $R_{\kappa l}(r)Y_{lm}(\theta,\phi)$. This implies the degeneracy as $E=f(\kappa)$ and the wave function is $f(\kappa,l,m)$ with the degree of degeneracy being $2l+1$. Generally, in such a state, the momentum does not have a defined value. This implies we may expand it as a function of the momentum $\vec{p}$.

In Equation 12.50, we observe that

$$z=r\cos\theta \tag{12.51}$$

This implies that $\exp\{i\kappa z\}$ is independent of ϕ. Consequently, the left-hand side of Equation 12.50 is an exact formula. However, we are interested in the asymptotic function. So expanding Equation 12.50 by $\frac{1}{r}$ and conserving the term where Equation 12.50 achieves the minimum value, we have

$$\overline{\exp\{i\kappa z\}}=\sum_{l=0}^{\infty}(-i)^l(2l+1)P_l(\cos\theta)\frac{1}{\kappa r}\sin\left(\kappa r-\frac{l\pi}{2}\right) \tag{12.52}$$

The quantity, $\frac{l\pi}{2}$, is the phase for the free particle. So in Equation 12.48, the quantity δ_l is an additional phase (**phase shift of the l partial wave**) that appears as a result of quasi-classical approach. This implies that δ_l should be the difference between the asymptotic forms of the actual radial function and the radial function in the absence of the scattering potential, $U = 0$. Equation 12.52 is indicative that a well-defined state of linear momentum is formed from the superposition of states consistent to all possible angular momenta.

From Equation 12.48, the quantity A_l can be defined from boundary conditions. So

$$\tilde{\psi} - \widetilde{\exp\{i\kappa z\}} = \sum_{l=0}^{\infty} \frac{1}{\kappa r} P_l(\cos\theta)\left[A_l \sin\left(\kappa r - \frac{l\pi}{2} + \delta_l\right) - (-i)^l (2l+1)\sin\left(\kappa r - \frac{l\pi}{2}\right)\right] \tag{12.53}$$

or

$$\tilde{\psi} - \widetilde{\exp\{i\kappa z\}} = \frac{1}{2i\kappa r}\sum_{l=0}^{\infty} P_l(\cos\theta)\left(F_{kl}(r) - \tilde{F}_{kl}^*(r)\right) \tag{12.54}$$

Here

$$F_{kl}(r) = \exp\left\{i\left(\kappa r - \frac{l\pi}{2}\right)\right\}\left[A_l \exp\{i\delta_l\} - (-i)^l (2l+1)\right] \tag{12.55}$$

and

$$\tilde{F}_{kl}^*(r) = \exp\left\{-i\left(\kappa r - \frac{l\pi}{2}\right)\right\}\left[A_l \exp\{-i\delta_l\} - (-i)^l (2l+1)\right] \tag{12.56}$$

We observe from Equation 12.54 that the partial wave $\tilde{\psi}$ like a free spherical wave stems from the **superposition of an incoming spherical wave and an outgoing spherical wave** as earlier indicated.

If $\frac{\exp\{i\kappa z\}}{r}$ is a divergent wave, then $\frac{\exp\{-i\kappa z\}}{r}$ is convergent. However, we do not need that since it does not correspond to the boundary conditions. So we have to delete it from the second term in Equation 12.54. So

$$A_l = (-i)^l (2l+1)\exp\{i\delta_l\}, \; (-i)^l = \exp\left\{-i\frac{l\pi}{2}\right\} \tag{12.57}$$

and

$$\tilde{\psi} - \widetilde{\exp\{i\kappa z\}} = \frac{1}{2i\kappa r}\exp\{i\kappa r\}\sum_{l=0}^{\infty}(-1)^l (2l+1)(\eta_l - 1)P_l(\cos\theta), \; \eta_l = \exp\{2i\delta_l\} \tag{12.58}$$

What should be the significance of the phase shift, $2\delta_l$? Initially, there exist the same incoming wave as in the case of a free particle. When the given incoming wave approaches the domain of influence of the potential, it becomes more perturbed by the given potential. On its return, it is transformed into an outgoing wave. In that case, there arose an **accumulated phase shift** $2\delta_l$ in relation to the free outgoing wave consistent with the case of a potential, $U(r)$, achieving the value zero. So the quantity η_l is an indication of the total effect of the potential on a particle with angular momentum l.

From Equation 12.58, we find the partial wave expansion of the scattering amplitude:

$$f(\theta) = \frac{1}{2i\kappa}\sum_{l=0}^{\infty}(2l+1)(\eta_l - 1)P_l(\cos\theta) \tag{12.59}$$

It is obvious from here that the problem of calculating the scattering amplitude $f(\theta)$ is reduced to the calculation of the scattering phase δ_l. Equation 12.59 is the exact formula for the **scattering amplitude** (of the particle in a central symmetric field) expressed in terms of the phase shift δ_l. The principal parameter of the function $f(\theta)$ is therefore δ_l. The method of solution of the problem with the help of Equation 12.59 is called the **phase method** or **method of partial waves**. What should be the method of partial waves? Any term with a defined l corresponds to some spherical wave that is called the partial wave, as earlier indicated.

From Equation 12.59, we can define the **total elastic scattering cross section** evaluated over a sphere:

$$d\sigma = 2\pi \sin\theta \, d\theta \left| f(\theta) \right|^2 \tag{12.60}$$

So there exists such an integral

$$\int_0^{\pi} P_l(\cos\theta) P_{l'}(\cos\theta) \sin\theta \, d\theta = \delta_{ll'} \frac{2}{2l+1} \tag{12.61}$$

and

$$d\sigma = \frac{4\pi}{\kappa^2} \sum_{l=0}^{\infty} (2l+1) \sin^2 \delta_l \tag{12.62}$$

From here, it follows that the partial cross section σ_l corresponding to the momentum l is not greater than its upper bound, $\frac{4\pi}{\kappa^2}(2l+1)$. In such cases, is it convenient to use Equation 12.62? If in the given problems, we could solve the Schrödinger equation. This implies we could define exactly the quantity δ_l. This phase shift δ_l as seen from Equation 12.62, completely determines the scattering and also the scattering cross section that achieves the value zero when δ_l achieves the value 0 or π.

In the majority of cases, to solve the Schrödinger equation exactly is not possible. So it is necessary to think of such cases when the given series faster converges. The example of a slowly moving particle. In the given case (scattering), the principal role is played by small r where the perturbation is maximal. We know that $R_{\kappa l}(r) \approx r^l$ and so as l is small, the effect of the field on the particle's trajectory is small. If $l = 0$, then we have the S-scattering or the S-wave for very low energy, and if $l = 1$, then we have the p-scattering or the p-wave for slightly greater energies and so on.

EXAMPLE

We consider the example of a particle subjected to the potential $U(r) = \frac{\alpha}{r^2}$. The wave function for such a particle:

$$\Psi_{n_r lm}(r,\theta,\phi) = R_{n_r l}(r) Y_{lm}(\theta,\phi) \tag{12.63}$$

The radial function, $R_{n_r l}(r)$, satisfies the following Schrödinger equation:

$$-\frac{\hbar^2}{2\mu r^2} \frac{d}{dr}\left(r^2 \frac{d}{dr} R_{\kappa l} \right) + \frac{\hbar^2 l(l+1)}{2\mu r^2} R_{\kappa l} + U(r) R_{\kappa l} = E R_{\kappa l} \tag{12.64}$$

We select the solution of Equation 12.64 in the form:

$$R_{\kappa l}(r) = \frac{X_{\kappa l}(r)}{\sqrt{r}} \tag{12.65}$$

then the Schrödinger Equation 12.64 becomes

$$\frac{d^2}{dr^2}X_{\kappa l}+\frac{1}{r}\frac{d}{dr}X_{\kappa l}+\left(\kappa^2-\frac{\nu^2}{r^2}\right)X_{\kappa l},\ \nu^2=\left(l+\frac{1}{2}\right)^2+\frac{2\mu\alpha}{\hbar^2},\ \kappa^2=\frac{2\mu E_{\kappa l}}{\hbar^2} \tag{12.66}$$

The solution of this equation has the form:

$$X_{\kappa l}=C_1 j_\nu(\kappa r)+C_2 n_\nu(\kappa r) \tag{12.67}$$

Since

$$\lim_{r\to 0} n_\nu(\kappa r)\to-\infty \tag{12.68}$$

and from the condition of the problem, $X_{\kappa l}(\kappa r)$ cannot achieve $-\infty$ for $r\to 0$. So $C_2=0$ for $r\to 0$ and the function X_κ now takes the following form:

$$X_{\kappa l}=Cj_\nu(\kappa r) \tag{12.69}$$

and so the boundary condition:

$$X_{\kappa l}(0)=0 \tag{12.70}$$

For $r\to\infty$, the function has the asymptotic form:

$$X_{\kappa l}(\kappa r)\cong C\sqrt{\frac{2}{\pi\kappa r}}\sin\left(\kappa r-\frac{\pi\nu}{2}+\frac{\pi}{4}\right)\equiv C\sqrt{\frac{2}{\pi\kappa r}}\sin\left(\kappa r-\frac{\pi l}{2}+\delta_l\right) \tag{12.71}$$

The asymptotic function permits us to calculate the phase shift, δ_l:

$$\delta_l=-\frac{\pi\nu}{2}+\frac{\pi}{2}\left(l+\frac{1}{2}\right) \tag{12.72}$$

This permits to calculation the scattering amplitude, $f(\theta)$:

$$f(\theta)=\frac{1}{2i\kappa}\sum_{l=0}^{\infty}(2l+1)\left(\exp\{2i\delta_l\}-1\right)P_l(\cos\theta) \tag{12.73}$$

Suppose

$$\frac{\mu\alpha}{\hbar^2}\ll 1 \tag{12.74}$$

then the phase shift approximates

$$|\delta_l|\cong\left|-\frac{\pi\mu\alpha}{(2l+1)\hbar^2}\right|\ll 1 \tag{12.75}$$

and so Taylor series expansion of the exponential in Equation 12.73, the scattering amplitude:

$$f(\theta)=-\frac{\pi\mu\alpha}{\kappa\hbar^2}\sum_{l=0}^{\infty}P_l(\cos\theta)=-\frac{\pi\mu\alpha}{2\kappa\hbar^2\sin\frac{\theta}{2}} \tag{12.76}$$

Consequently, the elastic scattering cross section evaluated over a sphere:

$$d\sigma=\left|f(\theta)\right|^2 d\Omega=\left(\frac{\pi\mu\alpha}{\hbar^2}\right)^2\frac{1}{4\kappa^2\sin^2\frac{\theta}{2}}d\Omega=\frac{\mu\pi^2\alpha^2}{8\hbar^2 E\sin^2\frac{\theta}{2}}d\Omega \tag{12.77}$$

For $E\to 0$, the total scattering cross section achieves an infinite value due to the slow decrease of the potential at large distances, $r\to\infty$.

12.3 S-Scattering of Slow Particles

It is instructive to note that at low energies, the centrifugal potential $\frac{l(l+1)}{r^2}$ suppresses the wave function at short distances. In this case, there appears slightly or no effect of the scattering potential. However, for the s-waves that equal zero, the centrifugal potential is zero. So we examine the case, when the particle scatters, some spherically square well potential U acting for a short moment where for $r\le a$, we have $U=U_0$, and for $r>a$, we have $U=0$. For $U_0>0$ and $U_0<0$, we have, respectively, a repulsive and attractive potential (see Figure 12.4).

We assume the particle to be slow if the de Broglie wavelength is exceedingly greater than the range a, i.e.,

$$\lambda\gg a \tag{12.78}$$

This implies that

$$\kappa a\ll 1 \tag{12.79}$$

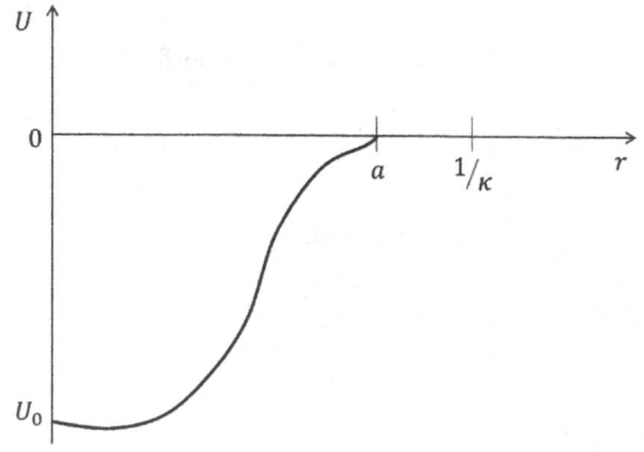

FIGURE 12.4 The spherically square well potential of depth $|U_0|$ and radius a.

We examine the solution of the radial Schrödinger equation in the following domain:

$$0 < r < a \tag{12.80}$$

We remember that $l = 0$ (s-wave) and so we have to solve the Schrödinger equation:

$$\frac{d^2}{dr^2}\mathrm{X} + \kappa^2\mathrm{X} = 0 \tag{12.81}$$

Now, we examine the domain:

$$a < r \ll \frac{1}{\kappa} \tag{12.82}$$

In this domain, Equation 12.81 becomes

$$\frac{d^2}{dr^2}\mathrm{X} = 0 \tag{12.83}$$

This implies

$$\mathrm{X} = C_1 r + C_2 \tag{12.84}$$

For $r > a$, the general solution considering the phase shift $\delta_{l=0} = \delta_0$ and $\frac{l\pi}{2} = 0$:

$$\mathrm{X} = C\cos\delta_0 \sin\kappa r + C\sin\delta_0 \cos\kappa r \tag{12.85}$$

From Equation 12.79, we have

$$\sin\kappa r \approx \kappa r,\ \cos\kappa r \approx 1 \tag{12.86}$$

So from Equations 12.79, 12.84, 12.85 and 12.86, we have

$$\mathrm{X} = C_1 = C\kappa\cos\delta_0,\ C_2 = C\sin\delta_0 \tag{12.87}$$

From here:

$$\frac{C_2}{C_1} = \frac{\sin\delta_0}{\kappa\cos\delta_0} \tag{12.88}$$

This implies that

$$\tan\delta_0 = \kappa\frac{C_2}{C_1} \tag{12.89}$$

We now have to find $\frac{C_2}{C_1}$ where we have to solve the Schrödinger equation from the left and right sides of the point a. So for that, we write

$$\mathrm{X}(a)=C_1 a+C_2,\ \mathrm{X}'(a)=C_1 \tag{12.90}$$

This implies

$$\frac{C_2}{C_1}=\frac{\mathrm{X}(a)}{\mathrm{X}'(a)}-a \tag{12.91}$$

We substitute Equation 12.91 into Equation 12.89 and have

$$\tan\delta_0=\kappa\left(\frac{\mathrm{X}(a)}{\mathrm{X}'(a)}-a\right) \tag{12.92}$$

Considering Equation 12.79, Equation 12.92 becomes

$$\tan\delta_0\cong\kappa\frac{\mathrm{X}(a)}{\mathrm{X}'(a)} \tag{12.93}$$

The most important supposition is that $\frac{\mathrm{X}(a)}{\mathrm{X}'(a)}$ is independent of the wave vector κ since in the domain of the slow-moving particle, we may neglect it in the Schrödinger equation compared to the potential, i.e.,

$$-\frac{\hbar^2}{2\mu}\frac{d^2}{dr^2}\mathrm{X}(r)+U(r)\mathrm{X}(r)=0 \tag{12.94}$$

We have now

$$\tan\delta_0\cong\sin\delta_0\cong\delta_0 \tag{12.95}$$

Substituting this in Equation 12.62 of the previous heading, we have

$$\sigma_0=4\pi a^2=\text{const} \tag{12.96}$$

So the s-scattering cross section of the particle is independent of the energy and four times the geometrical cross section πa^2. The scattering amplitude is independent of the scattering angle.

12.4 Resonance Scattering

In this problem, we are going to examine the scattering of slowly moving particles. This implies a short-lived potential. For the satisfaction of Equation 12.79, the de Broglie wave vector κ in this potential well has small discrete energy levels. The energy, E, of the particle has the same order as the scattering energy:

$$\mathrm{E}_\kappa=\frac{\hbar^2\kappa^2}{2\mu}>0 \tag{12.97}$$

and here

$$E_0 = -\frac{\hbar^2\chi^2}{2\mu} \tag{12.98}$$

with $\kappa \approx \chi$. So the system is at resonance and leads to a considerable increase in the effective cross section.

We examine *s*-scattering, i.e.,

$$f(\theta) = \frac{1}{2i\kappa}\sum_{l=0}^{\infty}(-1)^l(2l+1)(\eta_l - 1)P_l(\cos\theta) \tag{12.99}$$

We consider only the term with $l = 0$. In the domain $r > a$, we have a free motion of the particle and

$$X_\kappa(r) = C\sin(\kappa r + \delta_0) \tag{12.100}$$

Here, δ_0 is the phase of the *s*-wave.

We examine the domain at the neighborhood of *a* so considering Equation 12.79:

$$\kappa r \ll 1 \tag{12.101}$$

So from Equations 12.79, 12.100, 12.86, 12.87 and 12.88, we have

$$\frac{C_2}{C_1} = \frac{1}{\kappa}\tan\delta_0 \tag{12.102}$$

In order to find δ_0, it is necessary to solve the problem in the domain $r > 0$ and match with the solution at the point *a*, and so considering Equation 12.90, we have the relation in Equation 12.91:

$$\frac{X(a)}{X'(a)} = a + \frac{C_2}{C_1} \tag{12.103}$$

So to find $\frac{X(a)}{X'(a)}$, we examine the domain $r < a$ and the Schrödinger equation will then be

$$-\frac{\hbar^2}{2\mu}\frac{d^2}{dr^2}X(r) + U(r)X(r) = EX(r) \tag{12.104}$$

For the case of a slow-moving particle:

$$r < 0 \tag{12.105}$$

and

$$|E| \ll |U| \tag{12.106}$$

This implies that the right-hand side of Equation 12.104 should be zero. So in the domain Equation 12.105, the functions $X_0(r)$ and $X_\kappa(r)$ satisfy one and the same Schrödinger equation (without the energy). They also satisfy the first boundary condition:

$$X(0) = 0 \tag{12.107}$$

This means at the point *a*, the functions $X_0(r)$ and $X_\kappa(r)$ coincide to the approximation of a factor. So if we take the ratio in Equation 12.103, then the said factor has to fall out and implies that

$$\frac{X_\kappa(a)}{X'_\kappa(a)} \to \frac{X_0(a)}{X'_0(a)} \tag{12.108}$$

Here, $X_0(a)$ is the solution of Equation 12.104.

Let us examine again domain $r > a$, then $U = 0$ and we have the energy $E = -\frac{\hbar^2\chi^2}{2\mu}$. This implies Equation 12.104 and

$$\frac{d^2}{dr^2}X_0 - \chi^2 X_0 = 0 \tag{12.109}$$

The solution of Equation 12.109 yields

$$X_0(r) = \exp\{-\chi r\}, \chi > 0 \tag{12.110}$$

Form here:

$$\frac{X_0(a)}{X'_0(a)} = -\frac{1}{\chi} = a + \frac{1}{\kappa}\tan\delta_0 \tag{12.111}$$

or

$$\tan\delta_0 = -\frac{\kappa}{\chi} - \kappa a \tag{12.112}$$

As

$$\kappa a \ll 1 \tag{12.113}$$

then

$$\tan\delta_0 = -\frac{\kappa}{\chi} \tag{12.114}$$

This implies that the problem is solved. So from Equation 12.59, we have

$$f_0(\theta) = \frac{1}{2i\kappa}(\eta_0 - 1) \tag{12.115}$$

The total scattering cross section is then

$$\sigma = 4\pi|f_0(\theta)|^2 \tag{12.116}$$

This implies that

$$|f_0(\theta)|^2 = \frac{\sin^2\theta}{\kappa^2} = \frac{1}{\kappa^2(1 + \cot an^2\delta_0)} = \frac{1}{\kappa^2 + \chi^2} \tag{12.117}$$

So the **Wigner formula**:

$$\sigma_0 = \frac{4\pi}{\kappa^2 + \chi^2} = \frac{2\pi\hbar^2}{\mu\left(E_\kappa + |E_0|\right)} \tag{12.118}$$

Equation 12.118 is that of the resonance scattering. In this case, we obtain the gigantic scattering cross section, i.e., $\sigma_0 > \sigma$ if there does not exist the discrete level. For the existence of the discrete level:

$$\sigma_0 \cong \pi a^2 \tag{12.119}$$

Such a resonance scattering often is seen in nuclear physics, physics of elementary particles, solid state physics where the electrical resistance arises due to the scattering of charge carriers. An example is the scattering of phonons on defects. Equation 12.119 suggests the angular dependence of the diffraction scattering when the absorber is a sphere of radius a. Equation 12.117 may easily be found in the form:

$$f_0 = \frac{1}{\kappa + i\chi} \tag{12.120}$$

This expression may be examined as some complex function of the variable κ and has a pole at $\kappa = -i\chi$. If we substitute this pole in the expression of the energy in Equation 12.97, it follows that

$$\frac{\hbar^2\chi^2}{2\mu} = 0 \tag{12.121}$$

This implies we find the energy of the discrete levels. So the discrete energy corresponds to the pole of the scattering amplitude.

12.5 The Unitary Scattering Conditions

We examine some important relations such as the reciprocity theorem, the generalized optical theorem and the optical theorem. We derive them by considering once more the asymptotic behavior of the scattering wave function. We examine again the asymptotic function, $\tilde{\psi}$, and consider collision to give rise to several different reactions or scattering. So in the evaluation of the total cross section, we consider the sum of all the cross sections corresponding to all the given processes.

12.5.1 Optical Theorems

We consider $\vec{n}$ and $\vec{n}'$ to represent, respectively, unit vectors in the direction of the incoming and scattered waves as indicated in Figure 12.5.

We can then represent the z coordinate:

$$z = r\cos\theta = r\vec{n}\vec{n}' \tag{12.122}$$

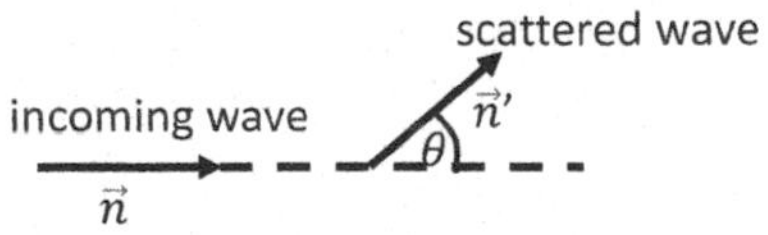

FIGURE 12.5 $\vec{n}$ and $\vec{n}'$ representing, respectively, unit vectors in the direction of the incoming and scattered waves with θ being the scattering angle.

and the asymptotic wave function:

$$\tilde{\psi} = \exp\{i\kappa r\vec{n}\vec{n}'\} + f(\vec{n},\vec{n}')\frac{\exp\{i\kappa r\}}{r} \tag{12.123}$$

The scattering amplitude $f(\vec{n},\vec{n}')$ corresponds to the scattering from $\vec{n}$ to $\vec{n}'$. The first summand in Equation 12.123 is due to the plane wave alone, while the second is due to the radially scattered wave.

We select an arbitrary function $F(\vec{n})$ and multiply it by the asymptotic wave function $\tilde{\psi}$ and the result integrated over the volume Ω in the entire space:

$$\int F(\vec{n})\tilde{\psi}\,d\Omega = \int F(\vec{n})\exp\{i\kappa r\vec{n}\vec{n}'\}\,d\Omega + \frac{\exp\{i\kappa r\}}{r}\int F(\vec{n})f(\vec{n},\vec{n}')d\Omega \tag{12.124}$$

This is a linear combination of functions. If the variable r is very large, $\kappa r \gg 1$, then the function $\exp\{i\kappa r\vec{n}\vec{n}'\}$ is an oscillatory one and the value of the integral in Equation 12.124 is determined when $\vec{n}$ is parallel to $\vec{n}'$:

$$\vec{n} = \vec{n}' \tag{12.125}$$

In this case, we have a **forward scattering** where the scattered spherical wave is aligned with the incomming plane wave in a way that they can and will interfer. Particularly, when there is a solid object in the path, we obtain shadow of no or low intensity. This implies, the interference is distructive. Generally, interference leads to a reduction of the total forward flux. This is because, side scattering can no longer propagate in the forward direction. We have now to find the relation between the forward scattering amplitude, $f(\vec{n},\vec{n}')$, and the total scattering cross section, σ_{total}, in the so-called **optical theorem**. We derive this from Equation 12.124:

$$\int F(\vec{n})\tilde{\psi}\,d\Omega = 2\pi\int_0^{\pi}\sin\theta\,d\theta F(\vec{n})\exp\{i\kappa r\cos\theta\} + \frac{\exp\{i\kappa r\}}{r}\int F(\vec{n})f(\vec{n},\vec{n}')d\Omega \tag{12.126}$$

or

$$\int F(\vec{n})\tilde{\psi}\,d\Omega = \frac{2\pi i}{\kappa}\left[\frac{1}{r}\exp\{-i\kappa r\}F(-\vec{n}') - \frac{1}{r}\exp\{i\kappa r\}\hat{S}F(\vec{n}')\right] \tag{12.127}$$

Here, the **scattering matrix (operator)** $\hat{S}$:

$$\hat{S} = 1 + 2i\kappa\hat{f} \tag{12.128}$$

and the integral operator $\hat{f}$ is defined as

$$\hat{f}F(\vec{n}') = \frac{1}{4\pi}\int F(\vec{n})f(\vec{n},\vec{n}')d\Omega \tag{12.129}$$

The first summand in Equation 12.127 is the wave going into the center and the second is the wave going out of the center. The conservation of the number of particles is guaranteed by the equality of the total fluxes of the particles in the incoming and outgoing waves. This implies that the two waves must be normalized the same, and for this to be achieved, the scattering matrix should have the property of unitarity:

$$\hat{S}\hat{S}^{\dagger}=\hat{\mathbb{I}} \tag{12.130}$$

This implies the conservation of particles. From the unitary property of the scattering matrix, we have

$$\hat{S}\hat{S}^{\dagger}=\left(1+2i\kappa\hat{f}\right)\left(1-2i\kappa\hat{f}^{\dagger}\right)=\hat{\mathbb{I}}+2i\kappa\left(\hat{f}-\hat{f}^{\dagger}\right)+4\kappa^2\hat{f}\hat{f}^{\dagger} \tag{12.131}$$

From the unitary Equation 12.130 of the scattering matrix $\hat{S}$, it follows that

$$\hat{f}-\hat{f}^{\dagger}=-\frac{4\kappa^2}{2i\kappa}\hat{f}\hat{f}^{\dagger}=2i\kappa\hat{f}\hat{f}^{\dagger} \tag{12.132}$$

So

$$f(\vec{n},\vec{n}')-f^*(\vec{n},\vec{n}')=\frac{2i\kappa}{4\pi}\int f(\vec{n},\vec{n}'')f^*(\vec{n}',\vec{n}'')d\Omega \tag{12.133}$$

Indeed,

$$\hat{f}\mathrm{F}(\vec{n}')=\frac{1}{4\pi}\int \mathrm{F}(\vec{n})f(\vec{n},\vec{n}')d\Omega,\ \hat{f}^{\dagger}\mathrm{F}(\vec{n}')=\frac{1}{4\pi}\int \mathrm{F}(\vec{n})f^*(\vec{n},\vec{n}')d\Omega \tag{12.134}$$

then

$$\left[\hat{f}-\hat{f}^{\dagger}\right]\mathrm{F}(\vec{n}')=\frac{1}{4\pi}\int \mathrm{F}(\vec{n})\left[f(\vec{n},\vec{n}')-f^*(\vec{n},\vec{n}')\right]d\Omega \tag{12.135}$$

If $\vec{n}=\vec{n}'$, then

$$\hat{f}-\hat{f}^{\dagger}=2i\,\mathrm{Im}\,f(\vec{n},\vec{n}') \tag{12.136}$$

Also,

$$\hat{f}\hat{f}^{\dagger}\mathrm{F}(\vec{n}')=\hat{f}\frac{1}{4\pi}\int \mathrm{F}(\vec{n}'')\hat{f}^*(\vec{n}',\vec{n}'')d\Omega''=\frac{1}{4\pi}\int \hat{f}\mathrm{F}(\vec{n}'')\hat{f}^*(\vec{n}',\vec{n}'')d\Omega'' \tag{12.137}$$

or

$$\hat{f}\hat{f}^{\dagger}\mathrm{F}(\vec{n}')=\frac{1}{(4\pi)^2}\int \hat{f}(\vec{n},\vec{n}'')\mathrm{F}(\vec{n})\hat{f}^*(\vec{n}',\vec{n}'')d\Omega d\Omega'' \tag{12.138}$$

and

$$\left[\hat{f}-\hat{f}^{\dagger}\right]\mathrm{F}(\vec{n}')=\frac{1}{4\pi}\int\left[\hat{f}(\vec{n},\vec{n}')-\hat{f}^*(\vec{n}',\vec{n})\right]\mathrm{F}(\vec{n})d\Omega=\frac{2i\kappa}{(4\pi)^2}\int \hat{f}(\vec{n},\vec{n}'')\mathrm{F}(\vec{n})\hat{f}^*(n',\vec{n}'')d\Omega d\Omega'' \tag{12.139}$$

So

$$\hat{f}(\vec{n},\vec{n}')-\hat{f}^*(\vec{n}',\vec{n})=\frac{2i\kappa}{4\pi}\int \hat{f}(\vec{n},\vec{n}'')\hat{f}^*(\vec{n}',\vec{n}'')d\Omega'' \tag{12.140}$$

Equation 12.133 or 12.140 is the **so-called generalized optical theorem**. These equations are valid only if U is real. If, $\vec{n}=\vec{n}'$, then

$$2i\,\mathrm{Im}\, f(0)=\frac{2i\kappa}{4\pi}\int\left|\hat{f}(\vec{n},\vec{n}'')\right|^2 d\Omega'' \tag{12.141}$$

Hence, the total scattering cross section, the **so-called optical theorem**:

$$\sigma_{\text{total}}=\frac{4\pi}{\kappa}\mathrm{Im}\,\hat{f}(0) \tag{12.142}$$

This relation shows that in order for the scattering to take place, particles must be removed from the incident beam in an amount proportional to σ_{total} so that its intensity is smaller behind the scattering region $(\theta\approx 0)$ than in front of the beam. This is possible only by interference between the two terms in the assymptotic expression in Equation 12.123. As such an interference term should be a linear function of the forward-scattered amplitude, a general form of Equation 12.142 is expected. The exact calculation of this intereference term shows that Equation 12.142 is generally satisfied where $\hat{f}$ is dependent on $\vec{n},\vec{n}''$ and when σ_{total} takes into consideration the inelastic scattering and also absorption together with the elastic scattering. From Equation 12.142, we have

$$\mathrm{Im}\,\hat{f}(0)=\frac{\kappa}{4\pi}\sigma_{\text{total}} \tag{12.143}$$

The famous **optical scattering theorem** is therefore represented in the form:

$$\mathrm{Im}\,\hat{f}(0)=\mathrm{Im}\,\hat{f}(\vec{n},\vec{n}') \tag{12.144}$$

This formula is more general, since it can be derived without reference to a potential say, U. The optical theorem established for the first time by Niels H.D. Bohr, Rudolf Peierls and George Placzek has a historical precursor an analog in classical optics. The optical scattering theorem is also applicable when special relativity is considered.

We examine again the example of a particle subjected to the potential $U(r)=\frac{\alpha}{r^2}$ where the scattering amplitude, $f(\theta)$, is obtained:

$$f(\theta)=\frac{1}{2i\kappa}\sum_{l=0}^{\infty}(2l+1)\left(\exp\{2i\delta_l\}-1\right)\mathrm{P}_l(\cos\theta) \tag{12.145}$$

The total scattering cross section considering the **so-called optical theorem**:

$$\sigma_{\text{total}}=\frac{4\pi}{\kappa}\mathrm{Im}\, f(0)=\frac{2\pi}{\kappa^2}\sum_{l=0}^{\infty}(2l+1)(1-\cos 2\delta_l) \tag{12.146}$$

12.6 Time-Reversal Symmetry

We consider another property of the scattering amplitude from the symmetry via the time reversal property. We consider the fact that the Hamiltonian $\hat{H}$ is Hermitian and so the eigenvalue is real valued:

$$\hat{H} = \hat{H}^{\dagger}, E^{*} = E \tag{12.147}$$

So the following function also corresponds to some scattering process:

$$\tilde{\psi}^{*} = \frac{1}{r}\exp\{i\kappa r\}F^{*}(-\vec{n}') - \frac{1}{r}\exp\{-i\kappa r\}\hat{S}^{*}F^{*}(\vec{n}') \tag{12.148}$$

Here, $\hat{S}$ is some transformation matrix whose properties will be shown under this heading.

Let us denote by

$$-\hat{S}^{*}F^{*}(\vec{n}') = \Phi(-\vec{n}') \tag{12.149}$$

then

$$F^{*}(\vec{n}') = -\left[\hat{S}^{*}\right]^{-1}\Phi(-\vec{n}') = \hat{\tilde{S}}\Phi(-\vec{n}') \tag{12.150}$$

$$\hat{\tilde{S}} = \left[\hat{S}^{*}\right]^{-1} \tag{12.151}$$

12.6.1 Inversion Operator and Reciprocity Theorem

We consider the action of the space inversion operator $\hat{P}$ on the function $F^{*}(\vec{n}')$ that possesses space inversion symmetry:

$$F^{*}(-\vec{n}') = \hat{P}F^{*}(\vec{n}') = \hat{I}\hat{\tilde{S}}\hat{I}\Phi(\vec{n}') \tag{12.152}$$

where $\hat{I}$ is also a space inversion operator. This implies that all vectors in the scattering problem may be inverted leaving the physical situation invariant. So the function in Equation 12.148 can be rewritten as follows:

$$\tilde{\psi}^{*} = \frac{1}{r}\exp\{-i\kappa r\}\Phi(-\vec{n}') - \frac{1}{r}\exp\{i\kappa r\}\hat{I}\hat{\tilde{S}}\hat{I}\Phi(\vec{n}') \tag{12.153}$$

Hence,

$$\hat{I}\hat{\tilde{S}}\hat{I} = \hat{S},\ \hat{\tilde{S}} = \left[\hat{S}^{*}\right]^{-1} = \hat{S}^{T} \tag{12.154}$$

So for unitarity,

$$\hat{S}\hat{S}^{\dagger} = \hat{S}\left[\hat{S}^{*}\right]^{T} = \hat{S}^{*}\hat{S}^{T} = \hat{\mathbb{I}} \tag{12.155}$$

and

$$\left[\hat{S}^{*}\right]^{T} = \hat{S}^{-1},\ \hat{S}^{T} = \left[\hat{S}^{*}\right]^{-1} \tag{12.156}$$

Also,

$$\hat{I}\hat{S}\hat{I} = \hat{I}\hat{S}^T\hat{I} = \hat{S} \tag{12.157}$$

so the **reciprocity theorem**:

$$\hat{S}(\vec{n},\vec{n}') = \hat{S}(-\vec{n},-\vec{n}'),\ \hat{f}(\vec{n},\vec{n}') = \hat{f}(-\vec{n},-\vec{n}') \tag{12.158}$$

This shows that the amplitude for the scattering from $\vec{n}$ to $\vec{n}'$ is equal to that for the scattering from the reversed final direction $-\vec{n}'$ to the reversed initial direction $-\vec{n}$. The thereom is valid for real or complex potential.

12.7 Schrödinger Equation Green's Function

We examine the time-dependent Schrödinger equation:

$$i\hbar\frac{\partial}{\partial t}\Psi = \hat{H}\Psi,\ \hat{H} = -\frac{\hbar^2}{2m}\Delta + V(\vec{r}) \tag{12.159}$$

The solution of the time-dependent Schrödinger equation in Equation 12.159:

$$\Psi = \exp\left\{-i\frac{E}{\hbar}t\right\}\psi(\vec{r}) \tag{12.160}$$

Substituting Equation 12.160 into 12.159, we have the stationary Schrödinger equation:

$$\left(-\frac{\hbar^2}{2m}\Delta + V(\vec{r})\right)\psi(\vec{r}) = \epsilon\,\psi(\vec{r}) \tag{12.161}$$

We consider the following equation for Green's function $G(\vec{r},t,\vec{r}',t')$:

$$\left(i\hbar\frac{\partial}{\partial t} + \frac{\hbar^2}{2m}\Delta - V(\vec{r})\right)G(\vec{r},t,\vec{r}',t') = \delta(t-t')\delta(\vec{r}-\vec{r}') \tag{12.162}$$

For a free particle, $V(\vec{r}) = 0$ and the free particle's Green's function $G_0(\vec{r},t,\vec{r}',t')$ can be obtained from the following equation:

$$\left(i\hbar\frac{\partial}{\partial t} + \frac{\hbar^2}{2m}\Delta\right)G_0(\vec{r},t,\vec{r}',t') = \delta(t-t')\delta(\vec{r}-\vec{r}') \tag{12.163}$$

Substituting the Fourier transform

$$G_0(\vec{r},t,\vec{r}',t') = \int G_0(\vec{p},\epsilon)\exp\left\{i\frac{\vec{p}}{\hbar}(\vec{r}-\vec{r}')\right\}\exp\left\{i\frac{E}{\hbar}(t-t')\right\}\frac{dE}{2\pi}\frac{d\vec{p}}{(2\pi\hbar)^3} \tag{12.164}$$

into Equation 12.163, we have

$$\left(E-\frac{p^2}{2m}\right)\mathrm{G}_0\left(\vec{p},E\right)=1 \tag{12.165}$$

or the causality or retarded Green's function:

$$\mathrm{G}_0\left(\vec{p},E\right)=\frac{1}{E-\frac{p^2}{2m}+i\delta} \tag{12.166}$$

Here, δ is an infinitesimal quantity. If we consider the interaction, then Green operator:

$$\hat{\mathrm{G}}=\left(i\hbar\frac{\partial}{\partial t}-\frac{\hat{p}^2}{2m}-V(\vec{r})\right)^{-1}=\left(\mathrm{G}_0^{-1}-V\right)^{-1}=\left(\mathrm{G}_0^{-1}\left(1-\mathrm{G}_0V\right)\right)^{-1}=\left(1-\mathrm{G}_0V\right)^{-1}\mathrm{G}_0 \tag{12.167}$$

If V is small, then

$$\left(1-\mathrm{G}_0V\right)^{-1}=\mathrm{G}_0+\mathrm{G}_0V+\mathrm{G}_0V\mathrm{G}_0V+\cdots \tag{12.168}$$

or the following Green operator that shows all the scattering mechanisms:

$$\hat{\mathrm{G}}=\mathrm{G}_0+\mathrm{G}_0V\mathrm{G}_0+\mathrm{G}_0V\mathrm{G}_0V\mathrm{G}_0+\cdots \tag{12.169}$$

where

$$\mathrm{G}_0\left(\epsilon,\vec{r},\vec{r}'\right)=\int\frac{\exp\left\{i\frac{\vec{p}}{\hbar}\left(\vec{r}-\vec{r}'\right)\right\}}{E-\frac{p^2}{2m}+i\delta}\frac{d\vec{p}}{\left(2\pi\hbar\right)^3}=\int_0^\infty\frac{p^2dp}{\left(2\pi\hbar\right)^3}\int_0^\pi\sin\theta\,d\theta\int_0^{2\pi}\frac{d\phi\exp\left\{i\frac{p}{\hbar}\left|\vec{r}-\vec{r}'\right|\cos\theta\right\}}{E-\frac{p^2}{2m}+i\delta} \tag{12.170}$$

or

$$\mathrm{G}_0\left(\epsilon,\vec{r},\vec{r}'\right)=\frac{1}{i\left(2\pi\hbar\right)^2}\frac{1}{\left|\vec{r}-\vec{r}'\right|}\int_0^\infty\frac{pdp\left[\exp\left\{i\frac{p}{\hbar}\left|\vec{r}-\vec{r}'\right|\right\}-\exp\left\{-i\frac{p}{\hbar}\left|\vec{r}-\vec{r}'\right|\right\}\right]}{E-\frac{p^2}{2m}+i\delta}$$

$$=-\frac{i}{\left(2\pi\hbar\right)^2}\frac{1}{\left|\vec{r}-\vec{r}'\right|}\int_{-\infty}^\infty\frac{pdp\exp\left\{i\frac{p}{\hbar}\left|\vec{r}-\vec{r}'\right|\right\}}{E-\frac{p^2}{2m}+i\delta} \tag{12.171}$$

Consider

$$E=\frac{\hbar^2\kappa^2}{2m} \tag{12.172}$$

then from here:

$$\mathrm{G}_0\left(\epsilon,\vec{r},\vec{r}'\right)=-\frac{m}{2\pi\hbar^2}\frac{\exp\left\{i\kappa\left|\vec{r}-\vec{r}'\right|\right\}}{\left|\vec{r}-\vec{r}'\right|} \tag{12.173}$$

12.8 Born Approximation

Until now, we examined the scattering of slowly moving particles. We now examine fast moving particles when the total energy is exceedingly greater than the potential energy. So the potential energy may be examined as a perturbation.

12.8.1 Scattering of Fast Charged Particles on Atoms

Suppose we have a system of atomic nuclei and atomic shells having electrons where each electron has the charge e. Suppose the electron of charge e_1 is projected toward and scatters on an atom of charge, Ze, the potential coupling the nucleus to the electrons on the electronic shells:

$$U(\vec{r}_1,\vec{r}_2,\ldots)=\frac{Zee_1}{r}-\sum_i \frac{ee_1}{|\vec{r}-\vec{r}_i|} \tag{12.174}$$

Further, this potential energy will be examined as a perturbation compared to the kinetic energy in the Born approximation that will be examined under Coulomb potential Type 3 of the current chapter.

12.8.1.1 Scattering Amplitude in Momentum Representation

We examine the time-independent Schrödinger equation:

$$\hat{\mathrm{H}}\psi(\vec{r})=E\psi(\vec{r}),\ \hat{\mathrm{H}}=-\frac{\hbar^2}{2\mu}\Delta+U(\vec{r}) \tag{12.175}$$

We write the Fourier transform of the wavefunction:

$$\psi(\vec{r})=\int\psi(\vec{q})\exp\{i\vec{q}\vec{r}\}\frac{d\vec{q}}{(2\pi)^3} \tag{12.176}$$

and so the wave equation:

$$\left(\frac{\hbar^2\vec{q}^{\,2}}{2\mu}-E\right)\psi(\vec{q})+\int U(\vec{q}-\vec{q}\,')\psi(\vec{q}\,')\frac{d\vec{q}\,'}{(2\pi)^3}=0 \tag{12.177}$$

$$\psi_{\vec{\kappa}}(\vec{r})=\exp\{i\vec{\kappa}\vec{r}\}+\mathrm{X}_{\vec{\kappa}}(\vec{r})\equiv\exp\{i\kappa r\vec{n}\vec{n}'\}+\frac{f(\theta)}{r}\exp\{i\kappa r\} \tag{12.178}$$

The Fourier transform:

$$\psi_{\vec{\kappa}}(\vec{q})=(2\pi)^3\delta(\vec{q}-\vec{\kappa})+\mathrm{X}_{\vec{\kappa}}(\vec{q}) \tag{12.179}$$

We substitute $\psi_{\vec{\kappa}}(\vec{q})$ into the Schrödinger equation and considering

$$E_{\vec{\kappa}}=\frac{\hbar^2\vec{\kappa}^2}{2\mu} \tag{12.180}$$

So

$$\left(\frac{\hbar^2\vec{q}^2}{2\mu}-\frac{\hbar^2\vec{\kappa}^2}{2\mu}\right)\left((2\pi)^3\delta(\vec{q}-\vec{\kappa})+X_{\vec{\kappa}}(\vec{q})\right)+\int U(\vec{q}-\vec{q}')\left((2\pi)^3\delta(\vec{q}'-\vec{\kappa})+X_{\vec{\kappa}}(\vec{q}')\right)\frac{d\vec{q}'}{(2\pi)^3}=0 \tag{12.181}$$

Considering the first term being zero:

$$\frac{\hbar^2}{2\mu}\left(\vec{\kappa}^2-\vec{q}^2\right)X_{\vec{\kappa}}(\vec{q})=U(\vec{q}-\vec{\kappa})+\int U(\vec{q}-\vec{q}')X_{\vec{\kappa}}(\vec{q}')\frac{d\vec{q}'}{(2\pi)^3} \tag{12.182}$$

We consider

$$X_{\vec{\kappa}}(\vec{q})=-\frac{2\mu}{\hbar^2}\frac{F(\vec{\kappa},\vec{q})}{\vec{\kappa}^2-\vec{q}^2+i\delta} \tag{12.183}$$

But

$$\frac{1}{\frac{\hbar^2\vec{\kappa}^2}{2\mu}-\frac{\hbar^2\vec{q}^2}{2\mu}+i\delta}=G_0(\in,\vec{q}) \tag{12.184}$$

So

$$X_{\vec{\kappa}}(\vec{q})=-G_0(\in,\vec{q})F(\vec{\kappa},\vec{q}) \tag{12.185}$$

and the **Lippmann-Schwinger equation**:

$$F(\vec{\kappa},\vec{q})=-V(\vec{q}-\vec{\kappa})-\frac{2\mu}{\hbar^2}\int U(\vec{q}-\vec{q}')\frac{F(\vec{\kappa},\vec{q}')}{\vec{q}'^2-\vec{\kappa}^2-i\delta}\frac{d\vec{q}'}{(2\pi)^3} \tag{12.186}$$

This is an important starting point for most modern studies of the quantum mechanical scattering theory.

We verify if this has something to do with scattering:

$$X_{\vec{\kappa}}(\vec{r})=-\frac{2\mu}{\hbar^2}\int\frac{d\vec{q}}{(2\pi)^3}\frac{F(\vec{\kappa},\vec{q})\exp\{i\vec{q}\vec{r}\}}{\vec{\kappa}^2-\vec{q}^2+i\delta},\; d\vec{q}=q^2dq\sin\theta\, d\theta d\phi \tag{12.187}$$

then

$$\begin{aligned}X_{\vec{\kappa}}(\vec{r})&=-i\frac{2\mu 2\pi}{(2\pi)^3\hbar^2}\int_0^\infty q^2dq\frac{F(\vec{\kappa},q\vec{n}')\exp\{iqr\}-F(\vec{\kappa},-q\vec{n})\exp\{-iqr\}}{\left(\vec{\kappa}^2-\vec{q}^2+i\delta\right)qr}\\&=-\frac{i\mu}{2\pi^2\hbar^2 r}\int_{-\infty}^{\infty}qdq\frac{F(\vec{\kappa},q\vec{n}')\exp\{iqr\}}{\vec{\kappa}^2-\vec{q}^2+i\delta}\end{aligned} \tag{12.188}$$

or

$$X_{\vec{\kappa}}(\vec{r})=\frac{\mu}{2\pi^2\hbar^2}\frac{\exp\{i\kappa r\}}{r}F(\kappa\vec{n},\kappa\vec{n}') \tag{12.189}$$

So

$$\hat{f}(\vec{n},\vec{n}')=\frac{\mu}{2\pi\hbar^2}F(\kappa\vec{n},\kappa\vec{n}') \tag{12.190}$$

or the following integral equation:

$$\hat{f}(\vec{n},\vec{n}')=-\frac{\mu}{2\pi\hbar^2}U(\vec{\kappa}'-\vec{\kappa})-\frac{2\mu}{\hbar^2}\int\frac{d\vec{\kappa}''}{(2\pi)^3}U(\vec{\kappa}-\vec{\kappa}'')\frac{\hat{f}(\vec{\kappa},\vec{\kappa}'')}{\vec{\kappa}^2-\vec{\kappa}''^2+i\delta} \tag{12.191}$$

Hence, the **first Born approxiamtion**:

$$\hat{f}(\vec{n},\vec{n}')=-\frac{\mu}{2\pi\hbar^2}U(\vec{\kappa}'-\vec{\kappa}'') \tag{12.192}$$

So the Born scattering amplitude is observed to be proportional to the spatial Fourier transform of the scattering potential with respect to $\vec{q}=\vec{\kappa}'-\vec{\kappa}''$. Here, $\hbar\vec{q}$ is the momentum transfer from the incident particle to the scattering potential within the collision.

For the **second Born approximation**:

$$\hat{f}(\vec{n},\vec{n}')=-\frac{\mu}{2\pi\hbar^2}\left[U(\vec{\kappa}'-\vec{\kappa})+\frac{2\mu}{\hbar^2}\int\frac{d\vec{\kappa}''}{(2\pi)^3}\frac{U(\vec{\kappa}-\vec{\kappa}'')U(\vec{\kappa}'-\vec{\kappa}'')}{\vec{\kappa}''^2-\vec{\kappa}^2+i\delta}\right],\ \vec{\kappa}=\kappa\vec{n},\ \vec{\kappa}'=\kappa\vec{n}' \tag{12.193}$$

This is continued to have other terms. For the applicability of the Born approximation, we have

$$|U|\ll\frac{\hbar^2}{\mu a^2} \tag{12.194}$$

$$|U|\ll\frac{\hbar v}{a}=\frac{\hbar^2}{\mu a^2}(\kappa a)^2 \tag{12.195}$$

The last inequality is for fast particles.

Let us examine the case of forward scattering, i.e., when $\theta=0$ and $\vec{\kappa}'=\vec{\kappa}$:

$$\hat{f}(\vec{n},\vec{n}')=-\frac{\mu}{2\pi\hbar^2}\left[U(\vec{\kappa}'-\vec{\kappa})+\frac{2\mu}{\hbar^2}\int\frac{d\vec{\kappa}''}{(2\pi)^3}\frac{|U(\vec{\kappa}-\vec{\kappa}'')|^2}{\vec{\kappa}''^2-\vec{\kappa}^2+i\delta}\right]\equiv\hat{f}^{(1)}(\vec{\kappa}',\vec{\kappa})+\hat{f}^{(2)}(\vec{\kappa},\vec{\kappa}) \tag{12.196}$$

where

$$\hat{f}^{(1)}(\vec{\kappa}',\vec{\kappa})\equiv-\frac{\mu}{2\pi\hbar^2}U(\vec{\kappa}'-\vec{\kappa}),\ \hat{f}^{(2)}(\vec{\kappa},\vec{\kappa})\equiv-\frac{\mu}{2\pi\hbar^2}\frac{2\mu}{\hbar^2}\int\frac{d\vec{\kappa}''}{(2\pi)^3}\frac{|U(\vec{\kappa}-\vec{\kappa}'')|^2}{\vec{\kappa}''^2-\vec{\kappa}^2+i\delta} \tag{12.197}$$

In Equation 12.196, we consider the fact that $U(\vec{q})=U^*(-\vec{q})$. From Equation 12.196 or 12.197, it follows that

$$\mathrm{Im}\,\hat{f}^{(2)}(\vec{\kappa},\vec{\kappa})\equiv\frac{\mu}{2\pi\hbar^2}\frac{2\mu}{\hbar^2}\int\frac{d\vec{\kappa}''}{(2\pi)^3}\frac{\delta}{(\vec{\kappa}^2-\vec{\kappa}''^2)+\delta^2}|U(\vec{\kappa}-\vec{\kappa}'')|^2 \tag{12.198}$$

Considering δ to be an infinitesimal quantity:

$$\frac{\delta}{(\vec{\kappa}^2-\vec{\kappa}''^2)+\delta^2}\to\pi\delta(\vec{\kappa}^2-\vec{\kappa}''^2) \tag{12.199}$$

Here, $\delta(x)$ is the Dirac delta function with argument x.

So

$$\operatorname{Im}\hat{f}^{(2)}(\vec{\kappa},\vec{\kappa}) \equiv \frac{\mu}{2\pi\hbar^2}\frac{2\mu}{\hbar^2}\int\frac{d\vec{\kappa}''}{(2\pi)^3}\pi\delta\left(\vec{\kappa}^2-\vec{\kappa}''^2\right)\left|U\left(\vec{\kappa}-\vec{\kappa}''\right)\right|^2 \tag{12.200}$$

From

$$\vec{q}=\vec{\kappa}-\vec{\kappa}',\ \vec{\kappa}=\kappa\vec{n},\ \vec{\kappa}'=\kappa\vec{n}' \tag{12.201}$$

$$\vec{q}^2=\vec{\kappa}^2+\vec{\kappa}'^2-2\vec{\kappa}\vec{\kappa}'=\kappa^2+\kappa^2-2\kappa^2\cos\theta=2\kappa^2(1-\cos\theta)=4\kappa^2\sin^2\frac{\theta}{2} \tag{12.202}$$

So

$$q=2\kappa\sin\frac{\theta}{2} \tag{12.203}$$

relates the scattering angle θ in accordance with Figure 12.5. It is instructive to note that κ is the absolute value of the incident wave vector and θ is the scattering angle. The second Born approximation gives

$$\operatorname{Im}\hat{f}^{(2)}(\vec{\kappa},\vec{\kappa}) \equiv \frac{\mu}{2\pi\hbar^2}\frac{2\mu}{\hbar^2}\pi\frac{1}{(2\pi)^3}\frac{1}{2}\int\delta\left(\vec{\kappa}^2-\vec{\kappa}''^2\right)\left|U\left(\kappa\vec{n}-\vec{\kappa}\right)\right|^2\kappa''d\kappa''^2\,d\Omega_n,\ d\vec{\kappa}''=\frac{\kappa''}{2}d\kappa''^2\,d\Omega_n \tag{12.204}$$

Equation 12.204 can be simplified as follows:

$$\operatorname{Im}\hat{f}^{(2)}(\vec{\kappa},\vec{\kappa}) \equiv \frac{\mu}{2\pi\hbar^2}\frac{2\mu}{\hbar^2}\pi\frac{1}{(2\pi)^3}\frac{1}{2}\kappa\int\left|U\left(\kappa\vec{n}-\vec{\kappa}\right)\right|^2 d\Omega_n \tag{12.205}$$

or

$$\operatorname{Im}\hat{f}^{(2)}(\vec{\kappa},\vec{\kappa})=\operatorname{Im}\hat{f}^{(2)}(E,\theta=0)=\frac{\mu^2\kappa}{16\pi^3\hbar^4}\int\left|U\left(\kappa\vec{n}-\vec{\kappa}\right)\right|^2 d\Omega_n=\frac{\kappa}{4\pi}\int\left|\hat{f}\left(\vec{\kappa}',\vec{\kappa}\right)\right|^2 d\Omega_\kappa \tag{12.206}$$

or

$$\operatorname{Im}\hat{f}^{(2)}(E,\theta=0)\equiv\frac{\kappa}{4\pi}\sigma(E) \tag{12.207}$$

or

$$\sigma(E)=\frac{4\pi}{\kappa}\operatorname{Im}\hat{f}^{(2)}(E,\theta=0) \tag{12.208}$$

This is the **so-called optical scattering theorem in the second Born approximation** of the perturbation theory and $\sigma(E)$ is the scattering cross section in the second Born approximation and

$$\operatorname{Im}\hat{f}^{(2)}(\vec{\kappa},\vec{\kappa})\equiv\frac{\mu^2\kappa}{16\pi^3\hbar^4}\int\left|U\left(\kappa\vec{n}-\vec{\kappa}\right)\right|^2 d\Omega_n\equiv\frac{\kappa}{4\pi}\int\left|\hat{f}\left(\vec{\kappa}',\vec{\kappa}\right)\right|^2 d\Omega_\kappa \tag{12.209}$$

The first Born approximation gives

$$\hat{f}(\vec{n},\vec{n}') = \hat{f}(\theta) = -\frac{\mu}{2\pi\hbar^2}U(\vec{q}) = -\frac{\mu}{2\pi\hbar^2}\int d\vec{r}\,U(\vec{r})\exp\{-i\vec{q}\vec{r}\} \tag{12.210}$$

Suppose the potential $U(\vec{r})$ is sphererically symmetric then expression may first be integrated over the angles θ and

$$\hat{f}(\theta) = -\frac{2\mu}{q\hbar^2}\int_0^\infty dr\,U(r)r\sin qr \tag{12.211}$$

From Equation 12.210, we can calculate the scattering cross section:

$$d\sigma = \left|\hat{f}(\theta)\right|^2 d\Omega = \left(\frac{\mu}{2\pi\hbar^2}\right)^2 \left|\int d\vec{r}\,U(\vec{r})\exp\{-i\vec{q}\vec{r}\}\right|^2 d\Omega \tag{12.212}$$

This is the Born formula that was proposed for fast moving particles. This formula may as well be used for slow moving particles when

$$|\vec{r} - \vec{r}'| \cong a \tag{12.213}$$

then as $\kappa a \ll 1$, we have

$$\bar{U} \ll \frac{\hbar^2}{\mu a^2} \tag{12.214}$$

This is the condition of applicability of Equation 12.212 for the case of slowly moving particles.

An example may be found for the elastic scattering of an electron by a neutral atom represented by a screened Coulomb potential, the so-called **Yukawa potential** named after Hideki Yukawa:

$$U(\vec{r}) = -\frac{Ze^2}{r}\exp\left\{-\frac{r}{a}\right\} \tag{12.215}$$

This imitates the nuclear Coulomb potential for atomic number Z when r is small. This potential imitates also a screening Coulomb potential with the range a. This potential decrease rapidly for large r compared to the atomic electronic cloud radius a screening the nucleus. From the Thomas-Fermi atomic statistical theory, for moderately heavy atoms, we have

$$a \cong \frac{\hbar^2}{\mu e^2 Z^{\frac{1}{3}}} \tag{12.216}$$

So considering the given potential in Equation 12.211, we have the scattering amplitude, $\hat{f}(\theta)$:

$$\hat{f}(\theta) = \frac{2\mu Ze^2}{\hbar^2\left(q^2 + a^{-2}\right)} \tag{12.217}$$

When the magnitude of the momentum transfer q is large enough so that a^{-2} can be safely neglected compared to q^2, the incident electron passes close enough to the nucleus. In this case, the screening of the nuclear potential by the atomic electrons is redundant.

We examine additional symmetry with the amplitude of the direct and reverse scattering processes:

$$\hat{f}(\vec{\kappa},\vec{\kappa}')=\hat{f}^*(\vec{\kappa}',\vec{\kappa}) \tag{12.218}$$

This implies we have processes differing in the initial and final momenta without a change of sign. We examine the optical scattering theorem by calculating the following scattering amplitude for forward scattering:

$$\hat{f}(\vec{\kappa},\vec{\kappa}')=\hat{f}(\theta=0)=-\frac{\mu}{2\pi\hbar^2}\int d\vec{r}\,U(\vec{r}) \tag{12.219}$$

Hence, the total scattering cross section (**optical scattering theorem**):

$$\sigma_{\text{total}}=\frac{4\pi}{\kappa}\text{Im}\,\hat{f}(\vec{\kappa},\vec{\kappa}') \tag{12.220}$$

From here:

$$\text{Im}\,\hat{f}(\vec{\kappa},\vec{\kappa}')=\frac{\kappa}{4\pi}\sigma_{\text{total}} \tag{12.221}$$

It can be observed from here that Equation 12.141 may be used in conjunction with the Born approximation. The Born approximation is observed to give an extremely simple expression for the scattering amplitude $\hat{f}$ that is of the first order in U where $\hat{f}$ is real when U is real. So Equation 12.141 permits us to calculate the imaginary part of $\hat{f}$ that is the lowest order in U. This is evident that it is of the second order.

12.8.2 Perturbation Theory Method Approach for Born Approximation

We consider again an elastic scattering of particles on some center, $U(r)$, and assume the particles interact with the given center. We consider the potential, $U(r)$, to be a small perturbation. We write for this case the Schrödinger equation in the form:

$$\Delta\psi(r)+\kappa^2\psi(r)=\frac{2\mu}{\hbar^2}U(r)\psi(r),\ \kappa^2=\frac{2\mu E}{\hbar^2} \tag{12.222}$$

We examine the state of a continuous spectrum where the perturbation theory may not be applicable, except we use the Born approximation:

$$\psi(r)=\psi^{(1)}(r)+\psi^{(0)}(r) \tag{12.223}$$

and

$$\Delta\psi^{(0)}(r)+\kappa^2\psi^{(0)}(r)=0 \tag{12.224}$$

Consequently,

$$\Delta\psi^{(1)}(r)+\kappa^2\psi^{(1)}(r)=\frac{2\mu}{\hbar^2}U(r)\psi^{(0)}(r) \tag{12.225}$$

This is a non-homogenous equation of the second order that can be solved with the help of Green's functions method. So we introduce the auxiliary function:

$$\Phi=\psi^{(1)}\exp\left\{-i\frac{E}{\hbar}t\right\} \tag{12.226}$$

where $\frac{E}{\hbar}$ is some auxiliary frequency and the wave vector for this case is $\kappa=\frac{E}{c\hbar}$, and c has the sense of the speed of light. We multiply Equation 12.225 by $\exp\left\{-i\frac{E}{\hbar}t\right\}$ and write the result as follows:

$$\Delta\Phi-\frac{1}{c^2}\frac{\partial^2}{\partial t^2}\Phi=\frac{2\mu}{\hbar^2}U(r)\psi^{(0)}\exp\left\{-i\frac{E}{\hbar}t\right\} \tag{12.227}$$

This is a well-known equation in electrodynamics called the D'Alembert equation, where

$$\frac{2\mu}{\hbar^2}U(r)\psi^{(0)}\exp\left\{-i\frac{E}{\hbar}t\right\}\to-4\pi\rho(r,t) \tag{12.228}$$

In electrodynamics, Φ in Equation 12.227 will have the sense of a retarded potential. So the solution of Equation 12.227 imitates that of the retardation potential solution:

$$\Phi(\vec{r},t)=\int\frac{\rho\left(\vec{r}',t-\frac{|\vec{r}-\vec{r}'|}{c}\right)d\vec{r}'}{|\vec{r}-\vec{r}'|}=-\frac{\mu}{2\pi\hbar^2}\int\frac{U(\vec{r}')\psi^{(0)}(\vec{r}')\exp\left\{-i\frac{\in}{\hbar}\left(t-\frac{|\vec{r}-\vec{r}'|}{c}\right)\right\}}{|\vec{r}-\vec{r}'|}d\vec{r}' \tag{12.229}$$

This implies from Equation 12.226 that the scattering wave has the form:

$$\psi^{(1)}(\vec{r})=-\frac{\mu}{2\pi\hbar^2}\int\frac{U(\vec{r}')\psi^{(0)}(\vec{r}')\exp\{i\kappa|\vec{r}-\vec{r}'|\}}{|\vec{r}-\vec{r}'|}d\vec{r}' \tag{12.230}$$

Equation 12.230 is the exact solution of Equation 12.225 for the retardation potential.

From the correct scattering theory, it is necessary to let

$$\psi^{(0)}=\exp\{i\kappa z\} \tag{12.231}$$

For $\psi^{(1)}$, we find its asymptotic solution. We find the description where z-axis coincides with the direction of the particle before scattering. Suppose $\vec{r}'$ is the coordinate of the source from where the integral is taken and $\vec{\kappa}'$ is the wave vector of the particle after scatter:

$$|\vec{\kappa}'|=|\vec{\kappa}|,\ |\vec{r}-\vec{r}'|=\sqrt{\vec{r}^2+\vec{r}'^2-2\vec{r}\,\vec{r}'\cos\theta} \tag{12.232}$$

Considering the following condition:

$$|\vec{r}| \gg |\vec{r}'| \tag{12.233}$$

then

$$|\vec{r} - \vec{r}'| \cong r - r'\cos\theta \tag{12.234}$$

and so

$$\exp\{i\kappa z\}\exp\{i\vec{\kappa}\vec{r} - i\kappa r'\cos\theta\} = \exp\{i\vec{\kappa}\vec{r}\}\exp\{i(\vec{\kappa} - \vec{\kappa}', \vec{r}')\} \tag{12.235}$$

Here, we consider that

$$\kappa r'\cos\theta = \vec{\kappa}'\vec{r}' \tag{12.236}$$

and κ' is in the direction of spread of the scattering wave. From Equation 12.233 and considering the denominator in Equation 12.234, we neglect r' compared to r and we have the following asymptotic wave function in the first-order Born approximation:

$$\psi^{(1)}(\vec{r}) = -\frac{\mu}{2\pi\hbar^2}\frac{\exp\{i\vec{\kappa}\vec{r}\}}{r}\int U(\vec{r}')\exp\{i(\kappa - \kappa', r')\}d\vec{r}' \tag{12.237}$$

If we take the solution described by the given potential, then we have $\frac{\exp\{-i\vec{\kappa}\vec{r}\}}{r}$. This is a convergent wave that we neglect due to boundary conditions. From Equation 12.237, we have

$$f(\theta) \equiv f(\kappa', \kappa) = -\frac{\mu}{2\pi\hbar^2}U(q),\ U(q) = \int U(\vec{r})\exp\{i\vec{q}\vec{r}\}d\vec{r},\ q = \kappa - \kappa' \tag{12.238}$$

From here:

$$q = \sqrt{\kappa^2 + \kappa'^2 - 2\kappa\kappa'\cos\theta} \tag{12.239}$$

So the scattering wave in Equation 12.237 can also be written as follows:

$$\psi^{(1)}(\vec{r}) = \frac{\exp\{i\vec{\kappa}\vec{r}\}}{r}f(\theta) \tag{12.240}$$

From Equation 12.239, since $\kappa = \kappa'$, we have Equation 12.203. The **effective cross section** is then

$$d\sigma = |f(\theta)|^2 d\Omega = \left(\frac{\mu}{2\pi\hbar^2}\right)^2\left|\int U(\vec{r})\exp\{i\vec{q}\vec{r}\}d\vec{r}\right|^2 d\Omega \tag{12.241}$$

This is the Born formula that we proposed for fast moving particles. This formula may also use for slow moving particles when

$$|\vec{r} - \vec{r}'| \cong a \tag{12.242}$$

then as $\kappa a \ll 1$ and from Equation 12.237, we have

$$\psi^{(1)} \cong \bar{U}\psi^{(0)} \frac{\mu}{2\pi\hbar^2} \int d\vec{r}' = \bar{U}\psi^{(0)} \frac{\mu}{2\pi\hbar^2 a} \frac{4\pi}{3} a^3 \tag{12.243}$$

From here:

$$\left|\frac{\psi^{(1)}}{\psi^{(0)}}\right| \cong \bar{U} \frac{\mu}{2\pi\hbar^2 a} \frac{4\pi}{3} a^3 \ll 1 \tag{12.244}$$

and implies

$$\bar{U} \ll \frac{\hbar^2}{\mu a^2} \tag{12.245}$$

This is the condition of applicability of Equation 12.241 for the case of slowly moving particles.

Let us examine the condition at the boundary for a continuous spectrum:

$$\psi^{(0)} + \psi^{(1)}, \left|\psi^{(1)}\right| \ll 1, \psi^{(0)} \neq 0 \tag{12.246}$$

This implies that the ground-state wave function does not have zeros. Our system cannot have discrete levels. This implies the Born approximation is applicable to the scattering of slowly moving particles if, at the field of the scattering center, there are no discrete levels. This implies that Equation 12.245 shows that in the field of the potential energy $\bar{U}$, there are no discrete levels.

For the case of a central symmetric field, we evaluate the integral for the effective cross section in Equation 12.241. For this, the potential, $U(\vec{r})$, is independent of the angle θ and so

$$\int U(\vec{r}) \exp\{i\vec{q}\vec{r}\} d\vec{r} = \int_0^\infty r^2 U(\vec{r})\, dr \int_0^\pi \sin\theta\, d\theta \exp\{iqr\cos\theta\} \int_0^{2\pi} d\phi \tag{12.247}$$

or

$$\int U(\vec{r}) \exp\{i\vec{q}\vec{r}\}\, d\vec{r} = 2\pi \int_0^\infty r^2 U(\vec{r})\, dr \frac{\exp\{iqr\} - \exp\{-iqr\}}{iqr} = \frac{4\pi}{q} \int_0^\infty r \sin qr\, U(\vec{r}) dr \tag{12.248}$$

or

$$\int U(\vec{r}) \exp\{i\vec{q}\vec{r}\} d\vec{r} = \frac{4\pi}{q} \int_0^\infty r \sin qr\, U(\vec{r}) dr = 4\pi \int_0^\infty r^2 \frac{\sin qr}{qr} U(\vec{r}) dr \tag{12.249}$$

and

$$f(\theta) = -\frac{\mu}{2\pi\hbar^2} U(\vec{q}) \tag{12.250}$$

Here, $U(\vec{q})$ is the Fourier transform of the potential, $U(\vec{r})$:

$$U(\vec{q}) = 4\pi \int_0^\infty U(\vec{r}) r^2 \frac{\sin qr}{qr} dr \tag{12.251}$$

Considering that

$$\frac{\sin qr}{qr}=\frac{\sin\sqrt{2\kappa^2r^2-22\kappa^2r^2\cos\theta}}{\sqrt{2\kappa^2r^2-22\kappa^2r^2\cos\theta}}=\frac{\pi}{2\kappa r}\sum_{l=0}^{\infty}(2l+1)\left[\mathrm{j}_{l+\frac{1}{2}}(\kappa r)\right]^2\mathrm{P}_l(\cos\theta) \quad (12.252)$$

then

$$f(\theta)=\sum_{l=0}^{\infty}(2l+1)\left(-\frac{\pi\mu}{\kappa\hbar^2}\int_0^{\infty}U(\vec{r})\left[\mathrm{j}_{l+\frac{1}{2}}(\kappa r)\right]^2 rdr\right)\mathrm{P}_l(\cos\theta) \quad (12.253)$$

The total effective cross section is then

$$\sigma_{\text{total}}=\int|f(\theta)|^2 d\Omega \quad (12.254)$$

Considering that

$$q^2=4\kappa^2\sin^2\frac{\theta}{2}=\frac{8\mu E}{\hbar^2}\sin^2\frac{\theta}{2} \quad (12.255)$$

then, for $E=0$, we have $q=0$ and the total effective cross section has its maximum value:

$$\sigma_{\text{total}}(E=0)=\int|f(0)|^2 d\Omega=|f(0)|^2\int d\Omega=4\pi|f(0)|^2 \quad (12.256)$$

Consider the inequality:

$$|U(\vec{q})|\equiv\left|\int U(\vec{r})\exp\{i\vec{q}\vec{r}\}d\vec{r}\right|\le\int|U(\vec{r})|d\vec{r}=\left|\int U(\vec{r})d\vec{r}\right|=|U(0)| \quad (12.257)$$

It is instructive to note that for $E\neq 0$, which corresponds to the scattering angle $\theta\neq 0$, $q^2\neq 0$ and

$$\sigma_{\text{total}}(E\neq 0)\le\sigma_{\text{total}}(E=0) \quad (12.258)$$

It is instructive to note that for a state with a defined angular momentum, l, we have

$$\hat{U}_{\text{exc}}\Psi(\vec{r})=U(\vec{r})\Psi(-\vec{r})=(-1)^l U(\vec{r})\Psi(\vec{r}) \quad (12.259)$$

This implies that the interaction with even l has an attractive character that with odd l has a repulsive character and $\hat{U}_{\text{exc}}$ is an exchange potential and exc denote exchange. Let us show that in the Born approxamination, the scattering amplitudes for the exchange and ordinary potential are related through the relation:

$$f_{\text{exc}}(\kappa',\kappa)\equiv f_{\text{exc}}(\theta)=f(\pi-\theta) \quad (12.260)$$

where,

$$f_{\text{exc}}(\kappa',\kappa)\equiv f_{\text{exc}}(\Delta)=-\frac{\mu}{2\pi\hbar^2}U(\Delta) \quad (12.261)$$

and

$$U(\Delta)=\int U(\vec{r})\exp\{i\vec{\Delta}\vec{r}\}d\vec{r},\ \vec{\Delta}=\vec{\kappa}+\vec{\kappa}' \tag{12.262}$$

with

$$\Delta^2=2\kappa^2(1+\cos\theta) \tag{12.263}$$

It is instructive to note that the scattering by fast particles, $a\gg 1$, on the ordinary potential $U(\vec{r})$, we expect basically forward scattering for the angles, $\theta\sim(\kappa a)^{-1}\ll 1$. Here, a is the range of the potential. From Equation 12.126, for the exchange potential, $U_{\text{exc}}(\vec{r})$, the scattering is basically backwards. This implies the scattering angles are at the neighborhood of π. If the potential is the sum of the ordinary and exchange potentials, then the scattering cross section for the fast particles will have pronounced maxima in the forward and backward scattering where we have respectively, $\theta\approx 0$ and $\theta\approx\pi$.

12.8.2.1 Phase Shift

Let us compare the scattering amplitude in Equation 12.253 with that of the partial waves in Equation 12.99 for small values of the phase shift:

$$|\delta_l(\kappa)|\ll 1 \tag{12.264}$$

So from Equation 12.99, we have

$$f(\theta)=\frac{1}{2l\kappa}\sum_{l=0}^{\infty}(2l+1)(\eta_l-1)\mathrm{P}_l(\cos\theta)\cong\frac{1}{\kappa}\sum_{l=0}^{\infty}(2l+1)\delta_l(\kappa)\mathrm{P}_l(\cos\theta) \tag{12.265}$$

Then comparing the equality in Equation 12.265 and 12.253, considering Equation 12.264, shows the equality of the phase shift for the Born and partial wave approximations:

$$\delta_l(\kappa)=-\frac{\pi\mu}{\hbar^2}\int_0^{\infty}U(\vec{r})\left[\mathrm{j}_{l+\frac{1}{2}}(\kappa r)\right]^2 rdr \tag{12.266}$$

This expression is the first-order term in the expansion of the phase shift relative to the interaction or potential. Considering partial waves, we use Equation 12.253 and find that for Equation 12.260 $\mathrm{P}_l(\cos\theta)$ is transformed to $(-1)^l\mathrm{P}_l(\cos\theta)$. Consequently, the phase shift for this case becomes

$$\delta_{\text{exc},l}(\kappa)=-(-1)^l\frac{\pi\mu}{\hbar^2}\int_0^{\infty}U(\vec{r})\left[\mathrm{j}_{l+\frac{1}{2}}(\kappa r)\right]^2 rdr \tag{12.267}$$

Let us consider the case for low scattering energy, i.e., when $E\to 0$ or $\sqrt{\frac{2\mu E}{\hbar^2}}\to 0$. Setting $z=\kappa r$ in Equation 12.266, for $z\ll 1$, we have

$$\mathrm{j}_{l+\frac{1}{2}}(z)\cong\frac{1}{\Gamma\left(l+\frac{3}{2}\right)}\left(\frac{z}{2}\right)^{l+\frac{1}{2}},\ z^2\ll l+\frac{1}{2} \tag{12.268}$$

and

$$\delta_l(\theta)=\lim_{E\to 0}\delta_l(\kappa)\cong\frac{\pi\mu}{\hbar^2\Gamma\left(l+\frac{3}{2}\right)}\left(\frac{\kappa}{2}\right)^{2l+1}\int_0^\infty U(\vec{r})r^{2l+2}dr \tag{12.269}$$

This formula is satsified only for the potentials of the type $U(\vec{r})\cong r^{-\nu}$ only when $l<\frac{\nu-3}{2}$ and not satisfied for the potentials that do not respect the condition for the Born approximation in Equation 12.245.

Let us examine the case of high scattering energy, i.e., when $E\to\infty$, or $\kappa=\sqrt{\frac{2\mu E}{\hbar^2}}\to\infty$. So setting $z=\kappa r$ in Equation 12.266, for $z\gg 1$, we have

$$j_{l+\frac{1}{2}}(z)\cong\left(\frac{2}{\pi z}\right)^{\frac{1}{2}}\sin\left(z-\frac{\pi l}{2}\right),\ z\gg\left(l+\frac{1}{2}\right)^2 \tag{12.270}$$

and

$$\delta_l(\theta)=\lim_{E\to\infty}\delta_l(\kappa)\cong\frac{2\mu}{\kappa\hbar^2}\int_0^\infty U(\vec{r})\sin^2\left(\kappa r-\frac{\pi l}{2}\right)dr\cong\frac{\mu}{\kappa\hbar^2}\int_0^\infty U(\vec{r})dr \tag{12.271}$$

The function, $\sin^2\left(\kappa r-\frac{\pi l}{2}\right)$, in the integrand faster oscillates and can be replaced by its mean value $\frac{1}{2}$. It is instructive to note that if the potential, $U(\vec{r})$, is bounded for all values of r, then for exceedingly large energies of the particles, this potential can be examined as a perturbation.

We consider the case of the s-wave scattering in the Born approximation:

$$\delta_l(\kappa)=-\frac{\pi\mu}{\hbar^2}\int_0^\infty rU(\vec{r})J^2_{\frac{1}{2}}(\kappa r)\,dr=-\frac{2\mu}{\kappa\hbar^2}\int_0^\infty U(\vec{r})\sin^2\kappa r\,dr \tag{12.272}$$

12.8.2.2 Spherical Potential Well

We apply this to the case of a spherical potential well:

$$U(\vec{r})=\begin{cases}-U_0, & r\le a\\ 0, & r\ge a\end{cases} \tag{12.273}$$

$$U(\vec{q})=\int U(\vec{r})\exp\{i\vec{q}\vec{r}\}d\vec{r}=\frac{4\pi}{q}\int_0^\infty r\sin qr\,U(\vec{r})\,dr=\frac{4\pi}{q}\int_0^a r\sin qr\,U_0\,dr \tag{12.274}$$

or

$$U(\vec{q})=\frac{4\pi U_0}{q^3}(-qa\cos qa+\sin qa) \tag{12.275}$$

Then the scattering cross section:

$$d\sigma=\left|\hat{f}(\theta)\right|^2 d\Omega=\left(\frac{\mu}{2\pi\hbar^2}\right)^2\left|\int d\vec{r}\,U(\vec{r})\exp\{-i\vec{q}\vec{r}\}\right|^2 d\Omega=\left(\frac{\mu}{2\pi\hbar^2}\right)^2\left|\frac{4\pi U_0}{q^3}\left[\sin qa-qa\cos qa\right]\right|^2 d\Omega \tag{12.276}$$

or

$$d\sigma = 4a^2\left(\frac{\mu U_0 a^2}{\hbar^2}\right)^2\left|\frac{\sin qa - qa\cos qa}{(qa)^3}\right|^2 d\Omega \tag{12.277}$$

The condition for applicability of the Born approximation:

$$\frac{\mu U_0 a^2}{\hbar^2} \ll 1 \tag{12.278}$$

Total Cross Section

We now find the total cross section by considering

$$q = 2\kappa\sin\frac{\theta}{2},\ dq = \kappa\cos\frac{\theta}{2}d\theta,\ qdq = \kappa^2\sin\theta\, d\theta,\ d\Omega = 2\pi\sin\theta\, d\theta = \frac{2\pi q dq}{\kappa^2} \tag{12.279}$$

$$\sigma_{\text{total}} = 4a^2\left(\frac{\mu U_0 a^2}{\hbar^2}\right)^2\int_0^{2\kappa}\frac{2\pi q dq}{\kappa^2}\left|\frac{\sin qa - qa\cos qa}{(qa)^3}\right|^2 = 8\pi a^2\left(\frac{\mu U_0 a^2}{\hbar^2}\right)^2\frac{1}{a^2\kappa^2}\int_0^{2\kappa a}\frac{xdx}{x^6}|\sin x - x\cos x|^2$$

$$= \left(\frac{\mu U_0 a^2}{\hbar^2}\right)^2\frac{2\pi}{\kappa^2}\left(1 - \frac{1}{(2\kappa a)^2} + \frac{\sin(4\kappa a)}{(2\kappa a)^3} - \frac{\sin^2(2\kappa a)}{(2\kappa a)^4}\right) \tag{12.280}$$

Considering that $E = \frac{\hbar^2\vec{\kappa}^2}{2\mu}$, for slow and fast scattering, we have, respectievely,

$$\lim_{E\to 0}\sigma_{\text{total}} \cong \pi\left(\frac{4\mu U_0 a^3}{3\hbar^2}\right)^2,\ \lim_{E\to\infty}\sigma_{\text{total}} \cong \left(\frac{U_0 a^2}{\hbar}\right)^2\frac{\mu\pi}{E} \tag{12.281}$$

Phase Shift

Considering the solution under Chapter 6.4.5 for the particle in a spherical symmetric well potential of finite depth, the radial wave function is selected in the form:

$$R(r) = \frac{X(r)}{r} \tag{12.282}$$

For s-wave scattering, $l = 0$. If $R(r)$ has to be finite at $r = 0$, then

$$X(r) = \theta(r \le a)A\sin\chi r + \theta(r \ge a)B\sin(\kappa r + \delta_0(\kappa)) \tag{12.283}$$

where

$$\chi^2 = \frac{2\mu}{\hbar^2}(E + U_0),\ \kappa^2 = -\frac{2\mu}{\hbar^2}E \tag{12.284}$$

From the condition of continuity of the wave function, $X(r)$, and its derivatives at the point $r = a$:

$$\chi\cot\chi a = \kappa\cot(\kappa a + \delta_0(\kappa)) \tag{12.285}$$

and the phase shift:

$$\delta_0(\kappa) = -\kappa a + \cot^{-1}\left(\frac{\chi}{\kappa}\cot\chi a\right) \tag{12.286}$$

For slow and fast scattering, we have, respectively,

$$\lim_{E\to 0}\delta_0(\kappa) \cong \kappa\left(-a + \frac{1}{\chi_0}\tan(\chi_0 a)\right),\ \lim_{E\to\infty}\delta_0(\kappa) \cong -E,\ \chi_0^2 = \frac{2\mu U_0}{\hbar^2} \tag{12.287}$$

12.8.2.3 Coulomb Interaction and Rutherford's Formula

Coulomb potential Type 1

We examine the following Coulomb interaction with the potential, $U(\vec{r})$, experienced by an electron of charge, $-e$, in an electrostatic field of a nucleus of charge, Ze:

$$U(\vec{r}) = -\frac{\alpha}{r},\ \alpha \equiv Ze^2 \tag{12.288}$$

The Fourier transform of this potential:

$$U(\vec{q}) = \frac{4\pi\alpha}{q^2} = \frac{4\pi\alpha}{4\kappa^2\sin^2\frac{\theta}{2}} = \frac{\pi\alpha}{\kappa^2}\frac{1}{\sin^2\frac{\theta}{2}} \tag{12.289}$$

The scattering cross section:

$$d\sigma = \left|\hat{f}(\theta)\right|^2 d\Omega = \left(\frac{\mu}{2\pi\hbar^2}\right)^2\left|\frac{\pi\alpha}{\kappa^2}\frac{1}{\sin^2\frac{\theta}{2}}\right|^2 d\Omega = \left(\frac{\mu}{2\pi\hbar^2}\right)^2\left|\frac{\pi\alpha}{\kappa^2}\frac{1}{\sin^2\frac{\theta}{2}}\right|^2 2\pi\sin\theta\, d\theta \tag{12.290}$$

or

$$d\sigma = \left(\frac{\mu\alpha}{\hbar^2}\right)^2\frac{2\pi\sin\theta\, d\theta}{\left(2\kappa^2\sin^2\frac{\theta}{2}\right)^2} = \left(\frac{\mu\alpha}{\hbar^2}\right)^2\frac{d\Omega}{\left(2\kappa^2\sin^2\frac{\theta}{2}\right)^2} = \frac{1}{a_B^2}\frac{d\Omega}{\left(2\kappa^2\sin^2\frac{\theta}{2}\right)^2},\ \frac{\mu\alpha}{\hbar^2} = \frac{1}{a_B} \tag{12.291}$$

Equation 12.291 is the Coulomb scattering cross section (**Rutherford's formula**) derived first by Ernest Rutherford, Lord Rutherford from classical mechanics, and was experimentally verified for collisions of alpha particles (helium nuclei) with heavier nuclei. The Rutherford formula gives also the relative probability of finding a particle scattered through an angle θ. For small values of θ, the total Coulomb scattering cross section becomes infinite as the corresponding integral diverges for small θ. This is due to an infinite range of the Coulomb potential. In this case, even if the particle passes very far from the origin, it will be affected by the potential and determines why the scattering cross section is infinite. It is instructive to note that usually it is difficult to achieve a rigorously pure Coulomb potential within an infinite range since the potential due to a charged particle is modified by the screening effect of the neighbouring charges.

Coulomb potential Type 2

Let us extend this example to the case of the **Yukawa potential (screened or shielded Coulomb potential)** seen in Equation 12.215 and named after **Hideki Yukawa**:

$$U(\vec{r}) = -\frac{\alpha}{r}\exp\{-\chi r\},\ \chi \equiv \frac{1}{a} \tag{12.292}$$

This potential imitates a screening Coulomb potential with the range a. The quantity, χ, is the inverse screening radius.

From Equation 12.250, the scattering amplitude:

$$f(\theta) = -\frac{2\mu}{\hbar^2}\int_0^\infty U(\vec{r})\frac{r\sin qr}{q}dr \tag{12.293}$$

Here, the momentum transfer, q, is defined in Equation 12.279. We now substitute Equation 12.292 into 12.293 and have

$$f(\theta) = \frac{2\mu}{\hbar^2}\int_0^\infty \frac{\alpha}{r}\exp\{-\chi r\}\frac{r\sin qr}{q}dr = \frac{2\mu\alpha}{\hbar^2 q}\frac{1}{2i}\left(\int_0^\infty \exp\{-(\chi - iq)r\}dr - \int_0^\infty \exp\{-(\chi + iq)r\}dr\right) \tag{12.294}$$

or

$$f(\theta) = \frac{\mu\alpha}{\hbar^2 qi}\left(\frac{1}{\chi - iq} - \frac{1}{\chi + iq}\right) = \frac{2\mu\alpha}{\hbar^2}\frac{1}{\chi^2 + q^2} \tag{12.295}$$

Considering Equation 12.203, the scattering amplitude, $f(\theta)$:

$$f(\theta) = \frac{2\mu\alpha}{\hbar^2}\frac{1}{\chi^2 + 4\kappa^2\sin^2\frac{\theta}{2}} \tag{12.296}$$

It is instructive to note that this relation is independent of the Azimuthal angle ϕ, indicating that the scattering by a central field is symmetric with respect to rotation about the incident direction of the wave. The scattering amplitude, $f(\theta)$, is dependent on the scattering angle θ and on the energy through the fixed absolute value of the incident wave vector, κ. The scattering amplitude, $f(\theta)$, in the forward direction, i.e., for $\theta = 0$, is greater than the scattering amplitude in the backward direction, i.e., for $\theta = \pi$. The scattering amplitude, $f(\theta)$, for fixed θ is a decreasing function of the energy.

The scattering cross section, $d\sigma$, can be evaluated as follows:

$$d\sigma = \left|\hat{f}(\theta)\right|^2 d\Omega = \left(\frac{2\mu\alpha}{\hbar^2}\right)^2\left|\frac{1}{\chi^2 + q^2}\right|^2 d\Omega = \left(\frac{2\mu\alpha}{\hbar^2}\right)^2\left|\frac{1}{\chi^2 + q^2}\right|^2 2\pi\sin\theta\, d\theta \tag{12.297}$$

The total scattering cross section, σ_{total}:

$$\sigma_{\text{total}} = 2\pi\left(\frac{2\mu\alpha}{\hbar^2}\right)^2\int_0^\pi d\theta\left|\frac{1}{\chi^2 + q^2}\right|^2\sin\theta = 2\pi\left(\frac{2\mu\alpha}{\hbar^2}\right)^2\int_0^\pi d\theta\frac{\sin\theta}{\left(\chi^2 + 2\kappa^2(1 - \cos\theta)\right)^2} \tag{12.298}$$

or

$$\sigma_{\text{total}} = 2\pi\left(\frac{2\mu\alpha}{\hbar^2}\right)^2 \frac{1}{2\kappa^2}\left(\frac{1}{\chi^2} - \frac{1}{\chi^2+4\kappa^2}\right) = \left(\frac{2\mu\alpha}{\chi\hbar^2}\right)^2 \frac{4\pi}{\chi^2+4\kappa^2} = \left(\frac{2\mu\alpha}{\chi^2\hbar^2}\right)^2 \frac{4\pi}{1+\dfrac{8E\mu}{\hbar^2\chi^2}} \tag{12.299}$$

then

$$\lim_{E\to 0}\sigma_{\text{total}} \cong 4\pi\left(\frac{2\mu\alpha}{\chi^2\hbar^2}\right)^2, \quad \lim_{E\to\infty}\sigma_{\text{total}} \cong \frac{2\pi\mu\alpha^2}{E\chi^2\hbar^2} \tag{12.300}$$

It is obvious from these limiting values that, for large E or large kinetic energy of the scattered particles, the total scattering cross section is small as physically previewed. So it is more difficult to deflect a fast-moving particle than a slow-moving one. The condition of applicability of the Born approximation:

$$\frac{\mu\alpha}{\chi^2\hbar^2} \ll 1 \tag{12.301}$$

Coulomb potential Type 3

Suppose we have a system of atomic nuclei and atomic shells having electrons where each electron has the charge e. Suppose a particle of charge e_1 is projected toward the atom. The atom is examined as a fixed center with charge Ze. The potential coupling the nucleus to the electrons on the electronic shells:

$$U(\vec{r}_1,\vec{r}_2,\ldots) = \frac{Zee_1}{r} - \sum_i \frac{ee_1}{|\vec{r}-\vec{r}_i|} \tag{12.302}$$

Suppose the wave function of the electronic shells is $\Psi(\vec{r}_1,\vec{r}_2,\ldots,\vec{r}_{i,\ldots})$, the electronic charge density on the electronic shells, $e\rho(r)$:

$$\rho(\vec{r}) = \sum_i \rho(\vec{r}_i),\ \rho(\vec{r}_i) = \int |\Psi(\vec{r}_1,\vec{r}_2,\ldots,\vec{r}_i)|^2 \prod_i d\vec{r}_i \tag{12.303}$$

This implies the potential coupling the nucleus to the electrons on the electronic shells can also be represented as follows:

$$\text{U}(r) = \frac{Zee_1}{r} - ee_1\int \frac{\rho(\vec{r}')d\vec{r}'}{|\vec{r}-\vec{r}'|} \tag{12.304}$$

Considering the screening effect of the electrons on the electronic shells on the nucleus, the function $\rho(r)$ can be selected in the following form:

$$\rho(r) = \rho_0 \exp\{-\chi r\},\ \chi \equiv \frac{1}{a} \tag{12.305}$$

From Equation 12.291, then we have the scattering amplitude, $f(\theta)$:

$$f(\theta) = -\frac{\mu Zee_1}{2\pi\hbar^2}\Phi(q) + \frac{\mu ee_1}{2\pi\hbar^2}\text{X}(q) \tag{12.306}$$

Here

$$\Phi(q) \equiv \int \frac{\exp\{i\vec{q}\vec{r}\}}{r} d\vec{r},\ X(q) \equiv \int \frac{\exp\{i\vec{q}\vec{r}\}}{|\vec{r}-\vec{r}'|} \rho(\vec{r}')d\vec{r}\,d\vec{r}',\ q = \kappa - \kappa' \tag{12.307}$$

Considering

$$\Delta I(\vec{r}') = \Delta \int \frac{\exp\{i\vec{q}\vec{r}\}}{|\vec{r}-\vec{r}'|} d\vec{r} = -4\pi \int \exp\{i\vec{q}\vec{r}\}\delta(\vec{r}-\vec{r}')d\vec{r} = -4\pi \exp\{i\vec{q}\vec{r}'\} \tag{12.308}$$

then letting

$$I(\vec{r}') = A\exp\{i\vec{q}\vec{r}'\} \tag{12.309}$$

we have

$$-q^2 A \exp\{i\vec{q}\vec{r}'\} = -4\pi \exp\{i\vec{q}\vec{r}'\} \tag{12.310}$$

So

$$A = \frac{4\pi}{q^2} \tag{12.311}$$

and

$$\Phi(q) = \frac{4\pi}{q^2},\ X(q) = \frac{4\pi}{q^2} \int \exp\{i\vec{q}\vec{r}'\}\rho(\vec{r}')d\vec{r}' \tag{12.312}$$

The scattering amplitude is then

$$f(\theta) = -\frac{\mu z e e_1}{2\pi\hbar^2}\frac{4\pi}{q^2} + \frac{\mu e e_1}{2\pi\hbar^2}\frac{4\pi}{q^2} \int \exp\{i\vec{q}\vec{r}'\}\rho(\vec{r}')d\vec{r}' = -\frac{\mu e e_1}{2\pi\hbar^2}\frac{4\pi}{q^2}\left(z - F(\theta)\right) \tag{12.313}$$

Here, the atomic form factor, $F(\theta)$, of the perturbation theory:

$$F(\theta) = \int \exp\{i\vec{q}\vec{r}'\}\rho(\vec{r}')d\vec{r}' = 4\pi \int_0^\infty \rho(r')\frac{\sin(qr')}{qr'} r'^2 dr' \tag{12.314}$$

This form factor can also be obtained from experimentation. Considering Equation 12.305 and the following normalization condition:

$$\int_0^\infty \rho(\vec{r})d\vec{r} = 4\pi\rho_0 \int_0^\infty \exp\{-\chi r\}r^2 dr = \frac{4\pi\rho_0 2}{\chi^3} = z \tag{12.315}$$

we have

$$\rho_0 = \frac{z\chi^3}{8\pi} \tag{12.316}$$

We can now evaluate the form factor, $F(\theta)$:

$$F(\theta)=4\pi\rho_0\int_0^{\infty}\exp\{-\chi r\}\frac{\exp\{iqr\}-\exp\{-iqr\}}{2iqr}r^2dr=\frac{z\chi^3}{4i}\left(\frac{1}{(\chi-iq)^2}-\frac{1}{(\chi+iq)^2}\right)=\frac{z}{\left(1+a^2q^2\right)^2} \tag{12.317}$$

So the scattering amplitude, $f(\theta)$:

$$f(\theta)=-\frac{\mu ee_1}{2\pi\hbar^2}\frac{4\pi}{q^2}\left(z-\frac{z}{\left(1+a^2q^2\right)^2}\right) \tag{12.318}$$

Considering that $q=2\kappa\sin\frac{\theta}{2}$, the scattering cross section:

$$d\sigma=\left|\hat{f}(\theta)\right|^2 d\Omega=\left(\frac{2z\mu ee_1}{\hbar^2q^2}\right)^2\left|1-\frac{1}{\left(1+a^2q^2\right)^2}\right|^2 d\Omega \tag{12.319}$$

For fast particles, the scattering through big angles, θ, when $aq\gg 1$, and we have the scattering cross section which is **Rutherford's formula**:

$$d\sigma=\left|\hat{f}(\theta)\right|^2 d\Omega=\left(\frac{2z\mu ee_1}{\hbar^2q^2}\right)^2 d\Omega \tag{12.320}$$

and for small $aq\ll 1$, the scattering amplitude:

$$f(\theta)=-\frac{\mu ee_1}{2\pi\hbar^2}\frac{4\pi}{q^2}z2a^2q^2 \tag{12.321}$$

The scattering cross section is then a constant

$$d\sigma=\left(\frac{4za^2\mu ee_1}{\hbar^2}\right)^2 d\Omega \tag{12.322}$$

12.8.2.4 Lippman Schwinger Equation, 1D Delta Potential

We consider again the **Lippman Schwinger equation**:

$$F(\vec{\kappa},\vec{\kappa}')=-U(\vec{\kappa}'-\vec{\kappa})+\frac{2\mu}{\hbar^2}\int\frac{d\vec{q}}{(2\pi)^3}\frac{U(\vec{\kappa}-\vec{q})F(\vec{\kappa},\vec{q})}{\vec{\kappa}^2-\vec{q}^2+i\delta} \tag{12.323}$$

$$U(r)=-U_0a\delta(r),\ U(q)=\int U(r)\exp\{-iqr\}dr=-U_0a \tag{12.324}$$

$$F(\vec{\kappa},\vec{\kappa}')=U_0a-\frac{2\mu U_0a}{\hbar^2}\int\frac{dq}{2\pi}\frac{F(\vec{\kappa},\vec{\kappa}')}{\vec{\kappa}^2-\vec{q}^2+i\delta} \tag{12.325}$$

We observe from here that there is no dependence on $\vec{\kappa}'$ and so

$$F(\vec{\kappa}) = U_0 a - \frac{2\mu U_0 a}{\hbar^2} F(\vec{\kappa}) \int \frac{dq}{2\pi} \frac{1}{\vec{\kappa}^2 - \vec{q}^2 + i\delta} = U_0 a + \frac{2\mu U_0 a}{\hbar^2} F(\vec{\kappa}) \frac{\pi i}{2\pi\kappa} \tag{12.326}$$

From here:

$$F(\vec{\kappa}) = \frac{U_0 a}{1 - i\frac{U_0\mu a}{\kappa\hbar^2}} = \frac{U_0 a\left(1 + i\frac{U_0\mu a}{\kappa\hbar^2}\right)}{1 + \left(\frac{U_0\mu a}{\kappa\hbar^2}\right)^2} \tag{12.327}$$

$$F(\vec{\kappa})\left(1 - i\frac{U_0\mu a}{\kappa\hbar^2}\right) = U_0 a \tag{12.328}$$

$$F(\vec{\kappa})\left(1 + \left(\frac{U_0\mu a}{\kappa\hbar^2}\right)^2\right) = U_0 a\left(1 + i\frac{U_0\mu a}{\kappa\hbar^2}\right) \tag{12.329}$$

$$F(\vec{\kappa}) = \frac{U_0 a\left(\frac{\hbar^2\kappa^2}{2\mu} + \frac{i}{2}U_0 a\kappa\right)}{\frac{\hbar^2\kappa^2}{2\mu} + \frac{\mu}{2\hbar^2}(U_0 a)^2} = \frac{\alpha\left(\in + \frac{i}{2}\alpha\kappa\right)}{\in + \frac{\mu\alpha^2}{2\hbar^2}} \tag{12.330}$$

where

$$E \equiv \frac{\hbar^2\kappa^2}{2\mu}, \quad \alpha \equiv U_0 a \tag{12.331}$$

and the energy of the bound state:

$$E = -\frac{\mu\alpha^2}{2\hbar^2} \tag{12.332}$$

Also,

$$\operatorname{Im} f = \frac{\mu\alpha^2\kappa}{2\left(E + \frac{\mu\alpha^2}{2\hbar^2}\right)}, \quad \operatorname{Re} f = \frac{\alpha \in}{E + \frac{\mu\alpha^2}{2\hbar^2}} \tag{12.333}$$

12.9 Elastic and Inelastic Collisions

12.9.1 Fast and Slow Particle Total Cross Section

Slow particles

For the case of slow particles:

$$qr \ll 1 \tag{12.334}$$

and

$$f(\theta)=\frac{\mu}{2\hbar^2}\int U(r)\exp\{-i\vec{q}\vec{r}\}d\vec{r}=\lim_{q\to 0}\frac{\mu}{2\hbar^2}\int U(r)\exp\{-i\vec{q}\vec{r}\}d\vec{r}=\frac{2\mu}{\hbar^2}\int_0^{\infty}U(r)r^2dr \tag{12.335}$$

$$d\sigma=|f(\theta)|^2\,d\Omega \tag{12.336}$$

The scattering is isotropic and is independent of κ.

We apply this to the following potential:

$$U(r)=\frac{\alpha}{r^n} \tag{12.337}$$

The scattering amplitude:

$$f(\theta)=-\frac{\mu}{2\pi\hbar^2}\int d\vec{r}\,U(\vec{r})\exp\{-i\vec{q}\vec{r}\} \tag{12.338}$$

For large distances, $r>a$, and small

$$q=2\kappa\sin\frac{\theta}{2}\to 0 \tag{12.339}$$

considering the potential in Equation 12.371, the scattering amplitude diverges. The contribution of the domain $r<a$ and for small $q\to 0$ is infinitesimally small.

We evaluate the scattering amplitude, $f(\theta)$:

$$f(\theta)\cong-\frac{\mu}{2\pi\hbar^2}\int_{r>a}^{\infty}\frac{\alpha}{r^n}\exp\{-i\vec{q}\vec{r}\}d\vec{r}=-\frac{2\mu\alpha}{q\hbar^2}\int_a^{\infty}\frac{\sin qr}{r^{n-1}}dr \tag{12.340}$$

Letting $z=qr$, we have

$$f(\theta)\cong-\frac{2\mu\alpha}{q^{3-n}\hbar^2}\int_{qa}^{\infty}\frac{\sin z}{z^{n-1}}dz \tag{12.341}$$

For $n<3$ and $qa\ll 1$, we set the lower limit in Equation 12.341 to zero:

$$A_n=\int_0^{\infty}\frac{\sin z}{z^{n-1}}dz=\Gamma(2-n)\sin\frac{\pi n}{2}=-\frac{\pi\sin\frac{\pi n}{2}}{\Gamma(n-1)\sin\pi n} \tag{12.342}$$

The scattering amplitude is then

$$f(\theta)\cong-\frac{2\mu\alpha}{q^{3-n}\hbar^2}A_n \tag{12.343}$$

We consider the case of slow moving particles when $E \to 0$:

$$\sigma = \int |f(\theta)|^2 d\Omega = 2\pi\left(\frac{2\mu\alpha}{\hbar^2}A_n\right)^2 \int_0^\pi \frac{\sin\theta}{\left(4\kappa^2\sin^2\frac{\theta}{2}\right)^{3-n}} d\theta = 2\pi\left(\frac{2\mu\alpha}{\hbar^2}A_n\right)^2\left(\frac{\hbar^2}{4\mu E}\right)^{3-n} \int_{-1}^{1} \frac{dx}{(1-x)^{3-n}} \to \infty \quad (12.344)$$

Suppose in Equation 12.337 $n = 3$, it will be convenient to proceed from Equation 12.341 so that for small q, we have

$$f(\theta) \cong -\frac{2\mu\alpha}{\hbar^2}\int_{qa}^{\infty} \frac{\sin z}{z^2} dz = -\frac{2\mu\alpha}{\hbar^2}\left(\int_{qa}^{b\ll 1} \frac{dz}{z^2} + \int_b^\infty \frac{\sin z}{z^2} dz\right) \cong \frac{2\mu\alpha}{\hbar^2}\ln(qa) \quad (12.345)$$

Considering $q = 2\kappa\sin\frac{\theta}{2}$, the total scattering cross section:

$$\sigma_{\text{total}} = \int |f(\theta)|^2 d\Omega \cong \left(\frac{2\mu\alpha}{\hbar^2}\right)^2 \int \ln^2(\kappa a) \cong 4\pi\left(\frac{2\mu\alpha}{\hbar^2}\right)^2 \ln^2(\kappa a) \to \infty \quad (12.346)$$

Fast particles

For the case of fast particles:

$$qr \gg 1 \quad (12.347)$$

and the function $\exp\{-i\vec{q}\vec{r}\}$ is rapidly oscillating except at the neighbourhood of $\theta = 0$

$$\sigma_{\text{total}} = \int |f(\theta)|^2 d\Omega \approx \frac{1}{\kappa^2} \quad (12.348)$$

The transport cross section can be evaluated as follows:

$$\sigma_{\text{tr}} = \int (1-\cos\theta) d\Omega \approx \frac{1}{E^2} \quad (12.349)$$

Two-dimensional delta potential

We consider again the **Lippman Schwinger equation:**

$$F(\vec{\kappa},\vec{\kappa}') = -U(\vec{\kappa}'-\vec{\kappa}) + \frac{2\mu}{\hbar^2}\int \frac{d\vec{q}}{(2\pi)^3} \frac{U(\vec{\kappa}-\vec{q})F(\vec{\kappa},\vec{q})}{\vec{\kappa}^2 - \vec{q}^2 + i\delta} \quad (12.350)$$

$$U(\vec{r}) = -U_0 a^2 \delta(\vec{r}),\ U(q) = \int U(r)\exp\{-iqr\} dr = -U_0 a^2 \quad (12.351)$$

$$F(\vec{\kappa},\vec{\kappa}') = U_0 a^2 - \frac{2\mu U_0 a^2}{\hbar^2}\int \frac{2\pi q dq}{(2\pi)^2} \frac{F(\vec{\kappa},\vec{\kappa}')}{\vec{\kappa}^2 - \vec{q}^2 + i\delta} \quad (12.352)$$

We observe again no dependence on $\vec{\kappa}'$ and so

$$F(\vec{\kappa}) = U_0 a^2 - \frac{2\mu U_0 a^2}{\hbar^2} F(\vec{\kappa}) \int_0^\infty \frac{2\pi q dq}{(2\pi)^2} \frac{1}{\vec{\kappa}^2 - \vec{q}^2 + i\delta} \quad (12.353)$$

Consider the cut-off length, $\Lambda = \frac{1}{a^2}$:

$$\int_0^{\Lambda} \frac{dq^2}{\vec{q}^2 - \vec{\kappa}^2 - i\delta} = \int_0^{\Lambda} \frac{dx}{x - \vec{\kappa}^2 - i\delta} = \mathrm{P}\frac{1}{x - \vec{\kappa}^2} + i\pi\delta\left(x - \vec{\kappa}^2\right) = \ln\left[\vec{q}^2 - \vec{\kappa}^2 - i\delta\right]\Big|_0^{\Lambda} \tag{12.354}$$

So

$$\int_0^{\Lambda} \frac{dq^2}{\vec{q}^2 - \vec{\kappa}^2 - i\delta} = \ln\left|\frac{\Lambda - \vec{\kappa}^2}{\vec{\kappa}^2}\right| + i\pi \underset{\kappa a \ll 1}{=} \ln\frac{1}{\kappa^2 a^2} + i\pi \tag{12.355}$$

and

$$\mathrm{F}(E) = \frac{U_0 a^2}{1 + \frac{2\mu U_0 a^2}{4\pi\hbar^2}\ln\left(-\kappa^2 a^2\right)} = \frac{U_0 a^2}{\frac{\mu U_0 a^2}{2\pi\hbar^2}\left(\frac{2\pi\hbar^2}{\mu U_0 a^2} + \ln\left(-\frac{2E\mu a^2}{\hbar^2}\right)\right)} = \frac{2\pi\hbar^2}{\mu}\frac{1}{\frac{2\pi\hbar^2}{\mu U_0 a^2} + \ln\left(-\frac{2E\mu a^2}{\hbar^2}\right)} \tag{12.356}$$

The pole of this function is

$$\ln\left(-\frac{2E\mu a^2}{\hbar^2}\right) = -\frac{2\pi\hbar^2}{\mu U_0 a^2} \tag{12.357}$$

From here, the energy

$$E = -\frac{\hbar^2}{2\mu a^2}\exp\left\{-\frac{2\pi\hbar^2}{\mu U_0 a^2}\right\} \tag{12.358}$$

12.10 Wentzel-Kramer-Brillouin (WKB) Method

12.10.1 Motion in a Central Symmetric Field

We consider motion in a central symmetric field where, in addition to the potential $V(r)$, there is the centrifugal term:

$$U_{\text{centrifugal}}(r) = \frac{\hbar^2 l(l+1)}{2m r^2} \tag{12.359}$$

The solution of the Schrödinger equation for the motion in a central symmetric field has the form:

$$\Psi(r,\theta,\phi) = R(r)\Theta_l(\theta)\exp\{im\phi\} \tag{12.360}$$

where the function, $\Theta_l(\theta)$, satisfies the equation:

$$\frac{1}{\sin\theta}\frac{d}{d\theta}\left(\sin\theta\frac{d}{d\theta}\Theta_l\right) + \left[l(l+1) - \frac{m^2}{\sin^2\theta}\right]\Theta_l = 0 \tag{12.361}$$

The solution of this equation has the form:

$$\Theta(\theta) = P_l^m(\cos\theta) \tag{12.362}$$

For quasi-classical approximation $l \gg 1$, $l\theta \gg 1$ and $l(\pi - \theta) \gg 1$ is everywhere fulfilled except at the small neighborhood of $\theta = 0$ and π, and $\frac{1}{l\theta} \ll 1$ and

$$P_l(\theta) \approx \sqrt{\frac{2}{\pi l}} \frac{1}{\sin^{\frac{1}{2}}\theta} \sin\left(\left(l + \frac{1}{2}\right)\theta + \frac{\pi}{4}\right) \tag{12.363}$$

and

$$\Theta_l = i^l \left(l + \frac{1}{2}\right)^{\frac{1}{2}} P_l(\cos\theta) \approx i^l \sqrt{\frac{2}{\pi}} \frac{1}{\sin^{\frac{1}{2}}\theta} \sin\left(\left(l + \frac{1}{2}\right)\theta + \frac{\pi}{4}\right) \tag{12.364}$$

For the quasi-classical approximation where $l \gg 1$, the effective potential, $U_{\text{eff}}(r)$:

$$U_{\text{eff}}(r) = U(r) + \frac{\hbar^2}{2\mu r^2}\left(l + \frac{1}{2}\right)^{\frac{1}{2}} \tag{12.365}$$

The phase of the wave function for $r < r_0$:

$$\int_{r_0}^{r} \sqrt{\kappa^2 - \frac{2\mu}{\hbar^2} U_{\text{eff}}(r)}\, dr + \frac{\pi}{4} - \left(\int_{r_0}^{r} \sqrt{\kappa^2 - \frac{1}{r^2}\left(l + \frac{1}{2}\right)^{\frac{1}{2}}}\, dr + \frac{\pi}{4} \right) = \delta_l \tag{12.366}$$

or

$$\delta_l = -\int_{r_0}^{r} \frac{\mu}{\hbar^2} \frac{U(r)}{\sqrt{\kappa^2 - \frac{1}{r^2}\left(l + \frac{1}{2}\right)^{\frac{1}{2}}}} dr \approx \frac{\mu r_0 U(r_0)}{\kappa\hbar^2},\ r_0 = \frac{1}{\kappa}\left(l + \frac{1}{2}\right) \tag{12.367}$$

The total scattering cross section, σ_{total}:

$$\sigma_{\text{total}} = \frac{4\pi}{\kappa^2} \sum_{l=0}^{\infty} (2l + 1)\sin^2\delta_l \tag{12.368}$$

Suppose

$$U(r) = \frac{\alpha}{r^n},\ n > 1 \tag{12.369}$$

then δ_l is finite.

If $n < 1$, then δ_l is infinite. We consider the total cross section σ_{total} for $l \gg 1$. This implies that $\delta_l \ll 1$:

$$\sin^2\delta_l \approx \delta_l^2 \tag{12.370}$$

and

$$\sigma_{\text{total}} \approx \int l\delta_l^2\, dl \approx \int r_0^3 U_0^2(r)\, dr \tag{12.371}$$

If $n > 2$, then σ_{total} is finite, and if $n \leq 2$, then σ_{total} is infinite. The small angle scattering is very large. If $n = 1$, then $\mathrm{U}(r) = \dfrac{\alpha}{\mathrm{r}}$ and σ_{total} is finite.

Suppose

$$\mathrm{U}(r) = \frac{\alpha}{r}\exp\left\{-\frac{r}{\mathrm{R}}\right\} \tag{12.372}$$

then the total scattering amplitude, σ_{total}:

$$\sigma_{\text{total}} = \frac{2\pi\mu\alpha^2 R^2}{\in \hbar^2} \approx \frac{1}{E} \tag{12.373}$$

So

$$\lim_{R\to\infty} \sigma_{\text{total}} = \infty \tag{12.374}$$

We find the scattering amplitude considering

$$\mathrm{U}(r) = \frac{\alpha}{r^n} \tag{12.375}$$

$$f(\theta) = \frac{1}{2i\kappa}\sum_{l=0}^{\infty} \mathrm{P}_l(\cos\theta)(-1)^l(2l+1)\left(\exp\{2i\delta_l\} - 1\right) \tag{12.376}$$

Considering

$$\exp\{2i\delta_l\} - 1 \approx \delta_l \tag{12.377}$$

then

$$f(\theta) \approx \int_0^{\infty} U(r_0) r_0^2\, dr_0 = \int_0^{\infty} \frac{\alpha}{r_0^{n-2}}\, dr_0,\ n-2>1 \tag{12.378}$$

and if $n < 3$, the $f(\theta)$ is infinite. So the total cross section σ_{total} is infinite.

12.11 Scattering of Indistinguishable Particles

We examine the symmetrization of the wave function:

$$\psi(r,\theta) = \exp\{i\kappa z\} \pm \exp\{-i\kappa z\}\frac{1}{r}\left(f(\theta) \pm f(\pi - \theta)\right)\exp\{i\kappa r\} \tag{12.379}$$

The plus sign is for the symmetric wave function with odd intgral total spin and the minus sign for the anti-symmetric wave function with even integral total spin. The scattering cross section:

$$d\sigma = |f|^2 \, d\Omega = d\sigma_s + d\sigma_a \tag{12.380}$$

where the symmetric $d\sigma_s$ and antisymmetric $d\sigma_a$ cross sections are, respectively, as follows:

$$d\sigma_s = |f(\theta) + f(\pi - \theta)|^2 \, d\Omega, \; d\sigma_a = |f(\theta) - f(\pi - \theta)|^2 \, d\Omega \tag{12.381}$$

For the point out, there is interference described by the term:

$$\pm\left[f(\theta) f^*(\pi - \theta) + f^*(\theta) f(\pi - \theta) \right] \tag{12.382}$$

For the distinguishability in the classicla sense, there is no interfence and we have

$$d\sigma = \left[|f(\theta)|^2 + |f(\pi - \theta)|^2 \right] d\Omega \tag{12.383}$$

For half-integral spins s, the scattering cross section:

$$d\sigma = \left[|f(\theta)|^2 + |f(\pi - \theta)|^2 \right] d\Omega - \frac{1}{2s+1} \left[f(\theta) f^*(\pi - \theta) + f^*(\theta) f(\pi - \theta) \right] d\Omega \tag{12.384}$$

and for integral spins s, the scattering cross section:

$$d\sigma = \left[|f(\theta)|^2 + |f(\pi - \theta)|^2 \right] d\Omega + \frac{1}{2s+1} \left[f(\theta) f^*(\pi - \theta) + f^*(\theta) f(\pi - \theta) \right] d\Omega \tag{12.385}$$

Consider two particles with spin $s = \frac{1}{2}$, for the singlet state, $s_{\text{total}} = 0$, and for the triplet state, $s_{\text{total}} = 1$. So we have one symmentric state and three antisymmetric states such that the total cross section:

$$d\sigma_{\text{total}} = \frac{1}{4} d\sigma_s + \frac{3}{4} d\sigma_a \tag{12.386}$$

13

Polaron Theory

We examine under this heading the motion in a polar crystal of a Bloch conduction electron together with its accompanying nonradiative polarization field in the phonon cloud. This state results into a quasi-particle called the **polaron**. From Fröhlich description, the crystal is treated as a continuous medium, and in the absence of the polarization field, the conduction electron is assumed to move as a free particle with a Bloch mass, say m. According to Fröhlich, only the long wavelength longitudinal optical modes interact significantly with the electron.

Polarons may be classified as weak-, intermediates- and strong-coupling. So, this problem involves two subsystems, i.e., the electron together with its polarization and the phonon cloud. Different techniques are used to solve the polaron problem, depending on the strength of the electron-phonon coupling. The Feynman polaron have been extensively studied in our books [6, 17] applying the Feynman variational technique. This Feynman technique considers all coupling polarons.

13.1 Lee-Low-Pines (LLP) Technique

We are going to single out the weak- and intermediate-coupling polarons under this heading applying the Lee-Low-Pines (LLP) canonical transformation [24]. This is a variational method that eliminates the electron coordinate, followed by the displacement transformation of oscillator amplitudes. The latter serves to account for that part of the lattice polarization which follows the electron instantaneously. The problem of determining the self-energy and effective mass of the polaron is of well-known interest as it serves as an ideal test for some types of field theoretical methods of approximation. The LLP canonical transformation is applied not only to the one-electron (polaron) problems but also to exciton problems that we see further under this heading as well as other problems. This LLP canonical transformation method involves the exact transformation of the Hamiltonian by a variational method. This is a very rigorous method and can be realized in so many variants. We will be examining under this heading a variant that unites strong- and weak- coupling regimes.

13.1.1 Lee-Low-Pines (LLP) Bulk Polaron

We consider the bulk polaron described by the following Fröhlich Hamiltonian, $\hat{\mathrm{H}}$:

$$\hat{\mathrm{H}} = \frac{\hat{\vec{p}}^2}{2m} + \sum_{\vec{\kappa}} \hbar\omega_{\vec{\kappa}} \hat{b}^{\dagger}_{\vec{\kappa}} \hat{b}_{\vec{\kappa}} + \sum_{\vec{\kappa}} \left(V_{\vec{\kappa}} \hat{b}_{\vec{\kappa}} + V^*_{\vec{\kappa}} \hat{b}^{\dagger}_{\vec{\kappa}} \right) \tag{13.1}$$

Here, the first summand is the quantum mechanical kinetic energy operator of the electron, the second represents the polarization field energy (phonon energy) resulting from the long wavelength longitudinal optical crystal modes in the absence of interaction with the electron and the third represents the interaction energy between the electron and the phonons; $\omega_{\vec{\kappa}}$ and $V_{\vec{\kappa}}$ are, respectively, the phonon

DOI: 10.1201/9781003273073-14

frequency and the amplitude of the interaction of the electron with the phonons numbered by the wave vector $\vec{\kappa}$; $\hat{b}_{\vec{\kappa}}$ and $\hat{b}^{\dagger}_{\vec{\kappa}}$ are phonon creation and destruction operators numbered by the wave vector, $\vec{\kappa}$. Consider the following transformation of the annihilation operator, $\hat{b}$:

$$\hat{b}(\beta)=\exp\left\{i\beta\hat{b}^{\dagger}\hat{b}\right\}\hat{b}\exp\left\{-i\beta\hat{b}^{\dagger}\hat{b}\right\} \tag{13.2}$$

and

$$\frac{\partial}{\partial\beta}\hat{b}(\beta)=\exp\left\{i\beta\hat{b}^{\dagger}\hat{b}\right\}i\hat{b}\left[\hat{b}^{\dagger},\hat{b}\right]\exp\left\{-i\beta\hat{b}^{\dagger}\hat{b}\right\}=-i\exp\left\{i\beta\hat{b}^{\dagger}\hat{b}\right\}\hat{b}\exp\left\{-i\beta\hat{b}^{\dagger}\hat{b}\right\}=-i\hat{b}(\beta) \tag{13.3}$$

then

$$\hat{b}(\beta)=A\exp\{-i\beta\} \tag{13.4}$$

From

$$\hat{b}(0)=\hat{b}=A \tag{13.5}$$

then

$$\hat{b}(\beta)=\hat{b}\exp\{-i\beta\} \tag{13.6}$$

Also,

$$\tilde{\hat{b}}(f)=\exp\left\{i\left(f\hat{b}^{\dagger}+f^{*}\hat{b}\right)\right\}\hat{b}\exp\left\{-i\left(f\hat{b}^{\dagger}+f^{*}\hat{b}\right)\right\} \tag{13.7}$$

and

$$\frac{\partial}{\partial f}\tilde{\hat{b}}(f)=\exp\left\{i\left(f\hat{b}^{\dagger}+f^{*}\hat{b}\right)\right\}\left[\hat{b},\hat{b}^{\dagger}\right]\exp\left\{-i\left(f\hat{b}^{\dagger}+f^{*}\hat{b}\right)\right\}=-i \tag{13.8}$$

then

$$\tilde{\hat{b}}(f)=C-if \tag{13.9}$$

From here, it follows that

$$\tilde{\hat{b}}(0)=C=\hat{b} \tag{13.10}$$

So

$$\tilde{\hat{b}}(f)=\hat{b}-if \tag{13.11}$$

Similarly, we do for the creation operator, $\hat{b}^{\dagger}$:

$$\frac{\partial}{\partial f^{*}}\tilde{\hat{b}}^{\dagger}(f)=i \tag{13.12}$$

and

$$\tilde{\hat{b}}^{\dagger}(f) = \hat{b}^{\dagger} + if^{*} \tag{13.13}$$

We apply the LLP technique to the bulk polaron that applies to the weak localization of the electron where the wave function for the canonical transformation may be written as follows:

$$\Psi = C\psi(\vec{r})\exp\{\hat{S}_1\}\exp\{\hat{S}_2\}|0\rangle \tag{13.14}$$

Here, $|0\rangle$ is the phonon vacuum state wave function and $\psi(\vec{r})$ is the electron wave function, $\exp\{\hat{S}_1\}$, a unitary transformation that eliminates the coordinate $\vec{r}$ from the electron-phonon interaction Hamiltonian with

$$\hat{S}_1 = i\sum_{\vec{\kappa}} \hat{b}^{\dagger}_{\vec{\kappa}}\hat{b}_{\vec{\kappa}}\vec{\kappa}\vec{r} \tag{13.15}$$

The second transformation, $\exp\{\hat{S}_2\}$, displaces the amplitude of the oscillators:

$$\hat{S}_2 = \sum_{\vec{\kappa}}\left(f_{\vec{\kappa}}\hat{b}_{\vec{\kappa}} + f^{*}_{\vec{\kappa}}\hat{b}^{\dagger}_{\vec{\kappa}}\right) \tag{13.16}$$

Here, $f_{\vec{\kappa}}$ is a variational function to be found after finding the extremal value of the polaron energy.

We consider the transformation of the momentum $\hat{p}$ that yields the total conserved momentum $\hat{\vec{P}}$:

$$\exp\{\hat{S}_1\}\hat{\vec{p}}\exp\{-\hat{S}_1\} = \hat{\vec{p}} - \sum_{\vec{\kappa}}\hbar\vec{\kappa}\hat{b}^{\dagger}_{\vec{\kappa}}\hat{b}_{\vec{\kappa}} \equiv \hat{\vec{P}} \tag{13.17}$$

The eigenfunction obtained from an energy eigenvalue equation will also be the eigenfunction of the total conserved momentum $\hat{\vec{P}}$. The total Hamiltonian in Equation 13.1 after the first unitary transformation yields the following reduced polaron Hamiltonian:

$$\tilde{\hat{H}} \equiv \exp\{\hat{S}_1\}\hat{H}\exp\{-\hat{S}_1\} \tag{13.18}$$

or

$$\tilde{\hat{H}} = \frac{\hat{\vec{P}}^2}{2m} + \sum_{\vec{\kappa}}\hbar\omega_{\vec{\kappa}}\hat{b}^{\dagger}_{\vec{\kappa}}\hat{b}_{\vec{\kappa}} + \sum_{\vec{\kappa}}\left(V_{\vec{\kappa}}\hat{b}_{\vec{\kappa}} + V^{*}_{\vec{\kappa}}\hat{b}^{\dagger}_{\vec{\kappa}}\right) \tag{13.19}$$

This Hamiltonian characterizes the dependence of the energy of the free polaron upon its exactly conserved total momentum $\hat{\vec{P}}$. This is therefore directly relevant to the description of various processes in which polarons are involved.

We perform the second unitary transformation that displaces the amplitude of the oscillators:

$$\tilde{\tilde{\hat{H}}} \equiv \exp\{\hat{S}_2\}\tilde{\hat{H}}\exp\{-\hat{S}_2\} \tag{13.20}$$

or

$$\tilde{\tilde{\mathrm{H}}}=\frac{1}{2m}\left[\hat{\vec{p}}-\sum_{\vec{\kappa}}\hbar\vec{\kappa}\left(f_{\vec{\kappa}}^{*}+\hat{b}_{\vec{\kappa}}^{\dagger}\right)\left(f_{\vec{\kappa}}+\hat{b}_{\vec{\kappa}}\right)\right]^{2}+\sum_{\vec{\kappa}}\hbar\omega_{\vec{\kappa}}\left(f_{\vec{\kappa}}^{*}+\hat{b}_{\vec{\kappa}}^{\dagger}\right)\left(f_{\vec{\kappa}}+\hat{b}_{\vec{\kappa}}\right)+\sum_{\vec{\kappa}}\left(V_{\vec{\kappa}}\left(f_{\vec{\kappa}}+\hat{b}_{\vec{\kappa}}\right)+V_{\vec{\kappa}}^{*}\left(f_{\vec{\kappa}}^{*}+\hat{b}_{\vec{\kappa}}^{\dagger}\right)\right) \quad (13.21)$$

We average this total Hamiltonian by the phonon vacuum state $|0\rangle$ considering

$$\left\langle 0\left|\hat{b}_{\vec{\kappa}}\right|0\right\rangle=\left\langle 0\left|\hat{b}_{\vec{\kappa}}^{\dagger}\right|0\right\rangle=0 \quad (13.22)$$

$$\left\langle 0\left|\left[\hat{\vec{p}}-\sum_{\vec{\kappa}}\hbar\vec{\kappa}\left(f_{\vec{\kappa}}^{*}+\hat{b}_{\vec{\kappa}}^{\dagger}\right)\left(f_{\vec{\kappa}}+\hat{b}_{\vec{\kappa}}\right)\right]^{2}\right|0\right\rangle$$

$$=\left\langle 0\left|\left[\hat{\vec{p}}^{2}-2\hat{\vec{p}}\sum_{\vec{\kappa}}\hbar\vec{\kappa}\left(f_{\vec{\kappa}}^{*}+\hat{b}_{\vec{\kappa}}^{\dagger}\right)\left(f_{\vec{\kappa}}+\hat{b}_{\vec{\kappa}}\right)+\sum_{\vec{\kappa}\vec{\kappa}'}\hbar\vec{\kappa}\left(f_{\vec{\kappa}}^{*}+\hat{b}_{\vec{\kappa}}^{\dagger}\right)\left(f_{\vec{\kappa}}+\hat{b}_{\vec{\kappa}}\right)\hbar\vec{\kappa}'\left(f_{\vec{\kappa}'}^{*}+\hat{b}_{\vec{\kappa}'}^{\dagger}\right)\left(f_{\vec{\kappa}'}+\hat{b}_{\vec{\kappa}'}\right)\right]\right|0\right\rangle \quad (13.23)$$

$$\left\langle 0\left|\sum_{\vec{\kappa}}\hbar\vec{\kappa}\left(f_{\vec{\kappa}}^{*}+\hat{b}_{\vec{\kappa}}^{\dagger}\right)\left(f_{\vec{\kappa}}+\hat{b}_{\vec{\kappa}}\right)\right|0\right\rangle=\sum_{\vec{\kappa}}\hbar\vec{\kappa}\left|f_{\vec{\kappa}}\right|^{2} \quad (13.24)$$

For $\vec{\kappa}=\vec{\kappa}'$, we have

$$\left\langle 0\left|\left[\sum_{\vec{\kappa}\vec{\kappa}'}\hbar\vec{\kappa}\left(f_{\vec{\kappa}}^{*}+\hat{b}_{\vec{\kappa}}^{\dagger}\right)\left(f_{\vec{\kappa}}+\hat{b}_{\vec{\kappa}}\right)\hbar\vec{\kappa}'\left(f_{\vec{\kappa}'}^{*}+\hat{b}_{\vec{\kappa}'}^{\dagger}\right)\left(f_{\vec{\kappa}'}+\hat{b}_{\vec{\kappa}'}\right)\right]\right|0\right\rangle=\sum_{\vec{\kappa}}\hbar^{2}\vec{\kappa}^{2}\left|f_{\vec{\kappa}}\right|^{2}+\left(\sum_{\vec{\kappa}}\hbar\vec{\kappa}\left|f_{\vec{\kappa}}\right|^{2}\right)^{2} \quad (13.25)$$

Also,

$$\left\langle 0\left|\left[\sum_{\vec{\kappa}}\hbar\omega_{\vec{\kappa}}\left(f_{\vec{\kappa}}^{*}+\hat{b}_{\vec{\kappa}}^{\dagger}\right)\left(f_{\vec{\kappa}}+\hat{b}_{\vec{\kappa}}\right)+\sum_{\vec{\kappa}}\left(V_{\vec{\kappa}}\left(f_{\vec{\kappa}}+\hat{b}_{\vec{\kappa}}\right)+V_{\vec{\kappa}}^{*}\left(f_{\vec{\kappa}}^{*}+\hat{b}_{\vec{\kappa}}^{\dagger}\right)\right)\right]\right|0\right\rangle=\sum_{\vec{\kappa}}\hbar\omega_{\vec{\kappa}}\left|f_{\vec{\kappa}}\right|^{2}+\sum_{\vec{\kappa}}\left(V_{\vec{\kappa}}f_{\vec{\kappa}}+V_{\vec{\kappa}}^{*}f_{\vec{\kappa}}^{*}\right) \quad (13.26)$$

So

$$\left\langle 0\left|\tilde{\tilde{\mathrm{H}}}\right|0\right\rangle=\frac{\hat{\vec{p}}^{2}}{2m}-\frac{\hat{\vec{p}}}{m}\sum_{\vec{\kappa}}\hbar\vec{\kappa}\left|f_{\vec{\kappa}}\right|^{2}+\frac{1}{2m}\sum_{\vec{\kappa}}\hbar^{2}\vec{\kappa}^{2}\left|f_{\vec{\kappa}}\right|^{2}+\frac{1}{2m}\left(\sum_{\vec{\kappa}}\hbar\vec{\kappa}\left|f_{\vec{\kappa}}\right|^{2}\right)^{2}+\sum_{\vec{\kappa}}\hbar\omega_{\vec{\kappa}}\left|f_{\vec{\kappa}}\right|^{2}+\sum_{\vec{\kappa}}\left(V_{\vec{\kappa}}f_{\vec{\kappa}}+V_{\vec{\kappa}}^{*}f_{\vec{\kappa}}^{*}\right) \quad (13.27)$$

or

$$\left\langle 0\left|\tilde{\tilde{\mathrm{H}}}\right|0\right\rangle=\frac{\hat{\vec{p}}^{2}}{2m}+\sum_{\vec{\kappa}}\left(V_{\vec{\kappa}}f_{\vec{\kappa}}+V_{\vec{\kappa}}^{*}f_{\vec{\kappa}}^{*}\right)+\frac{\hbar^{2}}{2m}\left(\sum_{\vec{\kappa}'}{}^{\prime}\,\vec{\kappa}'\left|f_{\vec{\kappa}'}\right|^{2}\right)^{2}+\sum_{\vec{\kappa}}\left|f_{\vec{\kappa}}\right|^{2}\left(\hbar\omega_{\vec{\kappa}}-\frac{\hat{\vec{p}}\hbar\vec{\kappa}}{m}+\frac{\hbar^{2}\vec{\kappa}^{2}}{2m}\right) \quad (13.28)$$

The variational functions $f_{\vec{\kappa}}$ and $f_{\vec{\kappa}}^{*}$ can be excluded from the variational energy by taking the extremum:

$$\frac{\partial}{\partial f_{\vec{\kappa}}}\tilde{\tilde{\mathrm{H}}}=V_{\vec{\kappa}}+f_{\vec{\kappa}}^{*}\left(\hbar\omega_{\vec{\kappa}}-\frac{\vec{p}\hbar\vec{\kappa}}{m}+\frac{\hbar^{2}\vec{\kappa}^{2}}{2m}\right)+\frac{\hbar^{2}\vec{\kappa}}{m}f_{\vec{\kappa}}^{*}\left(\sum_{\vec{\kappa}'}\vec{\kappa}'\left|f_{\vec{\kappa}'}\right|^{2}\right) \quad (13.29)$$

and

$$f_{\vec{\kappa}}^{*} = -\frac{V_{\vec{\kappa}}}{\hbar\omega_{\vec{\kappa}} - \frac{\vec{p}\hbar\vec{\kappa}}{m} + \frac{\hbar^2\vec{\kappa}^2}{2m} + \frac{\hbar^2\vec{\kappa}}{m}\sum_{\vec{\kappa}'}\vec{\kappa}'\left|f_{\vec{\kappa}'}\right|^2} \quad (13.30)$$

We note in this problem that the only preferred direction is that of $\vec{p}$ and consider the slow motion of the electron where we introduce the parameter η that imitates a drag parameter:

$$\sum_{\vec{\kappa}}\hbar\vec{\kappa}\left|f_{\vec{\kappa}}\right|^2 \equiv \eta\vec{p} \quad (13.31)$$

and

$$f_{\vec{\kappa}}^{*} = -\frac{V_{\vec{\kappa}}}{\hbar\omega_{\vec{\kappa}} + \frac{\hbar^2\vec{\kappa}^2}{2m} - \frac{\vec{p}\hbar\vec{\kappa}}{m}(1-\eta)} \quad (13.32)$$

The parameter η imitates the drag parameter, since the above mechanism is very similar to the drag effect in uniform systems on a conserved polaron's total momentum.

Taylor expanding the expression for $f_{\vec{\kappa}}^{*}$ by the parameter $\vec{p}$ and limiting ourselves to the linear term in $\vec{p}$ since other terms are infinitesimally small result in:

$$f_{\vec{\kappa}}^{*} \cong -\frac{V_{\vec{\kappa}}}{\hbar\omega_{\vec{\kappa}} + \frac{\hbar^2\vec{\kappa}^2}{2m}} - \frac{V_{\vec{\kappa}}}{\left(\hbar\omega_{\vec{\kappa}} + \frac{\hbar^2\vec{\kappa}^2}{2m}\right)^2}\frac{\vec{p}\hbar\vec{\kappa}}{m}(1-\eta) \quad (13.33)$$

From here and considering Equation 13.31, we have

$$\eta\vec{p} = \sum_{\vec{\kappa}}\hbar\vec{\kappa}\left|f_{\vec{\kappa}}\right|^2 = \sum_{\vec{\kappa}}\hbar\vec{\kappa}\left|V_{\vec{\kappa}}\right|^2\left|\frac{1}{\hbar\omega_{\vec{\kappa}} + \frac{\hbar^2\vec{\kappa}^2}{2m}} + \frac{\frac{\vec{p}\hbar\vec{\kappa}}{m}(1-\eta)}{\left(\hbar\omega_{\vec{\kappa}} + \frac{\hbar^2\vec{\kappa}^2}{2m}\right)^2}\right|^2 = 2\sum_{\vec{\kappa}}\hbar\vec{\kappa}\left|V_{\vec{\kappa}}\right|^2\frac{\frac{\vec{p}\hbar\vec{\kappa}}{m}(1-\eta)}{\left(\hbar\omega_{\vec{\kappa}} + \frac{\hbar^2\vec{\kappa}^2}{2m}\right)^3} \quad (13.34)$$

This implies from here that

$$\eta = \frac{\alpha/6}{1+\alpha/6} \quad (13.35)$$

Here, α is the Fröhlich electron-phonon coupling constant that depends on the properties of the crystal. To evaluate Equation 13.28 in terms of the momentum $\vec{p}$ and considering dispersionless phonon modes with frequency ω_0, we have

$$\sum_{\vec{\kappa}}\left(V_{\vec{\kappa}}f_{\vec{\kappa}} + V_{\vec{\kappa}}^{*}f_{\vec{\kappa}}^{*}\right) = -2\sum_{\vec{\kappa}}\left|V_{\vec{\kappa}}\right|^2\left(\frac{1}{\hbar\omega_{\vec{\kappa}} + \frac{\hbar^2\vec{\kappa}^2}{2m}} + \frac{1}{\left(\hbar\omega_{\vec{\kappa}} + \frac{\hbar^2\vec{\kappa}^2}{2m}\right)^2}\frac{\vec{p}\hbar\vec{\kappa}}{m}(1-\eta)\right) = -2\alpha\hbar\omega_0 \quad (13.36)$$

where integrating over the polar angle θ, we find that

$$\sum_{\vec{\kappa}} |V_{\vec{\kappa}}|^2 \frac{\frac{\vec{p}\hbar\vec{\kappa}}{m}(1-\eta)}{\left(\hbar\omega_{\vec{\kappa}} + \frac{\hbar^2\vec{\kappa}^2}{2m}\right)^2} = 0 \tag{13.37}$$

Also,

$$\frac{\hbar^2}{2m}\left(\sum_{\vec{\kappa}} \vec{\kappa}|f_{\vec{\kappa}}|^2\right)^2 = \frac{\eta^2\vec{p}^2}{2m} \tag{13.38}$$

$$\sum_{\vec{\kappa}} |f_{\vec{\kappa}}|^2 \left(\hbar\omega_{\vec{\kappa}} - \frac{\hat{\vec{p}}\hbar\vec{\kappa}}{m} + \frac{\hbar^2\vec{\kappa}^2}{2m}\right) = \sum_{\vec{\kappa}} |f_{\vec{\kappa}}|^2 \left(\hbar\omega_{\vec{\kappa}} + \frac{\hbar^2\vec{\kappa}^2}{2m}\right) - \sum_{\vec{\kappa}} |f_{\vec{\kappa}}|^2 \frac{\hat{\vec{p}}\hbar\vec{\kappa}}{m} \tag{13.39}$$

and

$$\sum_{\vec{\kappa}} |f_{\vec{\kappa}}|^2 \left(\hbar\omega_{\vec{\kappa}} + \frac{\hbar^2\vec{\kappa}^2}{2m}\right) = \sum_{\vec{\kappa}} \frac{|V_{\vec{\kappa}}|^2}{\left(\hbar\omega_{\vec{\kappa}} + \frac{\hbar^2\vec{\kappa}^2}{2m}\right)} + 2\sum_{\vec{\kappa}} \frac{|V_{\vec{\kappa}}|^2 \frac{\vec{p}\hbar\vec{\kappa}}{m}(1-\eta)}{\left(\hbar\omega_{\vec{\kappa}} + \frac{\hbar^2\vec{\kappa}^2}{2m}\right)^2} + \sum_{\vec{\kappa}} \frac{|V_{\vec{\kappa}}|^2 \left(\frac{\vec{p}\hbar\vec{\kappa}}{m}\right)^2 (1-\eta)^2}{\left(\hbar\omega_{\vec{\kappa}} + \frac{\hbar^2\vec{\kappa}^2}{2m}\right)^3} \tag{13.40}$$

$$\frac{(1-\eta)^2}{m} \sum_{\vec{\kappa}} \frac{|V_{\vec{\kappa}}|^2 \frac{(\vec{p}\hbar\vec{\kappa})^2}{m}}{\left(\hbar\omega_{\vec{\kappa}} + \frac{\hbar^2\vec{\kappa}^2}{2m}\right)^3} = \frac{(1-\eta)}{2m} 2(1-\eta) \sum_{\vec{\kappa}} \frac{|V_{\vec{\kappa}}|^2 \frac{(\vec{p}\hbar\vec{\kappa})^2}{m}}{\left(\hbar\omega_{\vec{\kappa}} + \frac{\hbar^2\vec{\kappa}^2}{2m}\right)^3} \tag{13.41}$$

or

$$\frac{(1-\eta)^2}{m} \sum_{\vec{\kappa}} \frac{|V_{\vec{\kappa}}|^2 \frac{(\vec{p}\hbar\vec{\kappa})^2}{m}}{\left(\hbar\omega_{\vec{\kappa}} + \frac{\hbar^2\vec{\kappa}^2}{2m}\right)^3} = \frac{(1-\eta)\eta\vec{p}^2}{2m} \tag{13.42}$$

From Equation 13.34 and considering the second summand of Equation 13.40, we have

$$\eta\vec{p}^2 = 2(1-\eta) \sum_{\vec{\kappa}} \frac{|V_{\vec{\kappa}}|^2 \frac{(\vec{p}\hbar\vec{\kappa})^2}{m}}{\left(\hbar\omega_{\vec{\kappa}} + \frac{\hbar^2\vec{\kappa}^2}{2m}\right)^3} \tag{13.43}$$

$$-\sum_{\vec{\kappa}} |f_{\vec{\kappa}}|^2 \frac{\hat{\vec{p}}\hbar\vec{\kappa}}{m} = -2\frac{(1-\eta)}{m}\sum_{\vec{\kappa}} \frac{|V_{\vec{\kappa}}|^2 \frac{(\vec{p}\hbar\vec{\kappa})^2}{m}}{\left(\hbar\omega_{\vec{\kappa}} + \frac{\hbar^2\vec{\kappa}^2}{2m}\right)^3} = -\frac{\eta\vec{p}^2}{m} \tag{13.44}$$

The variational energy is then

$$E(\vec{p}) = \left\langle 0 \middle| \tilde{\tilde{\hat{H}}} \middle| 0 \right\rangle = \frac{\vec{p}^2}{2m} - 2\alpha\hbar\omega_0 + \frac{\eta^2\vec{p}^2}{2m} + \alpha\hbar\omega_0 + \frac{(1-\eta)\eta\vec{p}^2}{2m} - 2\frac{\eta\vec{p}^2}{2m} = \frac{\vec{p}^2}{2m}(1-\eta) - \alpha\hbar\omega_0 \tag{13.45}$$

or

$$E(\vec{p}) = \frac{\vec{p}^2}{2m\left(1+\alpha/6\right)} - \alpha\hbar\omega_0 \tag{13.46}$$

This is the weak- and intermediate-coupling polaron energy for the bulk with the effective polaron mass, m^*:

$$m^* = m\left(1+\alpha/6\right) \tag{13.47}$$

These results have been possible for the weak- and intermediate-coupling polarons by using a trial wave function in Equation 13.14 that is an exact eigenfunction of the total momentum vector and describes the ground state of a polaron with momentum vector, $\vec{p}$, accurately at least for small $\vec{p}$.

13.1.2 Lee-Low-Pines (LLP) Surface and Slow Moving Polaron

To construct a solution with the total angular momentum in the z-direction conserved where we use first the unitary operator, $\hat{U}_1$:

$$\hat{U}_1 = \exp\left\{\hat{S}_1\right\},\ \hat{U}_1^{-1} = \exp\left\{-\hat{S}_1\right\} \tag{13.48}$$

where

$$\hat{S}_1 = -i\vec{\rho}\left[\sum_{\vec{\eta}} \hbar\vec{\eta}\hat{b}_{\vec{\eta}}^{\dagger}\hat{b}_{\vec{\eta}} + \sum_{\vec{Q}} \hbar\vec{\eta}\hat{b}_{\vec{Q}}^{\dagger}\hat{b}_{\vec{Q}}\right],\ \rho^2 = x^2 + y^2 \tag{13.49}$$

The unitary operator, $\hat{U}_1$, transforms as follows:

$$\hat{U}_1^{-1}\hat{b}_{\vec{\eta}}\hat{U}_1 = \hat{b}_{\vec{\eta}}\exp\{-i\vec{\eta}\vec{\rho}\},\ \hat{U}_1^{-1}\hat{b}_{\vec{\eta}}^{\dagger}\hat{U}_1 = \hat{b}_{\vec{\eta}}^{\dagger}\exp\{i\vec{\eta}\vec{\rho}\} \tag{13.50}$$

$$\hat{U}_1^{-1}\hat{b}_{\vec{Q}}\hat{U}_1 = \hat{b}_{\vec{Q}}\exp\{-i\vec{\eta}\vec{\rho}\},\ \hat{U}_1^{-1}\hat{b}_{\vec{Q}}^{\dagger}\hat{U}_1 = \hat{b}_{\vec{Q}}^{\dagger}\exp\{i\vec{\eta}\vec{\rho}\} \tag{13.51}$$

This unitary transformation through the unitary operator, $\hat{U}_1$, will permit the elimination of the electronic coordinate $\vec{\rho}$ from the electron-polarization interaction. The transformed Hamiltonian after the first unitary transformation reads

$$\tilde{\hat{\mathrm{H}}} \equiv \hat{U}_1^{-1}\hat{\mathrm{H}}\hat{U}_1 \tag{13.52}$$

So

$$\begin{aligned}\tilde{\hat{\mathrm{H}}} = \frac{\hat{\vec{\mathrm{P}}}_\perp^2}{2m_\perp} + \frac{\hat{\vec{p}}_\parallel^2}{2m_\parallel} + \sum_{\vec{\eta}} \hbar\Omega_{\vec{\eta}}\hat{b}_{\vec{\eta}}^\dagger \hat{b}_{\vec{\eta}} + \sum_{\vec{Q}} \hbar\omega_{\vec{Q}}\hat{b}_{\vec{Q}}^\dagger \hat{b}_{\vec{Q}} + \sum_{\vec{\eta}}\left(V_{\vec{\eta}}g_{\vec{\eta}}(z)\hat{b}_{\vec{\eta}} + \text{c.c.}\right)\\ + \sum_{\vec{Q}}\left(V_{\vec{Q}}g_{\vec{Q}}(z)\exp\{iqz\} + \text{c.c.}\right) + V(z)\end{aligned} \tag{13.53}$$

Here, due to the translational symmetry in the xoy-plane, the total transversal polaron momentum, $\hat{\vec{\mathrm{P}}}_\perp$:

$$\hat{U}_1^{-1}\hat{\vec{p}}_\perp\hat{U}_1 \equiv \hat{\vec{\mathrm{P}}}_\perp = \hat{\vec{p}}_\perp - \sum_{\vec{\eta}} \hbar\vec{\eta}\hat{b}_{\vec{\eta}}^\dagger \hat{b}_{\vec{\eta}} - \sum_{\vec{\eta}} \hbar\vec{\eta}\hat{b}_{\vec{Q}}^\dagger \hat{b}_{\vec{Q}} \tag{13.54}$$

The transformation in Equation 13.54 is a very convenient property when we describe the system with the total momentum of $\hat{\vec{\mathrm{P}}}_\perp$. The integral of motion for this transformation should be the transversal component of the momentum, $\hat{\vec{p}}_\perp$, that for now can be assumed to be a c-number. The cost we have to pay in the transformation in Equation 13.54 is that a new interaction among bosons appears in the transformed Hamiltonian $\tilde{\hat{\mathrm{H}}}$.

Similarly, we do the second transformation, $\hat{U}_2 = \exp\{\hat{S}_2\}$, dependent on z by taking the expectation value of $\tilde{\hat{\mathrm{H}}}$. This displaces the amplitude of the oscillators and gives the total Hamiltonian that reads

$$\tilde{\tilde{\hat{\mathrm{H}}}} \equiv \hat{U}_2^{-1}\tilde{\hat{\mathrm{H}}}\hat{U}_2 \tag{13.55}$$

where

$$\hat{S}_2 = \sum_{\vec{Q}}\left(\mathrm{F}_{\vec{Q}}\hat{b}_{\vec{Q}} - \mathrm{F}_{\vec{Q}}^*\hat{b}_{\vec{\kappa}}^\dagger\right) + \sum_{\vec{\eta}}\left(f_{\vec{\eta}}\hat{b}_{\vec{\eta}} - f_{\vec{\eta}}^*\hat{b}_{\vec{\eta}}^\dagger\right) \tag{13.56}$$

and

$$\hat{U}_2^{-1}\hat{b}_{\vec{Q}}\hat{U}_2 = \hat{b}_{\vec{Q}} - \mathrm{F}_{\vec{Q}}^*,\ \hat{U}_2^{-1}\hat{b}_{\vec{Q}}^\dagger\hat{U}_2 = \hat{b}_{\vec{Q}}^\dagger - \mathrm{F}_{\vec{Q}},\ \mathrm{F}_{\vec{Q}} = \exp\{iqz\}f_{\vec{Q}},\ \mathrm{F}_{\vec{Q}}^* = \exp\{-iqz\}f_{\vec{Q}}^* \tag{13.57}$$

We find that the displacement of the oscillators is dependent on the electronic coordinate z (the approximation of a slow motion in the z-axis direction) as well as the quantum state of the electron. In the above, $\mathrm{F}_{\vec{Q}}$ and $f_{\vec{\eta}}$ (or their complex conjugates $\mathrm{F}_{\vec{Q}}^*$ and $f_{\vec{\eta}}^*$) are variational parameters. Their physical meanings are, respectively, the probability amplitudes for bulk and surface phonons for a given state.

The elimination of the z electronic coordinate is done as well by the unitary operator $\hat{U}_2$ in the following unitary transformation:

$$\mathrm{A} \equiv \hat{U}_2^{-1}\hat{\vec{\mathrm{P}}}_\perp^2\hat{U}_2 = \left[\hat{p}_\perp - \sum_{\vec{\eta}} \hbar\vec{\eta}\left(\hat{b}_{\vec{\eta}}^\dagger - f_{\vec{\eta}}\right)\left(\hat{b}_{\vec{\eta}} - f_{\vec{\eta}}^*\right) - \sum_{\vec{Q}} \hbar\vec{\eta}\left(\hat{b}_{\vec{Q}}^\dagger - \mathrm{F}_{\vec{Q}}\right)\left(\hat{b}_{\vec{Q}} - \mathrm{F}_{\vec{Q}}^*\right)\right]^2 \tag{13.58}$$

or

$$\hat{\mathrm{A}} \equiv \left[\hat{p}_\perp - \sum_{\vec{\eta}} \hbar\vec{\eta}\left(\hat{b}_{\vec{\eta}}^\dagger \hat{\mathrm{b}}_{\vec{\eta}} - \hat{b}_{\vec{\eta}}^\dagger f_{\vec{\eta}}^* + f_{\vec{\eta}}\hat{b}_{\vec{\eta}} - f_{\vec{\eta}} f_{\vec{\eta}}^*\right) - \sum_{\vec{Q}} \hbar\vec{\eta}\left(\hat{b}_{\vec{Q}}^\dagger \hat{b}_{\vec{Q}} - \hat{b}_{\vec{Q}}^\dagger \mathrm{F}_{\vec{Q}}^* + \mathrm{F}_{\vec{Q}}\hat{b}_{\vec{Q}} - \mathrm{F}_{\vec{Q}}\mathrm{F}_{\vec{Q}}^*\right)\right]^2 \tag{13.59}$$

We average the operator $\hat{\mathrm{A}}$ by the phonon vacuum state $\langle 0|$ and we have

$$\begin{aligned}\langle 0|\hat{\mathrm{A}}|0\rangle = \hat{p}_\perp^2 - 2\hat{p}_\perp\left[\sum_{\vec{\eta}} \hbar\vec{\eta}|f_{\vec{\eta}}|^2 + \sum_{\vec{Q}} \hbar\vec{\eta}|f_{\vec{Q}}|^2\right] + \sum_{\vec{Q}} (\hbar\vec{\eta})^2|f_{\vec{Q}}|^2 + \left(\sum_{\vec{Q}} \hbar\vec{\eta}|f_{\vec{Q}}|^2\right)^2 \\ + \sum_{\vec{\eta}} (\hbar\vec{\eta})^2|f_{\vec{\eta}}|^2 + \left(\sum_{\vec{\eta}} \hbar\vec{\eta}|f_{\vec{\eta}}|^2\right)^2\end{aligned} \tag{13.60}$$

Considering

$$\hat{\mathrm{B}} = \sum_{\vec{\eta}} \hbar\Omega_{\vec{\eta}}\hat{b}_{\vec{\eta}}^\dagger \hat{b}_{\vec{\eta}} + \sum_{\vec{Q}} \hbar\omega_{\vec{Q}}\hat{b}_{\vec{Q}}^\dagger \hat{b}_{\vec{Q}} \tag{13.61}$$

then

$$\langle 0|\hat{U}_2^{-1}\hat{\mathrm{B}}\hat{U}_2|0\rangle = \left\langle 0\left|\hat{U}_2^{-1}\left[\sum_{\vec{\eta}} \hbar\Omega_{\vec{\eta}}\hat{b}_{\vec{\eta}}^\dagger \hat{b}_{\vec{\eta}} + \sum_{\vec{Q}} \hbar\omega_{\vec{Q}}\hat{b}_{\vec{Q}}^\dagger \hat{b}_{\vec{Q}}\right]\hat{U}_2\right|0\right\rangle = \sum_{\vec{\eta}} \hbar\Omega_{\vec{\eta}}|f_{\vec{\eta}}|^2 + \sum_{\vec{Q}} \hbar\omega_{\vec{Q}}|f_{\vec{Q}}|^2 \tag{13.62}$$

Also, considering

$$\hat{\mathrm{M}} = \sum_{\vec{\eta}}\left(V_{\vec{\eta}}g_{\vec{\eta}}(z)\hat{b}_{\vec{\eta}} + \text{c.c.}\right) + \sum_{\vec{Q}}\left(V_{\vec{Q}}g_{\vec{Q}}(z)\exp\{iqz\} + \text{c.c.}\right) \tag{13.63}$$

then

$$\begin{aligned}\langle 0|\hat{U}_2^{-1}\hat{\mathrm{M}}\hat{U}_2|0\rangle &= \left\langle 0\left|\hat{U}_2^{-1}\left[\sum_{\vec{\eta}}\left(V_{\vec{\eta}}g_{\vec{\eta}}(z)\hat{b}_{\vec{\eta}} + \text{c.c.}\right) + \sum_{\vec{Q}}\left(V_{\vec{Q}}g_{\vec{Q}}(z)\exp\{iqz\} + \text{c.c.}\right)\right]\hat{U}_2\right|0\right\rangle \\ &= -\sum_{\vec{\eta}}\left(V_{\vec{\eta}}g_{\vec{\eta}}(z)f_{\vec{\eta}}^* + \text{c.c.}\right) - \sum_{\vec{Q}}\left(V_{\vec{Q}}g_{\vec{Q}}(z)f_{\vec{Q}}^* + \text{c.c.}\right)\end{aligned} \tag{13.64}$$

We examine now

$$\left(\hat{U}_2^{-1}\hat{\tilde{p}}_{\|}^2\hat{U}_2\right)\phi(z)=\hat{U}_2^{-1}\left(\hat{\tilde{p}}_{\|}^2\hat{U}_2\right)\phi(z)+2\hat{U}_2^{-1}\left(\hat{\tilde{p}}_{\|}\hat{U}_2\right)\left(\hat{\tilde{p}}_{\|}\phi(z)\right)+\hat{U}_2^{-1}\hat{U}_2\hat{\tilde{p}}_{\|}^2\phi(z) \tag{13.65}$$

and

$$\hat{\tilde{p}}_{\|}\hat{U}_2=-i\hbar\sum_{\vec{Q}}\left(\hat{b}_{\vec{Q}}\frac{\partial}{\partial z}\mathrm{F}_{\vec{Q}}-\hat{b}_{\vec{\kappa}}^{\dagger}\frac{\partial}{\partial z}\mathrm{F}_{\vec{Q}}^{*}\right)\hat{U}_2-i\hbar\sum_{\vec{\eta}}\left(\hat{b}_{\vec{\eta}}\frac{\partial}{\partial z}f_{\vec{\eta}}-\hat{b}_{\vec{\eta}}^{\dagger}\frac{\partial}{\partial z}f_{\vec{\eta}}^{*}\right)\hat{U}_2 \tag{13.66}$$

From here, we have the effective Hamiltonian as

$$\begin{aligned}\hat{\tilde{\tilde{\mathrm{H}}}}=&\frac{\hat{p}_{\perp}^2}{2m_{\perp}}-\frac{\hat{p}_{\perp}}{m_{\perp}}\left[\sum_{\vec{\eta}}\hbar\vec{\eta}\left|f_{\vec{\eta}}\right|^2+\sum_{\vec{Q}}\hbar\vec{\eta}\left|f_{\vec{Q}}\right|^2\right]+\frac{1}{2m_{\perp}}\sum_{\vec{Q}}(\hbar\vec{\eta})^2\left|f_{\vec{Q}}\right|^2+\frac{1}{2m_{\perp}}\left(\sum_{\vec{Q}}\hbar\vec{\eta}\left|f_{\vec{Q}}\right|^2\right)^2\\&+\frac{1}{2m_{\perp}}\sum_{\vec{\eta}}(\hbar\vec{\eta})^2\left|f_{\vec{\eta}}\right|^2+\frac{1}{2m_{\perp}}\left(\sum_{\vec{\eta}}\hbar\vec{\eta}\left|f_{\vec{\eta}}\right|^2\right)^2+\sum_{\vec{\eta}}\hbar\Omega_{\vec{\eta}}\left|f_{\vec{\eta}}\right|^2+\sum_{\vec{Q}}\hbar\omega_{\vec{Q}}\left|f_{\vec{Q}}\right|^2-\sum_{\vec{\eta}}\left(V_{\vec{\eta}}g_{\vec{\eta}}(z)f_{\vec{\eta}}^{*}+\text{c.c.}\right)\\&-\sum_{\vec{Q}}\left(V_{\vec{Q}}g_{\vec{Q}}(z)f_{\vec{Q}}^{*}+\text{c.c.}\right)+\frac{\hbar^2}{2m_{\|}}\sum_{\vec{Q}}q^2\left|f_{\vec{Q}}\right|^2+\frac{\hbar^2}{2m_{\|}}\left(\sum_{\vec{Q}}q\left|f_{\vec{Q}}\right|^2\right)^2-\frac{\hbar\hat{p}_{\|}}{m_{\|}}\sum_{\vec{Q}}q\left|f_{\vec{Q}}\right|^2+\frac{\hat{\tilde{p}}_{\|}^2}{2m_{\|}}+V(z)\end{aligned} \tag{13.67}$$

We now find the variational functions $f_{\vec{Q}}^{*}$ and $f_{\vec{Q}}$ by taking the variation of the Hamiltonian, $\hat{\tilde{\tilde{\mathrm{H}}}}$, for the saddle-point condition as

$$\begin{aligned}\frac{\partial}{\partial f_{\vec{Q}}^{*}}\hat{\tilde{\tilde{\mathrm{H}}}}=&-\frac{\hat{p}_{\perp}}{m_{\perp}}\hbar\vec{\eta}f_{\vec{Q}}-\frac{\hbar\hat{p}_{\|}}{m_{\|}}qf_{\vec{Q}}+\frac{1}{2m_{\perp}}(\hbar\vec{\eta})^2f_{\vec{Q}}+\frac{2}{2m_{\perp}}\left(\sum_{\vec{Q}'}\hbar\vec{\eta}\left|f_{\vec{Q}'}\right|^2\right)\hbar\vec{\eta}f_{\vec{Q}}+\hbar\omega_{\vec{Q}}f_{\vec{Q}}\\&-V_{\vec{Q}}g_{\vec{Q}}(z)+\frac{\hbar^2}{2m_{\|}}q^2f_{\vec{Q}}+2\frac{\hbar^2}{2m_{\|}}\left(\sum_{\vec{Q}'}q'\left|f_{\vec{Q}'}\right|^2\right)qf_{\vec{Q}}=0\end{aligned} \tag{13.68}$$

we obtain a variational solution for the phonon probability amplitude:

$$f_{\vec{Q}}=\frac{V_{\vec{Q}}g_{\vec{Q}}(z)}{\hbar\omega_{\vec{Q}}+\frac{1}{2m_{\perp}}(\hbar\vec{\eta})^2+\frac{\hbar^2}{2m_{\|}}q^2-\frac{\hat{p}_{\perp}}{m_{\perp}}\hbar\vec{\eta}-\frac{\hbar\hat{p}_{\|}}{m_{\|}}q+\frac{2}{2m_{\perp}}\left(\sum_{\vec{Q}'}\hbar\vec{\eta}\left|f_{\vec{Q}'}\right|^2\right)\hbar\vec{\eta}+2\frac{\hbar^2}{2m_{\|}}\left(\sum_{\vec{Q}'}q'\left|f_{\vec{Q}'}\right|^2\right)q} \tag{13.69}$$

Let us consider

$$m_{\perp}=m_{\|}=m,\; Q^2=q^2+\eta^2 \tag{13.70}$$

then

$$f_{\vec{Q}}=\frac{V_{\vec{Q}}g_{\vec{Q}}(z)}{\hbar\omega_{\vec{Q}}+\frac{\hbar^2}{2m}Q^2-\frac{\hat{p}_{\perp}}{m}\hbar\vec{\eta}-\frac{\hbar\hat{p}_{\|}}{m}q+\frac{1}{m}\left(\sum_{\vec{Q}}\hbar\vec{\eta}\left|f_{\vec{Q}}\right|^2\right)\hbar\vec{\eta}+\frac{\hbar^2}{m}\left(\sum_{\vec{Q}'}q'\left|f_{\vec{Q}'}\right|^2\right)q} \tag{13.71}$$

The only preferred direction is that of $\vec{p}$ and we consider the slow motion of the electron where we introduce the parameters ξ_v and ξ_s:

$$\sum_{\vec{Q}} \hbar\vec{\eta}\left|f_{\vec{Q}}\right|^2 \equiv \xi_v \vec{p}_\perp, \quad \sum_{\vec{Q}} \hbar\vec{\eta}\left|f_{\vec{\eta}}\right|^2 \equiv \xi_s \vec{p}_\perp, \tag{13.72}$$

$$\sum_{\vec{\eta}} \hbar\vec{\eta}\left(\left|f_{\vec{Q}}\right|^2 + \left|f_{\vec{\eta}}\right|^2\right) \equiv \xi_\perp \vec{p}_\perp, \quad \sum_{\vec{Q}} \hbar q\left|f_{\vec{Q}}\right|^2 \equiv \xi_\| \vec{p}_\| \tag{13.73}$$

So

$$f_{\vec{Q}} = \frac{V_{\vec{Q}} g_{\vec{Q}}(z)}{\hbar\omega_{\vec{Q}} + \frac{\hbar^2}{2m}Q^2 - \frac{\vec{p}}{m}\hbar\vec{Q} + \frac{1}{m}\xi_v \vec{p}\hbar\vec{\eta} + \frac{\hbar^2}{m}q\sum_{\vec{Q}'} q'\left|f_{\vec{Q}'}\right|^2} \tag{13.74}$$

or

$$f_{\vec{Q}} = \frac{\left\langle V_{\vec{Q}} g_{\vec{Q}}(z)\right\rangle}{\hbar\omega_{\vec{Q}} - \frac{1-\xi_\perp}{m}\hbar\vec{\eta}\vec{p}_\perp + \frac{\hbar^2}{m}\xi_\| q\vec{p}_\| + \frac{\hbar^2}{2m}Q^2}, \quad Q^2 = q^2 + \eta^2 \tag{13.75}$$

We now find the variational functions $f_{\vec{\eta}}^*$ and $f_{\vec{\eta}}$ by taking the variation of the Hamiltonian, $\tilde{\tilde{\hat{\mathrm{H}}}}$, for the saddle-point condition as

$$\frac{\partial}{\partial f_{\vec{\eta}}}\tilde{\tilde{\hat{\mathrm{H}}}} = -\frac{\hat{p}_\perp}{m_\perp}\hbar\vec{\eta}f_{\vec{\eta}}^* + +\frac{1}{2m_\perp}(\hbar\vec{\eta})^2 f_{\vec{\eta}}^* + \frac{1}{m_\perp}\left(\sum_{\vec{\eta}}(\hbar\vec{\eta})^2\left|f_{\vec{\eta}}\right|^2 + \sum_{\vec{Q}}(\hbar\vec{\eta})^2\left|f_{\vec{Q}}\right|^2\right) f_{\vec{\eta}}^* + \hbar\Omega_{\vec{\eta}} f_{\vec{\eta}}^* + V_{\vec{\eta}} g_{\vec{\eta}}(z) = 0 \tag{13.76}$$

or

$$f_{\vec{\eta}}^* = -\frac{V_{\vec{\eta}} g_{\vec{\eta}}(z)}{\hbar\Omega_{\vec{\eta}} - \frac{\hat{p}_\perp}{m_\perp}\hbar\vec{\eta} + \frac{1}{2m_\perp}(\hbar\vec{\eta})^2 + \frac{1}{m_\perp}\left(\sum_{\vec{\eta}}(\hbar\vec{\eta})^2\left|f_{\vec{\eta}}\right|^2 + \sum_{\vec{Q}}(\hbar\vec{\eta})^2\left|f_{\vec{Q}}\right|^2\right)} \tag{13.77}$$

From the above, and letting

$$V_{\vec{\eta}} g_{\vec{\eta}}(z) = V_{\vec{\eta}}', \quad V_{\vec{Q}} g_{\vec{Q}}(z) = V_{\vec{Q}}' \tag{13.78}$$

then

$$f_{\vec{\eta}}^* = -\frac{V_{\vec{\eta}} g_{\vec{\eta}}(z)}{\hbar\Omega_{\vec{\eta}} + \frac{1}{2m_\perp}(\hbar\vec{\eta})^2 - \frac{\hat{p}_\perp}{m_\perp}\hbar\vec{\eta}(1-\xi_\perp)} \tag{13.79}$$

or

$$f_{\vec{\eta}}^* \cong \frac{V_{\vec{\eta}} g_{\vec{\eta}}(z)}{\hbar\Omega_{\vec{\eta}} + \frac{1}{2m_\perp}(\hbar\vec{\eta})^2} - \frac{V_{\vec{\eta}} g_{\vec{\eta}}(z)}{\left(\hbar\Omega_{\vec{\eta}} + \frac{1}{2m_\perp}(\hbar\vec{\eta})^2\right)^2}\frac{\hat{p}_\perp}{m_\perp}\hbar\vec{\eta}(1-\xi_\perp) \tag{13.80}$$

$$f_{\vec{Q}}^* = -\frac{V_{\vec{\eta}} g_{\vec{\eta}}(z)}{\hbar\omega_{\vec{Q}} + \frac{1}{2m_\perp}(\hbar\vec{Q})^2 - \frac{\hat{p}_\perp}{m_\perp}\hbar\vec{\eta}(1-\xi_\perp)} \tag{13.81}$$

or

$$f_{\vec{Q}}^* \cong \frac{V_{\vec{\eta}} g_{\vec{\eta}}(z)}{\hbar\omega_{\vec{Q}} + \frac{1}{2m_\perp}(\hbar\vec{Q})^2} - \frac{V_{\vec{\eta}} g_{\vec{\eta}}(z)}{\left(\hbar\omega_{\vec{Q}} + \frac{1}{2m_\perp}(\hbar\vec{Q})^2\right)^2} \frac{\hat{p}_\perp}{m_\perp}\hbar\vec{\eta}(1-\xi_\perp) \tag{13.82}$$

In the above relations, we have done the Taylor series expansion over the parameter $\hat{p}_\perp$.

So the polaron energy:

$$E = \frac{\hat{p}_\perp^2}{2m_\perp}(1-\xi_\perp) - \sum_{\vec{Q}} \frac{|V_{\vec{Q}} g_{\vec{Q}}(z)|^2}{\hbar\omega_{\vec{Q}} + \frac{1}{2m_\perp}(\hbar\vec{Q})^2} - \sum_{\vec{\eta}} \frac{|V_{\vec{\eta}} g_{\vec{\eta}}(z)|^2}{\hbar\Omega_{\vec{\eta}} + \frac{1}{2m_\perp}(\hbar\vec{\eta})^2} + \langle V(z)\rangle \tag{13.83}$$

where the polaron effective mass:

$$m_\perp^* = \frac{m}{1-\xi_\perp} \tag{13.84}$$

13.1.3 Lee-Low-Pines (LLP) Surface and Fast Moving Polaron

We examine the case of fast moving polaron in the direction of the oz-axis. For this case, the displacement of the oscillators will not be defined by the coordinate of the electron but only defined by the quantum state of the electron. The elimination of the electron coordinate $\vec{\rho}$ will still be affected by the unitary operator $\hat{U}_1$. This time, we will average the result of the first unitary transformation by the wave function $\psi_\nu(z)$, where ν is the set of quantum numbers:

$$\hat{\tilde{\mathrm{H}}} \equiv \hat{U}_1^{-1}\hat{\mathrm{H}}\hat{U}_1 \tag{13.85}$$

Similarly, introduce the second transformation, $\hat{U}_2 = \exp\{\hat{S}_2\}$, dependent on z by taking the expectation value of $\hat{\tilde{\mathrm{H}}}$ after which we average by the phonon vacuum state. We calculate the ground-state energy considering that $\vec{p}_\perp = 0$ and have

$$E = \left\langle \frac{\hat{p}_\parallel^2}{2m_\parallel} \right\rangle - \sum_{\vec{Q}} \frac{|\langle V_{\vec{Q}} \langle g_{\vec{Q}}(z)\rangle\rangle|^2}{\hbar\omega_{\vec{Q}} + \frac{1}{2m}(\hbar\vec{Q})^2} - \sum_{\vec{\eta}} \frac{|\langle V_{\vec{\eta}} \langle g_{\vec{\eta}}(z)\rangle\rangle|^2}{\hbar\Omega_{\vec{\eta}} + \frac{1}{2m}(\hbar\vec{\eta})^2} + \langle V(z)\rangle \tag{13.86}$$

Here

$$\langle g_{\vec{Q}}(z)\rangle = \langle \psi_\nu(z)|g_{\vec{Q}}(z)|\psi_\nu(z)\rangle,\ \langle g_{\vec{\eta}}(z)\rangle = \langle \psi_\nu(z)|g_{\vec{\eta}}(z)|\psi_\nu(z)\rangle \tag{13.87}$$

$$\langle V(z)\rangle = \langle \psi_\nu(z)|V(z)|\psi_\nu(z)\rangle,\ \left\langle \frac{\hat{p}_\parallel^2}{2m_\parallel} \right\rangle = \left\langle \psi_\nu(z) \left| \frac{\hat{p}_\parallel^2}{2m_\parallel} \right| \psi_\nu(z) \right\rangle \tag{13.88}$$

The case of the fast oscillatory electron in the direction of the oz-axis can be achieved when the system is subjected to an electric field in the perpendicular direction to the surface. This case can also be applied even for small electron-phonon interaction.

We can have the case of strong coupling in the xoy-plane and weak coupling on the surface. In this case, instead of applying the first unitary transformation $\hat{U}_1$, we average the Hamiltonian by the wave function in the transversal direction by $\psi(\rho)$, where $\rho^2 = x^2 + y^2$. The field operators in the resultant Hamiltonian can be eliminated with help of the procedure indicated at the beginning of this chapter.

For the case of strong coupling and strong localization of the electron, we average the Hamiltonian by the wave function, $\psi(r)$, after which we apply the unitary transformation via the unitary operator, $\hat{U}_2$. We apply the example of a polaron in a quantum wire where the electron is subjected to a transversal parabolic potential.

13.2 Polaron in a Quantum Wire

We examine the case of a polaron in a quantum wire where the electron is confined in the transversal direction by a transversal parabolic potential while the motion in the longitudinal direction (i.e., oz-axis direction) is free. The Hamiltonian for the polaron is written in the form:

$$\hat{H} = \frac{\hat{\vec{P}}_\perp^2}{2m} + \frac{\hat{\vec{p}}_\parallel^2}{2m} + \frac{m\Omega^2\rho^2}{2} + \hat{H}_{ph} + \hat{H}_{e-ph} \tag{13.89}$$

Here, $\rho^2 = x^2 + y^2$ and $\vec{r} = (x, y, z)$ is the radius vector; $\omega_{\vec{q}}$ is the phonon frequency numbered by the wave vector $\vec{q}$; the first term is the operator of the kinetic energy of the electron in the transversal:

$$\frac{\hat{\vec{P}}_\perp^2}{2m} = -\frac{\hbar^2}{2m}\left(\frac{1}{\rho}\frac{\partial}{\partial\rho}\left(\rho\frac{\partial}{\partial\rho}\right) + \frac{\partial^2}{\partial z^2}\right) \tag{13.90}$$

and the second term is the operator of the kinetic energy of the electron in the longitudinal direction; the third term is the transversal parabolic potential; the fourth term is the phonon energy:

$$\hat{H}_{ph} = \sum_{\vec{q}} \hbar\omega_{\vec{q}} \hat{b}_{\vec{q}}^\dagger \hat{b}_{\vec{q}} \tag{13.91}$$

and the fifth term is the energy of interaction of the electron and the phonon:

$$\hat{H}_{e-ph} = \sum_{\vec{q}} \left(V_{\vec{q}} \exp\{i\vec{q}\vec{r}\} \hat{b}_{\vec{q}} + V_{\vec{q}}^* \exp\{-i\vec{q}\vec{r}\} \hat{b}_{\vec{q}}^\dagger\right) \tag{13.92}$$

Here, $V_{\vec{q}}$ is the amplitude of the electron-phonon coupling.

The variational wave function of the polaron is selected in the form:

$$\Psi(r) = \psi(r)\hat{U}|0\rangle \tag{13.93}$$

Here

$$\hat{U} = \exp\{\hat{S}\},\ \hat{S} = \sum_{\vec{q}} \left(f_{\vec{q}}^* \hat{b}_{\vec{q}} - f_{\vec{q}} \hat{b}_{\vec{q}}^\dagger\right)|0\rangle \tag{13.94}$$

$$\psi(r) = \mathrm{A}\exp\left\{-\frac{1}{2}(\beta_{\parallel}z)^2 - \frac{1}{2}(\beta_{\perp}\rho)^2\right\} \tag{13.95}$$

and A is the normalization constant that can be obtained from the normalization condition of the wave function:

$$\mathrm{A}^2\int_{-\infty}^{+\infty}dz\int_0^{2\pi}d\phi\int_0^{+\infty}\rho d\rho\exp\left\{-(\beta_{\parallel}z)^2 - (\beta_{\perp}\rho)^2\right\} = 1 \tag{13.96}$$

This yields

$$\mathrm{A}^2 = \frac{\beta_{\parallel}\beta_{\perp}^2}{\pi^{\frac{3}{2}}} \tag{13.97}$$

But

$$\left\langle\psi(r)\left|\frac{\mathrm{m}\Omega^2\rho^2}{2}\right|\psi(r)\right\rangle = \frac{\mathrm{m}\Omega^2}{2\beta_{\perp}^2} \tag{13.98}$$

$$\left\langle\psi(r)\left|\frac{\hat{\vec{\mathrm{P}}}_{\perp}^2}{2m}\right|\psi(r)\right\rangle = \frac{\hbar^2}{2m}\left(\frac{\beta_{\parallel}^2}{2} + \beta_{\perp}^2\right), \left\langle\psi(r)\left|\frac{\hat{p}_{\parallel}^2}{2m}\right|\psi(r)\right\rangle = -\frac{\beta_{\parallel}^2}{4m} \tag{13.99}$$

Also,

$$\left\langle\hat{\mathrm{U}}\left|\hat{b}_{\vec{q}}\right|\hat{\mathrm{U}}\right\rangle = \hat{b}_{\vec{q}} - f_{\vec{q}}^*, \left\langle\hat{\mathrm{U}}\left|\hat{b}_{\vec{q}}^{\dagger}\right|\hat{\mathrm{U}}\right\rangle = \hat{b}_{\vec{q}}^{\dagger} - f_{\vec{q}} \tag{13.100}$$

$$\left\langle 0\left|\hat{b}_{\vec{q}}^{\dagger}\hat{b}_{\vec{q}}\right|0\right\rangle = \left\langle 0\left|\hat{b}_{\vec{q}}^{\dagger}\right|0\right\rangle\left\langle 0\left|\hat{b}_{\vec{q}}\right|0\right\rangle = \left\langle 0\left|\hat{b}_{\vec{q}}^{\dagger}\right|0\right\rangle = \left\langle 0\left|\hat{b}_{\vec{q}}\right|0\right\rangle = 0 \tag{13.101}$$

$$\left\langle\hat{\mathrm{U}}\left|\hat{\mathrm{H}}_{\mathrm{ph}}\right|\hat{\mathrm{U}}\right\rangle = \sum_{\vec{q}}\hbar\omega_{\vec{q}}\left\langle\hat{\mathrm{U}}\left|\hat{b}_{\vec{q}}^{\dagger}\hat{b}_{\vec{q}}\right|\hat{\mathrm{U}}\right\rangle = \sum_{\vec{q}}\hbar\omega_{\vec{q}}\left\langle\hat{\mathrm{U}}\left|\hat{b}_{\vec{q}}^{\dagger}\right|\hat{\mathrm{U}}\right\rangle\left\langle\hat{\mathrm{U}}\left|\hat{b}_{\vec{q}}\right|\hat{\mathrm{U}}\right\rangle = \sum_{\vec{q}}\hbar\omega_{\vec{q}}\left(\hat{b}_{\vec{q}}^{\dagger}\hat{b}_{\vec{q}} - \hat{b}_{\vec{q}}^{\dagger}f_{\vec{q}}^* - f_{\vec{q}}\hat{b}_{\vec{q}} + \left|f_{\vec{q}}\right|^2\right) \tag{13.102}$$

and

$$\left\langle\Psi(r)\left|\hat{\mathrm{H}}_{\mathrm{ph}}\right|\Psi(r)\right\rangle = \sum_{\vec{q}}\hbar\omega_{\vec{q}}\left|f_{\vec{q}}\right|^2 \tag{13.103}$$

then also

$$\left\langle\hat{\mathrm{U}}\left|\hat{\mathrm{H}}_{e-\mathrm{ph}}\right|\hat{\mathrm{U}}\right\rangle = \sum_{\vec{q}}\hbar\omega_{\vec{q}}\left(V_{\vec{q}}\exp\{i\vec{q}\vec{r}\}\left(\hat{b}_{\vec{q}} - f_{\vec{q}}^*\right) + V_{\vec{q}}^*\exp\{-i\vec{q}\vec{r}\}\left(\hat{b}_{\vec{q}}^{\dagger} - f_{\vec{q}}\right)\right) \tag{13.104}$$

with

$$\left\langle \psi(r) \middle| \hat{\mathrm{H}}_{e-ph} \middle| \psi(r) \right\rangle = -\sum_{\vec{q}} \left(V_{\vec{q}} f_{\vec{q}}^{*} \left\langle \psi(r) \middle| \exp\{i\vec{q}\vec{r}\} \middle| \psi(r) \right\rangle + V_{\vec{q}}^{*} f_{\vec{q}} \left\langle \psi(r) \middle| \exp\{-i\vec{q}\vec{r}\} \middle| \psi(r) \right\rangle \right) \tag{13.105}$$

The resultant variational Hamiltonian of the phonon and electron-phonon interaction, $\tilde{\hat{\mathrm{H}}}_{e,ph}$, takes the form:

$$\widetilde{\hat{\mathrm{H}}_{e,ph}} = \sum_{\vec{q}} \hbar\omega_{\vec{q}} \left| f_{\vec{q}} \right|^2 - \sum_{\vec{q}} \left(V_{\vec{q}} f_{\vec{q}}^{*} \left\langle \psi(r) \middle| \exp\{i\vec{q}\vec{r}\} \middle| \psi(r) \right\rangle + V_{\vec{q}}^{*} f_{\vec{q}} \left\langle \psi(r) \middle| \exp\{-i\vec{q}\vec{r}\} \middle| \psi(r) \right\rangle \right) \tag{13.106}$$

The variational functions $f_{\vec{q}}$ and $f_{\vec{q}}^{*}$ can be obtained from the variational Hamiltonian, $\widetilde{\hat{\mathrm{H}}_{e,ph}}$, by taking the extremum:

$$\frac{\partial}{\partial f_{\vec{q}}} \widetilde{\hat{\mathrm{H}}_{e,ph}} = \sum_{\vec{q}} \left(\hbar\omega_{\vec{q}} f_{\vec{q}}^{*} - V_{\vec{q}}^{*} \left\langle \psi(r) \middle| \exp\{-i\vec{q}\vec{r}\} \middle| \psi(r) \right\rangle \right) = 0 \tag{13.107}$$

with

$$f_{\vec{q}}^{*} = \frac{V_{\vec{q}}^{*}}{\hbar\omega_{\vec{q}}} \left\langle \psi(r) \middle| \exp\{-i\vec{q}\vec{r}\} \middle| \psi(r) \right\rangle \tag{13.108}$$

then

$$\frac{\partial}{\partial f_{\vec{q}}^{*}} \widetilde{\hat{\mathrm{H}}_{e,ph}} = \sum_{\vec{q}} \left(\hbar\omega_{\vec{q}} f_{\vec{q}} - V_{\vec{q}} \left\langle \psi(r) \middle| \exp\{i\vec{q}\vec{r}\} \middle| \psi(r) \right\rangle \right) = 0 \tag{13.109}$$

with

$$f_{\vec{q}} = \frac{V_{\vec{q}}}{\hbar\omega_{\vec{q}}} \left\langle \psi(r) \middle| \exp\{i\vec{q}\vec{r}\} \middle| \psi(r) \right\rangle \tag{13.110}$$

So the resultant variational Hamiltonian of the phonon and electron-phonon interaction, $\tilde{\hat{\mathrm{H}}}$, now takes the form:

$$\widetilde{\hat{\mathrm{H}}_{e,ph}} = -\sum_{\vec{q}} \frac{\left| V_{\vec{q}} \right|^2}{\hbar\omega_{\vec{q}}} \left| \left\langle \psi(r) \middle| \exp\{i\vec{q}\vec{r}\} \middle| \psi(r) \right\rangle \right|^2 \tag{13.111}$$

For the strong coupling regime, we have

$$\left| V_{\vec{q}} \right|^2 = \frac{4\pi\alpha}{V} \left(\frac{\hbar\omega_{LO}}{q} \right)^2 R_p,\ R_p = \left(\frac{\hbar}{2m\omega_{LO}} \right)^{\frac{1}{2}} \tag{13.112}$$

Here, R_p is the polaron radius, ω_{LO} is the dispersionless longitudinal phonon frequency, V is the volume of the crystal and α is the dimensionless electron-phonon coupling constant.

To find the total resultant variational Hamiltonian, $\tilde{\hat{H}}$, we find first the resultant variational Hamiltonian of the phonon and electron-phonon interaction, $\widetilde{\hat{H}_{e,ph}}$, in the continuum integral:

$$\widetilde{\hat{H}_{e,ph}} = -\sum_{\vec{q}} \frac{|V_{\vec{q}}|^2}{\hbar\omega_{\vec{q}}} \left|\langle \psi(r)| \exp\{i\vec{q}\vec{r}\} |\psi(r)\rangle\right|^2 = -\frac{V}{(2\pi)^3}\int d\vec{q}\, \frac{|V_{\vec{q}}|^2}{\hbar\omega_{LO}} \left|\langle \psi(r)| \exp\{i\vec{q}\vec{r}\} |\psi(r)\rangle\right|^2 \quad (13.113)$$

$$\langle \psi(r)| \exp\{i\vec{q}\vec{r}\} |\psi(r)\rangle = A^2 \int_{-\infty}^{+\infty} dz \int_0^{2\pi} d\phi \int_0^{+\infty} \rho d\rho \exp\left\{-(\beta_{\|}z)^2 - (\beta_{\perp}\rho)^2\right\} \exp\left\{i\left(q_{\|}z + q_{\perp}\rho\cos\phi\right)\right\} \quad (13.114)$$

$$\int_0^{2\pi} d\phi \exp\{iq_{\perp}\rho\cos\phi\} = 2\pi J_0(q_{\perp}\rho) \quad (13.115)$$

Here, $J_0(x)$ is the Bessel function of zero order and argument.

Also,

$$\int_0^{+\infty} d\rho\, \rho J_0(q_{\perp}\rho) \exp\left\{-(\beta_{\perp}\rho)^2\right\} = \frac{1}{2\beta_{\perp}^2} \exp\left\{-\frac{q_{\perp}^2}{4\beta_{\perp}^2}\right\} \quad (13.116)$$

$$\int_{-\infty}^{+\infty} dz \exp\left\{-(\beta_{\|}z)^2 + iq_{\|}z\right\} = \frac{\sqrt{\pi}}{\beta_{\|}} \exp\left\{-\frac{q_{\|}^2}{4\beta_{\|}^2}\right\} \quad (13.117)$$

So

$$\widetilde{\hat{H}_{e,ph}} = -\frac{4\pi\alpha}{(2\pi)^3} \hbar\omega_{LO} R_p \int_0^{2\pi} d\phi \int_0^{\pi} d\theta \sin\theta \int_0^{+\infty} dq\, q^2 \frac{1}{q^2} \exp\left\{-q^2\left(\frac{\sin^2\theta}{2\beta_{\perp}^2} + \frac{\cos^2\theta}{2\beta_{\|}^2}\right)\right\} \quad (13.118)$$

We do a change of variables to facilitate the computation of the integrals:

$$k = R_p q,\ \beta_{\|}R_p = b_{\|},\ \beta_{\perp}R_p = b_{\perp},\ \cos\theta = x \quad (13.119)$$

So

$$\widetilde{\hat{H}_{e,ph}} = -\frac{\alpha\hbar\omega_{LO}}{\pi} \int_{-1}^{1} dx \int_0^{+\infty} dk \exp\left\{-k^2\left(\frac{1-x^2}{2\beta_{\perp}^2} + \frac{x^2}{2\beta_{\|}^2}\right)\right\} = -\frac{\alpha\hbar\omega_{LO} b_{\perp}}{\sqrt{2\pi}} \int_{-1}^{1} \frac{dx}{\sqrt{1 + \left(\frac{b_{\perp}^2}{b_{\|}^2} - 1\right)x^2}} \quad (13.120)$$

or

$$\widetilde{\hat{H}_{e,ph}} - \frac{\alpha\hbar\omega_{LO} b_{\perp}}{\sqrt{2\pi}} \int_{-1}^{1} \frac{dx}{\sqrt{1 + \left(\frac{b_{\perp}^2}{b_{\|}^2} - 1\right)x^2}} \quad (13.121)$$

or

$$\widetilde{\hat{H}_{e,ph}} \equiv -\alpha\hbar\omega_{LO}F(b_\perp, b_\parallel) \tag{13.122}$$

Here

$$F(b_\perp, b_\parallel) = \sqrt{\frac{2}{\pi}} b_\perp \begin{cases} \dfrac{\sinh^{-1}\sqrt{\dfrac{b_\perp^2}{b_\parallel^2}-1}}{\sqrt{\dfrac{b_\perp^2}{b_\parallel^2}-1}} & b_\perp > b_\parallel \\ \dfrac{\sin^{-1}\sqrt{1-\dfrac{b_\perp^2}{b_\parallel^2}}}{\sqrt{1-\dfrac{b_\perp^2}{b_\parallel^2}}} & b_\perp < b_\parallel \end{cases} \tag{13.123}$$

The total resultant variational Hamiltonian $\tilde{\hat{H}}$ is obtained:

$$\tilde{\hat{H}} = \hbar\omega_{LO} F(b_\perp, b_\parallel, \Omega) - \alpha\hbar\omega_{LO} F(b_\perp, b_\parallel) \tag{13.124}$$

Here, Ω is dimensionless and is in units of the dispersionless longitudinal phonon frequency ω_{LO} and

$$F(b_\perp, b_\parallel, \Omega) = \left(\frac{b_\parallel^2}{2} + b_\perp^2\right) + \frac{\Omega^2}{4b_\perp^2} - \frac{b_\parallel^2}{2} \tag{13.125}$$

13.3 Polaronic Exciton and Haken Exciton

We write the Haken polaronic exciton Hamiltonian [25, 26]:

$$\hat{H} = \frac{\hat{P}_1^2}{2m_1} + \frac{\hat{P}_2^2}{2m_2} + W(r_{12}) + \hbar\omega\sum_{\vec{q}} \hat{b}_{\vec{q}}^\dagger \hat{b}_{\vec{q}}$$
$$- i\hbar\omega\left(\frac{4\pi\alpha}{uV}\right)^{\frac{1}{2}} \sum_{\vec{q}} \frac{1}{q}\left(\hat{b}_{\vec{q}}\left[\exp\{i\vec{q}\vec{r}_1\} - \exp\{i\vec{q}\vec{r}_2\}\right] - \hat{b}_{\vec{q}}^\dagger\left[\exp\{-i\vec{q}\vec{r}_1\} - \exp\{-i\vec{q}\vec{r}_2\}\right]\right) \tag{13.126}$$

and the eigenvalue and eigenfunction equation:

$$\hat{H}\Psi = E\Psi \tag{13.127}$$

We introduce the coordinate of relative motion and that of the center of mass:

$$\vec{r} = \vec{r}_1 - \vec{r}_2,\ \vec{R} = \mu_1\vec{r}_1 + \mu_2\vec{r}_2,\ M = m_1 + m_2,\ \frac{1}{M'} = \frac{1}{m_1} + \frac{1}{m_2},\ \mu_i = \frac{m_i}{M} \tag{13.128}$$

$$V_{\vec{q}}(\vec{r}) = -i\hbar\omega\left(\frac{4\pi\alpha}{uV}\right)^{\frac{1}{2}}\frac{1}{q}\left(\exp\{i\vec{q}\mu_2\vec{r}\} - \exp\{-i\vec{q}\mu_1\vec{r}\}\right) \quad (13.129)$$

$$u = \left(\frac{2M\omega}{\hbar}\right)^{\frac{1}{2}}, \alpha = \frac{e^2 u}{2\hbar\omega}\left(\frac{1}{\varepsilon_\infty} - \frac{1}{\varepsilon_0}\right) \quad (13.130)$$

The wave function of the polaronic exciton has the form:

$$\Psi = \exp\{i\vec{K}\vec{R}\}\psi(\vec{r})\exp\left\{\sum_{\vec{q}}\left[\hat{b}_{\vec{q}}\mathrm{F}_{\vec{q}}^{*} - \hat{b}_{\vec{q}}^{\dagger}\mathrm{F}_{\vec{q}}\right]\right\}|0\rangle \quad (13.131)$$

Here

$$\mathrm{F}_{\vec{q}} = \exp\{-iq\vec{R}\}f_{\vec{q}}, \mathrm{F}_{\vec{q}}^{*} = \exp\{iq\vec{R}\}f_{\vec{q}}^{*} \quad (13.132)$$

This wave function is assumed normalized. The exciton function in a deformed lattice is selected in this form because the electron, due to its motion, behaves like a polarization wave and displaces ions from their equilibrium position. So in this case, the wave function has this form due to LLP [24]:

$$\chi = \exp\left\{\sum_{\vec{q}}\left[\hat{b}_{\vec{q}}\hat{b}_{\vec{q}}^{0} - \hat{b}_{\vec{q}}\hat{b}_{\vec{q}}^{\dagger}\right]\right\}|0\rangle \quad (13.133)$$

Here, the operator, $\hat{b}_{\vec{q}}^{0}$, describes the new equilibrium position and may be defined by a variational method. In the case of the exciton, there is displacement of ions from their equilibrium position in addition to the Coulomb interaction of the electron and hole. For this Coulomb interaction, there should be a correlation between the position of the electron and hole that should now be $\hat{b}_{\vec{q}}^{0}(\vec{r}_e - \vec{r}_h)$ which is dependent on the positions of the electron $\vec{r}_e$ and hole $\vec{r}_h$. For this reason, we introduce the following variational function:

$$\Psi = \exp\{i\vec{K}\vec{R}\}\psi(\vec{r})\exp\left\{\sum_{\vec{q}}\left[\hat{b}_{\vec{q}}\mathrm{F}_{\vec{q}}^{*} - \hat{b}_{\vec{q}}^{\dagger}\mathrm{F}_{\vec{q}}\right]\right\}|0\rangle \quad (13.134)$$

Here, $f_{\vec{q}}$ is a variational function independent of r.

We find the eigenvalue:

$$\int d\vec{r}_1\, d\vec{r}_2 \left\langle \Psi \middle| \hat{\mathrm{H}} \middle| \Psi \right\rangle = E \quad (13.135)$$

Here

$$\hat{\mathrm{H}} = -\frac{\hbar^2\Delta_{\vec{R}}}{2M} - \frac{\hbar^2\Delta_{\vec{r}}}{2M'} + W(r) + \hbar\omega\sum_{\vec{q}}\hat{b}_{\vec{q}}^{\dagger}\hat{b}_{\vec{q}} + \sum_{\vec{q}}\left(\hat{b}_{\vec{q}}\mathrm{M}_{\vec{q}} + \hat{b}_{\vec{q}}^{\dagger}\mathrm{M}_{\vec{q}}^{*}\right) \quad (13.136)$$

$$\mathrm{M}_{\vec{q}} = \exp\left\{iq\vec{R}\right\} V_{\vec{q}}(\vec{r}),\ \mathrm{M}_{\vec{q}}^{*} = \exp\left\{-iq\vec{R}\right\} V_{\vec{q}}^{*}(\vec{r}) \tag{13.137}$$

To facilitate the evaluation of the energy, E, we find first the following:

$$\nabla_{\vec{R}}\Psi = i\vec{\mathrm{K}}\exp\left\{i\vec{\mathrm{K}}\vec{R}\right\}\psi(\vec{r})\mathrm{F} + \exp\left\{i\vec{\mathrm{K}}\vec{R}\right\}\psi(\vec{r})\sum_{\vec{q}}\mathrm{M}_{\vec{q}}^{+} i\vec{q}\mathrm{F} \tag{13.138}$$

$$\begin{aligned}\Delta_{\vec{R}}\Psi = &-\vec{\mathrm{K}}^{2}\exp\left\{i\vec{\mathrm{K}}\vec{R}\right\}\psi(\vec{r})\mathrm{F} - 2\vec{\mathrm{K}}\exp\left\{i\vec{\mathrm{K}}\vec{R}\right\}\psi(\vec{r})\sum_{\vec{q}}\mathrm{M}_{\vec{q}}^{+}\vec{q}\,\mathrm{F} - \exp\left\{i\vec{\mathrm{K}}\vec{R}\right\}\psi(\vec{r})\sum_{\vec{q}}\mathrm{M}_{\vec{q}}^{-}\vec{q}^{2}\,\mathrm{F}\\ &-\exp\left\{i\vec{\mathrm{K}}\vec{R}\right\}\psi(\vec{r})\sum_{\vec{q}\vec{q}'}\mathrm{M}_{\vec{q}}^{+}\mathrm{M}_{\vec{q}'}^{+}\vec{q}'\vec{q}\,\mathrm{F}\end{aligned} \tag{13.139}$$

Here

$$\mathrm{M}_{\vec{q}}^{\pm} = \hat{b}_{\vec{q}}\mathrm{F}_{\vec{q}}^{*} \pm \hat{b}_{\vec{q}}^{\dagger}\mathrm{F}_{\vec{q}},\ \mathrm{F} = \exp\left\{\sum_{\vec{q}}\mathrm{M}_{\vec{q}}^{-}\right\}|0\rangle \tag{13.140}$$

Considering

$$\exp\left\{\hat{a}+\hat{b}\right\} = \exp\left\{\hat{a}\right\}\exp\left\{\hat{b}\right\}\exp\left\{-\frac{1}{2}\left[\hat{a},\hat{b}\right]\right\} \tag{13.141}$$

then

$$\mathrm{F} = \exp\left\{\sum_{\vec{q}}\mathrm{M}_{\vec{q}}^{-}\right\}|0\rangle = \exp\left\{-\sum_{\vec{q}}\hat{b}_{\vec{q}}^{\dagger}\mathrm{F}_{\vec{q}}\right\}\exp\left\{-\frac{1}{2}\sum_{\vec{q}}\left|f_{\vec{q}}\right|^{2}\right\}|0\rangle \tag{13.142}$$

Also,

$$\Delta_{\vec{r}}(\psi\mathrm{F}) = \mathrm{F}\Delta_{\vec{r}}\psi + 2\nabla_{\vec{r}}\psi\nabla_{\vec{r}}\mathrm{F} + \psi\Delta_{\vec{r}}\mathrm{F} \tag{13.143}$$

Considering

$$\left\langle \mathrm{F}^{\dagger}\middle|\mathrm{F}\right\rangle = 1 \tag{13.144}$$

and multiplying Equation 13.143 from the left-hand side by $\psi^{*}\mathrm{F}^{\dagger}$, we have

$$\int d\vec{r}\left\langle\psi^{*}\mathrm{F}^{\dagger}\Delta_{\vec{r}}\psi\mathrm{F}\right\rangle = \int d\vec{r}\psi^{*}\Delta_{\vec{r}}\psi + 2\int d\vec{r}\left\langle\psi^{*}\mathrm{F}^{\dagger}\nabla_{\vec{r}}\psi\nabla_{\vec{r}}\mathrm{F}\right\rangle + \int d\vec{r}\left\langle\psi^{*}\mathrm{F}^{\dagger}\psi\Delta_{\vec{r}}\mathrm{F}\right\rangle \tag{13.145}$$

For further calculations, we consider that

$$|\psi\rangle = \exp\left\{-\frac{x^{2}}{2}\right\}\exp\left\{x\hat{a}^{\dagger}\right\}|0\rangle,\ \exp\left\{x\hat{a}\right\}|0\rangle = |0\rangle \tag{13.146}$$

and

$$\exp\{x\hat{a}\}|0\rangle = |0\rangle \tag{13.147}$$

where x, in our case, is

$$x = -F_{\bar{q}} \tag{13.148}$$

Also,

$$\langle\hat{a}\rangle = \langle\psi|\hat{a}|\psi\rangle = \exp\{-x^2\}\langle 0|\exp\{x\hat{a}\}\,\hat{a}\exp\{x\hat{a}^\dagger\}|0\rangle \tag{13.149}$$

$$\left[\hat{a}, \exp\{x\hat{a}^\dagger\}\right] = \frac{\partial}{\partial\hat{a}^\dagger}\exp\{x\hat{a}^\dagger\} = x\exp\{x\hat{a}^\dagger\} \tag{13.150}$$

$$\hat{a}\exp\{x\hat{a}^\dagger\} = \exp\{x\hat{a}^\dagger\}\hat{a} + x\exp\{x\hat{a}^\dagger\} \tag{13.151}$$

Since

$$\hat{a}|0\rangle = 0 \tag{13.152}$$

then

$$\exp\{-x^2\}\langle 0|\exp\{x\hat{a}\}\hat{a}\exp\{x\hat{a}^\dagger\}|0\rangle x = x\exp\{-x^2\}\exp\{x^2\}\langle 0|\exp\{x\hat{a}^\dagger\}\exp\{x\hat{a}\}|0\rangle = x \tag{13.153}$$

Since

$$\langle 0|\exp\{x\hat{a}^\dagger\} = \langle 0| \tag{13.154}$$

$$\exp\{x\hat{a}\}|0\rangle = 0 \tag{13.155}$$

$$\exp\{x\hat{a}\}\exp\{x\hat{a}^\dagger\} = \exp\{x\hat{a}\}\exp\{x\hat{a}^\dagger\}\exp\left\{-\frac{1}{2}\left[x\hat{a}, x\hat{a}^\dagger\right]\right\}\exp\left\{-\frac{x^2}{2}\right\} \tag{13.156}$$

or

$$\exp\{x\hat{a}\}\exp\{x\hat{a}^\dagger\} = \exp\{x\hat{a}^\dagger\}\exp\{x\hat{a}\}\exp\left\{\frac{1}{2}\left[x\hat{a}, x\hat{a}^\dagger\right]\right\}\exp\left\{\frac{x^2}{2}\right\} \tag{13.157}$$

or

$$\exp\{x\hat{a}\}\exp\{x\hat{a}^\dagger\} = \exp\{x^2\}\exp\{x\hat{a}^\dagger\}\exp\{x\hat{a}\} \tag{13.158}$$

$$\langle\hat{a}^\dagger\rangle = \langle\psi|\hat{a}|\psi\rangle^* = \langle\psi|\hat{a}^\dagger|\psi\rangle = x \tag{13.159}$$

But we have

$$x = -F_{\vec{q}}, x^* = -F_{\vec{q}}^* \tag{13.160}$$

Similarly,

$$\left\langle F^\dagger \hat{b}_{\vec{q}}^\dagger F \right\rangle = -F_{\vec{q}}^* \tag{13.161}$$

$$\left\langle F^\dagger \hat{b}_{\vec{q}}^\dagger \hat{b}_{\vec{q}'}^\dagger F \right\rangle_{\vec{q} \neq \vec{q}'} = F_{\vec{q}}^* F_{\vec{q}'}^* \tag{13.162}$$

$$\left\langle \int d\vec{r} \psi^* F^\dagger \nabla \psi \nabla F \right\rangle = \left\langle \int \psi^* F^\dagger \nabla \psi \left(-\sum_{\vec{q}} \left(\nabla F_{\vec{q}} \hat{b}_{\vec{q}}^\dagger \right) - \frac{1}{2} \sum_{\vec{q}} \nabla \left| f_{\vec{q}} \right|^2 \right) F d\vec{r} \right\rangle \tag{13.163}$$

or

$$\left\langle \int d\vec{r} \psi^* F^\dagger \nabla \psi \nabla F \right\rangle = \int d\vec{r} \sum_{\vec{q}} \left(\nabla f_{\vec{q}} \right) f_{\vec{q}}^* \nabla \psi - \frac{1}{2} \int d\vec{r} \sum_{\vec{q}} \nabla \left| f_{\vec{q}} \right|^2 \psi^* \nabla \psi \tag{13.164}$$

or

$$\left\langle \int d\vec{r} \psi^* F^\dagger \nabla \psi \nabla F \right\rangle = \frac{1}{2} \int d\vec{r} \psi^* \nabla \psi \sum_{\vec{q}} \left(\nabla f_{\vec{q}} f_{\vec{q}}^* - f_{\vec{q}} \nabla f_{\vec{q}}^* \right) \tag{13.165}$$

Also,

$$\Delta_{\vec{r}} F = \left(-\sum_{\vec{q}} \hat{b}_{\vec{q}}^\dagger \Delta_{\vec{r}} F_{\vec{q}} - \frac{1}{2} \sum_{\vec{q}} \Delta_{\vec{r}} \left| f_{\vec{q}} \right|^2 \right) F + \sum_{\vec{q}\vec{q}'} \left[\hat{b}_{\vec{q}}^\dagger \nabla F_{\vec{q}} + \frac{1}{2} \nabla \left| f_{\vec{q}} \right|^2 \right] \left[\hat{b}_{\vec{q}'}^\dagger \nabla F_{\vec{q}'} + \frac{1}{2} \nabla \left| f_{\vec{q}'} \right|^2 \right] F \tag{13.166}$$

$$\Delta_{\vec{r}} \psi F = F \Delta_{\vec{r}} \psi + 2\nabla \psi \nabla F + \psi \Delta_{\vec{r}} F \tag{13.167}$$

We now find $\Delta_{\vec{R}}$:

$$\nabla_{\vec{R}} \Psi = i\vec{K} \exp\left\{ i\vec{K}\vec{R} \right\} \psi(\vec{r}) F + \exp\left\{ i\vec{K}\vec{R} \right\} \psi(\vec{r}) i \sum_{\vec{q}} \vec{q} M_{\vec{q}}^\pm F \tag{13.168}$$

$$\Delta_{\vec{R}} \Psi = -\vec{K}^2 \exp\left\{ i\vec{K}\vec{R} \right\} \psi(\vec{r}) F - 2\vec{K} \exp\left\{ i\vec{K}\vec{R} \right\} \psi(\vec{r}) \sum_{\vec{q}} \vec{q} M_{\vec{q}}^\pm F - \exp\left\{ i\vec{K}\vec{R} \right\} \psi(\vec{r}) \sum_{\vec{q}} \vec{q}^2 M_{\vec{q}}^- F$$

$$- \vec{K} \exp\left\{ i\vec{K}\vec{R} \right\} \psi(\vec{r}) \sum_{\vec{q}\vec{q}'} \vec{q} M_{\vec{q}}^\pm \vec{q}' M_{\vec{q}'}^+ F \tag{13.169}$$

So

$$\left\langle F^\dagger \Delta_{\vec{r}} F \right\rangle = \frac{1}{2} \sum_{\vec{q}} \left[f_{\vec{q}}^* \Delta_{\vec{r}} f_{\vec{q}} - f_{\vec{q}} \Delta_{\vec{r}} f_{\vec{q}}^* - 2\nabla_{\vec{r}} f_{\vec{q}}^* \nabla_{\vec{r}} f_{\vec{q}} \right] - \frac{1}{2} \sum_{\vec{q}\vec{q}'} f_{\vec{q}} f_{\vec{q}'}^* \nabla_{\vec{r}} f_{\vec{q}}^* \nabla_{\vec{r}} f_{\vec{q}'}$$

$$+ \frac{1}{4} \sum_{\vec{q}} \left[f_{\vec{q}}^* f_{\vec{q}'}^* \nabla_{\vec{r}} f_{\vec{q}'} \nabla_{\vec{r}} f_{\vec{q}} + f_{\vec{q}} f_{\vec{q}'} \nabla_{\vec{r}} f_{\vec{q}'}^* \nabla_{\vec{r}} f_{\vec{q}}^* \right] \tag{13.170}$$

$$\frac{1}{N}\sum_{\vec{R}}\int d\vec{r}\langle\Psi|\hat{\mathrm{H}}|\Psi\rangle=\int d\vec{r}\psi^*(\vec{r})\,\psi(\vec{r})\left\{\frac{\hbar^2\vec{\mathrm{K}}^2}{2M}-\frac{\hbar^2}{M}\sum_{\vec{q}}\vec{q}\vec{\mathrm{K}}|f_{\vec{q}}|^2+\frac{\hbar^2}{2M}\sum_{\vec{q}}\vec{q}^2|f_{\vec{q}}|^2+\frac{\hbar^2}{2M}\sum_{\vec{q}\vec{q}'}|f_{\vec{q}}|^2|f_{\vec{q}'}|^2\vec{q}\vec{q}'\right\}$$

$$-\frac{\hbar^2}{2M'}\int d\vec{r}\psi^*(\vec{r})\,\Delta_{\vec{r}}\psi(\vec{r})+\frac{\hbar^2}{2M'}2\int d\vec{r}\psi^*(\vec{r})\,\nabla_{\vec{r}}\psi(\vec{r})\frac{1}{2}\sum_{\vec{q}}\left[f_{\vec{q}}\nabla_{\vec{r}}f_{\vec{q}}^*-f_{\vec{q}}^*\nabla_{\vec{r}}f_{\vec{q}}\right]-\frac{\hbar^2}{2M'}\int d\vec{r}\psi^*(\vec{r})\,\psi(\vec{r})$$

$$\left\{\frac{1}{2}\sum_{\vec{q}}\left[f_{\vec{q}}^*\Delta_{\vec{r}}f_{\vec{q}}-f_{\vec{q}}\Delta_{\vec{r}}f_{\vec{q}}^*-2\nabla_{\vec{r}}f_{\vec{q}}^*\nabla_{\vec{r}}f_{\vec{q}}\right]-\frac{1}{2}\sum_{\vec{q}\vec{q}'}f_{\vec{q}}f_{\vec{q}'}^*\nabla_{\vec{r}}f_{\vec{q}}^*\nabla_{\vec{r}}f_{\vec{q}'}+\frac{1}{4}\sum_{\vec{q}}\left[f_{\vec{q}}^*f_{\vec{q}'}^*\nabla_{\vec{r}}f_{\vec{q}'}\nabla_{\vec{r}}f_{\vec{q}}+f_{\vec{q}}f_{\vec{q}'}\nabla_{\vec{r}}f_{\vec{q}'}^*\nabla_{\vec{r}}f_{\vec{q}}^*\right]\right\}$$

$$+\int d\vec{r}\psi^*(\vec{r})W(\vec{r})\,\psi(\vec{r})+\hbar\omega\int d\vec{r}\psi^*(\vec{r})\,\psi(\vec{r})\sum_{\vec{q}}|f_{\vec{q}}|^2+\int d\vec{r}\psi^*(\vec{r})\,\psi(\vec{r})\sum_{\vec{q}}\left[f_{\vec{q}}V_{\vec{q}}(\vec{r})+f_{\vec{q}}^*V_{\vec{q}}^*(\vec{r})\right]$$

We write this equation in another form except the last term:

$$\frac{1}{N}\sum_{\vec{R}}\int d\vec{r}\langle\Psi|\hat{\mathrm{H}}|\Psi\rangle=\int d\vec{r}\psi^*(\vec{r})\,\psi(\vec{r})\left\{\frac{\hbar^2\vec{\mathrm{K}}^2}{2M}+\frac{\hbar^2}{2M}\sum_{\vec{q}\vec{q}'}|f_{\vec{q}}|^2|f_{\vec{q}'}|^2\vec{q}\vec{q}'+\sum_{\vec{q}}|f_{\vec{q}}|^2\left(\frac{\hbar^2\vec{q}^2}{2M}-\frac{\hbar^2\vec{\mathrm{K}}\vec{q}}{M}\right)\right\}$$

$$-\frac{\hbar^2}{2M'}\left[\int d\vec{r}\psi^*(\vec{r})\,\Delta_{\vec{r}}\psi(\vec{r})+2\int d\vec{r}\psi^*(\vec{r})\,\nabla_{\vec{r}}\psi(\vec{r})\frac{1}{2}\sum_{\vec{q}}\left[f_{\vec{q}}\nabla_{\vec{r}}f_{\vec{q}}^*-f_{\vec{q}}^*\nabla_{\vec{r}}f_{\vec{q}}\right]-\int d\vec{r}\psi^*(\vec{r})\,\psi(\vec{r})\right.$$

$$\left.\left\{\frac{1}{2}\sum_{\vec{q}}\left[f_{\vec{q}}^*\Delta_{\vec{r}}f_{\vec{q}}-f_{\vec{q}}\Delta_{\vec{r}}f_{\vec{q}}^*-2\nabla_{\vec{r}}f_{\vec{q}}^*\nabla_{\vec{r}}f_{\vec{q}}\right]-\frac{1}{2}\sum_{\vec{q}\vec{q}'}f_{\vec{q}}f_{\vec{q}'}^*\nabla_{\vec{r}}f_{\vec{q}}^*\nabla_{\vec{r}}f_{\vec{q}'}+\frac{1}{4}\sum_{\vec{q}}\left[f_{\vec{q}}^*f_{\vec{q}'}^*\nabla_{\vec{r}}f_{\vec{q}'}\nabla_{\vec{r}}f_{\vec{q}}+f_{\vec{q}}f_{\vec{q}'}\nabla_{\vec{r}}f_{\vec{q}'}^*\nabla_{\vec{r}}f_{\vec{q}}^*\right]\right\}\right]$$

$$+\int d\vec{r}\psi^*(\vec{r})W(\vec{r})\,\psi(\vec{r})+\hbar\omega\int d\vec{r}\psi^*(\vec{r})\,\psi(\vec{r})\sum_{\vec{q}}|f_{\vec{q}}|^2+\int d\vec{r}\psi^*(\vec{r})\,\psi(\vec{r})\sum_{\vec{q}}\left[f_{\vec{q}}V_{\vec{q}}(\vec{r})+f_{\vec{q}}^*V_{\vec{q}}^*(\vec{r})\right] \quad (13.171)$$

This equation describes the self-consistent motion of the electron and the hole in the given crystal lattice.

If we let $\vec{\mathrm{K}}=0$, then we have 1*s* function of the relative motion and so $f_{\vec{q}}$ will be of the 1*s* type. So $\sum_{\vec{q}}\left[f_{\vec{q}}\nabla_{\vec{r}}f_{\vec{q}}^*-f_{\vec{q}}^*\nabla_{\vec{r}}f_{\vec{q}}\right]$ and $\sum_{\vec{q}}|f_{\vec{q}}|^2\vec{q}$ will disappear.

Hence,

$$\overline{\mathrm{H}}=\int d\vec{r}\psi^*(\vec{r})\,\psi(\vec{r})\left\{\frac{\hbar^2}{2M}\sum_{\vec{q}\vec{q}'}|f_{\vec{q}}|^2|f_{\vec{q}'}|^2\vec{q}\vec{q}'+\sum_{\vec{q}}|f_{\vec{q}}|^2\frac{\hbar^2\vec{q}^2}{2M}\right\}-\frac{\hbar^2}{2M'}\left[\int d\vec{r}\psi^*(\vec{r})\,\Delta_{\vec{r}}\psi(\vec{r})-\int d\vec{r}\psi^*(\vec{r})\,\psi(\vec{r})\right.$$

$$\left.\left\{\frac{1}{2}\sum_{\vec{q}}\left[f_{\vec{q}}^*\Delta_{\vec{r}}f_{\vec{q}}-f_{\vec{q}}\Delta_{\vec{r}}f_{\vec{q}}^*-2\nabla_{\vec{r}}f_{\vec{q}}^*\nabla_{\vec{r}}f_{\vec{q}}\right]-\frac{1}{2}\sum_{\vec{q}\vec{q}'}f_{\vec{q}}f_{\vec{q}'}^*\nabla_{\vec{r}}f_{\vec{q}}^*\nabla_{\vec{r}}f_{\vec{q}'}+\frac{1}{4}\sum_{\vec{q}}\left[f_{\vec{q}}^*f_{\vec{q}'}^*\nabla_{\vec{r}}f_{\vec{q}'}\nabla_{\vec{r}}f_{\vec{q}}+f_{\vec{q}}f_{\vec{q}'}\nabla_{\vec{r}}f_{\vec{q}'}^*\nabla_{\vec{r}}f_{\vec{q}}^*\right]\right\}\right]$$

$$+\int d\vec{r}\psi^*(\vec{r})W(\vec{r})\,\psi(\vec{r})+\hbar\omega\int d\vec{r}\psi^*(\vec{r})\,\psi(\vec{r})\sum_{\vec{q}}|f_{\vec{q}}|^2+\int d\vec{r}\psi^*(\vec{r})\,\psi(\vec{r})\sum_{\vec{q}}\left[f_{\vec{q}}V_{\vec{q}}(\vec{r})+f_{\vec{q}}^*V_{\vec{q}}^*(\vec{r})\right] \quad (13.172)$$

In order to find the excitonic state, it is necessary to find $f_{\vec{q}}$ and $f_{\vec{q}}^*$ through the extremal value of the exciton energy:

$$\frac{\partial}{\partial f_{\vec{q}}^*}\bar{H}=\int d\vec{r}\,\psi^*(\vec{r})\,\psi(\vec{r})\left\{f_{\vec{q}}\frac{\hbar^2\vec{q}^2}{2M}-\frac{\hbar^2}{2M'}\Delta_{\vec{r}}f_{\vec{q}}+\hbar\omega f_{\vec{q}}-V_{\vec{q}}^*(\vec{r})+\frac{1}{2}\frac{\hbar^2}{2M'}\frac{\Delta_{\vec{r}}\left[\psi^*(\vec{r})\,\psi(\vec{r})\right]}{\psi^*(\vec{r})\,\psi(\vec{r})}f_{\vec{q}}\right\}=0 \quad (13.173)$$

From here:

$$-\frac{\hbar^2}{2M'}\Delta_{\vec{r}}f_{\vec{q}}+\left\{\hbar\omega+\frac{\hbar^2\vec{q}^2}{2M}+\frac{1}{2}\frac{\hbar^2}{2M'}\frac{\Delta_{\vec{r}}\left[\psi^*(\vec{r})\,\psi(\vec{r})\right]}{\psi^*(\vec{r})\,\psi(\vec{r})}\right\}f_{\vec{q}}-V_{\vec{q}}^*(\vec{r})=0 \quad (13.174)$$

In addition, differentiating by ψ^* and with the help of Equations 13.173 and 13.172, we have:

$$\sum_{\vec{q}}\left|f_{\vec{q}}\right|^2\left(\frac{\hbar^2\vec{q}^2}{2M}+\hbar\omega\right)\psi(\vec{r})-\frac{\hbar^2}{2M'}\Delta_{\vec{r}}\psi(\vec{r})-\frac{1}{2}\frac{\hbar^2}{2M'}\sum_{\vec{q}}\left[f_{\vec{q}}^*\Delta_{\vec{r}}f_{\vec{q}}+f_{\vec{q}}\Delta_{\vec{r}}f_{\vec{q}}^*-\Delta_{\vec{r}}\left|f_{\vec{q}}\right|^2\right]$$
$$+W(\vec{r})\,\psi(\vec{r})-\psi(\vec{r})\sum_{\vec{q}}\left[f_{\vec{q}}V_{\vec{q}}(\vec{r})+f_{\vec{q}}^*V_{\vec{q}}^*(\vec{r})\right]=\lambda\psi(\vec{r}) \quad (13.175)$$

Equation 13.175 describes the relative motion in the field of an effective potential. This effective potential in addition to $W(\vec{r})$ includes the field created by the displacement of the oscillators. In order to find this potential, we have to solve the nonhomogeneous Equation 13.174 with the displacement function of the oscillators $f_{\vec{q}}$ that is expressed through $V_{\vec{q}}^*$. It can be seen from Equation 13.174 that the displacement function of the oscillators $f_{\vec{q}}$ is dependent on M', $\hbar\omega$ and $\psi(\vec{r})$. Equation 13.174 for lattice vibrations and Equation 13.175 for the relative motion of the electron and hole are coupled to each other. So in a sense, this problem is that of a self-consistent field.

To solve this problem, we consider that the relative motion of the electron and hole is hydrogen-type and the radius will be a variational parameter. So $\psi(\vec{r})$ will be selected in the form:

$$\psi(\vec{r})=\sqrt{\frac{\beta^3}{\pi}}\exp\{-\beta r\} \quad (13.176)$$

Here, β is a variational parameter. So, from Equation 13.174, we have

$$-\frac{\hbar^2}{2M'}\Delta_{\vec{r}}f_{\vec{q}}+\left\{\hbar\omega+\frac{\hbar^2\vec{q}^2}{2M}+\frac{1}{2}\frac{\hbar^2}{2M'}4\beta^2\exp\{-2\beta r\}\left(1-\frac{1}{\beta r}\right)\right\}f_{\vec{q}}+V_{\vec{q}}^*(\vec{r})=0 \quad (13.177)$$

However, integrating this equation results to a very complex equation expressed through the effective potential. Hence, the term $4\beta^2-\frac{4\beta}{r}$ should be omitted for a long series. On the basis of Equation 13.172, all the terms that contain $f_{\vec{q}}$ are weighted by

$$d\vec{r}\,\psi^*(\vec{r})\,\psi(\vec{r})\cong r^2\exp\{-2\beta r\} \quad (13.178)$$

We can now find the approximation of Equation 13.177:

$$-\frac{\hbar^2}{2M'}\Delta_{\vec{r}}f_{\vec{q}}+\left(\hbar\omega+\frac{\hbar^2\vec{q}^2}{2M}\right)f_{\vec{q}}+V_{\vec{q}}^*(\vec{r})=0 \quad (13.179)$$

We can solve this equation for

$$V_{\vec{q}}(\vec{r}) = \gamma_{\vec{q}}\left(\exp\{i\vec{q}\mu_1\vec{r}\} - \exp\{-i\vec{q}\mu_2\vec{r}\}\right) \tag{13.180}$$

$$f_{\vec{q}} = \gamma^*_{\vec{q}}\left[-\frac{\exp\{-iq\mu_2 r\cos\theta\}}{\hbar\omega + \frac{\hbar^2\vec{q}^2}{2m_1}} + \frac{\exp\{iq\mu_1 r\cos\theta\}}{\hbar\omega + \frac{\hbar^2\vec{q}^2}{2m_2}}\right] \tag{13.181}$$

Then considering Equation 13.177, we have

$$\left(\hbar\omega + \frac{\hbar^2\vec{q}^2}{2M}\right)\exp\{iq\mu_1 r\cos\theta\}\frac{}{\hbar\omega + \frac{\hbar^2\vec{q}^2}{2m_2}} + \left(\exp\{-iq\mu_2 r\cos\theta\} - \exp\{iq\mu_1 r\cos\theta\}\right) = 0 \tag{13.182}$$

Now we consider a shorten equation for $f_{\vec{q}}$ in Equation 13.179 and we find from Equation 13.175:

$$-\frac{\hbar^2}{2M'}\Delta_{\vec{r}}\psi(\vec{r}) + \left\{W(\vec{r}) - \frac{1}{2}\frac{\hbar^2}{2M'}\sum_{\vec{q}}\left(f_{\vec{q}}\Delta_{\vec{r}}f^*_{\vec{q}} - \Delta_{\vec{r}}\left|f_{\vec{q}}\right|^2\right) + \sum_{\vec{q}} f_{\vec{q}}V_{\vec{q}}(\vec{r})\right\}\psi(\vec{r}) = 0 \tag{13.183}$$

Equation 13.175 can be written in the form:

$$\begin{aligned}
&-\frac{\hbar^2}{2M'}\Delta_{\vec{r}}\psi(\vec{r}) + W(\vec{r})\,\psi(\vec{r}) + \frac{1}{2}\left\{\sum_{\vec{q}} f^*_{\vec{q}}\left[-\frac{\hbar^2}{2M'}\Delta_{\vec{r}}f_{\vec{q}} + \left(\hbar\omega + \frac{\hbar^2\vec{q}^2}{2M}\right)f_{\vec{q}} + V^*_{\vec{q}}(\vec{r})\right]\right\}\psi(\vec{r})\\
&+\frac{1}{2}\left\{\sum_{\vec{q}} f_{\vec{q}}\left[-\frac{\hbar^2}{2M'}\Delta_{\vec{r}}f^*_{\vec{q}} + \left(\hbar\omega + \frac{\hbar^2\vec{q}^2}{2M}\right)f^*_{\vec{q}} + V_{\vec{q}}(\vec{r})\right]\right\}\psi(\vec{r}) + \frac{\hbar^2}{2M'}\sum_{\vec{q}}\frac{1}{2}\Delta_{\vec{r}}\left|f_{\vec{q}}\right|^2\\
&+\frac{1}{2}\sum_{\vec{q}}\left[f_{\vec{q}}V_{\vec{q}}(\vec{r}) + f^*_{\vec{q}}V^*_{\vec{q}}(\vec{r})\right] = \lambda\psi(\vec{r})
\end{aligned} \tag{13.184}$$

So the correction to the potential energy has the form:

$$\delta_{\vec{q}} = \frac{1}{2}\sum_{\vec{q}}\left[f_{\vec{q}}V_{\vec{q}}(\vec{r}) + f^*_{\vec{q}}V^*_{\vec{q}}(\vec{r})\right] + \frac{\hbar^2}{2M'}\frac{1}{2}\sum_{\vec{q}}\Delta_{\vec{r}}\left|f_{\vec{q}}\right|^2 \tag{13.185}$$

Considering the first summand in Equation 13.181 and integrating Equation 13.185 over q, we have

$$\sum_{\vec{q}}\left[f_{\vec{q}}V_{\vec{q}}(\vec{r}) + f^*_{\vec{q}}V^*_{\vec{q}}(\vec{r})\right] = \frac{e^2}{2\varepsilon^*}\left[\frac{1}{r}\left(2 - \exp\{-u_1 r\} - \exp\{-u_2 r\}\right) - (u_1 + u_2)\right] \tag{13.186}$$

Here

$$\frac{1}{\varepsilon^*} = \frac{1}{\varepsilon_\infty} - \frac{1}{\varepsilon_0},\; u_i = \left(\frac{2m_i\omega}{\hbar}\right)^{\frac{1}{2}} \tag{13.187}$$

The quantity, u_i, is the inverse length of the polarization wave. So we have the correction to the ionic polarization which is the potential energy of the interaction of the electron and the hole in the given field.

The second term in Equation 13.185 can be calculated and found to be

$$\delta_2 = \frac{\hbar^2}{2M'}\frac{1}{2}\sum_{\vec{q}}\Delta_{\vec{r}}\left|f_{\vec{q}}\right|^2 = \frac{e^2}{\varepsilon^* r}\frac{19}{2(m_1+m_2)}\left(\exp\{-u_2 r\}-\exp\{-u_1 r\}\right) \tag{13.188}$$

We consider the case of a large radius:

$$r > \frac{1}{u_i} \tag{13.189}$$

For this case, we observe the Coulomb interaction of the electron and the hole of the type $\frac{e^2}{\varepsilon^* r}$. If now

$$r < \frac{1}{u_i} \tag{13.190}$$

then we can expand the exponential function in Equation 13.188 by $u_1 r$ and $u_2 r$. We see that the potential energy of interaction compensates with the eigenenergy of the particles. This implies that the exciton does not feel the field. That is, it behaves like an electrical neutral particle. However, the correction δ_2 behaves otherwise. For larger distances r, it disappears, and for small distances r, it has a finite positive value. We can improve the variational problem if we introduce an additional variational parameter, say δ. For this, we substitute the resultant function $f_{\vec{q}}$ in the expression for $\psi(\vec{r})$. If $f_{\vec{q}}$ is independent of r, then the polarization accompanies the general exciton motion, i.e., the motion of the center of mass.

We examine the case, $\vec{K}=0$, which implies Equation 13.188 disappears. We come to this conclusion from the average energy in Equation 13.172 and now $f_{\vec{q}}$ becomes a constant. The summands containing the following terms disappear:

$$\sum_{\vec{q}}\left(f_{\vec{q}}^*\Delta_{\vec{r}}f_{\vec{q}} - \frac{1}{2}\Delta_{\vec{r}}\left|f_{\vec{q}}\right|^2\right) \tag{13.191}$$

So

$$\begin{aligned}\bar{\mathrm{H}} = &\int d\vec{r}\,\psi^*(\vec{r})\,\psi(\vec{r})\sum_{\vec{q}}\left|f_{\vec{q}}\right|^2\frac{\hbar^2\vec{q}^2}{2M} - \frac{\hbar^2}{2M'}\int d\vec{r}\,\psi^*(\vec{r})\,\Delta_{\vec{r}}\psi(\vec{r}) + \int d\vec{r}\,\psi^*(\vec{r})W(\vec{r})\,\psi(\vec{r})\\ &+\hbar\omega\int d\vec{r}\,\psi^*(\vec{r})\,\psi(\vec{r})\sum_{\vec{q}}\left|f_{\vec{q}}\right|^2 + \int d\vec{r}\,\psi^*(\vec{r})\,\psi(\vec{r})\sum_{\vec{q}}\left[f_{\vec{q}}V_{\vec{q}}(\vec{r}) + f_{\vec{q}}^*V_{\vec{q}}^*(\vec{r})\right]\end{aligned} \tag{13.192}$$

Hence, doing differentiation over $f_{\vec{q}}^*$, we have

$$\int d\vec{r}\,\psi^*(\vec{r})\,\psi(\vec{r})\left(\hbar\omega + \frac{\hbar^2\vec{q}^2}{2M}\right)f_{\vec{q}} + \int d\vec{r}\,\psi^*(\vec{r})\,V_{\vec{q}}^*(\vec{r})\psi(\vec{r}) = 0 \tag{13.193}$$

and

$$f_{\vec{q}} = -\frac{\int d\vec{r}\,\psi^*(\vec{r})\,V_{\vec{q}}^*(\vec{r})\psi(\vec{r})}{\left(\hbar\omega + \frac{\hbar^2\vec{q}^2}{2M}\right)} \tag{13.194}$$

II

Relativistic Theory

14

Case of an Electron

It is not possible to use the Schrödinger equation to describe the motion of a particle with relativistic effects since it is not invariant relative to the Lorentz transformation. However, the attempt to generalize quantum mechanics in the relativistic domain on the basis of the one particle theory is somehow difficult. So we find the appropriate equations on the basis of the relativistic invariance. To obtain the non-relativistic equations, we take the limit as the speed of light tends to infinity. This obviously recovers the Schrödinger equation. Under this heading, we extend the non-relativistic Schrödinger equation in the description of particles with speeds comparable with that of light. This extension should be consistent with the Lorentz transformation [12]. The characteristic feature of the relativistic wave equations is the spin which provides a useful gauge in the applicability of particular equations that describe particular kinds of particles. We have two types of relativistic equations, i.e., the equation with spin-zero due to Schrödinger and spin-half due to Dirac that describes the electron.

14.1 Spin Operators

A particle has an intrinsic angular momentum called the spin which is a representation introduced in quantum mechanics in 1925 by Goudsmit and Uhlenbeck [25, 27, 28] as a hypothesis to explain some spectroscopic phenomena which further enhance the explanation of several spectroscopic regularities. In 1927, Pauli and then Dirac introduced the non-relativistic formalism for the spin [29]. From the Dirac equations which we see later under this heading where a relativistic description of particle motion is inherent the spin. Note that the spin is not a consequence of relativity and has no classical analogue. The spin is completely an intrinsic quantum mechanical property of a particle.

Consider an explicit non-elementary particle, say the atomic nucleus. It may be displaced in space as a whole, i.e., it may execute an orbitary motion and have an angular momentum. But the nucleus is made up of protons p and neutrons n which are also in motion. This is an internal motion for which we may have its intrinsic angular momentum. It is right to consider the result of the spin as a general property of the particles, elementary or complex system of particles.

We revisit the material under the operator of the angular momentum and observe that the Azimuthal quantum number:

$$l = \frac{n}{2}, n = 0, 1, 2, \ldots \tag{14.1}$$

Here, l takes integral and half-integral values that essentially is new from what we saw earlier. This time the spin comes into play and the entire scenario changes also for the wave function.

We studied the angular momentum, $\hat{L}$, within the framework of the Schrödinger equation and found that projection of the angular momentum operator, $\hat{L}_z$, on the z-axis has the eigenvalues $m\hbar$ where m is the magnetic quantum number. The eigenvalues of the square of the operator of the operator of the angular momentum, $L^2 = \hbar^2 l(l+1)$. Here, l is the Azimuthal quantum number, which, from Equation 14.1,

DOI: 10.1201/9781003273073-16

takes multiple of integral and half-integer values. This introduces the existence of half-integer angular momentum in addition to the integer values which we have seen earlier. From experimentation such as the anomalous Zeeman effect, fine structure, the Stern-Gerlach experiment and so on, there is evidence of the half-integer values of the angular momentum associated with the electron:

Anomalous Zeeman Effect

We have seen earlier the rotational symmetry yields the degeneracy in the magnetic quantum number, m, or the Azimuthal quantum number, l. Here, $-l \le m \le l$. So the magnetic quantum number, m takes $2l+1$ values. Applying a magnetic field along the z-axis direction, the rotational symmetry is broken and the degeneracy in the magnetic quantum number, m, lifted. So due to the action of the magnetic field, each level is $(2l+1)$-fold degenerate. When the Azimuthal quantum number, l, has integer values, such levels are odd in the phenomenon which is the so-called normal Zeeman effect. But for atoms with an odd nuclear charge, the level splits up into an even number of closely spaced levels in a phenomenon known as the anomalous Zeeman effect which cannot be explained by an integer value of the Azimuthal quantum number, l.

Fine Structure

We observe further in the book that for the hydrogen atom, even without the application of an external magnetic field, the energy levels show fine structure which can be explained when a half-integer angular momentum is connected to an electron.

Stern-Gerlach Experiment

Passing a beam of silver atoms through a non-homogeneous magnetic field splits the beam into two showing that the electron should have a half-integer angular momentum associated with it. The magnitude of the splitting may be defined by the magnetic moment of the atom. It takes the value of the Bohr magneton:

$$\mathrm{M}_z = \pm\frac{e\hbar}{2\mathrm{m}c} \equiv \pm\mathrm{M}_B \tag{14.2}$$

Here, M_B is the Bohr magneton. What explains the magnetic moment of the atom? The silver atom is paramagnetic and has an odd nuclear charge. The Lorentz force is not experienced by a neutral silver atom. However, the silver atom possesses a permanent magnetic dipole moment due to its outer electron. When a magnetic field $\vec{\mathrm{H}}$ is applied, this gives rise to an interaction or potential energy:

$$\mathrm{U} = -\left(\vec{\mathrm{M}},\vec{\mathrm{H}}\right) \tag{14.3}$$

Here, $\vec{\mathrm{M}}$ is the magnetic dipole moment of the electron which relates the angular momentum of the electron. So the behavior of the atoms due to the influence of the magnetic field, $\vec{\mathrm{H}}$, gives information on the angular momentum of the electron. The Stern-Gerlach experiment involves the splitting of the atomic beam into two components in an inhomogeneous magnetic field and justifies that the electron to have half-integer angular momentum associated with it. For a revolving electron in a circular orbit, the magnetic moment and angular momentum are associated with the gyromagnetic ratio. This is the magnetic moment of a particle divided by the angular momentum associated with its revolution:

$$\frac{\mathrm{M}_z}{L_z} = \frac{e}{2\mathrm{m}c} \tag{14.4}$$

The equation $M_z = \pm M_B$ represents the intrinsic magnetic moment of an electron, i.e., M_{S_Z}. From the Pauli exclusion principle, the electron has an intrinsic moment the projection of which on the oz-axis takes two values: $s_z = \pm\frac{1}{2}$ and two values of $M_{S_Z} \equiv \pm M_B$. From here, it follows that

$$\frac{M_{S_Z}}{S_Z} = \frac{e}{mc} \tag{14.5}$$

This implies that for the spin of an electron, the gyromagnetic ratio is twice less than for the orbitary motion.

14.1.1 Spin and Spin Operator Commutation Relations

We examine again Equation 14.1 and consider $n = 1$:

$$l = \frac{1}{2} = s \tag{14.6}$$

Since l is an angular momentum value, s also is an intrinsic angular momentum value, the so-called spin. Let us denote the spin vector by $\vec{S}$ and its operator by $\hat{S}$:

$$\hat{S} = \left(\hat{S}_x, \hat{S}_y, \hat{S}_z\right) \tag{14.7}$$

Note that the total angular momentum, $\hat{J}$, of a particle consists of two parts, i.e., the part due to its orbital angular momentum described by the angular momentum operator, $\hat{L}$, and the part due to its intrinsic angular momentum (spin) described by the spin operator, $\hat{S}$:

$$\hat{J} = \hat{L} + \hat{S} \tag{14.8}$$

The operator $\hat{J}$ satisfies the same commutation relations as the operator $\hat{L}$

$$\left[\hat{J}_i, \hat{J}_j\right] = i\hbar \in_{ijk} \hat{J}_k, \left[\hat{J}^2, \hat{J}_j\right] = 0 \tag{14.9}$$

This implies that the spin operator, $\hat{S}$, will satisfy the same commutation relations as $\hat{J}$ and $\hat{L}$:

$$\left[\hat{S}_i, \hat{S}_j\right] = i\hbar \in_{ijk} \hat{S}_k, \left[\hat{S}^2, \hat{S}_j\right] = 0 \tag{14.10}$$

Here, the Hermitian operators $\hat{S}_x, \hat{S}_y$ and $\hat{S}_z$ should have

$$\hat{S}^2 = \hat{S}_x^2 + \hat{S}_y^2 + \hat{S}_z^2 \tag{14.11}$$

and $\hat{S}^2$ is also a Hermitian operator. In Equation 14.10, the quantity, $\in_{ijk}$, is the complete anti-symmetric (**Levi-Civita**) tensor:

$$\in_{ijk} = \begin{cases} 1, & \text{for even permutations of } 1,2,3 \\ 0, & \text{for 2 or more equal indices} \\ -1, & \text{for odd permutations of } 1,2,3 \end{cases} \tag{14.12}$$

With the help of this ϵ_{ijk} tensor, e.g., the cross product of two vectors $\vec{A}=\{A_i\}$ and $\vec{B}=\{B_i\}$ can be written as follows:

$$\left[\vec{A},\vec{B}\right]_k = \sum_{i,j} \epsilon_{ijk} A_i B_j \tag{14.13}$$

For this, we have the grounds to suppose that $\hat{S}$ is the consequence of the interior angular momentum. Note that an external magnetic field does leave the spin invariant.

The first commutation relation in Equation 14.10 shows the components of the spin, $\hat{S}$, to have mutually incompatible observables, while the second equation in Equation 14.10 shows there should exist simultaneous eigenstates of $\hat{S}^2$ and one of the components, $\hat{S}_j$. The operator, $\hat{S}$, quantizes according to the standard rule, i.e.,

$$S_z = m_s\hbar \tag{14.14}$$

where m_s is a quantum number defined as

$$-s \le m_s \le s \tag{14.15}$$

The quantum number, s, can be defined from the eigenvalue of the square of the spin operator, $\hat{S}^2$, that quantizes according to the standard rule:

$$S^2 = \hbar^2 s(s+1) \tag{14.16}$$

The quantum number s also relates the Azimuthal quantum number, l. It is obvious from Equation 14.15 that m_s takes the values, $s-1, s-2,\ldots,-s$. This implies we have $2s+1$ values running from $-s$ to s in steps of unity. If this number is even, then $s=\frac{1}{2},\frac{3}{2},\ldots$, and if odd, then $s=0,1,2,\ldots$. So the quantum number, s, takes either integer or half-integral values and a particle with half-integral spin, s, is called a Fermion and obeys Fermi-Dirac statistics while that with an integral spin, s, is called a Boson and obeys Bose-Einstein statistics. The electron is a Fermion with $s=\frac{1}{2}$ and obeys Fermi-Dirac statistics. If $s=\frac{1}{2}$, then

$$S_z = \pm\frac{\hbar}{2} \tag{14.17}$$

and from Equation 14.16, we have

$$S^2 = \frac{3\hbar^2}{4} \tag{14.18}$$

Note that the spin operators for the electron act on a $2s+1=2$ dimensional vector space and the basis states in this two-dimensional space are:

$$\left|s=\frac{1}{2}, s_z=\frac{1}{2}\right\rangle; \left|s=\frac{1}{2}, s_z=-\frac{1}{2}\right\rangle \tag{14.19}$$

We show further that

$$\hat{S}_z\left|s=\frac{1}{2}, s_z=\frac{1}{2}\right\rangle = \frac{\hbar}{2}\left|s=\frac{1}{2}, s_z=\frac{1}{2}\right\rangle;\ \hat{S}_z\left|s=\frac{1}{2}, s_z=-\frac{1}{2}\right\rangle = -\frac{\hbar}{2}\left|s=\frac{1}{2}, s_z=-\frac{1}{2}\right\rangle \tag{14.20}$$

14.1.2 Pauli Matrices

We represent the spin operator, $\hat{S}$, through a dimensionless quantity, $\hat{\sigma}$:

$$\hat{S} = \frac{\hbar}{2}\hat{\sigma} \tag{14.21}$$

The operator $\hat{\sigma}$ is yet to be defined. From the commutation relations in Equation 14.10:

$$[\hat{\sigma}_x, \hat{\sigma}_y] = 2i\hat{\sigma}_z, [\hat{\sigma}_y, \hat{\sigma}_z] = 2i\hat{\sigma}_x, [\hat{\sigma}_z, \hat{\sigma}_x] = 2i\hat{\sigma}_y \tag{14.22}$$

This can be written in a more compact form in the same manner as in Equation 14.10:

$$[\hat{\sigma}_i, \hat{\sigma}_j] = 2i \in_{ijk} \hat{\sigma}_k \tag{14.23}$$

Since the spin components, $\hat{S}_i$, from their possible orientations have only two eigenvalues $\pm\frac{\hbar}{2}$, the spin matrix operators $\hat{\sigma}_x, \hat{\sigma}_y, \hat{\sigma}_z$ must be 2×2 matrices:

$$\hat{\sigma}_x^2 = \hat{\sigma}_y^2 = \hat{\sigma}_z^2 = \hat{\mathbb{I}} \tag{14.24}$$

Here, $\hat{\mathbb{I}}$ represents the two-dimensional identity matrix.

How do we find the matrices $\hat{\sigma}_x$ and $\hat{\sigma}_y$ in the eigenrepresentations of $\hat{\sigma}_z$? Multiply the third equation in Equation 14.29 from the left and the right by $\hat{\sigma}_y$ and add both equations:

$$2i(\hat{\sigma}_x\hat{\sigma}_y + \hat{\sigma}_y\hat{\sigma}_x) = (\hat{\sigma}_y\hat{\sigma}_z - \hat{\sigma}_z\hat{\sigma}_y)\hat{\sigma}_y + \hat{\sigma}_y(\hat{\sigma}_y\hat{\sigma}_z - \hat{\sigma}_z\hat{\sigma}_y) = \hat{\sigma}_y^2\hat{\sigma}_z - \hat{\sigma}_z\hat{\sigma}_y^2 = 0 \tag{14.25}$$

From Equation 14.24:

$$\hat{\sigma}_x\hat{\sigma}_y + \hat{\sigma}_y\hat{\sigma}_x = 0 \tag{14.26}$$

We may also perform it for the first and second equations of Equation 14.29 and have

$$\{\hat{\sigma}_x, \hat{\sigma}_y\} = \{\hat{\sigma}_y, \hat{\sigma}_z\} = \{\hat{\sigma}_z, \hat{\sigma}_x\} = 0 \tag{14.27}$$

$$\hat{\sigma}_x\hat{\sigma}_y = i\hat{\sigma}_z, \hat{\sigma}_z\hat{\sigma}_x = i\hat{\sigma}_y, \hat{\sigma}_y\hat{\sigma}_z = i\hat{\sigma}_x \tag{14.28}$$

Equation 14.28 can be written in a compact form:

$$\hat{\sigma}_i\hat{\sigma}_j = i \in_{ijk} \hat{\sigma}_\text{k} \tag{14.29}$$

Equations 14.28 and 14.27 may be combined in a more compact form:

$$\{\hat{\sigma}_i, \hat{\sigma}_j\} = 2\delta_{ij} \tag{14.30}$$

The operators $\hat{\sigma}_x$, $\hat{\sigma}_y$ and $\hat{\sigma}_z$ are observed to be anti-commuting.

As the oz-axis is the quantization direction then the orientation of the spin is related to it. Mathematically, this implies that the spin functions are given as eigenfunctions of the operator $\hat{\sigma}_z$. The operator $\hat{\sigma}_z$ is diagonal in its eigenrepresentation and has the eigenvalues ± 1 as diagonal elements:

$$\hat{\sigma}_z = \begin{bmatrix} 1 & 0 \\ 0 & -1 \end{bmatrix} \tag{14.31}$$

Pauli in the form of 2×2 matrices represented these operators $\hat{\sigma}_x$, $\hat{\sigma}_y$ and $\hat{\sigma}_z$, the **so-called Pauli matrices**. From the representation where $\hat{\sigma}_z$ is a diagonal matrix, it may be shown also that in this representation, we have

$$\hat{\sigma}_x = \begin{bmatrix} 0 & 1 \\ 1 & 0 \end{bmatrix}, \quad \hat{\sigma}_y = \begin{bmatrix} 0 & -i \\ i & 0 \end{bmatrix} \tag{14.32}$$

14.1.3 Derivation of Pauli Matrices

Suppose we represent the Pauli matrices, $\hat{\sigma}_x$ and $\hat{\sigma}_y$:

$$\hat{\sigma}_x = \begin{bmatrix} \alpha_{11} & \alpha_{12} \\ \alpha_{21} & \alpha_{22} \end{bmatrix}, \quad \hat{\sigma}_y = \begin{bmatrix} \beta_{11} & \beta_{12} \\ \beta_{21} & \beta_{22} \end{bmatrix} \tag{14.33}$$

From the anti-commutation relation in Equation 14.27 of $\hat{\sigma}_x$ and $\hat{\sigma}_y$ with $\hat{\sigma}_z$, we have

$$\begin{bmatrix} \alpha_{11} & \alpha_{12} \\ -\alpha_{21} & -\alpha_{22} \end{bmatrix} = \begin{bmatrix} -\alpha_{11} & \alpha_{12} \\ -\alpha_{21} & \alpha_{22} \end{bmatrix} \tag{14.34}$$

So $\alpha_{11} = \alpha_{22} = 0$ and

$$\hat{\sigma}_x = \begin{bmatrix} 0 & \alpha_{12} \\ \alpha_{21} & 0 \end{bmatrix} \tag{14.35}$$

The matrices $\hat{\sigma}_x$ and $\hat{\sigma}_y$ with $\hat{\sigma}_z$ should be Hermitian:

$$\hat{\sigma}_x = \hat{\sigma}_x^{\dagger} = \hat{\sigma}_x^{*} \tag{14.36}$$

So

$$\alpha_{21} = \alpha_{12}^{*} \tag{14.37}$$

and

$$\hat{\sigma}_x = \begin{bmatrix} 0 & \alpha_{12} \\ \alpha_{12}^{*} & 0 \end{bmatrix}, \quad \hat{\sigma}_x^2 = \begin{bmatrix} |\alpha_{12}|^2 & 0 \\ 0 & |\alpha_{12}|^2 \end{bmatrix} \tag{14.38}$$

From Equation 14.24:

$$\hat{\sigma}_x^2 = \hat{\mathbb{I}} \tag{14.39}$$

and

$$|\alpha_{12}|^2 = 1 \tag{14.40}$$

From here, it is convenient to represent the matrix element, α_{12}:

$$\alpha_{12} = \exp\{i\alpha\} \tag{14.41}$$

Here, α is a real-valued quantity. We proceed with $\hat{\sigma}_y$ in a similar manner as $\hat{\sigma}_x$ and may write

$$\hat{\sigma}_x = \begin{bmatrix} 0 & \exp\{i\alpha\} \\ \exp\{-i\alpha\} & 0 \end{bmatrix}, \quad \hat{\sigma}_y = \begin{bmatrix} 0 & \exp\{i\beta\} \\ \exp\{-i\beta\} & 0 \end{bmatrix} \tag{14.42}$$

From the anti-commutation equation in Equation 14.27 for $\hat{\sigma}_x$ and $\hat{\sigma}_y$:

$$\begin{bmatrix} \exp\{i(\alpha-\beta)\} & 0 \\ 0 & \exp\{-i(\alpha-\beta)\} \end{bmatrix} = \begin{bmatrix} \exp\{-i(\alpha-\beta)\} & 0 \\ 0 & \exp\{i(\alpha-\beta)\} \end{bmatrix} \tag{14.43}$$

or

$$\exp\{i(\alpha-\beta)\} = -\exp\{-i(\alpha-\beta)\} \tag{14.44}$$

This implies

$$\exp\{2i(\alpha-\beta)\} = -1 \tag{14.45}$$

and so

$$\alpha - \beta = \frac{\pi}{2} \tag{14.46}$$

So for all the equations to be satisfied, we set

$$\alpha = 0, \quad \beta = -\frac{\pi}{2} \tag{14.47}$$

This yields the Pauli matrices in the $\hat{\sigma}_z$ representation:

$$\hat{\sigma}_x = \begin{bmatrix} 0 & 1 \\ 1 & 0 \end{bmatrix}, \quad \hat{\sigma}_y = \begin{bmatrix} 0 & -i \\ i & 0 \end{bmatrix}, \quad \hat{\sigma}_z = \begin{bmatrix} 1 & 0 \\ 0 & -1 \end{bmatrix} \tag{14.48}$$

It may be verified that each of the squares of the operators $\hat{\sigma}_x$, $\hat{\sigma}_y$ and $\hat{\sigma}_z$ give the unit matrix $\mathbb{I}$, say for example:

$$\hat{\sigma}_y^2 = \begin{bmatrix} 0 & -i \\ i & 0 \end{bmatrix}\begin{bmatrix} 0 & -i \\ i & 0 \end{bmatrix} = \begin{bmatrix} 1 & 0 \\ 0 & 1 \end{bmatrix} = \hat{\mathbb{I}} \tag{14.49}$$

The matrices in Equation 14.48 are called **Pauli matrices**.

From the above derivations, we observe that the Pauli matrices are traceless:

$$\mathrm{Tr}\,\hat{\sigma}_i = 0 \tag{14.50}$$

We prove this from

$$\mathrm{Tr}\left(\hat{\sigma}_i\hat{\sigma}_j\right) = \mathrm{Tr}\left(\hat{\sigma}_j\hat{\sigma}_i\right) = -\mathrm{Tr}\left(\hat{\sigma}_i\hat{\sigma}_j\right),\ i \neq j \tag{14.51}$$

and from here:

$$\mathrm{Tr}\left(\hat{\sigma}_i\hat{\sigma}_j\right) = 0,\ i \neq j \tag{14.52}$$

From Equation 14.50:

$$\mathrm{Tr}\left(\hat{\sigma}_i\hat{\sigma}_j\right) = i \in_{ijk} \mathrm{Tr}\left(\hat{\sigma}_k\right) = 0 \tag{14.53}$$

So

$$\mathrm{Tr}\left(\hat{\sigma}_k\right) = 0 \tag{14.54}$$

Question:
Let us use the property of the Pauli matrix $\hat{\sigma}_x$ and simplify the exponential operator, $\exp\left\{i\lambda\hat{\sigma}_x\right\}$.

Solution:
We Taylor series expand the exponential operator:

$$\exp\left\{i\lambda\hat{\sigma}_x\right\} = 1 + i\lambda\hat{\sigma}_x + \frac{\left(i\lambda\hat{\sigma}_x\right)^2}{2!} + \frac{\left(i\lambda\hat{\sigma}_x\right)^3}{3!} + \cdots \tag{14.55}$$

From Equation 14.24:

$$\exp\left\{i\lambda\hat{\sigma}_x\right\} = 1 + i\lambda\hat{\sigma}_x - \frac{\lambda^2}{2!} + \frac{i\lambda^3}{3!}\hat{\sigma}_x + \cdots = \left(1 - \frac{\lambda^2}{2!} + \cdots\right) + i\hat{\sigma}_x\left(\lambda - \frac{\lambda^3}{6} + \cdots\right) \tag{14.56}$$

Here

$$1 - \frac{\lambda^2}{2!} + \cdots = \cos\lambda \tag{14.57}$$

and

$$\lambda - \frac{\lambda^3}{6} + \cdots = \sin\lambda \tag{14.58}$$

So

$$\exp\{i\lambda\hat{\sigma}_x\} = \cos\lambda + i\hat{\sigma}_x \sin\lambda \quad (14.59)$$

14.2 Spinors

We earlier observed a **spinor** to be a mathematical object which can be Lorentz-transformed while allowing a more general treatment of the notion of invariance under rotation. So spinors arise while treating the Lorentz group. Spinors play a more general role in physics when seeing electrons and other particles to have an intrinsic angular momentum (spin). Pauli formalized this equation by representing the electron spin using a two-component complex vector via Pauli spin matrices. Seeking a quantum mechanical description of the electron consistent with the requirements of Lorentz covariance, **Paul Dirac introduced a four-component complex vector, called a Dirac spinor, the so-called bispinor** that we examine further. Interesting properties of the spin angular momentum can be achieved easily with the use of spinors. Rotations of two-component spinors describe rotations in three dimensions in addition to the Lorentz transformations in 3+1-dimensions. A rotation about a given axis does affect the spinor but the overall sign of the spinor in this case, though more subtle has no consequence when spinors are examined one at a time. However, this sign change can be relevant when one spinor is compared with another.

We examine the wave function of an electron considering the spin, S. In the general case, the state of an electron should depend on the direction of the spin S. So we have

$$\Psi = \Psi(x,y,z,\sigma_Z) \quad (14.60)$$

where σ_Z is the fourth independent variable and x,y,z continuously change. The variable σ_Z takes only two variables, i.e., ± 1 and, as a consequence, when considering the spin, the state of an electron should be described by a two-component function, Ψ, called a **spinor**:

$$\Psi = \begin{bmatrix} \Psi_1 \\ \Psi_2 \end{bmatrix}, \Psi_1 = \Psi(x,y,z,1),\ \Psi_2 = \Psi(x,y,z,-1) \quad (14.61)$$

The spin operators, $\hat{S}$, for the case of the electron act on the spin variables on a two-dimensional Hilbert space and so the operators transform the functions Ψ_1 and Ψ_2 through each other. The total electron wave function is dependent on both position as well as spin variables. As the angular momentum commutes with both the coordinates and momenta, it is obvious that the entire Hilbert space for the electron should be a direct product of the given two spaces and so the total wave function Ψ:

$$\Psi(\vec{r},\sigma_Z) = \Psi(\vec{r})\mathbb{X}(\sigma_Z) \quad (14.62)$$

where $\mathbb{X}(\sigma_Z)$ is the spinor:

$$\mathbb{X} = \begin{bmatrix} X_1 \\ X_2 \end{bmatrix} \quad (14.63)$$

and introducing the normalization condition, we have

$$|X_1|^2 + |X_2|^2 = 1 \quad (14.64)$$

The coordinate and the spinor function are normalized separately.

We find the eigenspinors of the operator $\hat{\sigma}_Z$ through the eigenvalue equation:

$$\hat{\sigma}_Z \mathbb{X} = \sigma_Z \mathbb{X} \tag{14.65}$$

or

$$\begin{bmatrix} 1 & 0 \\ 0 & -1 \end{bmatrix} \begin{bmatrix} X_1 \\ X_2 \end{bmatrix} = \pm \begin{bmatrix} X_1 \\ X_2 \end{bmatrix} \tag{14.66}$$

If $\sigma_Z = +1$, then

$$\begin{bmatrix} X_1 \\ -X_2 \end{bmatrix} = \begin{bmatrix} X_1 \\ X_2 \end{bmatrix} \tag{14.67}$$

From here, it follows that $X_2 = 0$. So from the normalization condition, we have $X_1 = 1$. The eigenvalue of $\sigma_Z = 1$ corresponds to the spinor (spin up):

$$\mathbb{X}_{\frac{1}{2}} = \begin{bmatrix} 1 \\ 0 \end{bmatrix} \tag{14.68}$$

then for $\sigma_z = -1$, we have (spin down)

$$\mathbb{X}_{-\frac{1}{2}} = \begin{bmatrix} 0 \\ 1 \end{bmatrix} \tag{14.69}$$

It is easily seen that these functions are not eigenfunctions of $\hat{\sigma}_x$ and $\hat{\sigma}_y$ but that of $\hat{\sigma}_z$. The spinor $\mathbb{X}$ describing the spin operator $\hat{\sigma}$ has with certainty an eigenvalue +1 in the (θ,ϕ)-direction. Suppose an observation of the spin component in the z-axis direction with outcome, σ_Z, the result of observation performed with the help of a Stern-Gerlach experiment with the magnetic field in the z-axis direction leads to either the state $\mathbb{X}_{\frac{1}{2}} = \begin{bmatrix} 1 \\ 0 \end{bmatrix}$ or the state $\mathbb{X}_{-\frac{1}{2}} = \begin{bmatrix} 0 \\ 1 \end{bmatrix}$. This implies that σ_Z is observed to have either the value of $\sigma_Z = 1$ or -1.

14.2.1 Lorentz Transformation and Spinor Transformation

Consider the system of coordinates of a particle at a fixed reference frame, K, where the particle is described by the spinor:

$$\Psi = \begin{bmatrix} \Psi_1 \\ \Psi_2 \end{bmatrix} \tag{14.70}$$

We examine the particle in a moving reference frame, K′, moving with velocity, $\vec{V}$, relative to the fixed reference frame, K, and the spinor representation of the wave function seems not to be sufficient. We examine the following coordinate, x, in a fixed reference frame, K, relative to x' in a moving reference frame, K′ [12–14]:

$$x = x'\cos\psi - \tau'\sin\psi,\ \tau = \tau'\cos\psi + x'\sin\psi \tag{14.71}$$

Here, $\tau' \equiv ct'$ where x' and t' are, respectively, the coordinate and time in the moving reframe frame, K′; c is the speed of light and ψ is the rotation angle. From the transformations $\psi \to i\psi$ and $\tau = i\tau$, Equation 14.71 becomes

$$x = x'\cosh\psi + \tau'\sinh\psi,\ \tau = \tau'\cosh\psi + x'\sinh\psi \tag{14.72}$$

If we consider the origin of the reference frame, K′, then $x' = 0$ and $x = Vt$. From Equation 14.72:

$$Vt = \tau'\sinh\psi,\ ct = \tau'\cosh\psi \tag{14.73}$$

From here, we formulate the relativistic theory in the four-dimensional Minkowski space [12] where an event is defined by a four-radius vector, $(x, y, z, \tau \equiv ct)$, and denote components of the four-radius vector:

$$\tau \equiv x^0,\ x \equiv x^1,\ y \equiv x^2,\ z \equiv x^3 \tag{14.74}$$

This can also be represented, respectively, by the covariant, x_μ, and contravariant, x^μ, vectors:

$$x_\mu = \{\tau, -\vec{r}\},\ x^\mu = \{\tau, \vec{r}\},\ \mu = 0,1,2,3 \tag{14.75}$$

The interval, s, which is an invariant quantity can then be represented as follows:

$$x^\mu x_\mu = c^2t^2 - \vec{r}^{\,2} \equiv s^2 \tag{14.76}$$

Here, $\vec{r}$ is a three-dimensional radius vector. The Lorentz transformation of the four-radius vector.

So from here and Equation 14.76, we have the Lorentz transform for the coordinate and time:

$$x^0 = \frac{x'^0 + \frac{V}{c}x'^1}{\sqrt{1 - \frac{V^2}{c^2}}},\ x^1 = \frac{x'^1 + \frac{V}{c}x'^0}{\sqrt{1 - \frac{V^2}{c^2}}},\ x^2 = x'^2,\ x^3 = x'^3 \tag{14.77}$$

Similarly, with the rotation operator, $\exp\left\{-\frac{i\phi\hat{\sigma}_x}{2}\right\}$, we define the Lorentz transformation operator as $\exp\left\{\frac{\phi\hat{\sigma}_x}{2}\right\}$ where we have considered $\phi \to i\phi$. The spinor

$$\Psi = \begin{bmatrix} \Psi_1 \\ \Psi_2 \end{bmatrix} \tag{14.78}$$

can be transformed:

$$\Phi = \begin{bmatrix} \Psi_1' \\ \Psi_2' \end{bmatrix} = \exp\left\{\frac{\phi\hat{\sigma}_x}{2}\right\}\begin{bmatrix} \Psi_1 \\ \Psi_2 \end{bmatrix} \tag{14.79}$$

For the inversion of the coordinate, the spinor, $\begin{bmatrix} \Psi_1' \\ \Psi_2' \end{bmatrix}$, is transformed:

$$\begin{bmatrix} \Psi_1'' \\ \Psi_2'' \end{bmatrix} = \exp\left\{-\frac{\phi\hat{\sigma}_x}{2}\right\}\begin{bmatrix} \Psi_1 \\ \Psi_2 \end{bmatrix} \tag{14.80}$$

How should we find an equation with a solution where applying inversion of the coordinate is self-transformed? For this, it is necessary to introduce two component spinors, $\begin{bmatrix} \Psi_1' \\ \Psi_2' \end{bmatrix}$ and $\begin{bmatrix} \Psi_1'' \\ \Psi_2'' \end{bmatrix}$, which implies we have to introduce a four-component wave function (bispinor):

$$\Psi = \begin{bmatrix} \Psi_1 \\ \Psi_2 \\ \Psi_3 \\ \Psi_4 \end{bmatrix} \tag{14.81}$$

14.2.2 Arbitrary Spinor Transformation

We write an arbitrary transformation spin matrix operator in the form:

$$\hat{\mathbb{S}} = \begin{bmatrix} S_{11} & S_{12} \\ S_{21} & S_{22} \end{bmatrix} \tag{14.82}$$

that transforms one spinor into the other:

$$\Phi = \hat{\mathbb{S}}\Psi \tag{14.83}$$

where

$$\Phi = \begin{bmatrix} \Phi_1 \\ \Phi_2 \end{bmatrix}, \quad \Psi = \begin{bmatrix} \Psi_1 \\ \Psi_2 \end{bmatrix} \tag{14.84}$$

We write Equation 14.83 in detail:

$$\begin{bmatrix} \Phi_1 \\ \Phi_2 \end{bmatrix} = \begin{bmatrix} S_{11} & S_{12} \\ S_{21} & S_{22} \end{bmatrix} \begin{bmatrix} \Psi_1 \\ \Psi_2 \end{bmatrix} = \begin{bmatrix} S_{11}\Psi_1 + S_{12}\Psi_2 \\ S_{21}\Psi_1 + S_{22}\Psi_2 \end{bmatrix} \tag{14.85}$$

and find the average value of the operator $\hat{\mathbb{S}}$:

$$\left\langle \hat{\mathbb{S}} \right\rangle = \int \Psi^{\dagger} \hat{\mathbb{S}} \Psi \, d\vec{r} \tag{14.86}$$

If the wave functions are spinors, then we use Hermitian-conjugate wave functions instead of complex-conjugate, i.e.,

$$\begin{bmatrix} \Psi_1 \\ \Psi_2 \end{bmatrix}^{\dagger} = \begin{bmatrix} \Psi_1^* & \Psi_2^* \end{bmatrix} \tag{14.87}$$

We find the average value:

$$\left\langle \hat{\mathbb{S}}(t) \right\rangle = \int \left\langle \hat{\mathbb{S}}(\vec{r},t) \right\rangle d\vec{r} \equiv \int \Psi^{\dagger} \hat{\mathbb{S}}(\vec{r},t) \Psi \, d\vec{r} \tag{14.88}$$

and

$$\left\langle \hat{\mathbb{S}}(\vec{r},t) \right\rangle = \begin{bmatrix} \Psi_1^* & \Psi_2^* \end{bmatrix} \begin{bmatrix} S_{11} & S_{12} \\ S_{21} & S_{22} \end{bmatrix} \begin{bmatrix} \Psi_1 \\ \Psi_2 \end{bmatrix} = \Psi_1^* S_{11} \Psi_1 + \Psi_1^* S_{12} \Psi_2 + \Psi_2^* S_{21} \Psi_1 + \Psi_2^* S_{22} \Psi_2 \quad (14.89)$$

In the above, $\left\langle \hat{\mathbb{S}}(\vec{r},t) \right\rangle$ is the average value of the spin operator at the location $\vec{r}$ and time t while $\left\langle \hat{\mathbb{S}}(t) \right\rangle$ is averaged over the spin directions with each location at time t. From the above, we evaluate the average values over both possible spin states of the Pauli matrices:

$$\left\langle \hat{\sigma}_x(\vec{r},t) \right\rangle = \Psi^\dagger \hat{\sigma}_x \Psi = \begin{bmatrix} \Psi_1^* & \Psi_2^* \end{bmatrix} \begin{bmatrix} 0 & 1 \\ 1 & 0 \end{bmatrix} \begin{bmatrix} \Psi_1 \\ \Psi_2 \end{bmatrix} = \Psi_1^* \Psi_2 + \Psi_2^* \Psi_1 \quad (14.90)$$

$$\left\langle \hat{\sigma}_y(\vec{r},t) \right\rangle = \begin{bmatrix} \Psi_1^* & \Psi_2^* \end{bmatrix} \begin{bmatrix} 0 & -i \\ i & 0 \end{bmatrix} \begin{bmatrix} \Psi_1 \\ \Psi_2 \end{bmatrix} = -i\Psi_1^* \Psi_2 + i\Psi_2^* \Psi_1 \quad (14.91)$$

$$\left\langle \hat{\sigma}_z(\vec{r},t) \right\rangle = \begin{bmatrix} \Psi_1^* & \Psi_2^* \end{bmatrix} \begin{bmatrix} 1 & 0 \\ 0 & -1 \end{bmatrix} \begin{bmatrix} \Psi_1 \\ \Psi_2 \end{bmatrix} = \Psi_1^* \Psi_1 - \Psi_2^* \Psi_2 \quad (14.92)$$

We find the eigenspinors of the spin operator $\hat{s}_x = \frac{\hat{\sigma}_x}{2}$ through the eigenvalue equation:

$$\hat{s}_x \mathbb{X} = s_x \mathbb{X} \quad (14.93)$$

From

$$\mathbb{X}_{s_x} = \begin{bmatrix} \mathrm{X}_1 \\ \mathrm{X}_2 \end{bmatrix} \quad (14.94)$$

then

$$\frac{1}{2} \begin{bmatrix} 0 & 1 \\ 1 & 0 \end{bmatrix} \begin{bmatrix} \mathrm{X}_1 \\ \mathrm{X}_2 \end{bmatrix} = s_x \begin{bmatrix} \mathrm{X}_1 \\ \mathrm{X}_2 \end{bmatrix} \quad (14.95)$$

From here:

$$\mathrm{X}_2 = 2s_x \mathrm{X}_1, \ \mathrm{X}_1 = 2s_x \mathrm{X}_2 \quad (14.96)$$

The non-trivial solution of the system of equation in Equation 14.69 exists when

$$4s_x^2 = 1 \quad (14.97)$$

This gives the possible eigenvalues:

$$s_x = \pm \frac{1}{2} \quad (14.98)$$

From Equation 14.96, considering $s_x = +\frac{1}{2}$, $X_1 = X_2 \equiv X$ and for $s_x = -\frac{1}{2}$, we have $X_1 = -X_2$:

$$\begin{cases} s_x = +\frac{1}{2} \quad \rightarrow \quad X_1 = X_2 \equiv X \\ s_x = -\frac{1}{2} \quad \rightarrow \quad X_1 = -X_2 \equiv -X \end{cases} \tag{14.99}$$

So

$$\mathbb{X}_{s_x=+\frac{1}{2}} = X\begin{bmatrix} 1 \\ 1 \end{bmatrix}, \mathbb{X}_{s_x=-\frac{1}{2}} = X\begin{bmatrix} 1 \\ -1 \end{bmatrix} \tag{14.100}$$

From the normalization condition:

$$\mathbb{X}^{\dagger}_{s_x}\mathbb{X}_{s_x} = |X_1|^2 + |X_2|^2 = 1 \tag{14.101}$$

so

$$\mathbb{X}_{s_x=+\frac{1}{2}} = \frac{1}{\sqrt{2}}\begin{bmatrix} 1 \\ 1 \end{bmatrix}, \mathbb{X}_{s_x=-\frac{1}{2}} = \frac{1}{\sqrt{2}}\begin{bmatrix} 1 \\ -1 \end{bmatrix} \tag{14.102}$$

Similarly, for the operator $\hat{\sigma}_y$, we have

$$\frac{1}{2}\begin{bmatrix} 0 & -i \\ i & 0 \end{bmatrix}\begin{bmatrix} X_1 \\ X_2 \end{bmatrix} = s_y\begin{bmatrix} X_1 \\ X_2 \end{bmatrix} \tag{14.103}$$

and

$$-iX_2 = 2s_yX_1, \; iX_1 = 2s_yX_2 \tag{14.104}$$

from where:

$$s_y = \pm\frac{1}{2} \quad \rightarrow \quad X_2 = \pm\frac{i}{2}X_1 \equiv \pm\frac{i}{2}X \tag{14.105}$$

So

$$\mathbb{X}_{s_y=+\frac{1}{2}} = \frac{1}{\sqrt{2}}\begin{bmatrix} 1 \\ i \end{bmatrix}, \mathbb{X}_{s_y=-\frac{1}{2}} = \frac{1}{\sqrt{2}}\begin{bmatrix} 1 \\ -i \end{bmatrix} \tag{14.106}$$

We examine two particles with spin, $s = \frac{1}{2}$, and find the eigenkets, $|s, s_z\rangle$, of the total spin. The functions $|1,0\rangle$ and $|0,0\rangle$ can be obtained from the ket when $s_z = 0$:

$$|s, s_z = 0\rangle = c_1\begin{bmatrix} 1 \\ 0 \end{bmatrix}_1\begin{bmatrix} 0 \\ 1 \end{bmatrix}_2 + c_2\begin{bmatrix} 0 \\ 1 \end{bmatrix}_1\begin{bmatrix} 1 \\ 0 \end{bmatrix}_2 \tag{14.107}$$

The operators $\hat{\mathbb{S}}^2$ and $\hat{\mathbb{S}}_\pm$ can always be useful as well as the symmetry properties of the eigenkets, $|s,s_z\rangle$. It is obvious from Equations 4.161 and 4.167 in Chapter 4 that the spin eigenkets of the two particles for the projection in z-axis direction:

$$|1,1\rangle = \begin{bmatrix} 1 \\ 0 \end{bmatrix}_1 \begin{bmatrix} 0 \\ 1 \end{bmatrix}_2, \quad |1,-1\rangle = \begin{bmatrix} 0 \\ 1 \end{bmatrix}_1 \begin{bmatrix} 1 \\ 0 \end{bmatrix}_2 \tag{14.108}$$

From

$$\hat{\mathbb{S}}^2|0,0\rangle = 0 \tag{14.109}$$

then

$$\langle 0,0|\hat{\mathbb{S}}^2|0,0\rangle = \langle 0,0|\hat{\mathbb{S}}_x^2|0,0\rangle + \langle 0,0|\hat{\mathbb{S}}_y^2|0,0\rangle + \langle 0,0|\hat{\mathbb{S}}_z^2|0,0\rangle = 0 \tag{14.110}$$

and

$$\begin{aligned} \hat{\mathbb{S}}_x|0,0\rangle &= \frac{1}{2}(\hat{\sigma}_{x1} + \hat{\sigma}_{x2})\left(c_1 \begin{bmatrix} 1 \\ 0 \end{bmatrix}_1 \begin{bmatrix} 0 \\ 1 \end{bmatrix}_2 + c_2 \begin{bmatrix} 0 \\ 1 \end{bmatrix}_1 \begin{bmatrix} 1 \\ 0 \end{bmatrix}_2 \right) \\ &= (c_1 + c_2)\left(\begin{bmatrix} 1 \\ 0 \end{bmatrix}_1 \begin{bmatrix} 1 \\ 0 \end{bmatrix}_2 + \begin{bmatrix} 0 \\ 1 \end{bmatrix}_1 \begin{bmatrix} 0 \\ 1 \end{bmatrix}_2 \right) = 0 \end{aligned} \tag{14.111}$$

From here, for the case of $|0,0\rangle$, we have

$$c_1 = -c_2 \tag{14.112}$$

Since the ket $|1,0\rangle$ is orthogonal to $|0,0\rangle$, for the ket $|0,0\rangle$, we have

$$c_1 = c_2 \tag{14.113}$$

So the normalized kets $|1,0\rangle$ and $|0,0\rangle$ that are, respectively, symmetrical and anti-symmetrical have the form:

$$|s=1,0\rangle = \frac{1}{\sqrt{2}}\left(\begin{bmatrix} 1 \\ 0 \end{bmatrix}_1 \begin{bmatrix} 0 \\ 1 \end{bmatrix}_2 + \begin{bmatrix} 0 \\ 1 \end{bmatrix}_1 \begin{bmatrix} 1 \\ 0 \end{bmatrix}_2 \right), \quad |s=0,0\rangle = \frac{1}{\sqrt{2}}\left(\begin{bmatrix} 1 \\ 0 \end{bmatrix}_1 \begin{bmatrix} 0 \\ 1 \end{bmatrix}_2 - \begin{bmatrix} 0 \\ 1 \end{bmatrix}_1 \begin{bmatrix} 1 \\ 0 \end{bmatrix}_2 \right) \tag{14.114}$$

The eigenket, $|1,0\rangle$, describes the state where with equal probability, the first particle has the spin projection $\frac{1}{2}$ and the second $-\frac{1}{2}$ and vice versa and implies spin delocalization which also applies to $|0,0\rangle$.

15

Klein-Gordon Equation

Before examining the Dirac equation that gives a relativistic quantum mechanical description of the electron, we will examine the Klein-Gordon equation. Compared to the Schrödinger equation, the Dirac equation should have an appreciable modification in the quantum description of the properties of the electron. In the non-relativistic quantum mechanical formulation, we begin with the classical relation for the energy in the absence of interaction, i.e., the kinetic energy:

$$\mathrm{H} = \frac{\mathrm{p}^2}{2\mathrm{m}} \tag{15.1}$$

In the non-relativistic Schrödinger equation, we raise both H and p to operator forms:

$$\hat{\mathrm{H}} \to i\hbar\frac{\partial}{\partial t} \tag{15.2}$$

and

$$\mathrm{p} \to \hat{\mathrm{p}} = -i\hbar\nabla \tag{15.3}$$

So for a free particle, we have the wave equation:

$$i\hbar\frac{\partial}{\partial t}\Psi = -\frac{\hbar^2}{2\mathrm{m}}\Delta\Psi \tag{15.4}$$

This non-relativistic Schrödinger equation cannot be used to describe the motion of a particle considering relativistic effects since it is not invariant relative to the Lorentz transformation. So the attempt to adapt quantum mechanics to the relativistic domain for a single particle theory has some difficulties. We investigate this character by considering the example of a spinless single particle. This non-relativistic character is inferred from the time variable in Equation 15.4. Note that a relativistic character of a dynamical variable must treat both space and time symmetrically. So a wave equation which has a relativistic description of the system must be invariant relative to Lorentz transformation. In the said equation, we put the constraint where in the non-relativistic limit, the light speed, $c \to \infty$, yields the Schrödinger equation. For this, we generalize the relativistic by first generalizing the Lorentz transformation in Equation 14.77 by introducing, respectively, the contravariant, A^{μ}, and covariant, A_{μ}, four-vectors:

$$A_{\mu} = \left\{A_0, -\vec{A}\right\},\ A^{\mu} = \left\{A^0, \vec{A}\right\},\ A_0 = A^0,\ A_{\mu} = -A^{\mu},\ \mu = 0,1,2,3 \tag{15.5}$$

DOI: 10.1201/9781003273073-17

Here, $\vec{A}$ is a three-dimensional vector in a three-dimensional Euclidean space and the scalar, A^0, imitates the time. So the invariant quantity should be

$$A^{\mu}A_{\mu} \equiv A_{\mu}A^{\mu} = \left(A^0\right)^2 - \vec{A}^2 \tag{15.6}$$

Further, we denote $\frac{\partial}{\partial x_{\mu}}$ or $\frac{\partial}{\partial x^{\mu}}$, respectively, by ∂_{μ} or ∂^{μ}.

We examine the relativistic classical Hamiltonian, H, for a free particle [12]:

$$\mathrm{H}^2 = \mathrm{p}^2c^2 + \mathrm{m}^2c^4 \tag{15.7}$$

where H includes the rest-mass energy, mc^2. From Equation 15.7, the relativistic invariant relation follows:

$$\mathrm{H}^2 - \mathrm{p}^2c^2 = \mathrm{m}^2c^4 \tag{15.8}$$

or

$$\frac{\mathrm{H}^2}{c^2} - \mathrm{p}^2 = \mathrm{m}^2c^2 \tag{15.9}$$

This is the so-called Einstein relation and is satisfied in all Lorentz frames.

Raising the relativistic Equation 15.7 to operator form, we arrive at a wave equation:

$$-\hbar^2\frac{\partial^2}{\partial t^2}\Psi \equiv \left(\hat{\mathrm{p}}^2c^2 + \mathrm{m}^2c^4\right)\Psi = -\hbar^2c^2\Delta\Psi + \mathrm{m}^2c^4\Psi \tag{15.10}$$

or

$$\Delta\Psi - \frac{1}{c^2}\frac{\partial^2}{\partial t^2}\Psi = \frac{\mathrm{m}^2c^2}{\hbar^2}\Psi \tag{15.11}$$

or

$$\partial_{\mu}\partial^{\mu}\Psi \equiv \frac{\mathrm{m}^2c^2}{\hbar^2}\Psi,\ \partial_{\mu}\partial^{\mu} \equiv \Delta - \frac{1}{c^2}\frac{\partial^2}{\partial t^2} \tag{15.12}$$

where $\partial_{\mu}\partial^{\mu}$ is the Laplacian in the four-dimensional Monkowski space. Equation 15.12 was also obtained by Schrödinger but was not published and is the so-called **Klein-Gordon equation**. The quantity $\frac{\hbar}{\mathrm{m}c}$ on the right-hand side of Equation 15.12 is the Compton wavelength. If $\mathrm{m} = 0$, then the particle is massless and the equation represents the wave equation describing traveling waves.

Equation 15.12 appears to be difficult to interpret. This difficulty can be overcome if we consider Equation 15.12 to describe a many-particle system and Ψ examined as a field operator for the many-particle system. So to talk of the creation of a particle with a negative energy, we imply the annihilation of anti-particles with positive energies and for the annihilation of particles with negative energies with talk of the creation of the anti-particles with positive energies. So for that, Equation 15.12 has a solution in the superposition form:

$$\Psi(\vec{r},t) = \Psi^{(+)}(\vec{r},t) + \Psi^{(-)}(\vec{r},t) \tag{15.13}$$

where the wave packet is the partial solution of the Klein-Gordon equation in Equation 15.12:

$$\Psi^{(\pm)}(\vec{r},t)=\int d\vec{\kappa}\mathrm{a}^{(\pm)}(\vec{\kappa})\Psi_{\vec{\kappa}}^{(\pm)}(\vec{r},t) \tag{15.14}$$

Here, $\mathrm{a}^{(\pm)}(\vec{\kappa})$ is examined as the wave function of the particle or anti-particle in the momentum representation. The solution of the Klein-Gordon equation has a formal sense since it does not describe any single-particle state. The single-particle state can only be described either by the wave function, $\Psi^{(+)}(\vec{r},t)$ or $\Psi^{(-)}(\vec{r},t)$. It is in this wise that the Klein-Gordon equation has a physical realizable solution. The partial solution of Equation 15.11 or 15.12 forms the complete system where

$$\Psi_{\vec{\kappa}}^{(\pm)}(\vec{r},t)=\exp\left\{\pm i\left(\vec{\kappa}\vec{r}-\omega(\kappa)t\right)\right\},\ \omega(\kappa)=\sqrt{\vec{\kappa}^2c^2+\frac{\mathrm{m}^2c^4}{\hbar^2}}>0 \tag{15.15}$$

Here, $\Psi_{\vec{\kappa}}^{(+)}(\vec{r},t)$ are eigenfunctions of the operators $\hat{\mathrm{p}}$ and $\hat{\mathrm{H}}$ with eigenvalues, respectively, as follows:

$$\vec{\mathrm{p}}=\hbar\vec{\kappa} \tag{15.16}$$

and

$$\epsilon=\hbar\omega\geq\mu c^2 \tag{15.17}$$

while $\Psi_{\vec{\kappa}}^{(-)}(\vec{r},t)$ are eigenfunctions also of the operators $\hat{\mathrm{p}}$ and H with eigenvalues, respectively, as follows:

$$\vec{\mathrm{p}}=-\hbar\vec{\kappa} \tag{15.18}$$

and

$$\epsilon'=-\hbar\omega\leq-\mu c^2 \tag{15.19}$$

The sign "±" shows the character of the time dependence of the wave function corresponding to the positive "+" and negative "–" energy and frequency. So the Klein-Gordon equation has solutions with positive as well as negative energies. This stems from the fact that the Klein-Gordon equation, unlike the Schrödinger equation, is a second-order differential equation over time.

We examine the invariance of the Klein-Gordon equation in Equation 15.12 for the free particle relative to the transformation operator, $\hat{\mathrm{C}}$. This implies that if $\Psi(\vec{r},t)$ is the solution of Equations 15.11 or 15.12, then the following is also a solution:

$$\Psi_{\mathrm{c}}=\hat{\mathrm{C}}\Psi \tag{15.20}$$

From the complex conjugate of Equation 15.12, its invariance can be confirmed and so $\hat{\mathrm{C}}^2=1$. For the scalar function, we have $\hat{\mathrm{C}}=+1$, and for the pseudoscalar function, $\hat{\mathrm{C}}=-1$. So we write the Klein-Gordon equation solution in the form:

$$\Psi(\vec{r},t)=\Psi^{(+)}(\vec{r},t)+\Psi^{(-)}(\vec{r},t)=\Psi^{(+)}(\vec{r},t)+\hat{\mathrm{C}}\Psi_{\mathrm{c}}^{(+)}(\vec{r},t) \tag{15.21}$$

The solutions $\Psi^{(+)}(\vec{r},t)$ and $\Psi_{\mathrm{c}}^{(+)}(\vec{r},t)$ have the same character relative to the inversion of the coordinate and describe separately the particle and anti-particle states. This implies, each solution is either scalar or pseudoscalar. This refers as well to the functions $\Psi^{(+)}$ and $\Psi_{\mathrm{c}}^{(+)}$. So the solution in Equation 15.21 being in a superpositional state has no physical sense since one of the solutions in the superpositional state is that of an anti-particle.

From Equation 15.20, it is also obvious that

$$\Psi_{c,\epsilon} = \Psi_{-\epsilon}, \Psi_{c,p} = \Psi_{-p}, \Psi_{c,m} = \Psi_{-m}, \Psi_{c,t} = \Psi_{-t} \tag{15.22}$$

So if

$$\Psi \sim \exp\{i\vec{\kappa}\vec{r}\} \tag{15.23}$$

then the eigenfunction of the operator, $\hat{\vec{p}}$, corresponds to the eigenvalue, $\hbar\vec{\kappa}$. The corresponding complex conjugate

$$\Psi_c \sim \exp\{-i\vec{\kappa}\vec{r}\} \tag{15.24}$$

corresponds to the eigenvalue, $-\hbar\vec{\kappa}$, and so forth.

Note that in the relativistic case, the internal parity of bosons can be defined from the law of parity conservation, considering bosonic creation and annihilation. In the non-relativistic case, during the interaction of particles, the particle number is invariant. So the internal parity of the entire system at all stages of the process is unique and is not an experimental observable quantity. So it is impossible to differentiate the character of the wave function relative to the inversion of the coordinate, and from brevity, the wave function selected may be scalar. For particles of arbitrary spin, the internal parity of the particle and anti-particle is the same for bosons and opposite in sign for fermions. So there is a need for a new relativistic wave equation.

15.1 Probability and Charge Densities

From the Klein-Gordon equation 15.12, the continuity equation implies the conservation of the quantity, $Q^{(\pm)}$, over time for the particle and anti-particle:

$$Q^{(\pm)} = \int d\vec{r}\, \rho^{(\pm)}(\vec{r},t) = \frac{i\hbar}{2mc^2}\int d\vec{r}\left(\Psi^{(\pm)*}(\vec{r},t)\frac{\partial}{\partial t}\Psi^{(\pm)}(\vec{r},t) - \Psi^{(\pm)}(\vec{r},t)\frac{\partial}{\partial t}\Psi^{(\pm)*}(\vec{r},t)\right) \tag{15.25}$$

This quantity, $Q^{(\pm)}$, is used for the normalization of the wave function, $\Psi^{(\pm)}(\vec{r},t)$, where from

$$\int d\vec{r} \exp\{\pm i(\vec{\kappa}-\vec{\kappa}')\vec{r}\} = (2\pi)^3\delta(\vec{\kappa}-\vec{\kappa}') \tag{15.26}$$

we obtain

$$Q^{(\pm)} = \pm\frac{(2\pi)^3}{mc^2}\int d\vec{\kappa}\omega(\vec{\kappa})\left|a^{(\pm)}(\vec{\kappa})\right|^2 \tag{15.27}$$

If e is the charge and $Q^{(\pm)}$ relates to the integrand, $\rho^{(\pm)}$ in Equation 15.27 relates to the bulk charge density $e\rho^{(\pm)}$ for the particle or anti-particle normalized as follows:

$$Q^{(\pm)} = \pm 1 \tag{15.28}$$

The plus and negative signs relate, respectively, to the particle and anti-particle states. Note that $Q^{(+)}$ and $Q^{(-)}$ individually are conserved, and if both are conserved simultaneously, then we assume the particle and anti-particle are pairwise created and annihilated. If the particle has the electronic charge, then from Equation 15.12, the anti-particle has the charge of opposite sign. So in this case, the process of creation or annihilation of the pair of particle-anti-particle will satisfy the law of conservation of the total charge.

For neutral particles, there arises some difficulty in interpreting $\rho^{(\pm)}$ unlike the case of the Schrödinger equation. In the present case, $\rho^{(\pm)}$ cannot be thought of as a probability density since it has to be strictly non-negative and, in addition, the local conservation law is already violated:

$$\mathrm{div}\vec{j} + \frac{\partial}{\partial t}\rho^{(\pm)} = 0 \tag{15.29}$$

Here, $\vec{j}(\vec{r},t)$ imitates the usual probability current density. From the theoretical point of view, the description of the particle or anti-particle states by Equation 15.29 presents some serious difficulty since it does not possess any defined Lorentz transformation properties. So its interpenetration in the relativistic domain has no physical sense. In addition, the localization of a particle in a small domain of space requires very strong fields which provoke the creation of new particles and renders our problem no more a single particle one. So in the relativistic theory, the wave function regarded as a probability amplitude of particles is conserved in the momentum representation unlike the coordinate representation. So in not a very small volume of space, if we average the quantity, $\rho^{(\pm)}$, then it can be interpreted as the probability density of the corresponding particle or anti-particle.

15.2 Motion in an Electromagnetic Field

For a particle moving in the potential field, $\hat{U}$, Equation 15.11 becomes

$$-\hbar^2\frac{\partial^2}{\partial t^2}\Psi(\vec{r},t) = \left(-\hbar^2c^2\Delta + m^2c^4 + 2mc^2\hat{U}\right)\Psi(\vec{r},t) \tag{15.30}$$

From Equation 15.22, Equation 15.30 can be written as follows:

$$-\hbar^2\frac{\partial^2}{\partial t^2}\Psi_c(\vec{r},t) = \left(-\hbar^2c^2\Delta + m^2c^4 + 2mc^2\hat{U}\right)\Psi_c(\vec{r},t) \tag{15.31}$$

From here, we observe that the Klein-Gordon equation for the particle in a scalar field is invariant relative to its complex conjugate. We observe also in Equations 15.30 and 15.31 that the scalar potential, $\hat{U}$, is invariant apart from the fact that Equations 15.30 and 15.31 describe the physical realizable particle and anti-particle. So $\hat{U}$ has the same character in the field of the particle and anti-particle.

The vector $\vec{A}$ and scalar ϕ potentials have the same Lorentz-transformation properties as $\hat{p}$ and H and for a particle of charge, e, in an electromagnetic field. So the Klein-Gordon equation for this case can be written as follows:

$$\left(i\hbar\frac{\partial}{\partial t} - e\phi\right)^2\Psi(\vec{r},t) = \left(c^2\left(\hat{p} - \frac{e}{c}\vec{A}\right)^2 + m^2c^4\right)\Psi(\vec{r},t) \tag{15.32}$$

The complex conjugate yields the Klein-Gordon equation for a particle of charge, $-e$, in an electromagnetic field:

$$\left(i\hbar\frac{\partial}{\partial t}+e\phi\right)^2\Psi_c(\vec{r},t)=\left(c^2\left(\hat{\vec{p}}+\frac{e}{c}\vec{A}\right)^2+m^2c^4\right)\Psi_c(\vec{r},t) \quad (15.33)$$

Equation 15.33 can be obtained from Equation 15.32 with the help of Equation 15.22 by the following correspondence:

$$\vec{A}\to\vec{A}_c,\ \phi\to\phi_c \quad (15.34)$$

We return to Equation 15.32 when

$$\vec{A}_c=-\vec{A},\ \phi_c=-\phi \quad (15.35)$$

If we consider a constant electromagnetic field, i.e., $\vec{A}$ and ϕ are independent of time, then Equation 15.33 has the solution:

$$\Psi_\in(\vec{r},t)=\exp\left\{-\frac{i\in t}{\hbar}\right\}\Psi(\vec{r}) \quad (15.36)$$

This describes the particle moving with positive as well negative energy, $\in$. We can imitate this solution to the case of a free particle when the field is switched on adiabatically. When $\in>0$, the solution imitates that of a free particle which has a physical sense and describes the possible states of the particle and also corresponds to the arbitrary superposition of such solutions like $\Psi^{(+)}$. The solution, when $\in<0$ corresponds to the arbitrary superposition of such solutions like $\Psi^{(-)}$, has no physical sense. In this case, we have

$$\Psi_c^{(+)}(\vec{r},t)=\hat{C}\Psi^{(-)}(\vec{r},t)\equiv\Psi^{(-)*}(\vec{r},t) \quad (15.37)$$

This is the wave function of an anti-particle and has the right time dependence which also satisfies the Klein-Gordon Equation 15.33 of a particle with charge, $-e$, opposite in sign to that of the particle with wave function, $\Psi^{(+)}$. From here, we observe that the solution of the Klein-Gordon equation for the particle in an electromagnetic field has the same form as that of the free particle:

$$\Psi(\vec{r},t)=\Psi^{(+)}(\vec{r},t)+\Psi^{(-)}(\vec{r},t)=\Psi^{(+)}(\vec{r},t)+\hat{C}\Psi_c^{(+)}(\vec{r},t) \quad (15.38)$$

We observe from here a particle and anti-particle. So any state of a particle described by the wave function, $\Psi^{(+)}(\vec{r},t)$, corresponds exactly to such a state of the anti-particle with the wave function:

$$\Psi_c^{(+)}(\vec{r},t)\equiv\Psi^{(+)}(\vec{r},t) \quad (15.39)$$

So to make a transition to an anti-particle, the potential of the external electromagnetic field should change sign and is consistent with the physical sense of the complex conjugate.

15.3 Spinless Charge Particle in a Coulombic Field

We find the energy levels of the discrete spectrum of a spinless charged particle with charge, $-e$, in a Coulombic field of the nucleus with point charge e. The energy levels and the corresponding wave function of the stationary states can be described by the Klein-Gordon equation:

$$\left(-\hbar^2c^2\Delta+\mathrm{m}^2c^4\right)\Psi=\left(E+\frac{Ze^2}{r}\right)^2\Psi \tag{15.40}$$

Since the problem has spherical symmetry, the wave function is selected in the form:

$$\Psi(\vec{r})=R_l(r)\mathrm{Y}_{lm}(\theta,\phi) \tag{15.41}$$

So from Equation 15.40, we have

$$\left(-\frac{\hbar^2}{2\mathrm{m}r^2}\frac{d}{dr}\left(r^2\frac{d}{dr}\right)+\frac{\hbar^2\left(\left(l+\frac{1}{2}\right)^2-\frac{1}{4}\right)}{2\mathrm{m}r^2}-\frac{Ze^2E}{\mathrm{m}c^2r}-\frac{Z^2e^4E}{2\mathrm{m}c^2r^2}\right)R=\frac{E^2-\mathrm{m}^2c^4}{2\mathrm{m}c^2}R \tag{15.42}$$

This problem is easily solvable by letting

$$\alpha=\frac{e^2}{\hbar c},\ Z\to\frac{ZE}{\mathrm{m}c^2},\ \left(l+\frac{1}{2}\right)^2\to\left(l+\frac{1}{2}\right)^2-Z^2\alpha^2,\ E_{n_rl}\to\frac{E^2-\mathrm{m}^2c^4}{2\mathrm{m}c^2} \tag{15.43}$$

and rewriting the equation for a hydrogen-like atom:

$$\left(-\frac{\hbar^2}{2\mathrm{m}r^2}\frac{d}{dr}\left(r^2\frac{d}{dr}\right)+\frac{\hbar^2\left(\left(l+\frac{1}{2}\right)^2-\frac{1}{4}\right)}{2\mathrm{m}r^2}-\frac{Ze^2}{r}\right)\tilde{R}_{n_rl}=E_{n_rl}\tilde{R}_{n_rl} \tag{15.44}$$

Use the energy levels of the non-relativistic hydrogen-like atom:

$$E_{n_rl}\equiv E_n=-\frac{\mathrm{m}\left(Ze^2\right)^2}{2\hbar^2n^2}=-\frac{\mathrm{m}\left(Ze^2\right)^2}{2\hbar^2\left(n_r+\frac{1}{2}+l+\frac{1}{2}\right)^2} \tag{15.45}$$

From here and Equation 15.43, we have

$$\left(E^2-\mathrm{m}^2c^4\right)\left(n_r+\frac{1}{2}+\sqrt{\left(l+\frac{1}{2}\right)^2-Z^2\alpha^2}\right)^2=-Z^2\alpha^2E^2 \tag{15.46}$$

and

$$E_{n_rl}=\pm\frac{\mathrm{m}c^2}{\sqrt{1+\dfrac{Z^2\alpha^2}{\left(n_r+\frac{1}{2}+\sqrt{\left(l+\frac{1}{2}\right)^2-Z^2\alpha^2}\right)^2}}} \tag{15.47}$$

Note that the minus sign is not applicable since it gives rise to unwanted energy levels which are not solutions of Equation 15.42. This is confirmed by Equation 15.45 which conserves its form for a sign change of Z to $-Z$. This does not describe any spectrum as the potential becomes repulsive and shows the none existence of a discrete spectrum. Note that the Coulomb potential in Equation 15.42 is repulsive for $E<0$ and so

$$E_{n_r l} = \frac{mc^2}{\sqrt{1+\frac{Z^2\alpha^2}{\left(n_r+\frac{1}{2}+\sqrt{\left(l+\frac{1}{2}\right)^2 - Z^2\alpha^2}\right)^2}}} \tag{15.48}$$

15.4 Non-Relativistic Limiting Equation

We find from the spinless free particle Klein-Gordon equation the non-relativistic limiting case of the Schrödinger equation and the first-order correction to this equation. This will be examined again in Chapter 20 on second-order relativistic correction. The stationary Klein-Gordon equation takes the form:

$$\left(c^2\left(\hat{p}-\frac{e}{c}\vec{A}\right)^2 + m^2c^4\right)\Psi = (\varepsilon - U)^2\Psi,\ U = e\phi \tag{15.49}$$

Here, we consider a particle of charge e in an electromagnetic field described by the vector $\vec{A}$ and scalar ϕ potentials. For the limiting case,

$$|U| \ll mc^2,\ |E| \ll mc^2,\ \varepsilon = mc^2 + E \tag{15.50}$$

we have

$$\left(\frac{\left(\hat{p}-\frac{e}{c}\vec{A}\right)^2}{2m} + U - E\right)\Psi = \frac{(\varepsilon - U)^2}{2mc^2}\Psi \tag{15.51}$$

The right-hand side of this equation is exceedingly greater than the left-hand side. Neglecting the right-hand side in the zero approximation, we have the non-relativistic Schrödinger equation. To find the first-order correction to the Schrödinger equation, select the wave function from Equation 15.51 in the form:

$$\Psi = \Psi_0 + \Psi_1 \tag{15.52}$$

where Ψ_0 is the solution of the Schrödinger equation, while Ψ_1 is the change to the given solution due to the relativistic effect since

$$|\Psi_1| \ll |\Psi_0| \tag{15.53}$$

This permits us to transform Equation 15.51:

$$\frac{(\varepsilon - e\phi)^2}{2mc^2}(\Psi_0 + \Psi_1) \approx \frac{(\varepsilon - e\phi)^2}{2mc^2}\Psi_0 = \frac{1}{2mc^2}\left(\frac{\left(\hat{p} - \frac{e}{c}\vec{A}\right)^2}{2m}\right)^2 \Psi_0 \approx \frac{\left(\hat{p} - \frac{e}{c}\vec{A}\right)^4}{8m^3c^2}(\Psi_0 + \Psi_1) \quad (15.54)$$

From Equation 15.51, we have

$$\left(\frac{\left(\hat{p} - \frac{e}{c}\vec{A}\right)^2}{2m} - \frac{\left(\hat{p} - \frac{e}{c}\vec{A}\right)^4}{8m^3c^2} + U\right)\Psi = E\Psi \quad (15.55)$$

This implies that it is the equation of the eigenfunction and eigenvalue of the Hamiltonian, $\hat{H}$:

$$\hat{H} = \sqrt{\left(\hat{p} - \frac{e}{c}\vec{A}\right)^2 c^2 + m^2c^4} + U - mc^2 \approx \frac{\left(\hat{p} - \frac{e}{c}\vec{A}\right)^2}{2m} - \frac{\left(\hat{p} - \frac{e}{c}\vec{A}\right)^4}{8m^3c^2} + U \quad (15.56)$$

This is the generalized quantum mechanical expression of the classical theory.

16

Dirac Equation

We re-examine the square of the Hamiltonian, H, in Equation 15.7 and also the Dirac approach in finding the relativistic Hamiltonian. The classical Hamiltonian for the free particle should be the positive square root of Equation 15.7 which is not a linear function of the momenta. Applying this to the problem, say for example in Equation 15.10, we have the resultant wave equation that is asymmetrical with respect to space and time derivatives and so the necessity to linearize the equation in the momentum operators. So for Equation 15.10, it is impossible to include the Pauli spin matrices without destroying the invariance of the theory since the spin matrices transform the components of a three-dimensional rather than a four-dimensional vector. In addition, the function, Ψ, has one component rather than two components like spinors and so the Schrödinger relativistic equation represents a particle with no spin. So for a potential to be added in Equation 15.10, the Lorentz-transformation properties of such an addition should be guaranteed by matrix algebra which is the requirement for the Hamiltonian $\hat{H}$ to be linear in the momentum operators and guarantees relativistic invariance. This matrix algebra which permits the linearization of the Hamiltonian in the momentum operator was introduced by Dirac with the help of the Pauli matrices.

For the Pauli matrix, we consider

$$(\hat{p},\hat{\sigma})^2 = (\hat{p}_x\hat{\sigma}_x + \hat{p}_y\hat{\sigma}_y + \cdots)(\hat{p}_x\hat{\sigma}_x + \hat{p}_y\hat{\sigma}_y + \cdots) = \hat{p}_x^2\hat{\sigma}_x^2 + \hat{p}_y^2\hat{\sigma}_y^2 + \hat{p}_z^2\hat{\sigma}_z^2 + \hat{p}_x\hat{p}_y(\hat{\sigma}_x\hat{\sigma}_y + \hat{\sigma}_y\hat{\sigma}_x) + \cdots = \hat{p}^2 \quad (16.1)$$

and

$$\hat{H}^2 = \hat{p}^2c^2 + m^2c^4,\ \hat{p} = -i\hbar\nabla \quad (16.2)$$

We ask ourselves whether the rule for quantization of p in the relativistic domain is considered. We can say yes, as the only supposition in the evaluation is the invariance of space. For $\sqrt{\hat{H}^2}$, we find the relativistic Hamiltonian of a particle written for the first time by Dirac that is linear in the momentum and mass term:

$$\hat{H} = c\hat{\alpha}\hat{p} + \beta mc^2 \quad (16.3)$$

where $\hat{\alpha}$, β are not numbers but matrices independent of coordinates and momenta which should have the following properties:

$$\hat{\alpha}_x^2 = \hat{\alpha}_y^2 = \hat{\alpha}_z^2 = \beta^2 = \mathbb{I} \quad (16.4)$$

and

$$\hat{\alpha}_x\hat{\alpha}_y + \hat{\alpha}_y\hat{\alpha}_x = \hat{\alpha}_y\hat{\alpha}_z + \hat{\alpha}_z\hat{\alpha}_y = \hat{\alpha}_z\hat{\alpha}_x + \hat{\alpha}_x\hat{\alpha}_z = 0 \quad (16.5)$$

DOI: 10.1201/9781003273073-18

$$\hat{\alpha}_x\beta+\beta\hat{\alpha}_x=\hat{\alpha}_y\beta+\beta\hat{\alpha}_y=\hat{\alpha}_z\beta+\beta\hat{\alpha}_z=0 \tag{16.6}$$

The four quantities in Equation 16.4 are said to anti-commute in pairs as seen from Equations 16.5 and 16.6. As $\hat{\alpha}$ and β anti-commute rather than commute with each other, they cannot be numbers. Equation 16.3 is the simplest Hamiltonian that is linear in the momentum and mass term.

If we square Equation 16.3, then we have Equation 16.2 (this should be done independently by the reader). What should be the form of the matrices $\hat{\alpha}$ and β? What should be their rank? We construct $\hat{\alpha}$ in terms of the Pauli matrices, $\hat{\sigma}_i$ and so the representation space must have even dimension. We take the least number of rows and columns (which should be four) in order that these properties should be satisfied. So $\hat{\alpha}$ and β are 4×4 matrices. How can we write them? The Dirac representation should be used:

$$\hat{\alpha}_i=\begin{bmatrix}\mathbb{O}_2 & \hat{\sigma}_i\\ \hat{\sigma}_i & \mathbb{O}_2\end{bmatrix},\quad \beta=\begin{bmatrix}\mathbb{I}_2 & \mathbb{O}_2\\ \mathbb{O}_2 & -\mathbb{I}_2\end{bmatrix},\mathbb{O}_2=\begin{bmatrix}0 & 0\\ 0 & 0\end{bmatrix} \tag{16.7}$$

From these matrices, we have

$$\hat{\alpha}_i^2=\begin{bmatrix}\mathbb{O}_2 & \hat{\sigma}_i\\ \hat{\sigma}_i & \mathbb{O}_2\end{bmatrix}\begin{bmatrix}\mathbb{O}_2 & \hat{\sigma}_i\\ \hat{\sigma}_i & \mathbb{O}_2\end{bmatrix}=\begin{bmatrix}\hat{\sigma}_i^2 & \mathbb{O}_2\\ \mathbb{O}_2 & \hat{\sigma}_i^2\end{bmatrix}=\begin{bmatrix}\mathbb{I}_2 & \mathbb{O}_2\\ \mathbb{O}_2 & \mathbb{I}_2\end{bmatrix}=\mathbb{I}_4 \tag{16.8}$$

$$\beta^2=\begin{bmatrix}\mathbb{I}_2 & \mathbb{O}_2\\ \mathbb{O}_2 & -\mathbb{I}_2\end{bmatrix}\begin{bmatrix}\mathbb{I}_2 & \mathbb{O}_2\\ \mathbb{O}_2 & -\mathbb{I}_2\end{bmatrix}=\begin{bmatrix}\mathbb{I}_2 & \mathbb{O}_2\\ \mathbb{O}_2 & \mathbb{I}_2\end{bmatrix}=\mathbb{I}_4 \tag{16.9}$$

$$\begin{aligned}\hat{\alpha}_i\hat{\alpha}_j+\hat{\alpha}_j\hat{\alpha}_i&=\begin{bmatrix}\mathbb{O}_2 & \hat{\sigma}_i\\ \hat{\sigma}_i & \mathbb{O}_2\end{bmatrix}\begin{bmatrix}\mathbb{O}_2 & \hat{\sigma}_j\\ \hat{\sigma}_j & \mathbb{O}_2\end{bmatrix}+\begin{bmatrix}\mathbb{O}_2 & \hat{\sigma}_j\\ \hat{\sigma}_j & \mathbb{O}_2\end{bmatrix}\begin{bmatrix}\mathbb{O}_2 & \hat{\sigma}_i\\ \hat{\sigma}_i & \mathbb{O}_2\end{bmatrix}=\begin{bmatrix}\hat{\sigma}_i\hat{\sigma}_j+\hat{\sigma}_j\hat{\sigma}_i & \mathbb{O}_2\\ \mathbb{O}_2 & \hat{\sigma}_i\hat{\sigma}_j+\hat{\sigma}_j\hat{\sigma}_i\end{bmatrix}\\ &=\begin{bmatrix}2\left[\delta_{ij}\right] & \mathbb{O}_2\\ \mathbb{O}_2 & 2\left[\delta_{ij}\right]\end{bmatrix}=2\delta_{ij}\begin{bmatrix}\mathbb{I}_2 & \mathbb{O}_2\\ \mathbb{O}_2 & \mathbb{I}_2\end{bmatrix}\end{aligned} \tag{16.10}$$

$$\hat{\alpha}_i\beta+\beta\hat{\alpha}_i=\begin{bmatrix}\mathbb{O}_2 & \hat{\sigma}_i\\ \hat{\sigma}_i & \mathbb{O}_2\end{bmatrix}\begin{bmatrix}\mathbb{I}_2 & \mathbb{O}_2\\ \mathbb{O}_2 & -\mathbb{I}_2\end{bmatrix}+\begin{bmatrix}\mathbb{I}_2 & \mathbb{O}_2\\ \mathbb{O}_2 & -\mathbb{I}_2\end{bmatrix}\begin{bmatrix}\mathbb{O}_2 & \hat{\sigma}_i\\ \hat{\sigma}_i & \mathbb{O}_2\end{bmatrix}=\begin{bmatrix}\mathbb{O}_2 & \mathbb{O}_2\\ \mathbb{O}_2 & \mathbb{O}_2\end{bmatrix}=\mathbb{O}_4,\ i,j=1,2,3\to x,y,z \tag{16.11}$$

The Dirac time dependent equation gives

$$i\hbar\frac{\partial}{\partial t}\Psi=\left(c\hat{\alpha}\hat{p}+\beta mc^2\right)\Psi \tag{16.12}$$

This equation is symmetrical in space and time. It is a matrix equation or, equivalently, four simultaneous partial differential equations. Suppose Equation 16.12 describes a free particle, there are no terms dependent on space coordinates or time since such terms would have space-time-energy dependent properties that produce forces. In addition, the space and time derivatives appear only in the momentum, $\hat{p}$, and energy, E, and not in $\hat{\alpha}$ and β. As $\hat{\alpha}$ and β are independent of the coordinates and momenta, they commute with all such operators. Note that $\hat{\alpha}$ and β should be Hermitian so that the Hamiltonian should be Hermitian.

We imitate the solution of Equation 16.12 as for case of the time Schrödinger equation, and the stationary Dirac equation is:

$$\left(c\hat{\alpha}\hat{p}+\beta mc^2\right)\Psi = E\Psi \tag{16.13}$$

where $\hat{p}$ is a differential operator. But in $\hat{H}$ we find $\hat{\alpha}$ and β and so it follows from here that $\hat{H}$ is a 4×4 matrix operator and, in the matrix representation, the wave function is a four-component column matrix, Ψ, called a **bispinor**:

$$\Psi \equiv \begin{bmatrix} \phi \\ \phi' \end{bmatrix}, \phi = \begin{bmatrix} \Psi_1 \\ \Psi_2 \end{bmatrix}, \ \phi' = \begin{bmatrix} \Psi_3 \\ \Psi_4 \end{bmatrix} \tag{16.14}$$

A relativistic electron should have a four-component wave function-bispinor. The Dirac equation in Equation 16.12 or 16.13 is a system of 4×4 equations relative to the components of the bispinor. If it is written in the explicit form, then we have

$$\hat{\alpha}_x p_x \begin{bmatrix} \phi \\ \phi' \end{bmatrix} = p_x \begin{bmatrix} \mathbb{O}_2 & \hat{\sigma}_x \\ \hat{\sigma}_x & \mathbb{O}_2 \end{bmatrix} \begin{bmatrix} \phi \\ \phi' \end{bmatrix} = \hat{\sigma}_x p_x \begin{bmatrix} \phi' \\ \phi \end{bmatrix} \tag{16.15}$$

It follows that $\hat{\alpha}_x$ rotates the bispinor. The Dirac matrix is a compact matrix form of these systems of equations. Since Ψ is a four-component wave function, it relates a particle with a non-trivial spin.

17

Probability and Current Densities

The Dirac equation has both positive and negative energy solutions and so we find a way to define a meaningful probability density. To find the probability current density and probability density of the corresponding Dirac equation, it is necessary to write the conjugate equation of Equation 16.12:

$$-i\hbar\frac{\partial}{\partial t}\Psi^{*} = -\hat{p}\left(\hat{\alpha}\Psi\right)^{\dagger} + \left(\beta\Psi\right)^{\dagger} mc^{2},\quad \left(\hat{p}\to -\hat{p}\right) \tag{17.1}$$

or

$$-i\hbar\frac{\partial}{\partial t}\Psi^{\dagger} = -c\hat{p}\Psi^{\dagger}\hat{\alpha} + mc^{2}\Psi^{\dagger}\beta \tag{17.2}$$

We use the fact that

$$\left(AB\right)^{\dagger} = B^{\dagger}A^{\dagger} \tag{17.3}$$

and the fact that $\hat{\alpha}$ and β are Hermitian operators:

$$\hat{\alpha}^{\dagger} = \hat{\alpha},\quad \beta^{\dagger} = \beta \tag{17.4}$$

Suppose we multiply Equation 16.12 from the left by $\Psi^{\dagger}$ and Equation 17.2 from the right by Ψ and we take the difference of the results, we have

$$i\hbar\frac{\partial}{\partial t}\left(\Psi^{\dagger}\Psi\right) = c\left\{\Psi^{\dagger}\hat{\alpha}\hat{p}\Psi + \left(\hat{p}\Psi^{\dagger}\right)\hat{\alpha}\Psi\right\} \tag{17.5}$$

The right-hand side of which may be written as

$$-i\hbar\operatorname{div}\left(\Psi^{\dagger}\hat{\alpha}\Psi\right) = \operatorname{div}\hat{\alpha} = 0 \tag{17.6}$$

This follows that the components are independent of the coordinates and

$$-i\hbar\left[\Psi^{\dagger}\operatorname{div}\hat{\alpha}\Psi + \left(\hat{\alpha},\operatorname{grad}\left(\Psi^{\dagger}\Psi\right)\right)\right] = -i\hbar\left[\hat{\alpha}\left(\Psi\operatorname{grad}\Psi^{\dagger} + \Psi^{\dagger}\operatorname{grad}\Psi\right)\right] - i\hbar\operatorname{div}\left(\Psi^{\dagger}\hat{\alpha}\Psi\right) \tag{17.7}$$

from where the continuity equation:

$$\frac{\partial}{\partial t}\left(\Psi^{\dagger}\Psi\right) + c\operatorname{div}\left(\Psi^{\dagger}\hat{\alpha}\Psi\right) \equiv \frac{\partial}{\partial t}\rho + \operatorname{div}\vec{j} = 0 \tag{17.8}$$

DOI: 10.1201/9781003273073-19

where

$$\rho = \Psi^{\dagger}\Psi = \sum_{i=1}^{4} |\Psi_i|^2 \tag{17.9}$$

imitates the generalized quantum mechanical probability density for the case of a multi-component function, and the probability current density imitates

$$\vec{j} = c\Psi^{\dagger}\hat{\alpha}\Psi \tag{17.10}$$

This leads us to some paradoxical result:

$$\vec{j} = \rho\vec{v},\ \vec{v} = c\hat{\alpha} \tag{17.11}$$

If the eigenvalue of $\hat{\alpha}$ is ± 1, then it follows that $v = c$, which is the velocity of Dirac electron that is the velocity of light in vacuum. But an electron may move with any velocity and may not move with the velocity $v = c$.

We have seen earlier that $\vec{j}(\vec{r},t)$ imitates the usual probability current density. The description of particle or anti-particle states by the functions $\vec{j}$ and ρ is paradoxical since they do not possess any defined Lorentz transformation properties. So their interpenetration in the relativistic domain has no physical sense as earlier indicated. So it is suggestive to define the current density as in reference [12]:

$$j^{\mu} = \left\{ j^0, \vec{j} \right\} \equiv \left\{ \Psi^{\dagger}\Psi, c\Psi^{\dagger}\hat{\alpha}\Psi \right\} \tag{17.12}$$

This permits the continuity equation to be written in the covariant form which obeys the Lorentz transformation:

$$\partial_{\mu} j^{\mu} = 0 \tag{17.13}$$

This is an indication that the probability density, ρ, is the time component of j^{μ} and can then imitate the time coordinate under Lorentz transformation. It is evident that j^{μ} transforms like a four-vector. The time component, j^0, of the continuity equation 17.13 is always positive. As earlier seen above, the Lorentz property of ρ is used for the normalization of the wave function.

18

Electron Spin in the Dirac Theory

In non-relativistic quantum mechanics, the angular momentum, $\hat{L}$, is an integral of motion. The spin is added as a supplementary hypothesis. What should be the integral of motion in the Dirac theory for a particle described by the Hamiltonian, $\hat{\mathrm{H}}$?:

$$\hat{\mathrm{H}} = c\hat{\alpha}\hat{\mathrm{p}} + \beta \mathrm{m}c^2 \tag{18.1}$$

In the relativistic quantum mechanics, the angular momentum, $\hat{L}$, does not commute with $\hat{\mathrm{H}}$. From

$$\left[\hat{\mathrm{H}}, \hat{L}_z\right] = c\hat{\alpha}_x\left[\hat{\mathrm{p}}_x, \hat{L}_z\right] + c\hat{\alpha}_y\left[\hat{\mathrm{p}}_y, \hat{L}_z\right] = c\hat{\alpha}_x\left[\hat{\mathrm{p}}_x, \hat{L}_z\right] + c\hat{\alpha}_y\left[\hat{\mathrm{p}}_y, \hat{L}_z\right] \tag{18.2}$$

and considering

$$\left[\hat{L}_i, \hat{L}_j\right] = i\hbar \in_{ijk} \hat{L}_k \tag{18.3}$$

where $\in_{ijk}$ is the Levi Civitas tensor:

$$\left[\hat{\mathrm{H}}, \hat{L}\right] = i\hbar c\left[\hat{\alpha}, \hat{\mathrm{p}}\right] \tag{18.4}$$

Here, now $\left[\hat{\alpha}, \hat{\mathrm{p}}\right]$ is a vector or cross product and $\left[\hat{\mathrm{H}}, \hat{L}\right]$ is a commutator. So from Equation 18.4, the angular momentum, $\hat{L}$, is not an integral of motion.

From the matrix vector:

$$\hat{\Sigma} = \begin{bmatrix} \hat{\sigma} & \mathbb{O}_2 \\ \mathbb{O}_2 & \hat{\sigma} \end{bmatrix} \tag{18.5}$$

it is easily seen that

$$\left[\hat{\Sigma}_x, \hat{\alpha}_x\right] = \left[\hat{\Sigma}_x, \beta\right] = \mathbb{O}_2 \tag{18.6}$$

We find

$$\left[\hat{\Sigma}_x, \hat{\alpha}_y\right] = \begin{bmatrix} \hat{\sigma}_x & \mathbb{O}_2 \\ \mathbb{O}_2 & \hat{\sigma}_x \end{bmatrix}\begin{bmatrix} \mathbb{O}_2 & \hat{\sigma}_y \\ \hat{\sigma}_y & \mathbb{O}_2 \end{bmatrix} - \begin{bmatrix} \mathbb{O}_2 & \hat{\sigma}_y \\ \hat{\sigma}_y & \mathbb{O}_2 \end{bmatrix}\begin{bmatrix} \sigma_x & \mathbb{O}_2 \\ \mathbb{O}_2 & \sigma_x \end{bmatrix} = \begin{bmatrix} \mathbb{O}_2 & \hat{\sigma}_x\hat{\sigma}_y \\ \hat{\sigma}_x\hat{\sigma}_y & \mathbb{O}_2 \end{bmatrix} - \begin{bmatrix} \mathbb{O}_2 & \hat{\sigma}_y\hat{\sigma}_x \\ \hat{\sigma}_y\hat{\sigma}_x & \mathbb{O}_2 \end{bmatrix}$$

$$= 2i\begin{bmatrix} \mathbb{O}_2 & \hat{\sigma}_z \\ \hat{\sigma}_z & \mathbb{O}_2 \end{bmatrix} = 2i\hat{\alpha}_z \tag{18.7}$$

DOI: 10.1201/9781003273073-20

This can be proven to involve all components:

$$\left[\hat{\Sigma}_i,\hat{\alpha}_k\right]=\begin{bmatrix}\hat{\sigma}_i & \mathbb{O}_2\\ \mathbb{O}_2 & \hat{\sigma}_i\end{bmatrix}\begin{bmatrix}\mathbb{O}_2 & \hat{\sigma}_k\\ \hat{\sigma}_k & \mathbb{O}_2\end{bmatrix}-\begin{bmatrix}\mathbb{O}_2 & \hat{\sigma}_k\\ \hat{\sigma}_k & \mathbb{O}_2\end{bmatrix}\begin{bmatrix}\hat{\sigma}_i & \mathbb{O}_2\\ \mathbb{O}_2 & \hat{\sigma}_i\end{bmatrix}=\begin{bmatrix}\mathbb{O}_2 & \hat{\sigma}_i\hat{\sigma}_k\\ \hat{\sigma}_i\hat{\sigma}_k & \mathbb{O}_2\end{bmatrix}-\begin{bmatrix}\mathbb{O}_2 & \hat{\sigma}_k\hat{\sigma}_i\\ \hat{\sigma}_k\hat{\sigma}_i & \mathbb{O}_2\end{bmatrix}$$

$$=2i\begin{bmatrix}\mathbb{O}_2 & 2i\in_{ikl}\hat{\sigma}_l\\ 2i\in_{ikl}\hat{\sigma}_l & \mathbb{O}_2\end{bmatrix}=2i\in_{ikl}\hat{\alpha}_l \tag{18.8}$$

We consider the fact that

$$\hat{\sigma}_x\hat{\sigma}_y=i\hat{\sigma}_z \tag{18.9}$$

So

$$\left[\hat{\mathrm{H}},\hat{\Sigma}_x\right]=c\hat{\mathrm{p}}_y\left[\hat{\alpha}_y,\hat{\Sigma}_x\right]+c\hat{\mathrm{p}}_z\left[\hat{\alpha}_z,\hat{\Sigma}_x\right]=-2ic\left(\hat{\alpha}_z\hat{\mathrm{p}}_y-\hat{\alpha}_y\hat{\mathrm{p}}_z\right)=2ic\left[\hat{\alpha},\hat{\mathrm{p}}\right]_x \tag{18.10}$$

where $\left[\hat{\mathrm{H}},\hat{\Sigma}_x\right]$ is a commutator and $\left[\hat{\alpha},\hat{\mathrm{p}}\right]$ is a vector product:

$$\left[\hat{\mathrm{H}},\hat{\Sigma}\right]=2ic\left[\hat{\alpha},\hat{\mathrm{p}}\right] \tag{18.11}$$

We introduce the spin, $\hat{S}$:

$$\hat{S}=\frac{\hbar}{2}\hat{\Sigma} \tag{18.12}$$

Multiply Equation 18.11 by $\frac{\hbar}{2}$ and take the sum of the result with Equation 18.4, and we have

$$\left[\hat{\mathrm{H}},\hat{L}+\hat{S}\right]\equiv\left[\hat{\mathrm{H}},\hat{J}\right]=0 \tag{18.13}$$

This follows that the integral of motion in the Dirac theory should be $\hat{J}$, where $\hat{S}$ is the operator of the intrinsic angular momentum (spin) which is independent of the coordinates and momenta. This shows a considerable difference with the non-relativistic theory where, for the free particle, the angular momentum, $\hat{L}$, without the spin, $\hat{S}$, is an integral of motion. In the relativistic theory, the angular momentum $\hat{L}$ and the spin $\hat{S}$ are not conserved separately. It is their sum, $\hat{J}$, that is conserved and so it is an integral of motion in the Dirac theory. **We observe that in the Dirac theory, the spin of an electron is obtained very elegantly without any supplementary postulate as for the Schrödinger theory.**

From the commutation relation, $\left[\hat{J}_i,\hat{J}_j\right]=i\hbar\in_{ijk}\hat{J}_k$, the components of the angular momentum should have mutually incompatible observables, while $\left[\hat{J}^2,\hat{J}_j\right]=0$ shows the existence of simultaneous eigenstates of $\hat{J}^2$ and one of the components, $\hat{J}_j$. The operator, $\hat{J}$, quantizes according to the standard rule for the spin one-half particle:

$$J_z=m_j\hbar \tag{18.14}$$

where

$$m_j=m+m_s=m\pm\frac{1}{2} \tag{18.15}$$

The operator $\hat{J}^2$ quantizes as

$$J^2 = \hbar^2 j(j+1), \quad j = l+s,\ldots,|l-s| \tag{18.16}$$

But in the given case, $s = \frac{1}{2}$ for an electron, and if $l = 0$, then $s = \frac{1}{2}$, and if $l \neq 0$, then

$$j = l \pm \frac{1}{2} \tag{18.17}$$

This emphasizes the existence of half-integer values of the angular momentum associated with the electron as confirmed by the Stern-Gerlach experiment.

The appearance of the spin, S, is not a hypothesis but the result of Dirac theory and we should note:

- The spin S of an electron is a relativistic effect.
- The spin S is a real quantum effect and has eigenvalue of its projection on some direction equal to $\frac{\hbar}{2}$ or 0. If $\hbar \to 0$, i.e., for the transition to classical mechanics, then the spin S vanishes. So it is not correct to say the spin relates the electron about some axis.

19

Free Electron State with Defined Momentum-Positronium Motion

We examine a **positronium**, a so-called atom made up of a bound state of an electron, e^-, and a **positron**, e^+. This imitates a hydrogen atom, except that a positron replaces the proton. The positronium like the hydrogen atom has many states. In addition, like the hydrogen, the ground state of the positronium is split into a hyperfine structure by the interaction with magnetic moments. The electron and positron have each spin one-half that can be either parallel or anti-parallel to any given axis. It is instructive to note that in the ground state, there is no other angular momentum due to orbital motion and so there should be four states, i.e., three sub-states with spin-one system all having the same energy, and the other one is a state of spin zero with a different energy.

19.1 Stationary Dirac Equation

Our problem is to find the solution of the stationary Dirac equation for the positronium:

$$\hat{H}\Psi = E\Psi \tag{19.1}$$

or

$$\left(c\hat{\alpha}\hat{p} + \beta mc^2\right)\Psi = E\Psi \tag{19.2}$$

We find the state with a defined p. The eigenfunctions of the operators $\hat{p}_x, \hat{p}_y, \hat{p}_z$ are de Broglie plane waves $\Psi \approx \exp\left\{i\frac{\vec{p}\vec{r}}{\hbar}\right\}$ that are the same as in relativistic quantum mechanics as $\hat{p}$ has the same form. It follows that Ψ satisfying Equation 19.1 should have the same factor where Ψ is a bispinor:

$$\Psi = \mathfrak{u} \exp\left\{i\frac{\vec{p}\vec{r}}{\hbar}\right\} \tag{19.3}$$

and $\mathfrak{u}$ is a **numerical bispinor (four-component spinor)** that is coordinate-independent:

$$\mathfrak{u} = \begin{bmatrix} u_1 \\ u_2 \\ u_3 \\ u_4 \end{bmatrix} \tag{19.4}$$

DOI: 10.1201/9781003273073-21

If now in Equation 19.1, we substitute Equation 19.3 then the exponent is cancelled. It should be noted that the operator $\hat{p}$ acts on the free positron to give p its eigenvalue:

$$\left(c\hat{\alpha}\hat{p}+\beta mc^2\right)\mathbb{u}=\mathbb{u}E \tag{19.5}$$

We solve Equation 19.5. Suppose that

$$\mathbb{u}=\begin{bmatrix} \mathbb{w} \\ \mathbb{w}' \end{bmatrix}, \mathbb{w}=\begin{bmatrix} u_1 \\ u_2 \end{bmatrix}, \quad \mathbb{w}'=\begin{bmatrix} u_3 \\ u_4 \end{bmatrix} \tag{19.6}$$

We find

$$\left\{c\vec{p}\begin{bmatrix} \mathbb{O} & \hat{\sigma} \\ \hat{\sigma} & \mathbb{O} \end{bmatrix}+mc^2\begin{bmatrix} \mathbb{I} & \mathbb{O} \\ \mathbb{O} & -\mathbb{I} \end{bmatrix}\right\}\begin{bmatrix} \mathbb{w} \\ \mathbb{w}' \end{bmatrix}=E\begin{bmatrix} \mathbb{w} \\ \mathbb{w}' \end{bmatrix} \tag{19.7}$$

or

$$c\vec{p}\hat{\sigma}\mathbb{w}'+mc^2\mathbb{w}=E\mathbb{w},\ c\vec{p}\hat{\sigma}\mathbb{w}-mc^2\mathbb{w}'=E\mathbb{w}' \tag{19.8}$$

This system has a solution different from zero if its determinant is equal to zero:

$$\begin{vmatrix} mc^2-E & c\vec{p}\hat{\sigma} \\ c\vec{p}\hat{\sigma} & -mc^2-E \end{vmatrix}=0 \tag{19.9}$$

or

$$E^2-m^2c^4-c^2p^2=0 \tag{19.10}$$

from where:

$$E^2=m^2c^4+c^2p^2 \tag{19.11}$$

This is the relativistic energy of a free particle.

Thus,

$$E=\pm\sqrt{m^2c^4+c^2p^2}=\pm\varepsilon \tag{19.12}$$

where ε is the absolute value of the energy. The positive $+\varepsilon$ and negative $-\varepsilon$ energies may correspond, respectively, to the energies of the particle and anti-particle. It is instructive to note from here that the positronium is short-lived and this is a significant difference with the hydrogen atom. The positron should be the anti-particle of the electron since they annihilate each other. When the two particles completely disappear, they transform their rest energy into radiation, which appears as γ-rays or photons. During disintegration, two particles with a finite rest mass go into two or more objects which have zero rest mass.

We find the eigenspinors, $\mathbb{v}$ and $\mathbb{w}$. If we consider Equations 19.8, 19.12 and the fact that $E=+\varepsilon$, then we have

$$\mathbb{w}=\frac{c\vec{p}\hat{\sigma}}{E-mc^2}\mathbb{w}' \tag{19.13}$$

Let us introduce a supplementary condition: The particle is moving with respect to the oz-axis:

$$\vec{p}\hat{\sigma}=p\sigma_z \tag{19.14}$$

We suppose that $\mathbb{w}$ and $\mathbb{w}'$ are the eigenspinors of the operator $\hat{\sigma}_z$:

$$\mathbb{w}=\begin{bmatrix}1\\0\end{bmatrix}, \mathbb{w}'=\begin{bmatrix}0\\1\end{bmatrix} \tag{19.15}$$

For which, in the first case, we have $\sigma_z=1$, and in the second, $\sigma_z=-1$. So we have the solution of Dirac equation:

$$\mathbb{u}^{(1)}\equiv\mathbb{u}_{\uparrow}^{(+)}=C_1\begin{bmatrix}1\\0\\\dfrac{cp}{\varepsilon+mc^2}\\0\end{bmatrix}, \quad \mathbb{u}^{(2)}\equiv\mathbb{u}_{\downarrow}^{(+)}=C_2\begin{bmatrix}0\\1\\0\\-\dfrac{cp}{\varepsilon+mc^2}\end{bmatrix} \tag{19.16}$$

The coefficients C_1 and C_2 can be obtained from the normalization condition:

$$\sum_{i=1}^{4}\left|\mathbb{u}^{(i)}\right|^2=1 \tag{19.17}$$

The expressions in Equation 19.16 are two solutions of Dirac equation that correspond to the plus sign in Equation 19.12 (positive energy E). There are two because for the given value of the energy, the two values take the projection of the spin in the direction of the motion ($\sigma_z=\pm1$). The projection of the spin in the direction of motion is the polarization of the electron.

Let us move to the non-relativistic case:

$$m^2c^4 \gg p^2c^2 \text{ or } m^2c^2 \gg p^2 \tag{19.18}$$

then

$$\varepsilon \approx mc^2 \tag{19.19}$$

From Equation 19.13, we have

$$\frac{\mathbb{w}}{\mathbb{w}'}=\frac{cp}{2mc^2}=\frac{mv}{2mc}=\frac{1}{2}\frac{v}{c}\rightarrow 0 \tag{19.20}$$

then

$$\mathbb{W} \to 0 \tag{19.21}$$

This follows that the bispinor tends to the spinor.

Consider the second solution: If we consider Equation 19.8, then we have for $-\varepsilon$:

$$\mathbb{W}' = \frac{c\vec{p}\hat{\sigma}}{-\varepsilon + mc^2}\mathbb{W} \tag{19.22}$$

Similarly, with $\mathbb{W}'$ being the eigenspinor of σ_z:

$$\mathbb{u}^{(3)} \equiv \mathbb{u}_{\uparrow}^{(-)} = C_3 \begin{bmatrix} -\frac{cp}{-\varepsilon + mc^2} \\ 0 \\ 1 \\ 0 \end{bmatrix}, \quad \mathbb{u}^{(4)} \equiv \mathbb{u}_{\downarrow}^{(-)} = C_4 \begin{bmatrix} 0 \\ \frac{cp}{-\varepsilon + mc^2} \\ 0 \\ 1 \end{bmatrix} \tag{19.23}$$

So we have four states from the secular equation for which we have two values of the energy E with each sign which corresponds to a solution as expected. This implies the electron has two polarizations. We can also talk about a particle and anti-particle.

In the non-relativistic mechanics, the energy ε is defined to the approximation of a constant summand in relativistic mechanics (the absolute value). The negative value of the energy ε implies the negative mass m. These states are paradoxical. It is instructive to note that due to negative energies, the number of components in the relativistic theory doubles compared with the non-relativistic counterpart. This is because each solution is possible for positive as well as negative energy eigenvalues.

Consider the energy spectrum of a free Dirac equation (Figure 19.1):

$$E = \pm\sqrt{m^2c^4 + c^2p^2} \tag{19.24}$$

then we have the gap separating the positive and negative energies:

$$\Delta E = 2mc^2 \tag{19.25}$$

There is no lower limit in the side of the negative value. It follows that the electron does not have the ground state. This contradicts the law of conservation of energy. This implies that the supposition of the electron having a negative energy is not possible.

The classical electron may not overcome the gap ΔE through quantum mechanics, as transitions occur between discrete states. The negative energy solutions are lower in energy eigenvalue is lower and our system will prefer to **jump** down into the lowest energy state and completely counter intuitive and would lead to the collapse of the known models like the hydrogen atom. If this is the case, why not just rule out the negative energy as being unphysical. It is instructive to note that in quantum, mechanics, our system leaves in the Hilbert space which contains states with positive as well as negative energies. So restricting ourselves to only positive energy states will limit ourselves to only a sub-space of the complete Hilbert space. So for a relativistic theory, positive as well as negative energies are needed as

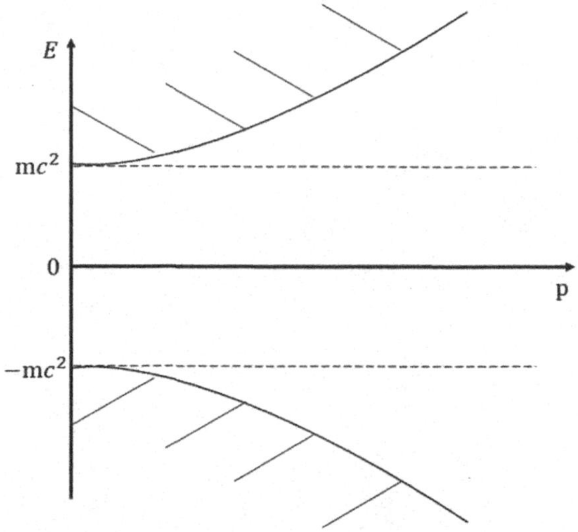

FIGURE 19.1 Positive as well as negative energies for a relativistic theory.

depicted in Figure 19.1. For negative energy states, there arises the problem of stability of matter where Dirac found a hypothesis for it.

19.1.1 Dirac Hypothesis-Hole Theory

For in the Dirac hypothesis, the **vacuum state** is that where negative energy ($-E$) states are filled with electrons. It is instructive to note that these are all passive electrons since they do not produce any observable electromagnetic field say. From this vacuum state definition, there is no stability problem since all the negative energy states are filled with electrons. In addition, as electrons are fermions, the Pauli exclusion principle prohibits any positive energy ($+E$) electron to cascade down.

There exist far reaching consequences of the Dirac hypothesis like when the electron with a negative energy is given the energy $E > 2mc^2$, which excites the negative energy electron into a positive energy state. As the negative energy states are all filled, the absence of the electron would appear as a hole with opposite charge and positive energy. So we have a system with charge q and energy E for the state when all the negative levels are occupied. The absence of an electron with a negative mass ($-$m) implies the appearance of an electron with a positive mass (+m). The absence of a particle with a negative charge ($-q$) implies the appearance of a particle with a positive charge ($+q$). This follows the appearance of a hole with a positive charge ($+q$) and a positive charge (**positron**) with mass (+m). This process is called **pair production**. So the **condition for the appearance of a hole (positron)** is as follows – In order that an electron with a negative energy in an impurity, say, should be in interaction with a pure state with energy $E \geq 2mc^2$, there should be the energy quantization $\hbar\omega \geq 2mc^2$ that is accompanied by the creation of an electron-positron pair. In the zone with the negative energy, we have a hole with energies of the order $\hbar\omega \approx Mev$. A similar examination may be done for a proton. Here arises a proton-anti-proton pair. So the Dirac hypothesis predicts that for every particle, there must exist an **anti-particle** of identical mass and opposite charge.

It is instructive to note that the Dirac theory is not really a particle on its own but an infinitely many-particles one as the vacuum is redefined to contain infinitely many particles. So any physical wave function must have this information. This general feature combines relativity with quantum mechanics. So dealing with many particle states cannot be avoided and may lead us to the study of quantum field theory in an elegant manner [23]. Unlike the Dirac equation where the vacuum is so unsymmetrical in the charges, the Dirac field leads to a charge symmetrical description of the vacuum.

20

Dirac Equation

20.1 Electron Motion in an External Electromagnetic Field

Consider the motion of an electron in an external electromagnetic field. The four-dimensional vector potential should characterize the electromagnetic field $A_i = \{\phi, \vec{A}\}$, where $\vec{A}$ is the three-dimensional vector potential and ϕ is the scalar potential.

From the relativistic mechanics, it follows that the transition from free motion to the motion in an electromagnetic field follows the change:

$$\hat{p}_i \rightarrow \hat{p}_i - \frac{e}{c} A_i \tag{20.1}$$

Here, e is the charge on the particle and c is the speed of light. In the Dirac equation, we have

$$\hat{\alpha}\hat{p} \rightarrow \hat{\alpha}\left(\hat{p} - \frac{e}{c}\vec{A}\right) \tag{20.2}$$

This is not sufficient and so it is necessary to make the transformation using the Hamiltonian, $\hat{H}$. From the time Schrödinger equation, it follows that for $x_4 = ict$, then

$$\hat{H} \rightarrow i\hbar \frac{\partial}{\partial t} = -\hbar c \frac{\partial}{\partial x_0} \tag{20.3}$$

On the other hand,

$$\hat{p}_i = -i\hbar \frac{\partial}{\partial x_i} \tag{20.4}$$

and

$$\hat{p}_0 = -i\hbar \frac{\partial}{\partial x_0} \tag{20.5}$$

Hence,

$$-\hbar \frac{\partial}{\partial x_0} = i\hat{p}_0 \tag{20.6}$$

DOI: 10.1201/9781003273073-22

and

$$\hat{H} \to ic\hat{p}_0 \tag{20.7}$$

If we consider Equation 20.1, then we have

$$\hat{H} \to ic\left(\hat{p}_0 - \frac{e}{c}A_0\right) = i\hbar\frac{\partial}{\partial t} + e\phi = \hat{H} + e\phi = \hat{H} + \hat{U} \tag{20.8}$$

Thus, effecting the transformation for $\hat{H}$, it is necessary to add $\hat{U}$ and do the change in Equation 20.2.
Finally,

$$\hat{H} = c\hat{\alpha}\left(\hat{p} - \frac{e}{c}\vec{A}\right) + \beta mc^2 + \hat{U} \tag{20.9}$$

This is the Dirac Hamiltonian in an external electromagnetic field.

It may be used for the solution analogous stationary problems in the Schrödinger quantum mechanics. In particular, Kepler's problem – the motion of an electron in a Coulomb field – is solved exactly and also in Dirac theory.

20.1.1 Quasi-Relativistic Approximation-Pauli Equation

We have in this case two limiting expressions for the Hamiltonian $\hat{H}$: The Schrödinger for the non-relativistic case and the Dirac for the relativistic case. Between them, there should exist a limiting transformation. If we tend the velocity of light to infinity, i.e., $c \to \infty$, then the Hamiltonian $\hat{H}$ for the Dirac case is transformed to that of Schrödinger.

Consider a general problem: We series expand the Dirac Hamiltonian with respect to $\frac{v}{c}$. This gives us a quasi-relativistic case (there is a relativistic effect to a small degree).

Consider the stationary Dirac equation:

$$\hat{H}\Psi = E\Psi \tag{20.10}$$

where $\hat{H}$ is obtained from Equation 20.9. In Equation 20.10, Ψ is a bispinor:

$$\Psi = \begin{bmatrix} \phi \\ \chi \end{bmatrix} \tag{20.11}$$

where ϕ, χ are spinors. It follows that we have the system:

$$c\left(\hat{p} - \frac{e}{c}\vec{A}\right)\hat{\sigma}\chi + mc^2\phi + U\phi = E\phi,\ c\left(\hat{p} - \frac{e}{c}\vec{A}\right)\hat{\sigma}\phi - mc^2\chi + U\chi = E\chi,\ E = mc^2 + \varepsilon \tag{20.12}$$

From here:

$$c\left(\hat{p} - \frac{e}{c}\vec{A}\right)\hat{\sigma}\chi + U\chi = \varepsilon\chi \tag{20.13}$$

and

$$\chi = \frac{c\left(\hat{p} - \frac{e}{c}\vec{A}\right)\hat{\sigma}}{2mc^2 + \varepsilon - U}\phi \tag{20.14}$$

For now, we have no approximation.

Suppose that

$$|\varepsilon|,|U| \ll mc^2 \tag{20.15}$$

In the quasi-classical case, we limit ourselves to the terms proportional to $\frac{v}{c}$ neglecting ε and U in Equation 20.14, and so we have

$$\frac{\chi}{\phi} \approx \frac{p}{2mc} \approx \frac{1}{2}\frac{v}{c} \tag{20.16}$$

If

$$\frac{v}{c} \to 0 (c \to \infty) \tag{20.17}$$

then

$$\chi \to 0 \tag{20.18}$$

and so the bispinor in Equation 20.11 tends to a spinor and

$$\frac{\left[\left(\hat{p} - \frac{e}{c}\vec{A}\right)\hat{\sigma}\right]^2}{2m}\phi + U\phi = \varepsilon\phi \tag{20.19}$$

It should be noted that

$$(\hat{p}\hat{\sigma})^2 = \hat{p}^2 \tag{20.20}$$

We should be very careful here as the components of $\hat{p}$ commute.

Let us find the equivalent of the operator:

$$\left(\hat{\vec{A}},\hat{\sigma}\right)\left(\hat{\vec{B}},\hat{\sigma}\right) \tag{20.21}$$

where $\hat{\vec{A}}$ and $\hat{\vec{B}}$ are two non-commutable arbitrary vectors acting on the electron subsystem which, however, commute with the Pauli matrices:

$$\left(\hat{\vec{A}},\hat{\sigma}\right)\left(\hat{\vec{B}},\hat{\sigma}\right) = \left(\hat{A}_x\hat{\sigma}_x + \hat{A}_y\hat{\sigma}_y + \cdots\right)\left(\hat{B}_x\hat{\sigma}_x + \hat{B}_y\hat{\sigma}_y + \cdots\right) \tag{20.22}$$

or

$$\left(\hat{\vec{A}},\hat{\sigma}\right)\left(\hat{\vec{B}},\hat{\sigma}\right) = \hat{A}_x\hat{\sigma}_x\hat{B}_x\hat{\sigma}_x + \hat{A}_x\hat{B}_x + \hat{A}_y\hat{B}_y + \cdots + \hat{A}_x\hat{B}_y\hat{\sigma}_x\hat{\sigma}_y + \hat{A}_y\hat{B}_x\hat{\sigma}_y\hat{\sigma}_x \tag{20.23}$$

or

$$\left(\hat{\vec{A}},\hat{\sigma}\right)\left(\hat{\vec{B}},\hat{\sigma}\right) = \left(\hat{\vec{A}},\hat{\vec{B}}\right)_x + \hat{\sigma}_x\hat{\sigma}_y\left(\hat{A}_x\hat{B}_y - \hat{A}_y\hat{B}_y\right) + \cdots \tag{20.24}$$

or

$$\left(\hat{\vec{A}},\hat{\sigma}\right)\left(\hat{\vec{B}},\hat{\sigma}\right)=\left(\hat{\vec{A}},\hat{\vec{B}}\right)_x+\hat{\sigma}_x\hat{\sigma}_y\left[\hat{\vec{A}},\hat{\vec{B}}\right]_z+\cdots \tag{20.25}$$

So,

$$\left(\hat{\vec{A}},\hat{\sigma}\right)\left(\hat{\vec{B}},\hat{\sigma}\right)=\left(\hat{\vec{A}},\hat{\vec{B}}\right)+i\hat{\sigma}_z\left[\hat{\vec{A}},\hat{\vec{B}}\right]_z+\cdots=\left(\hat{\vec{A}},\hat{\vec{B}}\right)+i\left(\hat{\sigma},\left[\hat{\vec{A}},\hat{\vec{B}}\right]\right) \tag{20.26}$$

For our case, we consider

$$\hat{\vec{A}}=\hat{\vec{B}}=\hat{p}-\frac{e}{c}\vec{A} \tag{20.27}$$

then

$$\left[\hat{\vec{A}},\hat{\vec{A}}\right]\neq 0 \tag{20.28}$$

This shows that the operator $\hat{\vec{A}}$ is non-commutable with itself.

In the given case,

$$\left[\hat{p}-\frac{e}{c}\hat{\vec{A}},\hat{p}-\frac{e}{c}\hat{\vec{A}}\right]=-\frac{e}{c}\left\{\left[\hat{p},\hat{\vec{A}}\right]+\left[\hat{\vec{A}},\hat{p}\right]\right\} \tag{20.29}$$

where the quantity that stands on the right-hand side (left-hand side) are vector products and for

$$\hat{p}=-i\hbar\nabla \tag{20.30}$$

we have

$$\left[\hat{p}-\frac{e}{c}\hat{\vec{A}},\hat{p}-\frac{e}{c}\hat{\vec{A}}\right]=\frac{ie\hbar}{c}\left\{\left[\nabla,\hat{\vec{A}}\right]+\left[\hat{\vec{A}},\nabla\right]\right\}=\frac{ie\hbar}{c}\mathrm{curl}\hat{\vec{A}} \tag{20.31}$$

It should be noted that

$$\left[\hat{p},\hat{p}\right]=0,\quad \left[\vec{A},\vec{A}\right]=0 \tag{20.32}$$

and from the magnetic field strength:

$$\mathrm{curl}\vec{A}=\vec{H} \tag{20.33}$$

then we have

$$\frac{1}{2m}\left(\hat{p}-\frac{e}{c}\vec{A}\right)^2\phi+U\phi-\frac{e\hbar}{2mc}\left(\hat{\sigma},\vec{H}\right)\phi=\varepsilon\phi \tag{20.34}$$

which is the **Pauli equation** that was originally postulated by him. From electrodynamics, it is known that from the Hamilton equation for the motion in a magnetic field, we do the transformation:

$$\hat{p} \to \hat{p} - \frac{e}{c}\vec{A} \tag{20.35}$$

From the magnetic moment of an electron, $\vec{M}$:

$$\vec{M} = \frac{e\hbar}{2mc}\hat{\sigma} \tag{20.36}$$

Then in the Pauli equation when a magnetic field $\vec{H}$ is applied, this gives rise to an interaction energy:

$$\frac{e\hbar}{2mc}\left(\hat{\sigma},\vec{H}\right) = \left(\vec{M},\vec{H}\right) \tag{20.37}$$

which is the potential energy of a magnetic dipole moment $\vec{M}$ in the magnetic field $\vec{H}$. It follows that $\vec{M}$ is the operator of the intrinsic magnetic moment of an electron. So the electron behaves as a particle with a magnetic dipole moment in an external magnetic field. Note that here the magnetic moment of an electron is automatically obtained from the theory, unlike the non-relativistic case where it is obtained from experiment.

If

$$\hat{S} = \frac{\hbar}{2}\hat{\sigma} \tag{20.38}$$

then

$$\frac{M_z}{S_z} = \frac{e}{mc} \tag{20.39}$$

which is a gyromagnetic ratio. In the classical theory, it gives us $\frac{e}{2mc}$.

It follows that

a. An electron in addition to the spin has an intrinsic magnetic moment. Considering Equation 20.36 its numerical value:

$$M_z = \pm\mu_B, \quad \mu_B = \frac{e\hbar}{2mc} \tag{20.40}$$

– is the **so-called Bohr magneton.**

b. For a free magnetic moment, an analogous gyromagnetic relation is true that gives Equation 20.39. It is twice more than that given by the classical theory.

Both results correspond to that foreseen by the experiments. Equation 20.39 was experimentally obtained in 1913 in the experiment of Einstein-de Haas.

It follows that ferromagnetism is connected with the spin for which on that basis Heisenberg constructed the theory on ferromagnetism. For spontaneous magnetization, the spin of all electrons is directed to one direction. Ferromagnetism is a quantum macroscopic effect. This can be seen in superconductivity, superfluidity and lasers.

The Stern-Gerlach experiment confirms Equation 20.34. The Pauli equation differs from that of Schrödinger by the spinor, ϕ.

20.1.2 Second-Order Relativistic Correction

20.1.2.1 Spin-Orbital Interaction

During the evaluation of Pauli equation, we made a limitation to the terms of the first order with respect to $\frac{v}{c}$. We consider that there exists only the external electric field strength and so for that reason, we have

$$\vec{H}=0, \quad \vec{A}=0 \tag{20.41}$$

Considering Equation 20.12, we have

$$c(\hat{p}\hat{\sigma})\chi=(\varepsilon-U)\phi,\ c(\hat{p}\hat{\sigma})\phi=(\varepsilon-U+2mc^2)\chi=,\ \varepsilon=E-mc^2 \tag{20.42}$$

and the second equation relative to χ:

$$\chi=\frac{c(\hat{p}\hat{\sigma})}{2mc^2+\varepsilon-U}\phi=\frac{(\hat{p}\hat{\sigma})}{2mc\left(1+\frac{\varepsilon-U}{2mc^2}\right)}\phi \tag{20.43}$$

We now write to the approximation of $\frac{\varepsilon-U}{2mc^2}$ up to the first order:

$$\chi\cong\frac{1}{2mc}\left(1-\frac{\varepsilon-U}{2mc^2}\right)(\hat{p}\hat{\sigma})\phi \tag{20.44}$$

Substitute χ in the first equation of Equation 20.42, we have the equation for ϕ:

$$c(\hat{p}\hat{\sigma})\frac{1}{2mc}\left(1-\frac{\varepsilon-U}{2mc^2}\right)(\hat{p}\hat{\sigma})\phi=(\varepsilon-U)\phi \tag{20.45}$$

and as

$$(\hat{p}\hat{\sigma})^2=\hat{p}^2 \tag{20.46}$$

then

$$\frac{\hat{p}^2}{2m}\phi+U\phi-\frac{\varepsilon}{4m^2c^2}\hat{p}^2\phi+\frac{(\hat{p}\hat{\sigma})U(\hat{p}\hat{\sigma})}{4m^2c^2}\phi=\varepsilon\phi \tag{20.47}$$

The operator $\hat{p}$ does not commute with U. The operators $\hat{\sigma}$ and $\hat{p}$ commute and from $\hat{p}$:

$$U-U\hat{p}=-i\hbar\nabla U \tag{20.48}$$

then

$$(\hat{\sigma}\hat{p})U(\hat{p}\hat{\sigma})=U\hat{p}^2-i\hbar(\hat{\sigma}\nabla U)(\hat{\sigma}\hat{p})=U\hat{p}^2-i\hbar(\nabla U,\hat{p})+\hbar(\hat{\sigma},[\nabla U,\hat{p}]) \tag{20.49}$$

Substituting Equation 20.49 into 20.47:

$$\hat{H}=\frac{\hat{p}^2}{2m}+U-\frac{\varepsilon-U}{4m^2c^2}\hat{p}^2-i\hbar\frac{(\nabla U,\hat{p})}{4m^2c^2}+\frac{\hbar}{4m^2c^2}(\hat{\sigma},[\nabla U,\hat{p}]) \tag{20.50}$$

and, consequently, the Schrödinger equation:

$$\hat{H}\phi = E\phi \tag{20.51}$$

If in Equation 20.50 we move to the non-relativistic case, then we obtain the non-relativistic Schrödinger equation. Thus, the expression:

$$\hat{H}' = -\frac{\varepsilon - U}{4m^2c^2}\hat{p}^2 - i\hbar\frac{(\nabla U, \hat{p})}{4m^2c^2} + \frac{\hbar}{4m^2c^2}(\hat{\sigma}, [\nabla U, \hat{p}]) \tag{20.52}$$

is the **second-order correction** to the Hamiltonian $\hat{H}$ due to the quantity $\frac{1}{c^2}$. The operator in Equation 20.50 acting on ϕ may be examined as a relativistic Hamiltonian considering the correction of the second order.

The wave function should be normalized up to the term of the order of the correction considering the perturbation theory. Thus, χ should be normalized to the approximation of the term of the second order and

$$\int \chi^{\dagger}\chi dq = 1 \tag{20.53}$$

This is the condition of normalization.

20.1.2.2 Fine Structure Levels

Let us denote by

$$\hat{H}_0 = \frac{\hat{p}^2}{2m} + U \tag{20.54}$$

the Hamiltonian of the zero approximation and

$$\frac{\varepsilon - U}{4m^2c^2}\hat{p}^2 = (\varepsilon - U)\frac{1}{4m^2c^2}\hat{p}^2 \cong \frac{1}{2m}\hat{p}^2\frac{1}{4m^2c^2}\hat{p}^2 = \frac{\hat{p}^4}{8m^3c^2} \tag{20.55}$$

is the relativistic correction to the Hamiltonian $\hat{H}$ which is not related to the spin $\hat{S}$ but rather has the dependence of the kinetic energy on the momentum. The fourth summand in Equation 20.55 is also independent of the spin. The fifth summand depends on $\hat{\sigma}$ as well as on $\hat{p}$ and follow the operator of the **spin-orbit interaction** which is a consequence of the Dirac equation:

$$\hat{U}_{so} = \frac{\hbar}{4m^2c^2}(\hat{\sigma}, [\nabla U, \hat{p}]) \tag{20.56}$$

Consider $\nabla U = -\vec{E}$, where $\vec{E}$ is the electric field strength. So Equation 20.56 can be considered as the interaction energy of a moving magnetic moment with the electric field. The term with div $\vec{E}$ should be different from zero only at points of location of the charges serving as the source of the field. The quantity $(\nabla U, \hat{p})$ is different from zero only at the points of location of the charges acting at sources of the field.

Consider the special case when the electron moves in a central symmetric field (or case of a spherical symmetric field), i.e., the potential energy U is dependent on the absolute value of the radius vector $\vec{r}$ and it follows that

$$\nabla U = \frac{dU}{dr}\nabla r = \frac{dU}{dr}\frac{\vec{r}}{r} \tag{20.57}$$

If we substitute Equation 20.57 into 20.56, then we have

$$[\vec{r},\hat{p}]=\hat{L} \text{ such that } \frac{\hbar}{2}\hat{\sigma}=\hat{S}.$$

So the **spin-orbit interaction operator**:

$$\hat{U}_{so}=\frac{1}{2m^2c^2r}\frac{dU}{dr}(\hat{S},\hat{L})=\lambda(r)(\hat{S},\hat{L}) \tag{20.58}$$

From Equation 20.58, we observe that the spin-orbit interaction is due to the torque exerted on the magnetic dipole moment of the spinning electron by the magnetic field of the proton in the electron's instantaneous rest frame. In the presence of the spin-orbit coupling in Equation 20.58, the Hamiltonian no longer commutes with $\hat{S}$ as well as $\hat{L}$. Hence, the spin and orbital angular momenta are not separately conserved. However, the spin-orbit interaction operator $\hat{U}_{so}$ commutes with $\hat{L}^2$ and $\hat{S}^2$ as well as the total angular momentum operator $\hat{J}$ and so the quantities are conserved. This implies that the eigenstates of $\hat{L}_z$ as well as $\hat{S}_z$ will not be good states to use in the perturbation theory, but the eigenstates of $\hat{L}^2$, $\hat{S}^2$, $\hat{J}^2$ and $\hat{J}_z$ can be used.

Considering

$$\hat{J}=\hat{L}+\hat{S},\ \hat{J}^2=\hat{L}^2+\hat{S}^2+2(\hat{L},\hat{S}) \tag{20.59}$$

then

$$(\hat{L},\hat{S})=\frac{1}{2}\left\{\hat{J}^2-\hat{L}^2-\hat{S}^2\right\} \tag{20.60}$$

If $\hat{L}$ and $\hat{S}$ have defined values, then they commute with $\hat{J}$ which also has a defined value. It follows also that $(\hat{L},\hat{S})$ has a defined value simultaneously with $\hat{L}$, $\hat{S}$ and $\hat{J}$. If $\hat{U}_{so}$ should have been examined as a perturbation, then the levels should be degenerate. As a result of the perturbation, there is splitting of the degenerate levels. The perturbation removes the degeneracy. But

$$\hat{J}^2=\hat{L}^2+\hat{S}^2+2(\hat{L},\hat{S}) \tag{20.61}$$

From where:

$$(\hat{L},\hat{S})=\frac{1}{2}\left\{\hat{J}^2-\hat{L}^2-\hat{S}^2\right\} \tag{20.62}$$

Considering this relation, we confirm that from the theory of moments $\hat{L}^2$ and $\hat{S}^2$ commute with $\hat{J}^2$ and so $(\hat{S},\hat{L})$ commutes with $\hat{J}^2$ and, consequently, $\hat{J}^2$ commutes with $\hat{U}_{SO}$. So from Equation 20.62, the eigenvalues of $(\hat{L},\hat{S})$ are therefore as follows:

$$\frac{\hbar^2}{2}\left(j(j+1)-l(l+1)-s(s+1)\right) \tag{20.63}$$

In the given problem, the quantity j is such that $j=l\pm\frac{1}{2}$.

As a result of the spin-orbitary interaction, the degeneracy with respect to the magnetic quantum number l is partially removed and the levels for which $l \neq 0$ split into two. This is the so-called **fine structure** of levels.

If we consider the spin-orbitary interaction, then the orbitary moment itself and the spin itself are not integrals of motion. Considering Equations 20.58 and 20.60, it follows that the integral of motion (as for the free Dirac electron) is the total angular momentum $\hat{J}$. In every state l and s, there arise as many levels as j, i.e., two. These results are true not only for one electron but also for many electron atoms. But for them, $\hat{L}$ and $\hat{S}$ are the contribution of all electrons and we may define j from the rule of addition of momenta.

We can apply the above result for the evaluation of the relativistic correction to the energy levels of the electron in the Coulomb field, $-\frac{Ze}{r}$. In this case, the relativistic terms in Equation 20.50 become

$$\hat{\mathrm{H}}'' = -\frac{\mathrm{p}^4}{8\mathrm{m}^3 c^2} + \frac{Ze^2}{2\mathrm{m}^2 c^2 r^3}\left(\hat{S}, \hat{L}\right) + \frac{Ze^2\hbar^2}{2\mathrm{m}^2 c^2}\pi\delta(r) \tag{20.64}$$

This can be regarded as a perturbation and evaluated in the first order of the perturbation theory as a correction to the non-relativistic energy level in the Coulomb field.

20.1.2.3 Fine Structure Effect

We examine the case of a Coulomb field with the corrections to the energy levels that are the so-called **fine structure corrections** having two sources, i.e., the kinetic energy and spin-orbit interaction sources.

20.1.2.3.1 Kinetic Energy Source

For brevity, we consider the Hamiltonian:

$$\hat{\mathrm{H}} = \hat{\mathrm{H}}_0 - \frac{\hat{\mathrm{p}}^4}{8\mathrm{m}^3 c^2} \equiv \hat{\mathrm{H}}_0 + \hat{\mathrm{H}}'',\ Ze^2 \equiv \tilde{\alpha} \tag{20.65}$$

where

$$\hat{\mathrm{H}}_0 = \frac{\hat{\mathrm{p}}^2}{2\mathrm{m}} - \frac{\tilde{\alpha}}{r} \tag{20.66}$$

and

$$\frac{\hat{\mathrm{p}}^2}{2\mathrm{m}} = \hat{\mathrm{H}}_0 + \frac{\tilde{\alpha}}{r} \tag{20.67}$$

So

$$\hat{\mathrm{p}}^2 = 2\mathrm{m}\left(\hat{\mathrm{H}}_0 + \frac{\tilde{\alpha}}{r}\right) \tag{20.68}$$

and

$$\hat{\mathrm{p}}^4 = 2\mathrm{m}\hat{\mathrm{p}}^2\left(\hat{\mathrm{H}}_0 + \frac{\tilde{\alpha}}{r}\right) = 4\mathrm{m}^2\left(\hat{\mathrm{H}}_0 + \frac{\tilde{\alpha}}{r}\right)^2 \tag{20.69}$$

The first-order correction to the energy:

$$E_n^{(1)} = \frac{1}{8\mathrm{m}^3c^2}\langle n,l,m|\hat{\mathrm{p}}^4|n,l,m\rangle \tag{20.70}$$

From Equation 20.69:

$$E_n^{(1)} = \frac{1}{2\mathrm{m}c^2}\langle n,l,m|\left(\hat{\mathrm{H}}_0+\frac{\tilde{\alpha}}{r}\right)^2|n,l,m\rangle = \frac{1}{2\mathrm{m}c^2}\left(\left(E_n^{(0)}\right)^2+\tilde{\alpha}^2\left\langle\frac{1}{r^2}\right\rangle_{nlm}+2\tilde{\alpha}E_n^{(0)}\left\langle\frac{1}{r}\right\rangle_{nlm}\right) \tag{20.71}$$

We calculate the averages via the virial theorem:

$$\langle T\rangle_{nlm} = \left\langle -\frac{1}{2}\mathrm{U}\right\rangle_{nlm} \tag{20.72}$$

then

$$\left\langle\hat{\mathrm{H}}_0\right\rangle_{nlm} = \langle T+\mathrm{U}\rangle_{nlm} = \left\langle -\frac{1}{2}\mathrm{U}+\mathrm{U}\right\rangle_{nlm} = \left\langle\frac{1}{2}\mathrm{U}\right\rangle_{nlm} = -\frac{\tilde{\alpha}}{2}\left\langle\frac{1}{r}\right\rangle_{nlm} \tag{20.73}$$

and

$$\left\langle\frac{1}{r}\right\rangle_{nlm} = -\frac{2}{\tilde{\alpha}}\left\langle\hat{\mathrm{H}}_0\right\rangle_{nlm} = -\frac{2}{\tilde{\alpha}}E_n^{(0)} = -\frac{2}{\tilde{\alpha}}\left(-\frac{\tilde{\alpha}}{2\mathrm{a}_0n^2}\right) = \frac{1}{\mathrm{a}_0n^2},\ \mathrm{a}_0 = \frac{\hbar^2}{\mathrm{m}\tilde{\alpha}} \tag{20.74}$$

We calculate $\left\langle\frac{1}{r^2}\right\rangle_{nlm}$ by adding the perturbation $\frac{\lambda}{r^2}$ to the Hamiltonian, $\hat{\mathrm{H}}$:

$$\hat{\mathrm{H}} = \hat{\mathrm{H}}_0+\hat{\mathrm{H}}_1 = -\frac{\hbar^2}{2\mathrm{m}}\left(\frac{1}{r^2}\frac{\partial}{\partial r}\left(r^2\frac{\partial}{\partial r}\right)-\frac{l(l+1)}{r^2}\right)-\frac{\tilde{\alpha}}{r}+\frac{\lambda}{r^2} = -\frac{\hbar^2}{2\mathrm{m}}\frac{1}{r^2}\frac{\partial}{\partial r}\left(r^2\frac{\partial}{\partial r}\right)+\frac{\hbar^2}{2\mathrm{m}}\frac{l'(l'+1)}{r^2}-\frac{\tilde{\alpha}}{r} \tag{20.75}$$

Here

$$l'(l'+1) = l(l+1)+\frac{2\mathrm{m}\lambda}{\hbar^2},\ l' = l'(\lambda) \tag{20.76}$$

The eigenenergy values are obtained as follows:

$$E_n \equiv E_n(\lambda) = \frac{\tilde{\alpha}}{2\mathrm{a}_0n^2},\ n = k+l'+1 \tag{20.77}$$

We Taylor series expand $E_n(\lambda)$ relative to λ:

$$E_n(\lambda) = E_n(0)+\lambda\frac{d}{d\lambda}E_n\bigg|_{\lambda=0}+\frac{\lambda^2}{2!}\frac{d^2}{d\lambda^2}E_n\bigg|_{\lambda=0}+\cdots \tag{20.78}$$

Since $E_n(0)$ is the unperturbed energy:

$$\left\langle \hat{H}_1 \right\rangle_{nlm} = \lambda \left\langle \frac{1}{r^2} \right\rangle_{nlm} \equiv \lambda \frac{d}{d\lambda} E_n \bigg|_{\lambda=0} = -\lambda \frac{\tilde{\alpha}}{2a_0 n^3} (-2) \frac{d}{d\lambda} l' \bigg|_{l'=l} \tag{20.79}$$

But

$$\frac{d}{d\lambda} l' = \frac{2m}{\hbar^2 (2l+1)} \tag{20.80}$$

So

$$\left\langle \frac{1}{r^2} \right\rangle_{nlm} = \frac{4n}{\tilde{\alpha}^2 \left(l + \frac{1}{2} \right)} \left(E_n^{(0)} \right)^2 \tag{20.81}$$

We evaluate $\left\langle \frac{1}{r^3} \right\rangle_{nlm}$ via the Hermitian radial momentum, p_r:

$$p_r = -i\hbar \left(\frac{\partial}{\partial r} + \frac{1}{r} \right) \tag{20.82}$$

from where:

$$p_r^2 = -\hbar^2 \frac{1}{r^2} \frac{\partial}{\partial r} \left(r^2 \frac{\partial}{\partial r} \right) \tag{20.83}$$

and the Hamiltonian, $\hat{H}_0$:

$$\hat{H}_0 = \frac{1}{2m} \left(p_r^2 + \frac{\hat{L}^2}{r^2} \right) - \frac{\tilde{\alpha}}{r} \tag{20.84}$$

So

$$\left[\hat{H}_0, p_r \right] = -i\hbar \left(\frac{\tilde{\alpha}}{r^2} - \frac{\hat{L}^2}{mr^3} \right) \tag{20.85}$$

For an energy eigenbasis, we have

$$\langle n,l,m | \left[\hat{H}_0, p_r \right] | n,l,m \rangle = \left\langle \left(\frac{\tilde{\alpha}}{r^2} - \frac{\hbar^2 l(l+1)}{mr^3} \right) \right\rangle_{nlm} = 0 \tag{20.86}$$

From here:

$$\left\langle \frac{1}{r^3} \right\rangle_{nlm} = \frac{1}{a_0^3 n^3 l(l+1) \left(l + \frac{1}{2} \right)} \tag{20.87}$$

The first-order correction to the energy becomes

$$E_n^{(1)} = \frac{mc^2 Z^4 \alpha^4}{2n^3}\left(\frac{1}{l+\frac{1}{2}} - \frac{3}{4n}\right), \alpha = \frac{e^2}{\hbar c} \approx \frac{1}{137} \tag{20.88}$$

Here, α is the **fine structure constant.**

20.1.2.3.2 Spin-Orbit Interaction Source

So, the **spin-orbit interaction operator** for the Coulomb field has the form:

$$\hat{U}_{so} = \frac{\tilde{\alpha}}{2m^2c^2r^3}\left(\hat{S},\hat{L}\right) \tag{20.89}$$

From Equations 20.63 to 20.60 and 20.63, we average the operator $\hat{U}_{so}$ in the $|jm,ls\rangle$ basis where $s = \frac{1}{2}$ to find the first-order change in the spin-orbit interaction energy:

$$E_{n,so}^{(1)} = \left\langle jm,ls\right|\hat{U}_{so}\left|jm,ls\right\rangle = \frac{\hbar^2\tilde{\alpha}}{4m^2c^2}\left\langle\frac{1}{r^3}\right\rangle_{nl}\left(j(j+1) - l(l+1) - \frac{3}{4}\right) \tag{20.90}$$

Since

$$j = l \pm \frac{1}{2} \tag{20.91}$$

then

$$E_{n,so}^{(1)} = \frac{\hbar^2\tilde{\alpha}}{4m^2c^2}\left\langle\frac{1}{r^3}\right\rangle_{nl}\begin{cases} l, & l+\frac{1}{2} \\ -(l+1), & l-\frac{1}{2}\end{cases} = \frac{mc^2Z^4\alpha^4}{4n^3 l(l+1)\left(l+\frac{1}{2}\right)}\begin{cases} l, & l+\frac{1}{2} \\ -(l+1), & l-\frac{1}{2}\end{cases} \tag{20.92}$$

The total correction to the energy level of the first order for $j = l \pm \frac{1}{2}$:

$$\Delta E = E_n^{(1)} + E_{n,so}^{(1)} = -\frac{mc^2Z^4\alpha^4}{2n^3}\left(\frac{1}{j+\frac{1}{2}} - \frac{3}{4n}\right) \tag{20.93}$$

From this relation, the degeneracy of the Coulomb level relative to l is partially lifted. The splitting of the levels for given n yielding a group near energy levels is called the fine structure. The set of levels considering the fine structure are as follows:

$$1s_{\frac{1}{2}}, \left(2s_{\frac{1}{2}}, 2p_{\frac{1}{2}}\right), 2p_{\frac{3}{2}}, \ldots \tag{20.94}$$

where, $s, p, d, \ldots$ correspond, respectively, to $l = 0, 1, 2, \ldots$. The indices are the values of j. The brackets in Equation 20.94 implies levels that are degenerate. This degeneracy is conserved in the exact solution of

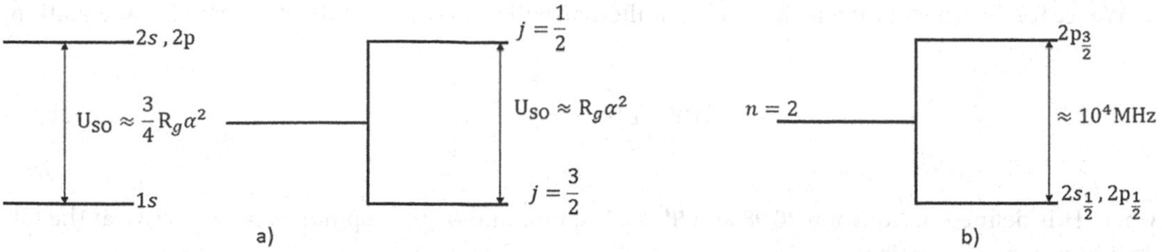

FIGURE 20.1 (20.1a) The hyperfine structure of the $n = 2$ level with = 0; (20.1b) the hyperfine structure of the $n = 2$ level with $l = 1$.

the Dirac equation for the Coulomb field. Note that from experimentation, even in the absence of an external field, a fluctuating electromagnetic field exist in space, a phenomenon related to the quantum mechanical nature of the electromagnetic field. The coupling of the electron to the fluctuating electromagnetic field removes the degeneracy between the $2p_{\frac{1}{2}}$ and $2s_{\frac{1}{2}}$ levels. The $2s_{\frac{1}{2}}$ level is raised with respect to $2p_{\frac{1}{2}}$ by the **so-called Lamb shift** of order 1,060 MHz and gives the divergence of Dirac theory and experimentation for the single particle in the given field.

We examine in a crude way of evaluating the spin-orbit interaction operator in Equation 20.58:

$$|U_{so}| \approx \frac{e^2\hbar^2}{2m^2c^2r^2 2} = \frac{e^2}{4a} \frac{\hbar^2}{m^2c^2a^2} \approx \frac{e^2}{4a} \frac{e^4}{\hbar^2c^2} \approx R_g\alpha^2 \tag{20.95}$$

where r is the first Bohr orbit and α characterizes the fine structure compared to the breadth of the energy level with $n = 2$ (Figure 20.1a):

$$l = 0: 2s \to j = \frac{1}{2}, \quad 2p \to j = \frac{1}{2}, \frac{3}{2} \tag{20.96}$$

From Equation 20.88, the level with $n = 2$ splits into two levels of the fine structure and the breadth between the levels is of the order of $R_g\alpha^2$. We see that the level with $n = 2$ splits into two levels if we consider the spin-orbitary interaction (Figure 20.1b):

$$2p_{\frac{1}{2}}: \quad n = 2, \quad l = 1, \quad j = \frac{1}{2} \tag{20.97}$$

The energy level depends on m_j and not on l, and this result is related to the full solution of Dirac equation which of course relates the fine structure.

20.2 Bound Electronic States in a Coulombic Field

We write the Dirac Hamiltonian, $\hat{H}$, for a particle in a Coulomb field, $-\frac{Ze}{r}$:

$$\hat{H} = c\hat{\alpha}\hat{p} + \beta mc^2 + \hat{U}, \; \hat{U} = -\frac{Ze^2}{r} \equiv -\frac{\tilde{\alpha}}{r}, \; Ze^2 \equiv \tilde{\alpha} \tag{20.98}$$

We see further the solution with such a Hamiltonian will be exact. From the stationary Dirac equation:

$$\hat{\mathrm{H}}\Psi = E\Psi,\ \Psi = \begin{bmatrix} \phi \\ \chi \end{bmatrix} \tag{20.99}$$

where $\hat{\mathrm{H}}$ is defined in Equation 20.98 and Ψ is a bispinor and ϕ, χ are spinors, and we arrive at the following system of equations:

$$c\hat{p}\hat{\alpha}\chi + mc^2\phi + U\phi = E\phi,\ c\hat{p}\hat{\alpha}\phi - mc^2\chi + U\chi = E\chi \tag{20.100}$$

Letting

$$E - mc^2 = \varepsilon,\ E + mc^2 = \varepsilon' \tag{20.101}$$

then

$$c\hat{p}\hat{\sigma}\chi = (\varepsilon - U)\phi,\ c\hat{p}\hat{\sigma}\phi = (\varepsilon' - U)\chi \tag{20.102}$$

Since the functions χ and ϕ have opposite parity, using the method of separation of variables, we have

$$\phi = F(r)\Omega_{jlM},\ \chi = i\mathcal{F}(r)\Omega_{jl'M} \tag{20.103}$$

where, for $l' = 2j - l$, Ω_{jlM} and $\Omega_{jl'M}$ are spherical spinors defined in Equation 4.95 (Chapter 4, Section 4.4). Substitute Equation 20.103 into 20.102, considering

$$\Omega_{jl'M} = -\hat{\sigma}\hat{n}\Omega_{jlM} = -\hat{\sigma}\frac{\vec{r}}{r}\Omega_{jlM} = -\hat{\sigma}\vec{r}\frac{1}{r}\Omega_{jlM} \tag{20.104}$$

then we rewrite Equation 20.36 in the form:

$$-c\hat{p}\hat{\sigma}i\mathcal{F}\hat{\sigma}\vec{r}\frac{1}{r}\Omega_{jlM} = (\varepsilon - U)F\,\Omega_{jlM} \tag{20.105}$$

But

$$(\hat{p},\hat{\sigma})(\vec{r},\hat{\sigma}) = (\hat{p},\vec{r}) + i(\hat{\sigma},[\hat{p},\vec{r}]) \tag{20.106}$$

and

$$[\hat{p},\vec{r}] = \hbar\hat{i},\ \nabla\left(\vec{r},\mathcal{F}\frac{1}{r}\right) = \frac{d}{dr}\mathcal{F} + \frac{2}{r}\mathcal{F} \tag{20.107}$$

So Equation 20.105 becomes

$$-c\hbar\left(\frac{d}{dr}\mathcal{F} + \frac{2}{r}\mathcal{F} + \frac{1}{r}\mathcal{F}\hat{\sigma}\hat{i}\right)\Omega_{jlM} = (\varepsilon - U)F\,\Omega_{jlM} \tag{20.108}$$

Letting $\mathcal{F}c\hbar \equiv \text{G}$, Equation 20.108 becomes

$$-\left(\frac{d}{dr}\text{G}+\frac{2}{r}\text{G}+\frac{1}{r}\text{G}\hat{\sigma}\hat{\text{i}}\right)\Omega_{jlM}=(\varepsilon-\text{U})\text{F}\,\Omega_{jlM} \tag{20.109}$$

From

$$\hat{\sigma}\hat{\text{i}} \equiv 2\hat{\text{s}}\hat{\text{i}} = \hat{\text{j}}^2-\hat{\text{i}}^2-\hat{\text{s}}^2 \tag{20.110}$$

then the eigenvalues of the operator, $\hat{\sigma}\hat{\text{i}}$:

$$j(j+1)-l(l+1)-\frac{3}{4} \tag{20.111}$$

This implies $j-\frac{1}{2}$ for $l=j-\frac{1}{2}$ and $-j-\frac{3}{2}$ for $=j+\frac{1}{2}$. We denote the eigenvalues of the operator, $\hat{\sigma}\hat{\text{i}}$, through $-(1+\nu)$, where

$$\nu=\begin{cases} j+\frac{1}{2}, & l=j+\frac{1}{2} \\ -\left(j+\frac{1}{2}\right), & l=j-\frac{1}{2} \end{cases} \tag{20.112}$$

Here, ν takes all half integer values except zero. So

$$\hat{\sigma}\hat{\text{i}}\Omega_{jlM}=-(1+\nu)\Omega_{jlM} \tag{20.113}$$

The spherical spinor, Ω_{jlM}, falls off in Equation 20.109 and we have

$$\frac{d}{dr}\text{G}+\frac{1-\nu}{r}\text{G}+(\varepsilon-\text{U})\text{F}=0 \tag{20.114}$$

Similarly,

$$\frac{d}{dr}\text{F}+\frac{1+\nu}{r}\text{F}-(\varepsilon'-\text{U})\text{G}=0 \tag{20.115}$$

Neglecting ε and ε' and from the solution for small r, we find F and G in the form:

$$\text{F}=c_1r^{\gamma-1},\ \text{G}=c_2r^{\gamma-1} \tag{20.116}$$

Substituting these into Equations 20.114 and 20.115, we have

$$c_1(\gamma+\nu)-c_2\tilde{\alpha}=0,\ c_1\tilde{\alpha}+c_2(\gamma-\nu)=0 \tag{20.117}$$

The condition for the non-trivial solution of the system of equations:

$$\gamma^2=\nu^2-\tilde{\alpha}^2 \tag{20.118}$$

So the regular solution for $r \to 0$:

$$F = \frac{\tilde{\alpha}}{\gamma + \nu} G = \text{const} \times r^{-1 \pm \sqrt{\nu^2 - \tilde{\alpha}^2}} \tag{20.119}$$

If $\nu^2 > \tilde{\alpha}^2$, then in Equation 20.119, we take the "+" sign since the "–" sign will give a divergent normalized integral. If $\nu^2 < \tilde{\alpha}^2$, then $\sqrt{\nu^2 - \tilde{\alpha}^2}$ is imaginary and the functions F and G for $r \to 0$ behave as $\frac{1}{r}\cos(|\gamma| \ln r)$. This implies it has an infinitely number of nodes. In the non-relativistic case, this corresponds to the ground-state energy extending to $-\infty$ and also the falling into the force center of the particle. But in the non-relativistic theory, such a situation for the Coulombic field does not exist in general. However, there appears the field, $-\frac{C}{r^2}$, for sufficiently large C. In the relativistic theory, we find that for $\tilde{\alpha}^2 > 1$ and $\nu^2 = 1$, the ground-state energy becomes imaginary. So the pure Coulombic field can be examined in the relativistic theory when $\tilde{\alpha} < 1$. This implies that $Z < 137$.

We find now the functions and the energy levels in the form:

$$F = \sqrt{\varepsilon}(Q_1 + Q_2)\exp\left\{-\frac{\rho}{2}\right\}\rho^{\gamma-1}, \; G = -\sqrt{-\varepsilon'}(Q_1 - Q_2)\exp\left\{-\frac{\rho}{2}\right\}\rho^{\gamma-1} \tag{20.120}$$

Here

$$\rho = 2\lambda r, \; \lambda = \sqrt{\varepsilon\varepsilon'} \tag{20.121}$$

We substitute the Equation 20.120 into the Equations 20.114 and 20.115 considering Equation 20.121:

$$\rho\frac{d}{d\rho}Q_1 + (\gamma - a)Q_1 + (\nu - b)Q_2 = 0, \; \rho\frac{d}{d\rho}Q_2 + (\gamma + a - \rho)Q_2 + (\nu + b)Q_1 = 0, \; a = \frac{\tilde{\alpha}E}{\lambda}, \; b = \frac{\tilde{\alpha}mc^2}{\lambda} \tag{20.122}$$

These equations are transformed to the following system of equations:

$$\rho\frac{d^2}{d\rho^2}Q_1 + (2\gamma + 1 - \rho)\frac{d}{d\rho}Q_1 - (\gamma - a)Q_2 = 0, \; \rho\frac{d^2}{d\rho^2}Q_2 + (2\gamma + 1 - \rho)\frac{d}{d\rho}Q_2 - (\gamma + 1 - a)Q_2 = 0 \tag{20.123}$$

Here

$$\gamma^2 - a^2 = \nu^2 - b^2 \tag{20.124}$$

The solution of the system of Equations 20.123 and 20.135 are expressed via the confluent hypergeometric functions, $F(\alpha, \beta, \rho)$:

$$Q_1 = C_1 F(\gamma - a, 2\gamma + 1, \rho), \; Q_2 = C_2 F(\gamma + 1 - a, 2\gamma + 1, \rho) \tag{20.125}$$

Substituting these in Equations 20.122 and 20.133 for $\rho = 0$, we have

$$C_2 = -C_1(\gamma - a)(\nu - b)^{-1} \tag{20.126}$$

Both confluent hypergeometric functions in Equation 20.125 have to tend to a polynomial and, conversely, Q_1 and Q_2 will exponentially diverge for $\rho \to \infty$. The function, Q_1, tends to a polynomial for

$$\gamma - a = -n_r,\ n_r = 0,1,2,\ldots \tag{20.127}$$

The function, Q_2, tends to a polynomial for positive integral n_r except zero. For $n_r = 0$, from Equations 20.127 and 20.124, we have

$$\gamma = a,\ \nu = b \tag{20.128}$$

If $\nu < 0$, then from Equation 20.126, $C_2 = 0$, and, consequently, $Q_2 = 0$. So

$$n_r = \begin{cases} 0,1,2,\ldots, & \nu < 0 \\ 1,2,\ldots, & \nu > 0 \end{cases} \tag{20.129}$$

The energy spectrum can be obtained from Equation 20.127:

$$\frac{E}{mc^2} = 1 + \left[\frac{\tilde{\alpha}^2}{\left(\sqrt{\nu^2 - \tilde{\alpha}^2} + n_r\right)^2} \right]^{-\frac{1}{2}} \tag{20.130}$$

The ground-state energy is obtained when $\nu = 1$ and $n_r = 0$:

$$E = mc^2\sqrt{1 - \tilde{\alpha}^2} \tag{20.131}$$

From Equation 20.130, the state for a given j and two possible values of l have the same energies. So the ground state is non-degenerate. For a point charge, the inequality, $\tilde{\alpha} > 1$ for large Z is over approximation. For a realistic relativistic approximation, the nucleus is a finite volume.

We observe that for the solution of Dirac equation for the given model, bound states exist for $\tilde{\alpha} > 1$. For an increase in Z, the energy levels increase, and for a certain critical value, Z_c, the ground-state energy becomes $-mc^2$. Further increase in Z leads to the possibility of the creation of electron and positron pair. So Z_c should be the maximum charge achieved by the nucleus (sphere) of radius, $r_0 = 1.2 \times 10^{-12}$ cm, where the nuclear charge is uniformly distributed.

21

Motion in a Magnetic Field

21.1 Landau Levels

The motion of a charged particle in a constant homogenous magnetic field leads to the motion of a harmonic oscillator. Examine first the motion of a charged particle in a constant magnetic field described by the **Pauli Hamiltonian**:

$$\hat{H} = \frac{1}{2m}\left(\hat{p} - \frac{e}{c}\vec{A}\right)^2 + U - \frac{e\hbar}{2mc}\left(\hat{\sigma}, \vec{H}\right) \tag{21.1}$$

where the first summand is the kinetic energy, the second is the potential energy and the third is the energy due to the interaction of the intrinsic magnetic moment of the electron with the external magnetic field. It should be noted that the operators of the momentum $\hat{p}$ and the vector potential $\vec{A}$ are non-commutable operators:

$$\left(\hat{p} - \frac{e}{c}\vec{A}\right)^2 = \hat{p}^2 - \frac{e}{c}\left(\hat{p}\vec{A} + \vec{A}\hat{p}\right) + \frac{e^2}{c^2}\vec{A}^2 \tag{21.2}$$

From

$$\left(\hat{p}, \vec{A}\right)\Psi = -i\hbar\left(\nabla, \vec{A}\right)\Psi = -i\hbar\Psi \mathrm{div}\vec{A} - i\hbar\left(\nabla\Psi, \vec{A}\right) = -i\hbar\Psi \mathrm{div}\vec{A} + \left(\vec{A}, \hat{\vec{p}}\right)\Psi \tag{21.3}$$

then

$$\hat{H} = \frac{1}{2m}\hat{p}^2 + U - \frac{e}{mc}\left(\vec{A}, \hat{p}\right) - i\frac{\hbar}{mc}\mathrm{div}\vec{A} + \frac{e^2}{2mc^2}\vec{A}^2 - \frac{e\hbar}{2mc}\left(\hat{\sigma}, \vec{H}\right) \tag{21.4}$$

Consider the motion of a free electron in a constant homogenous magnetic field, $\vec{H}$. In equation 21.4, no dependence on time t of the magnetic field and the potential is considered and if there exists a time dependence then there exists the Lorentz relation:

$$\mathrm{div}\vec{A} + \frac{1}{c}\frac{dA_0}{dt} = 0 \tag{21.5}$$

where A_0 is the scalar potential:

$$U = eA_0 \tag{21.6}$$

DOI: 10.1201/9781003273073-23

But if the magnetic field is constant, then

$$A_0 \neq f(t) \tag{21.7}$$

From Equation 21.5, we have the **Coulomb gauge**:

$$\text{div}\vec{A} = 0 \tag{21.8}$$

For this, it is necessary that the quantity should only be constant with time t and not necessary in space. From $\text{div}\vec{A} = 0$, $\vec{A}$ and $\hat{p}$ commute.

For the case of a constant homogenous magnetic field in the direction of the oz-axis:

$$H_x = H_y = 0, \quad H_z = H, \quad U = 0 \tag{21.9}$$

then

$$\vec{H} = \begin{vmatrix} \vec{e}_x & \vec{e}_y & \vec{e}_z \\ \frac{\partial}{\partial x} & \frac{\partial}{\partial y} & \frac{\partial}{\partial z} \\ A_x & A_y & A_z \end{vmatrix} \tag{21.10}$$

We set $\vec{A}$ such that

$$H_z = H \neq 0 \tag{21.11}$$

and select the gauge:

$$A_x = -Hy, \quad A_y = A_z = 0 \tag{21.12}$$

Substitute into Equation 21.10 and we have

$$H_x = 0, \quad H_y = 0, \quad H_z = H \tag{21.13}$$

This selection of $\vec{A}$ relates the magnetic field directed toward the oz-axis. For

$$A_x = -\frac{1}{2}Hy, \quad A_y = -\frac{1}{2}Hx \tag{21.14}$$

we obtain the same results.

From the principle of gauge invariance:

$$\vec{H} = \text{curl}\left(\vec{A} + \text{grad}\chi\right) \tag{21.15}$$

then

$$\hat{H} = \frac{1}{2m}\hat{p}^2 + \frac{e}{mc}H\hat{p}_x y + \frac{e^2}{2mc^2}H^2 y^2 - \frac{e\hbar}{2mc}H\hat{\sigma}_z \tag{21.16}$$

where the variables are separable and there is no spin-orbit interaction. Also, from Equation 21.16, we observe the wave function of the full system is an eigenfunction of the operator σ_z. The Hamiltonian in Equation 21.16 commutes with p_x and p_z and does not commute with p_y:

$$\Psi(x,y,z) = \exp\left\{\frac{i}{\hbar}(p_x x + p_z z)\right\}\chi(y) \tag{21.17}$$

If we substitute Equation 21.17 into 21.16, then we have the linear harmonic equation:

$$\left\{\frac{1}{2m}\hat{p}_x^2 + \frac{1}{2m}\hat{p}_y^2 + \frac{1}{2m}\hat{p}_z^2 - \frac{e}{mc}Hp_x y + \frac{e^2}{2mc^2}H^2y^2 + \frac{e\hbar}{2mc}H\right\}\chi(y) = E\chi(y) \tag{21.18}$$

from where

$$\left\{\frac{1}{2m}\hat{p}_y^2 + \frac{M\omega_c^2}{2}(y-y_0)^2 + \frac{1}{2m}\hat{p}_z^2 + \frac{\omega_c}{2}\right\}\chi(y) = E\chi(y),\ \omega_c = \frac{eH}{mc},\ y^2\frac{m\omega_c^2}{2} = \frac{p_x^2}{2m},\ y_0 = \frac{p_x}{m\omega_c} \tag{21.19}$$

and ω_c is called the **cyclotron frequency.**

Suppose the electron moves in a uniform magnetic field $\vec{H}$ which is in the oz-axis direction (Figure 21.1). If such a motion is a classical one, then the electron moving with a constant velocity describes a circular helix in the oz-axis direction. This circular motion has the frequency which imitates a cyclotron motion

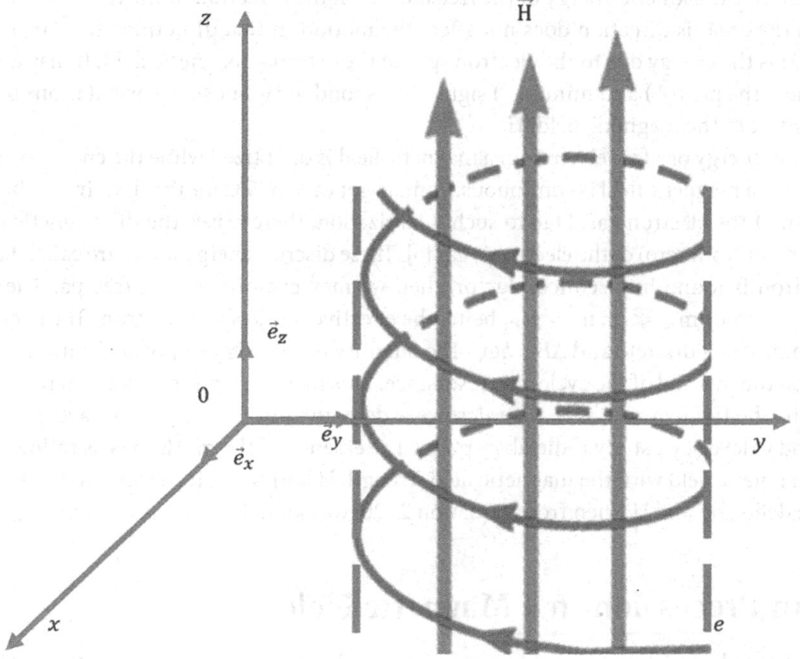

FIGURE 21.1 The electron moving in a uniform magnetic field, $\vec{H}$, in the direction of the oz-axis. The motion is examined to be classical with the electron, e, moving at a constant velocity and describes a circular helix in the direction of the oz-axis.

given in Equation 21.20. In classical physics, the frequency which imitates the cyclotron frequency is the Larmor precession, $\Omega_L = \frac{\omega_c}{2}$.

From Equation 21.19, the Hamiltonian describes motion of the particle in the oy-axis direction:

$$\hat{H} = \frac{1}{2m}\hat{p}_y^2 + \frac{m\omega_c^2}{2}(y - y_0)^2 + \frac{1}{2m}\hat{p}_z^2 + \frac{\omega_c}{2} \tag{21.20}$$

Here the last two summands are constants, and so in the total Hamiltonian in Equation 21.20, we can consider only the first two summands representing a linear harmonic oscillator moving in the oy-axis direction with frequency ω_c and with the equilibrium position being y_0. The projection of the angular momentum, $\vec{p}_x$, is an integral of motion. The Schrödinger equation for the linear harmonic oscillator has a well-known solution:

$$E = \hbar\omega_c\left(n + \frac{1}{2}\right) + \frac{1}{2m}p_z^2 \pm \frac{\hbar\omega_c}{2},\ n = 0,1,2,\ldots \tag{21.21}$$

which is the energy of a free electron in a homogenous external magnetic field. The first summand in Equation 21.21 is the quantized energy of the transversal oscillatory motion with frequency ω_c. The multiplicity of degeneracy is infinite. So the energy is independent of $\hat{p}_y$ which takes arbitrary values:

$$-\infty < p_y < +\infty \tag{21.22}$$

For the given stationary states, the particle is not localized in a bounded domain of space (in the transversal direction) notwithstanding the discrete character of the energy spectrum. The second summand of Equation 21.21 is the kinetic energy of the free motion in the direction of the magnetic field. The magnetic field in the oz-axis direction does not affect the motion in that direction. The third summand in Equation 21.21 is the energy due to the electron spin in the external magnetic field. It may also be written as $\pm\mu_B H$, where the plus (+) and minus (−) signs correspond to two possible orientations of the electron spin with respect to the magnetic field, $\vec{H}$.

Note that the energy of a free electron in a magnetic field is quantized while the energy of a free particle in the absence of a magnetic field is continuous. Landau got this in 1932 for the first time in his work on the diamagnetism of the electron gas. Due to such quantization, there arises the diamagnetic susceptibility (which is different from zero) of the electronic gas [6]. These discrete energy levels are called **Landau levels.**

If an electron is found in a semiconductor, then we may examine it as a free particle with a mass $m \to m^* = m_{eff} < m$ or $m_{eff} \ll m$, $m^* = m_{eff}$ being the effective mass of the electron. The energy spectrum of the electronic gas is discrete and $\Delta E = \hbar\omega_c$. The quantity m^* is a very important parameter and can be obtained from the method of the cyclotron resonance. This method for a free electron relates radio spectroscopy while the frequency ω_c is in the microwave domain and relates a radio wave. So in the investigation of Landau levels, we study radio absorption. **The essence of the method is as follows**: A specimen placed in a magnetic field with the magnetic field strength $\vec{H}$ and for a given radio wave changes $\vec{H}$ when $\omega = \omega_c$. If we define ω_c and $\vec{H}$, then from Equation 21.20, we can define m^*.

21.2 Spin Precession in a Magnetic Field

A spinning charged particle subjected to a magnetic field, $\vec{H}$, constitutes a magnetic dipole $\vec{M}$ proportional to its spin angular momentum, $\vec{s}$:

$$\vec{M} = \gamma\vec{s} \tag{21.23}$$

The proportionality constant, γ, is the so-called **gyromagnetic ratio**. A magnetic dipole placed in a magnetic field, $\vec{H}$, experiences a torque, $\left[\vec{M},\vec{H}\right]$, which tends to align it up parallel to the field. The interaction Hamiltonian associated with this torque being

$$\hat{H}_s = -\left(\vec{M},\vec{H}\right) \tag{21.24}$$

So the Hamiltonian of the spinning charged particle at rest in the presence of a magnetic field $\vec{H}$ is:

$$\hat{H}_s = -\frac{e\hbar}{2mc}\left(\hat{\sigma},\vec{H}\right) \tag{21.25}$$

where $\hat{\sigma}$ is the Pauli matrix.

If the projection of the spin in the direction of the magnetic field has a defined value, then the state of the particle placed in the magnetic field will not be stationary. For a neutral spin $\frac{1}{2}$ particle, for example, a spin $\frac{1}{2}$ neutral atom where we consider an exceedingly weak field. For that, we neglect the field influence on the atomic structure.

In presence of the magnetic field $\vec{H}$, the magnetic moment operator interacting with the field is independent of the coordinate. So the wave function is examined as the product of the spin and coordinate parts. For the neutral particle, the coordinate part of the wave function is independent of the magnetic field and so describes simply the free motion of the particle. Interest here will only be on the effect related to the spin part of the wave function. So for spin $\frac{1}{2}$, we consider the Hamiltonian in Equation 21.25.

The unit normal vector $\hat{n}$ dependent on the polar angles, θ and ϕ, is considered to be in the direction of the magnetic field, $\vec{H}$:

$$\hat{n} \equiv \frac{\vec{H}}{H} = \left(n_x, n_y, n_z\right) \tag{21.26}$$

where

$$n_x = \sin\theta\,\cos\phi,\; n_y = \sin\theta\,\sin\phi,\; n_z = \cos\theta \tag{21.27}$$

This permits us to rewrite Equation 21.25 now in the form:

$$\hat{H}_s = -\frac{e\hbar}{2mc}H\hat{\sigma}\hat{n} \tag{21.28}$$

The evolution (unitary) operator of the spin part can be written as follows:

$$\hat{U} = \exp\left\{i\hbar\Omega_L\hat{\sigma}\hat{n}t\right\} \tag{21.29}$$

where the Larmor precession frequency, Ω_L:

$$\Omega_L = \frac{eH}{2mc} \tag{21.30}$$

From Equations 4.177 and 4.178, the unitary operator relating state picture and operator picture:

$$\hat{U} = \exp\left\{i\Omega_L\hat{\sigma}\hat{n}t\right\} \equiv \cos(\Omega_L t) + i\hat{\sigma}\hat{n}\sin(\Omega_L t) \tag{21.31}$$

This represents a spin rotation through the angle, $2\Omega_L t$, around $\hat{n}$, considering the chapter on finite rotation of the operator for spin $\frac{1}{2}$. If at the initial moment the spin of the particle has a projection of $\frac{1}{2}$ on the z-axis, then the probability amplitude has a projection of $-\frac{1}{2}$ on the z-axis at time moment t equal:

$$\mathbb{d}_{-\frac{1}{2},\frac{1}{2}} = \left\langle -\frac{1}{2}\middle|\hat{U}\middle|\frac{1}{2}\right\rangle \tag{21.32}$$

From the action of the Pauli matrices, $\hat{\sigma}_x$, $\hat{\sigma}_y$ and $\hat{\sigma}_z$ on the state $\left|\frac{1}{2}\right\rangle$:

$$\mathbb{d}_{-\frac{1}{2},\frac{1}{2}} = i\exp\{i\phi\}\sin\theta\sin(\Omega_L t) \tag{21.33}$$

From the scattering component of the magnetic field:

$$H_\perp = \sqrt{H_x^2 + H_y^2} \tag{21.34}$$

then

$$\sin\theta = \frac{H_\perp}{H} = \frac{H_\perp}{\sqrt{H_z^2 + H_\perp^2}} \tag{21.35}$$

The probability of the spin $-\frac{1}{2}$ rotation can then be written as follows:

$$W_{-\frac{1}{2},\frac{1}{2}} = \left|\mathbb{d}_{-\frac{1}{2},\frac{1}{2}}\right|^2 = \frac{H_\perp^2}{H_z^2 + H_\perp^2}\sin^2(\Omega_L t) \tag{21.36}$$

This probability is zero when $t = 0$ and varies sinusoidally with respect to time, t, between the values 0 and $\frac{H_\perp^2}{H_z^2 + H_\perp^2}$.

To get a complete picture of what is happening, we calculate the expectation value of $\hat{\sigma}$ as a function of time, the so-called polarization vector (time-dependent polarization vector characterizing the spin):

$$\hat{P}(t) = \mathrm{Tr}\{\hat{\rho}(t)\hat{\sigma}\} \tag{21.37}$$

Here, $\hat{\rho}$ is the density matrix operator satisfying the equation of motion:

$$i\hbar\frac{\partial}{\partial t}\hat{\rho}(t) = \left[\hat{H}_s, \hat{\rho}(t)\right] \tag{21.38}$$

From Equation 21.37:

$$i\hbar\frac{\partial}{\partial t}\hat{P}(t) = i\hbar\frac{\partial}{\partial t}\mathrm{Tr}\{\hat{\rho}(t)\hat{\sigma}\} = \mathrm{Tr}\left\{\hat{\sigma}\left[\hat{H}_s\hat{\rho}(t)\right]\right\} \tag{21.39}$$

and

$$\mathrm{Tr}\{\hat{\sigma}_x^2\} = \mathrm{Tr}\{\hat{\sigma}_y^2\} = \mathrm{Tr}\{\hat{\sigma}_z^2\} = 2 \tag{21.40}$$

then Equations 21.30 and 21.37:

$$\frac{\partial}{\partial t}\hat{P}(t) = 2\Omega_L H\left[\hat{P}(t), \hat{n}\right] \tag{21.41}$$

This is the classical equation describing the precession of the vector, $\hat{P}(t)$, around the direction of the vector $\hat{n}$ (or magnetic field, $\vec{H}$) with angular frequency, $2\Omega_L$. Visualizing the spin via processing polarization vector, $\hat{P}$, is permissible and does not contradict the probability aspect of the Stern-Gerlach experiment. Equation 21.41 is also satisfied for a mixed state when there is no defined vector state. In particular, when there is such a mixed state, then $\rho_{12} = 0$ and so

$$P_z \neq 0, P_x = P_y = 0 \tag{21.42}$$

From Equation 21.41, P_z is a conservable quantity.

We examine the behavior of the spin on a rotating magnetic field:

$$H_x = H\sin\theta\cos(\omega t), H_y = H\sin\theta\sin(\omega t), H_z = H\cos\theta \tag{21.43}$$

We examine the equation for the evolution operator:

$$i\hbar\frac{\partial}{\partial t}\hat{U}(t) = \hat{H}_s\hat{U}(t) \tag{21.44}$$

and move to the system of rotating coordinates:

$$\hat{U}(t) = \exp\{-i\omega s_z t\}\hat{U}'(t) \tag{21.45}$$

Letting

$$\hat{s} = \frac{1}{2}\hat{\sigma} \tag{21.46}$$

then from the evolution operator in Equation 21.45, we have

$$\exp\{-i\omega\hat{s}_z t\} \equiv \exp\left\{-i\frac{\omega t}{2}\hat{\sigma}_z\right\} \equiv \cos\left(\frac{\omega t}{2}\right) - i\hat{\sigma}_z\sin\left(\frac{\omega t}{2}\right) \tag{21.47}$$

This can also be put in the form of a diagonal matrix:

$$\exp\{-i\omega\hat{s}_z t\} \equiv \begin{bmatrix} \exp\left\{-i\frac{\omega t}{2}\right\} & 0 \\ 0 & \exp\left\{i\frac{\omega t}{2}\right\} \end{bmatrix} \tag{21.48}$$

From Equation 21.46, Equation 21.44 becomes

$$i\hbar\frac{\partial}{\partial t}\hat{U}'(t) = -\hbar\left[\omega\hat{s}_z + 2\Omega_L H\exp\{i\omega\hat{s}_z t\}\hat{s}\hat{n}\exp\{-i\omega\hat{s}_z t\}\right]\hat{U}'(t) \tag{21.49}$$

Here

$$\hat{s}\hat{n} = \hat{s}_z \cos\theta + \frac{1}{2}\sin\theta\left(\hat{s}_+ \exp\{-i\omega t\} + \hat{s}_- \exp\{i\omega t\}\right) \tag{21.50}$$

and

$$\hat{s}_\pm(t) = \hat{s}_x \pm i\hat{s}_y \tag{21.51}$$

By virtue of the commutation relations of the Pauli matrices:

$$\left[\hat{\sigma}_y, \hat{\sigma}_z\right] = 2i\hat{\sigma}_x, \left[\hat{\sigma}_z, \hat{\sigma}_x\right] = 2i\hat{\sigma}_y \tag{21.52}$$

then

$$\hat{s}_x(t) = \exp\{-i\omega\hat{s}_z t\}\hat{s}_x \exp\{i\omega\hat{s}_z t\} = \frac{1}{2}\left(\hat{\sigma}_x + \frac{1}{1!}\left(-\frac{1}{2}i\omega t\right)\left[\hat{\sigma}_z, \hat{\sigma}_x\right] + \frac{1}{2!}\left(-\frac{1}{2}i\omega t\right)^2\left[\hat{\sigma}_z, \left[\hat{\sigma}_z, \hat{\sigma}_x\right]\right] + \cdots\right) \tag{21.53}$$

or

$$\hat{s}_x(t) = \frac{1}{2}\left(\hat{\sigma}_x + \frac{1}{1!}\omega t\hat{\sigma}_y - \frac{1}{2!}(\omega t)^2\hat{\sigma}_x + \cdots\right) = \hat{s}_x \cos(\omega t) + \hat{s}_y \sin(\omega t) \tag{21.54}$$

Similarly,

$$\hat{s}_y(t) = \hat{s}_y \cos(\omega t) - \hat{s}_x \sin(\omega t) \tag{21.55}$$

Here, $\hat{s}_x$ and $\hat{s}_y$ are spin operators for the spin-$\frac{1}{2}$ particle acting in a space of two states $|0\rangle$ and $|1\rangle$.
So

$$\exp\{i\omega\hat{s}_z t\}\hat{s}_\pm \exp\{-i\omega\hat{s}_z t\} = \exp\{\pm i\omega t\}\hat{s}_\pm \tag{21.56}$$

and

$$\exp\{i\omega\hat{s}_z t\}\hat{s}\hat{n} \exp\{-i\omega\hat{s}_z t\} = \hat{s}_z \cos\theta + \hat{s} \sin\theta \tag{21.57}$$

From the above formulae, Equation 21.49 becomes

$$i\frac{\partial}{\partial t}\hat{U}'(t) = -\left[(\omega + 2\Omega_L \mathrm{H}\cos\theta)\hat{s}_z + 2\Omega_L \mathrm{H}\sin\theta\hat{s}_x\right]\hat{U}'(t) \tag{21.58}$$

So

$$\hat{U}'(t) = \exp\left\{it\left[(\omega + 2\Omega_L \mathrm{H}\cos\theta)\hat{s}_z + 2\Omega_L \mathrm{H}\sin\theta\hat{s}_x\right]\right\} \tag{21.59}$$

We select $\theta > \frac{\pi}{2}$ for $\Omega_L > 0$ and $\theta < \frac{\pi}{2}$ for $\Omega_L < 0$ and also

$$\omega_0 = -2\Omega_L \mathrm{H}\cos\theta,\ \omega_1 = 2\Omega_L \mathrm{H}\sin\theta,\ \Omega = \sqrt{(\omega_0 - \omega_1)^2 + \omega_1^2} \tag{21.60}$$

We introduce the unit vector, $\hat{n}'$, with the components:

$$n'_x = \frac{\omega_1}{\Omega},\ n'_z = \frac{\omega_0 - \omega_1}{\Omega} \tag{21.61}$$

and also express the operators s_x and s_z through the Pauli matrices:

$$\hat{U}'(t) = \exp\left\{i\frac{\Omega t}{2}\hat{\sigma}\hat{n}'\right\} \tag{21.62}$$

or

$$\hat{U}'(t) = \exp\left\{i\frac{\Omega t}{2}\hat{\sigma}\hat{n}'\right\} \equiv \cos\left(\frac{\Omega t}{2}\right) + i\hat{\sigma}\hat{n}'\sin\left(\frac{\Omega t}{2}\right) \tag{21.63}$$

This form permits to calculate the action of the operator $\hat{U}'(t)$ on any spin state. So from Equation 21.63, the transition amplitude of $\hat{U}'(t)$ and $\hat{U}(t)$ are observed to differ by a phase difference:

$$\mathbb{d}_{-\frac{1}{2},\frac{1}{2}} = \left\langle -\frac{1}{2}\middle|\hat{U}'\middle|\frac{1}{2}\right\rangle = \exp\left\{i\frac{\pi}{2}\right\}\frac{\omega_1}{\Omega}\sin\left(\frac{\Omega t}{2}\right) \tag{21.64}$$

while the transition probability for a spin-flip or the probability for finding at time, t, the state in which the spin has the value, $-\frac{1}{2}$:

$$W_{-\frac{1}{2},\frac{1}{2}} = \left|\mathbb{d}_{-\frac{1}{2},\frac{1}{2}}\right|^2 = \frac{\omega_1^2}{(\omega_0 - \omega_1)^2 + \omega_1^2}\sin^2\left(\frac{\Omega t}{2}\right) \tag{21.65}$$

This coincides with the probability of the **resonance spin rotation** in the rotating magnetic field. The amplitude, $\frac{\omega_1^2}{(\omega_0 - \omega_1)^2 + \omega_1^2}$, has the form of a peak with height located at the **resonance frequency** $\omega_1 = \omega_0$ and breadth, $2\omega_1$. At **resonance** ($\omega_1 = \omega_0$), the transition probability for a spin-flip is observed to have the form:

$$W_{-\frac{1}{2},\frac{1}{2}} = \sin^2\left(\frac{\omega_1 t}{2}\right) \tag{21.66}$$

The given resonance can be exploited to measure the frequency, ω_0, and, consequently, the gyromagnetic ratio. It is obvious that at resonance (**resonance spin rotation**), the transition probability, $W_{-\frac{1}{2},\frac{1}{2}}$, periodically achieves the value unity. This implies that the achievement of a total blow-up of the probability of arrival from the first level to the second level with period, $T = \frac{2\pi}{\omega_1}$.

For the vector

$$\hat{P} = \hat{\sigma} \tag{21.67}$$

then the rotating system of coordinates precess about the direction of $\hat{n}'$. For $\omega_1 = \omega_0$, the vector $\hat{n}'$ is in the ox-axis direction of the rotating system of coordinate. For these conditions, the vector $\hat{P}$ rotates on the yz-plane about the ox-axis and periodically is directed in the oz-axis direction.

From Equation 21.65, we observe that for $\omega_0 \gg \omega_1$, the resonance appears to be very sharp. Considering Equation 21.61, the resonance frequency ω_0 is proportional to the Lamour frequency Ω_L that tailors the resonance spin rotation. This effect of resonance spin rotation lies on the basis of solid-state radio-spectroscopy, liquid bodies and the magnetic resonance methods in the investigation of atomic rays. That is why the problem of the motion of the spin in a rotating field is very important.

21.3 Theory of the Zeeman Effect

We have studied so far the free hydrogen atom which constitutes a system of an electron and a proton exerting an electrostatic attraction for each other but not subjected to an external field. We consider under this heading the problem of the hydrogen atom subjected to an external homogenous magnetic field $\vec{\mathrm{H}}$. Though we already have knowledge of the interaction of an electron with an electric field alone or a magnetic field alone, the energy levels of an electron subjected both to the influence of the internal electric field of the atom and to an external magnetic field are lifted in different fashions. The present study will show that when an atom is placed in a uniform external magnetic field $\vec{\mathrm{H}}$, the energy levels are shifted in a phenomenon known as the **Zeeman effect**.

To simplify discussions, we have the following assumptions:

- **Neglect the nuclear finite mass effect**
- **Effect of the external magnetic field is much smaller than that of the internal electric field of the atom**
- **Atomic level shifts due to the magnetic field are much smaller than the energy separations in a zero field**

We consider the Hamiltonian in Equation 21.4:

$$\hat{\mathrm{H}} = \frac{1}{2\mathrm{m}}\hat{\mathrm{p}}^2 + \mathrm{U} - \frac{e}{\mathrm{m}c}\left(\vec{\mathrm{A}},\hat{\mathrm{p}}\right) - \frac{ie\hbar}{2\mathrm{m}c}\mathrm{div}\vec{\mathrm{A}} + \frac{e^2}{2\mathrm{m}c^2}\vec{\mathrm{A}}^2 - \frac{e\hbar}{2\mathrm{m}c}\left(\hat{\sigma},\vec{\mathrm{H}}\right) \tag{21.68}$$

As the magnetic field is constant:

$$\mathrm{div}\vec{\mathrm{A}} = 0 \tag{21.69}$$

and in this case, we select

$$\vec{\mathrm{A}} = \frac{1}{2}\left[\vec{r},\vec{\mathrm{H}}\right] \tag{21.70}$$

then we simplify the Hamiltonian in Equation 21.68 by calculating the quantities:

$$-\frac{e}{\mu c}\left(\vec{\mathrm{A}},\hat{\mathrm{p}}\right) = -\frac{e}{2\mu c}\left(\left[\vec{r},\vec{\mathrm{H}}\right],\hat{\mathrm{p}}\right) = \frac{e}{2\mu c}\left(\left[\vec{r},\hat{\mathrm{p}}\right],\vec{\mathrm{H}}\right) = \frac{e}{2\mu c}\left(\hat{L},\vec{\mathrm{H}}\right) \tag{21.71}$$

$$\mathrm{curl}\vec{\mathrm{A}} = \left[\nabla,\vec{\mathrm{A}}\right] = \frac{1}{2}\left[\nabla,\left[\vec{\mathrm{H}},\vec{r}\right]\right] = \frac{1}{2}\left\{\vec{\mathrm{H}}\nabla\vec{r} - \left(\nabla,\vec{\mathrm{H}}\right)\vec{r}\right\} = \vec{\mathrm{H}} \tag{21.72}$$

So

$$\hat{\mathrm{H}} = \frac{1}{2\mathrm{m}}\hat{\mathrm{p}}^2 + \mathrm{U} + \frac{e}{2\mathrm{m}c}\left(\hat{L},\vec{\mathrm{H}}\right) - \frac{ie\hbar}{2\mathrm{m}c}\mathrm{div}\vec{\mathrm{A}} + \frac{e^2}{2\mathrm{m}c^2}\vec{\mathrm{A}}^2 - \frac{e\hbar}{2\mathrm{m}c}\left(\hat{\sigma},\vec{\mathrm{H}}\right) \tag{21.73}$$

Consider again

$$\hat{H} = \hat{H}_0 + \hat{U}' + \hat{U}'' \tag{21.74}$$

where $\hat{H}_0$ is the Hamiltonian of the atom in the absence of the magnetic field:

$$\hat{H}_0 = \frac{1}{2m}\hat{p}^2 + U + U_{so} \tag{21.75}$$

and we have added U_{so} which is independent of the external magnetic field and $\hat{U}'$ with $\hat{U}''$ are linear and quadratic with respect to external magnetic field $\vec{H}$ correspondingly. In the linear term, we have Equation 21.71 as well as the Pauli equation for the interaction of the intrinsic magnetic moment with the external magnetic field:

$$-\frac{e\hbar}{2mc}\left(\hat{\sigma},\vec{H}\right) = -\frac{e}{2mc}\left(2\hat{S},\vec{H}\right) \tag{21.76}$$

So from Equations 21.71 and 21.76, the linear term relative to the magnetic field strength vector $\vec{H}$ is

$$\hat{U}' = \frac{e}{2mc}\left(\hat{L},\vec{H}\right) + \frac{e}{2mc}\left(2\hat{S},\vec{H}\right) = \frac{e}{2mc}\left(\vec{H},\hat{L}+2\hat{S}\right) \equiv \frac{e}{2mc}\left(\vec{H},\hat{J}+\hat{S}\right) \tag{21.77}$$

From Equation 21.70, the quadratic term relative to the magnetic field strength vector $\vec{H}$:

$$\hat{U}'' = \frac{e^2}{2mc^2}\vec{A}^2 = \frac{e^2}{8mc^2}\left[\vec{H},\vec{r}\right]^2 \tag{21.78}$$

We assume that the field is weak and $\hat{U}'$ and $\hat{U}''$ are examined as small perturbations. We find the first correction to the energy and we neglect $\hat{U}''$ since it is quadratic relative to the field $\vec{H}$ which is small. So as a perturbation, we select $\hat{U}'$. We observe from Equations 21.77 and 21.78 that the nature of the Zeeman splitting will depend critically on the strength of the external magnetic field $\vec{H}_{ext} = \vec{H}$ compared to the internal field $\vec{H}_{int}$ which gives rise to the spin-orbit coupling. If the external field is much less than the internal field:

$$H_{ext} \ll H_{int} \tag{21.79}$$

then the fine structure dominates and, for this case, $\hat{U}'$ is treated as a perturbation and if

$$H_{ext} \gg H_{int} \tag{21.80}$$

then the Zeeman effect dominates and the fine structure becomes the perturbation. The good quantum numbers are then n,l,j and m_j but not m_l and m_s. This is because in the presence of the spin-orbit coupling, $\hat{L}$ and $\hat{S}$ are not separately conserved. They only precess about the fixed total angular momentum, $\hat{J}$, as shown in Figure 21.2.

For the intermediate regime, the two fields are comparable. In what follows, we evaluate each of the regimes for the hydrogen atom. We consider the magnetic field strength $\vec{H}$ in the oz-axis direction. So in the first-order perturbation theory, the Zeeman correction to the energy is

$$U' = \frac{eH}{2mc}\overline{\left(\hat{J}_z + \hat{S}_z\right)} \tag{21.81}$$

Here, $H_z = H$ and $\overline{x}$ denotes the average value of x.

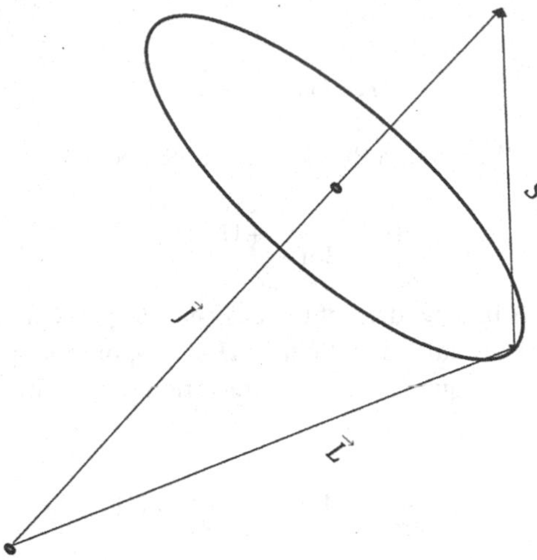

FIGURE 21.2 In the presence of the spin-orbit coupling, $\hat{L}$ and $\hat{S}$ are not separately conserved but only precess about the fixed total angular momentum, $\hat{J}$.

We find the correction to the energy in the first-order approximation as a result of $\hat{U}'$. As we have included in $\hat{H}_0$ the spin-orbitary interaction, the integral of motion is the total angular momentum $\hat{J}$. The wave function of the zero approximation can be written as $|j,m_j\rangle$, where j is the interior quantum number for which

$$J^2 = \hbar^2 j(j+1) \tag{21.82}$$

and m_j is the magnetic quantum number for which

$$J_z = \hbar m_j \tag{21.83}$$

The energy of the zero-order approximation depends only on j and not on m_j. For the zero approximation:

$$U_{so} = \lambda(r)\overline{(\hat{L},\hat{S})} \tag{21.84}$$

This implies that the state with a given j relative for m_j is $f = (2j+1)-$fold degenerate. If we want to apply rightly the perturbation theory, then we have to apply it to a degenerate level.

21.3.1 Russell-Saunders Coupling

We consider the Russell-Saunders coupling and write down the secular equation that permits to find E, the energy of the first approximation $E^{(1)}$ and the proper wave function of the zero approximation which is a linear combination for different m_j. We get a result that is less accurate though correct. Suppose we find these eigenfunctions of the zero approximation, we can get $E^{(1)}$ like in the non-degenerate case. We evaluate it from the formula:

$$\Delta E^{(1)} = \bar{U'} \tag{21.85}$$

If we assumed these wave functions to be the proper ones (without the linear combination), then J_z has a defined value and follows Equation 21.83. It is necessary to evaluate the mean value $\overline{S}_z$ since it does not have a defined value and does not commute with $\overline{U}_{so}$.

We suppose that the mean value $\overline{\hat{S}}$ is parallel to the mean value $\overline{\hat{J}}$ and so

$$\overline{S} = a\overline{J} \tag{21.86}$$

As the vectors $\overline{S}$ and $\overline{L}$ have a unique direction:

$$\overline{S}_z = a\overline{J}_z \tag{21.87}$$

Let us multiply Equation 21.86 scalarly by $\overline{J}$:

$$\overline{\hat{J}\hat{S}} = aJ^2 = a\hbar^2 j(j+1) \tag{21.88}$$

from where:

$$a = \frac{\overline{\hat{J}\hat{S}}}{\hbar^2 j(j+1)} \tag{21.89}$$

Considering Equation 21.82:

$$\hat{J} - \hat{S} = \hat{L},\ \hat{J}^2 + \hat{S}^2 - 2\hat{J}\hat{S} = \hat{L}^2 \tag{21.90}$$

from where:

$$\hat{J}\hat{S} = \frac{1}{2}\left(\hat{J}^2 + \hat{S}^2 - \hat{L}^2\right) \tag{21.91}$$

We examine the states where J^2, L^2 and S^2 have defined values:

$$L^2 = \hbar^2 l(l+1),\ S^2 = \hbar^2 s(s+1) \tag{21.92}$$

If we consider Equation 21.91, then the state, $\left(\hat{J},\hat{S}\right)$, also has a defined value where if we substitute into Equations 21.82 and 21.92, we have

$$a = \frac{\overline{\hat{J}\hat{S}}}{\hbar^2 j(j+1)} = \frac{\hbar^2\left[j(j+1)+s(s+1)-l(l+1)\right]}{2\hbar^2 j(j+1)} = \frac{\hbar^2\left[j(j+1)+s(s+1)-l(l+1)\right]}{2\hbar^2 j(j+1)} \tag{21.93}$$

So choosing the z-axis to be along the external field H, the energy:

$$\Delta E_{nljm_j} = \mu_B H g m_j \equiv \hbar\Omega_L g m_j \tag{21.94}$$

where the **Larmor precession frequency**, Ω_L, is written in Equation 21.30 and

$$\mu_B = \frac{e\hbar}{2mc} \tag{21.95}$$

being the **Bohr magneton** and m_j is the magnetic quantum number.

As

$$\overline{J_z + S_z} = \overline{J}_z(1+a) \equiv \overline{J}_z\left(1+\frac{j(j+1)+s(s+1)-l(l+1)}{2j(j+1)}\right) \equiv \overline{J}_z g \tag{21.96}$$

Here, g is called the **Landé g-factor**. The energy of an atom in the absence of the magnetic field H is independent of m but dependent on n,l,j. In the Equation 21.94, the energy levels split $2j+1$-fold in the presence of a magnetic field. The perturbation removes the $2j+1$-fold degeneracy of the angular part of the wave function: For each value of m_j, the energies of the orbitals are shifted by an amount shown in Equation 21.94 – **Zeeman effect**. This is **an anomaly of the Zeeman effect of the hydrogen atom and, at times, called** the **anomalous Zeeman effect. This invokes the Wigner-Eckhart theorem which shows the effect of the magnetic field,** H, **on the energy levels of an atom. The theorem very well predicts in a very general way that the magnetic field completely removes degeneracies of the multiplet, leading to the appearance of equidistant energy levels to the first order in the field** H. **This implies that the energy difference of the given states is proportional to** H **and to the constant,** g, **the Landé g-factor.**

In the zero approximation, the level is characterized by the quantum numbers n,l,j and follows that these levels are fine structures. For

$$l \neq 0;\ j = l \pm \frac{1}{2} \tag{21.97}$$

then in accordance with the general criterion of the perturbation theory:

$$|\Delta E| \ll \left|E_{l+\frac{1}{2}} - E_{l-\frac{1}{2}}\right| \tag{21.98}$$

which is the distance between energy levels. The criterion of the applicability of our theory

$$\left|U^{\overline{\gamma}}\right| \ll |U_{so}| \tag{21.99}$$

For $l = 0$, which is the s-state and

$$j = s = \frac{1}{2},\ g = 2,\ n = 2 \tag{21.100}$$

then the Zeeman splitting is independent of the principal quantum number.

Let us consider the special case of a spin-half particle via the spin operator $\hat{S}$ expressed through the Pauli matrix, $\hat{\sigma}$:

$$\hat{S} = \frac{\hbar}{2}\hat{\sigma} \tag{21.101}$$

From here, we can construct the following ladder operators:

$$\hat{S}_{\pm} = \hat{S}_x \pm i\hat{S}_y \tag{21.102}$$

Here

$$\hat{S}_x = \frac{\hbar}{2}\begin{bmatrix} 0 & 1 \\ 1 & 0 \end{bmatrix}, \hat{S}_y = \frac{\hbar}{2}\begin{bmatrix} 0 & -i \\ i & 0 \end{bmatrix}, \hat{S}_z = \frac{\hbar}{2}\begin{bmatrix} 1 & 0 \\ 0 & -1 \end{bmatrix}, \hat{S}_+ = \hbar\begin{bmatrix} 0 & 1 \\ 0 & 0 \end{bmatrix}, \hat{S}_- = \hbar\begin{bmatrix} 0 & 0 \\ 1 & 0 \end{bmatrix} \tag{21.103}$$

So

$$\hat{S}_z|\uparrow\rangle = \frac{\hbar}{2}|\uparrow\rangle,\ \hat{S}_z|\downarrow\rangle = -\frac{\hbar}{2}|\downarrow\rangle, |\downarrow\rangle = \begin{bmatrix} 0 \\ 1 \end{bmatrix}, |\uparrow\rangle = \begin{bmatrix} 1 \\ 0 \end{bmatrix} \quad (21.104)$$

$$\hat{S}_+|\downarrow\rangle = \hbar\begin{bmatrix} 0 & 1 \\ 0 & 0 \end{bmatrix}\begin{bmatrix} 0 \\ 1 \end{bmatrix} = \hbar\begin{bmatrix} 1 \\ 0 \end{bmatrix} = \hbar|\uparrow\rangle,\ \hat{S}_-|\uparrow\rangle = \hbar\begin{bmatrix} 0 & 0 \\ 1 & 0 \end{bmatrix}\begin{bmatrix} 1 \\ 0 \end{bmatrix} = \hbar\begin{bmatrix} 0 \\ 1 \end{bmatrix}$$

$$= \hbar|\downarrow\rangle,\ \hat{S}_+|\uparrow\rangle = 0,\ \hat{S}_-|\downarrow\rangle = 0,\ \hat{S}_+\left|\frac{1}{2}, -\frac{1}{2}\right\rangle = \hbar\sqrt{\frac{3}{4} + \frac{1}{4}}\left|\frac{1}{2}, \frac{1}{2}\right\rangle,\ \hat{S}_-\left|\frac{1}{2}, \frac{1}{2}\right\rangle = \hbar\sqrt{\frac{3}{4} + \frac{1}{4}}\left|\frac{1}{2}, -\frac{1}{2}\right\rangle \quad (21.105)$$

So for

$$\hat{L}\cdot\hat{S} = \left(\hat{L}_x\hat{S}_x + \hat{L}_y\hat{S}_y + \hat{L}_z\hat{S}_z\right) = \frac{1}{2}\left(\hat{L}_-\hat{S}_+ + \hat{L}_+\hat{S}_-\right) + \hat{L}_z\hat{S}_z \quad (21.106)$$

and

$$l = -1, 0, 1;\ n = 2;\ S = \frac{1}{2} \quad (21.107)$$

we have

$$\hat{L}_+|1\rangle = 0,\ \hat{L}_+|-1\rangle = \hbar\sqrt{2}|0\rangle,\ \hat{L}_+|0\rangle = \hbar\sqrt{2}|1\rangle,\ \hat{L}_-|1\rangle = \hbar\sqrt{2}|0\rangle,\ \hat{L}_-|0\rangle = \hbar\sqrt{2}|-1\rangle \quad (21.108)$$

$$\hat{L}\cdot\hat{S}|1,\downarrow\rangle = -\frac{\hbar^2}{2}|1,\downarrow\rangle + \frac{1}{2}\hbar^2\sqrt{2}|0,\uparrow\rangle,\ \hat{L}\cdot\hat{S}|0,\uparrow\rangle = \frac{1}{2}\hat{L}_+\hat{S}_-|0,\uparrow\rangle = \frac{1}{2}\hbar^2\sqrt{2}|1,\downarrow\rangle \quad (21.109)$$

$$\hat{L}\cdot\hat{S}|0,\downarrow\rangle = \frac{1}{2}\hat{L}_-\hat{S}_+|0,\downarrow\rangle = \frac{1}{2}\hbar^2\sqrt{2}|-1,\uparrow\rangle,\ \hat{L}\cdot\hat{S}|-1,\uparrow\rangle = -\frac{\hbar^2}{2}|-1,\uparrow\rangle + \frac{1}{2}\hbar^2\sqrt{2}|0,\downarrow\rangle,\ \hat{L}\cdot\hat{S}|-1,\downarrow\rangle = \frac{\hbar^2}{2}|-1,\downarrow\rangle \quad (21.110)$$

From here, we construct the following table:

	$\vert 1,\uparrow\rangle$	$\vert 1,\downarrow\rangle$	$\vert 0,\uparrow\rangle$	$\vert 0,\downarrow\rangle$	$\vert -1,\uparrow\rangle$	$\vert -1,\downarrow\rangle$
$\vert 1,\uparrow\rangle$	1/2	0	0	0	0	0
$\vert 1,\downarrow\rangle$	0	−1/2	$1/\sqrt{2}$	0	0	0
$\vert 0,\uparrow\rangle$	0	$1/\sqrt{2}$	0	0	0	0
$\vert 0,\downarrow\rangle$	0	0	0	0	$1/\sqrt{2}$	0
$\vert -1,\uparrow\rangle$	0	0	0	$1/\sqrt{2}$	−1/2	0
$\vert -1,\downarrow\rangle$	0	0	0	0	0	1/2

(21.111)

The matrix of the spin-orbit coupling from here can be written as follows:

$$\left(\hat{H}_{LS}\right)_{mm'} = \begin{bmatrix} a & 0 \\ 0 & b \end{bmatrix} \quad (21.112)$$

where

$$\mathbb{a}=\omega\begin{bmatrix}1 & 0 & 0\\ 0 & -1 & \sqrt{2}\\ 0 & \sqrt{2} & 0\end{bmatrix},\ \mathbb{b}=\omega\begin{bmatrix}0 & \sqrt{2} & 0\\ \sqrt{2} & -1 & 0\\ 0 & 0 & 1\end{bmatrix},\ \mathbb{O}=\begin{bmatrix}0 & 0 & 0\\ 0 & 0 & 0\\ 0 & 0 & 0\end{bmatrix},\ \omega=\frac{1}{2}\hbar^2\lambda \tag{21.113}$$

The non-trivial solution is obtained from the secular equation:

$$\det\left[\left(\hat{H}_{LS}\right)_{mm'}-\varepsilon\delta_{mm'}\right]=0 \tag{21.114}$$

or

$$\det\mathbb{M}=\det\mathbb{M}_1\times\det\mathbb{M}_2=0 \tag{21.115}$$

Here

$$\det\mathbb{M}_1=\det\begin{bmatrix}-\varepsilon+\omega & 0 & 0\\ 0 & -\varepsilon-\omega & \omega\sqrt{2}\\ 0 & \omega\sqrt{2} & -\varepsilon\end{bmatrix},\ \det\mathbb{M}_2=\det\begin{bmatrix}-\varepsilon & \omega\sqrt{2} & 0\\ \omega\sqrt{2} & -\varepsilon-\omega & 0\\ 0 & 0 & -\varepsilon+\omega\end{bmatrix} \tag{21.116}$$

So from

$$\det\mathbb{M}_1=0 \tag{21.117}$$

then

$$\det\begin{bmatrix}-\varepsilon+\omega & 0 & 0\\ 0 & -\varepsilon-\omega & \omega\sqrt{2}\\ 0 & \omega\sqrt{2} & -\varepsilon\end{bmatrix}=0 \tag{21.118}$$

or

$$-(\varepsilon-\omega)^2(\varepsilon+2\omega)=0 \tag{21.119}$$

with the energy levels being the following solutions:

$$\varepsilon_1=\omega,\ \varepsilon_2=-2\omega \tag{21.120}$$

The change in energy level:

$$\Delta E=\varepsilon_1-\varepsilon_2=3\omega \tag{21.121}$$

For

$$\det \mathbb{M}_2 = \det \begin{bmatrix} -\varepsilon & \omega\sqrt{2} & 0 \\ \omega\sqrt{2} & -\varepsilon-\omega & 0 \\ 0 & 0 & -\varepsilon+\omega \end{bmatrix} = 0 \tag{21.122}$$

then

$$-(\varepsilon-\omega)^2(\varepsilon+2\omega) = 0 \tag{21.123}$$

and the solution

$$\varepsilon_1 = \omega,\ \varepsilon_2 = -2\omega \tag{21.124}$$

So

$$\det \mathbb{M} = \det \mathbb{M}_1 \times \det \mathbb{M}_2 = (\varepsilon-\omega)^4(\varepsilon+2\omega)^2 \tag{21.125}$$

For $j = \frac{3}{2}$, the first factor in Equation 21.125 corresponds to $2j+1=4$ degeneracy, which gives the quartet and the second factor corresponds to $2j+1=2$ degeneracy that gives the doublet.

For

$$|U^7| \gg |U_{so}| \tag{21.126}$$

we neglect in the Hamiltonian $\hat{H}$ the spin-orbit interaction and U^7 introduces the quantity $\hat{H}_0$ that considers the presence of the magnetic field H expressed through Ω_L:

$$\hat{H}_0 = \frac{1}{2m}\hat{p}^2 + U - \Omega_L\left(\hat{L}_z + 2\hat{S}_z\right) \tag{21.127}$$

Here, the factor 2 in front of $\hat{S}_z$ comes from the electron spin gyromagnetic ratio. In addition, $\hat{S}_z$ commutes with the Hamiltonian $\hat{H}$ as there is no spin-orbit interaction. So considering the **Zeeman energy**, $\hbar\Omega_L$:

$$\left(\hat{L}_z + 2\hat{S}_z\right)|1,\uparrow\rangle = 2\Omega|1,\uparrow\rangle;\ \left(\hat{L}_z + 2\hat{S}_z\right)|1,\downarrow\rangle = 0\Omega|1,\downarrow\rangle \tag{21.128}$$

$$\left(\hat{L}_z + 2\hat{S}_z\right)|0,\uparrow\rangle = \Omega|0,\uparrow\rangle;\ \left(\hat{L}_z + 2\hat{S}_z\right)|0,\downarrow\rangle = -\Omega|0,\downarrow\rangle \tag{21.129}$$

$$\left(\hat{L}_z + 2\hat{S}_z\right)|-1,\uparrow\rangle = 0\Omega|-1,\uparrow\rangle;\ \left(\hat{L}_z + 2\hat{S}_z\right)|-1,\downarrow\rangle = -2\Omega|-1,\downarrow\rangle \tag{21.130}$$

then

$$\mathbb{M} = \begin{bmatrix} \mathbb{M}_1 & \mathbb{O} \\ \mathbb{O} & \mathbb{M}_2 \end{bmatrix},\ \mathbb{M}_1 = \begin{bmatrix} -\varepsilon+\omega+2\Omega & 0 & 0 \\ 0 & -\varepsilon-\omega & \omega\sqrt{2} \\ 0 & \omega\sqrt{2} & -\varepsilon+\Omega \end{bmatrix},\ \mathbb{M}_2 = \begin{bmatrix} -\varepsilon-\Omega & \omega\sqrt{2} & 0 \\ a\sqrt{2} & -\varepsilon-\omega & 0 \\ 0 & 0 & -\varepsilon+\omega-2\Omega \end{bmatrix} \tag{21.131}$$

and

$$\det \mathbb{M} = \det \mathbb{M}_1 \times \det \mathbb{M}_2 \tag{21.132}$$

with

$$\det \mathbb{M}_1 = (-\varepsilon + \omega + 2\Omega)\left((\varepsilon+\omega)(\varepsilon-\Omega) - 2\omega^2\right), \det \mathbb{M}_2 = (-\varepsilon + \omega - 2\Omega)\left((\varepsilon+\omega)(\varepsilon+\Omega) - 2\omega^2\right) \tag{21.133}$$

Hence,

$$\det \mathbb{M} = \left((\varepsilon-\omega)^2 - 4\Omega^2\right)\left((\varepsilon+\omega)(\varepsilon-\Omega) - 2\omega^2\right)\left((\varepsilon+\omega)(\varepsilon+\Omega) - 2\omega^2\right) = 0 \tag{21.134}$$

then

$$(\varepsilon-\omega)^2 - 4\Omega^2 = 0 \tag{21.135}$$

or

$$\varepsilon_{1,2} = \omega \pm 2\Omega \tag{21.136}$$

and

$$(\varepsilon+\omega)(\varepsilon-\Omega) - 2\omega^2 = \varepsilon^2 + \varepsilon(\omega-\Omega) - \left(2\omega^2 + \omega\Omega\right) = 0 \tag{21.137}$$

$$(\varepsilon+\omega)(\varepsilon+\Omega) - 2\omega^2 = \varepsilon^2 + \varepsilon(\omega+\Omega) - \left(2\omega^2 - \omega\Omega\right) = 0 \tag{21.138}$$

The solutions of Equations 21.137 and 21.138 give, respectively,

$$\varepsilon_{3,4} = \frac{1}{2}(\Omega-\omega) \pm \frac{1}{2}\sqrt{9\omega^2 + 2\omega\Omega + \Omega^2}, \ \varepsilon_{5,6} = -\frac{1}{2}(\Omega-\omega) \pm \frac{1}{2}\sqrt{9\omega^2 - 2\omega\Omega + \Omega^2} \tag{21.139}$$

We consider limiting cases for weak and strong magnetic fields.

21.3.2 Weak Field Limiting Case – Zeeman Effect

For the weak magnetic field, we consider the p-state when $l = 1$ where the spin-orbit coupling dominates:

$$\omega \gg \Omega \tag{21.140}$$

Here, the good quantum numbers will be n, l, j and m_j and not m_L and m_S. This is because in the presence of the spin-orbit coupling, $\vec{L}$ and $\vec{S}$ are not separately conserved. So in the first-order approximation, we have

$$\varepsilon_{3,4} = \frac{1}{2}(\Omega-\omega) \pm \frac{3\omega}{2}\sqrt{1 + \frac{2\Omega}{9\omega} + \frac{\Omega^2}{9\omega^2}} \cong \frac{1}{2}(\Omega-\omega) \pm \frac{3\omega}{2}\left(1 + \frac{\Omega}{9\omega} + \frac{\Omega^2}{18\omega^2}\right) \tag{21.141}$$

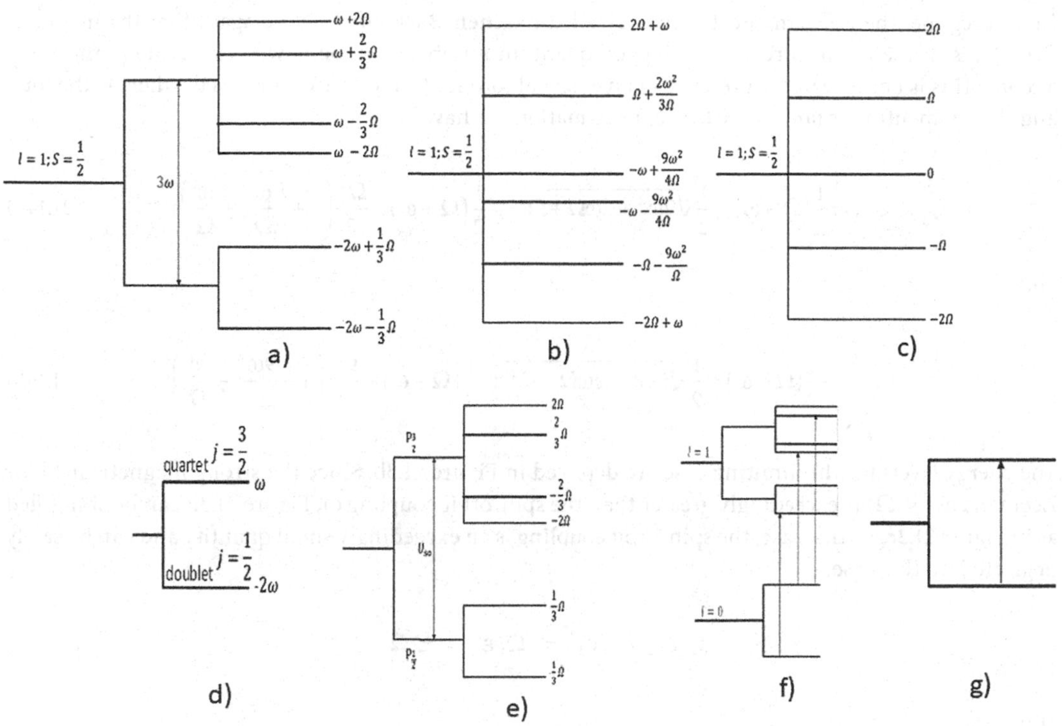

FIGURE 21.3 (21.3a) A quartet and a doublet state for the weak magnetic field (Zeeman effect); (21.3b) the splitting of the energy level for strong magnetic field for exceedingly small spin-orbit interaction; (21.3c) the splitting of the energy level for strong magnetic field with no spin-orbit interaction. There is splitting of equidistant levels with distance between two adjacent levels proportional to Ω; (21.3d) the quartet states for $j=\frac{3}{2}$ and the doublet states for $j=\frac{1}{2}$; (21.3e) shows in detail the quartet states for $j=\frac{3}{2}$ and the doublet states for $j=\frac{1}{2}$; (21.3f) show transitions between different degenerate levels; (21.3g) the transition between two levels of a doublet.

Similarly,

$$\varepsilon_{5,6}=-\frac{1}{2}(\Omega-\omega)\pm\frac{3\omega}{2}\sqrt{1-\frac{2\Omega}{9\omega}+\frac{\Omega^2}{9\omega^2}}\cong-\frac{1}{2}(\Omega-\omega)\pm\frac{3\omega}{2}\left(1-\frac{\Omega}{9\omega}+\frac{\Omega^2}{18\omega^2}\right) \tag{21.142}$$

We observe from these relations that the total energy is the sum of the spin-orbit coupling and the Zeeman contributions. The contributions from the spin-orbit coupling merely shows the shifts in the energy levels without modifying the Zeeman diagram. The energy levels for this limiting case are depicted in Figure 21.3a showing a quartet and a doublet state.

21.3.3 Strong Field for Exceedingly Small Spin-Orbit Interaction – Paschen-Back Effect†

For the weak magnetic field, we consider the p-state for $l=1$ where the Zeeman effect dominates:

$$\omega \ll \Omega \tag{21.143}$$

In this regime, the Zeeman effect is the so-called **Paschen-Back effect**. Note again that the magnetic field, $\vec{H}$, is in the z-axis direction. The good quantum numbers will be now n,l,m_L and m_S and not j and m_j. This is because in the presence of an external torque, L_z and S_z are conserved, whereas the total angular momentum is not. In the first approximation, we have:

$$\varepsilon_{3,4} = \frac{1}{2}(\Omega - \omega) \pm \frac{1}{2}\sqrt{9\omega^2 + 2\omega\Omega + \Omega^2} \cong \frac{1}{2}(\Omega - \omega) \pm \frac{\Omega^2}{2}\left(1 + \frac{9\omega^2}{2\Omega^2} + \frac{\omega}{\Omega}\right) \tag{21.144}$$

and

$$\varepsilon_{5,6} = -\frac{1}{2}(\Omega - \omega) \pm \frac{1}{2}\sqrt{9\omega^2 - 2\omega\Omega + \Omega^2} \cong \frac{1}{2}(\Omega - \omega) \pm \frac{\Omega^2}{2}\left(1 + \frac{9\omega^2}{2\Omega^2} - \frac{\omega}{\Omega}\right) \tag{21.145}$$

The energy levels for this limiting case are depicted in Figure 21.3b. Since the strong magnetic field (or Zeeman energy Ω) is exceedingly greater than the spin-orbit coupling ω, Figure 21.3b can be simplified as in Figure 21.3c. In this case, the spin-orbit coupling is an exceedingly small quantity and can be safely neglected. In this case,

$$\varepsilon_{1,2} = 0,\ \varepsilon_{3,4} = \pm\Omega,\ \varepsilon_{5,6} = \pm 2\Omega \tag{21.146}$$

and

$$\Delta E_{nljm_j} = -\Omega\left(m_j \pm 1\right) \tag{21.147}$$

For this limiting case, the Zeeman normal effect in a strong field (**Paschen-Back effect**) results in the splitting of the levels given simply by the number m_j and the distance between levels is equal to Ω.

21.3.4 Landau Case

For the Landau case, we consider only spin-orbit coupling:

$$\Omega \ll \omega,\ \Omega \to 0 \tag{21.148}$$

$$\Delta E_{J,J-1} = \lambda J = \lambda\left(l + \frac{1}{2}\right) \tag{21.149}$$

For $l = 1; s = \frac{1}{2}$, we have the quartet states for $j = \frac{3}{2}$ and the doublet states for $j = \frac{1}{2}$ as in Figure 21.3d and described by the following equation:

$$(\varepsilon - \omega)^2(\varepsilon - \omega)^2(\varepsilon + 2\omega)^2 = 0 \tag{21.150}$$

We consider the spin-orbit coupling for Equation 21.148 and

$$(\varepsilon - \omega)^2(\varepsilon - \omega)^2(\varepsilon + 2\omega)^2 = 0 \tag{21.151}$$

The relation in Equation 21.76 is obtained for the one electron atom where the electron is at the center of the field. This formula has a more general character and may be used for the many-electron atom. But

$$g=1+\frac{j(j+1)+s(s+1)-l(l+1)}{2j(j+1)} \tag{21.152}$$

where j, s and l are respectively total, spin and azimuthal quantum numbers of all electrons. The rest of the formulae are invariant. In particular, if s equal to zero (for the many-electron atom), then $j=l$ and $g=1$. Then we have the normal Zeeman effect. The splitting of levels in a magnetic field can be observed not only in the optical domain but also in the microwave domain. In the optical domain, the Zeeman effect differs by the type of the transitions. In the microwave domain, we have the transition between sub-levels of one of the levels. For any magnetic field strength H, the frequency of the field falls in the microwave domain, i.e., such a transition belongs to radio spectroscopy.

21.4 Atomic Paramagnetism and Diamagnetism

If an atom has an intrinsic magnetic moment, $\vec{M}$, and if its energy E in the magnetic field $\vec{H}$ is dependent on the field, then

$$M_z=-\frac{\partial E}{\partial H_z} \tag{21.153}$$

We apply it to Equation 21.94:

$$M_z=-m_B g m_j \tag{21.154}$$

It is independent of the external magnetic field H_z and is related to paramagnetism where in the absence of the field the dipole moments are chaotically oriented and in the presence of the magnetic field they are oriented in the direction of the magnetic field. The quantity in Equation 21.154 characterizes the paramagnetic moment of an atom. For one electron atom, that quantity may not be equal to zero. Equation 21.154 can also be applicable to many electron-atom (for the same condition). In this case, m_j may be equal to zero. For such atoms, paramagnetism does not exist and are diamagnetic. Atoms with odd number of electrons can be paramagnetic as well as diamagnetic for the case when the intrinsic moment of the atom is equal to zero. If the correction of the first order to the energy is equal to zero, then it is necessary to find the correction of the second order. In the definition of the Hamiltonian, we obtain the term $\hat{U}''$:

$$\hat{U}''=\frac{e^2}{8mc^2}\left[\vec{H},\vec{r}\right]^2 \tag{21.155}$$

We find the term of the second order with respect to H if we consider $\hat{U}$:

$$\hat{U}''=\frac{e^2}{8mc^2}\sum_i\left[\vec{H},\vec{r}_i\right]^2 \tag{21.156}$$

Here, the sum is taken with respect to all electron coordinates.

Suppose we have many electrons then $\hat{U}'$ and $\hat{U}''$ are additive with respect to these electrons. If we assume that the state is non-degenerate, then the energy correction:

$$\Delta E^{(2)} = \frac{e^2}{8mc^2}\sum_i \overline{\left[\vec{H}, \vec{r}_\iota\right]^2} = \frac{e^2}{8mc^2}\sum_i \overline{\left[Hr_\iota \sin\theta_\iota\right]^2} \tag{21.157}$$

If we assume that in the ground state, the wave function is spherically symmetric:

$$\overline{\sin^2\theta_\iota} = \frac{1}{4\pi}\int_0^\pi\int_0^{2\pi} \sin^2\theta_i \sin\theta_i d\theta_i d\phi = \frac{2}{3} \tag{21.158}$$

and

$$\Delta E^{(2)} = \frac{e^2 H^2}{12\mu c^2}\sum_i \overline{r_\iota^2} \tag{21.159}$$

and the diamagnetic moment

$$M_d = -\frac{\partial}{\partial H}\Delta E^{(2)} = -\frac{e^2 H}{6mc^2}\sum_i \overline{r_\iota^2} \tag{21.160}$$

If we consider the quadratic term of the Hamiltonian, then the atom has a magnetic moment proportional to H which is antiparallel to the field. This is a diamagnetic effect. If matter is placed in a magnetic field, then it is magnetized with the dipole moments being antiparallel to that field. The diamagnetic susceptibility is given as follows:

$$X_d = -\frac{e^2}{6\mu c^2}\sum_i \overline{r_i^2} \tag{21.161}$$

If the paramagnetic moment is different from zero, then

$$\left|\frac{M_d}{M_p}\right| = -\frac{e^2 H \overline{r^2}\, 2\mu c}{6mc^2 l\hbar} = \frac{\alpha}{3}/H/\frac{l}{\overline{r^2}},\ \alpha = \frac{e^2}{\hbar c} \approx \frac{1}{137} \tag{21.162}$$

Here, α is the fine structure and $\frac{l}{\overline{r^2}}$ is the mean electric field at the interior of an atom created by the nucleus. If

$$\frac{l}{\overline{r^2}} \cong 10^8 \text{ volts/cm} \tag{21.163}$$

then it follows that

$$\left|\frac{M_d}{M_p}\right| \ll 1 \tag{21.164}$$

This implies that if the paramagnetic moment is different from zero, then it overshadows diamagnetism.

III

Appendix: Special Functions

22

Gamma Functions

22.1 First Kind Euler Integral-Beta Function

Under this chapter, we consider the so-called Euler functions, namely the well-known Gamma and Beta functions. The Euler integral of the first kind, also known as the Beta function, is defined as follows:

$$B(\alpha,\beta)=\int_0^1 y^{\alpha-1}(1-y)^{\beta-1}\,dy \tag{22.1}$$

and converges for $\operatorname{Re}\alpha>0$ and $\operatorname{Re}\beta>0$. We show that for the $\alpha-\beta$ symmetry to be preserved there must be symmetry of the integrand y and $1-y$. So, we compute the value of the Beta function recursively by setting

$$1-y\equiv x \tag{22.2}$$

then

$$B(\alpha,\beta)=-\int_1^0 (1-x)^{\alpha-1}x^{\beta-1}dx=\int_0^1 x^{\beta-1}(1-x)^{\alpha-1}\,dx=B(\beta,\alpha) \tag{22.3}$$

This shows that the Beta function is a symmetric function relative to the permutation of its arguments.
Do integration by parts in Equation 22.1:

$$B(\alpha,\beta)=\int_0^1 y^{\alpha-1}(1-y)^{\beta-1}\,dy=\int_0^1 \frac{(\beta-1)y^{\alpha}(1-y)^{\beta-2}}{\alpha}dy \tag{22.4}$$

From

$$y^{\alpha}=y^{\alpha-1}-y^{\alpha-1}(1-y) \tag{22.5}$$

then Equation 22.4 becomes

$$B(\alpha,\beta)=\frac{\beta-1}{\alpha}B(\alpha,\beta-1)-\frac{\beta-1}{\alpha}B(\alpha,\beta) \tag{22.6}$$

DOI: 10.1201/9781003273073-25

from where:

$$B(\alpha,\beta)=\frac{\beta-1}{\alpha+\beta-1}B(\alpha,\beta-1)=\frac{\alpha-1}{\alpha+\beta-1}B(\alpha-1,\beta) \tag{22.7}$$

If $=n$, where n is an integer, then Equation 22.7 becomes

$$B(\alpha,n)=B(n,\alpha)=\frac{n-1}{\alpha+n-1}\frac{n-2}{\alpha+n-2}\cdots\frac{1}{\alpha+1}B(\alpha,1)=\frac{1\cdot 2\cdots(n-1)}{\alpha(\alpha+1)\cdots(\alpha+n-1)} \tag{22.8}$$

For the integers, $\alpha=m$, $\beta=n$, Equation 22.7 becomes

$$B(m,n)=B(n,m)=\frac{1\cdot 2\cdots(n-1)}{(m+1)\cdots(m+n-1)}B(m,1)=\frac{1\cdot 2\cdots(n-1)}{(m+1)\cdots(m+n-1)}\frac{m-1}{m}\frac{m-2}{m-1}\cdots\frac{1}{2}B(1,1) \tag{22.9}$$

Since

$$B(1,1)=1 \tag{22.10}$$

then

$$B(m,n)=B(n,m)=\frac{(n-1)!(m-1)!}{(m+n-1)!} \tag{22.11}$$

If $\alpha=\beta$, then Equation 22.1 becomes

$$B(\alpha,\alpha)=\frac{1}{2^{2\alpha-1}}B\left(\frac{1}{2},\alpha\right) \tag{22.12}$$

Letting in Equation 22.1, $y=\frac{x}{1+x}$, $x=\frac{y}{1-y}$ and

$$B(\alpha,\beta)=\int_0^\infty\frac{x^{\alpha-1}}{(1+x)^{\alpha+\beta}}dx=\int_0^1\frac{x^{\alpha-1}}{(1+x)^{\alpha+\beta}}dx+\int_1^\infty\frac{x^{\alpha-1}}{(1+x)^{\alpha+\beta}}dx \tag{22.13}$$

Doing the change of variable, $x=\frac{1}{y}$, in the second integral:

$$B(\alpha,\beta)=\int_0^1\frac{x^{\alpha-1}}{(1+x)^{\alpha+\beta}}dx+\int_0^1\frac{1}{y^2}\frac{y^{\alpha+\beta}}{y^{\alpha-1}(1+y)^{\alpha+\beta}}dy \tag{22.14}$$

22.2 Gamma Function (Second Kind Euler Integral)

The Gamma function, $\Gamma(\alpha)$, also known as Euler integral of the second kind, is the most widely used of all the special functions where we take for its definition the most common integral representation:

$$\Gamma(\alpha)=\int_0^\infty\exp\{-x\}x^{\alpha-1}dx,\ \mathrm{Re}\,\alpha>0 \tag{22.15}$$

and valid only in the right half-plane of the complex plane. Letting $x = ty$:

$$\Gamma(\alpha) = t^{\alpha} \int_0^{\infty} \exp\{-ty\} y^{\alpha-1} dy \tag{22.16}$$

Swap α for $\alpha+\beta$ and t for $t+1$:

$$\frac{\Gamma(\alpha+\beta)}{(t+1)^{\alpha+\beta}} = \int_0^{\infty} \exp\{-(t+1)y\} y^{\alpha+\beta-1} dy \tag{22.17}$$

Multiply this equality by $t^{\alpha-1}$ and integrate over t from 0 to ∞:

$$\Gamma(\alpha+\beta) \int_0^{\infty} \frac{t^{\alpha-1}}{(t+1)^{\alpha+\beta}} dt = \int_0^{\infty} dt \left(\int_0^{\infty} \exp\{-(t+1)y\} y^{\alpha+\beta-1} dy \right) t^{\alpha-1} \tag{22.18}$$

From Equation 22.13, we have

$$\int_0^{\infty} \frac{t^{\alpha-1}}{(t+1)^{\alpha+\beta}} dt = \mathrm{B}(\alpha,\beta) \tag{22.19}$$

So

$$\Gamma(\alpha+\beta)\mathrm{B}(\alpha,\beta) = \int_0^{\infty} \left(\int_0^{\infty} \exp\{-ty\} (ty)^{\alpha-1} d(ty) \right) y^{\beta-1} \exp\{-y\} dy = \Gamma(\alpha) \int_0^{\infty} y^{\beta-1} \exp\{-y\} dy = \Gamma(\alpha)\Gamma(\beta) \tag{22.20}$$

From here, we observed the Euler beta function relates the Gamma function:

$$\mathrm{B}(\alpha,\beta) = \frac{\Gamma(\alpha)\Gamma(\beta)}{\Gamma(\alpha+\beta)} \tag{22.21}$$

Letting $x = \mathrm{v}^2$, Equation 22.15 becomes

$$\Gamma(\alpha) = 2\int_0^{\infty} \exp\{-\mathrm{v}^2\} \mathrm{v}^{2\alpha-1} d\mathrm{v}, \ \mathrm{Re}\,\alpha > 0 \tag{22.22}$$

If $\alpha = \frac{1}{2}$, then

$$\Gamma\left(\frac{1}{2}\right) = \int_{-\infty}^{+\infty} \exp\{-\mathrm{v}^2\} d\mathrm{v} = \sqrt{\pi} \tag{22.23}$$

For integer values, n, then

$$\Gamma\left(n+\frac{1}{2}\right) = \int_{-\infty}^{+\infty} \exp\{-\mathrm{v}^2\} \mathrm{v}^n d\mathrm{v} = \Gamma\left(\frac{1}{2}\right) \frac{(2n-1)!!}{2^n} = \sqrt{\pi} \frac{(2n)!}{2^{2n} n!} \tag{22.24}$$

In Equation 22.15, swap α with $\alpha+1$ and take integration by parts:

$$\Gamma(\alpha+1)=\int_0^\infty \exp\{-x\}x^\alpha dx=-\exp\{-x\}x^\alpha\Big|_0^\infty+\alpha\int_0^\infty \exp\{-x\}x^{\alpha-1}dx=\alpha\Gamma(\alpha) \qquad (22.25)$$

For positive integer values, $\alpha=n$:

$$\Gamma(n+1)=n\Gamma(n)=n(n-1)\Gamma(n-1)=\cdots=n(n-1)\cdots 2\cdot 1\cdot\Gamma(1) \qquad (22.26)$$

From

$$\Gamma(1)=\int_0^\infty \exp\{-x\}xdx=1 \qquad (22.27)$$

then the Gamma function defines the factorial function for positive integer values, n:

$$\Gamma(n+1)=n! \qquad (22.28)$$

22.3 Gamma Function Analytic Continuation

Find the analytic continuation of the Gamma function, $\Gamma(z)$, in the complex domain when $\mathrm{Re}\,z>0$:

$$\Gamma(z)=\int_0^\infty \exp\{-t\}t^{z-1}dt=\int_0^\infty \exp\{(z-1)\ln t-t\}dt \qquad (22.29)$$

where t is real and positive, i.e., $\ln t$ is real. When t is a complex-valued variable, then the function, $\exp\{(z-1)\ln t-t\}$ has a singularity at $=0$. The given function is single-valued on the plane, (t), and defined on the real axis from point $t=0$ to $t=+\infty$. From the Cauchy integral:

$$\int_l \exp\{-t\}t^{z-1}dt=\int_l \exp\{(z-1)\ln t-t\}dt \qquad (22.30)$$

Here, the contour l with ends points leading to $+\infty$ and enclosing the point $t=0$ (Figure 22.1a).

The contour l can be represented in the form of the interval $(+\infty,\varepsilon)$ which is the upper part of the circle, C_ε, and the interval $(\varepsilon,+\infty)$ which is the lower part of the given circle (Figure 22.1b). At the upper part, the function $\ln t$ is real-valued, while at the lower part, it is complex-valued since $\ln t+2\pi i$:

$$\int_l \exp\{-t\}t^{z-1}dt=\int_{+\infty}^{\varepsilon} \exp\{-t\}t^{z-1}dt+\int_{C_\varepsilon} \exp\{-t\}t^{z-1}dt+\exp\{2(z-1)\pi i\}\int_{\varepsilon}^{+\infty} \exp\{-t\}t^{z-1}dt \qquad (22.31)$$

From the upper part of the circle, consider

$$t^{z-1}=\exp\{(z-1)\ln t\} \qquad (22.32)$$

and so

$$\exp\{(z-1)(\ln t+2\pi i)\}=\exp\{(z-1)\ln t\}\exp\{(z-1)2\pi i\}=t^{z-1}\exp\{(z-1)2\pi i\} \qquad (22.33)$$

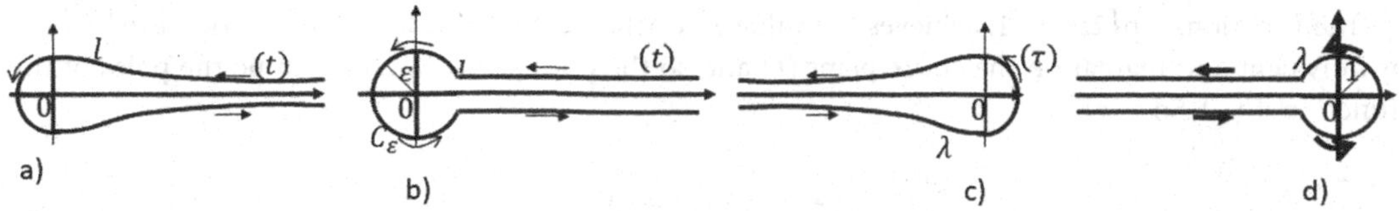

FIGURE 22.1 (22.1a) The contour l with ends points leading to $+\infty$ and enclosing the point $t=0$; (22.1b) the contour, l in the form of the interval, $(+\infty,\varepsilon)$ which is the upper part of the circle, C_ε and the interval $(\varepsilon,+\infty)$ that is the lower part of the given circle; (22.1c) the contour λ when rotating the plane (t) (with the positive cut of the real axis) through the angle $(-\pi)$ yields the plane, (τ), with the cut in the negative part of the real axis; (22.1d) the contour, λ, taking the lower edge of the cut of the negative part of the real axis from the point $t=-\infty$ to the point $t=-1$ of the circle, $|t|=1$, and the upper edge of the same cut.

On the circle, C_ε, we set $|t|=\varepsilon$ and from $z=x+iy$:

$$\left|t^{z-1}\right|=\left|\exp\{(x-1)\ln|t|-y\arg t\}\right|\times\left|\exp\{i((x-1)\arg t+y\ln|t|)\}\right|=|t|^{x-1}\exp\{-y\arg t\}=\varepsilon^{x-1}\exp\{-y\arg t\} \quad (22.34)$$

From

$$\max_{C_\varepsilon}\left|\exp\{-y\arg t-t\}\right|=M \quad (22.35)$$

then

$$\left|\int_{C_\varepsilon}\exp\{-t\}t^{z-1}dt\right|\le\varepsilon^{x-1}M2\pi\varepsilon \quad (22.36)$$

and

$$\lim_{\varepsilon\to 0}\int_{C_\varepsilon}\exp\{-t\}t^{z-1}dt=0 \quad (22.37)$$

This implies that when $\varepsilon\to 0$:

$$\int_l\exp\{-t\}t^{z-1}dt=\int_{+\infty}^{0}\exp\{-t\}t^{z-1}dt+\exp\{2(z-1)\pi i\}\int_0^{+\infty}\exp\{-t\}t^{z-1}dt \quad (22.38)$$

From

$$\exp\{-2\pi i\}=1 \quad (22.39)$$

and Equation 22.29:

$$\Gamma(z)=\left(\exp\{2z\pi i\}-1\right)^{-1}\int_l\exp\{-t\}t^{z-1}dt \quad (22.40)$$

The function, $\exp\{2z\pi i\}-1$, achieves the value zero when $z=0,\pm1,\pm2,\ldots$. The function $\exp\{-t\}t^{z-1}$ is univalent and regular on the entire plane (t) and so the points $z=1,2,\ldots$ cannot be the poles of the function, $\Gamma(z)$. So

$$\int_l \exp\{-t\}t^{z-1}dt=0 \tag{22.41}$$

For the poles, $z=0,-1,-2,\ldots$ of the function, $\Gamma(z)$:

$$\int_l \exp\{-t\}t^{z-1}dt\neq 0 \tag{22.42}$$

So from the analytic continuation, Equation 22.40 is defined on the entire (z)-plane. Swapping z with $1-z$:

$$\int_l \exp\{-t\}t^{-z}dt=\left(\exp\{-2z\pi i\}-1\right)\Gamma(1-z) \tag{22.43}$$

22.4 Hankel Integral Representations

From

$$t=\tau\exp\{\pi i\}=-\tau \tag{22.44}$$

then rotating the plane (t) (with the cut on positive part of the real axis) through the angle $(-\pi)$ yields the plane, (τ) (with the cut on the negative part of the real axis). Besides, the upper part of the cut of the plane (t) (where $\arg t=0$) matches with the lower part of the cut of the plane (τ) as on the lower part:

$$\tau\exp\{\pi i\}=0 \tag{22.45}$$

implying

$$\arg\exp\{\pi i\}+\arg\tau=\pi+\arg\tau=0 \tag{22.46}$$

In this case, the contour l is transformed to the contour λ as in Figure 22.1c. So

$$\int_l \exp\{-t\}t^{-z}dt=-\exp\{-z\pi i\}\int_\lambda \exp\{\tau\}\tau^{-z}d\tau \tag{22.47}$$

But

$$\int_\lambda \exp\{\tau\}\tau^{-z}d\tau=\left(\exp\{z\pi i\}-\exp\{-z\pi i\}\right)\Gamma(1-z)=2i\sin z\pi\,\Gamma(1-z)=\frac{2i\pi}{\Gamma(z)} \tag{22.48}$$

then we have the complex integral representation for the function, $\frac{1}{\Gamma(z)}$, provided by Hankel in 1864 and valid for unrestricted z:

$$\frac{1}{\Gamma(z)}=\frac{1}{2\pi i}\int_{\lambda}\exp\{\tau\}\tau^{-z}d\tau \tag{22.49}$$

Setting in Equation 22.48 the change of variable $1-z=z'$ and from

$$\sin(1-z')\pi=\sin z'\pi \tag{22.50}$$

then

$$\Gamma(z')=\frac{1}{2i\sin z'\pi}\int_{\lambda}\exp\{\tau\}\tau^{z'-1}d\tau \tag{22.51}$$

Letting $\tau=-\tau'$:

$$\Gamma(z')=-\frac{1}{2i\sin z'\pi}\int_{\lambda'}\exp\{-\tau'\}(-\tau')^{z'-1}d\tau' \tag{22.52}$$

Here, λ' denotes the Hankel path defined in Figure 22.1c with the contour starting at $\tau'=+\infty$ and goes to the positive direction of the point $\tau'=0$ on the real axis, encircles the origin in the counterclockwise direction and returns to the starting point $\tau'=+\infty$. The Hankel formula is satisfied for all values of z' except $z'=0,\pm1,\pm2,\ldots$.

Consider Equation 22.29, the Hankel formula can be transformed:

$$\int_{\lambda'}\exp\{-\tau'\}(-\tau')^{z'-1}d\tau'=-2i\sin z'\pi\int_{0}^{\infty}\exp\{-\tau'\}\tau'^{z'-1}d\tau' \tag{22.53}$$

22.5 Reflection or Complementary Formula

From Equation 22.48, we also have

$$\Gamma(1-z)\Gamma(z)=\frac{\pi}{\sin z\pi} \tag{22.54}$$

This is the **so-called reflection formula** which shows that the Gamma function has no zeros. Actually, the zeros cannot be $z=0,\pm1,\pm2,\ldots$. If the Gamma function vanishes at non-integer z, then from Equation 22.54, this zero would be a pole of $\Gamma(1-z)$, which cannot be true. This implies that $\frac{1}{\Gamma(z)}$ is an entire function. So we can imagine $\frac{1}{\Gamma(1-z)}$ collecting the positive zeros of $\sin z\pi$, while $\frac{1}{\Gamma(z)}$ collects the non-positive zeros.

23

Confluent Hypergeometric Functions

23.1 Classical Gauss Confluent Hypergeometric Function

The confluent hypergeometric equation (classical Gauss equation) is represented by the differential equation:

$$z(z-1)\frac{d^2}{dz^2}f+\left((\alpha+\beta+1)z-\gamma\right)\frac{d}{dz}f+\alpha\beta f=0 \tag{23.1}$$

This equation has a partial solution which is convergent on the disk $|z|<1$ and for the pole, $z=0$, we have

$$f_{0,1}\equiv \mathrm{F}(\alpha,\beta,\gamma;z) \tag{23.2}$$

This is the confluent hypergeometric function defined by the Gauss series:

$$\mathrm{F}(\alpha,\beta,\gamma;z)=1+\frac{\alpha\beta}{1!\gamma}z+\frac{\alpha(\alpha+1)\beta(\beta+1)}{2!\gamma(\gamma+1)}z^2+\cdots+\frac{\alpha(\alpha+1)\cdots\alpha(\alpha+k-1)\beta(\beta+1)\cdots(\beta+k-1)}{k!\gamma(\gamma+1)\cdots(\gamma+k-1)}z^k\cdots \tag{23.3}$$

For

$$\alpha=1,\beta=\gamma \tag{23.4}$$

then the confluent hypergeometric function yields a geometric progression leading to an elementary function:

$$\mathrm{F}(1,\beta,\beta;z)=1+z+z^2+\cdots+z^k\cdots=\frac{1}{1-z} \tag{23.5}$$

The second partial solution of Equation 23.1 is convergent on the disk, $|z|<1$, and for the pole, $z=0$, we have

$$f_{0,2}\equiv z^{1-\gamma}\mathrm{F}(\alpha-\gamma+1,\beta-\gamma+1,2-\gamma;z) \tag{23.6}$$

Here, we consider the condition:

$$2-\gamma\neq 0,-1,-2,\ldots \tag{23.7}$$

DOI: 10.1201/9781003273073-26

From the partial solution of Equation 23.1 and from the pole, $z=1$, we set $z'=1-z$ while the pole $z=0$ is shifted to $z'=1$ and the pole $z=1$ shifted to $z'=0$, while the point $z=\infty$ coincides with the point $z'=\infty$. So for convergence on the disk, $|z-1|<1$, we have

$$f_{1,1} \equiv \mathrm{F}(\alpha,\beta,1+\alpha+\beta-\gamma;1-z) \tag{23.8}$$

for the condition:

$$1+\alpha+\beta-\gamma \neq 0,-1,-2,\ldots \tag{23.9}$$

and

$$f_{1,2} \equiv (1-z)^{\gamma-\alpha-\beta}\,\mathrm{F}(\beta-\gamma,\alpha-\gamma,1+\gamma-\alpha-\beta;1-z) \tag{23.10}$$

for the condition:

$$1+\gamma-\alpha-\beta \neq 0,-1,-2,\ldots \tag{23.11}$$

The partial solution of Equation 23.1 when the pole $z=\infty$ is obtained by setting $z''=\frac{1}{z}$ when the point $z=0$ becomes $z''=\infty$ and $z=\infty$ becomes $z''=0$, while the point $z=1$ coincides with $z''=1$. So for convergence on the disk, $|z|>1$, we have

$$f_{\infty,1} \equiv \left(\frac{1}{z}\right)^{\alpha}\mathrm{F}\left(\alpha,1+\alpha-\gamma,1+\alpha-\beta;\frac{1}{z}\right) \tag{23.12}$$

for the condition:

$$1+\alpha-\beta \neq 0,-1,-2,\ldots \tag{23.13}$$

and

$$f_{\infty,2} \equiv \left(\frac{1}{z}\right)^{\beta}\mathrm{F}\left(\beta,1+\beta-\gamma,1+\beta-\alpha;\frac{1}{z}\right) \tag{23.14}$$

for the condition:

$$1+\beta-\alpha \neq 0,-1,-2,\ldots \tag{23.15}$$

We note, respectively, $f_{0,1}=f_{0,2}$, $f_{1,1}=f_{1,2}$ and $f_{\infty,1}=f_{\infty,2}$ for $\gamma=1$, $\gamma=\alpha+\beta$ and $\alpha=\beta$.

We rewrite the confluent hypergeometric equation in the form:

$$\frac{d}{dz}\left(z^{\gamma}(z-1)^{\alpha+\beta+1-\gamma}\frac{d}{dz}f\right)+\alpha\beta z^{\gamma-1}(z-1)^{\alpha+\beta-\gamma}f=0 \tag{23.16}$$

with the solution:

$$f \equiv \mathrm{F}(\alpha,\beta,\gamma;z) \tag{23.17}$$

Swapping α,β,γ for $\alpha+n,\beta+n,\gamma+n$:

$$f_n \equiv F(\alpha+n,\beta+n,\gamma+n;z) \tag{23.18}$$

and

$$\frac{d}{dz}\left(z^{\gamma+n}(z-1)^{\alpha+\beta+1-\gamma+n}\frac{d}{dz}f_n\right)+(\alpha+n)(\beta+n)z^{\gamma-1+n}(z-1)^{\alpha+\beta-\gamma+n}f_n=0 \tag{23.19}$$

So from Equation 23.3:

$$\frac{d^n}{dz^n}F(\alpha,\beta,\gamma;z)=\frac{\alpha(\alpha+1)\cdots(\alpha+n-1)\beta(\beta+1)\cdots(\beta+\mathrm{n}-1)}{\gamma(\gamma+1)\cdots(\gamma+n-1)}F(\alpha+n,\beta+n,\gamma+n;z) \tag{23.20}$$

If $n=1$, then

$$\frac{d}{dz}F(\alpha,\beta,\gamma;z)=\frac{\alpha\beta}{\gamma}F(\alpha+1,\beta+1,\gamma+1;z) \tag{23.21}$$

We find the confluent hypergeometric function, $F(\alpha,\beta,\gamma;z)$, satisfies the differential equation:

$$\begin{aligned}&\frac{d^k}{dz^k}\left(z^{\gamma+k-1}(z-1)^{\alpha+\beta-\gamma+k}F^{(k)}(\alpha,\beta,\gamma;z)\right)\\&=(-1)^k\alpha(\alpha+1)\cdots(\alpha+k-1)\beta(\beta+1)\cdots(\beta+k-1)z^{\gamma-1}(z-1)^{\alpha+\beta-\gamma}F(\alpha,\beta,\gamma;z),\ k=1,2,\ldots\end{aligned} \tag{23.22}$$

Subsequently, we show that if any of the parameters α and β in the confluent hypergeometric function, $F(\alpha,\beta,\gamma;z)$, is equal to a negative integer, $-n$, then the confluent hypergeometric function becomes a polynomial of order n. Also, if $\alpha=-n_1$ and $\beta=-n_2$, when the integer $n_1>0$ and $n_2>0$, then the confluent hypergeometric function tends to a polynomial of order less than the integers n_1 and n_2.

23.2 Euler Integral Representation: Mellin–Barnes Integral Representation

The series in Equation 23.3 within the disk $|z|<1$ can have an analytic prolongation on the entire z-plane and with a cut on the real axis at the point $z=1$ up to the point $=\infty$. When $\mathrm{Re}\gamma>\mathrm{Re}\beta>0$, then for the entire z-plane with the given cut, we can represent the confluent hypergeometric function, $F(\alpha,\beta,\gamma;z)$ in the Euler integral representation:

$$F(\alpha,\beta,\gamma;z)=\frac{1}{B(\beta,\gamma-\beta)}\int_0^1 t^{\beta-1}(1-t)^{\gamma-\beta-1}(1-tz)^{-\alpha}\,dt \tag{23.23}$$

Here, $(1-tz)^{-\alpha}=1$ for $z=0$ and

$$B(\beta,\gamma-\beta)=\frac{\Gamma(\beta)\Gamma(\gamma-\beta)}{\Gamma(\gamma)} \tag{23.24}$$

If

$$\mathrm{Re}(\gamma-\alpha-\beta)>0, \mathrm{Re}\gamma>\mathrm{Re}\beta>0 \quad (23.25)$$

then from Equation 23.23 follows the asymptotic formula for the confluent hypergeometric function near the branch point $z=1$:

$$\lim_{z\to 1}\mathrm{F}(\alpha,\beta,\gamma;z)=\frac{\Gamma(\gamma)\Gamma(\gamma-\alpha-\beta)}{\Gamma(\gamma-\alpha)\Gamma(\gamma-\beta)} \quad (23.26)$$

23.3 Confluent Hypergeometric Function – Kummer Function

In Equation 23.1, we swap z for $\frac{z}{\beta}$ and we have

$$z\left(\frac{z}{\beta}-1\right)\frac{d^2}{dz^2}f+\left(\left(\frac{\alpha}{\beta}+1+\frac{1}{\beta}\right)z-\gamma\right)\frac{d}{dz}f+\alpha f=0 \quad (23.27)$$

For $\beta\to\infty$, we arrive at the confluent hypergeometric or **Kummer function differential equation**:

$$z\frac{d^2}{dz^2}f+(\gamma-z)\frac{d}{dz}f-\alpha f=0 \quad (23.28)$$

The confluent hypergeometric function has two poles with one regular at $z=0$ and one irregular at $z=\infty$. For $\gamma\neq 0,-1,-2,\ldots$, the first partial solution is the confluent hypergeometric function, the **so-called Kummer function**:

$$f_1\equiv\mathrm{F}(\alpha,\gamma;z)=\lim_{\beta\to\infty}\left(1+\frac{\alpha\beta}{1!\gamma}\frac{z}{\beta}+\frac{\alpha(\alpha+1)\beta(\beta+1)}{2!\gamma(\gamma+1)}\frac{z^2}{\beta^2}+\cdots\right)=\lim_{\beta\to\infty}\mathrm{F}\left(\alpha,\beta,\gamma;\frac{z}{\beta}\right) \quad (23.29)$$

Here, $\mathrm{F}\left(\alpha,\beta,\gamma;\frac{z}{\beta}\right)$ is another confluent hypergeometric function.

For $2-\gamma\neq 0,-1,-2,\ldots$, the second partial solution of Equation 23.28 is linearly independent:

$$f_2\equiv z^{1-\gamma}\mathrm{F}(\alpha+1-\gamma,2-\gamma;z) \quad (23.30)$$

For $\gamma\neq 0,\pm 1,\pm 2,\ldots$, the general solution of the given equation is defined by the formula:

$$f\equiv C_1\mathrm{F}(\alpha,\gamma;z)+C_2z^{1-\gamma}\mathrm{F}(\alpha+1-\gamma,2-\gamma;z) \quad (23.31)$$

Here, C_1 and C_2 are arbitrary constants, and if $\gamma=1$, then $f_1=f_2$.

For

$$\mathrm{Re}\gamma>\mathrm{Re}\alpha>0 \quad (23.32)$$

then from Equation 23.32, the function, $F(\alpha,\gamma;z)$, can also be represented in the form:

$$F(\alpha,\gamma;z)=\frac{1}{B(\alpha,\gamma-\alpha)}\int_0^1 t^{\alpha-1}(1-t)^{\gamma-\alpha-1}\left[\sum_{k=0}^{\infty}\frac{(zt)^k}{k!}\right]dt \tag{23.33}$$

So for $\text{Re}\gamma > \text{Re}\alpha > 0$, we have the integral representation of the confluent hypergeometric function:

$$F(\alpha,\gamma;z)=\frac{1}{B(\alpha,\gamma-\alpha)}\int_0^1 t^{\alpha-1}(1-t)^{\gamma-\alpha-1}\exp\{-zt\}dt \tag{23.34}$$

Letting

$$t=1-s \tag{23.35}$$

then

$$F(\alpha,\gamma;z)=\frac{1}{B(\alpha,\gamma-\alpha)}\exp\{z\}\int_0^1 (1-s)^{\alpha-1}s^{\gamma-\alpha-1}\exp\{-zs\}ds \tag{23.36}$$

Comparing Equation 23.34 with 23.36, we find the following Kummer transformation:

$$F(\alpha,\gamma;z)=\exp\{z\}F(\gamma-\alpha,\gamma;-z) \tag{23.37}$$

In this case, the condition $\text{Re}\gamma > \text{Re}\alpha > 0$ is swapped with $\gamma \neq 0,-1,-2,\ldots$, since both sides are regular for all α and γ except $\gamma = 0,-1,-2,\ldots$.

If we differentiate the function in Equation 23.32 n-times, then

$$\frac{d^n}{dz^n}F(\alpha,\gamma;z)=\frac{\alpha(\alpha+1)\cdots(\alpha+n-1)}{\gamma(\gamma+1)\cdots(\gamma+n-1)}F(\alpha+n,\gamma+n;z) \tag{23.38}$$

If $n=1$, then

$$\frac{d}{dz}F(\alpha,\gamma;z)=\frac{\alpha}{\gamma}F(\alpha+1,\gamma+1;z) \tag{23.39}$$

24

Cylindrical Functions

24.1 Cylindrical Function of the First Kind

We rewrite the confluent hypergeometric differential Equation 23.1 in the form:

$$t(t-1)\frac{d^2}{dt^2}f+\left((\alpha+\beta+1)t-\gamma\right)\frac{d}{dt}f+\alpha\beta f=0 \tag{24.1}$$

Letting $t=-\frac{z^2}{4\alpha\beta}$ and dividing the resultant equation by $\alpha\beta$, we have

$$\left(\frac{z^2}{4\alpha\beta}+1\right)\frac{d^2}{dz^2}f+\left(\left(\frac{1}{\beta}+\frac{1}{\alpha}+\frac{1}{\alpha\beta}\right)\frac{z^2}{4}+\gamma\right)\frac{2}{z}\frac{d}{dz}f+f=0 \tag{24.2}$$

Taking the limits $\alpha\to\infty$ and $\beta\to\infty$, we have

$$z\frac{d^2}{dz^2}f+2\gamma\frac{d}{dz}f+zf=0 \tag{24.3}$$

For $\nu=\text{const}$ and letting $f=z^{-\nu}W$, the differential equation for the Bessel functions:

$$z^2\frac{d^2}{dz^2}W+2(\gamma-\nu)z\frac{d}{dz}W+\left(z^2+\nu(\nu+1)-2\gamma\nu\right)W=0 \tag{24.4}$$

Selecting γ such that $2(\gamma-\nu)=1$:

$$\gamma=\nu+\frac{1}{2},\ \nu(\nu+1)-2\gamma\nu=-\nu^2 \tag{24.5}$$

From here follows the Bessel function differential equation:

$$z^2\frac{d^2}{dz^2}W+z\frac{d}{dz}W+\left(z^2-\nu^2\right)W=0 \tag{24.6}$$

The solution is selected in the form of a power series:

$$W=\sum_{k=0}^{\infty}a_k z^{r+k} \tag{24.7}$$

DOI: 10.1201/9781003273073-27

and substituting into Equation 24.6:

$$z^2\sum_{k=0}^{\infty}a_k(r+k)(r+k-1)z^{r+k-2}+z\sum_{k=0}^{\infty}a_k(r+k)z^{r+k-1}+(z^2-\nu^2)\sum_{k=0}^{\infty}a_kz^{r+k}=0 \tag{24.8}$$

Equating coefficients of z^r, z^{r+1} and so on to zero, two partial solutions of Equation 24.6 follow:

$$J_\nu(z)=\sum_{k=0}^{\infty}\frac{(-1)^k}{\Gamma(k+1)\Gamma(\nu+k+1)}\left(\frac{z}{2}\right)^{\nu+2k},\ J_{-\nu}(z)=\sum_{k=0}^{\infty}\frac{(-1)^k}{\Gamma(k+1)\Gamma(-\nu+k+1)}\left(\frac{z}{2}\right)^{-\nu+2k} \tag{24.9}$$

The functions $J_\nu(z)$ and $J_{-\nu}(z)$ are the so-called cylindrical or the Bessel functions of the first kind and, respectively, of order ν and $-\nu$. We represent these functions via the confluent hypergeometric function through

$$\begin{aligned}J_{\pm\nu}(z)&=\frac{1}{\Gamma(\pm\nu+1)}\left(\frac{z}{2}\right)^{\pm\nu}\sum_{k=0}^{\infty}\frac{(-1)^k\Gamma(\pm\nu+1)}{\Gamma(k+1)\Gamma(\pm\nu+k+1)}\left(\frac{z}{2}\right)^{2k}\\&=\lim_{\alpha,\beta\to\infty}\frac{1}{\Gamma(\pm\nu+1)}\left(\frac{z}{2}\right)^{\pm\nu}\sum_{k=0}^{\infty}\frac{\alpha(\alpha+1)\cdots(\alpha+k-1)\beta(\beta+1)\cdots(\beta+k-1)}{1\cdot2\cdots k(\pm\nu+1)(\pm\nu+2)\cdots(\pm\nu+k)}\left(-\frac{z^2}{4\alpha\beta}\right)^k\\&=\lim_{\alpha,\beta\to\infty}\frac{1}{\Gamma(\pm\nu+1)}\left(\frac{z}{2}\right)^{\pm\nu}F\left(\alpha,\beta,\pm\nu+1;-\frac{z^2}{4\alpha\beta}\right)\end{aligned} \tag{24.10}$$

24.2 Neumann Function

The partial solution of the Bessel differential equation is the Neumann function (cylindrical function of the second kind), $N_\nu(z)$, which is a linear combination of the function $J_\nu(z)$ for all ν:

$$N_\nu(z)=\frac{1}{\sin\nu\pi}\left(J_\nu(z)\cos\nu\pi-J_{-\nu}(z)\right) \tag{24.11}$$

For integer value $\nu=n$, the right-hand side of Equation 24.11 is an indeterminacy of the type, $\frac{0}{0}$. So from l'Hôpital's rule, the Neumann function in Equation 24.11 when $\nu\to n$:

$$\begin{aligned}N_n(z)&=\lim_{\nu\to n}N_\nu(z)=\lim_{\nu\to n}\frac{1}{\sin\nu\pi}\left(J_\nu(z)\cos\nu\pi-J_{-\nu}(z)\right)\\&=\lim_{\nu\to n}\frac{1}{\pi\cos\nu\pi}\left(\cos\nu\pi\frac{\partial}{\partial\nu}J_\nu(z)-\pi J_\nu(z)\sin\nu\pi-\frac{\partial}{\partial\nu}J_{-\nu}(z)\right)\end{aligned} \tag{24.12}$$

or

$$N_n(z)=\frac{1}{\pi}\left(\frac{\partial}{\partial\nu}J_\nu(z)-(-1)^n\frac{\partial}{\partial\nu}J_{-\nu}(z)\right)\Bigg|_{\nu=n} \tag{24.13}$$

So the general solution of the Bessel Equation 24.6 can be represented as follows:

$$W=C_1J_\nu(z)+C_2N_\nu(z) \tag{24.14}$$

24.3 Hankel Functions

We examine the Hankel or the cylindrical function of the third kind defined as follows:

$$H_{\nu}^{(1)}(z)=J_{\nu}(z)+iN_{\nu}(z),\; H_{\nu}^{(2)}(z)=J_{\nu}(z)-iN_{\nu}(z) \tag{24.15}$$

which can be expressed through the cylindrical or Bessel function of the first kind, $J_{\nu}(z)$, of order ν:

$$H_{\nu}^{(1)}(z)=J_{\nu}(z)+\frac{i}{\sin\nu\pi}\left(J_{\nu}(z)\cos\nu\pi-J_{-\nu}(z)\right)=\frac{1}{i\sin\nu\pi}\left(J_{-\nu}(z)-\exp\{-i\nu\pi\}J_{\nu}(z)\right) \tag{24.16}$$

$$H_{\nu}^{(2)}(z)=J_{\nu}(z)-\frac{i}{\sin\nu\pi}\left(J_{\nu}(z)\cos\nu\pi-J_{-\nu}(z)\right)=\frac{1}{i\sin\nu\pi}\left(-J_{-\nu}(z)+\exp\{i\nu\pi\}J_{\nu}(z)\right) \tag{24.17}$$

These equations are true when ν do not take integral values, and for $\nu = n$, we have the indeterminacy, $\frac{0}{0}$. Applying l'Hôpital's rule when $\nu \to n$:

$$H_{n}^{(1)}(z)=J_{n}(z)+\frac{i}{\pi}\left(\frac{\partial}{\partial\nu}J_{\nu}(z)-(-1)^{n}\frac{\partial}{\partial\nu}J_{-\nu}(z)\right)\Bigg|_{\nu=n},$$
$$H_{n}^{(2)}(z)=J_{n}(z)-\frac{i}{\pi}\left(\frac{\partial}{\partial\nu}J_{\nu}(z)-(-1)^{n}\frac{\partial}{\partial\nu}J_{-\nu}(z)\right)\Bigg|_{\nu=n} \tag{24.18}$$

We swap in Equations 24.16 and 24.17 the value ν with $-\nu$:

$$H_{-\nu}^{(1)}(z)=\frac{1}{-i\sin\nu\pi}\left(J_{\nu}(z)-\exp\{i\nu\pi\}J_{-\nu}(z)\right)=-\frac{\exp\{i\nu\pi\}}{i\sin\nu\pi}\left(J_{-\nu}(z)-\exp\{-i\nu\pi\}J_{\nu}(z)\right) \tag{24.19}$$

$$H_{-\nu}^{(2)}(z)=\frac{1}{-i\sin\nu\pi}\left(-J_{\nu}(z)+\exp\{-i\nu\pi\}J_{-\nu}(z)\right)=\frac{\exp\{-i\nu\pi\}}{i\sin\nu\pi}\left(-J_{\nu}(z)+\exp\{i\nu\pi\}J_{\nu}(z)\right) \tag{24.20}$$

So

$$H_{-\nu}^{(1)}(z)=\exp\{i\nu\pi\}H_{\nu}^{(1)}(z),\; H_{-\nu}^{(2)}(z)=\exp\{-i\nu\pi\}H_{\nu}^{(2)}(z) \tag{24.21}$$

From Equation 24.15:

$$J_{\nu}(z)=\frac{1}{2}\left(H_{\nu}^{(1)}(z)+H_{\nu}^{(2)}(z)\right),\; N_{\nu}(z)=\frac{1}{2i}\left(H_{\nu}^{(1)}(z)-H_{\nu}^{(2)}(z)\right) \tag{24.22}$$

24.4 Modified Bessel Function

The modified Bessel function equation of the first kind of order ν can be obtained through the Bessel function equation of the first kind of order ν by setting $z \to iz$ in the differential Equation 24.6:

$$z^{2}\frac{d^{2}}{dz^{2}}W+z\frac{d}{dz}W-\left(z^{2}+\nu^{2}\right)W=0 \tag{24.23}$$

with the solution:

$$W = C_1 J_\nu(iz) + C_2 N_\nu(iz) \tag{24.24}$$

So the solution of Equation 24.23 is also the modified cylindrical function or the modified Bessel function, $I_\nu(z)$, of the first kind of order ν

$$I_\nu(z) = i^{-\nu} J_\nu(iz) = \exp\left\{-i\frac{\nu\pi}{2}\right\} J_\nu(iz) = \sum_{k=0}^{\infty} \frac{1}{\Gamma(k+1)\Gamma(\nu+k+1)} \left(\frac{z}{2}\right)^{\nu+2k}, -\pi < \arg z \le \frac{\pi}{2} \tag{24.25}$$

24.5 Modified Bessel Function with Imaginary Argument

We introduce the modified Bessel function of the second kind of order ν, the so-called McDonald function, which is a particular solution of Equation 24.23:

$$K_\nu(z) = \frac{i\pi}{2} \exp\left\{i\frac{\nu\pi}{2}\right\} H_\nu^{(1)}(iz) \tag{24.26}$$

Here, $H_\nu^{(1)}(iz)$ is the first Hankel function. The function, $K_\nu(z)$, can be expressed through $I_\nu(z)$ and $I_{-\nu}(z)$:

$$K_\nu(z) = \frac{\pi}{2} \frac{1}{\sin\nu\pi} \left(\exp\left\{i\frac{\nu\pi}{2}\right\} J_{-\nu}(iz) - \exp\left\{-i\frac{\nu\pi}{2}\right\} J_\nu(iz) \right) \tag{24.27}$$

And can be re-expressed as follows:

$$K_\nu(z) = \frac{\pi}{2} \frac{1}{\sin\nu\pi} \left(I_{-\nu}(z) - I_\nu(z) \right) \tag{24.28}$$

from where:

$$K_{-\nu}(z) = K_\nu(z) \tag{24.29}$$

If $\nu = n$ is an integer value, then

$$K_n(z) = \lim_{\nu \to n} K_\nu(z) \tag{24.30}$$

and from l'Hôpital's rule applied to Equation 24.28:

$$K_n(z) = \frac{(-1)^n}{2} \left(\frac{\partial}{\partial\nu} I_{-\nu}(z) - \frac{\partial}{\partial\nu} I_\nu(z) \right)\Bigg|_{\nu=n} \tag{24.31}$$

So the general solution of Equation 24.23:

$$W = C_1 I_\nu(z) + C_2 K_\nu(z) \tag{24.32}$$

24.6 Bessel Function of the First Kind Integral Formula

We rewrite the series for the Bessel function of the first kind in Equation 24.9:

$$J_\nu(z)=\sum_{k=0}^{\infty}\frac{(-1)^k}{\Gamma(k+1)\Gamma(\nu+k+1)}\left(\frac{z}{2}\right)^{\nu+2k} \tag{24.33}$$

and also the complex integral representation of the function, $\frac{1}{\Gamma(z)}$ provided by Hankel in 1864 and valid for unrestricted z:

$$\frac{1}{\Gamma(z)}=\frac{1}{2\pi i}\int_{\lambda}\exp\{\tau\}\tau^{-z}d\tau \tag{24.34}$$

From the series, we isolated the reciprocal of the Gamma function, $\frac{1}{\Gamma(\nu+k+1)}$, and then applied the integral representation in Equation 24.34:

$$\frac{1}{\Gamma(\nu+k+1)}=\frac{1}{2\pi i}\int_{\lambda}\exp\{\tau\}\tau^{-(\nu+k+1)}d\tau \tag{24.35}$$

We then substitute this for the corresponding expression in Equation 24.33:

$$J_\nu(z)=\frac{1}{2\pi i}\int_{\lambda}\exp\{\tau\}\tau^{-(\nu+1)}\left(\frac{z}{2}\right)^{\nu}\sum_{k=0}^{\infty}\frac{(-1)^k}{\Gamma(k+1)}\tau^{-k}\left(\frac{z}{2}\right)^{2k}d\tau \tag{24.36}$$

From

$$\sum_{k=0}^{\infty}\frac{(-1)^k}{\Gamma(k+1)}\tau^{-k}\left(\frac{z}{2}\right)^{2k}=\sum_{k=0}^{\infty}\frac{(-1)^k}{\Gamma(k+1)}\left(\frac{z^2}{4\tau}\right)^{k}=\exp\left\{-\frac{z^2}{4\tau}\right\} \tag{24.37}$$

then Equation 24.36 becomes

$$J_\nu(z)=\frac{1}{2\pi i}\int_{\lambda}\left(\frac{z}{2}\right)^{\nu}\tau^{-(\nu+1)}\exp\left\{\tau-\frac{z^2}{4\tau}\right\}d\tau \tag{24.38}$$

Considering $|\arg z|<\frac{\pi}{2}$ and letting $\tau=\frac{zt}{2}, \frac{z^2}{4\tau}=\frac{z}{2t}$ and Equation 24.38 becomes

$$J_\nu(z)=\frac{1}{2\pi i}\int_{\lambda}t^{-(\nu+1)}\exp\left\{\frac{z}{2}\left(t-\frac{1}{t}\right)\right\}dt \tag{24.39}$$

Here, we consider the contour, λ, which takes the lower edge of the cut of the negative part of the real axis from point $t=-\infty$ to point $t=-1$ of the circle, $|t|=1$ and the upper edge of the same cut as seen in Figure 22.1d. Letting $t=\exp\{\psi\}$:

$$\frac{1}{2}\left(t-\frac{1}{t}\right)=\sinh\psi,\ t^{-(\nu+1)}=\exp\{-\nu\psi\}\exp\{-\psi\} \tag{24.40}$$

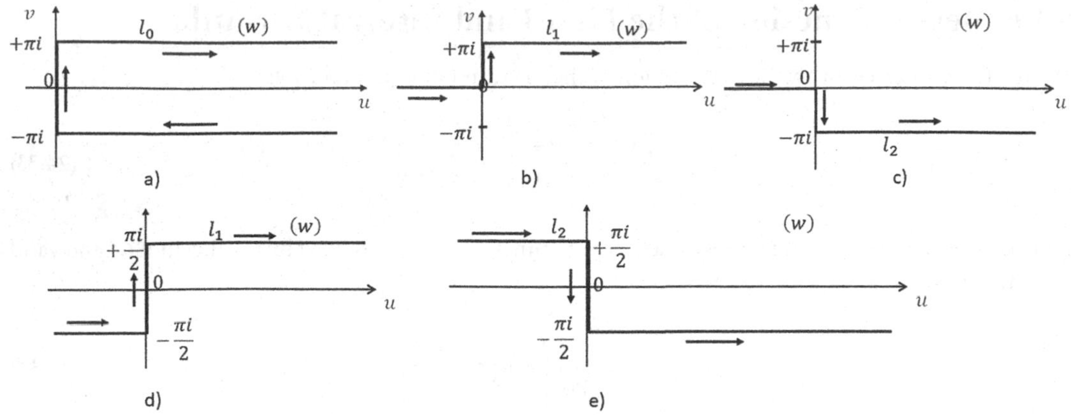

FIGURE 24.1 (24.1a) The contour, l_0 showing the segment of the imaginary axis on the plane (ψ) bounded by the points $-\pi i$ and $+\pi i$, where $\mathrm{Re}\,\psi = 0$ on the circle, $|t| = 1$; (24.1b) the contour, l_1, with the entire negative real and segment of the imaginary axes and semi-line parallel to the real axis (located at a distant π from it at the upper plane); (24.1c) the contour, l_2 with the entire negative real and segment of the imaginary axes and semi-line parallel to the real axis (located at a distant π from it at the lower plane); (24.1d) shows that for $\mathrm{Im}\,z > 0$, the first segment of l_1 constituting the segments from point $-\infty - \frac{\pi i}{2}$ to $-\frac{\pi i}{2}$ and from point $-\frac{\pi i}{2}$ to 0. The second segment of the contour, l_1, is from 0 to $+\frac{\pi i}{2}$ and point $+\frac{\pi i}{2}$ to $+\infty + \frac{\pi i}{2}$; (24.1e) shows that for $\mathrm{Im}\,z < 0$, the first segment of l_2 constituting the segments from point $-\infty + \frac{\pi i}{2}$ to $+\frac{\pi i}{2}$ and from point $+\frac{\pi i}{2}$ to 0. The second segment of the contour, l_2, is from 0 to $-\frac{\pi i}{2}$ and point $-\frac{\pi i}{2}$ to $+\infty - \frac{\pi i}{2}$.

The function, $t = \exp\{\psi\}$, transforms the line, $\mathrm{Im}\,\psi = c$, to the semi-line $\arg t = c$. The line segment of $\mathrm{Re}\,\psi = c'$ has a length 2π at the neighborhood of $|t| = \exp\{c'\}$. So the lines $\mathrm{Im}\,\psi = -\pi i$ and $\mathrm{Re}\,\psi = +\pi i$ are transformed to semi-lines, $\arg t = -\pi i$ and $\arg t = +\pi i$. This implies that at the lower and upper edge of the cut of the negative part of the real axis of the plane, (t). The segment of the imaginary axis on the plane (ψ) is bounded by the points $-\pi i$ and $+\pi i$, where $\mathrm{Re}\,\psi = 0$ on the circle, $|t| = \exp\{0\} = 1$. This permit constituting the contour, l_0, in Figure 24.1a.

So from here and Equation 24.39, we have

$$\mathrm{J}_\nu(z) = \frac{1}{2\pi i}\int_{l_0} \exp\{z\sinh\psi - \nu\psi\}\,d\psi \tag{24.41}$$

The contour, l_0, constitutes the following three-line segments, i.e.,

i. $(+\infty - \pi i, -\pi i)$ where $\psi = \phi - \pi i$ and $\sinh\psi = -\sinh\phi$
ii. $(-\pi i, +\pi i)$ where $\psi = \phi i$ and $\sinh\psi = i\sin\phi$
iii. $(+\pi i, +\infty + \pi i)$ where $\psi = \phi + \pi i$ and $\sinh\psi = -\sinh\phi$

Hence,

$$\mathrm{J}_\nu(z) = \frac{1}{2\pi i}\int_{+\infty-\pi i}^{-\pi i} \exp\{z\sinh\psi - \nu\psi\}\,d\psi + \frac{1}{2\pi i}\int_{-\pi i}^{+\pi i} \exp\{z\sinh\psi - \nu\psi\}\,d\psi$$
$$+\frac{1}{2\pi i}\int_{+\pi i}^{+\infty+\pi i} \exp\{z\sinh\psi - \nu\psi\}\,d\psi \tag{24.42}$$

or

$$J_\nu(z)=\frac{1}{2\pi i}\int_{+\infty}^{0}\exp\{-z\sinh\phi-\nu\phi+\nu\pi i\}d\phi+\frac{1}{2\pi}\int_{-\pi}^{+\pi}\exp\{iz\sin\phi-\nu\phi i\}d\phi$$
$$+\frac{1}{2\pi i}\int_{0}^{+\infty}\exp\{-z\sinh\phi-\nu\phi-\nu\pi i\}d\phi \tag{24.43}$$

With the help of the Euler formula, we have the generalized Bessel function of the order:

$$J_\nu(z)=\frac{1}{\pi}\int_{0}^{\pi}\cos(\nu\phi-z\sin\phi)d\phi-\frac{\sin\nu\pi}{\pi}\int_{0}^{\infty}\exp\{-\nu\phi-z\sinh\phi\}d\phi \tag{24.44}$$

From Equation 24.33 and Equations 22.1 and 22.21, we have

$$\frac{1}{\Gamma(\nu+k+1)}=\frac{1}{\Gamma\left(\nu+\frac{1}{2}+k+\frac{1}{2}\right)}=\frac{B\left(\nu+\frac{1}{2},k+\frac{1}{2}\right)}{\Gamma\left(\nu+\frac{1}{2}\right)\Gamma\left(k+\frac{1}{2}\right)}=\frac{1}{\Gamma\left(\nu+\frac{1}{2}\right)\Gamma\left(k+\frac{1}{2}\right)}\int_{0}^{1}y^{k-\frac{1}{2}}(1-y)^{\nu-\frac{1}{2}}dy \tag{24.45}$$

Setting $y=t^2$:

$$\frac{1}{\Gamma(\nu+k+1)}=\frac{1}{\Gamma\left(\nu+\frac{1}{2}\right)\Gamma\left(k+\frac{1}{2}\right)}\int_{-1}^{+1}t^{2k}\left(1-t^2\right)^{\nu-\frac{1}{2}}dt \tag{24.46}$$

and substituting this into Equation 24.33:

$$J_\nu(z)=\frac{1}{\Gamma\left(\nu+\frac{1}{2}\right)}\left(\frac{z}{2}\right)^{\nu}\int_{-1}^{+1}\left(1-t^2\right)^{\nu-\frac{1}{2}}\sum_{k=0}^{\infty}\frac{(-1)^k(zt)^{2k}}{2^{2k}\Gamma(k+1)\Gamma\left(k+\frac{1}{2}\right)}dt \tag{24.47}$$

From

$$\sqrt{\pi}\sum_{k=0}^{\infty}\frac{(-1)^k(zt)^{2k}}{2^{2k}\Gamma(k+1)\Gamma\left(k+\frac{1}{2}\right)}=\cos zt \tag{24.48}$$

then

$$J_\nu(z)=\frac{1}{\sqrt{\pi}\Gamma\left(\nu+\frac{1}{2}\right)}\left(\frac{z}{2}\right)^{\nu}\int_{-1}^{+1}\left(1-t^2\right)^{\nu-\frac{1}{2}}\cos zt\,dt,\ \mathrm{Re}\,\nu>-\frac{1}{2} \tag{24.49}$$

and setting $t=\cos\theta$:

$$J_\nu(z)=\frac{1}{\sqrt{\pi}\Gamma\left(\nu+\frac{1}{2}\right)}\left(\frac{z}{2}\right)^{\nu}\int_{-1}^{+1}\sin^{2\nu}\theta\cos(z\cos\theta)d\theta,\ \mathrm{Re}\,\nu>-\frac{1}{2} \tag{24.50}$$

Setting $z \to iz$ in Equation 24.49 and from Equation 24.25, we have the modified Bessel function:

$$I_\nu(z) = \frac{1}{\sqrt{\pi}\Gamma\left(\nu + \frac{1}{2}\right)}\left(\frac{z}{2}\right)^\nu \int_{-1}^{+1} \left(1 - t^2\right)^{\nu - \frac{1}{2}} \cosh zt\, dt, \left|\arg z\right| < \pi, \mathrm{Re}\,\nu > -\frac{1}{2} \tag{24.51}$$

24.7 Neumann Function Integral Formula

The Neumann function, $N_\nu(z)$, integral formula can be obtained by substituting the expression for $J_\nu(z)$ from Equation 24.44 into 24.11:

$$N_\nu(z) = \frac{1}{\pi \sin\nu\pi}\left(\cos\nu\pi \int_0^\pi \cos(\nu\phi - z\sin\phi)\,d\phi - \cos\nu\pi \sin\nu\pi \int_0^\infty \exp\{-\nu\phi - z\sin h\phi\}\,d\phi \right.$$
$$\left. - \int_0^\pi \cos(-\nu\phi - z\sin\phi)\,d\phi + \sin\nu\pi \int_0^\infty \exp\{\nu\phi - z\sin h\phi\}\,d\phi \right) \tag{24.52}$$

or

$$N_\nu(z) = \frac{1}{\pi}\left(\cot\nu\pi \int_0^\pi \cos(\nu\phi - z\sin\phi)\,d\phi - \cos\nu\pi \int_0^\infty \exp\{-\nu\phi - z\sin h\phi\}\,d\phi \right.$$
$$\left. - \frac{1}{\sin\nu\pi}\int_0^\pi \cos(-\nu\phi - z\sin\phi)\,d\phi + \int_0^\infty \exp\{\nu\phi - z\sin h\phi\}\,d\phi \right) \tag{24.53}$$

From the integral:

$$\int_0^\pi \cos(-\nu\phi - z\sin\phi)\,d\phi = \int_0^\pi \cos(\nu\phi + z\sin\phi)\,d\phi \tag{24.54}$$

we do the change of variable, $\phi = \pi - \phi'$, so

$$\int_0^\pi \cos(\nu\phi + z\sin\phi)\,d\phi = \cos\nu\pi \int_0^\pi \cos(\nu\phi' - z\sin\phi')\,d\phi' + \sin\nu\pi \int_0^\pi \sin(\nu\phi' - z\sin\phi')\,d\phi' \tag{24.55}$$

Then the Neumann function, $N_\nu(z)$, integral formula:

$$N_\nu(z) = \frac{1}{\pi}\int_0^\pi \sin(z\sin\phi - \nu\phi)\,d\phi - \frac{1}{\pi}\int_0^\infty \left(\exp\{\nu\phi\} + \exp\{-\nu\phi\}\cos\nu\pi\right)\exp\{-z\sin h\phi\}\,d\phi \tag{24.56}$$

24.8 Hankel Function Integral Formula

We find the Hankel functions, $H_\nu^{(1,2)}(z)$, integral formula by considering first $H_\nu^{(1)}(z)$ from Equation 24.15 where $J_\nu(z)$ is taken from Equation 24.44 and $N_\nu(z)$ from Equation 24.56:

$$H_\nu^{(1)}(z)=\frac{1}{\pi i}\int_0^\infty \exp\{\nu\phi - z\sinh\phi\}d\phi+\frac{1}{\pi}\int_0^\pi \exp\{(-\nu\phi+z\sin\phi)i\}d\phi$$
$$+\frac{1}{\pi i}\int_0^\infty \exp\{-\nu\pi i\}\exp\{-\nu\phi - z\sinh\phi\}d\phi \tag{24.57}$$

We apply the following, respectively, to the first, second and third integrals where integration is done on the entire negative real segment of the imaginary axes and then semi-line parallel to the real axis (located at a distant π from it at the upper plane). This constitutes the contour, l_1, of, respectively, the following three-line segments (see Figure 24.1b), i.e.,

i. $(-\infty,0)$ where $\phi=-\psi$ and $\sinh\phi=-\sinh\psi$
ii. $(0,+\pi i)$ where $\phi=i\psi$ and $\sin\phi=-i\sinh\psi$
iii. $(+\pi i,+\infty+\pi i)$ where $\phi=\psi-\pi i$ and $\sinh\phi=-\sinh\psi$

So these changes of variables permit to arrive at the following Hankel function integral formula:

$$H_\nu^{(1)}(z)=\frac{1}{\pi i}\int_{-\infty}^0 \exp\{-\nu\psi+z\sinh\psi\}d\psi+\frac{1}{\pi i}\int_0^{\pi i} \exp\{-\nu\psi+z\sinh\psi\}d\psi$$
$$+\frac{1}{\pi i}\int_{\pi i}^{+\infty+\pi i} \exp\{-\nu\psi+z\sinh\psi\}d\psi \tag{24.58}$$

Considering Figure 24.1b, we have

$$H_\nu^{(1)}(z)=\frac{1}{\pi i}\int_{l_1} \exp\{z\sinh\psi-\nu\psi\}d\psi \tag{24.59}$$

Similarly, considering Figure 24.1c, we have

$$H_\nu^{(2)}(z)=-\frac{1}{\pi i}\int_{l_2} \exp\{z\sinh\psi-\nu\psi\}d\psi \tag{24.60}$$

All the Bessel and Hankel functions integral formulae have been derived when $|\arg z|<\frac{\pi}{2}$. However, these formulae can equally be verified through the analytic continuation method.

We apply the following change of variables, respectively, to Equations 24.59 and 24.60 where

i. $\psi=t+\frac{\pi i}{2}$ and $\sinh\psi=i\cosh t$
ii. $\psi=t-\frac{\pi i}{2}$ and $\sinh\psi=-i\cosh t$ (see Figure 24.1e)

If $\text{Im}z > 0$, then the Cauchy theorem can be applied to the contour, l_1, in Figure 24.1d and for the first segment of l_1 constituting the segments from point $-\infty - \frac{\pi i}{2}$ to $-\frac{\pi i}{2}$ and from point $-\frac{\pi i}{2}$ to 0. For the segment from $-\infty$ to 0, i.e., the entire real axis for $t \to \infty$, the integral tends to zero on the vertical segment bounded by the points $-\frac{\pi i}{2}$ and 0 as a result of the invariance of the segment length and bounded modulus of the integrand. We consider similarly the segment of the contour, l_1, from 0 to $+\frac{\pi i}{2}$ and point $+\frac{\pi i}{2}$ to $+\infty + \frac{\pi i}{2}$. We exchange the entire positive part with the real axis.

So for $\text{Im}z > 0$, we have

$$\text{H}_\nu^{(1)}(z) = \frac{1}{\pi i}\exp\left\{-\frac{\nu\pi i}{2}\right\}\int_{-\infty}^{+\infty}\exp\{iz\cosh t - \nu t\}dt \tag{24.61}$$

Similarly, for $\text{Im}z < 0$, considering the contour, l_2, in Figure 24.1e, we have

$$\text{H}_\nu^{(2)}(z) = \frac{1}{\pi i}\exp\left\{\frac{\nu\pi i}{2}\right\}\int_{-\infty}^{+\infty}\exp\{-iz\cosh t - \nu t\}dt \tag{24.62}$$

We examine now the case when $\text{Re}\nu > -\frac{1}{2}$ and $\arg z = \frac{\pi}{2}$, i.e., z is pure imaginary. We consider again Equation 24.61 and set $\exp\{t\} = y$. So

$$t = \ln y,\ \cosh t = \cosh\ln y = \frac{1}{2}\left(y + \frac{1}{y}\right) \tag{24.63}$$

and

$$\begin{aligned}\text{H}_\nu^{(1)}(z) &= \frac{1}{\pi i}\exp\left\{-\frac{\nu\pi i}{2}\right\}\int_0^\infty \exp\left\{\frac{iz}{2}\left(y+\frac{1}{y}\right) - \nu\ln y\right\}\frac{dy}{y} \\ &= \frac{1}{\pi i}\exp\left\{-\frac{\nu\pi i}{2}\right\}\int_0^\infty \exp\left\{\frac{iz}{2}\left(y+\frac{1}{y}\right)\right\}y^{-\nu-\frac{1}{2}}y^{-\frac{1}{2}}dy\end{aligned} \tag{24.64}$$

From

$$y^{-\nu-\frac{1}{2}} = \frac{1}{\Gamma\left(\nu+\frac{1}{2}\right)}\int_0^\infty \exp\{-yx\}x^{\nu-\frac{1}{2}}dx \tag{24.65}$$

then

$$\text{H}_\nu^{(1)}(z) = \frac{1}{\pi i\Gamma\left(\nu+\frac{1}{2}\right)}\exp\left\{-\frac{\nu\pi i}{2}\right\}\int_0^\infty x^{-\nu-\frac{1}{2}}\left(\int_0^\infty \exp\left\{-y\left(x-\frac{iz}{2}\right)+\frac{iz}{2y}\right\}y^{-\frac{1}{2}}dy\right)dx \tag{24.66}$$

Letting $y = q^2$:

$$dy = 2q\,dq,\ \exp\left\{-y\left(x-\frac{iz}{2}\right)+\frac{iz}{2y}\right\}y^{-\frac{1}{2}} = \exp\left\{-q^2\left(x-\frac{iz}{2}\right)+\frac{iz}{2q^2}\right\}q^{-1} \tag{24.67}$$

and

$$\int_0^\infty \exp\left\{-y\left(x-\frac{iz}{2}\right)+\frac{iz}{2y}\right\}y^{-\frac{1}{2}}dy = 2\int_0^\infty \exp\left\{-q^2\left(x-\frac{iz}{2}\right)+\frac{iz}{2q^2}\right\}dq \tag{24.68}$$

or

$$2\int_0^\infty \exp\left\{-q^2\left(x-\frac{iz}{2}\right)+\frac{iz}{2q^2}\right\}dq = \sqrt{\frac{\pi}{x-\frac{iz}{2}}}\exp\left\{-2\sqrt{-\frac{iz}{2}\left(x-\frac{iz}{2}\right)}\right\} \tag{24.69}$$

then

$$\mathrm{H}_\nu^{(1)}(z) = \frac{1}{i\sqrt{\pi}\,\Gamma\left(\nu+\frac{1}{2}\right)}\exp\left\{-\frac{\nu\pi i}{2}\right\}\int_0^\infty \frac{1}{\sqrt{x-\frac{iz}{2}}}x^{\nu-\frac{1}{2}}\exp\left\{-2\sqrt{-\frac{iz}{2}\left(x-\frac{iz}{2}\right)}\right\}dx \tag{24.70}$$

Setting

$$x = -\frac{iz}{2}\left(t^2-1\right),\ t>0 \tag{24.71}$$

then

$$\sqrt{-\frac{iz}{2}\left(x-\frac{iz}{2}\right)} = -\frac{iz}{2}t,\ dx = -iztdt \tag{24.72}$$

and

$$\mathrm{H}_\nu^{(1)}(z) = \frac{2}{i\sqrt{\pi}\,\Gamma\left(\nu+\frac{1}{2}\right)}\exp\left\{-\frac{\nu\pi i}{2}\right\}\left(-\frac{iz}{2}\right)^\nu\int_1^\infty \left(t^2-1\right)^{\nu-\frac{1}{2}}\exp\{izt\}dt \tag{24.73}$$

Setting $t = 1-\frac{s}{iz}$ and substituting into Equation 24.73 while doing a series of transformations on the resultant function, we have

$$\mathrm{H}_\nu^{(1)}(z) = \sqrt{\frac{2}{\pi z}}\frac{1}{\Gamma\left(\nu+\frac{1}{2}\right)}\exp\left\{i\left(z-\frac{\nu\pi}{2}-\frac{\pi}{4}\right)\right\}\int_0^\infty s^{\nu-\frac{1}{2}}\left(1-\frac{s}{2iz}\right)^{\nu-\frac{1}{2}}\exp\{-s\}ds \tag{24.74}$$

We find in a similar manner the function:

$$H_\nu^{(2)}(z)=\sqrt{\frac{2}{\pi z}}\frac{1}{\Gamma\left(\nu+\frac{1}{2}\right)}\exp\left\{-i\left(z-\frac{\nu\pi}{2}-\frac{\pi}{4}\right)\right\}\int_0^\infty s^{\nu-\frac{1}{2}}\left(1+\frac{s}{2iz}\right)^{\nu-\frac{1}{2}}\exp\{-s\}ds \tag{24.75}$$

Setting $z\to iz$ in Equation 24.61 and considering Equation 24.26:

$$K_\nu(z)=\frac{1}{2}\int_{-\infty}^{+\infty}\exp\{-z\cosh t-\nu t\}dt=\int_0^{+\infty}\exp\{-z\cosh t\}\cosh\nu t\,dt,\ \mathrm{Re}\,z>0 \tag{24.76}$$

24.9 Airy Function

We define the Airy function, $A_i(z)$, with argument, z, by considering the integral:

$$\int_0^\infty dt\cos\left(t^3+zt\right)=\mathrm{Re}\int_0^\infty dt\exp\left\{i\left(t^3+zt\right)\right\},\ z>0 \tag{24.77}$$

From Figure 24.2, we examine the function, $\exp\left\{i\left(t^3+zt\right)\right\}$, that is regular on the entire t-plane. So the Cauchy theorem is valid on the closed contour, $OABO$:

$$\int_{OA}dt\exp\left\{i\left(t^3+zt\right)\right\}+\int_{AB}dt\exp\left\{i\left(t^3+zt\right)\right\}+\int_{BO}dt\exp\left\{i\left(t^3+zt\right)\right\}=0 \tag{24.78}$$

From the arc $\breve{AB}$:

$$t=R\exp\{i\phi\} \tag{24.79}$$

and

$$\left|\int_{AB}dt\exp\left\{i\left(t^3+zt\right)\right\}\right|=\left|Ri\int_0^{\frac{\pi}{6}}d\phi\exp\{i\phi\}\exp\left\{i\left(R^3\exp\{3i\phi\}+zR\exp\{i\phi\}\right)\right\}\right| \tag{24.80}$$

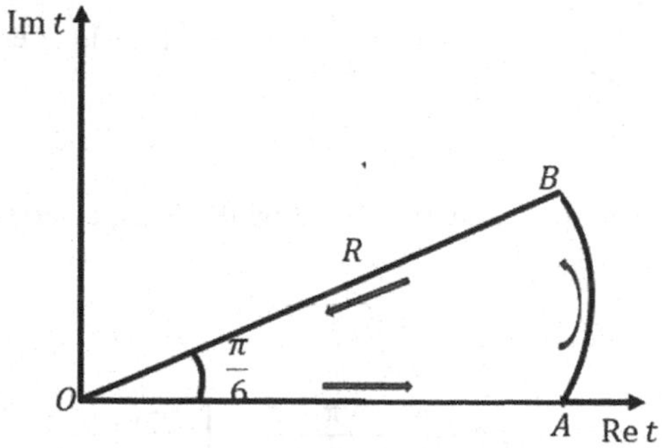

FIGURE 24.2 The closed contour, $OABO$, on the q-plane.

or

$$\left|\int_{AB} dt\exp\left\{i\left(t^3+zt\right)\right\}\right| \le R\int_0^{\frac{\pi}{6}} d\phi\exp\left\{-\left(R^3\sin(3\phi)+zR\sin\phi\right)\right\} = R\int_0^{\frac{\pi}{6}} d\phi\exp\left\{-\left(R^3\frac{6\phi}{\pi}+zR\frac{2\phi}{\pi}\right)\right\} \quad (24.81)$$

or

$$R\int_0^{\frac{\pi}{6}} d\phi\exp\left\{-\left(R^3\frac{6\phi}{\pi}+zR\frac{2\phi}{\pi}\right)\right\} = \frac{\pi}{2\left(3R^2+z\right)}\left(1-\exp\left\{-\frac{R\left(3R^2+z\right)}{3}\right\}\right) \quad (24.82)$$

This implies

$$\lim_{R\to\infty}\int_{AB} dt\exp\left\{i\left(t^3+zt\right)\right\}=0 \quad (24.83)$$

and

$$\lim_{R\to\infty}\int_{OA} dt\exp\left\{i\left(t^3+zt\right)\right\} = \lim_{R\to\infty}\int_{OB} dt\exp\left\{i\left(t^3+zt\right)\right\} \quad (24.84)$$

This implies that integration relative to the real axis can be swapped with the integration relative to the line, $\arg t=\frac{\pi}{6}$:

$$t=r\exp\left\{i\frac{\pi}{6}\right\} \quad (24.85)$$

and

$$\int_0^{\infty} dt\exp\left\{i\left(t^3+zt\right)\right\} = \exp\left\{i\frac{\pi}{6}\right\}\int_0^{\infty} dr\exp\left\{i\left(ir^3+zr\exp\left\{i\frac{\pi}{6}\right\}\right)\right\} \quad (24.86)$$

From

$$\exp\left\{izr\exp\left\{i\frac{\pi}{6}\right\}\right\} = \exp\left\{\exp\left\{i\frac{\pi}{2}\right\}zr\exp\left\{i\frac{\pi}{6}\right\}\right\} = \exp\left\{zr\exp\left\{i\frac{2\pi}{3}\right\}\right\} \quad (24.87)$$

then from Taylor series expansion of the last expression:

$$\exp\left\{zr\exp\left\{i\frac{2\pi}{3}\right\}\right\} = \sum_{m=0}^{n}\exp\left\{i\frac{2m\pi}{3}\right\}\frac{(zr)^m}{m!}+R_n(r) \quad (24.88)$$

where

$$R_n(r)=\theta\frac{(zr)^{n+1}}{(n+1)!}\exp\{zr\}, |\theta|<1 \quad (24.89)$$

So

$$\int_0^\infty dt \exp\{i(t^3+zt)\} = \exp\left\{i\frac{\pi}{6}\right\}\int_0^\infty dr \exp\{-r^3\}\left(\sum_{m=0}^{n}\exp\left\{i\frac{2m\pi}{3}\right\}\frac{(zr)^m}{m!}+R_n(r)\right) \tag{24.90}$$

When $n\to\infty$, then

$$\int_0^\infty d(r^3)\exp\{-r^3\}(r^3)^{\frac{m+1}{3}-1} = \frac{1}{3}\Gamma\left(\frac{m+1}{3}\right) \tag{24.91}$$

and

$$\int_0^\infty dt \exp\{i(t^3+zt)\} = \int_0^\infty dt\left(\cos(t^3+zt)+i\sin(t^3+zt)\right) \tag{24.92}$$

or

$$\int_0^\infty dt \exp\{i(t^3+zt)\} = \frac{1}{3}\sum_{m=0}^{\infty}\left(\cos\frac{\pi}{6}+i\sin\frac{\pi}{6}\right)\left(\cos\frac{2m\pi}{3}+i\sin\frac{2m\pi}{3}\right)\frac{z^m}{m!}\Gamma\left(\frac{m+1}{3}\right) \tag{24.93}$$

or

$$\int_0^\infty dt \cos(t^3+zt) = \frac{1}{3}\sum_{m=0}^{\infty}\sin\frac{2(m+1)\pi}{3}\frac{z^m}{m!}\Gamma\left(\frac{m+1}{3}\right) \tag{24.94}$$

Consider the sum when

$$m=3k,\ m=3k+1,\ m=3k+2,\ k=0,1,2,\ldots \tag{24.95}$$

and

$$\sin\frac{2(3k+1)\pi}{3}=\frac{\sqrt{3}}{2},\ \sin\frac{2(3k+2)\pi}{3}=-\frac{\sqrt{3}}{2},\ \sin\frac{2(3k+3)\pi}{3}=0 \tag{24.96}$$

so

$$\int_0^\infty dt \cos(t^3+zt) = \frac{1}{2\sqrt{3}}\left(\sum_{k=0}^{\infty}\frac{z^{3k}}{(3k)!}\Gamma\left(k+\frac{1}{3}\right)-\sum_{k=0}^{\infty}\frac{z^{3k+1}}{(3k+1)!}\Gamma\left(k+\frac{2}{3}\right)\right) \tag{24.97}$$

From

$$\Gamma(k)\Gamma\left(k+\frac{1}{3}\right)\Gamma\left(k+\frac{2}{3}\right) = 2\pi 3^{\frac{1}{2}-3k}\Gamma(3k) \tag{24.98}$$

then

$$\int_0^\infty dt\cos\left(t^3+zt\right)=\frac{\pi}{3}\left(\sum_{k=0}^{\infty}\frac{1}{\Gamma(k+1)\Gamma\left(k-\frac{1}{3}+1\right)}\left(\frac{z}{3}\right)^{3k}-\sum_{k=0}^{\infty}\frac{1}{\Gamma(k+1)\Gamma\left(k+\frac{1}{3}+1\right)}\left(\frac{z}{3}\right)^{3k+1}\right) \tag{24.99}$$

and

$$\frac{\pi}{3}\sqrt{\frac{z}{3}}\mathrm{I}_{-\frac{1}{3}}\left(\frac{2z\sqrt{z}}{3\sqrt{3}}\right)=\frac{\pi}{3}\sqrt{\frac{z}{3}}\sum_{k=0}^{\infty}\frac{1}{\Gamma(k+1)\Gamma\left(k-\frac{1}{3}+1\right)}\left(\frac{z}{3}\sqrt{\frac{z}{3}}\right)^{-\frac{1}{3}+2k} \tag{24.100}$$

$$\frac{\pi}{3}\sqrt{\frac{z}{3}}\mathrm{I}_{\frac{1}{3}}\left(\frac{2z\sqrt{z}}{3\sqrt{3}}\right)=\frac{\pi}{3}\sqrt{\frac{z}{3}}\sum_{k=0}^{\infty}\frac{1}{\Gamma(k+1)\Gamma\left(k+\frac{1}{3}+1\right)}\left(\frac{z}{3}\sqrt{\frac{z}{3}}\right)^{\frac{1}{3}+2k} \tag{24.101}$$

then

$$\int_0^\infty dt\cos\left(t^3+zt\right)=\frac{\pi}{3}\sqrt{\frac{z}{3}}\left(\mathrm{I}_{-\frac{1}{3}}\left(\frac{2z\sqrt{z}}{3\sqrt{3}}\right)-\mathrm{I}_{\frac{1}{3}}\left(\frac{2z\sqrt{z}}{3\sqrt{3}}\right)\right) \tag{24.102}$$

or

$$\int_0^\infty dt\cos\left(t^3+zt\right)=\frac{\sqrt{z}}{3}\mathrm{K}_{\frac{1}{3}}\left(\frac{2z\sqrt{z}}{3\sqrt{3}}\right) \tag{24.103}$$

Similarly,

$$\int_0^\infty dt\cos\left(t^3-zt\right)=\frac{\pi\sqrt{z}}{3\sqrt{3}}\left(\mathrm{J}_{-\frac{1}{3}}\left(\frac{2z\sqrt{z}}{3\sqrt{3}}\right)+\mathrm{J}_{\frac{1}{3}}\left(\frac{2z\sqrt{z}}{3\sqrt{3}}\right)\right) \tag{24.104}$$

From here and Equation 24.9:

$$\psi(z)=\frac{\pi\sqrt{z}}{3\sqrt{3}}\mathrm{J}_{-\frac{1}{3}}\left(\frac{2z\sqrt{z}}{3\sqrt{3}}\right)=\frac{\pi}{3}\frac{z}{3}\sum_{k=0}^{\infty}\frac{1}{\Gamma(k+1)\Gamma\left(k+\frac{4}{3}\right)}\left(-\frac{z}{3}\right)^{3k} \tag{24.105}$$

Letting

$$\frac{2z\sqrt{z}}{3\sqrt{3}}\equiv x \tag{24.106}$$

and

$$\varphi(x)=\frac{\pi}{3}\left(\frac{x}{2}\right)^{\frac{1}{3}}\mathrm{J}_{-\frac{1}{3}}(x),\ \psi(x)=\frac{\pi}{3}\left(\frac{x}{2}\right)^{\frac{1}{3}}\mathrm{J}_{\frac{1}{3}}(x) \tag{24.107}$$

then from the Bessell differential equation in Equation 24.6, we swap z, W, ν for $x, X, \frac{1}{3}$:

$$x^2 \frac{d^2}{dx^2}X + x\frac{d}{dx}X + \left(x^2 - \frac{1}{9}\right)X = 0 \tag{24.108}$$

so this yields the partial solutions:

$$X_1 = \frac{3}{\pi}\left(\frac{x}{2}\right)^{-\frac{1}{3}}\varphi(x),\ X_2 = \frac{3}{\pi}\left(\frac{x}{2}\right)^{-\frac{1}{3}}\psi(x) \tag{24.109}$$

with the functions $\varphi(x)$ and $\psi(x)$ being the solutions of the differential equation:

$$3x\frac{d^2}{dx^2}\Phi + \frac{d}{dx}\Phi + 3x\Phi = 0 \tag{24.110}$$

This relates Equation 24.108 by considering

$$X = \frac{3}{\pi}\left(\frac{x}{2}\right)^{-\frac{1}{3}}\Phi \tag{24.111}$$

where

$$\frac{d}{dx}X = -\frac{1}{2\pi}\left(\frac{x}{2}\right)^{-\frac{4}{3}}\Phi + \frac{3}{\pi}\left(\frac{x}{2}\right)^{-\frac{1}{3}}\Phi' \tag{24.112}$$

$$\frac{d^2}{dx^2}X = \frac{1}{3\pi}\left(\frac{x}{2}\right)^{-\frac{7}{3}}\Phi - \frac{1}{\pi}\left(\frac{x}{2}\right)^{-\frac{4}{3}}\Phi' + \frac{3}{\pi}\left(\frac{x}{2}\right)^{-\frac{1}{3}}\Phi'' \tag{24.113}$$

and

$$\frac{d}{dx}\Phi = \left(\frac{z}{3}\right)^{-\frac{1}{2}}\frac{d}{dz}\Phi,\ \frac{d^2}{dx^2}\Phi = -\frac{1}{6}\left(\frac{z}{3}\right)^{-2}\frac{d}{dz}\Phi + \left(\frac{z}{3}\right)^{-1}\frac{d^2}{dz^2}\Phi \tag{24.114}$$

then

$$\frac{d^2}{dz^2}\Phi + \frac{z}{3}\Phi = 0 \tag{24.115}$$

The partial solutions are the functions $\varphi(z)$ and $\psi(z)$ or their linear combinations leading to the so-called Airy function:

$$A_i(z) = \varphi(z) + \psi(z) \tag{24.116}$$

From

$$0<\varphi(-z)=\frac{\pi}{3}\left(\frac{|z|}{3}\right)^{\frac{1}{2}}\exp\left\{-i\frac{\pi}{6}\right\}J_{-\frac{1}{3}}\left(2\left(\frac{|z|}{3}\right)^{\frac{3}{2}}\exp\left\{-i\frac{\pi}{2}\right\}\right) \tag{24.117}$$

$$0>\psi(-z)=-\frac{\pi}{3}\left(\frac{|z|}{3}\right)^{\frac{1}{2}}\exp\left\{i\frac{\pi}{6}\right\}J_{\frac{1}{3}}(q) \tag{24.118}$$

then,

$$A_i(-z)=\varphi(-z)+\psi(-z) \tag{24.119}$$

or

$$A_i(-z)=\frac{\pi}{3}\left(\frac{|z|}{3}\right)^{\frac{1}{2}}\left(\exp\left\{-i\frac{\pi}{6}\right\}J_{-\frac{1}{3}}(t)-\exp\left\{i\frac{\pi}{6}\right\}J_{\frac{1}{3}}(t)\right) \tag{24.120}$$

where

$$t=2\left(\frac{|z|}{3}\right)^{\frac{3}{2}}\exp\left\{-i\frac{\pi}{2}\right\} \tag{24.121}$$

For $|z|\to\infty$, the Airy function decreases

$$A_i(-z)\cong\frac{1}{2}\left(\frac{\pi^2}{3|z|}\right)^{\frac{1}{4}}\exp\left\{-2\left(\frac{|z|}{3}\right)^{\frac{3}{2}}\right\} \tag{24.122}$$

This implies that $A_i(z)$ is bounded for $z\to-\infty$.

25

Orthogonal Polynomials

25.1 Orthogonal Polynomials General Properties

One of the central notions in mathematical physics is the complete orthogonal set of functions, $\psi_0(x), \psi_1(x), \psi_2(x),\ldots$, weighted by $\rho(x) \neq 0$ and orthogonal to each other within the interval, (a,b):

$$\int_a^b \rho(x)\psi_m(x)\psi_n(x)dx = \begin{cases} 0, & m<n \\ N_n \neq 0, & m=n \end{cases} \tag{25.1}$$

When $\rho(x) \neq 0$, the norm of the functions, $\psi_n(x)$ can be calculated as follows:

$$N_n = \int_a^b \rho(x)\left[\psi_n(x)\right]^2 dx, n=0,1,2,\ldots, \tag{25.2}$$

From the orthogonal functions, $\psi_k(x)$, we construct normalized functions:

$$\tilde{\psi}_k(x) = \frac{\psi_k(x)}{\sqrt{N_n}}, k=0,1,2,\ldots, \tag{25.3}$$

and so considering Equation 25.1:

$$\int_a^b \rho(x)\tilde{\psi}_m(x)\tilde{\psi}_n(x)dx = \begin{cases} 0, & m<n \\ 1, & m=n \end{cases} \tag{25.4}$$

Suppose $\psi_k(x)$ is a polynomial of order, k, weighted by $\rho(x)$ and orthogonal within the interval, (a,b), and if $f(x)$ is any continuous function defined on (a,b), then follows the series expansion:

$$f(x) = \sum_{k=0}^{\infty} c_k\psi_k(x) \tag{25.5}$$

Multiplying both sides of Equation 25.5 by $\rho(x)\psi_n(x)$ and integrating the resultant within the interval, (a,b):

$$\int_a^b \rho(x)f(x)\psi_n(x)dx = \sum_{k=0}^{\infty} c_k \int_a^b \rho(x)\psi_k(x)\psi_n(x)dx \tag{25.6}$$

DOI: 10.1201/9781003273073-28

From Equation 25.1:

$$\int_a^b \rho(x)f(x)\psi_n(x)dx = \sum_{k=0}^{\infty} c_k \int_a^b \rho(x)\psi_k(x)\psi_n(x)dx = \sum_{k=0}^{\infty} c_k\delta_{kn}N_k \tag{25.7}$$

or

$$\int_a^b \rho(x)f(x)\psi_n(x)dx = \sum_{k=0}^{\infty} c_k\delta_{kn}N_k \tag{25.8}$$

and

$$c_n = \frac{1}{N_n}\int_a^b \rho(x)f(x)\psi_n(x)dx,\ n = 0,1,2,\ldots, \tag{25.9}$$

If $f(x)$ is a polynomial of order, n, then

$$f(x) = \sum_{k=0}^{n} c_k\psi_k(x) \tag{25.10}$$

25.2 Transforming Confluent Hypergeometric Function into a Polynomial

We study orthogonal polynomials related to confluent hypergeometric functions when independent variables are real. The confluent hypergeometric series in Equation 23.3 is transformed into the polynomial, $\mathrm{R}_n^{(\lambda,\mu)}(t)$:

$$\begin{aligned}\mathrm{R}_n^{(\lambda,\mu)}(t) &= C_n\mathrm{F}(\lambda+\mu+n+1,-n,\lambda+1;t)\\ &= C_n\left(1-\frac{(\lambda+\mu+n+1)n}{1!(\lambda+1)}t+\frac{(\lambda+\mu+n+1)(\lambda+\mu+n+2)n(n-1)}{2!(\lambda+1)(\lambda+2)}t^2\right.\\ &\quad\left.-\cdots+(-1)^n\frac{(\lambda+\mu+n+1)\cdots(\lambda+\mu+2n)n!}{n!(\lambda+1)\cdots(\lambda+n)}t^n\right)\end{aligned} \tag{25.11}$$

where C_n is an arbitrary constant. From the differential equation in Equation 23.22:

$$\begin{aligned}&\frac{d^n}{dt^n}\left(\mathrm{f}_n^{(\lambda,\mu)}(t)C_n(-1)^n\frac{(\lambda+\mu+n+1)\cdots(\lambda+\mu+2n)n!}{(\lambda+1)\cdots(\lambda+n)}t^n\right)\\ &= (-1)^n(\lambda+\mu+n+1)\cdots(\lambda+\mu+2n)n!\mathrm{f}_0^{(\lambda,\mu)}(t)\mathrm{R}_n^{(\lambda,\mu)}(t)\end{aligned} \tag{25.12}$$

Here

$$\mathrm{f}_n^{(\lambda,\mu)}(t) \equiv t^{\lambda+n}(t-1)^{\mu+n} \tag{25.13}$$

and we find that

$$C_n = \frac{(\lambda+1)\cdots(\lambda+n)}{n!} \tag{25.14}$$

So

$$\mathrm{R}_n^{(\lambda,\mu)}(t) = \frac{(-1)^n}{n!} t^{-\lambda}(t-1)^{-\mu} \frac{d^n}{dt^n} \mathrm{f}_n^{(\lambda,\mu)}(t) \tag{25.15}$$

with

$$\mathrm{R}_0^{(\lambda,\mu)}(t) = 1 \tag{25.16}$$

$$\mathrm{R}_1^{(\lambda,\mu)}(t) = -(\lambda+\mu+2)t + (\lambda+1) \tag{25.17}$$

$$\mathrm{R}_2^{(\lambda,\mu)}(t) = \frac{1}{2}\left((\lambda+\mu+4)(\lambda+\mu+3)t^2 - 2(\lambda+2)(\lambda+\mu+3)t + (\lambda+2)(\lambda+1)\right) \tag{25.18}$$

and so on.

We also find within the interval, $(0,1)$, the polynomial $\mathrm{R}_n^{(\lambda,\mu)}(t)$ is the solution of the differential equation:

$$\frac{d}{dt}\left(\mathrm{f}_1^{(\lambda,\mu)}(t)\frac{dy}{dt}\right) - n(\lambda+\mu+n+1)\mathrm{f}_0^{(\lambda,\mu)}(t)y = 0 \tag{25.19}$$

For $m \le n$:

$$\int_0^1 \mathrm{f}_0^{(\lambda,\mu)}(t)\mathrm{R}_m^{(\lambda,\mu)}(t)\mathrm{R}_n^{(\lambda,\mu)}(t)dt = \frac{(-1)^{\lambda+\mu}}{n!}\int_0^1 \mathrm{f}_0^{(\lambda,\mu)}(t)\mathrm{R}_m^{(\lambda,\mu)}(t)\mathrm{R}_n^{(\lambda,\mu)}(t)\frac{d^n}{dt^n}\mathrm{f}_n^{(\lambda,\mu)}(t)dt \tag{25.20}$$

Integrating by parts n-times:

$$\int_0^1 \mathrm{f}_0^{(\lambda,\mu)}(t)\mathrm{R}_m^{(\lambda,\mu)}(t)\mathrm{R}_n^{(\lambda,\mu)}(t)dt = \frac{(-1)^{\mu}}{n!}\int_0^1 \mathrm{f}_n^{(\lambda,\mu)}(t)\frac{d^n}{dt^n}\mathrm{R}_m^{(\lambda,\mu)}dt \tag{25.21}$$

If $m < n$, then

$$\frac{d^n}{dz^n}\mathrm{R}_m^{(\lambda,\mu)} \equiv 0 \tag{25.22}$$

This implies that for $m \ne n$, the polynomials $\mathrm{R}_m^{(\lambda,\mu)}$ and $\mathrm{R}_n^{(\lambda,\mu)}$ weighted by $\mathrm{f}_0^{(\lambda,\mu)}(t)$ are orthogonal within the interval, $(0,1)$:

$$\int_0^1 \mathrm{f}_0^{(\lambda,\mu)}(t)\mathrm{R}_m^{(\lambda,\mu)}(t)\mathrm{R}_n^{(\lambda,\mu)}(t)dt = 0 \tag{25.23}$$

If $m=n$, then

$$\frac{d^n}{dt^n}\mathrm{R}_n^{(\lambda,\mu)}=(-1)^n\frac{\Gamma(\lambda+\mu+2n+1)}{\Gamma(\lambda+\mu+n+1)} \tag{25.24}$$

But

$$\int_0^1 \mathrm{f}_n^{(\lambda,\mu)}(t)dt=(-1)^{\mu+n}\frac{\Gamma(\lambda+n+1)\Gamma(\mu+n+1)}{\Gamma(\lambda+\mu+2n+2)} \tag{25.25}$$

So

$$\int_0^1 \mathrm{f}_0^{(\lambda,\mu)}(t)\left[\mathrm{R}_n^{(\lambda,\mu)}(t)\right]^2 dt=\frac{\Gamma(\lambda+n+1)\Gamma(\mu+n+1)}{n!\Gamma(\lambda+\mu+2n+1)\Gamma(\lambda+\mu+n+1)} \tag{25.26}$$

25.3 Jacobi Polynomials

From Equation 25.19 and doing the change of variables:

$$t=\frac{1-x}{2},\, t-1=-\frac{1+x}{2},\, \frac{dy}{dt}=-2\frac{dy}{dx},\, \frac{d^2y}{dt^2}=4\frac{d^2y}{dx^2} \tag{25.27}$$

then

$$\frac{d}{dx}\left(\mathcal{F}_1^{(\lambda,\mu)}(x)\frac{dy}{dx}\right)+n(\lambda+\mu+n+1)\mathcal{F}_0^{(\lambda,\mu)}(x)y=0 \tag{25.28}$$

So within the interval, $(-1,1)$, the polynomial $\mathrm{R}_n^{(\lambda,\mu)}\left(\frac{1-x}{2}\right)$ satisfies the equation:

$$\mathrm{R}_n^{(\lambda,\mu)}\left(\frac{1-x}{2}\right)\equiv \mathrm{P}_n^{(\lambda,\mu)}(x)=\frac{1}{2^n n!}\mathcal{F}_0^{(-\lambda,-\mu)}(x)\frac{d^n}{dx^n}\mathcal{F}_n^{(\lambda,\mu)}(x) \tag{25.29}$$

where

$$\mathcal{F}_n^{(\lambda,\mu)}(z)\equiv(z-1)^{\lambda+n}(z+1)^{\mu+n} \tag{25.30}$$

and $\mathrm{P}_n^{(\lambda,\mu)}(x)$ is the **so-called Jacobi polynomial** of order n with real roots different from each other within the interval, $(-1,1)$. Letting

$$\mathrm{F}_n^{(\lambda,\mu)}(z)\equiv\frac{1}{(z-x)^{n+1}}\mathcal{F}_n^{(\lambda,\mu)}(z) \tag{25.31}$$

then from the Cauchy and residue theorems, we have

$$\oint_l dz\mathrm{F}_n^{(\lambda,\mu)}(z)=2\pi i\operatorname*{Res}_{z=x}\mathrm{F}_n^{(\lambda,\mu)}(z)=\frac{2\pi i}{n!}\lim_{z\to x}\frac{d^n}{dx^n}\mathcal{F}_n^{(\lambda,\mu)}(z)=\frac{2\pi i}{n!}\frac{d^n}{dx^n}\mathcal{F}_n^{(\lambda,\mu)}(x) \tag{25.32}$$

Here, l is the contour encircling the point, $z=x$, while $z=+1$ and $z=-1$ are outside the given contour. So we represent the Jacobi polynomial through the contour integral:

$$P_n^{(\lambda,\mu)}(x)=\frac{1}{2^{n+1}\pi i}\mathcal{F}_0^{(-\lambda,-\mu)}(x)\oint_l dz F_n^{(\lambda,\mu)}(z) \tag{25.33}$$

and has the properties:

$$P_0^{(\lambda,\mu)}(x)=1,\ P_1^{(\lambda,\mu)}(x)=\frac{1}{2}(\lambda+\mu+2)x+(\lambda-\mu) \tag{25.34}$$

$$P_2^{(\lambda,\mu)}(x)=\frac{1}{8}\left((\lambda+\mu+4)(\lambda+\mu+3)x^2+2(\lambda-\mu)(\lambda+\mu+3)x+\left((\lambda-\mu)^2-(\lambda+\mu)-4\right)\right) \tag{25.35}$$

and so on.

We apply Leibnitz formula to Equation 25.29:

$$P_n^{(\lambda,\mu)}(x)=\frac{1}{2^n n!}\sum_{k=0}^{n}C_n^k\frac{\Gamma(\lambda+n+1)\Gamma(\mu+n+1)}{\Gamma(\lambda+k+1)\Gamma(\mu+n-k+1)}(x-1)^k(x+1)^{n-k} \tag{25.36}$$

So the Jacobi function expressed through the confluent hypergeometric function:

$$P_n^{(\lambda,\mu)}(x)=\frac{\Gamma(\lambda+n+1)}{n!\Gamma(\lambda+1)}F\left(\lambda+\mu+n+1,-n,\lambda+1;\frac{1-x}{2}\right) \tag{25.37}$$

Do the change of variables $t=\frac{1-x}{2}$ in Equations 25.23 and 25.26, then from

$$\rho(x)=(1-x)^{\lambda}(1+x)^{\mu}=(-1)^{\lambda}(x-1)^{\lambda}(1+x)^{\mu}\equiv(-1)^{\lambda}\mathcal{F}_0^{(\lambda,\mu)}(x) \tag{25.38}$$

and when $m\neq n$, then

$$\int_{-1}^{+1}\rho(x)P_m^{(\lambda,\mu)}(x)P_n^{(\lambda,\mu)}(x)dx=0 \tag{25.39}$$

while for $m=n$:

$$\int_{-1}^{+1}\rho(x)\left[P_n^{(\lambda,\mu)}(x)\right]^2dx=2^{\lambda+\mu+1}\frac{\Gamma(\lambda+n+1)\Gamma(\mu+n+1)}{n!\Gamma(\lambda+\mu+2n+1)\Gamma(\lambda+\mu+n+1)} \tag{25.40}$$

We find the Jacobi polynomial, $P_n^{(\lambda,\mu)}(x)$, weighted by $\rho(x)$ is orthogonal within the interval, $(-1,1)$. Now, examine the normalized Jacobi polynomial, $\tilde{P}_n^{(\lambda,\mu)}(x)$:

$$\tilde{P}_n^{(\lambda,\mu)}(x)=\sqrt{2^{-\lambda-\mu-1}\frac{n!\Gamma(\lambda+\mu+2n+1)\Gamma(\lambda+\mu+n+1)}{\Gamma(\lambda+n+1)\Gamma(\mu+n+1)}}P_n^{(\lambda,\mu)}(x) \tag{25.41}$$

and so

$$\int_{-1}^{+1}\rho(x)\tilde{P}_m^{(\lambda,\mu)}(x)\tilde{P}_n^{(\lambda,\mu)}(x)dx=\begin{cases}0, & m\neq n\\ 1, & m=n\end{cases} \tag{25.42}$$

25.4 Jacobi Polynomial Generating Function

We use Equation 25.33 to find the generating function, $\Phi(t,x)$, for the Jacobi polynomial:

$$\Phi(t,x)=\frac{1}{2\pi i}\mathcal{F}_0^{(-\lambda,-\mu)}(x)\oint_l F_0^{(\lambda,\mu)}(z)\sum_{n=0}^{\infty}\left(\frac{1}{2}\frac{z^2-1}{z-x}t\right)^n dz \tag{25.43}$$

From the points of the contour, l, which satisfy the inequality:

$$\frac{1}{2}\left|\frac{z^2-1}{z-x}t\right|<1 \tag{25.44}$$

then

$$\begin{aligned}\Phi(t,x)&=\frac{1}{2\pi i}\oint_l\frac{1}{(z-x)\left(1-\frac{t}{2}\frac{z^2-1}{z-x}\right)}\left(\frac{z-1}{x-1}\right)^{\lambda}\left(\frac{z+1}{x+1}\right)^{\mu}dz\\&=-\frac{1}{2\pi i}\oint_l\frac{1}{\frac{t}{2}(z-z_1)(z-z_2)}\left(\frac{z-1}{x-1}\right)^{\lambda}\left(\frac{z+1}{x+1}\right)^{\mu}dz\end{aligned} \tag{25.45}$$

where for small t, the points z_1 and z_2 are, respectively, at the exterior and interior of the contour, l:

$$\lim_{t\to0}z_1=\frac{1+\sqrt{1-2tx+t^2}}{t}=\infty,\ \lim_{t\to0}z_2=\frac{1-\sqrt{1-2tx+t^2}}{t}=x \tag{25.46}$$

So

$$\Phi(t,x)=-\operatorname*{Res}_{z=z_2}\frac{1}{\frac{t}{2}(z-z_1)(z-z_2)}\left(\frac{z-1}{x-1}\right)^{\lambda}\left(\frac{z+1}{x+1}\right)^{\mu}=\frac{1}{\frac{t}{2}(z_1-z_2)}\left(\frac{z_2-1}{x-1}\right)^{\lambda}\left(\frac{z_2+1}{x+1}\right)^{\mu} \tag{25.47}$$

or

$$\begin{aligned}\Phi(t,x)&=\frac{\left(1-\sqrt{1-2tx+t^2}-t\right)^{\lambda}\left(1-\sqrt{1-2tx+t^2}+t\right)^{\mu}}{t^{\lambda+\mu}(x-1)^{\lambda}(x+1)^{\mu}\sqrt{1-2tx+t^2}}\\&=\frac{2^{\lambda+\mu}}{\sqrt{1-2tx+t^2}\left(1-t+\sqrt{1-2tx+t^2}\right)^{\lambda}\left(1+t+\sqrt{1-2tx+t^2}\right)^{\mu}}\end{aligned} \tag{25.48}$$

So

$$P_n^{(\lambda,\mu)}(x)=\frac{1}{n!}\left[\frac{\partial^n}{\partial t^n}\Phi(t,x)\right]_{t=0} \tag{25.49}$$

and

$$P_n^{(\lambda,\mu)}(1)=\frac{\Gamma(\lambda+n+1)}{n!\Gamma(\lambda+1)},\ P_n^{(\lambda,\mu)}(-1)=(-1)^n\frac{\Gamma(\mu+n+1)}{n!\Gamma(\mu+1)} \tag{25.50}$$

with Equation 23.21 becoming

$$\frac{d}{dz}F(\alpha,\beta,\gamma;z)=\frac{\alpha\beta}{\gamma}F(\alpha+1,\beta+1,\gamma+1;z) \tag{25.51}$$

while z is real-valued and we set

$$\alpha=\lambda+\mu+n+1,\ \beta=-n,\ \gamma=\lambda+1,\ z=\frac{1-x}{2} \tag{25.52}$$

So from here and Equation 25.37:

$$\frac{d}{dx}P_n^{(\lambda,\mu)}(x)=\frac{\lambda+\mu+n+1}{2}P_{n-1}^{(\lambda+1,\mu+1)}(x),\ \frac{d^k}{dx^k}P_n^{(\lambda,\mu)}(x)=\frac{\Gamma(\lambda+\mu+n+k+1)}{2^k\Gamma(\lambda+\mu+n+1)}P_{n-k}^{(\lambda+k,\mu+k)}(x) \tag{25.53}$$

Setting $k=n$:

$$\frac{d^n}{dx^n}P_n^{(\lambda,\mu)}(x)=\frac{\Gamma(\lambda+\mu+2n+1)}{2^n\Gamma(\lambda+\mu+n+1)} \tag{25.54}$$

From Equation 25.29:

$$P_n^{(\lambda,\mu)}(-x)=(-1)^n P_n^{(\lambda,\mu)}(x) \tag{25.55}$$

We series expand the function $f(x)$ convergent in the interval, $(-1,+1)$, through the Jacobi polynomials:

$$f(x)=\sum_{n=0}^{\infty}c_nP_n^{(\lambda,\mu)}(x) \tag{25.56}$$

Multiply this expression by $\rho(x)P_m^{(\lambda,\mu)}(x)$ and then integrate the resultant within the interval, $(-1,+1)$:

$$\int_{-1}^{1}\rho(x)f(x)P_m^{(\lambda,\mu)}(x)dx=\sum_{n=0}^{\infty}c_n\int_{-1}^{1}\rho(x)P_m^{(\lambda,\mu)}(x)P_n^{(\lambda,\mu)}(x)dx \tag{25.57}$$

For $m=n$ and Equation 25.40:

$$\int_{-1}^{1}\rho(x)f(x)P_n^{(\lambda,\mu)}(x)dx=c_n 2^{\lambda+\mu+1}\frac{\Gamma(\lambda+n+1)\Gamma(\mu+n+1)}{n!\Gamma(\lambda+\mu+2n+1)\Gamma(\lambda+\mu+n+1)} \tag{25.58}$$

From here:

$$c_n=\frac{n!\Gamma(\lambda+\mu+2n+1)\Gamma(\lambda+\mu+n+1)}{2^{\lambda+\mu+1}\Gamma(\lambda+n+1)\Gamma(\mu+n+1)}\int_{-1}^{1}\rho(x)P_n^{(\lambda,\mu)}(x)f(x)dx \tag{25.59}$$

25.5 Gegenbauer Polynomials

From

$$\rho(x)=\left(1-x^2\right)^{\sigma-\frac{1}{2}}\equiv(-1)^{\sigma-\frac{1}{2}}\mathcal{F}_0^{\left(\sigma-\frac{1}{2},\sigma-\frac{1}{2}\right)}(x) \tag{25.60}$$

and Equation 25.29, we set

$$\lambda=\mu=\sigma-\frac{1}{2}>-1 \tag{25.61}$$

then follows the Gegenbauer polynomials, $P_n^{(\sigma)}(x)$:

$$P_n^{(\sigma)}(x)=\Lambda_n(\sigma)P_n^{\left(\sigma-\frac{1}{2},\sigma-\frac{1}{2}\right)}(x),\ \sigma>-\frac{1}{2} \tag{25.62}$$

where

$$P_n^{\left(\sigma-\frac{1}{2},\sigma-\frac{1}{2}\right)}(x)=\frac{1}{2^n n!}\left(x^2-1\right)^{-\sigma+\frac{1}{2}}\frac{d^n}{dx^n}\left(x^2-1\right)^{\sigma-\frac{1}{2}+n},\ \Lambda_n(\sigma)\equiv\frac{\Gamma\left(\sigma+\frac{1}{2}\right)\Gamma(2\sigma+n)}{\Gamma(2\sigma)\Gamma\left(\sigma+n+\frac{1}{2}\right)} \tag{25.63}$$

and

$$P_0^{(\sigma)}(x)=1,\ P_1^{(\sigma)}(x)=2\sigma x,\ P_2^{(\sigma)}(x)=\sigma\left(2(\sigma+1)x^2-1\right) \tag{25.64}$$

$$P_3^{(\sigma)}(x)=\frac{2\sigma(\sigma+1)}{3}\left(2(\sigma+1)x^3-3x\right) \tag{25.65}$$

and so on.

Apply Equation 25.61 to 25.60, and find the polynomial, $P_n^{(\sigma)}(x)$, weighted by $\rho(x)$ orthogonal within the interval, $(-1,+1)$:

$$\int_{-1}^{+1} dx\rho(x)P_m^{(\sigma)}(x)P_n^{(\sigma)}(x)=0,\ m\neq n \tag{25.66}$$

Also,

$$P_n^{(\sigma)}(-x)=(-1)^n P_n^{(\sigma)}(x),\ P_n^{(\sigma)}(\pm 1)=(\pm 1)^n \frac{2\sigma\cdots(2\sigma+n-1)}{n!} \tag{25.67}$$

and

$$P_n^{(0)}(x)\equiv 0,\ n\geq 1 \tag{25.68}$$

We observe from the first equation in Equation 25.67 that the parity of the Gegenbauer polynomial, $P_n^{(\sigma)}(x)$, is determined from the argument, x. From the orthogonality relation in Equation 25.66, we infer that all the roots of $P_n^{(\sigma)}(x)$ are real-valued, different and within the interval, $(-1,+1)$.

From Equation 25.61:

$$\frac{d}{dx}\left(\rho(x)\frac{dy}{dx}\right)+n(2\sigma+1)\rho(x)y=0 \tag{25.69}$$

We express the Gegenbauer polynomials through the confluent hypergeometric function considering Equation 25.37 for all σ:

$$P_n^{(\sigma)}(x)=\frac{2\sigma\cdots(2\sigma+n-1)}{n!}F\left(2\sigma+n,-n,\sigma+\frac{1}{2};\frac{1-x}{2}\right) \tag{25.70}$$

For $\sigma=-m$ and $n>2m$, $m=01,2,\ldots$:

$$P_n^{(\sigma)}(x)\equiv 0 \tag{25.71}$$

For $n=2m$ and Equation 25.70:

$$\lim_{\sigma\to -m}\frac{1}{\sigma+m}P_n^{(\sigma)}(x)=2\frac{(n-2m-1)!(2m)!}{n!}F\left(n-2m,-n,-m+\frac{1}{2};\frac{1-x}{2}\right) \tag{25.72}$$

In Equation 25.37, for $\lambda=\mu=-m-\frac{1}{2}$:

$$\lim_{\sigma\to -m}\frac{1}{\sigma+m}P_n^{(\sigma)}(x)=2\frac{(n-2m-1)!(2m)!2^{n+1}}{(-2m+1)\cdots(-2m+2n-1)}P_n^{\left(-m-\frac{1}{2},-m-\frac{1}{2}\right)}(x) \tag{25.73}$$

Also, for $n>2m$:

$$\lim_{\sigma\to -m}\frac{1}{\sigma+m}P_n^{(\sigma)}(x)=\frac{P_n^{(\sigma)}(x)-P_n^{(-m)}(x)}{\sigma-(-m)}=\left[\frac{d}{d\sigma}P_n^{(\sigma)}(x)\right]_{\sigma=-m} \tag{25.74}$$

Applying Equation 25.61 to 25.74:

$$\int_{-1}^{+1}\rho(x)\left[P_n^{(\sigma)}(x)\right]^2 dx=\frac{\pi\Gamma(2\sigma+n)}{2^{2\sigma-1}(\sigma+n)n!\Gamma^2(\sigma)},\ \sigma>-\frac{1}{2} \tag{25.75}$$

From Equation 25.53, we have

$$\frac{d}{dx}P_n^{(\sigma)}(x)=2\sigma P_{n-1}^{(\sigma+1)}(x),\ \frac{d^k}{dx^k}P_n^{(\sigma)}(x)=2^k\frac{\Gamma(\sigma+k)}{\Gamma(\sigma)}P_{n-k}^{(\sigma+k)}(x) \tag{25.76}$$

For $k=n$:

$$\frac{d^n}{dx^n}P_n^{(\sigma)}(x)=2^n\frac{\Gamma(\sigma+n)}{\Gamma(\sigma)} \tag{25.77}$$

To find the integral formula for the Gegenbauer Polynomial, we consider Equation 25.61:

$$P_n^{(\sigma)}(x)=\Lambda_n(\sigma)\frac{\left(x^2-1\right)^{-\sigma+\frac{1}{2}}}{2^{n+1}\pi i}\oint_l F_n^{\left(\sigma-\frac{1}{2},\sigma-\frac{1}{2}\right)}(z)dz \tag{25.78}$$

Here, l is the contour encircling the point $z=x$, while $z=+1$ and $z=-1$ are outside the given contour. The Gegenbauer polynomial has the following recurrence relation:

$$2(\sigma+n)xP_n^{(\sigma)}(x)=(n+1)P_{n+1}^{(\sigma)}(x)+(2\sigma+n-1)P_{n-1}^{(\sigma)}(x) \tag{25.79}$$

25.6 Gegenbauer Polynomial Generating Function

Use two methods to find the Gegenbauer polynomial generating function from Equations 25.48 and 25.61:

$$\Phi(t,x)=\sum_{n=0}^{\infty}\Lambda_n^{-1}(\sigma)P_n^{(\sigma)}(x)t^n=\frac{2^{\sigma-\frac{1}{2}}}{\sqrt{1-2tx+t^2}\left(1-tx+\sqrt{1-2tx+t^2}\right)^{\sigma-\frac{1}{2}}} \tag{25.80}$$

From here:

$$P_n^{(\sigma)}(x)=\Lambda_n(\sigma)\frac{1}{n!}\left[\frac{\partial^n}{\partial t^n}\Phi(t,x)\right]_{t=0} \tag{25.81}$$

The second method is via Equation 25.79 where we swap n with $n-1$, then the resultant is multiplied by t^{n-1} and the sum of the resultant is as follows:

$$\sum_{n=0}^{\infty}nP_n^{(\sigma)}(x)t^{n-1}-2xt\sum_{n=0}^{\infty}(n-1)P_{n-1}^{(\sigma)}(x)t^{n-2}+t^2\sum_{n=0}^{\infty}(n-2)P_{n-2}^{(\sigma)}(x)t^{n-3}$$
$$+2\sigma\left(t\sum_{n=0}^{\infty}P_{n-2}^{(\sigma)}(x)t^{n-3}-x\sum_{n=0}^{\infty}P_{n-1}^{(\sigma)}(x)t^{n-1}\right)=0 \tag{25.82}$$

Setting in the second and fifth term, $n-1=k$, and in the third and fourth terms, $n-2=l$, while considering

$$P_{-1}^{(\sigma)}(x)=P_{-2}^{(\sigma)}(x)=0 \tag{25.83}$$

and the result multiplied by $\left(1-2tx+t^2\right)^{\sigma-1}$:

$$\left(1-2tx+t^2\right)^{\sigma}\sum_{n=0}^{\infty} nP_n^{(\sigma)}(x)t^{n-1}-2\sigma(t-x)\left(1-2tx+t^2\right)^{\sigma-1}\sum_{n=0}^{\infty} P_n^{(\sigma)}(x)t^n=0 \tag{25.84}$$

So the generating function:

$$\Phi(t,x)=\sum_{n=0}^{\infty} P_n^{(\sigma)}(x)t^n \tag{25.85}$$

or

$$\Phi(t,x)=\left(1-2tx+t^2\right)^{-\sigma} \tag{25.86}$$

from where:

$$P_n^{(\sigma)}(x)=\frac{1}{n!}\left[\frac{\partial^n}{\partial t^n}\Phi(t,x)\right]_{t=0} \tag{25.87}$$

We series expand the function $f(x)$ convergent in the interval, $(-1,+1)$, via the Gegenbauer polynomial, $P_n^{(\sigma)}(x)$:

$$f(x)=\sum_{n=0}^{\infty} c_n P_n^{(\sigma)}(x) \tag{25.88}$$

From Equations 25.66 and 25.60:

$$c_n=\frac{2^{2\sigma-1}\Gamma^2(\sigma)(\sigma+n)n!}{\pi\Gamma(2\sigma+n)}\int_{-1}^{+1} dx\rho(x)P_n^{(\sigma)}(x)f(x) \tag{25.89}$$

25.7 First Kind Tschebycheff Polynomial

Examine the first kind Tschebycheff polynomial, $T_n(x)$, from Equations 25.73 and 25.29 for $m=0$:

$$T_n(x)=\frac{n}{2}\lim_{\sigma\to 0}\frac{1}{\sigma}P_n^{(\sigma)}(x)=\frac{2\cdot 4\cdots(2n)}{1\cdot 3\cdots(2n-1)}P_n^{\left(-\frac{1}{2},-\frac{1}{2}\right)}(x) \tag{25.90}$$

then

$$T_0(x)=1,\ T_1(x)=x,\ T_2(x)=2x^2-1,\ T_3(x)=4x^3-3x \tag{25.91}$$

and so on. The polynomial, $T_n(x)$, weighted by $\rho(x)=\frac{1}{\sqrt{1-x^2}}$ is orthogonal within the interval, $(-1,+1)$:

$$\int_{-1}^{+1}\rho(x)T_m(x)T_n(x)dx=0,\ m\neq n \tag{25.92}$$

Note

$$T_n(-x)=(-1)^n T_n(x),\ T_n(1)=1,\ T_n(-1)=(-1)^n \tag{25.93}$$

then

$$\int_{-1}^{+1}\rho(x)\left[T_n(x)\right]^2 dx=\left[\frac{2\cdot 4\cdots(2n)}{1\cdot 3\cdots(2n-1)}\right]^2\int_{-1}^{+1}\rho(x)\left[P_n^{\left(-\frac{1}{2},-\frac{1}{2}\right)}(x)\right]^2 dx \tag{25.94}$$

or

$$\int_{-1}^{+1}\rho(x)\left[T_n(x)\right]^2 dx=\left[\frac{2\cdot 4\cdots(2n)}{1\cdot 3\cdots(2n-1)}\right]^2\frac{\Gamma^2\left(n+\frac{1}{2}\right)}{n!2n\Gamma(n)}=\begin{cases}\frac{\pi}{2}, & n\neq 0\\ \pi, & n=0\end{cases} \tag{25.95}$$

From the orthogonality relation in Equation 25.92, we can infer that all the roots of $T_n(x)$ are real, different and are within the interval, $(-1,+1)$. When $\sigma=0$, then for the polynomial, $T_n(x)$, the differential Equation 25.69 takes the form:

$$\frac{1}{\rho(x)}\frac{d}{dx}\left(\frac{1}{\rho(x)}\frac{dy}{dx}\right)+n^2y=0 \tag{25.96}$$

Express the Gegenbauer polynomial, $T_n(x)$, through the confluent hypergeometric function considering Equations 25.90 and 25.37:

$$T_n(x)=F\left(n,-n,\frac{1}{2};\frac{1-x}{2}\right) \tag{25.97}$$

Dividing Equation 25.79 by σ and taking the limit, $\sigma\to 0$:

$$\lim_{\sigma\to 0}\frac{1}{\sigma}P_n^{(\sigma)}(x)=(n+1)\lim_{\sigma\to 0}\frac{1}{\sigma}P_{n+1}^{(\sigma)}(x)+(n-1)\lim_{\sigma\to 0}\frac{1}{\sigma}P_{n-1}^{(\sigma)}(x) \tag{25.98}$$

From Equation 25.90, the recurrent formula for the Tschebycheff polynomial, $T_n(x)$:

$$T_{n+1}(x)=2xT_n(x)-T_{n-1}(x),\ n=1,2,\ldots \tag{25.99}$$

From Equations 25.90 and 25.12 for $\lambda = \mu = -\frac{1}{2}$:

$$\frac{d^k}{dx^k}\mathrm{T}_n(x) = \frac{2^{2n-k}nn!(n+k-1)!}{(2n)!}\mathrm{P}_{n-k}^{\left(k-\frac{1}{2},k-\frac{1}{2}\right)}(x) \tag{25.100}$$

If $k=1$, then

$$\frac{d}{dx}\mathrm{T}_n(x) = \frac{2^{2n-1}n(n!)^2}{(2n)!}\mathrm{P}_{n-1}^{\left(\frac{1}{2},\frac{1}{2}\right)}(x) \tag{25.101}$$

and for $k=n$:

$$\frac{d^n}{dx^n}\mathrm{T}_n(x) = 2^{n-1}n! \tag{25.102}$$

and implies that the coefficient of $\mathrm{T}_n(x)$ for x^n is 2^{n-1}.

From Equations 25.33 and 25.90, consider the contour, l, encircling the point $z=x$, while $z=+1$ and $z=-1$ are outside the given contour. So we represent the Tschebycheff polynomial, $\mathrm{T}_n(x)$, through the contour integral of the form:

$$\mathrm{T}_n(x) = \frac{2^{n-1}(n!)^2}{(2n)!\pi i}\sqrt{1-x^2}\oint_l \mathrm{F}_n^{\left(-\frac{1}{2},-\frac{1}{2}\right)}(z)dz \tag{25.103}$$

Letting in Equation 25.96 $x=\cos\psi$:

$$\frac{d^2}{d\phi^2}y(\cos\psi) + n^2 y(\cos\psi) = 0 \tag{25.104}$$

From this solution, we arrive at the expression of the Tschebycheff polynomial, $\mathrm{T}_n(x)$:

$$\mathrm{T}_n(\cos\psi) = \cos n\psi \equiv \frac{1}{2}\left(\exp\{in\psi\} + \exp\{-in\psi\}\right) = \frac{1}{2}\left((\cos\psi + i\sin\psi)^n + (\cos\psi - i\sin\psi)^n\right) \tag{25.105}$$

or

$$\mathrm{T}_n(x) = \frac{1}{2}\left(\left(x+\sqrt{x^2-1}\right)^n + \left(x-\sqrt{x^2-1}\right)^n\right) \tag{25.106}$$

25.8 Generating Function of the First Kind Tschebycheff Polynomial

We write the Tschebycheff polynomial generating function:

$$\Phi(t,x) = \sum_{n=0}^{\infty}\mathrm{T}_n(x)t^n = \mathrm{Re}\sum_{n=0}^{\infty}\left(\exp\{i\psi\}t\right)^n = \frac{1-t\cos\psi}{1-2t\cos\psi+t^2} = \frac{1-tx}{1-2tx+t^2} \tag{25.107}$$

So

$$T_n(x)=\frac{1}{n!}\left[\frac{\partial^n}{\partial t^n}\Phi(t,x)\right]_{t=0} \tag{25.108}$$

We series expand the function $f(x)$ convergent within the interval, $(-1,+1)$ through the Tschebycheff polynomial, $T_n(x)$:

$$f(x)=\sum_{n=0}^{\infty}c_nT_n(x) \tag{25.109}$$

where

$$c_n=\frac{2}{\pi}\int_{-1}^{+1}\rho(x)T_n(x)f(x)dx,\ n=1,2,\ldots \tag{25.110}$$

25.9 Tschebycheff Polynomial of the Second Kind

We examine the second kind Tschebycheff polynomial, $U_n(x)$, by considering Equations 25.62 and 25.36 for $\sigma=1$ and $\lambda=\mu=\frac{1}{2}$:

$$U_n(x)=P_n^{(1)}(x)=C_nP_n^{\left(-\frac{1}{2},-\frac{1}{2}\right)}(x),\ C_n=\frac{\Gamma\left(\frac{3}{2}\right)\Gamma(n+2)}{\Gamma\left(n+\frac{3}{2}\right)\Gamma(2)} \tag{25.111}$$

From here:

$$U_0(x)=1,\ U_1(x)=2x,\ U_2(x)=4x^2-1,\ U_3(x)=8x^3-4x \tag{25.112}$$

and so on.

The polynomial, $U_n(x)$, weighted by $\rho(x)=\sqrt{1-x^2}$ is orthogonal within the interval, $(-1,+1)$:

$$\int_{-1}^{+1}\rho(x)U_m(x)U_n(x)dx=0,\ m\neq n \tag{25.113}$$

Note:

$$U_n(-x)=(-1)^nU_n(x),\ U_n(1)=n+1,\ U_n(-1)=(-1)^n(n+1) \tag{25.114}$$

We find the norm of $U_n(x)$ considering Equations 25.111 and 25.40:

$$\int_{-1}^{+1}\rho(x)\left[U_n(x)\right]^2dx=C_n^2\int_{-1}^{+1}\rho(x)\left[P_n^{\left(-\frac{1}{2},-\frac{1}{2}\right)}(x)\right]^2dx \tag{25.115}$$

or

$$\int_{-1}^{+1}\rho(x)\left[\mathrm{U}_n(x)\right]^2 dx = \mathrm{C}_n^2\frac{2^2\Gamma^2\left(n+\frac{3}{2}\right)}{n!(2n+1)\Gamma(n+2)} = \frac{\pi}{2} \tag{25.116}$$

From the orthogonality relation in Equation 25.113, we infer that all the roots of $\mathrm{U}_n(x)$ are real-valued, different and are within the interval, $(-1,+1)$. When $\sigma=1$, then for the polynomial, $\mathrm{U}_n(x)$, the differential Equation 25.69 takes the form:

$$\frac{d}{dx}\left((1-x^2)^{\frac{3}{2}}\frac{dy}{dx}\right)+n(n+2)(1-x^2)^{\frac{1}{2}}y=0 \tag{25.117}$$

We express the Tschebycheff polynomial, $\mathrm{U}_n(x)$, through the confluent hypergeometric function by considering Equations 25.111 and 25.37:

$$\mathrm{U}_n(x)=(n+1)\mathrm{F}\left(n+2,-n,\frac{3}{2};\frac{1-x}{2}\right) \tag{25.118}$$

We set in Equation 25.79 $\sigma=1$, so

$$\mathrm{U}_{n+1}(x)=2x\mathrm{U}_n(x)-\mathrm{U}_{n-1}(x),\ n=1,2,\ldots \tag{25.119}$$

From Equations 25.111 and 25.78, for $\lambda=\mu=\frac{1}{2}$:

$$\frac{d^k}{dx^k}\mathrm{U}_n(x)=\mathrm{C}_{nk}\mathrm{P}_{n-k}^{\left(k+\frac{1}{2},k+\frac{1}{2}\right)}(x),\ \mathrm{C}_{nk}=\frac{2^{2n-k}n!(n+k+1)!}{(2n+1)!} \tag{25.120}$$

If $k=1$, then

$$\frac{d}{dx}\mathrm{U}_n(x)=\mathrm{C}_{n1}\mathrm{P}_{n-1}^{\left(\frac{3}{2},\frac{3}{2}\right)}(x) \tag{25.121}$$

and for $k=n$:

$$\frac{d^n}{dx^n}\mathrm{U}_n(x)=2^n n! \tag{25.122}$$

implying that the coefficient of $\mathrm{U}_n(x)$ for x^n is 2^n.

We compare Equation 25.120 with 25.100 and swap, respectively, n and k for $n+1$ and $k+1$:

$$\frac{d^k}{dx^k}\mathrm{U}_n(x)=\frac{1}{n+1}\frac{d^{k+1}}{dx^{k+1}}\mathrm{T}_{n+1}(x) \tag{25.123}$$

and for $k=0$:

$$\mathrm{U}_n(x)=\frac{1}{n+1}\frac{d}{dx}\mathrm{T}_{n+1}(x) \tag{25.124}$$

From Equation 25.33, we consider the contour, l, encircling the point $z = x$, while $z = +1$ and $z = -1$ are outside the given contour. So we represent the Tschebycheff polynomial, $\mathrm{U}_n(x)$, through the contour integral:

$$\mathrm{U}_n(x) = \frac{\mathrm{C}_{n0}}{\pi i\sqrt{1-x^2}} \oint_l \mathrm{F}_n^{\left(\frac{1}{2},\frac{1}{2}\right)}(z)\,dz \tag{25.125}$$

In Equation 25.105, we swap n for $n+1$ considering Equation 25.124 and arrive at the expression of the Tschebycheff polynomial, $\mathrm{U}_n(x)$:

$$\mathrm{U}_n(x) = \frac{1}{\sqrt{1-x^2}} \sin\left((n+1)\cos^{-1} x\right) \tag{25.126}$$

From

$$\sin\left((n+1)\cos^{-1} x\right) = \sin\left((n+1)\psi\right) \tag{25.127}$$

or

$$\sin\left((n+1)\cos^{-1} x\right) = \frac{1}{2i}\left((\cos\psi + i\sin\psi)^{n+1} - (\cos\psi - i\sin\psi)^{n+1}\right) \tag{25.128}$$

then

$$\mathrm{U}_n(x) = \frac{1}{2\sqrt{x^2-1}}\left(\left(x+\sqrt{x^2-1}\right)^{n+1} - \left(x-\sqrt{x^2-1}\right)^{n+1}\right) \tag{25.129}$$

25.10 Generating Function of the Second Kind Tschebycheff Polynomial

We write the Tschebycheff polynomial, $\mathrm{U}_n(x)$, generating function considering Equation 25.126:

$$\Phi(t,x) = \sum_{n=0}^{\infty} \mathrm{U}_n(x) t^n = \frac{1}{t\sin\psi} \mathrm{Im} \sum_{n=0}^{\infty} \left(\exp\{i\psi\}t\right)^{n+1} = \frac{1}{1-2t\cos\psi + t^2} = \frac{1}{1-2tx+t^2} \tag{25.130}$$

So

$$\mathrm{U}_n(x) = \frac{1}{n!}\left[\frac{\partial^n}{\partial t^n}\Phi(t,x)\right]_{t=0} \tag{25.131}$$

We series expand the function $f(x)$ convergent within the interval, $(-1,+1)$, through the Tschebycheff polynomial, $\mathrm{U}_n(x)$:

$$f(x) = \sum_{n=0}^{\infty} c_n \mathrm{U}_n(x) \tag{25.132}$$

where

$$c_n = \frac{2}{\pi}\int_{-1}^{+1}\rho(x)U_n(x)f(x)dx, n=1,2,\ldots \tag{25.133}$$

25.11 Legendre Polynomials

We find the Legendre polynomials, $P_n(x)$, which is a complete set of polynomials within the finite interval $(-1,1)$. This Legendre polynomials, $P_n(x)$, are defined by the **Rodriguez recurrence formula** considering Equations 25.62 and 25.29 for, respectively, $\sigma = \frac{1}{2}$ and $\lambda = \mu = 0$:

$$P_n(x) = P_n^{\left(\frac{1}{2}\right)}(x) = P_n^{(0,0)}(x) = \frac{1}{2^n n!}\frac{d^n}{dx^n}(x^2-1)^n \tag{25.134}$$

When $\sigma = \frac{1}{2}$, then for the polynomial, $P_n(x)$, the differential Equation 25.69 takes the form:

$$\frac{d}{dx}\left((1-x^2)\frac{dy}{dx}\right) + n(n+1)y = 0 \tag{25.135}$$

From Equations 25.134 and 25.70, the Legendre polynomial, $P_n(x)$, can be expressed through the confluent hypergeometric function:

$$P_n(x) = F\left(n+1,-n,1;\frac{1-x}{2}\right) \tag{25.136}$$

From Equations 25.134, 25.77 and 25.78, considering, respectively, $\sigma = \frac{1}{2}$ and $\lambda = \mu = 0$:

$$\frac{d^k}{dx^k}P_n(x) = \frac{(n+1)\cdots(n+k)}{2^k}P_{n-k}^{(k,k)}(x) \tag{25.137}$$

If $k = n$, then the coefficient of x^n in the expansion of the polynomial, $P_n(x)$:

$$\frac{1\cdot 3\cdots(2n-1)}{n!} = \frac{(2n)!}{2^n (n!)^2} \tag{25.138}$$

From Equations 25.137 and 25.36, we also have

$$\left[\frac{d^k}{dx^k}P_n(x)\right]_{x=1} = \frac{(n+k)!}{2^k(n-k)!k!}, \left[\frac{d^k}{dx^k}P_n(x)\right]_{x=-1} = \frac{(-1)^{n-k}(n+k)!}{2^k(n-k)!k!} = (-1)^{n-k}\left[\frac{d^k}{dx^k}P_n(x)\right]_{x=1} \tag{25.139}$$

25.12 Legendre Polynomial Generating Function

We find the generating function, $\Phi(t,x)$, for the Legendre polynomial by considering Equation 25.80 when $\sigma = \frac{1}{2}$:

$$\Phi(t,x) = \sum_{n=0}^{\infty} P_n(x)t^n = \frac{1}{\sqrt{1-2tx+t^2}} \tag{25.140}$$

then

$$\mathrm{P}_n(x)=\frac{1}{n!}\left[\frac{\partial^n}{\partial t^n}\Phi(t,x)\right]_{t=0} \tag{25.141}$$

and

$$\mathrm{P}_{2n+1}(0)=0,\ \mathrm{P}_{2n}(0)=(-1)^n\frac{(2n)!}{2^{2n}(n!)^2} \tag{25.142}$$

From Equation 25.140, we have

$$\frac{\partial}{\partial t}\Phi(t,x)=\frac{x-t}{1-2tx+t^2}\Phi(t,x),\ \frac{\partial}{\partial x}\Phi(t,x)=\frac{t}{1-2tx+t^2}\Phi(t,x) \tag{25.143}$$

From here:

$$t\frac{\partial}{\partial t}\Phi(t,x)+(t-x)\frac{\partial}{\partial x}\Phi(t,x)=0 \tag{25.144}$$

But

$$\frac{\partial}{\partial t}\Phi(t,x)=\sum_{n=0}^{\infty}n\mathrm{P}_n(x)t^{n-1},\ \frac{\partial}{\partial x}\Phi(t,x)=\sum_{n=0}^{\infty}\mathrm{P}'_n(x)t^n \tag{25.145}$$

So

$$t\sum_{n=0}^{\infty}n\mathrm{P}_n(x)t^{n-1}+(t-x)\sum_{n=0}^{\infty}\mathrm{P}'_n(x)t^n=0 \tag{25.146}$$

and hence the recurrence Legendre equation:

$$n\mathrm{P}_n(x)-x\mathrm{P}'_n(x)+\mathrm{P}'_{n-1}(x)=0 \tag{25.147}$$

Applying Equation 25.145 to the equation:

$$\left(t\frac{\partial}{\partial t}t-t\left(t\frac{\partial}{\partial t}+1\right)\right)\Phi(t,x)=0 \tag{25.148}$$

then from Equations 25.145 and 25.140, we have

$$\left(t\frac{\partial}{\partial t}t+(tx-1)\frac{\partial}{\partial x}\right)\sum_{n=0}^{\infty}\mathrm{P}_n(x)t^n=0 \tag{25.149}$$

and so follows the third recurrence relation:

$$nP_{n-1}(x)-P_n'(x)+xP_{n-1}'(x)=0 \tag{25.150}$$

From Equation 25.150 minus Equation 25.147 and the resultant multiplied by x, the fourth recurrence relation:

$$\left(1-x^2\right)P_n'(x)+nxP_n(x)-nP_{n-1}(x)=0 \tag{25.151}$$

In Equation 25.147, we let $n\to n+1$ and sum the resultant with Equation 25.150:

$$(n+1)P_{n+1}(x)-xP_{n+1}'(x)+nP_{n-1}(x)+xP_{n-1}'(x)=0 \tag{25.152}$$

From here, we have

$$(2n+1)P_n(x)-P_{n+1}'(x)-P_{n-1}'(x)=0 \tag{25.153}$$

Considering $n=0,1,2,\ldots$:

$$P_{-1}'(x)=0,\ P_0'(x)=0,\ P_0(x)=1 \tag{25.154}$$

We series expand the function $f(x)$ convergent within the interval, $(-1,+1)$, through the Legendre polynomials:

$$f(x)=\sum_{n=0}^{\infty}c_nP_n(x) \tag{25.155}$$

and

$$c_n=\left(n+\frac{1}{2}\right)\int_{-1}^{+1}f(x)P_n(x)dx,\ n=0,1,2,\ldots \tag{25.156}$$

25.13 Legendre Polynomials Integral Representation

The Legendre polynomials can be represented in an integral form:

$$P_n(x)=\frac{1}{2^{n+1}\pi i}\oint_l\frac{\left(z^2-1\right)^n}{(z-x)^{n+1}}dz \tag{25.157}$$

Here, l is any contour enclosing the point $z=x$. Suppose, l is a circle with radius $\left|x^2-1\right|^{\frac{1}{2}}$ and center at $z=x$:

$$z=x+\sqrt{x^2-1}\exp\{i\psi\} \tag{25.158}$$

So

$$P_n(x)=\frac{1}{2^{n+1}\pi i}\int_{-\pi}^{+\pi}\frac{\left(2\sqrt{x^2-1}\exp\{i\psi\}\left(x+\sqrt{x^2-1}\cos\psi\right)\right)^n}{\left(\sqrt{x^2-1}\exp\{i\psi\}\right)^{n+1}}\sqrt{x^2-1}\exp\{i\psi\}id\psi \tag{25.159}$$

From here, we have

$$P_n(x)=\frac{1}{2\pi}\int_{-\pi}^{+\pi}\left(x+\sqrt{x^2-1}\cos\psi\right)^n d\psi=\frac{1}{\pi}\int_{0}^{+\pi}\left(x+\sqrt{x^2-1}\cos\psi\right)^n d\psi \tag{25.160}$$

and for $x=\cos\theta$:

$$P_n(\cos\theta)=\frac{1}{\pi}\int_{0}^{+\pi}(\cos\theta+i\sin\theta\cos\psi)^n d\psi \tag{25.161}$$

But

$$-1\leq x\leq +1 \tag{25.162}$$

and so

$$\left|x+\sqrt{x^2-1}\cos\psi\right|=\left|x+i\sqrt{1-\rho^2}\cos\psi\right|=\sqrt{x^2+\left(1-x^2\right)\cos^2\psi}=\sqrt{\cos^2\psi+x^2\sin^2\psi}\leq 1 \tag{25.163}$$

So, from Equation 25.160, we have

$$P_n(x)\leq\frac{1}{\pi}\int_{0}^{+\pi}d\psi=1 \tag{25.164}$$

This shows that the Legendre polynomials are bounded within the interval defined in Equation 25.162.

25.14 Associated Legendre Polynomials

We write the equation of the associated Legendre polynomials for $m>0$, $n>0$:

$$\left(1-x^2\right)\frac{d^2y}{dx^2}-2x\frac{dy}{dx}+\left(n(n+1)-\frac{m^2}{1-x^2}\right)y=0 \tag{25.165}$$

Setting

$$y=\left(1-x^2\right)^{\frac{m}{2}}\frac{d^mY}{dx^m} \tag{25.166}$$

then from Equation 25.165, we have

$$\left(1-x^2\right)\frac{d^{m+2}Y}{dx^{m+2}}-2(m+1)x\frac{d^{m+1}Y}{dx^{m+1}}+\left(n(n+1)-m(m+1)\right)\frac{d^mY}{dx^m}=0 \tag{25.167}$$

Swapping y for $P_n^m(x)$ and letting $Y = P_n(x)$ being the Legendre polynomial of order n, we find the equation of the associated Legendre polynomial of order n and rank m:

$$P_n^m(x)=(1-x^2)^{\frac{m}{2}}\frac{d^m}{dx^m}P_n(x)=(1-x^2)^{\frac{m}{2}}\frac{1}{2^n n!}\frac{d^{n+m}}{dx^{n+m}}(x^2-1)^n \tag{25.168}$$

Note that $P_n^0(x)=P_n(x)$ and $P_n^m(x)=0$ for $>n$. From the differential Equation 25.165, we have

$$\frac{d}{dx}\left((1-x^2)\frac{dy}{dx}\right)+\left(n(n+1)-\frac{m^2}{1-x^2}\right)y=0 \tag{25.169}$$

Using Equation 25.168, we express the associated Legendre polynomial, $P_n^m(x)$, through the confluent hypergeometric function:

$$\begin{aligned}P_n^m(x)&=(1-x^2)^{\frac{m}{2}}\frac{d^m}{dx^m}F\left(n+1,-n,1;\frac{1-x}{2}\right)\\&=(1-x^2)^{\frac{m}{2}}\frac{\Gamma(n+m+1)}{2^m m!\Gamma(n-m+1)}F\left(n+m+1,m-n,m+1;\frac{1-x}{2}\right)\end{aligned} \tag{25.170}$$

We express the associated Legendre polynomial, $P_n^m(x)$, through the Jacobi polynomial considering Equations 25.168 and 25.137:

$$P_n^m(x)=\frac{(n+1)\cdots(n+m)}{2^m}(1-x^2)^{\frac{m}{2}}P_{n-m}^{(m,m)}(x) \tag{25.171}$$

The associated Legendre polynomials weighted $\rho(x)=1$ are of different orders but of the same range and orthogonal within the interval, $(-1,+1)$:

$$\int_{-1}^{+1}\rho(x)P_n^m(x)P_k^m(x)dx=\frac{(n+1)\cdots(n+m)}{2^m}\frac{(k+1)\cdots(k+m)}{2^m}\int_{-1}^{+1}(1-x^2)^m P_{n-m}^{(m,m)}(x)P_{k-m}^{(m,m)}(x)dx=0 \tag{25.172}$$

The norm of $P_n^m(x)$ can be calculated as follows:

$$\int_{-1}^{+1}\rho(x)\left[P_n^m(x)\right]^2dx=\left[\frac{(n+1)\cdots(n+m)}{2^m}\right]^2\frac{2^{2m+1}\Gamma^2(n+1)}{(n-m)!(2n+1)\Gamma(n+m+1)} \tag{25.173}$$

or

$$\int_{-1}^{+1}\rho(x)\left[P_n^m(x)\right]^2dx=\frac{2}{2n+1}\frac{(n+m)!}{(n-m)!} \tag{25.174}$$

The recurrence formula for the associated Legendre polynomial can be obtained when $\lambda=\mu=m$ and swapping n for $n-m$ in the Jacobi polynomial while considering Equation 25.171:

$$(2n+1)xP_n^m(x)=(n-m+1)P_{n+1}^m(x)+(n+m)P_{n-1}^m(x) \tag{25.175}$$

For $n=m$, $P_{n-1}^{n}(x)=0$ and

$$(2n+1)xP_n^n(x)=P_{n+1}^n(x) \tag{25.176}$$

From Equation 25.168, it follows that the function, $P_n^m(x)$, has $n-m$ roots within the interval $(-1,+1)$.

25.15 Associated Legendre Polynomials Integral Representation

The associated Legendre polynomials can be represented in an integral form through

$$\oint_l \frac{(z^2-1)^n}{(z-x)^{n+m+1}}dz = 2\pi i \operatorname*{Res}_{z=x}\frac{(z^2-1)^n}{(z-x)^{n+m+1}} = \frac{2\pi i}{(n+m)!}\lim_{z\to x}\frac{d^{n+m}}{dx^{n+m}}(z^2-1)^n = \frac{2\pi i 2^n n!}{(n+m)!}(1-x^2)^{-\frac{m}{2}}P_n^m(x) \tag{25.177}$$

From where:

$$P_n^m(x)=\frac{(n+m)!}{\pi i 2^{n+1}n!}(1-x^2)^{\frac{m}{2}}\oint_l \frac{(z^2-1)^n}{(z-x)^{n+m+1}}dz \tag{25.178}$$

From Equation 25.160:

$$P_n^m(x)=\frac{(n+1)\cdots(n+m)}{2\pi i^m}\int_{-\pi}^{+\pi}\left(x+\sqrt{x^2-1}\cos\psi\right)^n \cos m\psi d\psi \tag{25.179}$$

and for $x=\cos\theta$:

$$P_n^m(x)=\frac{(n+1)\cdots(n+m)}{2\pi i^m}\int_{-\pi}^{+\pi}(\cos\theta+i\sin\theta\cos\psi)^n \cos m\psi d\psi \tag{25.180}$$

25.16 Spherical Functions

The Laplace equation:

$$\Delta u(x,y,z)=\frac{\partial^2 u}{\partial x^2}+\frac{\partial^2 u}{\partial y^2}+\frac{\partial^2 u}{\partial z^2} \tag{25.181}$$

has its solution in the form of a polynomial:

$$u_n(x,y,z)=\sum_{m=0}^{n}\varphi_m(x,y)z^{n-m} \tag{25.182}$$

where $\varphi_m(x,y)$ is a homogenous polynomial of order m with arguments x and y:

$$\varphi_m(x,y)=\sum_{k=0}^{m}C_{m,k}x^{m-k}y^k,\ k=0,1,\ldots,m,\ m=0,1,\ldots,n \tag{25.183}$$

Here, $C_{m,k}$ are constant coefficients. Since $\varphi_m(x,y)$ has $m+1$ coefficients, $u_n(x,y,z)$ has the coefficients:

$$1+2+\cdots+(n+1)=\frac{(n+1)(n+2)}{2} \quad (25.184)$$

For independent equations, the number of these coefficients relates to $\frac{(n-1)n}{2}$ equations, i.e., $2n+1$ coefficients of the polynomial, $u_n(x,y,z)$:

$$\frac{(n+1)(n+2)}{2}-\frac{(n-1)n}{2}=2n+1 \quad (25.185)$$

We easily show

$$u_n(x,y,z)=\sum_{m=0}^{n} A_{\alpha\beta\gamma}x^{\alpha}y^{\beta}z^{\gamma},\ \alpha+\beta+\gamma=n \quad (25.186)$$

where

$$A_{\alpha\beta\gamma}=\frac{1}{\alpha!\beta!\gamma!}\frac{\partial^{\alpha+\beta+\gamma}}{\partial x^{\alpha}\partial y^{\beta}\partial z^{\gamma}}u \quad (25.187)$$

has $2n+1$ independent coefficients. Note that any polynomial satisfying the Laplace equation is called a **harmonic polynomial**. So for any integer, n, there exist $2n+1$ linear independent homogenous harmonic polynomials. From the spherical system of coordinates:

$$x=r\sin\theta\cos\phi,\ y=r\sin\theta\sin\phi,\ z=r\cos\theta \quad (25.188)$$

we write the homogenous harmonic polynomial, $u_n(x,y,z)$, in the form:

$$u_n(x,y,z)=r^n Y_n(\theta,\phi) \quad (25.189)$$

which is called the **volume spherical function** of order n and $Y_n(\theta,\phi)$, **surface spherical harmonic** or, simply, **spherical function** of order n. So we have $2n+1$ linear independent spherical functions of order, n.

Note that for any function, $f(t,\tau)$, where

$$\tau=ix\cos t+iy\sin t+z \quad (25.190)$$

then the function:

$$F(x,y,z)=\int_{-\pi}^{+\pi} f(t,\tau)dt \quad (25.191)$$

satisfies the Laplace equation:

$$f_{n,m}^{(1)}(t,\tau)=(ix\cos t+iy\sin t+z)^n\cos mt,\ f_{n,m}^{(2)}(t,\tau)=(ix\cos t+iy\sin t+z)^n\sin mt,\ m=01,2,\ldots,n \quad (25.192)$$

So the homogenous harmonic polynomial of order:

$$F_{n,m}^{(1)}(t,\tau)=\int_{-\pi}^{+\pi}dt\left(ix\cos t+iy\sin t+z\right)^{n}\cos mt,\; F_{n,m}^{(2)}(t,\tau)$$
$$=\int_{-\pi}^{+\pi}dt\left(ix\cos t+iy\sin t+z\right)^{n}\sin mt,\; m=01,2,\ldots,n \tag{25.193}$$

move to spherical system of coordinates:

$$F_{n,m}^{(1)}(x,y,z)=r^{n}\overline{Y}_{n,m}^{(1)}(\theta,\phi),\; F_{n,m}^{(2)}(x,y,z)=r^{n}\overline{Y}_{n,m}^{(2)}(\theta,\phi) \tag{25.194}$$

where $\overline{Y}_{n,m}^{(1)}(\theta,\phi)$ and $\overline{Y}_{n,m}^{(2)}(\theta,\phi)$ are spherical functions:

$$\overline{Y}_{n,m}^{(1)}(\theta,\phi)=\int_{-\pi}^{+\pi}dt\left(\cos\theta+i\sin\theta\cos(t-\phi)\right)^{n}\cos mt,\; \overline{Y}_{n,m}^{(2)}(\theta,\phi)=\int_{-\pi}^{+\pi}dt\left(\cos\theta+i\sin\theta\cos(t-\phi)\right)^{n}\sin mt \tag{25.195}$$

Letting $t-\phi=\psi$:

$$\overline{Y}_{n,m}^{(1)}(\theta,\phi)=\int_{-\pi}^{+\pi}d\psi\left(\cos\theta+i\sin\theta\cos\psi\right)^{n}\cos m(\phi+\psi),\; \overline{Y}_{n,m}^{(2)}(\theta,\phi)$$
$$=\int_{-\pi}^{+\pi}d\psi\left(\cos\theta+i\sin\theta\cos\psi\right)^{n}\sin m(\phi+\psi) \tag{25.196}$$

or

$$\overline{Y}_{n,m}^{(1)}(\theta,\phi)=\cos m\phi\int_{-\pi}^{+\pi}d\psi\left(\cos\theta+i\sin\theta\cos\psi\right)^{n}\cos m\psi,\; \overline{Y}_{n,m}^{(2)}(\theta,\phi)$$
$$=\sin m\phi\int_{-\pi}^{+\pi}d\psi\left(\cos\theta+i\sin\theta\cos\psi\right)^{n}\sin m\psi \tag{25.197}$$

We relate these functions with the Legendre and associated Legendre polynomials:

$$\overline{Y}_{n,0}^{(1)}(\theta,\phi)=P_{n}(\cos\theta),\; \overline{Y}_{n,0}^{(2)}(\theta,\phi)=0 \tag{25.198}$$

$$\overline{Y}_{n,m}^{(1)}(\theta,\phi)=P_{n}^{m}(\cos\theta)\cos m\phi,\; \overline{Y}_{n,m}^{(2)}(\theta,\phi)=P_{n}^{m}(\cos\theta)\sin m\phi,\; m=1,2,\ldots,n \tag{25.199}$$

So from here, the solution of the Laplace equation is the linear combination:

$$Y_{n}(\theta,\phi)=\sum_{m=0}^{n}\left(A_{n,m}\cos m\phi+B_{n,m}\sin m\phi\right)P_{n}^{m}(\cos\theta) \tag{25.200}$$

where $A_{n,m}$ and $B_{n,m}$ are constant coefficients.

Examine the homogenous harmonic polynomial, $u_n(x,y,z)$ in Equation 25.189 and evaluate the integral:

$$\iint_S \left(u_k \frac{\partial u_l}{\partial r} - u_l \frac{\partial u_k}{\partial r}\right) dS = \iiint_V (u_k \Delta u_l - u_l \Delta u_k)\, dV \tag{25.201}$$

where S and V are, respectively, the surface and volume of the sphere. Since

$$\frac{\partial u_k}{\partial r} = kr^{k-1}\mathrm{Y}_k(\theta,\phi),\ \frac{\partial u_l}{\partial r} = kr^{l-1}\mathrm{Y}_l(\theta,\phi),\ \Delta u_l = \Delta u_k = 0 \tag{25.202}$$

then for $k \neq l$, we have

$$(l-k)\iint_S \mathrm{Y}_k(\theta,\phi)\mathrm{Y}_l(\theta,\phi)\, dS = 0,\ k \neq l \tag{25.203}$$

So the spherical functions of different orders are orthogonal:

$$\iint_S \mathrm{Y}_k(\theta,\phi)\mathrm{Y}_l(\theta,\phi)\, dS = 0,\ k \neq l \tag{25.204}$$

From Equation 25.198, we evaluate the surface integral:

$$\iint_S \left[\overline{\mathrm{Y}}_{n,0}^{(1)}(\theta,\phi)\right]^2 dS = \int_0^{2\pi}\left(\int_0^{\pi}\left[\mathrm{P}_n(\cos\theta)\right]^2 \sin\theta\, d\theta\right) d\phi \tag{25.205}$$

Letting $\cos\theta = x$:

$$\iint_S \left[\overline{\mathrm{Y}}_{n,0}^{(1)}(\theta,\phi)\right]^2 dS = \int_0^{2\pi}\left(\int_{-1}^{+1}\left[\mathrm{P}_n(x)\right]^2 dx\right) d\phi = \frac{2}{2n+1}\int_0^{2\pi} d\phi = \frac{4\pi}{2n+1} \tag{25.206}$$

Also, from Equation 25.199:

$$\iint_S \left[\overline{\mathrm{Y}}_{n,m}^{(1)}(\theta,\phi)\right]^2 dS = \int_0^{2\pi}\left(\int_0^{\pi}\left[\mathrm{P}_n^m(\cos\theta)\right]^2 \sin\theta\, d\theta\right)\cos^2 m\phi\, d\phi \tag{25.207}$$

then for $\cos\theta = x$, we have

$$\iint_S \left[\overline{\mathrm{Y}}_{n,m}^{(1)}(\theta,\phi)\right]^2 dS = \int_0^{2\pi}\left(\int_{-1}^{+1}\left[\mathrm{P}_n^m(x)\right]^2 dx\right)\cos^2 m\phi\, d\phi = \frac{2}{2n+1}\frac{(n+m)!}{(n-m)!}\int_0^{2\pi}\cos^2 m\phi\, d\phi \tag{25.208}$$

or

$$\iint_S \left[\overline{\mathrm{Y}}_{n,m}^{(1)}(\theta,\phi)\right]^2 dS = \frac{2\pi}{2n+1}\frac{(n+m)!}{(n-m)!},\ m = 1,2,\ldots,n \tag{25.209}$$

Similarly,

$$\iint_S \left[\overline{Y}^{(2)}_{n,m}(\theta,\phi)\right]^2 dS = \int_0^{2\pi}\left(\int_0^{\pi}\left[P_n^m(\cos\theta)\right]^2 \sin\theta d\theta\right)\sin^2 m\phi d\phi \tag{25.210}$$

then for $\cos\theta = x$, we have

$$\iint_S \left[\overline{Y}^{(2)}_{n,m}(\theta,\phi)\right]^2 dS = \int_0^{2\pi}\left(\int_{-1}^{+1}\left[P_n^m(x)\right]^2 dx\right)\sin^2 m\phi d\phi = \frac{2}{2n+1}\frac{(n+m)!}{(n-m)!}\int_0^{2\pi}\sin^2 m\phi d\phi \tag{25.211}$$

or

$$\iint_S \left[\overline{Y}^{(2)}_{n,m}(\theta,\phi)\right]^2 dS = \frac{2\pi}{2n+1}\frac{(n+m)!}{(n-m)!},\ m = 1,2,\ldots,n \tag{25.212}$$

We series expand the function $f(x)$ through the Legendre polynomials:

$$f(\theta,\phi) = c_{00} + \sum_{n=1}^{\infty}\left[c_{n0}P_n(\cos\theta) + \sum_{m=1}^{n}(c_{nm}\cos m\phi + b_{nm}\sin m\phi)P_n^m(\cos\theta)\right] \tag{25.213}$$

where the coefficients are obtained:

$$c_{n0} = \frac{2n+1}{4\pi}\iint_S P_n(\cos\theta) f(\theta,\phi) dS \tag{25.214}$$

$$c_{nm} = \frac{2n+1}{4\pi}\frac{(n-m)!}{(n+m)!}\iint_S P_n^m(\cos\theta)\cos m\phi f(\theta,\phi) dS \tag{25.215}$$

$$b_{nm} = \frac{2n+1}{4\pi}\frac{(n-m)!}{(n+m)!}\iint_S P_n^m(\cos\theta)\sin m\phi f(\theta,\phi) dS \tag{25.216}$$

25.17 Laguerre Polynomials

We investigate orthogonal polynomials related to confluent hypergeometric function equations. For this case, we assume α and γ are independent and real-valued variables. Suppose in Equation 23.33 we set $\gamma = \lambda + 1$ with $\lambda > -1$ and let $L_n^{(\lambda)}(x)$ be the so-called **associated Laguerre polynomials** when $\alpha = -n$ and n is a positive integer. The associated Laguerre polynomials, $L_n^{(\lambda)}(x)$, relate the confluent hypergeometric functions, $F(-n,\lambda+1;x)$:

$$L_n^{(\lambda)}(x) = C_n F(-n,\lambda+1;x) = C_n\left(1 - \frac{n}{1!(\lambda+1)}x + \frac{n(n-1)}{2!(\lambda+1)(\lambda+2)}x^2 + \cdots + (-1)^n\frac{n\cdots 1}{n!(\lambda+1)\cdots(\lambda+n)}x^n\right) \tag{25.217}$$

Here, C_n is an arbitrary constant obtained from

$$\frac{d^n}{dx^n}\left(x^n\rho_\lambda(x)C_n(-1)^n\frac{n!}{(\lambda+1)\cdots(\lambda+n)}\right)=(-1)^n n!\rho_\lambda(x)L_n^{(\lambda)}(x) \tag{25.218}$$

From here:

$$C_n=(\lambda+1)\cdots(\lambda+n)=\frac{\Gamma(\lambda+n+1)}{\Gamma(\lambda+1)} \tag{25.219}$$

and

$$L_n^{(\lambda)}(x)=x^{-\lambda}\exp\{x\}\frac{d^n}{dx^n}\left(x^n\rho_\lambda(x)\right),\ \rho_\lambda(x)\equiv x^\lambda\exp\{-x\} \tag{25.220}$$

25.18 Associated Laguerre Polynomial Generating Function

We Taylor series expand the associated Laguerre polynomials to find the productive function:

$$\Phi(t,x,\lambda)=\sum_{n=0}^{\infty}\frac{1}{n!}L_n^{(\lambda)}(x)t^n=\frac{\exp\left\{-\frac{tx}{1-t}\right\}}{(1-t)^{\lambda+1}},\ |t|<1 \tag{25.221}$$

from where:

$$L_n^{(\lambda)}(x)=\frac{\partial^n}{\partial t^n}\Phi(t,x,\lambda)\Big|_{t=0} \tag{25.222}$$

Considering Equation 25.221:

$$\frac{\partial}{\partial x}\Phi(t,x,\lambda)=-t\frac{\exp\left\{-\frac{tx}{1-t}\right\}}{(1-t)^{\lambda+2}},\ \frac{\partial}{\partial t}\Phi(t,x,\lambda)=\left(\lambda+1-\frac{x}{1-t}\right)\frac{1}{(1-t)^{\lambda+2}}\exp\left\{-\frac{tx}{1-t}\right\} \tag{25.223}$$

In Equation 25.223, we swap λ with $\lambda-1$:

$$(1-t)\frac{\partial}{\partial x}\Phi(t,x,\lambda-1)=\lambda\Phi(t,x,\lambda-1)-x\Phi(t,x,\lambda) \tag{25.224}$$

From Equations 25.223 to 25.224, we have

$$(1-t)\frac{\partial}{\partial x}\Phi+t\Phi=0,\ (1-t)^2\frac{\partial}{\partial t}\Phi+\left(x+(t-1)(\lambda+1)\right)\Phi=0 \tag{25.225}$$

From Equation 25.221:

$$\frac{\partial}{\partial x}\Phi=\sum_{n=0}^{\infty}\frac{1}{n!}t^n\frac{\partial}{\partial x}L_n^{(\lambda)}(x),\ \frac{\partial}{\partial t}\Phi=\sum_{n=0}^{\infty}\frac{1}{(n-1)!}t^{n-1}L_n^{(\lambda)}(x) \tag{25.226}$$

From here and considering Equations 25.221 and 25.225:

$$(1-t)\sum_{n=0}^{\infty}\frac{1}{n!}t^n\frac{d}{dx}L_n^{(\lambda)}(x)+t\sum_{n=0}^{\infty}\frac{1}{n!}L_n^{(\lambda)}(x)t^n=0 \tag{25.227}$$

$$(1-t)^2\sum_{n=0}^{\infty}\frac{1}{(n-1)!}t^{n-1}L_n^{(\lambda)}(x)+\big(x+(t-1)(\lambda+1)\big)\sum_{n=0}^{\infty}\frac{1}{n!}L_n^{(\lambda)}(x)t^n=0 \tag{25.228}$$

We derive from these equations the following recurrence relations:

$$\frac{d}{dx}L_n^{(\lambda)}(x)-n\frac{d}{dx}L_{n-1}^{(\lambda)}(x)+nL_{n-1}^{(\lambda)}(x)=0,\ L_{n+1}^{(\lambda)}(x)+(x-\lambda-2n-1)L_n^{(\lambda)}(x)+n(\lambda+n)L_{n-1}^{(\lambda)}(x)=0 \tag{25.229}$$

Eliminating $L_{n-1}^{(\lambda)}(x)$ from Equation 25.229:

$$\frac{d}{dx}L_{n+1}^{(\lambda)}(x)-L_{n+1}^{(\lambda)}(x)+(x-n-1)\frac{d}{dx}L_n^{(\lambda)}(x)+(\lambda+2n+2-x)L_n^{(\lambda)}(x)=0 \tag{25.230}$$

Eliminate $\frac{d}{dx}L_{n-1}^{(\lambda)}(x)$ from Equation 25.229 and using Equation 25.230 and swapping n with $n-1$:

$$x\frac{d}{dx}L_n^{(\lambda)}(x)-nL_n^{(\lambda)}(x)+n(\lambda+n)L_{n-1}^{(\lambda)}(x)=0 \tag{25.231}$$

Consider now Equation 25.229 and swap n with $n+1$:

$$L_n^{(\lambda)}(x)=\frac{d}{dx}L_n^{(\lambda)}(x)-\frac{1}{n+1}\frac{d}{dx}L_{n+1}^{(\lambda)}(x) \tag{25.232}$$

Considering Equation 25.217:

$$\int_0^t L_n^{(\lambda)}(x)dx=L_n^{(\lambda)}(t)-\frac{1}{n+1}L_{n+1}^{(\lambda)}(t)+\frac{\lambda}{n+1}\frac{\Gamma(\lambda+n+1)}{\Gamma(\lambda+1)} \tag{25.233}$$

Suppose $\lambda=0$:

$$\int_0^t L_n(x)dx=L_n(t)-\frac{1}{n+1}L_{n+1}(t) \tag{25.234}$$

Considering Equations 25.22–25.232:

$$L_n^{(\lambda-1)}(x) = L_n^{(\lambda)}(x) - nL_{n-1}^{(\lambda)}(x) = 0 \quad (25.235)$$

$$xL_n^{(\lambda)}(x) = (\lambda+n)L_n^{(\lambda-1)}(x) - L_{n+1}^{(\lambda)}(x) = 0 \quad (25.236)$$

We examine the following integral considering $m \leq n$ while applying m-times integration by parts:

$$\int_0^\infty \rho_\lambda(x)L_m^{(\lambda)}(x)L_n^{(\lambda)}(x)dx = \int_0^\infty L_m^{(\lambda)}(x)\frac{d^n}{dx^n}\left(x^n\rho_\lambda(x)\right)dx \quad (25.237)$$

or

$$\int_0^\infty L_m^{(\lambda)}(x)\frac{d^n}{dx^n}\left(x^n\rho_\lambda(x)\right)dx = (-1)^m\int_0^\infty \frac{d^m}{dx^m}L_m^{(\lambda)}(x)\frac{d^{n-m}}{dx^{n-m}}\left(x^n\rho_\lambda(x)\right)dx \quad (25.238)$$

If $m < n$ and integrating by parts once more:

$$\frac{d^{m+1}}{dx^{m+1}}L_m^{(\lambda)}(x) = 0 \quad (25.239)$$

So

$$\int_0^\infty \rho_\lambda(x)L_m^{(\lambda)}(x)L_n^{(\lambda)}(x)dx = 0,\ m \neq n \quad (25.240)$$

If $m = n$, then

$$\int_0^\infty \rho_\lambda(x)dx\left[L_n^{(\lambda)}(x)\right]^2 = \int_0^\infty \rho_\lambda(x)x^n dx = n!\Gamma(\lambda+n+1) \quad (25.241)$$

If $\lambda = 0$, then

$$\int_0^\infty \rho_0(x)L_m(x)L_n(x)dx = \begin{cases} 0, & m \neq n \\ n!\Gamma(n+1) = (n!)^2, & m = n \end{cases} \quad (25.242)$$

We series expand the function $f(x)$ convergent within the interval, $(0,\infty)$, through the associated Laguerre polynomials, $L_n^{(\lambda)}(x)$:

$$f(x) = \sum_{n=0}^\infty c_n L_n^{(\lambda)}(x),\ c_n = \frac{1}{n!\Gamma(\lambda+n+1)}\int_0^\infty \rho_\lambda(x)L_n^{(\lambda)}(x)f(x)dx \quad (25.243)$$

25.19 Hermite Polynomials

The Hermite function can be obtained from the following differential equation:

$$\frac{d^2y}{dx^2} - 2x\frac{dy}{dx} + 2(2n+1)y = 0 \tag{25.244}$$

This yields the **Hermite polynomial,** $y = \mathrm{H}_n(x)$, of order n:

$$\mathrm{H}_n(x) = (-1)^n \exp\{x^2\}\frac{d^n}{dx^n}\exp\{-x^2\} \tag{25.245}$$

From Equation 25.232, we swap n with $n+1$ for $\lambda = -\frac{1}{2}$:

$$\frac{1}{2x}\frac{d}{dx}\mathrm{L}_{n+1}^{\left(-\frac{1}{2}\right)}(x^2) = -(n+1)\mathrm{L}_n^{\left(\frac{1}{2}\right)}(x^2) \tag{25.246}$$

and

$$\mathrm{H}_{2n+1}(x) = \frac{(-1)^{n+1}2^{2n}}{n+1}\frac{d}{dx}\mathrm{L}_{n+1}^{\left(-\frac{1}{2}\right)}(x^2) \tag{25.247}$$

25.20 Hermite Polynomial Generating Function

Examine the generating function, $\Phi(t,x)$, for the Hermite polynomial:

$$\Phi(t,x) = \sum_{n=0}^{\infty}\frac{1}{n!}\mathrm{H}_n(x)t^n = \exp\{-t^2 + 2tx\} \tag{25.248}$$

then

$$\mathrm{H}_n(x) = \frac{1}{n!}\left[\frac{\partial^n}{\partial t^n}\Phi(t,x)\right]_{t=0} \tag{25.249}$$

From Equation 25.248:

$$\frac{\partial}{\partial t}\Phi(t,x) = 2(x-t)\Phi(t,x),\ \frac{\partial}{\partial x}\Phi(t,x) = 2t\Phi(t,x) \tag{25.250}$$

From here:

$$\frac{\partial}{\partial t}\Phi(t,x) + 2(x-t)\Phi(t,x) = 0,\ \frac{\partial}{\partial x}\Phi(t,x) - 2t\Phi(t,x) = 0 \tag{25.251}$$

But

$$\frac{\partial}{\partial t}\Phi(t,x)=\sum_{n=0}^{\infty}\frac{1}{(n-1)!}\mathrm{H}_n(x)t^{n-1},\ \frac{\partial}{\partial x}\Phi(t,x)=\sum_{n=0}^{\infty}\frac{1}{n!}\mathrm{H}_n'(x)t^n \tag{25.252}$$

So

$$\sum_{n=0}^{\infty}\frac{1}{n!}\mathrm{H}_n'(x)t^n-2t\sum_{n=0}^{\infty}\frac{1}{n!}\mathrm{H}_n(x)t^n=0 \tag{25.253}$$

$$\sum_{n=0}^{\infty}\frac{1}{(n-1)!}\mathrm{H}_n(x)t^{n-1}+2(t-x)\sum_{n=0}^{\infty}\frac{1}{n!}\mathrm{H}_n(x)t^n=0 \tag{25.254}$$

and hence the recurrence relations:

$$\mathrm{H}_n'(x)-2n\mathrm{H}_{n-1}(x)=0 \tag{25.255}$$

$$\mathrm{H}_{n+1}(x)-2x\mathrm{H}_n(x)+2\mathrm{H}_n(x)=0 \tag{25.256}$$

From these two recurrence relations, we have

$$\mathrm{H}_{n+1}(x)+\mathrm{H}_n'(x)-2x\mathrm{H}_n(x)=0 \tag{25.257}$$

Consider Equation 25.256 and swapping n with $n-1$, we represent the Hermite polynomial, $\mathrm{H}_n(x)$, in the form of a determinant:

$$\mathrm{H}_n(x)=\begin{vmatrix} 2x & 2(n-1) & 0 & \cdots & 0 & 0 & 0 \\ 1 & 2x & 2(n-1) & \cdots & 0 & 0 & 0 \\ 0 & 1 & 2x & \cdots & 0 & 0 & 0 \\ \cdots & \cdots & \cdots & \cdots & \cdots & \cdots & \cdots \\ 0 & 0 & 0 & \cdots & 2x & 2\cdot 2 & 0 \\ 0 & 0 & 0 & \cdots & 1 & 2x & 2\cdot 1 \\ 0 & 0 & 0 & \cdots & 0 & 1 & 2x \end{vmatrix} \tag{25.258}$$

Let us apply $(k-1)$-fold differentiation in Equation 25.255 and we have

$$\mathrm{H}_n^{(k)}(x)=2^k n(n-1)\cdots(n-k+1)\mathrm{H}_{n-k}(x)=0 \tag{25.259}$$

If we let $n=k$ in Equation 25.255, then

$$\mathrm{H}_n^{(n)}(x)=2^n n! \tag{25.260}$$

This implies that the coefficient of x^n is equal to 2^n.

References

1. L.D. Landau, E.M. Lifshitz, Quantum Mechanics, Non-relativistic Theory, Pergamon Press Ltd., United Kingdom, 1965.
2. Yuli V. Nazarov, Jeroen Danon, Advanced Quantum Mechanics, Cambridge University Press, United Kingdom, 2013.
3. Kyriakos Tamvakis, Basic Quantum Mechanics, Springer, Berlin, 2019.
4. Pereyra Pedro, Fundamentals of Quantum Physics, Springer-Verlag, Berlin Heidelberg, 2012.
5. Gabriel Popescu, Black body radiation, Principles of Biophotonics, Volume 2, IOP Publishing, Bristol, UK, 2019, pp. 6–14.
6. Lukong Cornelius Fai, Matthew Wysin, Statistical Thermodynamics: Understanding the Properties of Macroscopic Systems, CRC Press, United Kingdom, 2012.
7. L.D. Landau, E.M. Lifshitz, Statistical Physics, Pergamon Press Ltd., United Kingdom, 1980.
8. Leonard I. Schiff, Quantum Mechanics, 3rd ed., McGraw-Hill, Inc., New York, NY, 1968.
9. Maged Marghany, Chapter 3 - Quantization of Maxwell's equations, I M. Marghany (Ed.) Synthetic Aperture Radar Imaging Mechanism for Oil Spills, CRC Press, United Kingdom, 2020, pp. 41–60.
10. Reina Maruyama, Phenomenology of Gamma Ray and Charged Particles Interactions, In E. Greenspan (Ed.) Encyclopedia of Nuclear Energy, Elsevier, Oxford, 2021, pp. 117–124.
11. Michael F. L'Annunziata, Radioactivity Hall of Fame-Part III, In M.F. L'Annunziata (Ed.) Radioactivity, Elsevier Science B.V., Amsterdam, 2007, pp. 141–186.
12. Lukong Cornelius Fai, Cabisov Casbek Sardionivich, Ashok Mody, Relativistic Theory and Electrodynamics, 2nd ed., Shroff Publishers & Distributors Pvt Ltd, Navi Mumbai, India, 2010.
13. Lukong Cornelius Fai, Cabisov Casbek Sardionovich, Ashok Mody, Study Aid in Relativistic Theory and Electrodynamics, 2nd ed., Shroff Publishers & Distributors Pvt Ltd, Navi Mumbai, India, 2010.
14. L.D. Landau, E.M. Lifshitz, The Classical Theory of Fields, 3rd ed., Pergamon Press Ltd., United Kingdom, 1971.
15. C. Davisson, L.H. Germer, Diffraction of electrons by a crystal of nickel, Physical Review, 30 (1927) 705–740.
16. George Paget Thomson, Experiments on the diffraction of cathode rays, Proceedings of the Royal Society of London. Series A, containing papers of a mathematical and physical character, 117 (1928) 600–609.
17. Lukong Cornelius Fai, Feynman Path Integrals in Quantum Mechanics and Statistical Physics, 1st ed., CRC Press, United Kingdom, 2021.
18. Roger Bach, Damian Pope, Sy-Hwang Liou, Herman Batelaan, Controlled double-slit electron diffraction, New Journal of Physics, 15 (2013) 033018.
19. Dan C. Marinescu, Gabriela M. Marinescu, CHAPTER 1 - Preliminaries, In D.C. Marinescu, G.M. Marinescu (Eds.) Classical and Quantum Information, Academic Press, Boston, MA, 2012, pp. 1–131.

20. E Schrödinger, SCHRÖDINGER 1926C, Annalen der Physik, 79 (1926) 734.
21. E.U. Condon, G.H. Shortley, Theory of Atomic Spectra, Cambridge Univ. Press, London, 1957.
22. Res Jost, Abraham Pais, On the scattering of a particle by a static potential, Physical Review, 82 (1951) 840.
23. Lukong Cornelius Fai, Quantum Field Theory: Feynman Path Integrals and Diagrammatic Techniques in Condensed Matter, CRC Press, United Kingdom, 2019.
24. T. D. Lee, F. E. Low, D. Pines, The motion of slow electrons in a polar crystal, Physical Review, 90 (1953) 297–302.
25. Hermann Haken, On the theory of excitons in solids, Journal of Physics and Chemistry of Solids, 8 (1959) 166–171.
26. Beril S. I., E.P. Pokatilov, A.S. Zotov, M. Farah, Fai L. C., Screening Coulomb interaction and polaronic exciton in a metal-dielectric-semiconductor structure, Physics and Techniques of Semiconductors-Soviet Physics, 27 (1993) 12–20.
27. G. E. Uhlenbeck, S. Goudsmit, Spinning electrons and the structure of spectra, Nature, 117 (1926) 264–265.
28. L. H. Thomas, The motion of the spinning electron, Nature, 117 (1926) 514–514.
29. I. I. Guseinov, Dirac equation for particles with arbitrary half-integral spin, Philosophical Magazine, 91 (2011) 4063–4072.

Index

I

J

K

L

M

N

O

P

Q

R

Printed in the United States
by Baker & Taylor Publisher Services

Printed in the United States
by Baker & Taylor Publisher Services